Methods in Enzymology

Volume 296

NEUROTRANSMITTER TRANSPORTERS

METHODS IN ENZYMOLOGY

EDITORS-IN-CHIEF

John N. Abelson Melvin I. Simon

DIVISION OF BIOLOGY
CALIFORNIA INSTITUTE OF TECHNOLOGY
PASADENA, CALIFORNIA

FOUNDING EDITORS

Sidney P. Colowick and Nathan O. Kaplan

Methods in Enzymology

Volume 296

Neurotransmitter Transporters

EDITED BY

Susan G. Amara

VOLLUM INSTITUTE FOR ADVANCED BIOMEDICAL RESEARCH
AND HOWARD HUGHES MEDICAL INSTITUTE
OREGON HEALTH SCIENCES UNIVERSITY
PORTLAND, OREGON

ACADEMIC PRESS
San Diego London Boston New York Sydney Tokyo Toronto

This book is printed on acid-free paper. ♾

Academic Press
a division of Harcourt Brace & Company
525 B Street, Suite 1900, San Diego, California 92101-4495, USA
http://www.academicpress.com

Academic Press Limited
24-28 Oval Road, London NW1 7DX, UK
http://www.hbuk.co.uk/ap/

International Standard Book Number: 0-12-182197-8

PRINTED IN THE UNITED STATES OF AMERICA
98 99 00 01 02 03 MM 9 8 7 6 5 4 3 2 1

Table of Contents

Section I. Purification of Transporter Proteins and cDNAs

A. Plasma Membrane Carriers

B. Vesicular Carriers

Section II. Pharmacological Approaches and Binding Studies

Section III. Transport Assays and Kinetic Analyses

Section IV. Biochemical Approaches for Structure–Function Analyses

Section V. Expression Systems and Molecular Genetic Approaches

Section VI. Application of Electrophysiological Techniques to Neurotransmitter Carriers

Section VII. Microdialysis and Electrochemical Measurements

Contributors to Volume 296

Article numbers are in parentheses following the names of contributors.
Affiliations listed are current.

JINHI AHN (25), *Department of Cellular and Molecular Physiology, Yale University School of Medicine, New Haven, Connecticut 06510*

SUSAN G. AMARA (21, 22, 29, 32), *Vollum Institute for Advanced Biomedical Research and Howard Hughes Medical Institute, Oregon Health Sciences University, Portland, Oregon 97201*

BRIAN B. ANDERSON (46), *Department of Chemistry, Pennsylvania State University, University Park, Pennsylvania 16802*

CARMEN ARAGÓN (1), *Centro de Biología Molecular "Severo Ochoa," Facultad de Ciencias, Universidad Autónoma de Madrid, 28049 Madrid, Spain*

DAVID ATTWELL (41, 42), *Department of Physiology, University College London, London WC1E 6BT, England*

ERIC L. BARKER (33), *Department of Pharmacology and Center for Molecular Neuroscience, Vanderbilt University School of Medicine, Nashville, Tennessee 37232-6600*

FRANCISCO BEZANILLA (38), *Department of Physiology, University of California, Los Angeles, California 90024*

BRIAN BILLUPS (42), *Department of Physiology, University College London, London WC1E 6BT, England*

RANDY D. BLAKELY (24, 33), *Center for Molecular Neuroscience, Vanderbilt University School of Medicine, Nashville, Tennessee 37232-6600*

HEINZ BÖNISCH (18), *Institute of Pharmacology and Toxicology, University of Bonn, D-53113 Bonn, Germany*

JEAN-JACQUES BONNET (14), *Laboratoire de Neuropsychopharmacologie Expérimentale, UFR de Médecine et Pharmacie, 76800 Saint Etienne du Rouvray, France*

RICHARD J. BRIDGES (12), *Department of Pharmaceutical Sciences, University of Montana, Missoula, Montana 59812*

LYNN A. BRISTOL (35), *Department of Neurology, Johns Hopkins University, Baltimore, Maryland 21287*

DIETER BRUNS (40), *Howard Hughes Medical Institute, Boyer Center for Molecular Medicine, Yale University School of Medicine, New Haven, Connecticut 06510*

KARI JOHNSON BUCK (32), *Department of Behavioral Neuroscience and Portland Alcohol Research Center, Oregon Health Sciences University and Veterans Administration Medical Center, Portland, Oregon 97201*

MELISSA A. BUNIN (47), *Department of Chemistry, University of North Carolina, Chapel Hill, North Carolina 27599-3290*

YONGWEI CAO (37), *Monsanto Company, St. Louis, Missouri 63198*

MICHAEL J. CAPLAN (25), *Department of Cellular and Molecular Physiology, Yale University School of Medicine, New Haven, Connecticut 06510*

F. IVY CARROLL (17), *Chemistry and Life Sciences, Research Triangle Institute, Research Triangle Park, North Carolina 27709*

ALBERT CHA (38), *Department of Physiology, University of California, Los Angeles, California 90024*

A. RICHARD CHAMBERLIN (12), *Department of Chemistry, University of California, Irvine, California 92717*

F. A. CHAUDHRY (26), *Anatomical Institute, University of Oslo, N-0317 Oslo, Norway*

NIAN-HANG CHEN (17, 49), *Department of Chemistry, Emory University, Atlanta, Georgia 30322*

JANET A. CLARK (20), *Section on Genetics, National Institute of Mental Health, Bethesda, Maryland 20892-4094*

AMADOU T. CORERA (14), *Laboratoire de Neuropsychopharmacologie Expérimentale, UFR de Médecine et Pharmacie, 76800 Saint Etienne du Rouvray, France*

N. C. DANBOLT (26), *Anatomical Institute, University of Oslo, N-0317 Oslo, Norway*

KRISSTINA DANEK (44), *Department of Chemistry, Emory University, Atlanta, Georgia 30322*

GWYNN M. DANIELS (21), *Department of Cell and Developmental Biology, Vollum Institute for Advanced Biomedical Research, Oregon Health Sciences University, Portland, Oregon 97201*

DALIT E. DAR (31), *Molecular Neurobiology Branch, NIDA-IRP and Departments of Neurology and Neuroscience, Johns Hopkins University School of Medicine, Baltimore, Maryland 21224*

LOUIS J. DEFELICE (39), *Department of Pharmacology, Center for Molecular Neuroscience, Vanderbilt University, Nashville, Tennessee 37232-6600*

Y. DEHNES (26), *Anatomical Institute, University of Oslo, N-0317 Oslo, Norway*

SHELLY D. DICKINSON (48), *Department of Pharmacology and Neuroscience Program, University of Colorado Health Science Center, Denver, Colorado 80262*

JEAN-CLAUDE DO RÉGO (14), *Laboratoire de Neuropsychopharmacologie Expérimentale, UFR de Médecine et Pharmacie, 76800 Saint Etienne du Rouvray, France*

JANET S. DUERR (36), *Program in Molecular and Cell Biology, Oklahoma Medical Research Foundation, Oklahoma City, Oklahoma 73104*

CYNTHIA EARLES (45), *Department of Chemistry, Washington State University, Pullman, Washington 99164-4630*

ROBERT H. EDWARDS (10), *Departments of Neurology and Physiology, University of California School of Medicine, San Francisco, California 94143*

JEFFREY D. ERICKSON (6), *Neuroscience Center and Department of Pharmacology, Louisiana State University Medical Center, New Orleans, Louisiana 70112*

ANDREW G. EWING (46), *Department of Chemistry, Pennsylvania State University, University Park, Pennsylvania 16802*

ERIK FALCH (11), *Department of Medicinal Chemistry, Royal Danish School of Pharmacy, DK-2100 Copenhagen Ø, Denmark*

J. PATRICK FINN III (10), *Departments of Microbiology and Immunology, University of California School of Medicine, Los Angeles, California 90024*

DENNIS L. FRISBY (36), *Program in Molecular and Cell Biology, Oklahoma Medical Research Foundation, Oklahoma City, Oklahoma 73104*

BENTE F. FRØLUND (11), *Department of Medicinal Chemistry, Royal Danish School of Pharmacy, DK-2100 Copenhagen Ø, Denmark*

AURELIO GALLI (39), *Department of Pharmacology, Center for Molecular Neuroscience, Vanderbilt University, Nashville, Tennessee 37232-6600*

VADIVEL GANAPATHY (19), *Department of Biochemistry and Molecular Biology, Medical College of Georgia, Augusta, Georgia 30912*

BRUNO GASNIER (5), *Centre National de la Recherche Scientifique, Institut de Biologie Physico-Chimique, 75005 Paris, France*

GREG A. GERHARDT (48), *Departments of Psychiatry and Pharmacology, Neuroscience Program, and Rocky Mountain Center for Sensor Technology, University of Colorado Health Science Center, Denver, Colorado 80262*

MICHELLE L. GILMOR (27), *Department of Neurology, Emory University School of Medicine, Atlanta, Georgia 30322*

LAWRENCE M. GRACZ (7), *Department of Chemistry, Program in Biochemistry and Molecular Biology, Neuroscience Research Institute, University of California, Santa Barbara, California 93106*

MITCHELL GREEN (45), *Department of Chemistry, Washington State University, Pullman, Washington 99164-4630*

HIROMI GUNSHIN (2), *Renal Division, Department of Medicine, Brigham and Women's Hospital and Harvard Medical School, and Department of Biological Chemistry and Molecular Pharmacology, Harvard Medical School, Boston, Massachusetts 02115*

MATTHIAS A. HEDIGER (2), *Renal Division, Department of Medicine, Brigham and Women's Hospital and Harvard Medical School, and Department of Biological Chemistry and Molecular Pharmacology, Harvard Medical School, Boston, Massachusetts 02115*

JOHANNES W. HELL (8), *Department of Pharmacology, University of Wisconsin, Madison, Wisconsin 53706-1532*

JEAN-PIERRE HENRY (5), *Centre National de la Recherche Scientifique, Institut de Biologie Physico-Chimique, 75005 Paris, France*

MARIE-FRANÇOISE ISAMBERT (5), *Centre National de la Recherche Scientifique, Institut de Biologie Physico-Chimique, 75005 Paris, France*

REINHARD JAHN (8), *Howard Hughes Medical Institute, Boyer Center for Molecular Medicine, Yale University School of Medicine, New Haven, Connecticut 06536-0812*

JONATHAN A. JAVITCH (23), *Departments of Psychiatry and Pharmacology, Center for Molecular Recognition, College of Physicians and Surgeons, Columbia University, New York, New York 10032*

JAY B. JUSTICE, JR. (44), *Department of Chemistry, Emory University, Atlanta, Georgia 30322*

YOSHIKATSU KANAI (2), *Department of Pharmacology, Kyorin University School of Medicine, Mitaka, Tokyo, Japan*

MICHAEL KAVANAUGH (38), *Vollum Institute, Oregon Health Sciences University, Portland, Oregon 97201*

HANS P. KOCH (12), *Department of Pharmaceutical Sciences, University of Montana, Missoula, Montana 59812*

POVL KROGSGAARD-LARSEN (11), *Department of Medicinal Chemistry, Royal Danish School of Pharmacy, DK-2100 Copenhagen Ø, Denmark*

K. P. LEHRE (26), *Anatomical Institute, University of Oslo, N-0317 Oslo, Norway*

FREDERICK H. LEIBACH (19), *Department of Biochemistry and Molecular Biology, Medical College of Georgia, Augusta, Georgia 30912*

BARBARA H. LEIGHTON (22), *Howard Hughes Medical Institute and Vollum Institute for Advanced Biomedical Research, Oregon Health Sciences University, Portland, Oregon 97201*

HENRY A. LESTER (37), *Division of Biology, California Institute of Technology, Pasadena, California 91125*

ALLAN I. LEVEY (27), *Department of Neurology, Emory University School of Medicine, Atlanta, Georgia 30322*

L. M. LEVY (26), *Anatomical Institute, University of Oslo, N-0317 Oslo, Norway*

SCOTT M. LEWIS (9), *Brain Sciences Center, Department of Veterans Affairs Medical Center, and Department of Neurology, University of Minnesota Medical School, Minneapolis, Minnesota 55417*

HOLGER LILL (28), *Abteilung Biophysik, Fachberich Biologie/Chemie, Universitat Osnabruck, D-49069 Osnabruck, Germany*

ZHICHENG LIN (31), *Molecular Neurobiology Branch, NIDA-IRP and Departments of Neurology and Neuroscience, Johns Hopkins University School of Medicine, Baltimore, Maryland 21224*

QING-RONG LIU (3), *Molecular Neurobiology Branch, National Institute on Drug Abuse, Baltimore, Maryland 21224*

BEATRIZ LÓPEZ-CORCUERA (1), *Centro de Biología Molecular "Severo Ochoa," Facultad de Ciencias, Universidad Autónoma de Madrid, 28049 Madrid, Spain*

SELA MAGER (37), *Department of Physiology, University of North Carolina, Chapel Hill, North Carolina 27599-7545*

HALEY E. MELIKIAN (24), *Department of Pharmacology and Center for Molecular Neuroscience, Vanderbilt University School of Medicine, Nashville, Tennessee 37232-6600*

STEVEN MENNERICK (43), *Department of Psychiatry, Washington University School of Medicine, St. Louis, Missouri 63110*

ANDREW MERICKEL (10), *Interdepartmental Program in Neuroscience, University of California School of Medicine, Los Angeles, California 90024*

THOMAS METZGER (31), *Molecular Neurobiology Branch, NIDA-IRP and Departments of Neurology and Neuroscience, Johns Hopkins University School of Medicine, Baltimore, Maryland 21224*

GARY W. MILLER (27), *Department of Cell Biology, Howard Hughes Medical Institute, Duke University Medical Center, Durham, North Carolina 27710*

THEODORE R. MUTH (25), *Department of Cellular and Molecular Physiology, Yale University School of Medicine, New Haven, Connecticut 06510*

NATHAN NELSON (3, 28), *Department of Biochemistry, Tel Aviv University, 69978 Tel Aviv, Israel*

STANLEY M. PARSONS (7), *Department of Chemistry, Program in Biochemistry and Molecular Biology, Neuroscience Research Institute, University of California, Santa Barbara, California 93106*

GRAZIA PIETRINI (25), *C.N.R. Cellular and Molecular Pharmacology Center, Department of Pharmacology, University of Milan, Milan, Italy*

SUE L. POVLOCK (29), *Vollum Institute for Advanced Biomedical Research, Oregon Health Sciences University, Portland, Oregon 97201-3011*

PUTTUR D. PRASAD (19), *Department of Obstetrics and Gynecology, Medical College of Georgia, Augusta, Georgia 30912*

YAN QIAN (24), *Department of Pharmacology and Center for Molecular Neuroscience, Vanderbilt University School of Medicine, Nashville, Tennessee 37232-6600*

SAMMANDA RAMAMOORTHY (24), *Department of Pharmacology and Center for Molecular Neuroscience, Vanderbilt University School of Medicine, Nashville, Tennessee 37232-6600*

JAMES B. RAND (36), *Program in Molecular and Cell Biology, Oklahoma Medical Research Foundation, Oklahoma City, Oklahoma 73104*

MAARTEN E. A. REITH (17, 49), *Department of Biomedical and Therapeutic Sciences, University of Illinois, College of Medicine, Peoria, Illinois 61605*

MICHAEL B. ROBINSON (13), *Departments of Pediatrics and Pharmacology, Children's Hospital of Philadelphia, University of Pennsylvania, Philadelphia, Pennsylvania 19104-4318*

GARY A. ROGERS (7), *Department of Chemistry, Program in Biochemistry and Molecular Biology, Neuroscience Research Institute, University of California, Santa Barbara, California 93106*

MICHAEL F. ROMERO (2), *Department of Cellular and Molecular Physiology, Yale University School of Medicine, New Haven, Connecticut 06510*

DAVID ROSSI (42), *Department of Physiology, University College London, London WC1E 6BT, England*

JEFFREY D. ROTHSTEIN (35), *Department of Neurology, Johns Hopkins University, Baltimore, Maryland 21287*

GARY RUDNICK (16), *Department of Pharmacology, Yale University School of Medicine, New Haven, Connecticut 06510*

CORINNE SAGNÉ (5), *Inserm U288, Faculté de Médecine Pitié-Sal Pêtrère, 75013 Paris, France*

MONIQUE SARANTIS (41), *Department of Physiology, University College London, London WC1E 6BT, England*

JAMES O. SCHENK (45), *Departments of Chemistry, Biochemistry and Biophysics, and Programs in Pharmacology/Toxicology and Neuroscience, Washington State University, Pullman, Washington 99164-4630*

SHIMON SCHULDINER (4), *Alexander Silberman Institute of Life Sciences, Hebrew University of Jerusalem, Givat Ram, Jerusalem 91904, Israel*

REBECCA P. SEAL (22), *Program in Neuroscience, Vollum Institute for Advanced Biomedical Research, Oregon Health Sciences University, Portland, Oregon 97201*

DAVID SULZER (46), *Departments of Neurology and Psychiatry, Columbia University, and Department of Neuroscience, New York State Psychiatric Institute, New York, New York 10032*

MAREK SZATKOWSKI (42), *Department of Physiology and Biophysics, Imperial College School of Medicine at St. Mary's, London WC2 1PG, England*

MICHIKO TAKAHASHI (41), *Department of Physiology, University College London, London WC1E 6BT, England*

CHRISTOPHER G. TATE (30), *MRC Laboratory of Molecular Biology, Cambridge CB2 2QH, United Kingdom*

TETSUFUMI UEDA (9), *Departments of Psychiatry and Pharmacology, Mental Health Research Institute, University of Michigan, Ann Arbor, Michigan 48109-0720*

GEORGE UHL (31), *Molecular Neurobiology Branch, NIDA-IRP and Departments of Neurology and Neuroscience, Johns Hopkins University School of Medicine, Baltimore, Maryland 21224*

DAVID J. VANDENBERGH (34), *Center for Developmental and Health Genetics, Department of Biobehavioral Health, Pennsylvania State University, University Park, Pennsylvania 16802*

HELENE VAROQUI (6), *Neuroscience Center and Department of Pharmacology, Louisiana State University Medical Center, New Orleans, Louisiana 70112*

ROXANNE A. VAUGHAN (15), *Molecular Neurobiology Branch, National Institute on Drug Abuse, Intramural Research Program, Baltimore, Maryland 21224*

HOLLIE WAYMENT (45), *Department of Chemistry, Washington State University, Pullman, Washington 99164-4630*

R. MARK WIGHTMAN (47), *Department of Chemistry, University of North Carolina, Chapel Hill, North Carolina 27599-3290*

CEN XU (17), *Department of Neuroscience Drug Discovery, Bristol-Myers Squibb Company, Wallingford, Connecticut 06492*

RODRIGO YELIN (4), *Alexander Silberman Institute of Life Sciences, Hebrew University of Jerusalem, Givat Ram, Jerusalem 91904, Israel*

NANCY R. ZAHNISER (48), *Department of Pharmacology and Neuroscience Program, University of Colorado Health Science Center, Denver, Colorado 80262*

NOA ZERANGUE (38), *Neuroscience Graduate Program, University of California, San Francisco, California 94143*

CHARLES F. ZORUMSKI (43), *Department of Psychiatry, Washington University School of Medicine, St. Louis, Missouri 63110*

Preface

Although primary sequence data have been available for many families of ion channels, receptors, and transporters for nearly 15 years, the molecular biology of neurotransmitter transporters is a relatively new field. Since 1990 as many as five distinct new families of transport proteins have been shown to contribute to the transport of neurotransmitters in neurons and glial cells. Current work has just begun to examine the structural and mechanistic aspects of carrier function and to explore how transporters shape cellular communication in the nervous system. Initial studies of neurotransmitter uptake systems in brain slices, synaptosomes, and plasma membrane vesicles demonstrated a variety of activities that could be distinguished by kinetic properties, specific ionic requirements, and sensitivity to selective antagonists. However, recent advances in the techniques used to study transporter structure, function, and expression have offered surprising glimpses into the molecular mechanisms of neurotransmitter transport both at the plasma membrane and into vesicles.

This *Methods in Enzymology* volume focuses on biochemical, electrophysiological, pharmacological, molecular, and cell biological approaches used to study neurotransmitter transport systems. Although the series has a long history of producing highly informative volumes on topics related to membrane transport, this is the first volume to focus exclusively on the area of neurotransmitter transport and on the methods that have facilitated the rapid developments in this area. The articles provide detailed descriptions of procedures that should enable the reader to understand how they are accomplished and to repeat or adapt them for their own experimental requirements. Several other volumes in the series describe more general techniques that complement the topics covered. In particular, these include the *Methods in Enzymology* volumes on Biomembranes and the outstanding Volume 207 on Ion Channels.

This volume would not exist without the enthusiasm and dedication of the contributors, whom I thank for all their efforts. I am also delighted to acknowledge Gloria Ellis for her excellent organizational and administrative support and the Academic Press staff for their patience and advice. Valuable input also came from Mark Sonders, Geoff Murdoch, and the members of my laboratory, and I gratefully acknowledge their contributions.

SUSAN G. AMARA

METHODS IN ENZYMOLOGY

VOLUME I. Preparation and Assay of Enzymes
Edited by SIDNEY P. COLOWICK AND NATHAN O. KAPLAN

VOLUME II. Preparation and Assay of Enzymes
Edited by SIDNEY P. COLOWICK AND NATHAN O. KAPLAN

VOLUME III. Preparation and Assay of Substrates
Edited by SIDNEY P. COLOWICK AND NATHAN O. KAPLAN

VOLUME IV. Special Techniques for the Enzymologist
Edited by SIDNEY P. COLOWICK AND NATHAN O. KAPLAN

VOLUME V. Preparation and Assay of Enzymes
Edited by SIDNEY P. COLOWICK AND NATHAN O. KAPLAN

VOLUME VI. Preparation and Assay of Enzymes (*Continued*)
Preparation and Assay of Substrates
Special Techniques
Edited by SIDNEY P. COLOWICK AND NATHAN O. KAPLAN

VOLUME VII. Cumulative Subject Index
Edited by SIDNEY P. COLOWICK AND NATHAN O. KAPLAN

VOLUME VIII. Complex Carbohydrates
Edited by ELIZABETH F. NEUFELD AND VICTOR GINSBURG

VOLUME IX. Carbohydrate Metabolism
Edited by WILLIS A. WOOD

VOLUME X. Oxidation and Phosphorylation
Edited by RONALD W. ESTABROOK AND MAYNARD E. PULLMAN

VOLUME XI. Enzyme Structure
Edited by C. H. W. HIRS

VOLUME XII. Nucleic Acids (Parts A and B)
Edited by LAWRENCE GROSSMAN AND KIVIE MOLDAVE

VOLUME XIII. Citric Acid Cycle
Edited by J. M. LOWENSTEIN

VOLUME XIV. Lipids
Edited by J. M. LOWENSTEIN

VOLUME XV. Steroids and Terpenoids
Edited by RAYMOND B. CLAYTON

VOLUME XVI. Fast Reactions
Edited by KENNETH KUSTIN

VOLUME XXXVI. Hormone Action (Part A: Steroid Hormones)
Edited by BERT W. O'MALLEY AND JOEL G. HARDMAN

VOLUME XXXVII. Hormone Action (Part B: Peptide Hormones)
Edited by BERT W. O'MALLEY AND JOEL G. HARDMAN

VOLUME XXXVIII. Hormone Action (Part C: Cyclic Nucleotides)
Edited by JOEL G. HARDMAN AND BERT W. O'MALLEY

VOLUME XXXIX. Hormone Action (Part D: Isolated Cells, Tissues, and Organ Systems)
Edited by JOEL G. HARDMAN AND BERT W. O'MALLEY

VOLUME XL. Hormone Action (Part E: Nuclear Structure and Function)
Edited by BERT W. O'MALLEY AND JOEL G. HARDMAN

VOLUME XLI. Carbohydrate Metabolism (Part B)
Edited by W. A. WOOD

VOLUME XLII. Carbohydrate Metabolism (Part C)
Edited by W. A. WOOD

VOLUME XLIII. Antibiotics
Edited by JOHN H. HASH

VOLUME XLIV. Immobilized Enzymes
Edited by KLAUS MOSBACH

VOLUME XLV. Proteolytic Enzymes (Part B)
Edited by LASZLO LORAND

VOLUME XLVI. Affinity Labeling
Edited by WILLIAM B. JAKOBY AND MEIR WILCHEK

VOLUME XLVII. Enzyme Structure (Part E)
Edited by C. H. W. HIRS AND SERGE N. TIMASHEFF

VOLUME XLVIII. Enzyme Structure (Part F)
Edited by C. H. W. HIRS AND SERGE N. TIMASHEFF

VOLUME XLIX. Enzyme Structure (Part G)
Edited by C. H. W. HIRS AND SERGE N. TIMASHEFF

VOLUME L. Complex Carbohydrates (Part C)
Edited by VICTOR GINSBURG

VOLUME LI. Purine and Pyrimidine Nucleotide Metabolism
Edited by PATRICIA A. HOFFEE AND MARY ELLEN JONES

VOLUME LII. Biomembranes (Part C: Biological Oxidations)
Edited by SIDNEY FLEISCHER AND LESTER PACKER

VOLUME LIII. Biomembranes (Part D: Biological Oxidations)
Edited by SIDNEY FLEISCHER AND LESTER PACKER

VOLUME LIV. Biomembranes (Part E: Biological Oxidations)
Edited by SIDNEY FLEISCHER AND LESTER PACKER

Volume LV. Biomembranes (Part F: Bioenergetics)
Edited by Sidney Fleischer and Lester Packer

Volume LVI. Biomembranes (Part G: Bioenergetics)
Edited by Sidney Fleischer and Lester Packer

Volume LVII. Bioluminescence and Chemiluminescence
Edited by Marlene A. DeLuca

Volume LVIII. Cell Culture
Edited by William B. Jakoby and Ira Pastan

Volume LIX. Nucleic Acids and Protein Synthesis (Part G)
Edited by Kivie Moldave and Lawrence Grossman

Volume LX. Nucleic Acids and Protein Synthesis (Part H)
Edited by Kivie Moldave and Lawrence Grossman

Volume 61. Enzyme Structure (Part H)
Edited by C. H. W. Hirs and Serge N. Timasheff

Volume 62. Vitamins and Coenzymes (Part D)
Edited by Donald B. McCormick and Lemuel D. Wright

Volume 63. Enzyme Kinetics and Mechanism (Part A: Initial Rate and Inhibitor Methods)
Edited by Daniel L. Purich

Volume 64. Enzyme Kinetics and Mechanism (Part B: Isotopic Probes and Complex Enzyme Systems)
Edited by Daniel L. Purich

Volume 65. Nucleic Acids (Part I)
Edited by Lawrence Grossman and Kivie Moldave

Volume 66. Vitamins and Coenzymes (Part E)
Edited by Donald B. McCormick and Lemuel D. Wright

Volume 67. Vitamins and Coenzymes (Part F)
Edited by Donald B. McCormick and Lemuel D. Wright

Volume 68. Recombinant DNA
Edited by Ray Wu

Volume 69. Photosynthesis and Nitrogen Fixation (Part C)
Edited by Anthony San Pietro

Volume 70. Immunochemical Techniques (Part A)
Edited by Helen Van Vunakis and John J. Langone

Volume 71. Lipids (Part C)
Edited by John M. Lowenstein

Volume 72. Lipids (Part D)
Edited by John M. Lowenstein

VOLUME 91. Enzyme Structure (Part I)
Edited by C. H. W. HIRS AND SERGE N. TIMASHEFF

VOLUME 92. Immunochemical Techniques (Part E: Monoclonal Antibodies and General Immunoassay Methods)
Edited by JOHN J. LANGONE AND HELEN VAN VUNAKIS

VOLUME 93. Immunochemical Techniques (Part F: Conventional Antibodies, Fc Receptors, and Cytotoxicity)
Edited by JOHN J. LANGONE AND HELEN VAN VUNAKIS

VOLUME 94. Polyamines
Edited by HERBERT TABOR AND CELIA WHITE TABOR

VOLUME 95. Cumulative Subject Index Volumes 61–74, 76–80
Edited by EDWARD A. DENNIS AND MARTHA G. DENNIS

VOLUME 96. Biomembranes [Part J: Membrane Biogenesis: Assembly and Targeting (General Methods; Eukaryotes)]
Edited by SIDNEY FLEISCHER AND BECCA FLEISCHER

VOLUME 97. Biomembranes [Part K: Membrane Biogenesis: Assembly and Targeting (Prokaryotes, Mitochondria, and Chloroplasts)]
Edited by SIDNEY FLEISCHER AND BECCA FLEISCHER

VOLUME 98. Biomembranes (Part L: Membrane Biogenesis: Processing and Recycling)
Edited by SIDNEY FLEISCHER AND BECCA FLEISCHER

VOLUME 99. Hormone Action (Part F: Protein Kinases)
Edited by JACKIE D. CORBIN AND JOEL G. HARDMAN

VOLUME 100. Recombinant DNA (Part B)
Edited by RAY WU, LAWRENCE GROSSMAN, AND KIVIE MOLDAVE

VOLUME 101. Recombinant DNA (Part C)
Edited by RAY WU, LAWRENCE GROSSMAN, AND KIVIE MOLDAVE

VOLUME 102. Hormone Action (Part G: Calmodulin and Calcium-Binding Proteins)
Edited by ANTHONY R. MEANS AND BERT W. O'MALLEY

VOLUME 103. Hormone Action (Part H: Neuroendocrine Peptides)
Edited by P. MICHAEL CONN

VOLUME 104. Enzyme Purification and Related Techniques (Part C)
Edited by WILLIAM B. JAKOBY

VOLUME 105. Oxygen Radicals in Biological Systems
Edited by LESTER PACKER

VOLUME 106. Posttranslational Modifications (Part A)
Edited by FINN WOLD AND KIVIE MOLDAVE

VOLUME 107. Posttranslational Modifications (Part B)
Edited by FINN WOLD AND KIVIE MOLDAVE

VOLUME 108. Immunochemical Techniques (Part G: Separation and Characterization of Lymphoid Cells)
Edited by GIOVANNI DI SABATO, JOHN J. LANGONE, AND HELEN VAN VUNAKIS

VOLUME 109. Hormone Action (Part I: Peptide Hormones)
Edited by LUTZ BIRNBAUMER AND BERT W. O'MALLEY

VOLUME 110. Steroids and Isoprenoids (Part A)
Edited by JOHN H. LAW AND HANS C. RILLING

VOLUME 111. Steroids and Isoprenoids (Part B)
Edited by JOHN H. LAW AND HANS C. RILLING

VOLUME 112. Drug and Enzyme Targeting (Part A)
Edited by KENNETH J. WIDDER AND RALPH GREEN

VOLUME 113. Glutamate, Glutamine, Glutathione, and Related Compounds
Edited by ALTON MEISTER

VOLUME 114. Diffraction Methods for Biological Macromolecules (Part A)
Edited by HAROLD W. WYCKOFF, C. H. W. HIRS, AND SERGE N. TIMASHEFF

VOLUME 115. Diffraction Methods for Biological Macromolecules (Part B)
Edited by HAROLD W. WYCKOFF, C. H. W. HIRS, AND SERGE N. TIMASHEFF

VOLUME 116. Immunochemical Techniques (Part H: Effectors and Mediators of Lymphoid Cell Functions)
Edited by GIOVANNI DI SABATO, JOHN J. LANGONE, AND HELEN VAN VUNAKIS

VOLUME 117. Enzyme Structure (Part J)
Edited by C. H. W. HIRS AND SERGE N. TIMASHEFF

VOLUME 118. Plant Molecular Biology
Edited by ARTHUR WEISSBACH AND HERBERT WEISSBACH

VOLUME 119. Interferons (Part C)
Edited by SIDNEY PESTKA

VOLUME 120. Cumulative Subject Index Volumes 81–94, 96–101

VOLUME 121. Immunochemical Techniques (Part I: Hybridoma Technology and Monoclonal Antibodies)
Edited by JOHN J. LANGONE AND HELEN VAN VUNAKIS

VOLUME 122. Vitamins and Coenzymes (Part G)
Edited by FRANK CHYTIL AND DONALD B. MCCORMICK

VOLUME 123. Vitamins and Coenzymes (Part H)
Edited by FRANK CHYTIL AND DONALD B. MCCORMICK

VOLUME 124. Hormone Action (Part J: Neuroendocrine Peptides)
Edited by P. MICHAEL CONN

VOLUME 125. Biomembranes (Part M: Transport in Bacteria, Mitochondria, and Chloroplasts: General Approaches and Transport Systems)
Edited by SIDNEY FLEISCHER AND BECCA FLEISCHER

VOLUME 143. Sulfur and Sulfur Amino Acids
Edited by WILLIAM B. JAKOBY AND OWEN GRIFFITH

VOLUME 144. Structural and Contractile Proteins (Part D: Extracellular Matrix)
Edited by LEON W. CUNNINGHAM

VOLUME 145. Structural and Contractile Proteins (Part E: Extracellular Matrix)
Edited by LEON W. CUNNINGHAM

VOLUME 146. Peptide Growth Factors (Part A)
Edited by DAVID BARNES AND DAVID A. SIRBASKU

VOLUME 147. Peptide Growth Factors (Part B)
Edited by DAVID BARNES AND DAVID A. SIRBASKU

VOLUME 148. Plant Cell Membranes
Edited by LESTER PACKER AND ROLAND DOUCE

VOLUME 149. Drug and Enzyme Targeting (Part B)
Edited by RALPH GREEN AND KENNETH J. WIDDER

VOLUME 150. Immunochemical Techniques (Part K: *In Vitro* Models of B and T Cell Functions and Lymphoid Cell Receptors)
Edited by GIOVANNI DI SABATO

VOLUME 151. Molecular Genetics of Mammalian Cells
Edited by MICHAEL M. GOTTESMAN

VOLUME 152. Guide to Molecular Cloning Techniques
Edited by SHELBY L. BERGER AND ALAN R. KIMMEL

VOLUME 153. Recombinant DNA (Part D)
Edited by RAY WU AND LAWRENCE GROSSMAN

VOLUME 154. Recombinant DNA (Part E)
Edited by RAY WU AND LAWRENCE GROSSMAN

VOLUME 155. Recombinant DNA (Part F)
Edited by RAY WU

VOLUME 156. Biomembranes (Part P: ATP-Driven Pumps and Related Transport: The Na,K-Pump)
Edited by SIDNEY FLEISCHER AND BECCA FLEISCHER

VOLUME 157. Biomembranes (Part Q: ATP-Driven Pumps and Related Transport: Calcium, Proton, and Potassium Pumps)
Edited by SIDNEY FLEISCHER AND BECCA FLEISCHER

VOLUME 158. Metalloproteins (Part A)
Edited by JAMES F. RIORDAN AND BERT L. VALLEE

VOLUME 159. Initiation and Termination of Cyclic Nucleotide Action
Edited by JACKIE D. CORBIN AND ROGER A. JOHNSON

VOLUME 160. Biomass (Part A: Cellulose and Hemicellulose)
Edited by WILLIS A. WOOD AND SCOTT T. KELLOGG

VOLUME 196. Molecular Motors and the Cytoskeleton
Edited by RICHARD B. VALLEE

VOLUME 197. Phospholipases
Edited by EDWARD A. DENNIS

VOLUME 198. Peptide Growth Factors (Part C)
Edited by DAVID BARNES, J. P. MATHER, AND GORDON H. SATO

VOLUME 199. Cumulative Subject Index Volumes 168–174, 176–194

VOLUME 200. Protein Phosphorylation (Part A: Protein Kinases: Assays, Purification, Antibodies, Functional Analysis, Cloning, and Expression)
Edited by TONY HUNTER AND BARTHOLOMEW M. SEFTON

VOLUME 201. Protein Phosphorylation (Part B: Analysis of Protein Phosphorylation, Protein Kinase Inhibitors, and Protein Phosphatases)
Edited by TONY HUNTER AND BARTHOLOMEW M. SEFTON

VOLUME 202. Molecular Design and Modeling: Concepts and Applications (Part A: Proteins, Peptides, and Enzymes)
Edited by JOHN J. LANGONE

VOLUME 203. Molecular Design and Modeling: Concepts and Applications (Part B: Antibodies and Antigens, Nucleic Acids, Polysaccharides, and Drugs)
Edited by JOHN J. LANGONE

VOLUME 204. Bacterial Genetic Systems
Edited by JEFFREY H. MILLER

VOLUME 205. Metallobiochemistry (Part B: Metallothionein and Related Molecules)
Edited by JAMES F. RIORDAN AND BERT L. VALLEE

VOLUME 206. Cytochrome P450
Edited by MICHAEL R. WATERMAN AND ERIC F. JOHNSON

VOLUME 207. Ion Channels
Edited by BERNARDO RUDY AND LINDA E. IVERSON

VOLUME 208. Protein–DNA Interactions
Edited by ROBERT T. SAUER

VOLUME 209. Phospholipid Biosynthesis
Edited by EDWARD A. DENNIS AND DENNIS E. VANCE

VOLUME 210. Numerical Computer Methods
Edited by LUDWIG BRAND AND MICHAEL L. JOHNSON

VOLUME 211. DNA Structures (Part A: Synthesis and Physical Analysis of DNA)
Edited by DAVID M. J. LILLEY AND JAMES E. DAHLBERG

VOLUME 212. DNA Structures (Part B: Chemical and Electrophoretic Analysis of DNA)
Edited by DAVID M. J. LILLEY AND JAMES E. DAHLBERG

VOLUME 213. Carotenoids (Part A: Chemistry, Separation, Quantitation, and Antioxidation)
Edited by LESTER PACKER

VOLUME 214. Carotenoids (Part B: Metabolism, Genetics, and Biosynthesis)
Edited by LESTER PACKER

VOLUME 215. Platelets: Receptors, Adhesion, Secretion (Part B)
Edited by JACEK J. HAWIGER

VOLUME 216. Recombinant DNA (Part G)
Edited by RAY WU

VOLUME 217. Recombinant DNA (Part H)
Edited by RAY WU

VOLUME 218. Recombinant DNA (Part I)
Edited by RAY WU

VOLUME 219. Reconstitution of Intracellular Transport
Edited by JAMES E. ROTHMAN

VOLUME 220. Membrane Fusion Techniques (Part A)
Edited by NEJAT DÜZGÜNEŞ

VOLUME 221. Membrane Fusion Techniques (Part B)
Edited by NEJAT DÜZGÜNEŞ

VOLUME 222. Proteolytic Enzymes in Coagulation, Fibrinolysis, and Complement Activation (Part A: Mammalian Blood Coagulation Factors and Inhibitors)
Edited by LASZLO LORAND AND KENNETH G. MANN

VOLUME 223. Proteolytic Enzymes in Coagulation, Fibrinolysis, and Complement Activation (Part B: Complement Activation, Fibrinolysis, and Nonmammalian Blood Coagulation Factors)
Edited by LASZLO LORAND AND KENNETH G. MANN

VOLUME 224. Molecular Evolution: Producing the Biochemical Data
Edited by ELIZABETH ANNE ZIMMER, THOMAS J. WHITE, REBECCA L. CANN, AND ALLAN C. WILSON

VOLUME 225. Guide to Techniques in Mouse Development
Edited by PAUL M. WASSARMAN AND MELVIN L. DEPAMPHILIS

VOLUME 226. Metallobiochemistry (Part C: Spectroscopic and Physical Methods for Probing Metal Ion Environments in Metalloenzymes and Metalloproteins)
Edited by JAMES F. RIORDAN AND BERT L. VALLEE

VOLUME 227. Metallobiochemistry (Part D: Physical and Spectroscopic Methods for Probing Metal Ion Environments in Metalloproteins)
Edited by JAMES F. RIORDAN AND BERT L. VALLEE

VOLUME 228. Aqueous Two-Phase Systems
Edited by HARRY WALTER AND GÖTE JOHANSSON

VOLUME 229. Cumulative Subject Index Volumes 195–198, 200–227

Volume 230. Guide to Techniques in Glycobiology
Edited by William J. Lennarz and Gerald W. Hart

Volume 231. Hemoglobins (Part B: Biochemical and Analytical Methods)
Edited by Johannes Everse, Kim D. Vandegriff, and Robert M. Winslow

Volume 232. Hemoglobins (Part C: Biophysical Methods)
Edited by Johannes Everse, Kim D. Vandegriff, and Robert M. Winslow

Volume 233. Oxygen Radicals in Biological Systems (Part C)
Edited by Lester Packer

Volume 234. Oxygen Radicals in Biological Systems (Part D)
Edited by Lester Packer

Volume 235. Bacterial Pathogenesis (Part A: Identification and Regulation of Virulence Factors)
Edited by Virginia L. Clark and Patrik M. Bavoil

Volume 236. Bacterial Pathogenesis (Part B: Integration of Pathogenic Bacteria with Host Cells)
Edited by Virginia L. Clark and Patrik M. Bavoil

Volume 237. Heterotrimeric G Proteins
Edited by Ravi Iyengar

Volume 238. Heterotrimeric G-Protein Effectors
Edited by Ravi Iyengar

Volume 239. Nuclear Magnetic Resonance (Part C)
Edited by Thomas L. James and Norman J. Oppenheimer

Volume 240. Numerical Computer Methods (Part B)
Edited by Michael L. Johnson and Ludwig Brand

Volume 241. Retroviral Proteases
Edited by Lawrence C. Kuo and Jules A. Shafer

Volume 242. Neoglycoconjugates (Part A)
Edited by Y. C. Lee and Reiko T. Lee

Volume 243. Inorganic Microbial Sulfur Metabolism
Edited by Harry D. Peck, Jr., and Jean LeGall

Volume 244. Proteolytic Enzymes: Serine and Cysteine Peptidases
Edited by Alan J. Barrett

Volume 245. Extracellular Matrix Components
Edited by E. Ruoslahti and E. Engvall

Volume 246. Biochemical Spectroscopy
Edited by Kenneth Sauer

Volume 247. Neoglycoconjugates (Part B: Biomedical Applications)
Edited by Y. C. Lee and Reiko T. Lee

VOLUME 266. Computer Methods for Macromolecular Sequence Analysis
Edited by RUSSELL F. DOOLITTLE

VOLUME 267. Combinatorial Chemistry
Edited by JOHN N. ABELSON

VOLUME 268. Nitric Oxide (Part A: Sources and Detection of NO; NO Synthase)
Edited by LESTER PACKER

VOLUME 269. Nitric Oxide (Part B: Physiological and Pathological Processes)
Edited by LESTER PACKER

VOLUME 270. High Resolution Separation and Analysis of Biological Macromolecules (Part A: Fundamentals)
Edited by BARRY L. KARGER AND WILLIAM S. HANCOCK

VOLUME 271. High Resolution Separation and Analysis of Biological Macromolecules (Part B: Applications)
Edited by BARRY L. KARGER AND WILLIAM S. HANCOCK

VOLUME 272. Cytochrome P450 (Part B)
Edited by ERIC F. JOHNSON AND MICHAEL R. WATERMAN

VOLUME 273. RNA Polymerase and Associated Factors (Part A)
Edited by SANKAR ADHYA

VOLUME 274. RNA Polymerase and Associated Factors (Part B)
Edited by SANKAR ADHYA

VOLUME 275. Viral Polymerases and Related Proteins
Edited by LAWRENCE C. KUO, DAVID B. OLSEN, AND STEVEN S. CARROLL

VOLUME 276. Macromolecular Crystallography (Part A)
Edited by CHARLES W. CARTER, JR., AND ROBERT M. SWEET

VOLUME 277. Macromolecular Crystallography (Part B)
Edited by CHARLES W. CARTER, JR., AND ROBERT M. SWEET

VOLUME 278. Fluorescence Spectroscopy
Edited by LUDWIG BRAND AND MICHAEL L. JOHNSON

VOLUME 279. Vitamins and Coenzymes, Part I
Edited by DONALD B. MCCORMICK, JOHN W. SUTTIE, AND CONRAD WAGNER

VOLUME 280. Vitamins and Coenzymes, Part J
Edited by DONALD B. MCCORMICK, JOHN W. SUTTIE, AND CONRAD WAGNER

VOLUME 281. Vitamins and Coenzymes, Part K
Edited by DONALD B. MCCORMICK, JOHN W. SUTTIE, AND CONRAD WAGNER

VOLUME 282. Vitamins and Coenzymes, Part L
Edited by DONALD B. MCCORMICK, JOHN W. SUTTIE, AND CONRAD WAGNER

VOLUME 283. Cell Cycle Control
Edited by WILLIAM G. DUNPHY

VOLUME 284. Lipases (Part A: Biotechnology)
Edited by BYRON RUBIN AND EDWARD A. DENNIS

VOLUME 285. Cumulative Subject Index Volumes 263, 264, 266–289

VOLUME 286. Lipases (Part B: Enzyme Characterization and Utilization)
Edited by BYRON RUBIN AND EDWARD A. DENNIS

VOLUME 287. Chemokines
Edited by RICHARD HORUK

VOLUME 288. Chemokine Receptors
Edited by RICHARD HORUK

VOLUME 289. Solid Phase Peptide Synthesis
Edited by GREGG B. FIELDS

VOLUME 290. Molecular Chaperones
Edited by GEORGE H. LORIMER AND THOMAS BALDWIN

VOLUME 291. Caged Compounds
Edited by GERARD MARRIOTT

VOLUME 292. ABC Transporters: Biochemical, Cellular, and Molecular Aspects
Edited by Suresh V. Ambudkar and Michael M. Gottsman

VOLUME 293. Ion Channels (Part B)
Edited by P. MICHAEL CONN

VOLUME 294. Ion Channels (Part C) (in preparation)
Edited by P. MICHAEL CONN

VOLUME 295. Energetics of Biological Macromolecules (Part B)
Edited by GARY K. ACKERS AND MICHAEL L. JOHNSON

VOLUME 296. Neurotransmitter Transporters
Edited by SUSAN G. AMARA

VOLUME 297. Photosynthesis: Molecular Biology of Energy Capture (in preparation)
Edited by LEE MCINTOSH

VOLUME 298. Molecular Motors and the Cytoskeleton (Part B) (in preparation)
Edited by RICHARD B. VALLEE

VOLUME 299. Oxidants and Antioxidants (Part A) (in preparation)
Edited by LESTER PACKER

VOLUME 300. Oxidants and Antioxidants (Part B) (in preparation)
Edited by LESTER PACKER

VOLUME 301. Nitric Oxide: Biological and Antioxidant Activities (Part C) (in preparation)
Edited by LESTER PACKER

VOLUME 302. Green Fluorescent Protein (in preparation)
Edited by P. MICHAEL CONN

VOLUME 303. cDNA Preparation and Display (in preparation)
Edited by SHERMAN M. WEISSMAN

VOLUME 304. Chromatin (in preparation)
Edited by PAUL M. WASSERMAN AND ALAN P. WOLFFE

Section I

Purification of Transporter Proteins and cDNAs

A. Plasma Membrane Carriers
Articles 1 through 3

B. Vesicular Carriers
Articles 4 through 10

[1] Purification, Hydrodynamic Properties, and Glycosylation Analysis of Glycine Transporters

By CARMEN ARAGÓN and BEATRIZ LÓPEZ-CORCUERA

Introduction

Glycine acts as an inhibitory neurotransmitter in the central nervous system (CNS) of vertebrates, mainly in the spinal cord and the brain stem.[1] An additional role of glycine is the potentiation of glutamate excitatory action on postsynaptic *N*-methyl-D-aspartate (NMDA) receptors.[2] The reuptake of glycine into presynaptic nerve terminals or neighboring glial cells provides one way of clearing the neurotransmitter from the extracellular space, and constitutes an efficient mechanism by which its postsynaptic action can be terminated.[3,4] This process is carried out by a transport system that actively accumulates glycine and that is energized by the electrochemical gradient of sodium.[5,6] Two genes have been cloned encoding two different glycine transporters.[7,8] Their products are integral plasma membrane proteins which display similar functional properties, sodium and chloride dependence, electrogenicity, and kinetic parameters, although they are differentially inhibited by sarcosine (*N*-methylglycine) with the GLYT1 activity showing sensitivity to sarcosine and the GLYT2 activity being relatively resistant. Moreover, the distribution along the central nervous system of both proteins is different, and whereas GLYT1 presents a wide distribution, GLYT2 is more specifically located, being mainly found in the brain stem and the spinal cord, brain areas where glycine has an inhibitory physiological role.[9,10]

[1] R. W. Kerwin, and C. J. Pycock, *Eur. J. Pharmacol.* **54,** 93 (1979).
[2] J. W. Johnson and P. Ascher, *Nature* **325,** 529 (1987).
[3] G. A. R. Johnston and L. L. Iversen, *J. Neurochem.* **18,** 1951 (1971).
[4] M. J. Kuhar and M. A. Zarbin, *J. Neurochem.* **31,** 251 (1978).
[5] B. I. Kanner, *Biochim. Biophys. Acta* **726,** 293 (1983).
[6] B. I. Kanner and S. Schuldiner, *CRC Crit. Rev. Biochem.* **22,** 1 (1987).
[7] Q. R. Liu, H. Nelson, S. Mandiyan, B. López-Corcuera, and N. Nelson, *FEBS Lett.* **305,** 110 (1992).
[8] Q. R. Liu, B. López-Corcuera, S. Mandiyan, H. Nelson, and N. Nelson, *J. Biol. Chem.* **269,** 22802 (1993).
[9] F. Zafra, J. Gomeza, L. Olivares, C. Aragón, and C. Giménez, *Eur. J. Neurosci.* **7,** 1342 (1995).
[10] F. Zafra, C. Aragón, L. Olivares, N. C. Danbolt, C. Giménez, and J. Storm-Mathisen, *J. Neurosci.* **15,** 3952 (1995).

0076-6879/98 $25.00

This chapter describes a biochemical procedure to obtain a highly purified glycine transport activity with GLYT2-like properties from pig brain stem. The main advantage of this procedure is to provide the transporter in its native state, what readily permits the study of its biochemical, structural and functional properties.

Methods

Purification of Glycine Transporter

The purification of glycine transporter involves the solubilization of the transporter protein from plasma membrane vesicles followed by three chromatographic steps on phenyl-Sepharose, wheat germ agglutinin-Sepharose and hydroxylapatite columns and a final sucrose density gradient fractionation.[11] The whole procedure is performed at 4°, unless otherwise stated. The purification of the glycine transporter is monitored by sodium dodecyl sulfate–polyacrylamide gel electrophoresis (SDS–PAGE) and silver staining or immunoblotting as described later.[10,12] In every purification step glycine transport activity is checked after transporter reconstitution into liposomes following the procedure developed by Kanner, except for the use of higher ionic strength (0.6 *M*) in the reconstitution mixture.[11,13] The transport of glycine into the liposomes by the reconstituted protein is measured as described later. Protein content of each fraction is measured by the method of Peterson.[14]

Preparation of Crude Synaptic Plasma Membranes. Fresh pig brain stems, stored at −80°, are thawed and crude synaptic plasma membranes derived from the synaptosomal fraction are prepared as described.[13,15] Briefly, tissue is homogenized in 10 volumes of 0.3 *M* mannitol, 1 m*M* ethylenediaminetetraacetic acid (EDTA), 10 m*M* Tris–HCl, pH 7.4, and 0.1 m*M* phenylmethylsulfonyl fluoride (PMSF) using a Dounce homogenizer (all the chemicals are purchased from Sigma, St. Louis, MO). After centrifugation of the homogenate at 3000 *g* (5000 rpm in a Sorvall SS-34 rotor) for 10 min, a crude mitochondrial pellet is obtained by centrifugation of the supernatant for 20 min at 27,000 *g* (15,000 rpm in the same rotor). The pellet is subjected to osmotic shock by homogenation in a minimal volume of 5 m*M* Tris–HCl, 1 m*M* EDTA, pH 7.4, and stirred for 20 min in the same hypoosmotic solution. The suspension is then centrifuged at 27,000 *g* for 20 min, and

[11] B. López-Corcuera, J. Vázquez, and C. Aragón, *J. Biol. Chem.* **266,** 24809 (1991).
[12] B. López-Corcuera, R. Alcántara, J. Vázquez, and C. Aragón, *J. Biol. Chem.* **268,** 2239 (1993).
[13] B. López-Corcuera, and C. Aragón, *Eur. J. Biochem.* **181,** 519 (1989).
[14] G. L. Peterson, *Anal. Biochem.* **83,** 346 (1977).
[15] B. I. Kanner, *Biochemistry* **17,** 1207 (1978).

the pellet is resuspended in an isosmotic solution (145 mM NaCl, 1 mM $MgSO_4$, 0.5 mM EDTA, 10 mM HEPES–Tris, pH 7.4, 1% glycerol) to yield approximately 20 mg of protein per ml. Crude synaptic membranes are aliquoted and stored at −80° until used for purification.

Solubilization of Membranes. Membrane vesicles are thawed at 37° and immediately placed on ice. Ammonium sulfate (from a 100% saturated solutions adjusted to pH 7.4 with NH_4OH), sodium cholate [Sigma, St. Louis, MO, from a 20% (w/v) stock solution] and PMSF are added to a 15% (saturation), 2% (w/v) and 0.1 mM concentration, respectively. The mixture is stirred for 15 min on ice, and centrifuged at 135,000 g (45,000 rpm in a Beckman 70 Ti rotor) for 30 min. The resulting supernatant is supplemented with ammonium sulfate to yield a 30% final saturation and the cholate concentration is adjusted to 1.5%. The mixture is centrifuged at 27,000 g for 15 min. The latter supernatant (S_{30}) contains about 95% of the starting total glycine transport activity and 40% of the original proteins, which results in about 3-fold enrichment of glycine transport activity (Table I). The S_{30} fraction serves as a source to purify the glycine transporter as described later.

Note: The ability of different detergents to solubilize the glycine transporter in its active form has been tested using the reconstitution assay. Only sodium cholate and 3-[(3-cholamidopropyl)dimethylammonio]-1-propanesulfonic acid (CHAPS) [Sigma, used at 1.5% (w/v)] are able to yield active glycine transporter. The detergent/protein ratio (w/w) is a very critical parameter during the solubilization using cholate, being optimal at a value of 1. CHAPS can be used in a wider range of detergent/protein relationships.[13]

TABLE I
PURIFICATION OF GLYCINE TRANSPORTER[a]

Purification step	Activity (pmol/15 min)	Protein (mg)	Specific activity (pmol/mg protein/15 min)	Increase in specific activity[b] (-fold)
Membranes	13,997	832	17	1.0
S_{30}	13,477	277	49	2.9
Phenyl-Sepharose	3,530	40	88	5.2 (10)
Wheat germ agglutinin–Sepharose	827	2.5	331	19.5 (80)
Hydroxylapatite flow-through	603	0.83	727	42.8 (180)
Density gradient	23	0.025	920	54.1 (450)

[a] Reprinted with permission from B. López-Corcuera, J. Vázquez, and C. Aragón, *J. Biol. Chem.* **266,** 24809 (1991).

[b] Values in parentheses are the increase in specific activity corrected for the inactivation of transporter. Inactivation is estimated by calculating the fraction of activity recovered in the flow-through, washing, and eluted fractions, and normalizing for the protein recovery.

Phenyl-Sepharose Chromatography. The S_{30} fraction is applied to an 8-ml (3.1 × 1.8-cm) phenyl-Sepharose (PS) column (Pharmacia LKB Biotechnology Inc., Piscataway, NJ) equilibrated with 10 volumes of 30% saturated ammonium sulfate, 1.1% CHAPS, 0.1 m*M* PMSF, and 3.5% glycerol in 10 m*M* sodium phosphate, pH 6.6. Unbound proteins are removed by washing with 5 column volumes of equilibration buffer. Retained proteins are eluted with 1.1% CHAPS, 0.1 m*M* PMSF, and 20% glycerol in 10 m*M* sodium phosphate, pH 7.8 (flow rate: 4 ml/min). Under these high ionic strength conditions (30% ammonium sulfate saturation), the glycine transporter binds to the PS column and elutes after the removal of the salt within the first 2.5 elution volumes. The binding to the column partially inactivates the transporter since in the absence of glycerol only 20% of the applied activity is recovered (not shown). Thus, the presence of 20% (w/v) glycerol in the elution buffer helps minimize transporter inactivation and allows the recovery of 50–60% of the total activity loaded to the column. The PS eluate contains about 30% of the glycine transport activity and only 10–15% of the protein, which provides an additional 2- to 3-fold enrichment over the S_{30} fraction and a 5-fold purification over the starting material (Table I). During this chromatographic step, sodium cholate is exchanged for CHAPS, which better preserves the transport activity at higher detergent/protein ratios.[13] Active fractions of the PS eluate are pooled and concentrated by adding ammonium sulfate to a 50% saturation and centrifuged at 27,000 *g* for 15 min. The resulting pellet is resuspended in a minimal volume (6–10 times concentration) of buffer A (500 m*M* NaCl, 1.1% CHAPS, 0.1 m*M* sodium phosphate, pH 7.8) and used for the next purification step.

Note: The concentrative step provokes a 50% loss of transport activity, but it results in a higher yield in the subsequent chromatographic step.

Wheat Germ Agglutinin–Sepharose Affinity Chromatography. The pellet from the last purification step is loaded onto a 3-ml (2.7 × 1.2-cm) wheat germ agglutinin–Sepharose CL-4B (WGA) column (Pharmacia) equilibrated in buffer A. The elution is performed by using buffer A supplemented with 0.1 *M* *N*-acetylglucosamine (Sigma). Although the bulk of proteins (about 90%) is unretained by the WGA column, 40% of the activity is found in the eluate, resulting in an additional 4-fold increase in specific activity (Table I). A partial inactivation of the transporter also occurs during this step, although to a lesser extent compared to the PS column.

Hydroxylapatite Chromatography. Active fractions of the WGA column are pooled and applied to a 0.75-ml (0.7 × 1.2-cm) Bio-Gel HTP hydroxylapatite (HT) column (Bio-Rad, Richmond, CA) equilibrated in 100 m*M* NaCl, 1.1% CHAPS, 0.1 m*M* PMSF, 10 m*M* sodium phosphate, pH 6.6. The column is washed with 7 volumes of equilibration buffer. Glycine transport activity (about 75%) is recovered in the flow-through and washing fractions, with most of the proteins retained by the column (about 65%, as detected after elution using a 0.5 *M* sodium phosphate solution). This results in an additional 2-fold purification (Table I). By contrast to the earlier chromatographic steps, there is minimal inactivation of the transporter on the HT column, which does not bind the protein. Flow-through and washing fractions containing transport activity are pooled and concentrated (about 30-fold) in Centriprep 30 cartridges (Amicon, Danvers, MA), resulting in an approximately 60% recovery of protein and a 50% loss in transport activity.

Sucrose Density Gradient Fractionation. Concentrated fractions (250 μl) are layered on top of linear 5–20% sucrose density gradients (3 ml) in 1.1% CHAPS, 0.1 m*M* PMSF, 10 m*M* sodium phosphate, pH 6.6, and centrifuged at 372,000 *g* (95,000 rpm in a Beckman TL 100.3 rotor) for 3 h. Fractions of 250 μl are collected from the bottom of the tube. Glycine transport activity migrates as a single peak centered around the middle of the gradient, comprising about 70–75% of the applied activity. Regardless of the inactivation produced in the concentrative step, the density gradient fractionation yields a further 2-fold purification (Table I).

Glycine Transport Assay. Accumulation of glycine into liposomes after reconstitution is measured as previously described.[13,16] Briefly, 20 μl of proteoliposomes reconstituted in internal medium (145 m*M* potassium gluconate, 1 m*M* $MgSO_4$, 0.5 m*M* EDTA, 10 m*M* HEPES–Tris, pH 7.4, 1% glycerol), are incubated during 15 min at 25° with 280 μl of external solution (145 m*M* NaCl, 1 m*M* $MgSO_4$, 0.5 m*M* EDTA, 10 m*M* HEPES–Tris, pH 7.4, 1% glycerol), containing 2.5 μM valinomycin, and 0.3 μM [2-^{3}H]glycine.

Comment. Glycine transport activity in the reconstituted fractions of the final purification step shows functional properties identical to those displayed by the native transporter such as sodium and chloride dependence, electrogenicity and K_m (5 μM). The activity obtained in the final step is not sensitive to inhibition by sarcosine, a property that is displayed by the GLYT2 isoform.[8]

Global Yield of Purification Procedure. The final enrichment in specific glycine transport activity is between 50- and 60-fold (Table I). During the procedure, progressive inactivation of glycine transporter is observed, with the losses in activity mainly due to chemical interactions between transporter and chromatographic resins (PS and WGA steps) (Table I, values in parentheses). Taking into account the protein recovery in each purification step, it is estimated that the overall inactivation of glycine transporter accounts for an 8-fold loss in specific activity, and results in a purification

[16] F. Mayor, Jr., J. G. Marvizón, C. Aragón, C. Giménez, and F. Valdivieso, *Biochem. J.* **198,** 535 (1981).

factor of about 450-fold (Table I). The electrophoretic analysis of the different purification steps shows a progressive enrichment of a band with apparent molecular mass of 90–100 kDa.[11] Western blot of the purified fraction using sequence-directed antibodies shows that this band partially overlaps with the immunostained GLYT2 transporter, which indicates the presence of at least two proteins migrating with similar apparent molecular mass, one of which is the GLYT2 transporter (not shown). This preparation is free of GABA and glutamate transport activities,[11] but it is possible to immunodetect the GLYT1 transporter by using a sequence-specific antibody to this isoform (data not shown).[9] However, the glycine transport activity in the most purified fraction shows no sensitivity to inhibition by sarcosine, even when the inhibitor concentration is two or three orders of magnitude greater than the substrate concentration, and in this respect appears more similar to the GLYT2 isoform. This purification procedure yields a highly enriched GLYT2 transporter preparation (specific activity about 1 nmol glycine/mg protein/15 min) in a functionally active state.

Hydrodynamic Characterization of Glycine Transporter

The hydrodynamic characterization of the native glycine transporter involves the determination of some relevant physical properties such as size, density, sedimentation coefficient, and molecular weight. Partially purified glycine transporter is solubilized using the two detergents which are compatible with the functional state of the transporter: CHAPS and sodium cholate. The hydrodynamic properties of the glycine transporter are measured by a combination of size-exclusion chromatography and density gradient centrifugation.[12]

Preparation of Purified Glycine Transporter–Detergent Complexes. Purification of glycine transporter is performed in the presence of 1.1% CHAPS as described before,[11] except for the omission of the final sucrose density gradient fractionation. Purified transporter–cholate complexes are obtained using the same procedure, replacing CHAPS by 1.1% sodium cholate. Active fractions from the last chromatographic step (hydroxylapatite column, HT) are immediately used for the hydrodynamic characterization.

Note: The purification (fold) obtained using sodium cholate is similar to that attained with CHAPS for the corresponding step.[11]

Gel-Filtration Chromatography. Active HT fractions (300 μg of protein) are applied to a 1.5 × 40-cm column of Sepharose CL-6B (Pharmacia) preequilibrated with 10 volumes of 100 mM NaCl, 20 mM sodium phosphate, 0.1 mM PMSF, pH 6.6, and either 1.1% CHAPS or 1.1% cholate, and fractions (1 ml) are collected using the same buffer (flow rate: 0.5 ml/

min). Glycine transporter is identified by measuring glycine uptake after reconstitution,[11,13] and by immunoblotting as described later.[10,12] Protein concentration is determined in each fraction.[14] The column is calibrated using standard proteins of known Stokes radii that are chromatographed in identical experimental conditions. Urease (6.2 nm), catalase (5.2 nm), glucose oxidase (4.3 nm), bovine serum albumin (3.5 nm), and peroxidase (3.1 nm) are used for calibration. Blue dextran 2000 and bromphenol blue are used to determine void volume (V_o) and total volume (V_c) of the column, respectively. Retention of proteins in the matrix is evaluated by the partition coefficient (K_d) defined as:

$$K_d = (V_e - V_o)/(V_c - \mathrm{V}_o) \quad (1)$$

V_e being the elution volume of the protein. Calibration curves are constructed by plotting the Stokes radii versus the inverse error function of $1 - K_d[\mathrm{erf} - 1(1 - K_d]$, according to Ackers' method.[17]

Sucrose Density Gradient Centrifugation. Linear gradients (12 ml) of 5–20% (w/v) sucrose in H_2O and D_2O are prepared at room temperature in 20 m*M* sodium phosphate, 0.1 m*M* PMSF, pH 6.6, and either 1.1% CHAPS or 1.1% cholate with the aid of a gradient forming unit. The gradients are allowed to cool at 4° and the purified transporter (500 μl of the HT column fraction, about 500 μg protein) is layered on the top and centrifuged at 190,000 *g* (39,500 rpm in a Beckman SW 40 Ti rotor) for 18 h. The gradients are standardized by sedimenting 500 μg of standard proteins of known sedimentation coefficients and partial specific volumes on parallel gradients centrifuged in the same run. Malate dehydrogenase (4.32S), lactate dehydrogenase (6.95S), alcohol dehydrogenase (7.4S), and catalase (11.4S) are used as standards, all having the same partial specific volume (0.73 ml/g). Once the centrifugation is completed, fractions (350 μl) are collected from the bottom of the tubes. The glycine transporter is identified by measuring specific glycine uptake activity after reconstitution,[11,13] and by immunoblotting as described later.[10,12] Migration of marker proteins is measured by determining protein concentration in each fraction of the standard gradients.[14] The refractive index of each fraction is checked in every gradient to ensure linearity of the sucrose concentration. The density of each individual fraction is determined by the direct weighing of fixed volumes of sample.

Calculations of Hydrodynamic Parameters. The partial specific volumes, ν_c, and the sedimentation coefficients, $s_{20,w}$, of the transporter–detergent complexes are determined from the apparent sedimentation coefficients in H_2O (s_H) or D_2O (s_D) obtained on interpolation of their migration relative

[17] G. K. Ackers, *J. Biol. Chem.* **242,** 3237 (1976).

to the calibration curves. The calibration lines are constructed by plotting the migration of the marker proteins in the H_2O and D_2O gradients versus their known sedimentation coefficients. The partial specific volume, ν_c, and the sedimentation coefficient, $s_{20,w}$, are calculated according to the method of Clarke,[18] assuming that the partial specific volume of the protein component is 0.73 ml/g,[19] and that the same amount of detergent is bound to the transporter in H_2O and D_2O.[18] The following equation is used to calculate the partial specific volume, ν_c, of the protein–detergent complexes:

$$\nu_c = \frac{\left(\dfrac{s_H \eta_H}{s_D \eta_D}\right) - 1}{\rho_D \left(\dfrac{s_H \eta_H}{s_D \eta_D}\right) - \rho_H} \qquad (2)$$

where the subscripts H and D refer to sucrose density gradients in H_2O and D_2O, s_H and s_D are the experimentally determined apparent sedimentation coefficients, η is the viscosity of the solvent, and ρ is the density of the solvent, all calculuated as r_{average} as described by Clarke.[18] The values ν_c, η_H, and ρ_H for the transporter in detergent solutions are used to calculate the true sedimentation coefficient in water, $s_{20,w}$, for the transporter–detergent complexes. To make these calculations, the following relationship is used:

$$s_{20,w} = \frac{s_H \left(\dfrac{\eta_H}{\eta_{20,w}}\right)(1 - \nu_c \rho_{20,w})}{1 - \nu_c \rho_H}. \qquad (3)$$

The molecular weights of the glycine transporter–CHAPS and glycine transporter–cholate complexes (M_c) are calculated from their Stokes radii (r_s), partial specific volumes (ν_c), and sedimentation coefficients ($s_{20,w}$) using Svedverg's equation:

$$M_c = \frac{s_{20,w} r_s N_A 6\pi \eta_{20,w}}{1 - \nu_c \rho_{20,w}}, \qquad (4)$$

where N_A is Avogadro's number, $\eta_{20,w}$ is the viscosity of water at 20° (0.01002 g/cm · s), and $\rho_{20,w}$ is the density of water at 20° (0.99823 g/ml).

The weight fraction of protein in the complexes (X_p) is calculated from the partial specific volumes of detergent (ν_d), protein (ν_p) (0.73 ml/g),[19] and detergent–protein complex (ν_c) using the equation:

$$X_p = (\nu_c - \nu_d)/(\nu_p - \nu_d) \qquad (5)$$

[18] S. Clarke, *J. Biol. Chem.* **250,** 5459 (1975).

[19] R. G. Martin and B. W. Ames, *J. Biol. Chem.* **236,** 1372 (1961).

TABLE II
HYDRODYNAMIC PROPERTIES OF PURIFIED GLYCINE TRANSPORTER[a]

Parameter	CHAPS	Cholate
Stokes radius (r_s)	5.5 nm	6.0 nm
Sedimentation coefficient ($s_{20,\omega}$)	4.6 S	6.6 S
Partial specific volume of protein–detergent complex (ν_c)	0.75 ml/g	0.73–0.75 ml/g
Partial specific volume of detergent (ν_d)	0.81 ml/g[b]	0.75 ml/g[c]
Molecular weight of protein–detergent complex (M_c)	115,000	160,000[d]
Weight fraction of protein in the complex (X_p)	0.75	ND[e]
Molecular weight of protein (M_p)	86,000	ND

[a] Reprinted with permission from B. López-Corcuera, R. Alcántara, J. Vázquez, and C. Aragón, *J. Biol. Chem.* **268,** 2239 (1993).
[b] See L. M. Hjelmeland, D. W. Nebert, and J. C. Osborne, Jr., *Anal. Biochem.* **130,** 72 (1983).
[c] See D. M. Small, *Adv. Chem. Ser.* **84,** 31 (1968).
[d] Calculated assuming that $\nu_c = 0.73$ ml/g.
[e] ND, Not determined.

The ν_d values for cholate and CHAPS are taken from Hjelmeland *et al.*[20] and Small,[21] respectively.

Finally, the molecular weight of the protein alone (M_p) is calculated from that of the complex (M_c) by using

$$M_p = M_c X_p. \tag{6}$$

Hydrodynamic Parameters of Glycine Transporter. In the experimental conditions described for the gel filtration chromatography (see previous discussion), glycine transport activity is found in a well defined peak that roughly coincides with the protein peaks observed in the presence of either cholate or CHAPS. Stokes radii of 6.0 nm and 5.5 nm are found for the transporter–cholate and transporter–CHAPS complexes, respectively, by interpolation in the calibration plots (Table II). The higher Stokes radius of the transporter–cholate complex suggests a higher proportion of cholate in the protein/detergent particles.

The sedimenting behavior of the transporter–CHAPS complex in the described density gradient centrifugation (see earlier discussion) yields apparent sedimentation coefficients of 4.5S in H_2O and D_2O, respectively. These predictable differences due to the partial specific volume of CHAPS (0.81 ml/g),[20] which is different from that of the proteins used as standards (0.73 ml/g),[19] allow the determination of both the partial specific volume (0.75 ml/g) and the true sedimentation coefficient in water (4.6S) of the transporter–CHAPS complex (Table II). In contrast, the transporter–cholate complex gives indistinguishable apparent sedimentation coefficients

[20] L. M. Hjelmeland, D. W. Nebert, and J. C. Osborne, Jr., *Anal. Biochem.* **130,** 72 (1983).
[21] D. M. Small, *Adv. Chem. Ser.* **84,** 31 (1968).

in H_2O and D_2O (6.6S), because of the value of the partial specific volume of cholate (0.75 ml/g),[21] which is almost identical to that of the proteins used as standards (0.73 ml/g); therefore, it is not possible to determine the exact partial specific volume of transporter–cholate complex (we could only conclude that it lies between 0.73 and 0.75 ml/g, i.e., intermediate between the values of protein and detergent) (Table II). However, since the apparent sedimentation coefficient obtained by interpolation in the calibration curves corresponds, in this condition, to the true sedimentation coefficient in water,[18] this value (6.6S) is used to estimate the molecular weight of the complex (see later discussion) (Table II).

The molecular weight of the transporter–detergent complexes calculated by using Eq. (4) are 115,000 and 160,000 for transporter–CHAPS and transporter–cholate complexes, respectively (Table II). The differences in molecular weight between the two complexes could arise from conformational differences between the transporter in each detergent resulting in a distinct molecular shape and an altered hydrodynamic behavior, or, alternatively, from different numbers of detergent monomers bound per transporter molecule. Because of the difficulty in calculating the weight fraction of protein in the transporter–cholate complex (see later discussion), it is not possible to determine which of these two possibilities accounts for the observed molecular weight differences.

The weight fraction of protein in the transporter–CHAPS complex obtained from Eq. (5) (Table II) indicates that about 75% of the molecular weight of the complex is attributable to the protein moiety. The molecular mass of the glycine transporter (M_p), calculated from that of the complex by using Eq. (6), is 86 kDa (Table II). For the transporter–cholate complex it is not feasible to calculate the weight contributions of detergent and protein since the partial specific volume of the complex, ν_c, cannot be determined. However, a reasonable estimation can be made assuming that the protein component is responsible for 60–80% in weight of the complex. This would yield a molecular mass of the transporter of 96–128 kDa (Table II).

Comment. The calculated molecular mass for the glycine transporter is in good agreement with the size of the GLYT2 isoform of glycine transporter (80–100 kDa). This result also suggests that in its native state, the transporter does not act as an homooligomeric protein.

Deglycosylation of Glycine Transporter

The deglycosylation of the GLYT2 transporter allows the contribution of the carbohydrate moieties to the structure and function of the transporter to be assessed. The structure of the carbohydrate residues and their linkages are analyzed by taking advantage of the specificities of different glycosidases. Partially purified glycine transporter is treated with glycosidases under denaturing conditions and the apparent molecular masses of the deglycosylated proteins are analyzed by SDS–PAGE and Western blotting.[22] The contribution of the oligosaccharide chain to the functional state of the transporter is tested after enzymatic deglycosylation of the transporter reconstituted into liposomes and further glycine transport activity determination.[11,13]

Structural Analysis of Carbohydrate Moiety. Several glycosidic enzymes are used to assess the structural characteristics of the oligosaccharide bound to the glycine transporter. Endo-β-*N*-acetylglucosaminidase F (EC 3.2.1.96, Endo F) generally cleaves high mannose and certain complex carbohydrate structures.[23] *O*-Glycopeptide endo-D-galactosyl-*N*-acetyl-β-galactosaminohydrolase (EC 3.2.1.97, *O*-glycosidase) cleaves carbohydrates linked to the protein backbone through serine or threonine.[24] Sialidase (EC 3.2.1.18, neuraminidase), an exoglycosidase, is able to remove terminal sialic acid residues.[25] Peptide: *N*-glycosidase F (EC 3.5.1.52, PNGase F), hydrolyzes most types of *N*-linked carbohydrate groups from glycoproteins at a point between the di-*N*-acetylglucosamine core and the asparagine to which the carbohydrate is linked.[26] The removal of a carbohydrate moiety from the purified glycine transporter can be detected as a shift in the electrophoretical mobility of the glycine transporter protein either on SDS–PAGE or Western blot analysis.[22] The specificity of the enzymes that are able to yield a mobility shift of the transporter can provide insight into the structure of the sugar residues of the transporter.

Purification and Reconstitution of Glycine Transporter. The purification of the glycine transporter is performed in the presence of 1.1% CHAPS as described earlier, except for the omission of the final sucrose gradient fractionation. Active fractions from the hydroxylapatite column are immediately used for reconstitution into liposomes,[11,13,22] and for enzymatic deglycosylation, as described later.

Glycosidase Treatment. Active HT fractions (about 16 μg protein) are diluted with 1 volume of 2× denaturation buffer (50 m*M* sodium citrate, 100 m*M* sodium phosphate, 50 m*M* sodium chloride, 0.05% SDS, 100 μg/ml PMSF, and 1 μg/ml pepstatin A, pH 5.0) and denatured at 37 for 4 h.

[22] E. Núñez and C. Aragón, *J. Biol. Chem.* **269,** 16920 (1994).
[23] T. H. Plummer, Jr., J. H. Elder, S. Alexander, W. W. Phelan, and A. L. Tarentino, *J. Biol. Chem.* **259,** 10700 (1984).
[24] Y. Endo and A. Kobata, *J. Biochem.* **80,** 1 (1976).
[25] Y. Uchida, Y. Tsukada, and T. Sugimori, *J. Biochem.* **86,** 1573 (1979).
[26] A. L. Tarentino, C. M. Gómez, and T. H. Plummer, Jr., *Biochemistry* **24,** 4665 (1985).

Aliquots of the denatured transporter (about 3 μg protein) are incubated with 1 volume of reaction buffer and the indicated amount of the corresponding glycosidase. Endo F from *Flavobacterium meningosepticum* (Boehringer, Mannheim, Germany), 1.7 milliunits (mU), reaction buffer: 100 m*M* sodium phosphate, pH 6.1, 50 m*M* EDTA, 2% 2-mercaptoethanol and 1% Nonidet P-40 (NP-40). Neuraminidase from *Arthrobacter ureafaciens* (Oxford glycosystems, Abingdon, U.K.), 20 mU, reaction buffer: denaturation buffer. *O*-glycosidase from *Diplococcus pneumoniae* (Boehringer, Mannheim, Germany), 10 mU, reaction buffer: 100 m*M* sodium phosphate, pH 74, 50 m*M* EDTA, and 2% Nonidet P-40. PNGase F (recombinant enzyme, New England Biolabs, Beverly, MA), 2.2×10^{-3} mU, reaction buffer: 100 m*M* sodium phosphate, pH 7.4, 50 m*M* EDTA, 1% 2-mercaptoethanol, and 2% Nonidet P-40. Control preparations are treated in the same way, but omitting the enzyme. The samples are incubated overnight at 37° and the reactions are stopped by adding 1 volume of 2× SDS–PAGE sample buffer. The gels are analyzed by Western blot, as described later.

Note: Enzymatic units are defined as micromoles of substrate transformed per minute. Commercial suppliers define their own enzymatic units.

Electrophoresis and Blotting. Samples are subjected to SDS–PAGE according to the method of Laemmli,[27] using a 4% stacking gel and 10% resolving gel. The gels are run slowly (overnight) at constant current. Samples are electrotransferred onto a nitrocellulose membrane in a semidry electroblotting system (Life Technologies, Inc., Paisley, U.K.) at 1.2 mA/cm^2 for 2 h, in transfer buffer consisting of 192 m*M* glycine in 25 m*M* Tris–HCl, pH 7.5. After blocking nonspecific protein binding with 3% (w/v) nonfat milk protein in 10 m*M* Tris–HCl, pH 7.5, 150 m*M* NaCl for 4 h at 25°, the blot is probed either with a 1 : 100 dilution of a polyclonal antibody made against the purified transporter,[12] or with a 1 : 200–1 : 500 dilution of a GLYT2 sequence-specific antibody,[9,10] and incubated overnight at 4°. Filters are washed and bound antibodies detected with a peroxidase-linked anti-rabbit IgG. Bands are visualized with the ECL (enhanced chemiluminescence) detection method (Amersham Corp., Bucks, U.K.).

Glycan Detection. The glycan chains linked to the glycine transporter are detected by incorporation of digoxigenin into the sugars followed by Western blotting and visualization using a high-affinity antidigoxigenin antibody conjugated to alkaline phosphatase.[28] Partially purified glycine transporter is treated with the different glycosidases as described before, then transferred to nitrocellulose membranes, and the efficiency of the

[27] U. K. Laemmli, *Nature* **227,** 680 (1970).
[28] A. Haselbeck and W. Hosel, *Glycoconjugate J.* **7,** 63 (1990).

deglycosylation is monitored by using a commercial kit for *N*- and *O*-type glycan chain detection (Boehringer Mannheim), following the manufacturer's instructions.

Structure of Sugar Residues on the Glycine Transporter. The apparent molecular weight of the purified glycine transporter is reduced by 30% after the treatment with PNGase F (Fig. 1) and the glycan moiety of the transporter is not detected in PNGase F-treated samples (data not shown).[22] This indicates that the carbohydrates are *N*-linked to the glycine transporter. Furthermore, *O*-glycosidase treatment appears to have no effect on the migration of the carrier (Fig. 1). The neuraminidase-treated transporter is a partially glycosylated protein (as revealed by sugar detection, data not shown[22]) with a slightly decreased apparent molecular mass (5–10%) (Fig. 1). This indicates the presence of some sialic acid residues on the sugar chain of the transporter. Endo F is not able to deglycosylate the transporter, which suggests that high mannose carbohydrate groups are not associated with the glycine transporter. Combinations of different glycosidases do not have synergistic effects.

In conclusion, the carbohydrate moiety of the glycine transporter is most probably a tri- or tetraantennary complex structure containing terminal sialic acid residues. The presence of *N*-acetylglucosamine can also be

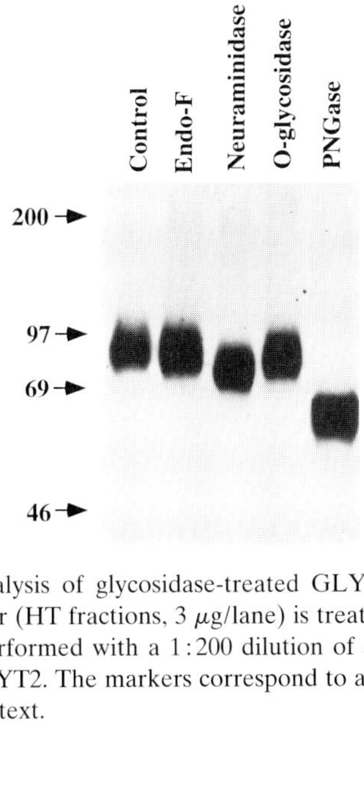

FIG. 1. Western blot analysis of glycosidase-treated GLYT2 transporter. Purified and denatured glycine transporter (HT fractions, 3 μg/lane) is treated with the indicated glycosidases. Immunoblotting is performed with a 1:200 dilution of an affinity purified sequence-specific antibody against GLYT2. The markers correspond to apparent molecular masses (in kilodaltons). For details see text.

inferred from the binding of the transporter to the WGA column (see earlier discussion).[11]

Glycosidase Treatment under Native Conditions. Active HT column-purified fractions are treated with glycosidases after reconstitution into liposomes.[11,13,22] Neuraminidase and recombinant PNGase F expressed in *Escherichia coli* are used for digestion of native glycine transporter at a 10-fold higher glycosidase/protein relationship than those used under denaturing conditions (see earlier discussion). No detergent or denaturing agent are used during the deglycosylation processes. The reaction mixture (110 μl, final volume), is incubated for 1, 3, and 5 h at 37°. After treatment, proteoliposomes are washed with 30 volumes of ice-cold liposome internal medium (145 mM potassium gluconate, 1 mM $MgSO_4$, 0.5 mM EDTA, 10 mM HEPES–Tris, 1% glycerol, pH 7.4) and centrifuged at 372,000 g (95,000 rpm in a Beckman TL 100.3 rotor) for 20 min. The resulting pellets are washed by three resuspension–centrifugation cycles and finally resuspended in internal medium (see earlier discussion). The proteoliposomes are immediately used for glycine transport assays or for SDS–PAGE and immunoblotting, as described previously.

Note: The recombinant PNGase F form is more convenient than the purified enzyme because it allows shorter incubation times, avoiding losses of reconstituted transporter activity. The same treatment is performed using purified transporter in solution in order to confirm that the accessibility to the cleavage sites is unchanged after reconstitution. Additionally, the efficiency of the protein insertion into liposomes can be reduced following deglycosylation; therefore, the enzymatric treatment is always performed on the reconstituted protein.

Functional Analysis of Carbohydrate Moiety. The removal of sialic acid residues from the glycine transporter by neuraminidase does not modify the glycine transport activity at any incubation time, indicating a limited involvement of these residues in the function of the transporter. However, PNGase F produces a time-dependent inhibition of the transport (not shown[22]) that parallels the time-dependent decrease in the apparent molecular mass of the transporter (Fig. 1, and data not shown). The maximal inhibition is reached after a 5-h treatment and accounts for a 70% decrease in glycine transport activity.[22] These results indicate a functional involvement of the carbohydrate moiety in glycine transport activity either by stabilizing the conformation of the transporter or by participating in ligand binding interactions.

Comment. A time-dependent impairment of the glycine uptake is also observed after deglycosylation of the reconstituted GLYT2 protein obtained by transient expression in mammalian cells (data not shown). Total deglycosylation of the expressed protein by PNGase F also produces a decrease in uptake activity (50%), the small differences being probably attributable to the specific glycosylation pattern of every cell type.[29]

Acknowledgments

The authors thank Dr. Jesús Vázquez and Enrique Núñez for valuable collaboration. This research was supported by the Spanish DGICYT, Boehringer Ingelheim España, S.A., the European Union BIOMED program (BMH1-CT93-1110 and BMH4-CT95-0571), and Fundación Ramón Areces.

[29] A. Patel, G. Uhl, and M. J. Kuhar, *J. Neurochem.* **61,** 496 (1993).

[2] Expression Cloning using *Xenopus laevis* Oocytes

By MICHAEL F. ROMERO, YOSHIKATSU KANAI, HIROMI GUNSHIN, and MATTHIAS A. HEDIGER

Introduction

How do cells move ions and nutrients from their environment to their interior, or vice versa? How does a cell respond to external stimuli such as hormones or changes in extracellular ion concentration? Many of these capabilities of cells are carried out by specialized membrane proteins such as transporters, channels, and receptors. Knowledge of the molecular structure and regulation of these proteins is extraordinarily valuable for understanding intact systems such as gastrointestinal absorption, hearing, renal filtration, and vision. The hydrophobic nature of integral membrane proteins has been a major hindrance, precluding their isolation in a form suitable for amino acid sequencing or immunization. Consequently, nucleic acid probes and antibodies for the screening of cDNA libraries were and are often not available. Expression cloning using *Xenopus laevis* oocytes[1] is a unique technique that has resolved many problems of isolating integral membrane proteins because cDNA clones are selected directly for their ability to induce the desired function in oocytes and no probes or antibodies are required for screening cDNA libraries.

For the experimental biologist, the *Xenopus* oocyte as an expression system offers several unique features.[2] The oocyte is a relatively large cell (~0.8–1.3 mm diameter) making it convenient for RNA microinjection

[1] M. A. Hediger, M. J. Coady, T. S. Ikeda, and E. M. Wright, *Nature* **330,** 379 (1987).
[2] B. K. Kay and H. B. Peng, *Methods in Cell Biology—Xenopus laevis: Practical Uses in Cell and Molecular Biology* **36,** 1 (1991).

0076-6879/98 $25.00

and two-electrode voltage clamp studies. In addition, the cell is in a relatively "silent state" and thus exhibits low levels of endogenous transport activity. Since it is primed for high volume protein synthesis before and following fertilization, the oocyte can readily synthesize membrane proteins from foreign RNAs microinjected into the cytoplasm.[3] There are several excellent reviews of *Xenopus*,[2] their oocytes,[4] and the properties of oocytes.[5] Additionally, there are many excellent *Xenopus* information resources available on the World Wide Web.[5a]

In many cases, this expression cloning strategy has worked very well and a variety of channel, enzymes, and transporter cDNAs have been isolated in this manner (see Table I, Refs. 6–19). Expression cloning with

[3] J. B. Gurdon, C. D. Lane, H. R. Woodland, and G. Marbaix, *Nature* **233,** 177 (1971).

[4] J. N. Dumont, *J. Morph.* **136,** 153 (1972).

[5] N. Dascal, *CRC Crit. Rev. Biochem.* **22,** 317 (1987).

[5a] *Xenopus* information resources on the World Wide Web:
http://vize222.zo.utexas.edu/
http://xenopus.com/
http://gto.ncsa.uiuc.edu/pingleto/xenopus.html
http://www.luc.edu/depts/biology/dev.htm
http://www.average.org/~pruss/Xlaevis.scripted.html
http://www.widomaker.com/~kea/edward/edward.html
http://www.nascofa.com/Science/Xenopus.html
http://www.wi.mit.edu/sive/home.html#400
http://www.bih.harvard.edu:80/sokol-lab/
http://sdb.bio.purdue.edu/Other/VL_DB_Organisms.html

[6] G. C. Frech, A. M. VanDongen, G. Schuster, A. M. Brown, and R. H. Joho, *Nature* **340,** 642 (1989).

[7] T. J. Jentsch, K. Steinmeyer, and G. Schwarz, *Nature* **348,** 510 (1990).

[8] B. Hagenbuch, B. Stieger, M. Foguet, H. Lübbert, and P. J. Meier, *Proc. Natl. Acad. Sci. USA* **88,** 10629 (1991).

[9] A. Werner, M. Moore, N. Mantei, J. Biber, G. Semenza, and H. Murer, *Proc. Natl. Acad. Sci. USA* **88,** 9608 (1991).

[10] Y. Kanai and M. A. Hediger, *Nature* **360,** 467 (1992).

[11] H. M. Kwon, A. Yamauchi, S. Uchida, A. S. Preston, A. Garcia-Perez, M. B. Burg, and J. S. Handler, *J. Biol. Chem.* **267,** 6297 (1992).

[12] G. You, C. P. Smith, Y. Kanai, W. S. Lee, M. Stelzner, and M. A. Hediger, *Nature* **365,** 844 (1993).

[13] G. Gamba, S. N. Saltzberg, M. Lombardi, A. Miyanoshita, J. Lytton, M. A. Hediger, B. M. Brenner, and S. C. Hebert, *Proc. Natl. Acad. Sci. USA* **90,** 2749 (1993).

[14] K. Ho, C. G. Nichols, W. J. Lederer, J. Lytton, P. M. Vassilev, M. V. Kanazirska, and S. C. Hebert, *Nature* **362,** 31 (1993).

[15] D. Markovich, J. Forgo, G. Stange, J. Biber, and H. Murer, *Proc. Natl. Acad. Sci. USA* **90,** 8073 (1993).

[16] Y-J. Fei, Y. Kanai, S. Nussberger, V. Ganapathy, F. H. Leibach, M. F. Romero, S. K. Singh, W. F. Boron, and M. A. Hediger, *Nature* **368,** 563 (1994).

[17] C. M. Canessa, L. Schild, G. Buell, B. Thorens, I. Gautschi, J-D. Horisberger, and B. C. Rossier, *Nature* **367,** 463 (1994).

[18] G. Dai, O. Levy, and N. Carrasco, *Nature* **379,** 458 (1996).

[19] M. F. Romero, M. A. Hediger, E. L. Boulpaep, and W. F. Boron, *FASEB J.* **10,** A89 (1996).

Xenopus oocytes has also led to the identification of several receptor cDNAs (see Table II), an approach particularly popular when ligand binding produces an electrophysiological response.

Selection of Organism and Tissue

When designing a new expression cloning project, an important consideration is the source of tissue, and the organism (Fig. 1, top). An organism and tissue type that has been extensively characterized is ideal. Another consideration is the accessibility of the tissue in question. For example, expression cloning was first used in oocytes to clone the Na^+/glucose cotransporter (SGLT1) from the rabbit small intestine.[1] Functionally, this cotransporter was well characterized in both the small intestine and the kidney cortex. However, the kidney cortex is a somewhat heterogeneous cell population, whereas the epithelia of the small intestine can be easily isolated as a pure population by scraping off the intestinal epithelia. Similar issues related to ease of tissue access or cell purity dictated the choice of starting material for several transporter or channel cDNAs: (i) The small intestine for the epithelial and neuronal high-affinity glutamate transporter EAAC1,[10] the H^+-coupled oligopeptide transporter PepT1,[16] and the epithelial Na^+ channel eNaC[17], (ii) the liver for the Na^+/bile acid cotransporter,[8] (iii) the kidney for the Na^+/PO_4^{2-} cotransporter NaPi-1,[9] the Na^+/SO_4^{2-} cotransporter NaSi-1,[15] and the urea transporter UT2,[12] and (iv) thyroid cells for the Na^+/I^- cotransporter NIS.[18]

Other circumstances necessitated a particular choice of organism. For example, the Ca^{2+}-sensing receptor, BoPCaR, was originally cloned from bovine parathyroid tissue[20] not only because the parathyroid was the only tissue thought to have such a receptor,[21] but also because bovine tissue was easily obtained and in high abundance. Similarly, the thiazide-sensitive NaCl cotransporter[13] was cloned from the bladder of the winter flounder, a tissue in which the cotransporter had been studied in great detail. More recently, *Xenopus* oocytes have even aided in cloning and expression of two silicon transporters from the marine diatom *Cylindrotheca fusiformis*.[22]

To clone the renal electrogenic Na^+/HCO_3^- cotransporter, NBC,[19,22a] it was necessary to have a particular tissue, kidney, and organism, the tiger salamander *Ambystoma tigrinum*. For unknown reasons, mammalian RNA for NBC did not express well in oocytes,[22b] whereas salamander RNA

[20] E. M. Brown, G. Gamba, D. Riccardi, M. Lombardi, R. Butters, O. Kifor, A. Sun, M. A. Hediger, J. Lytton, and S. C. Hebert, *Nature* **366,** 575 (1993).

[21] E. F. Nemeth and A. Scarpa, *FEBS Lett.* **203,** 15 (1986).

[22] M. Hildebrand, B. E. Volcani, W. Gassmann, and J. I. Schroeder, *Nature* **385,** 688 (1997).

[22a] M. F. Romero, M. A. Hediger, E. L. Boulpaep, and W. F. Boron, *Nature* **387,** 409 (1997).

[22b] M. F. Romero, P. Fong, U. V. Berger, M. A. Hediger, and W. F. Boron, *Am. J. Physiol.* **274,** F425 (1998).

TABLE I

cDNAs OBTAINED BY EXPRESSION CLONING USING OOCYTES

Channel/enzyme/transporter	cDNA	Tissue	Organism	Selection method	Reference
Na^+/glucose cotransporter	SGLT-1	Intestine	Rabbit	Na^+-dependent glucose uptake	1
Delayed rectifier K^+ channel	drk1	Brain	Rat	VC, K^+-selective I–V relationship	6
Cl^- channel	ClC-0	Electric organ	*Torpedo marmorata*	VC, Cl^- current	7
Type I iodothyronine deiodinase	G21	Liver	Rat	Deiodination of 3,3′,5′-triiodothyronine	25
Na^+/bile acid cotransporter	prLNaBA	Liver	Rat	Taurocholate uptake	8
Na^+/phosphate cotransporter	NaPi-1	Kidney	Rat	PO_4^{-2}	9
Cl^- channel	I_{Cln}	MDCK cells	Dog	VC, ±NPPB	43
High affinity glutamate transporter	EAAC1	Small intestine	Rabbit	Glutamate uptake	10
Na^+/*myo*-inositol cotransporter	SMIT	MDCK cells	Dog	Na^+-dependent, *myo*-inositol uptake	11
Urea transporter	UT2	Kidney	Rat	Urea uptake	12
Thiazide-sensitive NaCl cotransporter	TSC-1	Bladder	Flounder	Na uptake	13
ATP-sensitive K^+ channel	ROMK1 (Kir 1.0)	Renal outer medulla	Rat	VC, ATP-sensitive K^+ current	14
Na^+/sulfate cotransporter	NaSi-1	Intestine	Rat	Na-dependent SO_4^{2-}-uptake	15
H^+-oligopeptide cotransporter	PepT1	Intestine	Rat	GlySar uptake	16
Epithelial Na^+ channel	eNaC	Intestine	Rat	VC, amiloride-sensitive Na^+ current	17
Pyrimidine nucleoside transporter	cNT1	Jejunum	Rat	Pyrimidine nucleosides and adenosine	53

Organic cation transporter	OCT1	Kidney	Rat	TEA-uptake	44
Na^+ dependent dicarboxylate transporter 1	NaDC-1	Kidney	Rabbit	Na^+ dependent ^{14}C-succinate uptake	54
Na^+-iodide cotransporter	NIS	Thyroid	FRTL-5 cells	I^- uptake	18
Calcitonin gene–related peptide response protein	CGRP	Organ of Corti	Guinea pig	VC, activation of Cl^- current by calcitonin gene–related peptide	55
H^+-oligopeptide cotransporter 2	PepT2	Small intestine	Rabbit	pH-dependent ^3H-cefadroxil uptake	56
Na^+/bicarbonate cotransporter	NBC	Kidney	Salamander	V_m and pH_i	19, 19a
Divalent cation transporter	DCT1	Duodenum	Fe-deficient rat	Fe^{2+} uptake	22b
Na^+ dependent dicarboxylate transporter 2	NaDC-2	Intestine	*Xenopus*	Na^+ dependent ^{14}C-succinate uptake	57
Renal organic cation transporter 1, basolateral	ROAT1	Kidney	Rat	Probenecid-sensitive ^3H-*p*-amino-hippurate (PAH) uptake	58
Basolateral PAH/dicarboxylate exchanger	fROAT1	Kidney	Flounder	Probenecid-sensitive PAH uptake	59
P(i)-uptake stimulator	PiUS	Small intestine	Rabbit	Enhanced Na^+-independent PO_4^{-2} uptake	60
Stimulator of Fe transport	SFT	K562 cells (erythroleukemic)	Human	Transferrin-independent Fe uptake	61
"*Xenopus* cdc6"[a]	Cdc6	Oocytes	*Xenopus*	Sperm binding without meiotic maturation	62
K^+-coupled neutral amino acid transporter	KAAT1	Midgut	*Manduca sexta*	K^+-dependent, amino acid uptake	36b

[a] *Xenopus* cdc6 was named according to database homology rather than cDNA phenotype as expressed in *Xenopus* oocytes.

TABLE II
RECEPTOR cDNAs OBTAINED BY EXPRESSION CLONING USING OOCYTES[a]

Receptor	Tissue	Organism	Selection method	Reference
5-HT_{1C} serotonin receptor	Choroid plexus tumor	SV11 mouse	VC	45
Substance K receptor	Stomach	Bovine	VC	46
5-HT_{1C} serotonin receptor	Choroid plexus	Rat	VC	63
Glutamate receptor	Forebrain	Rat	VC	47
Substance P receptor	Brain	Rat	VC	30
Neurotensin receptor	Brain	Rat	VC	31
Histamine H1 receptor	Adrenal medulla	Bovine	VC	48
Platelet-activating factor receptor	Lung	Guinea pig	VC	49
Thrombin receptor	HEL, Dami cells	Human	VC	50
Ca^{2+}-sensing receptor	Parathyroid	Bovine	VC	20
ATP receptor	NG108-15 cells (neuroblastoma)	Mouse	VC	64

[a] Receptor cDNA expression cloning typically takes advantage of the normal transport mechanism inherent in the *Xenopus* oocyte. In most cases, the endogenous Ca^{2+} activated Cl^- channel has been used as the "read-out" while voltage clamping (VC) oocytes.[5,51,52]

expressed very well. Thus, one should keep in mind that expression cloning from a mammalian tissue may not always be the ideal strategy.

Finally, it is often necessary to maximize the expression of a certain transporter, such as through changes in the dietary content of nutrients or by hormonal treatment. For example, the Fe^{2+} transporter DCT1 cDNA, cloned by Gunshin *et al.*[22c] was isolated from the duodenum of iron-deficient rats using the *Xenopus* oocyte expression cloning approach. In this case, duodenal mRNA of normal rats did not stimulate iron uptake in oocytes, whereas duodenal mRNA from iron deficient rats greatly increased iron uptake in oocytes. Thus, the iron depletion approach was crucial for the cloning of DCT1.

Use of *Xenopus* Oocytes as Expression System

Like all experimental systems, *Xenopus* oocytes are not suited for every expression cloning project. Because the oocyte is a living cell, it has its own natively occurring repertoire of channels, enzymes, proteins, and transporters.[5] Thus, before embarking on an expression cloning project, one should evaluate native oocytes for the presence of the protein cDNA to be cloned by functional expression. However, if a similar activity exists in the oocyte, the use of oocytes as a tool for expression cloning is not necessarily dis-

[22c] H. Gunshin, B. Mackenzie, U. V. Berger, Y. Gunshin, M. F. Romero, W. F. Boron, S. Nussberger, J. L. Gollan, and M. A. Hediger, *Nature* **388,** 482 (1997).

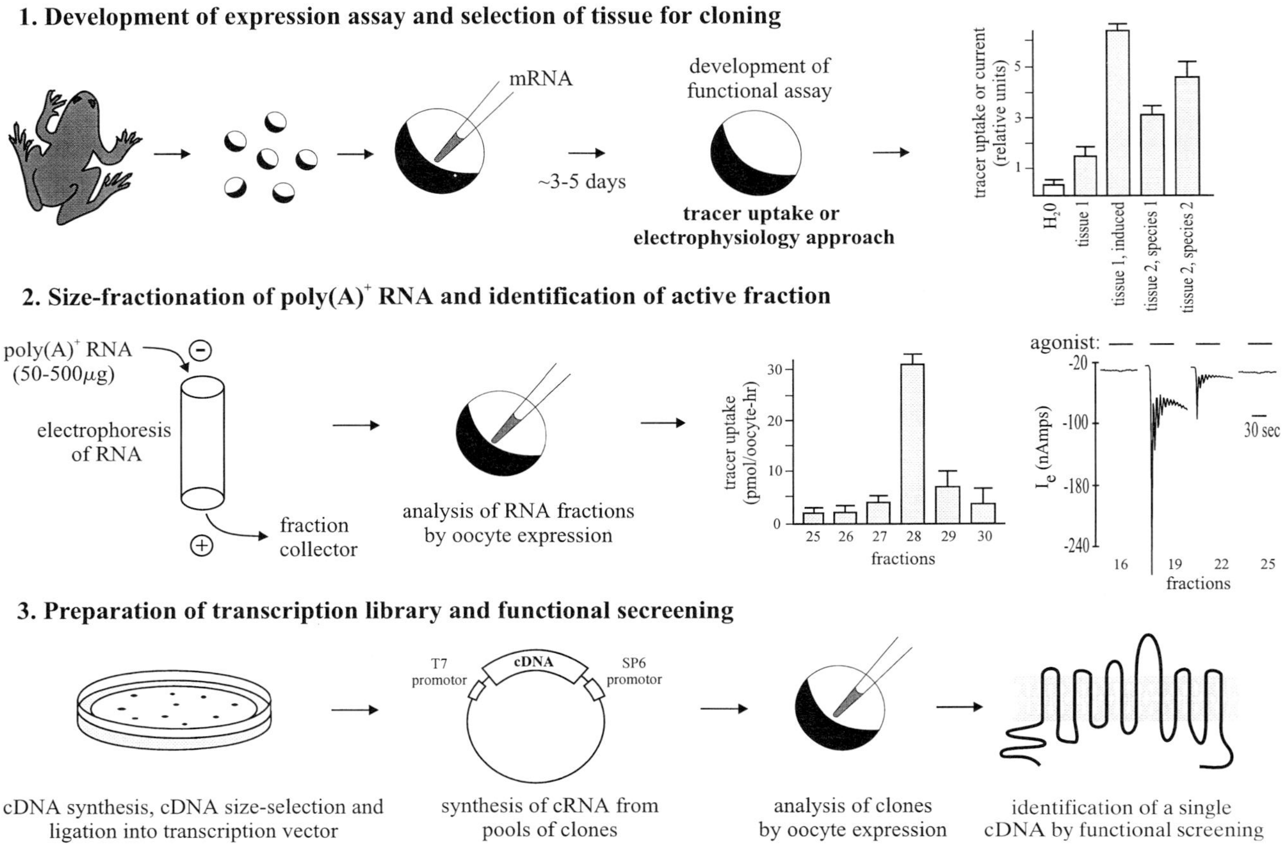

Fig. 1. Flow chart of expression cloning using *Xenopus* oocytes.

counted. For example, Jentsch and associates cloned the ClC-0 Cl^- channel though the oocyte has several endogenous Cl^- channels.[7] Gamba *et al.* successfully cloned a NaCl cotransporter though the *Xenopus* oocyte has an endogenous Na^+–K^+–$2Cl^-$ cotransporter.[13] By blocking the native channel or transporter, these investigators still maintained a null background to functionally test for the transporter of interest. Similarly, if the activity of interest exists in the oocyte, the investigator must be creative to distinguish the native oocyte activity from that of exogenous expression.

Isolation of RNA

Preparation of high-quality poly(A)$^+$ RNA is crucial for an expression cloning project. The RNA should be isolated from the freshest tissue possible. For large-scale tissue harvests it is convenient to quick freeze tissue in liquid nitrogen and then pulverize the tissue at a later date for RNA isolation. Alternatively, tissue culture cells can be harvested directly by treatment with guanidinium isothiocyanate. Poly(A)$^+$ RNA may be isolated by one of several methods. Most protocols require isolation of total RNA first, followed by poly(A)$^+$ RNA enrichment.

Total RNA Isolation

The method used in our laboratory is based on dissolution of tissue in guanidinium isothiocyanate,[23] followed by RNA extraction with cesium trifluoroacetate (CsTFA) density gradient ultracentrifugation (Pharmacia, Piscataway, NJ; see manufacturer's instructions).

Solutions

Guanidinium isothiocyanate (GITC) solution. 5.5 *M* guanidinium thiocyanate (Fluka), 0.5% sodium-*N*-laurylsarcosine (Sigma), 25 m*M* sodium citrate (Sigma), and 0.5% 2-mercaptoethanol (Sigma), pH 7.0, adjusted with NaOH. It is convenient to make 20-ml aliquots of this solution in 50-ml polypropylene tubes and to store them at −80°. To prevent excessive foaming during Polytron treatment (see below), add one drop (~50 μl) of antifoam agent (Sigma antifoam A) to 20 ml GITC solution just before use.

CsTFA solution. Mix 50 ml of CsTFA (Pharmacia, density of 2.0 g/ml) with 20 ml of 0.5 *M* EDTA, pH 8.0, and 30 ml of RNase-free H_2O (Gibco-BRL).

1. In a 50-ml polypropylene tube (Beckman) containing 20 ml GITC solution, homogenize ~2 g of fresh tissue using a Polytron homogenizer (Brinkmann/Kinematica model PT 3000). Homogenize at 25,000 rpm for

[23] J. Sambrook, E. F. Fritsch, and T. Maniatis, Molecular Cloning: A Laboratory Manual. Cold Spring Harbor Press, Cold Spring Harbor, New York (1989).

~30 sec. After homogenization, centrifuge at 5000*g* for 10 min at room temperature and transfer the supernatant to a fresh 50-ml polypropylene tube.

2. Add 6.2 ml of CsTFA solution into each Beckman polyallomer tube (14 × 95 mm) and carefully overlay ~7 ml of the RNA/GITC homogenate with a disposable transfer pipette. The tubes should be filled with this solution to ~2 mm below the top of the rim. Balance the tubes carefully by adding or removing GITC solution, and centrifuge at 27,000 rpm for 24 hr at 15° in a Beckman SW40-TI swinging bucket rotor.

3. After centrifugation, use disposable transfer pipettes to remove the supernatant down to ~1.5 cm from the bottom of the tube. Slowly overlay with diethyl pyrocarbonate (DEPC; see Ref. 23) treated RNase-free water and fill tube again to remove contaminating material. Using the transfer pipette, remove the supernatant again, this time down to ~1 cm from the bottom of the tube. Rapidly flip over the tube to pour off the remainder of the fluid and store the tube upside-down for a minute. Cut off the bottom of the tube (cut ~1 cm from the bottom) using a razor blade. The RNA will be evident at the bottom as a clear gel-like pellet (~50 μl). Suspend this pellet using a 200-μl pipetter. Suspend and rinse the tube bottom several times with small (50-μl) aliquots of RNase-free water and transfer RNA to a 1.5-ml microcentrifuge tube. It is best to cut off the pipette tip using a razor blade to prevent clogging by the viscous RNA solution during this step.

4. Heat RNA to 65° for 1 min, vortex, heat again, and repeat as needed until RNA is dissolved. Measure the total RNA concentration by UV spectroscopy. Dilute to 1 to 2 μg/μl and store at −80° until ready for poly(A)$^+$ RNA selection using oligo(dT)-cellulose chromatography. One gram of tissue should yield about 1 mg of total RNA.

Quantitation of RNA by UV Spectroscopy. To conserve RNA samples, a dilution of 1:50 is made. It is helpful to have a 50 μl quartz cuvette (e.g., Beckman). We have found that scanning from 220 to 320 nm is best. This scan not only provide A_{260} and A_{260}/A_{280}, but also will show abnormal peaks with contamination of the RNA. For example, phenol contamination is manifested as a peak at ~270 nm. For injection into oocytes, it is critical that RNA is free of chloroform, ethanol, phenol, and protein.

Poly(A)$^+$ RNA Selection

The following method is a modification of the oligo(dT)-cellulose chromatography described by Maniatis *et al.*[23]

Solutions

Elution buffer: 10 m*M* Tris–HCl, 0.05% SDS, 1 m*M* EDTA (pH 7.5)

2× binding buffer: 40 m*M* Tris–HCl, 1 *M* NaCl, 1% SDS, 2 m*M* EDTA, pH 7.5

Washing buffer: 10 m*M* Tris–HCl, 0.1 *M* NaCl, 0.1% SDS, 1 m*M* EDTA, pH 7.5

1. In a 15 ml-polypropylene tube (Corning), suspend 0.1 g of oligo(dT)-cellulose (type 3, Collaborative Research) in 5 ml of elution buffer. Let it soak for at least 15 min. Pour the slurry into a disposable plastic column (Evergreen Scientific, chromatography column, 70-mm filter) and allow buffer to drain through. Wash the column with 7 ml of 1× binding buffer.
2. Heat total RNA (~700 μg) for 3 min at 65° and quick chill in wet ice for ~3 min. Warm to room temperature. Add an equal volume of 2× binding buffer, apply RNA to the column, and collect the flow-through in microcentrifuge tubes. When nearly dry, add a few drops of 1× binding buffer and collect the flow-through again. Heat-treat the flow-through for 3 min at 65°, quick chill, allow to warm to room temperature, and reapply to the column. Wash the resin with 5 ml of 1× binding buffer, followed by 5 ml of washing buffer.
3. Elute with 6 aliquots of 400 μl elution buffer and collect the resulting eluants in separate microcentrifuge tubes. Identify RNA-containing fractions by UV spectroscopy. Ethanol-precipitate RNA: Add one-tenth of the volume of 3 *M* sodium acetate (pH 5.2) and 2.5 volumes of ethanol. After centrifugation, wash the pellets with one volume 70% (v/v) ethanol and dry for 5 min in a rotary evaporator. Dissolve the pellets in 10–20 μl of RNase-free water for at least 15 min on ice. Mix a few seconds using a vortex, rest the tubes on ice for 5 min, and vortex again to ensure that the RNA is dissolved completely. The yield of the poly(A)$^+$ RNA selection should be ~2–5%.

A less traditional method for RNA isolation involves direct isolation of poly(A)$^+$ RNA from tissue. This method has been popularized by the Ambion and Qiagen kits. The basic principle is to bind poly(A)$^+$ RNA to oligo(dT) carriers in the presence of RNase inhibitor and high salt. This solution, however, must be somewhat dilute and consequently is not useful. The method is cost prohibitive for large-scale RNA isolations poly(A)$^+$ RNA > 100 μg.

Isolation and Maintenance of *Xenopus* Oocytes

Xenopus has long been one of the favorite organisms of the developmental biologist.[2,4] In the past biochemists and physiologists have realized the advantage of this organism's oocytes for expression of exogenous RNAs.[3] Since the rearing and care of *Xenopus* has been reviewed,[24] we will focus here on the care and maintenance of these oocytes.

[24] M. Wu and J. Gerhart, *Meth. Cell Biol.* **36,** 3 (1991).

Vendors

Nasco (901 Janesville Ave., P.O. Box 901, Fort Atkinson, WI, 53538-0901, Tel: 800-558-9595); Xenopus 1 (716 Northside, Ann Arbor, MI, 48105, Tel: 313-426-2083); Xenopus Express (64 South Jackson St., Beverly Hills, FL, 34465, Tel: 800-936-6787); Hamamatsu Biological Research Service, Inc. (2370, Maisako, Maisako-Cho, Hamana-Gun, Shizuoka-Pref 431-02, Japan, Tel: 81-535-92-7822; Fax: 81-535-96-0522).

Isolation of Oocytes

Healthy oocytes are crucial for expression of foreign RNAs. Although many investigators focus on the first 3 days after RNA injection for expression, it is important to determine the optimal days for expression for each type of RNA. For example, in the expression cloning of the renal electrogenic Na^+/HCO_3^- cotransporter NBC, functional expression was not apparent until at least day 6 after RNA injection.[19] On the other hand, the expression of membrane proteins often declines 5 to 6 days postinjection. If treated properly, oocytes, injected or not, may be maintained for at least 1 week. The protocols detailed here are used to maintain oocyte health.

Solutions/Equipment

Anesthetic. 0.1% Tricaine (3-aminobenzoic acid ethyl ester, methane sulfonate salt, Sigma, St. Louis, MO) and 0.3% $KHCO_3$. Weigh out 1 g tricaine and 39 g $KHCO_3$ per liter nanopure Millipore (Bedford, MA) water. The tricaine solution should not be recycled.

Barth's solution. 300–400 ml per oocyte preparation, stored at 4°. 88 m*M* NaCl, 1 m*M* KCl, 0.33 m*M* $Ca(NO_3)_2$, 0.41 m*M* $CaCl_2$, 0.82 m*M* $MgSO_4$, 2.4 m*M* $NaHCO_3$, 10 m*M* HEPES, pH to 7.4. Sterile filter with 0.22- to 0.45-μm filter.

OR2. 300–400 ml per oocyte preparation, stored at 4°. 82.5 m*M* NaCl, 2 m*M* KCl, 1 m*M* $MgCl_2$, 5 m*M* HEPES, pH to 7.5. Sterile filter with 0.22- to 0.45-μm filter.

OR3 medium. For 2.0 liters, store at 4°. (This medium is an alternative to Barth's solution for oocyte storage.) One pack of powdered Leibovitz L-15 medium with L-glutamine (Gibco-BRL, Gaithersburg, MD), 100 ml penicillin/streptomycin solution in 0.9% NaCl (Sigma), 5 m*M* HEPES (final concentration), and pH to 7.5 with NaOH. Osmolarity should be 195–200 mOsm; if it is low, add 0.1–0.2 g NaCl. *Note:* For OR3 it is better to err on the high side of osmolarity (i.e., 200–210). Sterile filter with 0.22- to 0.45-μm filter and store at 4°.

Collagenase type A (Boehringer Mannheim). Prepare solution just before it is needed at a concentration of 1 mg/ml in OR2 buffer. Filter using a syringe filter (0.22 or 0.45 μm) and pour into a sterile 50-ml tube.

Autoclaved, cleaned dissecting instruments.

Small pair of scissors, scalpel, angled forceps, 2× #4 or #5 forceps, curved hemostats
5-0 chromic gut suture for muscle layer and 6-0 silk for skin.
Penicillin/streptomycin solution (Sigma). Dilute 50% using deionized water.
Frog tagging solution: 1 part concentrated HCl and 2 parts nanopure Millipore water.

Preparation of Xenopus Frogs for Partial Ovarectomy

Place a *Xenopus* frog in a beaker or plastic cage. Add enough tricaine solution to just cover *Xenopus*. Make sure that the nose remains above the fluid level. Keep *Xenopus* in tricaine solution for 3–5 min. This will "lightly" anesthetize the frog. Place *Xenopus* on ice and cover with ice. It is usually a good idea to have a container that has a cover since the frog may still move for a few minutes. Leave the frog on ice for at least 5 min. At this point the frog may be left longer if necessary. The purpose of the ice is to slow metabolism and place the frog in a state of hypothermia while anesthetized. This protocol allows the frog to recover more quickly from the anesthetic.

Marking/Numbering Xenopus. Knowing which *Xenopus* are "good expressors" requires record keeping, i.e., distinguishing one animal from another. The most reliable way to identify an animal is by tattooing or branding.[24] Using 5–10 μl of HCl solution (see above), trace out with a micropipetter a number or letter on the back of the unconscious frog. Once the skin begins to blister, wait about 5 sec and then completely rinse the area with the 50% penicillin/streptomycin solution. Keep a log book or computer database documenting surgeries and expression capacity of each animal.

Oocyte Removal. While lightly squeezing the sides of the frog together, make a diagonal cut about 1.5 cm above the leg. Cut through the skin and muscle layers; oocytes should be in view. Rinse the area with 50% penicillin/streptomycin solution. Using the angled forceps, reach into the abdomen and pull out part of an ovary lobe. Cut this lobe off and place oocytes into a 50-ml tube with about 10–15 ml of OR2. Repeat until enough oocytes are obtained.

Xenopus Recovery. When removal is finished, rinse the abdomen with 50% pencillin/streptomycin. Sew the muscle layer with 5-0 chromic gut and then sew the skin with 6-0 silk. Place the frog into a recovery tank with enough water to keep it wet, but do not allow the nose to submerge. The frog should recover in about 10–15 min. After recovery, replace water in the recovery tank several times to get rid of debris. Fill the recovery tank with 0.1 *M* NaCl and allow the frogs to recover overnight. The NaCl facilitates the frog recovery and apparently minimizes stress.[24]

Oocyte Treatment

This method is to be performed at room temperature, with solutions maintained at room temperature. It does not require manual defolliculation and thereby reduces injury of oocytes. The oocytes are dissociated in OR2 solution. First, the ovarian lobes should be cut open to create flat sheets of oocytes in membrane. Place oocytes into a 50-ml polypropylene tube or a petri dish and agitate them in ~20 ml OR2 solution using several solution changes. The most effective way to ensure proper agitation is to use a rotator or a shaker (slow speed!). Shake or rotate for 5–10 min after each solution change. Continue solution changes until cell/connective tissue debris in supernatant is greatly reduced (almost gone). This usually requires 5 or more solution changes over 1–1.5 hr.

Once oocytes have been "loosened" from the surrounding connective tissue, they are ready for a brief exposure to collagenase. Minimizing the collagenase treatment after a prolonged OR2 treatment helps to prevent oocyte membrane damage. Thus, oocytes in general are healthier and survive longer in culture. It should be noted that collagenase contains the proteolytic enzyme clostripain, which requires calcium as a cofactor. To minimize its activity, it is important to work with calcium-free OR2. Prepare and filter 1 mg/ml type A collagenase (Boehringer Mannheim) in OR2. Pour off the OR2 supernatant from the oocytes and pour the oocytes into a fresh 50-ml tube with the collagenase solution. About 5 to 10 ml of this solution is required per ovary lobe removed. Rotate or shake the oocytes gently for 15–20 min. Check the oocytes for dispersion and rinse with 2–3 changes of OR2. If necessary treat the oocytes for another 5–10 min with collagenase in OR2. Add 10–20 ml OR2 to the collagenase/oocyte solution, cap, and mix by inverting several times. Pour off the supernatant and small oocytes. Larger oocytes will fall to the bottom of the tube very quickly, so that part of the sorting process occurs by pouring off small oocytes in these later solution changes. Follicular membranes can be observed floating free at this point. Finally, add 20–30 ml Barth's solution (with calcium, filter solution) to the oocytes for 4–5 washes. Mix by inverting the tube, pour off the supernatant and the small oocytes. Oocytes are kept in 20–30 ml filtered Barth's solution with gentamicin (50 μg/ml) or OR3 for sorting.

Sorting Oocytes. Sort oocytes in either 100- or 150-mm petri dishes in filtered Barth's solution. Take a transfer pipette and get a group of oocytes (<100). Place oocytes in one of the petri dishes in a line. Sort oocytes by eye with good light perpendicular to the petri dish. Select oocytes based on (a) size (the largest oocytes are stage V/VI), (b) the color of the pigmented pole (it should be dark brown, not black), (c) a "healthy" look of the oocyte (do not choose damaged oocytes). Move good oocytes to one side of the dish. Once this group has been sorted, use the transfer pipette to move oocytes to a sterile dish with Barth's solution containing gentamycin or OR3 for storage at 17–19°. Repeat the process until all oocytes have been sorted.

Maintaining Oocytes. Oocytes are kept in an incubator at 17–19°. The Barth's/gentamicin medium or OR3 needs to be replaced every day. Check oocytes for health prior to changing the medium. If oocytes appear mottled, splotchy, have a white area on the pigmented (animal) pole, or have a "pasty" white nonpigmented (vegetal) pole, they should be discarded.

Development of Unique Expression Assay

Of critical importance is a good, reliable, and preferably unique experimental assay for the encoded protein to be cloned. Transporters can be tested by monitoring labeled substrate uptake or, if they are electrogenic, by monitoring substrate evoked uptake currents or membrane potential (V_m) changes. The use of specific inhibitors such as phlorizin (Na^+/glucose cotransporters) or phloretin (urea transporters), or the isolation of the ion-dependent uptakes can further contribute to the design of unique assays. Certain transporters require a more sophisticated experimental assay. An example is the Na^+/bicarbonate cotransporter, which depended on simultaneous recording intracellular pH and V_m using microelectrodes.[19] For channels, one typically would like to demonstrate ion specificity, a particular response to a second messenger system, sensitivity to a specific channel inhibitor, or a certain electrophysiological profile. Receptors can be easily analyzed in oocytes if ligand binding elicits an electrical response.[20] Enzymes such as the type-I iodothyronine deiodinase[25] have been cloned by monitoring enzymatic activity in oocytes.

Oocyte Behavior during Protein Expression

Sometimes expression of exogenous proteins in *Xenopus* oocytes results in unanticipated changes in oocytes' health, morphology, or physiology. The researcher should consider the potential effects for expressing the particular protein of interest in the oocyte. Many channels, when highly expressed, can be detrimental to the oocyte within a few days. Cl^- channels are therefore typically studied within 48 hr of RNA injection.[7] When the amiloride-sensitive epithelial Na^+ channel eNaC is expressed, oocytes must be maintained in a Na^+-free solution to prevent death due to high (60 m*M*) intracellular sodium.[17] Similarly, oocytes expressing the Na(K)Cl cotransporter are maintained in a Cl^--free solution for longevity.[26] In cloning the

[25] M. J. Berry, L. Banu, and P. R. Larsen, *Nature* **349,** 438 (1991).

[26] G. Gamba, A. Miyanoshita, M. Lombardi, J. Lytton, W. S. Lee, M. A. Hediger, and S. C. Hebert, *J. Biol. Chem.* **269,** 17713 (1994).

electrogenic Na^+/HCO_3^- cotransporter NBC, it was necessary to raise intracellular Na^+ from ~7.5 m*M* (Romero and Nakhoul, unpublished data, 1995). Intracellular Na^+ was raised by blocking the oocytes' Na^+ pump with K^+-free ND96 for 24 hr, thus allowing expression to be observed more easily.[19]

Injecting Oocytes

Oocytes are microinjected with usually 50 nl of fluid, with or without RNA. The oocytes are placed in a dish with sterile Barth's solution on a 0.5-mm nylon or polypropylene mesh. Oocytes are visualized under a dissecting microscope and injections are performed using micropipettes held by a micromanipulator. Pipettes should have tip diameters of 10–40 μ meter. Injection pipettes are filled with paraffin oil, and RNA is then pulled up into the pipette. The Drummond Nanoject pipetter is a very easy way to deliver 46–50 nl of solution to each oocyte. A microprocessor-controlled version of Nanoject (#3-000-203-XV) allows motorized injection of variable volumes (4.6 to 73.6 nl), and thereby improves the reproducibility and speed of injection. The site of injection in the oocyte should be in the transitional zone, where the animal and vegetal poles meet.

Size Fractionation of Poly(A)$^+$ RNA

Gel Electrophoresis

Size fractionation of RNA is an important step in expression cloning because it significantly reduces the number of clones to be screened. Poly(A)$^+$ RNA can be fractionated using an agarose gel with continuous elution at the cathode using an apparatus such as our GenePrep electrophoresis system[27] (see Fig. 2). This system allows efficient separation of RNAs from 0.5 to 15 kb. As shown in Fig. 2, it makes use of an externally and internally water-cooled, donut-shaped agarose gel chamber and a concentric elution chamber (Fig. 2, bottom). The elution chamber has a very small volume (~500 μl) and is separated from the gel chamber on the top by two RNA-permeable Millipore filters (GVWP 04700, 0.22-μm pore size, 45-mm diameter) with a reducing plate in between, and on the bottom by an RNA-impermeable dialysis membrane (Spectra/Por MWCO 6-8000, cut to 45-mm diameter circles). Before the unit is assembled, all parts, including the tubing and the filters must be soaked or rinsed in 3% hydrogen peroxide and then washed with DEPC-treated water. A 1.5% low-EEO agarose gel (<0.09%, Bio-Rad, Richmond, CA) is made in nondenaturing low salt buffer (10 m*M* MOPS, pH 7.0, 0.5 m*M* EDTA, 0.05% SDS). Then, 50–500

[27] M. A. Hediger, T. Ikeda, M. Coady, C. B. Gundersen, and E. M. Wright, *Proc. Natl. Acad. Sci. USA* **84,** 2634 (1987).

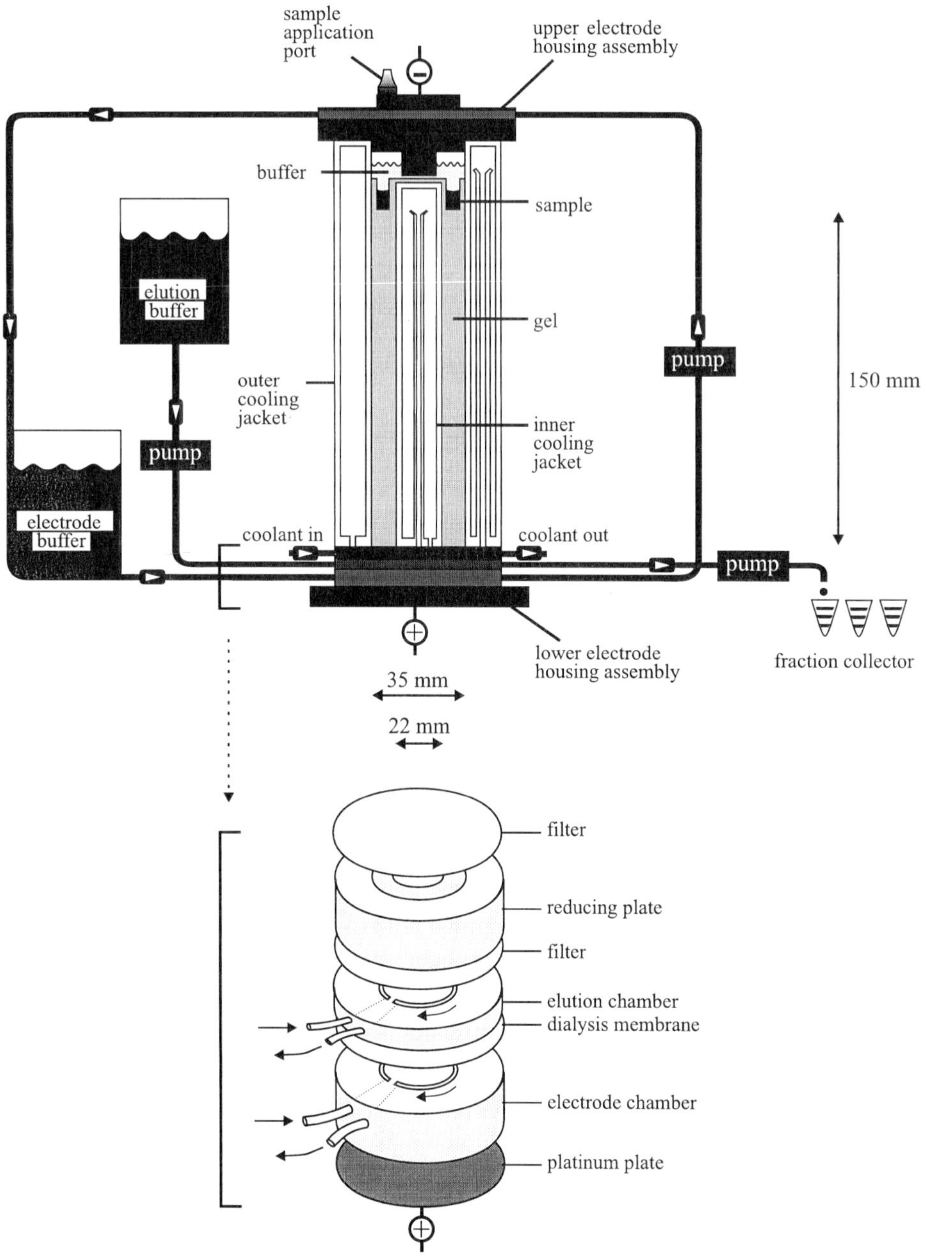

FIG. 2. Method for poly(A)$^+$ RNA fractionation. GenePrep electrophoresis device is shown (top). The design of the elution system is illustrated at the bottom. The outer cooling jacket and the inner cooling finger are made of glass, and the elution system and the upper electrode housing assembly are made of Plexiglas.

μg of RNA is carefully dissolved in 300–600 μl of Gibco water and one-tenth of the volume of 10× sample buffer (12.5 ml RNase-free water (Gibco-BRL), 12.5 ml glycerol, 0.1 g bromphenol blue, 0.1 g xylene cyanol, and 0.125 g SDS) is added. The mixture is heated to 70° for 1 min, mixed quickly using a vortex, incubated at 70° for another 2 min, quick-chilled on wet ice for a few minutes, and then loaded into a sample well formed on the upper side of the gel. The gel is cooled to 8° before sample application. Electrophoresis is then allowed to proceed at 400 V (~18 mA), 8° during 5–15 hr. The flow rate of electrode buffer circulating through upper and lower electrode chambers (see Fig. 2) is ~500 ml/hr. Elution and electrode buffers are the same as the gel buffer. The elution chamber at the bottom of the gel allows for continuous collection of separated RNA into a fraction collector. The flow rate for elution is 5 to 6 ml/hr, and ~900-μl fractions are collected every 10 min. About 50–100 μl from each fraction are moved after separation for minigel electrophoresis, as described below. Each fraction is then split into three subfractions of ~270 μl which are stored as ethanol precipitates: Per 270-μl fractions, 135 μl (0.5 volume) of 7.5 *M* ammonium acetate (Sigma, RNase-free) and 810 μl (2.0 volume) of 100% ethanol are added and the mixture is stored at −80°. For initial oocyte expression studies, ~10 RNA pools are prepared using aliquots of fractions from the entire size range (e.g., from 1.5–14 kb). A typical scenario for preparing suitable pools is as follows: For the 1.5- to 4-kb size range, mix 250-μl aliquots from ethanol precipitation mixes using 3 consecutive fractions for each pool. For the 4- to 8-kb size range, mix 150-μl aliquots of ethanol precipitation mixes using 6 consecutive fractions for each pool. After ethanol precipitation, dissolve each pool in 4 μl RNase-free water. After separating ~300 μg of poly(A)$^+$ RNA, the resulting concentrations of the pools should be between 0.3 and 0.8 μg/μl. The RNA is then ready for oocyte expression studies to identify the active fraction(s) (see Fig. 1).

The GenePrep system is not yet commercially available. A preparative electrophoresis device is available from Bio-Rad (Miniprep Cell Model 491), but its mRNA size fractions are not as well resolved as in the GenePrep (Fig. 3b).

Sucrose Density Gradient Centrifugation

An alternative method, usually more accessible, which can be used for size fractionation of RNA is the sucrose density gradient method.[23,28,29] This method works reasonably well and has >90% RNA recovery, though the separation is not as complete as that obtained using the GenePrep (Fig. 3c). Nevertheless, the separation may be satisfactory if the RNA of interest

[28] K. Geering, D. I. Meyer, M.-P. Paccolat, J-P. Kraehenbuhl, and B. C. Rossier, *J. Biol. Chem.* **260,** 5154 (1985).

[29] C. B. Gundersen and J. A. Umbach, *Neuron* **9,** 527 (1992).

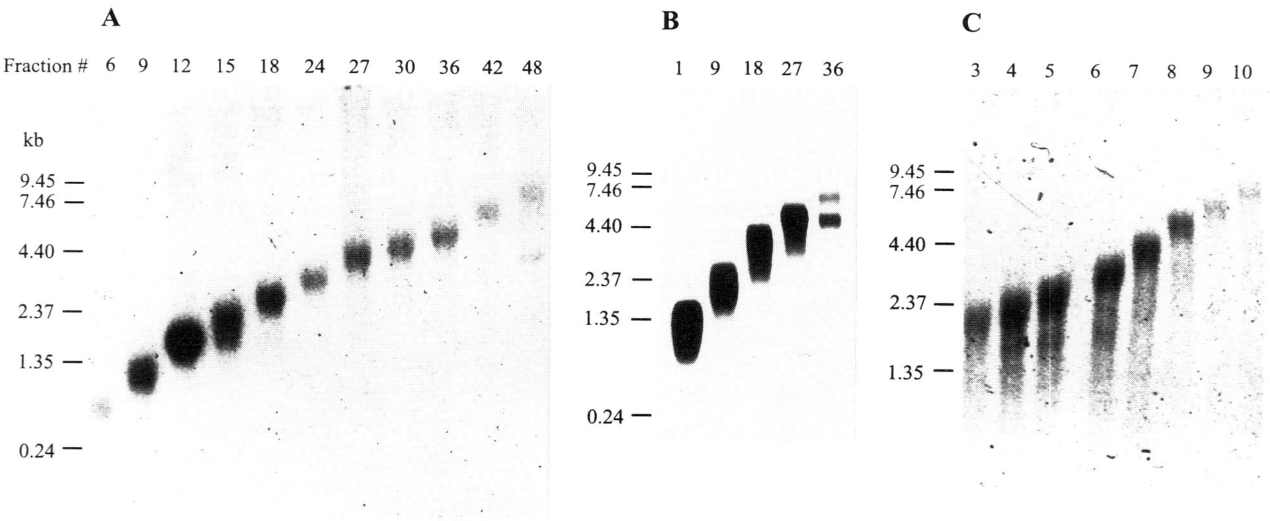

FIG. 3. Comparison of three different RNA separation techniques. (A) GenePrep, (B) Bio-Rad, (C) sucrose density gradient method. (Reproduced with permission from Gundersen and Umbach.[29]) Ethidium bromide staining of analytical gels performed on aliquots of fractions indicated.

is expressed at relatively high levels, as was the case for the ROMK1 K^+ channel[14] and the eNaC Na^+ channel.[17] Requirements are RNase-free reagents, a swinging bucket ultracentrifuge, a butterfly needle and connector, and graduated Eppendorf 1.5-ml tubes. Typically a 5 or 10 ml, 5–20% sucrose gradient is made in a swinging bucket ultracentrifuge. The gradient can be made by placing equal volumes of the 20% and the 5% sucrose in a gradient mixer (two flat-bottom cylinders connected at the bottom by tubing and another outlet tube) with stirrers at the bottom. A capillary glass pipette is connected to the outlet tubing and placed at the bottom of the 5- or 10-ml centrifuge tube. The flow between the two chambers is opened at the same time as the outlet in the centrifuge tube. The fluid will fill from the bottom to the top. Up to 100 μg of poly(A)$^+$ RNA may be layered on top of such a gradient. The sample is centrifuged for 4 hr at 45,000 RPM at 16° in an SW 50 rotor. After centrifugation, a butterfly needle with attached tubing is used to puncture the bottom of the tube. The flow to the sampling tubes is controlled by a clamp (hemostat or screw clamp). Fractions should be collected in graduated tubes at constant volume rather than counting drops, since drop size will vary with the density of the sucrose. An aliquot of each fraction is used to determine RNA concentration. Each fraction is precipitated with ammonium acetate and centrifuged to pellet. The pellet is washed with 70% ethanol (this step reduces the salt and solubilizes the sucrose) and dried in a rotary evaporator (usually 5 min). The RNA pellet is resuspended in RNase-free water to a final concentration of 0.5–1.0 μg/μl, and the fractionated RNA is ready for oocyte injection. A nondenaturing 7–35% sucrose gradient was used by Gundersen and Umbach,[29] and the result of their separation is shown in Fig. 3c.

Analysis of RNA Fractions by Minigel Electrophoresis

Denaturing RNA gels are quite sensitive for detection of small amounts of RNA, and 50 ng of poly(A)$^+$ RNA can be visualized with ethidium bromide staining and UV light. The following method, which is a modification of that described by Sambrook, Fritsch, and Maniatis,[23] is particularly suitable for detecting small amounts of RNA.

Solutions

5× MOPS buffer. 200 m*M* MOPS [3-(*N*-morpholino)propanesulfonic acid], 50 m*M* sodium acetate, 5 m*M* EDTA, pH 7.0 (NaOH), in nanopure Millipore water. Sterile filtered and stored at 4° in the dark for up to ~3 months. *Note:* This solution must be shielded from light (wrap with aluminum foil). Do not autoclave.

10× sample buffer. 50% (v/v) glycerol, 1 m*M* EDTA, 0.4% (w/v) bromphenol blue, 0.4% (w/v) xylene cyanol. Prepare in nanopure Millipore water. Mix well and store at room temperature.

Method

1. *Clean minigel apparatus.* Note that treatment with DEPC or H_2O_2 is not necessary since the denaturing agents serve to inhibit RNase activity. Wash the apparatus and associated parts with warm, soapy water. Rinse with deionized water, and dry. Set up the apparatus as well as a tray equipped with a 15- or 30-well comb in a fume hood, and level both of them carefully.

2. *Pour gel.* The amounts for a 30-ml, 1% denaturing RNA minigel are as follows:

Volume	Component	Notes
18.3 ml	Milli-Q–H_2O	
0.3 g	agarose	Use microwave to melt; make sure that there are no undissolved particles
6.2 ml	5× MOPS buffer	
5.3 ml	formaldehyde	Mix in fume hood

Solubilize the water and agarose in a microwave in a 125-ml screwcap bottle, and let it cool to ~60°. In a fume hood, immediately add 5× MOPS buffer and formaldehyde to the bottle, mix briefly, and pour into a gel tray. Allow 30–45 min for the gel to solidify.

3. *Prepare 300 ml 1× MOPS solution for electrode buffer.* Transfer the polymerized gel into an electrophoresis tank and pour electrode buffer into the device.

4. *Prepare samples while gel is solidifying.*

(a) *Prepare sample mix* (just before use): 100 μl deionized formamide (stored at −20° as aliquots), 40 μl formaldehyde, 20 μl 5× MOPS, 20 μl 10× sample buffer, and 4 μl ethidium bromide, 10 mg/ml.

(b) *Prepare RNA samples:* 0.5–2 μl RNA sample in water (Gibco/BRL) (or 1 μl RNA ladder from Gibco/BRL). Add 5 μl sample mix. Vortex briefly and heat mixture to 65° for 5 min, vortex again, quick spin, and heat for another 2 min. Quick spin again, and then load immediately (without cooling) onto the denaturing gel.

5. *Electrophorese* at 80 V during 1.5–2 hr.

6. *Wash gel.* After electrophoresis wash with tap water (several changes) during 20 to 60 min with shaking and proceed with photography on an UV transilluminator.

Synthesis of Size-Selected cDNA from the Positive RNA Fraction

Both plasmids and lambda (λ) phage can be used as vectors for expression cloning. Suitable vectors contain specific RNA polymerase promotors flanking the cloning site, for example, T3, T7, and SP6 RNA polymerase promoters (Fig. 4). This allows the synthesis of cRNA by *in vitro* transcription to be used for oocyte injection. Although λ vectors have been successfully used for expression cloning,[30,31] we prefer plasmid cDNA libraries because of the ease of handling bacterial colonies during filter replica preparation (see Fig. 5).

The choice of the cDNA synthesis method is critical to the success of expression cloning. cDNA libraries should be directional to ensure that only sense cRNA is obtained during *in vitro* transcription of clones. This reduces the number of clones to be screened and eliminates problems with antisense RNA inhibition. The SuperScript plasmid system for cDNA cloning (Gibco-BRL) is one of the more useful systems currently available. The protocol for cDNA synthesis using this system is outlined in Fig. 4. Oligo(dT) priming is performed using an oligo(dT) primer containing a 5′ adapter sequence with rare cutting cleavage sites such as *Not*I. This site can be used for plasmid linearization during *in vitro* cRNA synthesis later on (Fig. 4).

cDNA Synthesis using the SuperScript cDNA Synthesis System (Gibco-BRL)

Poly(A)$^+$ RNA (2–5 μg), dNTPs, and the oligo(dT) adapter (total volume of 9 μl) are heated to 65–70° for 10 min, quick-chilled on ice, and then reverse transcribed at 37° for 1 hr (total reaction volume is 20 μl). The RNA is next degraded by RNase H, and the second strand of cDNA is synthesized with the addition of dNTPs, *Escherichia coli* DNA polymerase I, and *E. coli* DNA ligase. Finally, a non-self-ligating blunt-end *Sal*I adapter is ligated to the cDNA, and product is cut with the *Not*I enzyme in 50 μl total volume (Fig. 4). Extract with phenol/chloroform and ethanol precipitate with ammonium acetate as described by the manufacturer. Dissolve the dry pellet in 30 μl water (Gibco-BRL).

The above described method will not work if the cDNA of interest has an internal *Not*I site. A suitable alternative method for directional cDNA synthesis is that of Stratagene (La Jolla, CA): The oligo(dT) primer contains an *Xho*I site, and first strand synthesis is performed in the presence of 5-methyl-dCTP, producing hemimethylated cDNA. *Eco*RI adapters are then ligated to the cDNA and the product is digested with *Xho*I. The hemimethylation prevents cleavage at internal *Xho*I sites. The directional cDNA is then ligated into pBluescript II (e.g., sk$^-$), which must be precut with *Eco*RI and *Xho*I and treated with calf intestinal phosphatase.[23] The product should be gel-purified before ligation with the cDNA.

[30] Y. Yokota, Y. Sasai, K. Tanaka, T. Fujiwara, K. Tsuchida, R. Shigemoto, A. Kakizuka, and S. Nakanishi, *J. Biol. Chem.* **264,** 17649 (1989).

[31] K. Tanaka, M. Masu, and S. Nakanishi, *Neuron* **4,** 847 (1990).

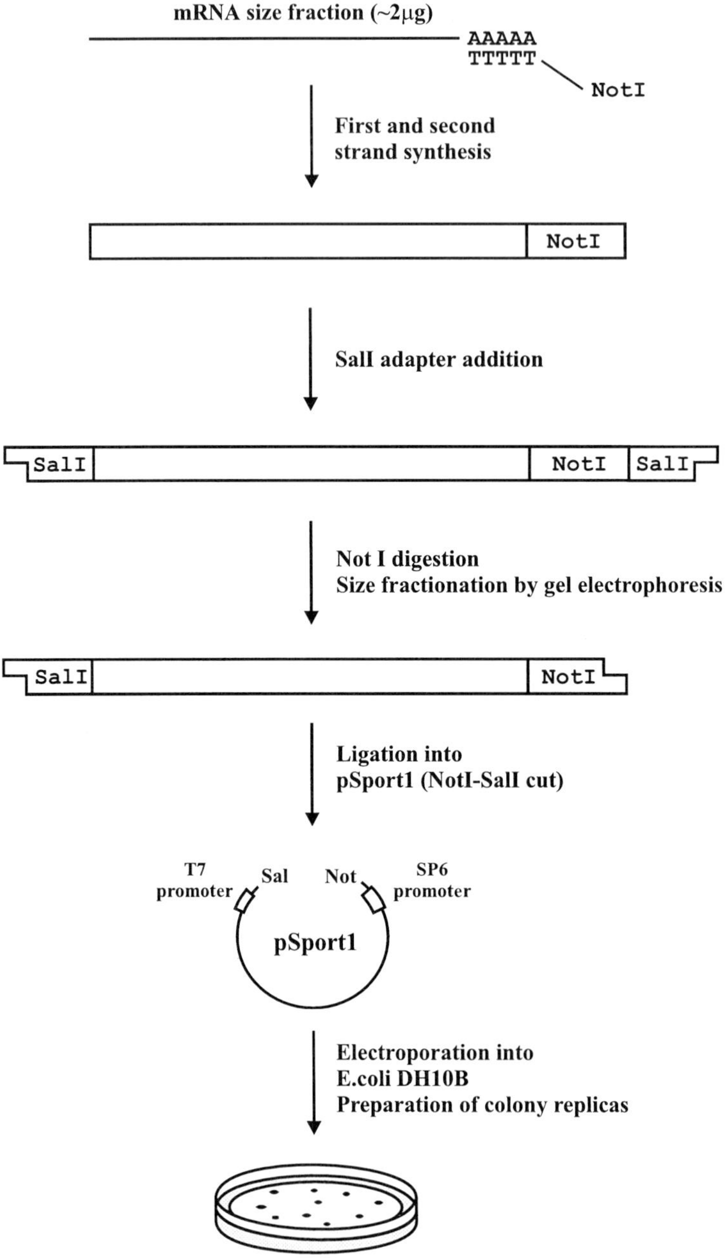

FIG. 4. Schematic diagram illustrating the strategy for directional cDNA synthesis.

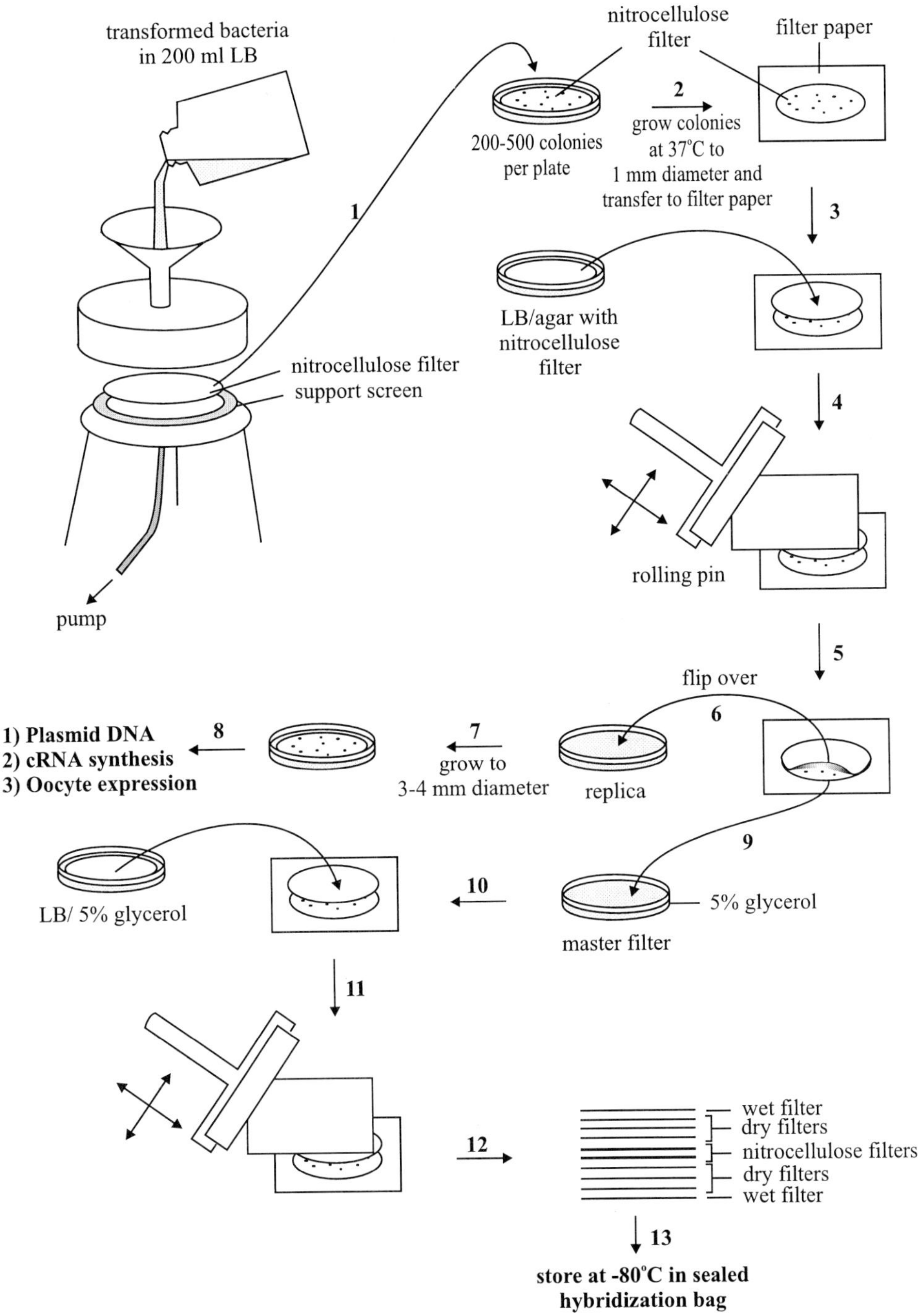

FIG. 5. Method for plating cDNA libraries for functional screening and for preparing filter replicas. On the top left, a filtration apparatus is shown which is used to distribute colonies evenly onto nitrocellulose filters.

Size Selection of cDNA

As important as it is to size-select poly(A)$^+$ RNA, it is equally important to size-select the cDNA from reverse transcription (RT). The purpose of this step is to ensure that the library is not filled with small fragments from incomplete RT reactions. The selected size of cDNAs must correspond to the size range of poly(A)$^+$ RNA used to make the library. This step will also remove oligomeric cDNA synthesis products.

One method of cDNA size fractionation is to use a small scale version of the GenePrep apparatus. We have used a 5-mm diameter cylindrical gel chamber (gel height ~75 mm) equipped with a cooling jacket. Although this device is not commercially available, it produces relatively high recovery of intact, size-fractionated DNA.

A readily available technique for cDNA fractionation is agarose minigel electrophoresis followed by extraction of DNA from gel slices using Geneclean Spin (BIO 101). This method works reasonably well if performed using the procedure outlined below, which is a modification of the manufacturer's instruction. The method minimizes the production of broken DNA and thereby increases the number of full-size clones.

1. Prepare a 0.8% minigel using an 8 teeth comb. Apply 10 μl of cDNA (one-third the total cDNA synthesis product) into the well of lane 4. Apply DNA ladder (2 μl/lane of Gibco-BRL) into the wells of lanes 1, 2, 6, and 7. Run gel at 80 V until bromphenol blue reaches the lower end of the gel. Carefully cut out cDNA lane 4 using a razor blade. Stain the remaining gel with ethidium bromide for 10 min, then wash for 10 min with water. Place the stained gel on a UV transilluminator, then place a transparency sheet on top. Mark the ladder and wells of the lanes with a pen or the transparency. Put the cut, unstained DNA lane back into the remaining gel and assemble the gel on top of the marked transparency. Cut out the desired cDNA size range which should correspond to the originally identified RNA size.

2. Weigh the gel slice (it should be less than 300 mg). Place into the spin tube (supplied by the Geneclean Spin kit). Shake glass milk well and immediately add 400–500 μl. Heat at 55° for 5 to 10 min or until the gel is completely dissolved. Spin out the liquid into the catch tube (provided by the kit).

3. Add 500 μl of "NEW Wash solution" (Geneclean kit) to the spin filter and spin for 30 s. Repeat the wash again. Transfer the spin filter to an elution catch tube (kit). Add 10–25 μl of elution solution (Geneclean kit) on top of the glass milk. In order to avoid shearing of DNA, it is critically important not to mix the elution buffer with the glass milk (flicking the tube or vortexing must be avoided). Simply let the mixture sit on ice for 5 min to allow DNA to elute from the glass milk by diffusion. Then

spin for 30 sec and collect the eluant with the cDNA in a catch tube. Repeat the elution twice and combine the eluants (yields ~200 μl).

4. Recover cDNA by ethanol precipitation in the presence of carrier RNA. Add 5 μl of yeast tRNA (1 μg/μl) to 200 μl of eluant. Add 0.5 volumes of 7.5 *M* ammonium acetate (100 μl) and 2.5 volumes (750 μl) of ethanol. Precipitate at −20° for at least 2 hr (or overnight). Recover DNA by centrifugation, and proceed with pellet washing and dry pellet with a Speed-Vac concentrator. Dissolve cDNA in 10 μl of RNase-free/DNAse-free water. The cDNA should be allowed to rehydrate by incubating the solution on ice for at least 10 min.

5. To estimate the concentration of the cDNA, run 2 μl of this cDNA on an agarose minigel along with different dilutions of pSport vector (e.g., 5 and 10 ng per lane, Gibco-BRL).

Ligation into Plasmid Vector

Since the library will be used for *in vitro* cRNA synthesis, it is important that it be ligated in a directional manner (see Fig. 4). For ligation, it is best to follow the instructions of the GibcoBRL SuperScript cDNA synthesis system. The ligation is typically performed in a 20-μl reaction containing 4 μl 5× DNA ligase buffer [250 m*M* Tris–HCl, pH 7.6, 50 m*M* $MgCl_2$, 5 m*M* ATP, 25% (w/v) polyethylene glycol (PEG) 8000, 5 m*M* dithiothreitol (DTT)], 1 μl (50 ng) pSport *Not*I/*Sal*I cut (kit), 4 μl DNA (10 ng), 1 μl (1 unit) T4 DNA ligase, and 10 μl water. The mixture is incubated at room temperature for 3 hr or overnight, and the cDNA is recovered by ethanol precipitation. To improve the efficiency of the precipitation, add 5 μl of yeast tRNA (kit) to the 20 μl ligation reaction. Then add 12.5 μl 7.5 *M* ammonium acetate and 70 μl ethanol (−20°), vortex the mixture thoroughly, and immediately centrifuge it at room temperature (20 min). Carefully remove the supernatant, proceed with a pellet wash using 0.5 ml 70% ethanol (−20°), centrifuge (2 min), remove the supernatant, and dry the sample at 37° for 10 min to evaporate residual ethanol. Dissolve the dry pellet in 10 μl sterile water and incubate on ice as before. Add 1 μl of the ligated DNA to 20 μl of electrotransformable cells and electroporate as described below.

Transformation of Escherichia coli

The cDNA library should be transfected into *E. coli* at the highest efficiency possible. Electroporation gives excellent transformation efficiencies of $>1 \times 10^{10}$ transformants per μg plasmid DNA. The 10- to 100-fold higher efficiency above the heat shock method increases the likelihood that low copy number messages are represented in the library and allows sparing use of the limited quantity of size-selected cDNA. The Gibco-BRL electroporation system with Electro MAX DH10B cells has been mostly used in our laboratory. In order to obtain maximum transformation efficiency, the DNA should not contain phenol, ethanol, salts, protein, or detergents. In general, we have followed the manufacturer's instructions. For best results, thaw each vial of electrocompetent cells only once. Only a small fraction if the cDNA (10–50 ng in 1 μl) is usually required for electroporation. After electroporation, the cells should be transferred to a tube containing 1 ml of S.O.C. medium [20 g Bacto-tryptone, 5 g Bacto-yeast extract, 0.5 g NaCl per liter, and 20 m*M* glucose (Gibco-BRL)]. The mixture is incubated for 1 hr at 37° with gentle agitation (225 rpm). This step helps the bacteria to recover and to express the antibiotic resistance marker encoded by the plasmid. Before filter replicas are prepared, (Fig. 5), the exact titer of the transformation mixture should be determined and the quality of the library must be evaluated as follows:

1. Make a series of dilutions of the transformation mixture in 100 μl S.O.C. medium (e.g., 1 : 10, 1 : 100, 1 : 1000). The rest of the transformation mixture can be stored at 4° for a limited time (preferably not longer than ~1–2 days).
2. Spread these 100-μl dilutions on 100-mm LB/ampicillin plates (final ampicillin concentration is 100 μg/ml). If blue/white screening is desired, spread 100 μl of a 40 m*M* isopropylthiogalactoside (IPTG) solution (in water) and 100 μl of a 20 mg/ml X-Gal solution (in dimethylformamide) on the plate 2 to 4 hr before plating the transformation. Incubate plates overnight at 37°.
3. To test the quality of the library, grow up 20 randomly picked colonies in 2 ml of LB ampicillin, prepare plasmid DNA using a minipreparation procedure, cut the DNA with *Not*I and *Sal*I, and run a minigel. Approximately 50% of the clones should contain inserts corresponding to the size range of interest.

Functional Screening of Library

Plating Master Filters of cDNA Library and Making Filter Replicas

Expression cloning should be executed in a systematic fashion. When using a plasmid cDNA library, one should be able to quickly screen the library. Typically, bacteria are plated at a density of 200–500 clones per 150-mm plate onto nitrocellulose filters (132-mm diameter, Schleicher and Schuell, BA85, 0.45-μm pore size, #20570) and then grown on 150-mm LB agar plates containing 100 μg/ml ampicillin (Fig. 5). By keeping the colony density low, colonies may be picked individually after the first round of screening to begin the second screen with the SubMaster filters. The agar plates must be dried for 2 hr at 37° (with lids off) before use. The master filters are labeled with a pencil or marking pen (Schleicher and Schuell, Keene, NH) to indicate the library and filter number. We usually perform the plating using a Millipore filtration device (316 stainless sanitary 142-mm filter holder) (see Fig. 5). The inlet of the device must be equipped with a top funnel which should be tightly sealed to the inlet port (e.g., using a rubber stopper with a hole), and the outlet should be connected to a diaphragm pump (e.g., VWR) via a vacuum flask, to produce a slight vacuum. The procedure for plating the library using the filtration device is as follows: Sterilize the filtration device using an autoclave. After titering the library, dilute 200–500 bacteria into 200 ml of sterile LB (without ampicillin) in a 500-ml screw cap bottle. With filter in place and filtration device clamped shut, the bacterial solution is poured into the funnel and the vacuum applied until air is pulled through the outlet. The pump is set to pump through the 200 ml solution in 1–2 min. The vacuum is broken at the outlet to prevent reflux of solution that would disturb the "stuck" bacteria. The master filter is transferred to an LB/agar/ampicillin plate (step 1 in Fig. 5) and colonies are grown only to ~1 mm in diameter (step 2). Next, a replica must be made for functional screening and the master filter is stored for later use (steps 3–13). To make a replica, the master filter is removed from the agar plate and placed on a clean filter paper (Schleicher and Schuell blotting paper, size 15 × 20 cm). A second, new nitrocellulose filter is labeled as above and moistened by placing the writing side down on a fresh LB/ampicillin plate and then on top of the master filter (step 3). Another filter paper on top completes the filter "sandwich" (step 4). The colonies are transferred to the second nitrocellulose filter by using a rolling pin, back and forth over the entire filter sandwich (step 5). The filters are then carefully pulled apart, the master filter is regrown (step 6), and the replica filter is grown on a fresh LB/ampicillin plate (step 9). To increase the amount of plasmid DNA isolated from the replica filters, the colonies are overgrown (colony sizes 3–4 mm) (step 7).

Prior to storage of the master filters, the colonies must be regrown on 5% glycerol/LB/ampicillin plates (step 10). Filters are grown for a few hours at 37° to bring colony size back to 1 mm with a rounded top appearance. The filter is then transferred on filter paper. Next, a labeled third nitrocellulose filter is moistened on a 5% glycerol plate and placed on top of the master filter (step 11). Another filter is placed on top to complete the filter "sandwich" and colonies are once again transferred using a rolling pin. Two more dry filter papers and one water-saturated filter paper are added to each side of the sandwich (step 12). The pile is placed into a hybridization bag (Gibco-BRL), sealed, and stored at −80° until needed (step 13).

When frozen filters are needed, they are thawed to room temperature and gently pulled apart. The master filter is grown as above on 5% glycerol/LB/ampicillin, whereas the third filter is grown on LB/amp. The master filter is sandwiched with a fourth filter and refrozen at −80°.

Isolating Plasmid DNA from Filters

A filter is removed from the LB/amp plate and placed into a clean 150-mm dish. Then, 2–3 ml of cold LB are added to the filter. Using a cell scraper (Falcon or Costar), the colonies are pooled toward the middle of the filter (Fig. 6, step 2). The bacteria are transferred from the filter to a clean 14-ml Falcon tube (#2059) using disposable transfer pipettes. The addition of LB and the scraping should be repeated at least once. Each filter will yield enough bacteria for a midi-size plasmid DNA preparation (40–60 μg). Filter scrapings of replicates will yield usually the equivalent of 50–100 ml of liquid bacterial growth, i.e., a midsized bacterial plasmid preparation of 40–150 μg. The commercially available plasmid purification kits (e.g., Qiagen) are quick and give high-quality plasmid DNA. At the end of the plasmid prep, dissolve DNA in an appropriate volume of DNase-free water or TE buffer (usually 100–200 μl). Since pipetting to resuspend the pellet can shear the plasmid DNA, the tube is incubated on wet ice for 15–30 min to allow the DNA to slowly rehydrate. The DNA solution is transferred to a labeled 1.5-ml tube and stored at $-20°$ until needed. The DNA concentration is quantitated by making a 1 : 100 dilution for UV spectroscopy (absorbance scan from 220 to 320 nm; see above).

Making cRNA

From each filter, 2–5 μg of plasmid are linearized with a restriction enzyme that cuts at the 3′ end of the cDNA inserts. If the SuperScript DNA synthesis system is used, the appropriate enzyme is *Not*I. The cut plasmid is purified by phenol–chloroform extraction[23] and the cDNA is precipitated. The linearized cDNA template is then ready to be in vitro transcribed. Several commercial kits are available to facilitate this step, for example, the Stratagene Capping Kit or Ambion mMessage mMachine. The protocol which is currently used is as follows:

1. Add into a 1.5-ml microcentrifuge tube: 5 μl RNase inhibitor (Promega, Madison, WI RNasin), 20 μl 5× transcription buffer (200 m*M* Tris, pH 7.5, 250 m*M* NaCl, 40 m*M* $MgCl_2$, 10 m*M* spermidine), 4 μl rNTP mix (10 m*M* rUTP, 10 m*M* rCTP, 10 m*M* rATP, 1 m*M* rGTP, in water, pH 7.5), 3 μl CAP analog (5′7meGppp5′G, 2.5 m*M*; Pharmacia), 3 μl 1 *M* DTT, linearized DNA (2–5 μg). Add RNase-free water (Gibco-BRL) to 98 μl. Gently mix, quick spin, and add 2.0 μl (100 U) of the appropriate RNA polymerase. T7 RNA polymerase is required when using the SuperScript cDNA synthesis system (Stratagene).
2. Incubate at 37° for 10 min. Add 1.0 μl of 10 m*M* rGTP. Incubate at 37° for 10 min. Add another 1.0 μl of 10 m*M* rGTP. Incubate 10 min at 37°. These additions favor the extension reaction after CAP analog incorporation.

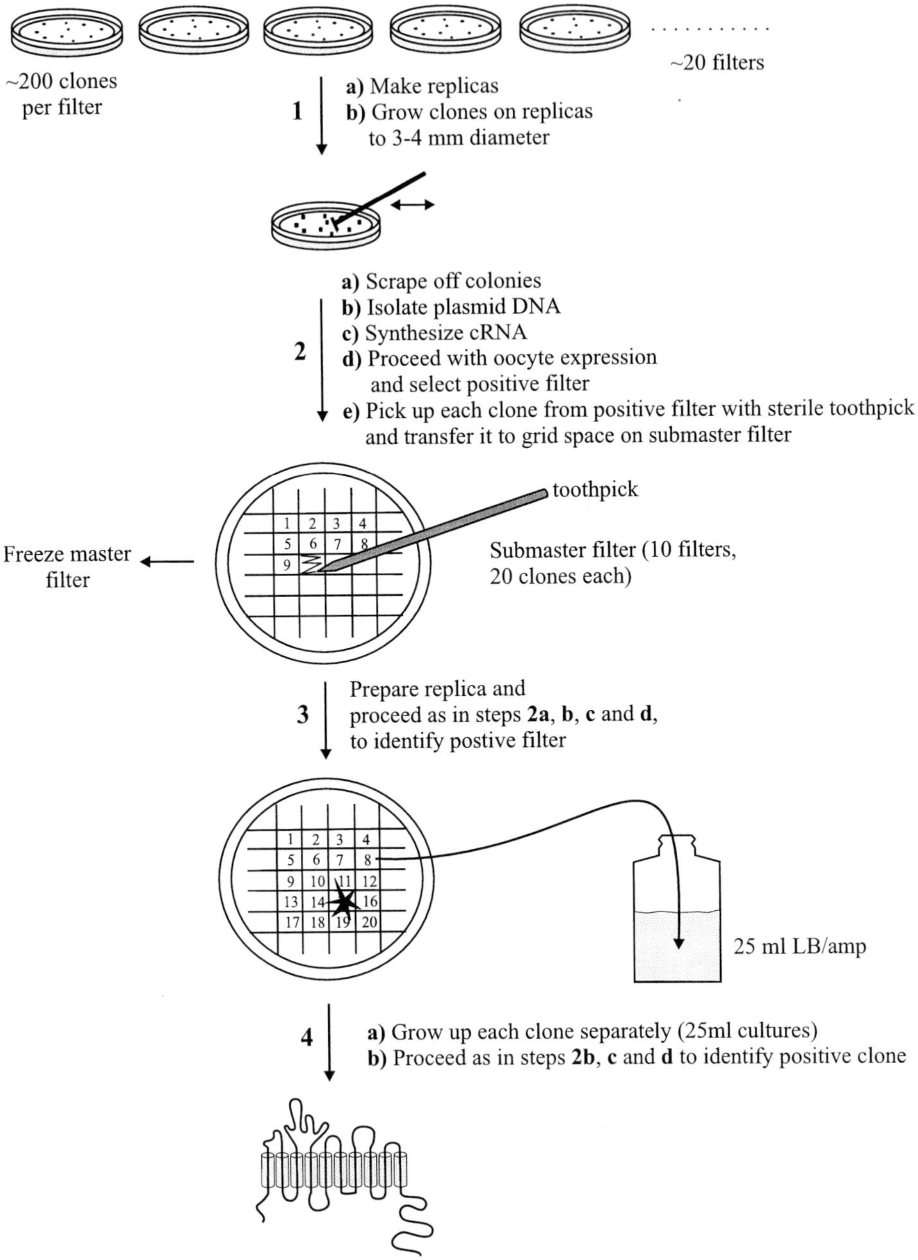

FIG. 6. Schematic diagram illustrating the procedures required for the functional screening of cDNA (colony) libraries.

3. Add 4.0 μl (40 U) of RNase-free DNase (Stratagene). Incubate 20 min at 37°.

4. Bring the final volume to 500 μl with RNase-free water and extract with 500 μl of Gibco-BRL phenol/chloroform/isoamyl alcohol (25:24:1, #15593-031). Vortex for 5 sec and spin in a microcentrifuge for 5 min at room temperature. Carefully transfer the top aqueous layer and transfer to another 1.5-ml tube. Reextract the cRNA with 1 volume of 24:1 chloroform:isoamyl alcohol. Vortex and spin again. Transfer the top aqueous layer to another 1.5-ml tube (tube B) and place it on ice. Precipitate cRNA with 0.5 volumes of 7.5 *M* ammonium acetate and 2.5 volumes (RNA + ammonium acetate) of ethanol at −20° for at least 2 hr. Pellet the cRNA by centrifugation, pour off the supernatant, and resuspend the wet pellet in 200 μl of RNase-free water (invert the tube several times). Ethanol precipitate with ammonium acetate at −20° as above. Wash the pellet with 70% ethanol and dry for 5 min in a rotary evaporator. Resuspend the pellet in 10–20 μl of RNase-free water (Gibco-BRL). Let the tube sit on ice for at least 15 min to hydrate the dry pellet and to ensure its complete dissolution. Precipitation of RNA with ammonium acetate is preferred, because other salts are not as easily removed in the 70% ethanol wash.[23] After cRNA synthesis, it is important to check the quality of the cRNA by UV spectroscopy and minigel electrophoresis. Use 0.5 μl for the minigel and 0.5 μl (diluted into 50 μl of water) for UV spectroscopy (see above).

Analyzing cRNA by Oocyte Expression

To analyze cRNA synthesized from pools of clones, inject 50 ng of cRNA (~1 mg/ml = 1 μg/μl) and proceed with the functional assay used to identify the positive RNA size fraction. If no positive pool can be identified, repeat the plating of the library and screen additional pools of clones.

Making Submaster Filters

Once a master plate is identified as having the activity of interest, the master filter is thawed to room temperature, a replica is made and grown on LB/ampicillin, and the master filter is grown on a 5% glycerol/LB/ampicillin plate, sandwiched with a fourth filter, and refrozen at −80° (steps 9–13 in Fig. 5). Colonies from that replica are then manually transferred to submaster filters (Fig. 6, step 2).

A grid pattern is drawn on the submaster filters as shown in Fig. 6, and one by one, colonies are transferred using sterile toothpicks from the master filter to the appropriate grid space. To obtain sufficient quantities of plasmid DNA, the picked colonies are streaked multiple times in their individual grid space (step 2, Fig. 6). After a filter's grid is filled, the filter is grown for 8–12 hr. The same procedure is executed for all the colonies of the identified master filter. Similar to the procedure with the master filters, submaster filters are next replicated and then frozen at $-80°$ after regrowth on 5% glycerol plates.

Screening Submaster Filters

As for the master filters (Fig. 5, step 7), the submaster replicas are grown until streaks are quite large and full. Filters are scraped and plasmid DNA is isolated as described above. The DNA yield will be ~100–150 μg per plate. As before, plasmid DNA from each submaster is linearized, cRNA synthesized, and cRNA injected into oocytes for functional expression. From this process a single submaster filter is identified.

Finding Single cDNA Clone

The submaster of interest is thawed, the replica grown on a LB/amp plate, and each of the 20 colonies is grown up separately in 25 ml LB/ampicillin (Fig. 6, step 4). DNA is isolated, cRNA is synthesized, and clones are analyzed by oocyte expression until a single clone is identified which encodes the activity sought.

Variations of Expression Cloning and Alternative Procedures

Multisubunit Proteins and Expression Cloning by Functional Complementation

Even when a single clone is identified, it is possible that there are additional subunits or accessory proteins that are necessary for complete function. ATPases,[32] as well as certain channels (e.g., GIRK K^+-channel subunits[33]) or the epithelial amiloride-sensitive sodium channel ENaC[17] (see proceeding text), have been shown to be composed of two or more heterologous subunits.

With a pure clone, the activity as expressed in oocytes should be severalfold larger than the activity of poly(A)$^+$ RNA, master filters, and submaster filters. If the activity is similar or lower than that of poly(A)$^+$ RNA, this is an indication that there is another component necessary for full activity. The library may be rescreened by coexpressing the single cRNA with each of the master filter cRNA pools. The coexpression is repeated for all the screening steps above until a second cDNA is identified. Canessa and coworkers[17] developed the functional complementation assay to rescreen their library, and thus they identified three distinct subunits of the epithelial Na^+ channel. These investigators assumed that the additional subunit is of the same poly(A)$^+$ RNA size as that from which the cDNA library was con-

[32] F. Jaisser, P. Jaunin, K. Geering, B. C. Rossier, and J. D. Horisberger, *J. Gen. Physiol.* **103,** 605 (1994).

[33] G. Krapivinsky, E. A. Gordon, K. Wickman, B. Velimirovic, L. Krapivinsky, and D. E. Clapham, *Nature* **374,** 135 (1995).

structed. However, if the RNA encoding the complementing subunit has a different size, another cDNA library will need to be constructed from the appropriate RNA size.

Other Limitations of Expressions Cloning using Xenopus Oocytes

For successful expression cloning, one of the most important requirements is that the endogenous transport activity of interest in non-RNA injected (i.e., control) oocytes be low. The low background uptake or electrical response can often be achieved using transporter-specific and/or nondegradable substrate analogs, e.g., α-methyl-D-glucopyranoside for the Na^+/glucose cotransporter SGLT1,[1] and glycylsarcosine for the oligopeptide transporter PepT1.[16] High levels of endogenous oocyte uptake are often encountered for amino acids, particularly neutral and basic amino acids. To clone amino acid transporters, it is therefore important to select the tissue with the highest possible expression level of the amino acid transporter of interest. The use of mRNA from rabbit small intestine rather than rat brain undoubtedly contributed to our success in isolating a high-affinity glutamate transporter cDNA (EAAC1) by *Xenopus* oocyte expression. Previous attempts to expression clone a rat-brain high-affinity glutamate transporter apparently failed, presumably because glutamate transporter mRNA is not as abundant in rat brain compared to rabbit intestine (see Ref. 34).

A common issue arising during attempts to clone dibasic and neutral amino acid transporters has been the existence of transporter activators such as D2/rBAT and 4F2 heavy chain.[35,36,36a] These are type II membrane glycoproteins with a single transmembrane domain. When expressed in oocytes, these proteins apparently induce the expression of endogenous oocyte transporters through a mechanism which is not yet clearly understood. Thus, many attempts to clone amino acid transporters have failed to yield the desired transporter cDNA and, instead, lead to the isolation of transporter activator cDNA. A possible strategy to circumvent this issue is to turn to another species that does not express transporter activators. Accordingly, using RNA from the midgut of *Manduca sexta*, we have isolated a system-B-like amino acid transporter cDNA (KAAT1).[36b]

A final consideration with expression cloning of substrate activated transport systems is the treatment of the oocyte after RNA injection. For

[34] R. D. Blakely, J. Clark, T. Pacholczyk, and S. G. Amara, *Neurochem,* **56,** 860 (1991).

[35] R. G. Wells, W. S. Lee, Y. Kanai, J. M. Leiden, and M. A. Hediger, *J. Biol. Chem.* **267,** 15285 (1992).

[36] R. G. Wells and M. A. Hediger, *Proc. Natl. Acad. Sci. USA* **89,** 5596 (1992).

[36a] J. Bertran, A. Werner, M. L. Moore, G. Stange, D. Markovich, J. Biber, X. Testar, A. Zorzano, M. Palacin, and H. Murer, *Proc. Natl. Acad. Sci. USA* **89,** 5601 (1992).

[36b] M. Castagna, C. Shayakul, D. Trotti, V. F. Sacchi, W. R. Harvey, and M. A. Hediger, *Proc. Natl. Acad. Sci. USA* **95,** 5395 (1998).

example, one of the determinants of using Barth's solution or OR3 to maintain injected oocytes is the composition of the culture solution. If one is interested in cloning an amino acid transport system, Barth's rather than OR3 should be used because OR3 is supplemented with most of the essential amino acids. On the other hand, if a channel, ion transporter, or enzyme is of expression cloning interest, OR3, which contains many nutrients, is likely the better choice of oocyte culture media.

Expression Cloning by Hybrid Depletion

One may encounter large messages (>5 kb) encoding the proteins of interest. In addition to efficient size fractionation of mRNA, careful size fractionation of cDNA before ligation into the plasmid vector will significantly facilitate the cloning of large cDNAs. Alternatively, this may be achieved by a hybrid-depletion screening approach in which cDNA size is not critical. Accordingly, a cDNA fragment is isolated by screening an antisense cDNA library for the ability of cDNA clones to decrease the expression of mRNA by hybrid depletion. This is achieved by preincubating mRNA with antisense cDNA before injection into oocytes. The cDNA fragment isolated in this manner is then used as a probe to obtain the full-length cDNA. Using this technique Jentsch and co-workers isolated the ClC-0 chloride channel from *Torpedo marmorata* electric organ, which is encoded by a 9- to 10-kb message.[7]

Expression Cloning Using *Xenopus* Embryos

A variant of oocyte expression cloning, injecting *Xenopus* embryos, has been used successfully to clone the cDNAs for proteins important in development. Lemaire and co-workers injected embryosin a ventral–vegetal blastomere with library cRNA and screened for secondary axis formation, isolating *siamois,* a novel *Xenopus* homeobox-containing gene.[36c] Kirschner's group has used this same approach to isolate *Xombi,* a T-box transcription factor involved in mesodermal patterning and blastopore lip formation.[36d] This same group cloned a second developmental factor, *sizzled* or *szl,* which encodes a secreted antagonist of wnt signaling apparently involved in ventral inhibition Xwnt8 and XmyoD, by looking for duplicated body axis.[36e] Thus, the *Xenopus* embryo may be used as a tool to clone novel factors that influence the developmental program.

Expression Cloning with Oocytes using Plasmid cDNA

Finally, another variant of expression cloning using *Xenopus* oocytes is to inject circular plasmid DNA into the nucleus of the oocyte for expression.

[36c] P. Lemaire, N. Garrett, and J. B. Gurdon, *Cell* **81,** 85 (1995).
[36d] K. D. Lustig, K. L. Kroll, E. E. Sun, and M. W. Kirschner, *Development* **122,** 4001 (1996).
[36e] A. N. Salic, K. L. Kroll, L. M. Evans, and M. W. Kirschner, *Development* **124,** 4739 (1997).

Coady and co-workers used nuclear injections to clone a rabbit renal neutral amino acid transporter, rBAT.[37] Rather than monitoring isotope uptake, these investigator alanine-elicited currents using either 2-electrode voltage clamping (VC) or the cut-open oocyte technique.[38] Although nuclear injections have a lower oocyte survival rate, this technique allows the investigator to inject picogram quantities of plasmid DNA and avoid the possibility of RNA degradation. Additionally, this technique forces the oocyte to make a *Xenopus* mRNA for protein synthesis, rather than relying on the oocyte to faithfully translate injected cRNA.

Expression Cloning with Cultured Cells

As an alternative approach to *Xenopus* oocyte expression cloning, one may use cultured cells as an expression cloning system. In general, this method works well in cases where a powerful screening approach is available to detect transfected cells, i.e., ligand binding to receptors or uptake of radiolabeled compounds by transporters. A review by Simonsen and Lodish describes suitable host cells for mammalian expression cloning and summarizes different classes of proteins cloned by this technique and different functional assays used.[39]

To clone the cocaine- and antidepressant-sensitive noradrenaline transporter, Amara and co-workers transfected a cDNA library into COS-1 cells.[40] Transfectants expressing the transporter were identified based on their ability to accumulate intracellular *m*-[^{125}I]iodobenzylguanidine, a noradrenaline analog. Positive cells were visualized by autoradiography. cDNA was then rescued from the positive COS-1 transfectants by Hirt lysis. The resulting plasmid pools were rescreened until a single clone was isolated. Using a similar strategy in COS cells, Wong and associates expression-cloned an small intestine (ileal) form of the Na^+-dependent bile acid transporter (IBAT) by monitoring taurocholate uptake.[41]

The vesicular monoamine transporter cDNA was also isolated using a cultured cell expression system.[42] The screening procedure relied on the

[37] M. J. Coady, F. Jahal, X. Chen, G. Lemay, A. Berteloot, and J-Y. Lapointe, *FEBS Lett.* **356,** 174 (1994).

[38] M. Taglialatela, L. Toro, and E. Stefani, *Biophys. J.* **61,** 78 (1992).

[39] H. Simonsen and H. F. Lodish, *Trends Pharmacol. Sci.* **15,** 437 (1994).

[40] T. Pacholczyk, R. D. Blakely, and S. G. Amara, *Nature* **350,** 350 (1991).

[41] M. H. Wong, P. Oelkers, A. L. Craddock, and P. A. Dawson, *J. Biol. Chem.* **269,** 1340 (1994).

[42] Y. Liu, D. Peter, A. Roghani, S. Schuldiner, G. G. Prive, D. Eisenberg, N. Brecha, and R. H. Edwards, *Cell* **70,** 539 (1992).

[43] M. Paulmichl, Y. Li, K. Wickman, M. Ackerman, E. Peralta, and D. Clapham, *Nature* **356,** 238 (1992).

[44] D. Grundemann, V. Gorboulev, S. Gambaryan, M. Veyhl, and H. Koepsell, *Nature* **372,** 549 (1994).

[45] H. Lübbert, B. Hoffman Jr., T. P. Snutch, T. Van Dyke, A. J. Levine, P. R. Hartig, H. A. Lester, and N. Davidson, *Proc. Natl. Acad. Sci. USA* **84,** 4332 (1987).

ability of the expressed vesicular transporter to protect cultured cells from the susceptibility to *N*-methyl-4-phenylpyridinium (MPP^+). The protection resulted from the toxin being sequestered into the intracellular vesicles. The investigators then rescued the cDNA clone that encodes the vesicular amine transporter.

Future Need for Expression Cloning

Although the sequencing projects for various genomes are rapidly progressing, and many disease genes have been identified by genetic linkage analysis, a universal procedure to determine the function of these genes has not yet been developed. In some cases, the function of genes can be predicted based on their similarity to others of known function. However, there are still many more gene families to be discovered, and the function of many gene families yet remains to be determined. Thus, expression cloning will remain, at least in the near future, an important approach to identifying genes and proteins responsible for specific functions.

---

[46] Y. Masu, K. Nakayama, H. Tamaki, Y. Harada, M. Kuno, and S. Nakanishi, *Nature* **329,** 836 (1987).
[47] M. Hollmann, A. O'Shea-Greenfield, S. W. Rogers, and S. Heinemann, *Nature* **342,** 643 (1989).
[48] M. Yamashita, H. Fukui, K. Sugama, Y. Horio, S. Ito, H. Mizuguchi, and H. Wada, *Proc. Natl. Acad. Sci. USA* **88,** 11515 (1991).
[49] Z-I. Honda, M. Nakamura, I. Miki, M. Minami, T. Watanabe, Y. Seyama, H. Okado, H. Toh, K. Ito, T. Miyamoto, and T. Shimizu, *Nature* **349,** 342 (1991).
[50] T-K. H. Vu, D. T. Hung, V. I. Wheaton, and S. R. Coughlin, *Cell* **64,** 1057 (1991).
[51] T. Takahashi, E. Neher, and B. Sakmann, *Proc. Natl. Acad. Sci. USA* **84,** 5063 (1987).
[52] T. Snutch, *Trends Neuro. Sci.* **11,** 250 (1988).
[53] Q. Q. Huang, S. Y. Yao, M. W. Ritzel, A. R. Paterson, C. E. Cass, and J. D. Young, *J. Biol. Chem.* **269,** 17757 (1994).
[54] A. M. Pajor, *J. Biol. Chem.* **270,** 5779 (1995).
[55] A. E. Luebke, G. P. Dahl, B. A. Roos, and I. M. Dickerson, *Proc. Natl. Acad. Sci. USA* **93,** 3455 (1996).
[56] M. Boll, M. Herget, M. Wagener, W. M. Weber, D. Markovich, J. Biber, W. Clauss, H. Murer, and H. Daniel, *Proc. Natl. Acad. Sci. USA* **93,** 284 (1996).
[57] L. Bai and A. M. Pajor, *Am. J. Physiol.* **273,** G267 (1997).
[58] D. H. Sweet, N. A. Wolff, and J. B. Pritchard, *J. Biol. Chem.* **272,** 30088 (1997).
[59] N. A. Wolff, A. Werner, S. Burkhardt, and G. Burckhardt, *FEBS Lett.* **417,** 287 (1997).
[60] F. Norbis, M. Boll, G. Stange, D. Markovich, F. Verrey, J. Biber, and H. Murer, *J. Memb. Biol.* **156,** 19 (1997).
[61] J. A. Gutierrez, J. Yu, S. Rivera, and M. Wessling-Resnick, *J. Cell. Biol.* **139,** 895 (1997).
[62] J. Tian, G. H. Thomsen, H. Gong, and W. J. Lennarz, *Proc. Natl. Acad. Sci. USA* **94,** 10729 (1997).
[63] D. Julius, A. B. MacDermott, R. Axel, and T. M. Jessell, *Science* **241,** 558 (1988).
[64] K. D. Lustig, A. K. Shiau, A. J. Brake, and D. Julius, *Proc. Natl. Acad. Sci. USA* **90,** 5113 (1993).

Acknowledgments

It is a particular pleasure to acknowledge Michela Castagna, Michael Coady, You-Jun Fei, Cameron Gundersen, Steven Hebert, Stephan Nussberger, Davide Trotti, Chairat Shayakul, Matthias Stelzner, Taro Tokui, Hiroyasu Tsukaguchi, Ernest Wright, Rebecca Wells, and Guofeng You, who have contributed to the development of the expression cloning procedures.

[3] Cloning of Genes or cDNAs Encoding Neurotransmitter Transporters and Their Localization by Immunocytochemistry

By NATHAN NELSON and QING-RONG LIU

The genes encoding most of the proteins in living cells can be grouped into gene families. The size of a gene family may vary in numbers of homologous proteins from hundreds to just two. The homology within a gene family may be as high as more than 90% and as low as less than 20%. However, there are clusters of highly homologous regions within a gene family. Degenerated oligonucleotide or cDNA fragments can be designed according to these homologous regions and utilized as radiolabeled probes to screen both cDNA and genomic libraries under low stringency conditions. Genomic libraries have the advantage of having equal distribution in the library regardless of their expression level and genomic clones can be used for designing probe to clone rarer cDNA that is expressed in specific tissues.[1] Additional members of a gene family can be thus isolated having similar structure and functions to the protein whose gene or cDNA was first cloned.[2] This technique is particularly useful for isolating homologs of neurotransmitter receptors and transporters because their function can be characterized by binding or uptake assays in various expression systems.[3,4]

Although the preferred technique for isolating homologous cDNAs is the polymerase chain reaction (PCR) for cloning new genes, low stringency screening may have some advantages in circumventing interference by long introns and finding additional homologs which may evade PCR detection.

[1] Q.-R. Liu, S. Mandiyan, H. Nelson, and N. Nelson, *Proc. Natl. Acad. Sci. USA* **89,** 6639 (1992).
[2] Q.-R. Liu, B. López-Corcuera, S. Mandiyan, H. Nelson, and N. Nelson, *J. Biol. Chem.* **268,** 2106 (1993).
[3] J. Guastella, N. Nelson, H. Nelson, L. Czyzyk, S. Keynan, M. C. Miedel, N. Davidson, H. A. Lester, and B. I. Kanner, *Science* **249,** 1303 (1990).
[4] T. Pacholczyk, R. D. Blakely, and S. G. Amara, *Nature* **350,** 350 (1991).

0076-6879/98 $25.00

Materials and Reagents

Libraries

Mouse cosmid (pWE15 vector) genomic library (Stratagene, La Jolla, CA). Mouse brain cDNA (Uni-ZAP XR vector) library (Stratagene). Mouse brain 5′-STRETCH PLUS cDNA (λgt 10 vector) library (Clontech, Palo Alto, CA). Rat brain cDNA (λZAP II vector) library (Stratagene).

Kits

Multiprime DNA labeling systems (Amersham, Arlington Heights, IL) for ^{32}P labeling of DNA probe. Geneclean II kit (BIO 101, Vista, CA) for purification of DNA fragment from agarose gel.

Materials

Nitrocellulose membranes (0.45 μm pore size, 137- and 82-mm diameter) for genomic and cDNA library screening (Schleicher and Schuell, Keene, NH).

Petri dishes (150 × 15 mm and 100 × 15 mm in sizes) for library plating (Fisher Scientific, Pittsburgh, PA). X-ray films (X-OMAT AR, 14 × 17 and 8 × 10 inches) for autoradiography (Eastman Kodak Co., Rochester, NY).

Seal-A-Meal bags (Dazey Corporation, Industrial Airport, KS) for hybridization.

Apparatus

Water bath shaker (Model G76D, New Brunswick, NJ) for hybridization. UV cross-linker (Stratalinker UV, Stratagene, La Jolla, CA) for fixing DNA on nitrocellulose membranes. Inoculating turntable (6-inch diameter, Fisher Scientific, Pittsburgh, PA) for spreading bacteria to agar plates. Heating block (Fisher Scientific) for denaturing probes and warming LB top agarose. Scotchpack Pouch Sealer (VWR Scientific, South Plainfield, NJ) for sealing hybridization bags. Dot-blot apparatus (Manifolds, Schleicher and Schuell). Autoradiography cassettes with intensifying screens (Sigma Chemical Co., St. Louis, MO) for film exposure.

Reagents

LB broth base (GIBCO-BRL, Gaithersburg, MD): Suspend 20 g LB powder in 1 liter of demineralized water and autoclave for 15 min at 121°.

LB agar (GIBCO-BRL, Gaithersburg, MD): Suspend 32 g of LB agar in 1 liter of demineralized water and autoclave 15 min at 121°.

LB agar containing 100 mg/ml of ampicillin: Cool the autoclaved LB agar to 55° and add 4 ml of 25 mg/ml ampicillin stock solution to 1 liter of LB agar.

LB top agarose: Microwave 100 ml of LB broth containing 0.6 g of agarose (GIBCO-BRL, Gaithersburg, MD) for 2 min. Cool to 55° and set on 60° heatblock.

SM buffer: One liter of SM buffer contains 5.8 g of NaCl, 2 g of $MgSO_4 \cdot 7H_2O$, 50 ml of 1 *M* Tris–HCl (pH 7.5), and 5 ml of 2% gelatin (w/v).

BLOTTO: Dissolve 5 g of powdered dry milk (Carnation brand, nonfat milk) in 100 ml phosphate-buffered saline (PBS) and 0.01% sodium azide.

20× SSC: Dissolve 175.3 g of NaCl and 88.2 g of sodium citrate in 800 ml of water. Adjust pH to 7.0 with NaOH and bring the volume to 1 liter.

Ampicillin stock (25 mg/ml): Dissolve 2.5 g of ampicillin to 100 ml of demineralized water and then filter through sterile 0.22-μm filter (Millipore, Bedford, MA). Store at 4° up to 1 month.

Low Stringency Screening

Colony Screening with ^{32}P-Labeled Oligonucleotides

1. *Escherichia coli* (XL1-blue MR) containing a mouse genomic cosmid library (constructed in pWE15 vector) is spread on six large LB agar/ampicillin agar plates (150 × 15 mm, prewarmed at 37°) to give approximately 10^4 colonies per plate. The plates are incubated at 37° for about 10 hr at the inverted position.
2. Check the colony growth after 8 hr incubation (no more than 12 hr) by holding the plates against the fluorescent light at different angles. When colonies appear as shiny speckles, transfer the plates to room temperature.
3. Number the plates with a marker and nitrocellulose membranes with a ballpoint pen.
4. Align and place nitrocellulose membrane on top of the plate evenly without trapping bubbles between the membrane and the agar.
5. Heat an 18-gauge needle on flame until red hot. Punch the membrane and agar asymmetrically. Mark the side of the plate where the number is on the nitrocellulose membrane.
6. Lift the nitrocellulose membrane immediately from the plate with forceps. Turn the membrane face up with attached bacteria on the top and carefully lay the membrane on top of a fresh LB/ampicillin agar plate.
7. Grow the plates and the replicas (bacteria grown on nitrocellulose membrane) at 37° overnight in the inverted position.
8. Next morning, wet three pieces of Whatman 3 MM (Clifton, NJ) filter paper completely with 10% (w/v) sodium dodecyl sulfate (SDS), dena-

ture buffer (0.2 *M* NaOH, 1.5 *M* NaCl) and neutralization buffer (0.5 *M* Tris–HCl, pH 8.0, 1.5 *M* NaCl), respectively.

9. Transfer the membranes face up from the plates to the prewet filter paper: SDS filter paper for 3 min, denature filter paper for 5 min, and neutralization filter paper for 5 min.

10. Dry the membrane on a piece of new filter paper and cross-link the bacteria DNA on nitrocellulose membrane with UV for two cycles (Stratalinker UV cross-linker).

11. Wash the membranes with 3× SSC at room temperature for 5 min, and wash again with 3× SSC/0.1% SDS at 65° for 1.5 hr.

12. Transfer the membranes to prehybridization solution (6× SSC, 0.1% SDS, and 0.25% dry milk) and incubate at 37° for 2–4 hr.

13. Transfer the membranes to a Seal-A-Meal bag containing hybridization solution (6× SSC and 0.25% dry milk) and ^{32}P-labeled oligonucleotide (10^6 cpm/ml). Drive out air bubbles and seal the bag with Scotchpack Pouch Sealer. Hybridize in a 65° waterbath with shaking overnight.

14. Next morning, cool down the waterbath gradually to 42°. Continue to incubate the membranes at 42° for 1–2 hr.

15. Wash the membrane first with 2× SSC/0.1% SDS three times, 10 min each time. Wash the membrane with 1× SSC/0.1% SDS twice, 30 min each time.

16. Wrap a piece of used X-ray film with a piece of plastic sheet. Put the nitrocellulose membranes face up on the top of the wrapped film (six membranes per film) and cover the membrane with another piece of plastic sheet.

17. Stick paper tapes on the corners of the wrapped film and put radioactive marks on the tapes which are then covered by Scotch tape to avoid radioactive contamination.

18. In a dark room, sandwich the membranes between X-ray film and intensify screen in a film cassette. Store the cassette in a −80° freezer for 12–24 hr.

19. Develop the X-ray film. Overlay the film on top of the wrapped membranes with the fluorescence light box as the background. Align the film according to the radioactive markers. Mark the needle holes on the film with marker and number the membranes.

20. Turn over the marked X-ray film on a fluorescence light box. Identify the hybridized colonies and mark the positive spots with a marker.

21. Overlay the plates with bacteria colonies on the top of the film. Use a sterile toothpick to pick up the colonies in the vicinity of the positive spot and inoculate in 5 ml LB/ampicillin medium.

22. Grow the bacteria at 37° overnight with shaking and save it for the bacteria stocks.

23. Dilute the stock 10^5 times and take 50 μl to spread on a small LB/ampicillin agar plate (100 × 15 mm). A few hundred colonies per plate are the appropriate titer for the secondary screening.

24. The rest of secondary screening is the same as the primary screening. A single positive colony can be picked up and grown for cosmid minipreparation.

25. Confirm the positive clones by dot-blot analysis as follows: Add 5 μl of plasmid prep to 300 μl of dot-blot denaturing solution (in 15 ml solution: 0.85 ml of 1 *M* Tris–HCl, 5 ml of 20× SSC, and 0.6 ml of 5 *M* NaOH). Heat samples at 80° for 10 min and chill on ice immediately for 5 min. Add 40 μl of 2 *M* Tris–HCl (pH 6.8) to the chilled samples. Apply denatured plasmid to nitrocellulose membrane using dot-blot apparatus. Cross-link DNA to the nitrocellulose membrane with UV and hybridize with ^{32}P-labeled oligonucleotide as described for library screening.

26. Digest the positive plasmids with *Sau*IIIA restriction enzyme and run parallel samples on an agarose gel.

27. Cut the agarose gel in half. One half will be used for the Southern blot and the other half will be used for the purification of DNA fragments.

28. Southern blot hybridization is performed at the same conditions as library screening.

29. The positive bands (usually below 1 kb) are excised for the agarose gel and DNA fragments purified by the Geneclean method (BIO 101, Vista, CA).

30. The *Sau*IIIA fragments are ligated with pBluescript (Stratagene) plasmid which is linearized with *Bam*HI and dephosphorylated by CIAP (calf intestinal alkaline phosphatase).

31. Plate the competent *E. coli* (DH5α) transformed with the subclones on a small LB/ampicillin plate (100 × 15 mm) for colony screening.

32. Screen positive subclones with the radiolabeled oligonucleotide as described above.

33. Purify the positive subclones and perform sequence analysis to identify the homolog of the neurotransmitter transporters.

Screening by Homology of cDNA Libraries

The library selected for screening will depend on the distribution of the clone of interest. The library should have a relatively large portion of long cDNA inserts. The source of the library may be human brain if medical applications are to be considered or mouse brain if one of the goals is to interrupt specific genes. One of the libraries utilized in our laboratory was Uni-ZAP XR cDNA library from neonatal mouse brain (Stratagene). The cDNA is directionally cloned into the *Eco*RI–*Xho*I site of the cloning vector. For obtaining new clones of the subfamily of GABA transporters, the library was screened with a ^{32}P-labeled cDNA encoding the taurine transporter.[1,3,5]

Plaque Screening with ^{32}P-Labeled DNA Probe

1. Streak *E. coli* (BB4, XL1-blue, or C600 Hf1 strains) on LB plates and grow at 37° overnight in an inverted position.
2. A single colony is picked and inoculated into 50 ml LB medium containing 0.2% maltose.
3. Incubate the bacteria culture at 37° overnight with shaking (<12 hr). Spin down the cells at 6000 *g* for 5 min at room temperature and discard the supernatant.
4. Resuspend the pellet in 25 ml of SM buffer and use appropriate volume of SM buffer to bring bacterial suspension to $OD_{600\ nm}$ of 1.0.
5. Dilute cDNA library 100 times [original titer 10^{10}–10^{12} plaque-forming units (pfu)/ml] with SM buffer and determine actual titer of the diluted suspension.
6. Add the diluted library (10^6 pfu) to 8 ml of the bacterial suspension. Incubate library and bacteria mixture at 37° for 20 min and then divide it into 12 culture tubes, each containing 0.6 ml.
7. At the same time, microwave 0.6% agarose in LB medium for 2 min. Cool to 55° and set on 65° heat block.
8. Dry 12 150 × 15 mm LB agar plates at 42° for 1 hr before plating. Add 7 ml of LB top agarose (55°) to each of the 12 culture tubes containing the library and bacterial mixture. Pour the top agarose evenly on the prewarmed LB agar plate.
9. After the agarose solidifies, incubate the plates at 37° in inverted position for 6–8 hr until clear plaques appear and become almost confluent.
10. Cool the plates to room temperature and then transfer them to 4° for at least 2 hr.
11. Number the plates with a marker and the nitrocellulose membranes with a ballpoint pen.
12. Align and place nitrocellulose membrane on top of the plate evenly without trapping bubbles between the membrane and the agar.
13. Heat an 18-gauge needle on flame until red hot. Punch the membrane and agar asymmetrically at three points. Mark the side of the plate where the number is on the nitrocellulose membrane.
14. Incubate at room temperature for 2 min and lift the nitrocellulose

[5] Q.-R. Liu, B. López-Corcuera, H. Nelson, S. Mandiyan, and N. Nelson, *Proc. Natl. Acad. Sci. USA* **89,** 12145 (1992).

membrane from the plates with forceps. Dry the membrane on a piece of filter paper for 10 min.

15. Denature plaque DNA by soaking the nitrocellulose membrane in denaturing solution (0.5 *M* NaOH, 1.5 *M* NaCl) for 2 min.

16. Neutralize plaque DNA by soaking the nitrocellulose membrane in neutralization solution (0.5 *M* Tris–HCl, pH 8.0, and 1.5 *M* NaCl) for 5 min.

17. Rinse the nitrocellulose membrane with 2× SSC for 5 min and transfer to a piece of filter paper for 5 min.

18. Cross-link plaque DNA to the nitrocellulose membrane with UV Stratalinker for two cycles.

19. Wet the nitrocellulose membrane with 5× SSC for 5 min and transfer them one by one into a plastic bag containing 30 ml of hybridization solution (50% formamide, 5× SSC, 14 m*M* Tris–HCl, pH 7.5, 1× Denhardt's, and 10% dextran sulfate).

20. Drive out air bubbles from the hybridization bag and seal the bag with Scotchpack Pouch Sealer. Prehybridize in a 42° water bath shaker for 2 hr.

21. Remove the nitrocellulose membrane from prehybridization bag and place them temporarily on a piece of plastic sheet.

22. Boil the ^{32}P-labeled DNA probe (labeled by multiprime labeling system from Amersham, England) for 5 min and chill on ice immediately for 5 min.

23. Add the denatured ^{32}P-labeled DNA probe to the hybridization solution (10^6 cpm/ml) in the plastic bag, and put the nitrocellulose membranes one by one back to the bag.

24. Drive out air bubbles and seal the bag with Scotchpack Pouch Sealer. Hybridize at 37° or 42° (depending on stringency) in a water bath shaker overnight.

25. Next morning, wash the nitrocellulose membranes three times at room temperature with 2× SSC and 0.1% SDS, 10 min each time.

26. Wash the nitrocellulose membrane twice at 37 or 42° (depending on stringency), 30 min each time.

27. Wrap a piece of used X-ray film with a piece of plastic sheet. Put the nitrocellulose membranes on top of the wrapped film (six membranes per film) and cover the membrane with another piece of plastic sheet.

28. Stick paper tapes on the corners of the wrapped film and put radioactive marks on the tapes which are then covered by Scotch tape to avoid radioactive contamination.

29. In a dark room, sandwich the membranes between X-ray film and intensify screen in a film cassette. Store the cassette in a −80° freezer for 12–24 hr.

30. Develop the X-ray film. Overlay the film on the top of the wrapped membranes with fluorescence light box as the background. Align the film according to the radioactive markers. Mark the needle holes on the film with marker and number the membranes.

31. Turn over the marked X-ray film on a fluorescence light box. Identify the hybridized plaques and mark the positive spots with a marker.

32. Overlay the plates with plaques on the top of the film. Use a cut-off 1 ml pipette tip and a pipetman to pick up the positive plaque with agar plug by applying gentle suction.

33. Add a single agar plug to 0.5 ml SM buffer with one drop of chloroform in an 1.5 ml of microtube. Store the phage stock at 4° overnight to release phages.

34. One microliter of the phage suspension from the phage stock of the primary screening is diluted to 1 ml of SM buffer from which 2 μl of the diluted suspension are used to infect *E. coli* host cells.

35. The secondary screening is performed same as the primary screening except 100 × 15 mm LB agar plates and 82-mm nitrocellulose membranes are used.

36. Use a sterile Pasteur pipette to pick up a single positive plaque as an agar plug by applying gentle suction.

37. Add the agar plug to 0.5 ml SM buffer with one drop of chloroform and store at 4° overnight to release phages.

38. The excision of plasmid from λZAP vector and the preparation of plasmids are performed according to the Stratagene protocol (Stratagene, La Jolla, CA).

39. Isolation of λgt10 DNA is performed according to the Clontech protocol (Clontech, Palo Alto, CA).

40. Sequence the positive subclones to identify neurotransmitter transporters.

Localization of Neurotransmitter Transporters by Immunocytochemistry

One of the questions asked after the cloning of a novel gene or cDNA is where is this gene expressed. Although Northern blots allow crude estimates of regional distribution of specific transporter mRNAs, *in situ* hybridization can localize the mRNA to specific neuronal populations or other cell types. Patterns of anatomic localization consistent with defined neuron populations may provide clues as to the nature and significance of the transporters encoded by cDNAs isolated in studies outlined above. The development of specific antibodies can provide greater insight into the neuroanatomical localization of transporter subtypes, and because these antibodies detect the protein product itself, data from this approach may indicate that the protein serves its function in a place far removed from the site of mRNA synthesis. Studies with specific antibodies are among the best ways to identify the site of action of the protein in question. Immunocytochemical localization of highly hydrophobic proteins may have several obstacles that have to be circumvented. First and foremost, it is difficult to obtain specific antibodies that will recognize hydrophobic membrane protein *in situ*. Second, membrane proteins are far more sensitive to the fixation method than soluble proteins. Therefore, special consideration should be given in choosing the sites to which antibodies are being raised, the assay of their specificity, and the appropriate immunocytochemical method to be used.[6]

Materials and Reagents

Animals: Rabbits or guinea pigs for antibody production, adult mice, adult rats, pregnant rats and mice.

Instruments: Cryostat microtome, microscope, peristaltic pump, gel electrophoresis apparatus, a pair of scissors, forceps, scalpels, syringes.

Materials: Protein Fusion and Purification System—pMAL-cRI (New England Biolabs), ECL kit and nitrocellulose membrane (Amersham). Affi-Gel 10 resin (Bio-Rad), Freund's complete adjuvant, diaminobenzidine (DAB) substrate kit (Vector), Triton X-100, Tween 20, zinc salicylate, paraformaldehyde, glutaraldehyde, goat serum, dry milk, ketamine, xylazine.

Preparation of Antibodies

Three domains of the neurotransmitter transporters that exhibit the lowest amino acid sequence homology have been chosen for the preparation of antibodies.[7] The domains are taken from the N- and C-terminal parts, and the large external loop of the transporters. We elect to use the system of pMAL-cRI (New England Biolabs) to express fusion proteins between the maltose-binding protein and the corresponding amino acid sequences of the individual transporters. The DNA fragments for constructing the recombinant fusion proteins are prepared by PCR (polymerase chain reaction) and cloned in frame usually into the *Eco*RI–*Hin*dIII restriction sites of the plasmid. A stop codon is placed in the oligonucleotide primer in front of the *Hin*dIII site. The inclusion of amino acid sequences encoding potential transmembrane helices is avoided. Although the inclusion of up to 10 hydrophobic amino acids has no consequence, we observe that inclusion of a single transmembrane helix could reduce the expression of the

[6] F. Jursky, S. Tamura, A. Tamura, S. Mandiyan, H. Nelson, and N. Nelson, *J. Exp. Biol.* **196,** 283 (1994).

[7] F. Jursky and N. Nelson, *J. Neurochem.* **64,** 1026 (1995).

fusion protein up to 100-fold. Expression and purification of fusion proteins are performed according to the manufacturer's instruction manual (Protein Fusion & Purification System, directions for expression and purification of proteins from cloned genes, New England Biolabs). Usually the fusion proteins are purified further by gel electrophoresis in the presence of SDS, followed by brief staining with Coomassie Brilliant Blue, and electroelution of the isolated protein bands.[8] In some instances the purification of the fusion proteins on the maltose column is avoided and the protein is purified only by SDS electrophoresis followed by electroelution.

The purified fusion proteins are mixed with two volumes of a complete Freund's adjuvant and the guinea pigs or rabbits are immunized by multisite injections into the skin as previously described.[8] The animals are subjected to up to 10 booster injections and approximately 10 days after each injection, serum samples are taken and assessed for their immunoreactivity on Western blots with membrane fractions isolated from various brain regions. For Western analysis rat brains are removed from the skull, and crude plasma membrane fractions are then isolated from the spinal cord, cerebellum, midbrain, cerebral cortex, and olfactory bulb as previously described.[9] The isolated membrane fractions are dissolved in SDS sample buffer and loaded onto 10% SDS–polyacrylamide gel (about 10 μg protein per lane). The proteins are electrotransferred from the gel onto an ECL (enhanced chemiluminescence) nitrocellulose membrane (Amersham). Membranes are processed with an ECL kit according to manufacturer's instructions in phosphate-buffered saline (PBS)/0.1% Tween 20. Prior to Western blot analysis, the serum containing antibodies is diluted 1:1000 and preabsorbed with excess of maltose binding protein (~50 μg protein in 20 ml solution) for 1 hr at room temperature.

For affinity purification, the fusion proteins are purified on maltose-binding columns. The purified fusion proteins are cross-linked onto Affi-Gel 10 resin according to manufacturer's instruction (Bio-Rad). The serum containing the corresponding antibodies is passed through the column and, following extensive washing, the affinity purified antibodies are eluted by a solution containing 0.1 *M* glycine hydrochloride (pH 3.5). Antibodies against the maltose-binding protein are removed on a Affi-Gel column to which the above protein was bound.

The quality and specificity of the antibodies are analyzed by several independent means, to be useful, an antibody is expected to fulfill several criteria. Antibodies give a single stained band at the correct position on Western blots and the intensity of the bands correspond to the expected distribution of the various neurotransmitter transporters in the different

[8] N. Nelson, *Methods Enzymol.* **97,** 510 (1983).

[9] M. K. Bennet, N. Calakos, T. Kreiner, and R. H. Scheller, *J. Cell Biol.* **116,** 761 (1992).

brain regions. The antibodies also should react exclusively with their specific fusion proteins and should not crossreact with any other proteins or with fusion proteins containing similar parts of other transporters. Preincubation with the corresponding fusion proteins should eliminate the reaction on the Western blots. Preincubation with purified maltose-binding protein or crude *E. coli* extract should have no effect. Although this procedure cannot exclude cross reactivity with an unknown and highly related transporter, these rigorous criteria eliminate mistaking one transporter with another known transporter.

Immunocytochemistry

Fixation Methods

Adult 60-day-old mice are deeply anesthetized with avertin and perfused intracardially through the left ventricle with a 20 ml solution containing 0.9% (w/v) NaCl followed by perfusion with a solution containing 0.9% (w/v) NaCl, 4% (w/v) paraformaldehyde, and 0.5% (w/v) zinc salicylate (pH 6.5). The perfusion is performed by a peristaltic pump at a flow rate of about 1 ml per min with a total volume of about 1 ml per 1 g of body weight. Alternatively, the perfusion is performed with PBS followed by a solution containing PBS, 4% (w/v) paraformaldehyde and 0.1% (w/v) glutaraldehyde (pH 7.5). The procedure used for immunocytochemistry with rat brain is the same as used for mouse, except that the volume of the perfusion solution is proportionally increased according to the body weight. A mixture of the ketamine (80 mg/ml) with xylazine (10 mg/ml) in physiological saline is used instead of avertin to anesthetize the rats. Adult 60-day-old rats and pregnant females are deeply anesthetized. The rats are perfused intracardially with PBS followed by perfusion with 4% (w/v) paraformaldehyde in PBS. Brains are removed from the skull, post fixed for 1 hr in the same fixative at room temperature, and cryoprotected in a solution containing 30% (w/v) sucrose in PBS (pH 6.5) for 2 days at 4°.

For developmental studies, pregnant mice are bred in-house.[10,11] The day of conception E0 is estimated by the presence of a vaginal plug. Birth occurs at E20 which is designated as P0. The pregnant females are deeply anesthetized with avertin. Embryos are anesthetized by hypothermia and

[10] F. Jursky and N. Nelson, *J. Neurochem.* **67,** 336 (1996).

[11] F. Jursky and N. Nelson, *J. Neurochem.* **67,** 857 (1996).

[12] F. Jursky and N. Nelson, *in* "Advances in Pharmacological Sciences: GABA: Receptors, Transporters and Metabolism" (C. Tanaka and N. G. Bowery, Eds.), p. 73. Birkhaeuser Verlag, Basel, 1996.

from E17 to adult, mice are perfused intracardially with paraformaldehyde or PBS followed by 4% paraformaldehyde in PBS. Brains are post fixed for 1 hr in the same fixative at room temperature and cryoprotected in 30% sucrose/PBS overnight at 4°. From E12 to E16, embryos are immersed in the same fixative for 2 hr at room temperature followed by cryoprotection at 4° in a sucrose solution.

Frozen sections (20 μm) are cut in a cryostat microtome and allowed to dry at room temperature for 2 days on microscope slides. The sections are then rehydrated by washing for 10 min three times by a solution of 0.5 *M* Tris–HCl (pH 7.6), blocked 1 hr in the same buffer containing 5% goat serum and 0.5% Triton X-100, and incubated with primary antibody 6 hr or overnight at room temperature in solution containing 0.5 *M* Tris–HCl (pH 7.6), 4% normal goat serum, and 0.1% Triton X-100. Sections are then incubated for 2 hr at room temperature with the secondary anti-guinea-pig antibody followed by incubation with avidin–biotin–horseradish peroxi-

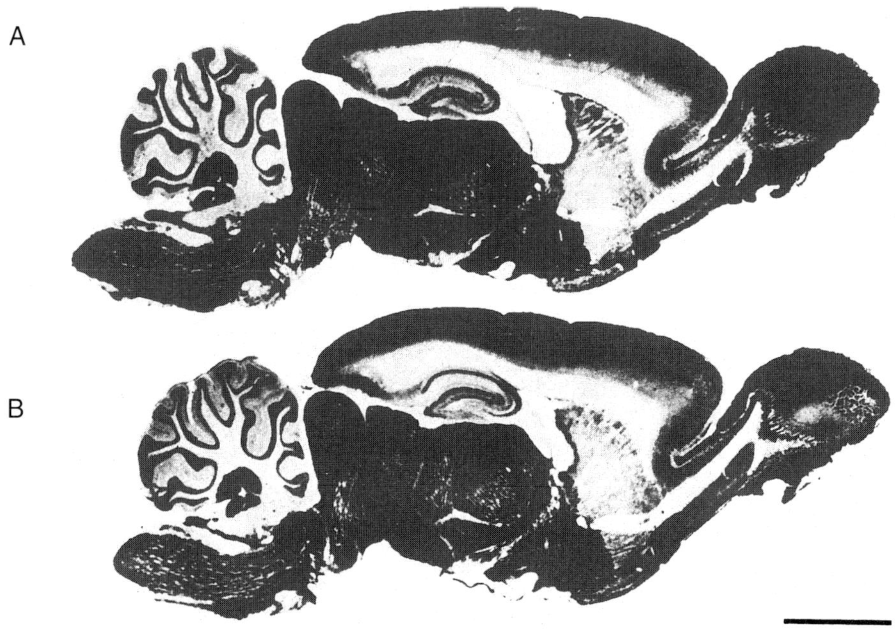

FIG. 1. Immunocytochemical staining of rat brain sagittal sections with antibodies against GAT4. (A) Antibody was raised against a pMAL fusion protein containing the C-terminal part of GAT4. (B) Antibody was raised against a pMAL fusion protein containing the extended loop of GAT4. (Published in F. Jursky and N. Nelson, *in* "Advances in Pharmacological Sciences: GABA: Receptors, Transporters and Metabolism" (C. Tanaka and N. G. Bowery, Eds.), p. 73. Birkhaeuser Verlag, Basel, 1996.)

dase complex in the same buffer, but without Triton X-100. The sections are washed 3 times for 10 min in a solution containing 0.5 *M* Tris–HCl (pH 7.6).

The sections are stained with a diaminobenzidine (DAB) substrate kit (Vector), washed in 0.9% NaCl, placed on slides (Superfrost Fisher), and let dry. The sections are dehydrated again by alcohol, cleared by hemo-D, and mounted with Permount (Fisher). For early embryonic stages, brain sections are placed directly on the microscope slides and processed as described above.

Figure 1 depicts an immunocytochemical staining of rat brain sagittal sections. The sections interacted with antibodies against fusion proteins containing the C-terminal part of the γ-aminobutyric acid (GABA) transporter GAT4 (A), and the extended loop of the same transporter (B). The staining pattern is essentially identical.

[4] Purification of Vesicular Monoamine Transporters: From Classical Techniques to Histidine Tags

By RODRIGO YELIN and SHIMON SCHULDINER

Molecular tools have greatly enhanced our ability to assign certain activities to given proteins. However, functional reconstitution of purified transport proteins into liposomes still provides the only unequivocal way to positively assign a given function to a single given polypeptide. In addition, reconstitution has provided important insights into transport mechanisms, specific lipid requirements, and bioenergetics. Several vesicular neurotransmitter transporters (VNTs) have been reconstituted. Bovine vesicular monoamine transporter (bVMAT) was reconstituted using a variety of techniques.[1–3] The driving force for accumulation of the monoamine was provided either by endogenous[2] or coreconstituted ATPase[3] and by artificially imposed pH gradients.[1,2] A major non-carrier-mediated component is observed in the presence of reserpine, a potent inhibitor of the vesicular monoamine transporter (VMAT). This component is due to passive permeation of monoamines across liposomes followed by equilibration with the

[1] Y. Stern-Bach, N. Greenberg-Ofrath, I. Flechner, and S. Schuldiner, *J. Biol. Chem.* **265,** 3961 (1990).
[2] R. Maron, H. Fishkes, B. Kanner, and S. Schuldiner, *Biochemistry* **18,** 4781 (1979).
[3] Y. Moriyama, A. Iwamoto, H. Hanada, M. Maeda, and M. Futai, *J. Biol. Chem.* **266,** 22141 (1991).

0076-6879/98 $25.00

pH gradient.[1,2] When the purified transporter and the proper protein–lipid ratio are used, the problem is ameliorated and the values of the background nonmediated transport are significantly reduced.[1] The vesicular γ-aminobutyric acid (GABA)[4] and glutamate[5,6] transporters were reconstituted as well. In the latter case, when coreconstituted with bacteriorhodopsin, light was used to generate the $\Delta\tilde{\mu}_{H^+}$ needed for accumulation of glutamate.[7]

Purification of Bovine VMAT

The only VNT purified in a functional form is bovine VMAT. The high stability of the complex [^{3}H]reserpine transporter has been used to label the transporter and follow its separation through a variety of procedures. In these experiments a small amount of Triton X-100 extracts from prelabeled membranes were mixed with a four- to fivefold higher amount of extract from unlabeled membranes (Fig. 1). Purification of the material labeled in this way has revealed the presence of two proteins that differ in p*I*, a very acidic one (3.5) and a moderately acidic one (5.0).[1] Reconstitution in proteoliposomes has shown that both catalyze monoamine transport with the expected properties. The highly acidic p*I* of the more acidic isoform facilitated its purification on DEAE cellulose at pH 4.0, conditions under which most proteins do not bind. Two additional purification steps were required to obtain a single glycoprotein of 80 kDa. The last step, binding to wheat germ agglutinin, was used also to exchange detergents from Triton X-100 to cholate. The latter was found to yield proteoliposomes which displayed higher and more reproducible transport activity. The purified VMAT catalyzes transport of serotonin with an apparent K_m of 2 μM and a V_{max} of 140 nmol/mg/min, about 200-fold higher than the one determined in the native system. Transport is inhibited by reserpine and tetrabenazine, ligands which bind to two distinct sites on the transporter. In addition, the reconstituted purified transporter binds reserpine with a biphasic kinetic behavior, typical of the native system. The results demonstrate that a single polypeptide is required for all the activities displayed by the transporter: reserpine- and tetrabenazine-sensitive, ΔpH-driven serotonin accumulation, and binding of reserpine in an energy-dependent and independent way. Based on these and additional findings it has been estimated that the transporter represents about 0.2–0.5% of the chromaffin granule membrane vesicle and it has a turnover of about 30 min^{-1}.

[4] J. W. Hell, P. R. Maycox, and R. Jahn, *J. Biol. Chem.* **265,** 2111 (1990).
[5] M. D. Carlson, P. E. Kish, and T. Ueda, *J. Biol. Chem.* **264,** 7369 (1989).
[6] P. R. Maycox, T. Deckwerth, J. W. Hell, and R. Jahn, *J. Biol. Chem.* **263,** 15423 (1988).
[7] P. Maycox, T. Deckwerth, and R. Jahn, *EMBO J.* **9,** 1465 (1990).

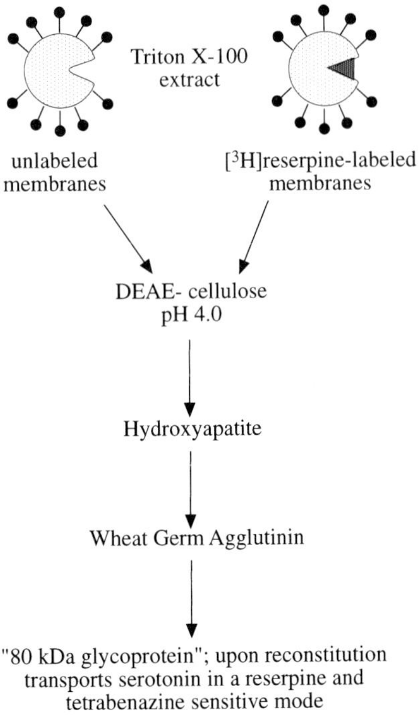

FIG. 1. Schematic representation of the purification of bovine VMAT2. Experimental details can be found in Ref. 1.

The assignment of the activity to the 80-kDa polypeptide and the localization of the tetrabenazine (TBZ) and reserpine binding sites in the same polypeptide are confirmed by several independent approaches, including direct sequencing of the purified protein[8,9] and cloning and analysis of the recombinant protein.[10-12] Vincent and Near purified a TBZ-binding glycoprotein from bovine adrenal medulla using a protocol identical to the one used to purify the functional transporter. The TBZ-binding protein displays an apparent M_r of 85 kDa.[13] In addition, Henry and collaborators

[8] Y. Stern-Bach, J. Keen, M. Bejerano, S. Steiner-Mordoch, M. Wallach, J. Findlay, and S. Schuldiner, *Proc. Natl. Acad. Sci. USA* **89,** 9730 (1992).

[9] E. Krejci, B. Gasnier, D. Botton, M. F. Isambert, C. Sagne, J. Gagnon, J. Massoulie, and J. P. Henry, *FEBS Lett.* **335,** 27 (1993).

[10] Y. Liu, D. Peter, A. Roghani, S. Schuldiner, G. Prive, D. Eisenberg, N. Brecha, and R. Edwards, *Cell* **70,** 539 (1992).

[11] S. Schuldiner, Y. Liu, and R. Edwards, *J. Biol. Chem.* **268,** 29 (1993).

[12] J. Erickson, L. Eiden, and B. Hoffman, *Proc. Natl. Acad. Sci. USA* **89,** 10993 (1992).

[13] M. Vincent and J. Near, *Molec. Pharmacol.* **40,** 889 (1991).

used 7-azido-8-[^{125}I]iodoketanserin, a photoactive derivative of ketanserin, which is thought to interact with the TBZ binding site. They labeled a polypeptide with an apparent M_r of 70,000 and a pI ranging from 3.8 to 4.6.[14] In all the cases broad diffuse bands, characteristic of membrane proteins, were detected so that the differences in the M_r values reported (70,000, 80,000 and 85,000) are probably due to a different analysis of the results and not to innate variations. Sequencing of the labeled protein confirmed that it is identical to the functional transporter.[9]

The basis of the difference between the two isoforms has not yet been studied and could be due to covalent modification (i.e., phosphorylation or different glycosylation levels) of the same polypeptide backbone, to limited proteolysis during preparation, or to a different polypeptide backbone. Two types of VMAT have been identified[10]: VMAT2 and VMAT1. They differ in their primary amino acid sequence, tissue localization, and pharmacology.[15–17] The functional purified transporter reconstituted from bovine adrenal is of the VMAT2 type as judged from direct sequencing of the N terminus of the protein.[8] The sequence of 26 N-terminal amino acids of the purified protein is practically identical to the predicted sequence of the bovine adrenal VMAT2.[9,18] Bovine VMAT2, the isoform with the low pI, accounts for 55 to 75% of the activity detected in membrane vesicles prepared from bovine adrenal chromaffin granules.[1] Antibodies raised against a synthetic peptide based on the VMAT2 sequences recognize the pure protein on Western blots and specifically immunoprecipitate reserpine binding activity.[8] It is still to be elucidated whether the isoform with a higher pI corresponds to VMAT1 which is also expressed in this tissue.[19]

Purification of Recombinant VMAT by His-tagging

To allow for rapid purification of recombinant membrane proteins, it is now possible to use tags with domains that allow for high affinity and rapid purification. For purification of bVMAT2 and rVMAT1 (a rat isoform[10]) we have chosen the polyhistidine tag because it has been successfully used to

[14] M. Isambert, B. Gasnier, D. Botton, and J. Henry, *Biochemistry* **31,** 1980 (1992).

[15] D. Peter, J. P. Finn, I. Klisak, Y. J. Liu, T. Kojis, C. Heinzmann, A. Roghani, R. S. Sparkes, and R. H. Edwards, *Genomics* **18,** 720 (1993).

[16] D. Peter, J. Jimenez, Y. J. Liu, J. Kim, and R. H. Edwards, *J. Biol. Chem.* **269,** 7231 (1994).

[17] E. Weihe, M.-H. Schafer, J. Erickson, and L. Eiden, *J. Molec. Neurosci.* **5,** 149 (1995).

[18] M. Howell, A. Shirvan, Y. Stern-Bach, S. Steiner-Mordoch, J. E. Strasser, G. E. Dean, and S. Schuldiner, *FEBS Lett.* **338,** 16 (1994).

[19] J. P. Henry, D. Botton, C. Sagne, M. F. Isambert, C. Desnos, V. Blanchard, R. Raisman Vozari, E. Krejci, J. Massoulie, and B. Gasnier, *J. Exp. Biol.* **196,** 251 (1994).

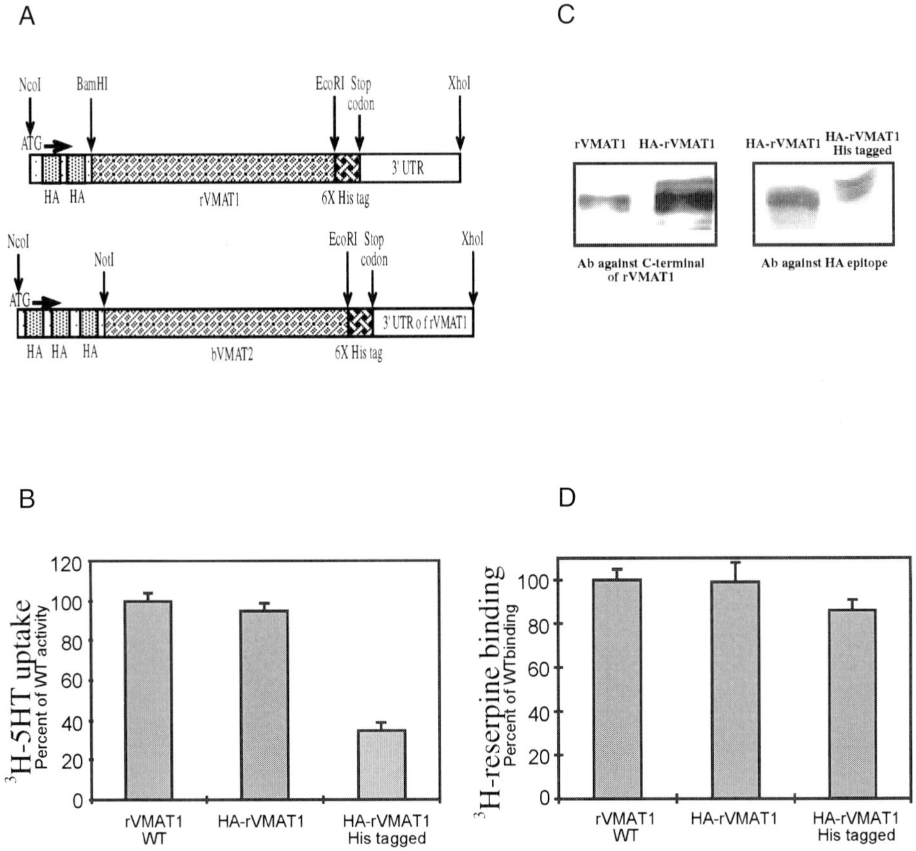

FIG. 2. Scheme showing the His-tagged rVMAT1, transport, expression, and [^{3}H]reserpine binding in CV-1 cells. (A) rVMAT1 and bVMAT2 were tagged with two and three copies of the HA epitope (YPYDVPDYA) at the N terminus and with 6 histidine residues at the C terminus. The genes were inserted in BFG-1, a yeast expression plasmid. BFG-1 (LEU2, 2μ, PGK promoter and terminator) contains an ATG site for start of translation, followed by a sequence encoding three copies of the HA epitope separated by cloning sites. The cDNA of rVMAT1 and bVMAT2 were cloned in BFG-1 at *Bam*H1 and *Not*I sites, respectively. As a result, rVMAT1 and bVMAT2 contain additional 24 and 38 amino acids having two and three copies of the HA, correspondingly. Addition of the 6$\times$ histidine tag (His$_6$ tag) at the C terminus was performed using PCR with a primer designed for that purpose. This primer inserts an *Eco*RI site followed by 6 codons for histidine immediately after the last residue of rVMAT1; it also introduces a new stop codon. Similarly, an *Eco*RI site was inserted before the stop codon of bVMAT2, which served to move the His$_6$ tag of rVMAT1 to bVMAT2. As a result, eight additional amino acids are appended to the C terminus: phenylalanine and glutamate corresponding to the *Eco*RI restriction site, and 6 histidine residues corresponding to the His$_6$ tag. For transient functional expression using the VTF7 expression system the above constructs were excised from BFG-1 and inserted in the high expression vector pTM1

purify a number of ion-coupled transporters.[20–22] To follow the protein during the purification process, immunological detection methods were used. The insertion of 6 histidines (His_6 tag) at the C terminus interfered with the recognition of the antibodies available (not shown), and therefore bVMAT2 and rVMAT1 were tagged inserting the hemagglutinin epitope (HA)[23] at the N terminus of the protein (Fig. 2A). To test whether the addition of these domains interferes with the activity of the transporter, the recombinant protein was transiently expressed in CV-1 (African green monkey kidney) cells. CV-1 cells expressing an HA-rVMAT1 His-tagged display transport activity lower than that detected in cells expressing wild-type rVMAT1 or HA-rVMAT1 (Fig. 2B). HA-bVMAT2 His-tagged displays 50% of the transport activity of rVMAT1 (not shown). Expression levels of the His-tagged and the untagged protein have been monitored with a monoclonal antibody that recognizes the HA epitope and they were found to be very similar (Fig. 2C, right). The untagged protein is expressed in this particular experiment at levels somewhat higher than the wild type (Fig. 2C, left). Expression levels and proper folding can be estimated also from the level of binding of [^{3}H]reserpine. HA-rVMAT1 His-tagged, HA-rVMAT1, and wild-type rVMAT1 bind [^{3}H]reserpine to practically identical levels (Fig. 2D). Therefore, we conclude that although the transport activity displayed by the His-tagged protein is lower than that of the wild type, it is still sufficiently high to guarantee purification and study of its properties. As previously reported for bVMAT2, the detergent solubilized protein retains the radiolabeled ligand if this is allowed to bind prior to solubilization, and therefore the complex [^{3}H]reserpine–protein can be

[20] G. G. Prive and H. R. Kaback, *J-Bioenerg-Biomembr* **28,** 29 (1996).

[21] T. Pourcher, S. Leclercq, G. Brandolin, and G. Leblanc, *Biochemistry* **34,** 4412 (1995).

[22] Y. Olami, A. Rimon, Y. Gerchman, A. Rothman, and E. Padan, *J. Biol. Chem.* **272,** 1761 (1997).

[23] I. A. Wilson, H. L. Niman, R. A. Houghten, A. R. Cherenson, M. L. Connolly, and R. A. Lerner, *Cell* **37,** 767 (1984).

[24] A. Shirvan, O. Laskar, S. Steiner-Mordoch, and S. Schuldiner, *FEBS Lett.* **356,** 145 (1994).

[25] S. Steiner-Mordoch, A. Shirvan, and S. Schuldiner, *J. Biol. Chem.* **271,** 13048 (1996).

plasmic,[18,24] using *Xho*I and partial digestion with *Nco*I. (B) [^{3}H]5HT (5-hydroxytryptamine) uptake assayed in CV-1 cells expressing rVMAT1, HA-rVMAT1 and HA-rVMAT1 His-tagged. Uptake was assayed for 10 min in cells permeabilized with digitonin as described by Erickson *et al.*[12] (C) Expression of rVMAT1, HA-rVMAT1 and HA-rVMAT1 His-tagged in CV-1 cells using the VTF7 expression system. *Left*: Western blot was exposed to antibody against the C terminus of rVMAT1. *Right*: Western blot was exposed to monoclonal antibody (12CA5) against the HA epitope.[23] (D) [^{3}H]Reserpine binding to lysates prepared from CV-1 cells, performed essentially as described in Ref. 25.

used in parallel with the use of the anti-HA antibody to monitor purification on a Ni-NTA resin (Qiagen). The results presented in Fig. 3 describe a typical purification protocol of bVMAT2 transiently expressed in CV-1 cells. Membranes were prepared from the above cells and a tenth of the preparation was allowed to bind [^{3}H]reserpine and was then mixed again with the original pool. Membrane proteins were extracted with 1.5% Triton X-100, a detergent previously shown to solubilize bVMAT2 in a functional state without causing substantial release of previously bound [^{3}H]reserpine. Imidazole (2.5 m*M*) was added to the Triton-solubilized extract and was then incubated with the Ni-NTA resin for 30 min. The void volume contained unbound [^{3}H]reserpine and a variety of other proteins, but no detectable bVMAT2 (Fig. 3, void). Wash of the column with binding buffer induces a small fraction of the transporter to elute (Fig. 3, wash). Washing with 10 and 20 m*M* imidazole was followed with 300 m*M*, which resulted in elution of the tagged bVMAT2 previously bound to the resin. The protein levels are too low for an estimation of the purification degree, but it is clear that the bulk of the immunologically reactive material is detected

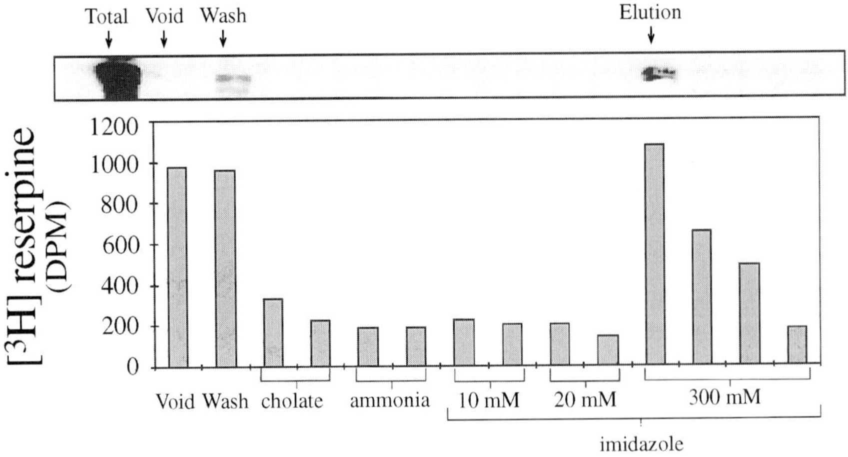

FIG. 3. Purification of His-rVMAT1 on a Ni-NTA column. The presence of the transporter in the different fractions was followed using immunoblots and measuring radioactivity. Membranes from CV-1 cells expressing HA-bVMAT2 His-tagged were labeled with [^{3}H]reserpine[25] and solubilized with 1.5% Triton X-100 in 0.3 *M* sucrose, 10 m*M* K-HEPES, pH 7.4, 5 m*M* $MgCl_2$, and 2.5 m*M* imidazole. Solubilized membranes were loaded into a Ni-NTA column and fractions collected were counted by liquid scintillation and tested by immunoblots. As shown here, it is possible to take advantage of the binding of VMAT to the column to exchange the binding buffer to 1% cholate in 150 m*M* NH_4Cl.

mainly in the elution fractions while very little is present in the other ones. Interestingly, the [^{3}H]reserpine seems to be displaced from bVMAT2 by higher concentrations of imidazole, a phenomenon that is now under investigation. On the other hand, rVMAT1 has a tendency to lose bound [^{3}H]re-

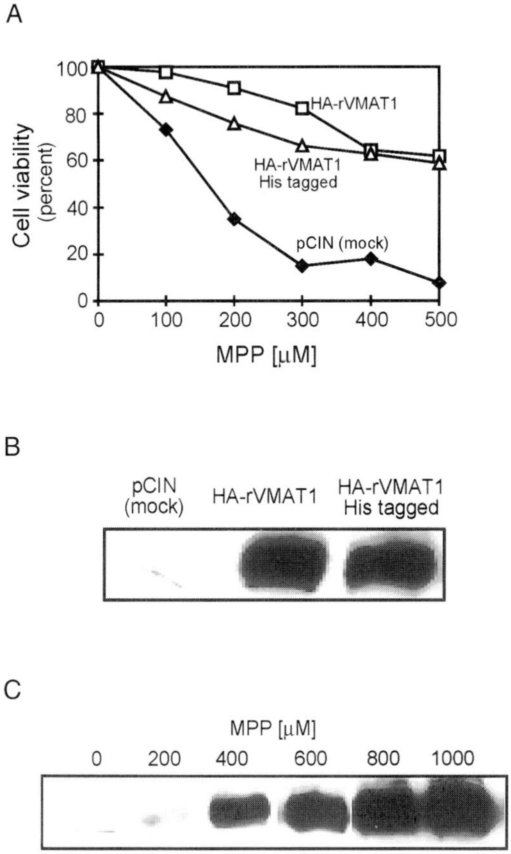

FIG. 4. Stable CHO cell lines expressing HA-rVMAT1 His-tagged are resistant to MPP^+. (A) Viability of cells in presence of MPP^+: CHO cells were seeded at low confluence; a day afterward MPP^+ was added at different concentrations. After 4 days, cell viability was checked using tetrazolium dye. After 2 hr in the presence of 0.3 μg/ml tetrazolium, the medium was removed and the tetrazolium precipitate was dissolved with 0.5 ml dimethyl sulfoxide (DMSO). The difference in absorbance between 540 and 650 nm was recorded. The values are expressed as percent of those measured in the absence of MPP^+. (B) Expression of HA-rVMAT1 and HA-rVMAT1 His-tagged in CHO cells. Membranes prepared from CHO cells were immunoblotted using monoclonal antibody against the HA epitope. (C) Cells were exposed to increasing concentrations of MPP^+ for 2 weeks; membranes were prepared and tested for HA-rVMAT1 His-tagged expression by Western blots.

serpine on solubilization. The purification profile of rVMAT1 is very similar to that of bVMAT2, as judged by the presence of immunoreactive material on the various fractions. In conclusion, this procedure using membranes from CV-1 cells yields a significant purification that can be used, for example, to follow specific radiolabeling of the protein, but does not yield large chemical amounts. Based on the levels of recovery of [^{3}H]reserpine we can estimate a yield of about 1 μg of rVMAT1 per 1.2×10^8 CV-1 cells. Transient expression of larger amounts of protein in CV-1 cells is not practical because of the labor and cost involved.

Therefore, we decided to generate stable cell lines expressing HA-rVMAT1 His-tagged that can be grown in larger quantities and do not require further manipulations for permanent expression of the protein. The rVMAT1 construct described in Fig. 2A was cloned into pCIN plasmid that allows for stable transfection of mammalian cells based on selection with neomycin.[26] After transfection of Chinese hamster ovary (CHO) cells, several neomycin-resistant colonies were isolated and tested for MPP^+ resistance. The results shown in Fig. 4A describe the cell viability in the presence of MPP^+ of cultures of CHO mock-transformed as compared to cells expressing either HA-rVMAT1 or HA-rVMAT1 His-tagged but that have not been previously exposed to the neurotoxin. The mock-transformed CHO cells show a decrease of 50% in their viability at around 150 μM of the neurotoxin. Cells expressing either construct display a significantly higher resistance (Fig. 4A). In both cases rVMAT1 can be detected immunologically (Fig. 4B). It has been previously shown that this resistance can be increased by chronic exposure of the cells to the toxin[10] and it was shown that, concomitantly, the amount of message increases.[11] In Fig. 4C it is shown that also the amount of protein in these cells dramatically increases on chronic exposure (2 weeks) to MPP^+. The levels of protein detected in the cells exposed to 1 mM MPP^+ are comparable to the levels detected in CV-1 cells. Moreover, further exposure to the toxin (1 month) improves expression even further (not shown). It is proposed, therefore, that these cells can be used as a good source for the purification of rVMAT1.

Acknowledgments

This work was supported by grants from the National Institute of Health (NS16708), the United States–Israel Binational Science Foundation (93-00051), and the National Institute of Psychobiology in Israel. We would like to thank Sonia Steiner Mordoch for assistance at various stages of this work.

[26] S. Rees, J. Coote, J. Stables, S. Goodson, S. Harris, and M. G. Lee, *Biotechniques* **20,** 102 (1996).

[5] Noncovalent and Covalent Labeling of Vesicular Monoamine Transporter with Tetrabenazine and Ketanserin Derivatives; Purification of Photolabeled Protein

By JEAN-PIERRE HENRY, CORINNE SAGNÉ, MARIE-FRANÇOISE ISAMBERT, and BRUNO GASNIER

Introduction

The vesicular monoamine transporter (VMAT) accumulates monoamines inside secretory granules or synaptic vesicles by catalyzing a H^+/monoamine antiport. Monoamine uptake is coupled to the generation of a proton electrochemical gradient by an ATP-dependent H^+ pump of the vacuolar type. The uptake is extremely efficient, and in the adrenal medulla chromaffin cells, the monoamine gradient between the cytosol and the secretory granules has been estimated to five orders of magnitude.[1,2] The transporter has a low specificity and will accept catecholamines, such as adrenaline, noradrenaline, or dopamine, the indoleamine serotonin (5-HT), histamine, and nonnatural substrates such as *m*-iodobenzylguanidine (MIBG) or the neurotoxin 1-methyl-4-phenylpyridinium (MPP^+).

The vesicular monoamine transporter from various origins (rat, human, and bovine) has been cloned and expressed in various cell types.[1,2] Two isoforms, VMAT1 and VMAT2, have been identified which have different localizations and slightly different properties. However, most of our knowledge regarding VMAT has been obtained on the protein from bovine chromaffin granules, where the major expressed isoform is VMAT2. An interesting pharmacology has been developed from this preparation.[3] Two different binding sites have been identified. The first one binds reserpine, whereas the second one binds tetrabenazine (TBZ) and ketanserin (KET). The H^+ electrochemical gradient generated in the presence of ATP greatly accelerates reserpine binding, but has no effect on TBZ and KET binding.

As for other transporters, the enzymology of VMAT is complicated by the fact that VMAT catalyzes a vectorial activity, which has to be coupled to a proton gradient generating system. It is possible to reconstitute the purified protein in liposomes and to impose a pH gradient on the vesi-

[1] S. Schuldiner, A. Shirvan, and M. Linial, *Physiol. Rev.* **75,** 369 (1995).
[2] Y. Liu and R. H. Edwards, *Am. Rev. Neurosc.* **20,** 125 (1997).
[3] J. P. Henry and D. Scherman, *Biochem. Pharmacol.* **38,** 2395 (1989).

0076-6879/98 $25.00

cles.[4] However, it is also possible to use nonvectorial ligand binding to assay VMAT. Because TBZ and KET derivative binding is not affected by the H^+ electrochemical gradient, these compounds are very useful ligands. Photosensitive derivatives of TBZ and KET can furthermore be used to label VMAT by photoaffinity techniques. An interesting aspect of covalent labeling is that the protein can be purified irrespective of its activity. During such purification studies, we introduced a new technique based on the selective aggregation of VMAT solubilized in sodium dodecyl sulfate (SDS) that results from the polytopic structure of the transporter. Consequently, it might be useful for the purification of other such transporters.

TBZ and KET Binding to Vesicular Monoamine Transporter

TBZ Binding

Various tritiated derivatives of TBZ have been used. The first one was α-[2-^{3}H]dihydrotetrabenazine (2-hydroxy-3-isobutyl-9,10-dimethoxy-1,2,3,4,6,7-hexahydro-11b*H*-benzo[*a*]quinolizine, [2-^{3}H]TBZOH), obtained by reduction of TBZ with KB^3H_4. This compound had a specific activity of about 15 Ci/mmol (0.5 TBq/mmol).[5] More recently, another tritiated derivative of α-dihydrotetrabenazine (α[*O-methyl*-^{3}H]dihydrotetrabenazine) has been commercialized that has a higher specific activity (4.44–6.48 TBq/mmol, 120–175 Ci/mmol, Amersham, Buckinghamshire, U.K.). α-Dihydrotetrabenazine contains two enantiomers and the binding is stereospecific, the (+) isomer being the active one.[6] [*O-methyl*-^{3}H]Tetrabenazine has also been tested successfully as a high specific activity ligand.[7]

For Scatchard analysis of α-[*O-methyl*-^{3}H]TBZOH binding to membrane preparations, various concentrations of the ligand at full specific activity (265–385 dpm/fmol) are incubated with membranes suspended in 0.2 ml of isosmotic medium pH 7.4 (saline or sucrose). When purified chromaffin granule membranes are used,[8,9] the membrane concentration is limited to about 10 μg of protein/ml to keep the site concentration lower than the equilibrium dissociation constant K_d.[5] Usually, duplicate incubations are performed for each concentration, and nonspecific binding is

[4] R. Yelin and S. Schuldiner, *Methods Enzymol.* **296,** [4], (1998) (this volume).

[5] D. Scherman, P. Jaudon, and J. P. Henry, *Proc. Natl. Acad. Sci. USA* **80,** 584 (1983).

[6] M. R. Kilbourn, L. C. Lee, T. M. Vander Borght, D. M. Jewett, and K. A. Frey, *Eur. J. Pharmacol.* **278,** 249 (1995).

[7] T. M. Vander Borght, A. A. F. Sima, M. R. Kilbourn, T. J. Desmond, D. E. Kuhl, and K. A. Frey, *Neuroscience* **68,** 955 (1995).

[8] A. D. Smith and H. Winkler, *Biochem. J.* **103,** 480 (1967).

[9] C. Sagné, M. F. Isambert, J. P. Henry, and B. Gasnier, *Biochem. J.* **316,** 825 (1996).

measured by adding 2 μM TBZ to a third series of incubations. The mixtures are incubated for at least 2 hr at 25°. Bound ligand is separated by filtration on HAWP filters (Millipore, Bedford, MA), pretreated with 0.3% PEI (polyethyleneimine) to decrease the background. The samples are diluted with 2 ml of ice-cold buffer containing 10 μM TBZ before filtration, and the radioactivity on the filter is measured by liquid scintillation counting in Ready Protein Plus mixture (Beckman, Fullerton, CA) or equivalent solutions. With purified chromaffin granule membranes, the signal-to-noise ratio at a ligand concentration equivalent to K_d is about 30.

In all cases tested, corresponding essentially to VMAT2 from various origins, the equilibrium dissociation constant K_d is in the 1–10 nM concentration range. With purified chromaffin granule membranes,[8,9] K_d and B_{max} (maximal concentration of bound ligand, corresponding to the density of sites) are 3 nM and 60 pmol/mg of protein, respectively. The assay of VMAT density in a given membrane preparation can be performed by measuring B_{max}, or more simply, if K_d is known, by measuring the bound ligand at a concentration equivalent to K_d.

This protocol can be applied to VMAT solubilized by cholate or octyl β-glucoside. Originally, the protein incubated with [^{3}H]TBZOH was precipitated with polyethylene glycol 6000 before filtration.[10] However, this is not necessary since the solubilized transporter is adsorbed on GF/B glass fiber filters (Whatman, Clifton, NJ) that have been pretreated by incubation for at least 3 hr with 0.3% PEI.[11] The incubation mixture is filtered under reduced pressure on these filters and the radioactivity is measured as previously.

[^{3}H]TBZOH binding has also been measured on bovine chromaffin cells in culture.[12] [^{3}H]TBZOH is a permeant molecule and specific binding can be measured on intact cells. The assay is performed in four plastic wells, each containing 500,000 cells. The culture medium is pipetted off and 3 nM [^{3}H]TBZOH is added in 1 ml of Locke medium. For nonspecific binding determination, 2 μM TBZ is added to four other wells. After 1 hr incubation at room temperature, the medium is removed and the cells are washed three times by incubating for 3 min with 2 ml of Locke solution. Finally, the cells are collected in 0.2% Triton X-100 for counting radioactivity. Under the conditions described and for 500,000 cells, specifically and nonspecifically bound [^{3}H]TBZOH are, respectively, 30–50 and 15 fmol. Specific binding does not change with the age of the culture, whereas nonspecific binding increases with the proliferation of nonchromaffin cells. The density

[10] D. Scherman and J. P. Henry, *Biochemistry* **22,** 2805 (1983).

[11] M. S. Vincent and J. A. Near, *Molec. Pharmacol.* **40,** 889 (1991).

[12] C. Desnos, M. P. Laran, and D. Scherman, *J. Neurochem.* **59,** 2105 (1992).

of [^{3}H]TBZOH binding sites, derived from B_{max}, is 1 pmol/mg of protein (K_d = 16 nM), which should be compared to the value obtained on adrenal medulla homogenates (B_{max} = 3 pmoles/mg of protein, K_d = 5 nM).

One difficulty in using this method is compromising the rate of dissociation of the ligand from its site during the washing of intact cells. An alternative would be to perform the assay on cells with membranes permeabilized by the bacterial toxin streptolysin O. In this case, cells are placed in a 2-cm diameter plastic well and are resuspended in a medium containing 150 mM potassium glutamate/0.5 mM EGTA/5 mM magnesium acetate/10 mM PIPES, pH 7.2 (KG medium) containing 50 U/ml streptolysin O (Biomerieux, Lyon, France)/2 mM dithiothreitol (DTT), and 0.1% bovine serum albumin (BSA). After incubation in this medium for 15 min at room temperature, the solution is pipetted off. Then the cells are incubated with 3 nM [^{3}H]TBZOH in KG medium for at least 1 hr at room temperature. At the end of the incubation, cells are rapidly washed two times with ice-cold KG and 0.3 ml of 0.1 N NaOH is added. After 15 min, the radioactivity of the suspension is counted by liquid scintillation. Nonspecific binding is measured by similar experiments to which 0.5 μM TBZ is added to the KG medium.

KET Binding

[^{3}H]KET, which is principally a 5-HT$_{2A}$ receptor ligand, is obtained commercially with a high specific activity (40–90 Ci/mmol, 1.5–3.33 Tbq/mmol), from NEN Life Science Products (Boston, MA). Binding to VMAT can be performed under the conditions used for [^{3}H]TBZOH, taking two precautions.[13] The first is to perform the filtrations rapidly in ice-cold buffer, because the rate of dissociation of [^{3}H]KET is extremely rapid (half-life of 40 sec at 0°). The incubation mixture is diluted 10 times in ice-cold buffer, rapidly filtered on GF/C glass filters (Whatman), and washed with 1 ml of ice-cold buffer. Binding is measured at 30° (K_d of 45 nM) or, more easily at 0° (K_d of 6 nM).

The second is to control pharmacologically the binding of [^{3}H]KET to VMAT and not to 5-HT$_{2A}$ receptors. [^{3}H]KET specifically bound to VMAT is obtained by subtracting the radioactivity measured in assays containing 2 μM TBZ, whereas that bound to 5-HT$_{2A}$ receptor was obtained by subtracting the radioactivity of incubations performed in the presence of 3 μM methysergide. In bovine chromaffin granule membranes, no 5-HT$_{2A}$ receptor binding can be detected.

Photoaffinity Labeling of Vesicular Monoamine Transporter using 7-azido[8-^{125}I]Iodoketanserin. 7-Azido[8-^{125}I]iodoketanserin (AZIK) can

[13] F. Darchen, D. Scherman, P. M. Laduron, and J. P. Henry, *Mol. Pharmacol.* **33,** 672 (1988).

be obtained from Amersham, but at the present time custom preparation is required. To decrease the cost, it is possible to order 7-amino[8-^{125}I]iodoketanserin and to use it in a one-step synthesis of AZIK.[14]

The carrier-free amino derivative solution (100 μCi) is dried under a flux of argon in the original vial to which an efflux active charcoal guard has been added. The compound is resuspended in 50 μl of 1 *N* HCl. Sodium nitrite (2 μl, 2.5 *M*) is added with a Hamilton syringe under magnetic stirring at 4°. The solution is stirred at 4° in the dark for 30 min. Sodium azide (2 μ, 2.5 *M*) is added at 4° and the mixture stirred for another 30 min in the dark.

The AZIK formed is separated from the mixture by high-performance liquid chromatography (HPLC) on a C_{18} μBondapak column (Waters, Milford, MA) with H_2O–CH_3OH–trifluoroacetic acid (TFA) (50:50:0.1) as the solvent and a UV detector operating at 315 nm. The active fractions are concentrated to dryness, resuspended in ethanol, and kept at −20°. The yield is 80%.

Photolabeling with [^{125}I]AZIK. For analytical purposes,[15] membrane samples with a concentration of about 3 n*M* of KET/TBZ binding sites are incubated with a one- to two-order-of-magnitude lower concentration of AZIK, in order to bind a large fraction of the probe. For instance, 50 μg of protein/ml of purified bovine chromaffin granule membranes[8,9] is incubated with 40 p*M* [^{125}I]AZIK (0.07 μCi/ml, 0.16×10^6 dpm/ml) in 5 ml of 0.3 *M* sucrose/10 m*M* HEPES, pH 7.5, for 1 hr at 0°, in a 3.5-cm diameter well of a cell culture plate. The incubated samples are irradiated for 10 min with a UV lamp operating at 350 nm at an average distance of 6 cm, under magnetic stirring. The conditions for irradiation are selected to flatten the 240 nm absorbance peak of 7-azidoketanserin and to minimize the labeling of phospholipids at 254 nm.

After irradiation, the membranes are washed twice by 4-fold dilution in sucrose buffer followed by centrifugation for 20 min at 140,000*g* at 4°. Samples (50–20 μg of protein) are analyzed by SDS–PAGE, using an acrylamide concentration of 10% (w/v) and a bisacrylamide:acrylamide ratio of 0.8:30 (w/w). Proteins are stained with silver nitrate, using sodium thiosulfate to sensitize the gels,[16] or Coomassie blue. Autoradiography is obtained by exposure of the dried gel to Kodak X-OMAT AR films with Lumix MR 800 screens (Agfa Gevaert, Rueil-Malmaison, France) for 2–30 days at −80°, or by using a Phosphorimager apparatus (Molecular Dynamics, Sunnyvale, CA). When quantitative data are required, the stained gel is cut into 2-mm slices and the radioactivity of the slices is measured with

[14] M. F. Isambert, B. Gasnier, D. Botton, and J. P. Henry, *Biochemistry* **31,** 1980 (1992).
[15] M. F. Isambert, B. Gasnier, P. M. Laduron, and J. P. Henry, *Biochemistry* **28,** 2265 (1989).
[16] T. Rabilloud, G. Carpentier, and P. Tarroux, *Electrophoresis* **9,** 288 (1988).

a γ-counter (Fig. 1). As described for [^{3}H]KET binding, the specificity of the labeling, i.e., labeling on VMAT and not on 5-HT_{2A} receptor, is tested pharmacologically: VMAT labeling is inhibited in the presence of 2 μM TBZ, whereas 2 μM methysergide inhibits 5-HT_{2A} receptor labeling.

In bovine chromaffin granules membranes, VMAT is labeled as a broad diffuse band with an apparent molecular mass centered at 73 kDa when 10% acrylamide concentration is used for SDS–PAGE. Increasing the acrylamide concentration increases the apparent molecular mass. In addition to the broad diffuse band corresponding to VMAT, another component is often visible as a sharp band with an apparent molecular mass of 40 kDa. This component is thought to be nonspecifically labeled since the corresponding band is not suppressed in the presence of TBZ. In some experiments, cytochrome *b*-561 (27 kDa), which is the major membrane protein, is also labeled. Bovine VMAT2 expressed in COS cells is labeled by the same technique as a component with an apparent molecular mass of 80 kDa.

[^{125}I]AZIK has also been used to photolabel a large stock of membranes,[14] for the purification of VMAT. Membranes (50 mg of protein) are thawed and suspended (2 mg/ml, final concentration) in 10 mM Tris–HCl buffer (pH 7.5) containing 1 mM EDTA, 6 μg/ml leupeptin, 5 μg/ml aprotinin, 10 μg/ml pepstatin, and 1 mM phenylmethylsulfonyl fluoride (PMSF), at 0°. [^{125}I]AZIK (80 μCi; 145 × 10^6 cpm) is added and incubated for 1 hr in the dark at 0°. The incubation medium is positioned at 6 cm from a 350-nm UV light source and irradiated for 12 min at 0°.

During these experiments, it has been noted that a radiolabeled pho-

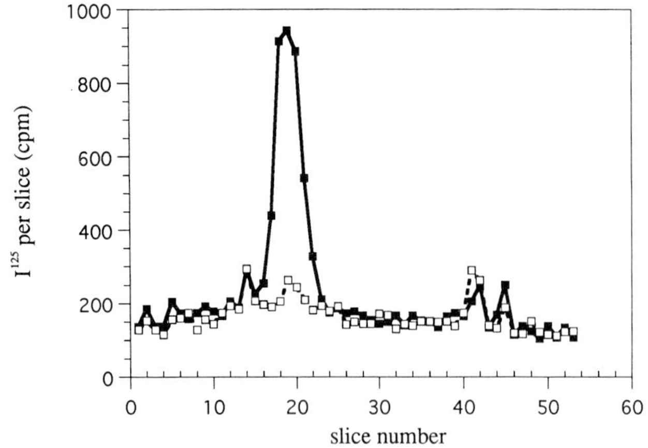

FIG. 1. Photoaffinity labeling of VMAT from purified bovine chromaffin granule membranes, using [^{125}I]AZIK. In this experiment, 60 μg of photolabeled protein, corresponding to 95,000 and 50,000 cpm for, respectively, the samples incubated in the absence (filled squares) and in the presence (open squares) of 0.2 μM TBZ, were analyzed by SDS–PAGE.

toproduct not covalently bound has a high affinity for the TBZ binding site, thus leading to a large overestimation of the protein covalently labeled (see Fig. 1). To minimize this effect,[14] TBZ is added to a final concentration of 2 μM after irradiation and the mixture is incubated for 2 hr at room temperature. After centrifugation for 20 min at 140,000*g* at 4°, the pellet is resuspended in about the same volume of TBZ-containing buffer and centrifuged again under the same conditions. The pellet is then resuspended in 5 ml of water containing leupeptin/aprotinin/pepstatin/PMSF at the concentration indicated above. The labeled membranes are stored in aliquots at −80°. The amount of photolabeled VMAT is determined by SDS–PAGE, gel slices, and γ counting of the 73-kDa peak. The specific activity of VMAT (typically 1000–2000 cpm/pmol) is calculated by dividing this amount by the [^{3}H]TBZOH binding site density of the membrane preparation.

Purification of Photolabeled VMAT

VMAT from bovine chromaffin granule membranes has been purified in an active or inactive form. In all cases, ligand binding has been used to monitor the purification, but this has been done in different ways. [^{3}H]TBZOH binding to the solubilized protein has allowed purification of an active protein in a 5-step protocol.[11] In a second procedure,[4] [^{3}H]RES (reserpine) is bound to membranes, and because RES dissociation is extremely slow, the protein can be solubilized without release of the ligand.[3]

Because the photolabeling of VMAT by AZIK is very specific, it is possible to purify the protein by simply measuring the specific activity of the ^{125}I-labeled material. Because the covalently modified protein is only a minor species, these measurements can be made to monitor the purification of not only the inactive but also the active protein. The inactive protein is purified by classical chromatographic techniques.[14] However, a new purification step, based on selective SDS-resistant thermal aggregation of VMAT, has been introduced[9] that may be of general use for the purification of polytopic membrane proteins. Two different procedures using this step will be described.

First Protocol

Bovine chromaffin granule VMAT partially purified by DEAE-cellulose chromatography at acid pH is used.[14] All steps are carried out at room temperature. Membranes (40 mg of protein) and photolabeled membranes (4 mg of protein) are suspended together in 10 ml of 10 mM sodium acetate/30 mM NaCl (pH 4.5) containing 1 mM EDTA, 6 μg/ml leupeptin, 5 μg/ml aprotinin, 10 μg/ml pepstatin, and 1 mM PMSF. They are solubilized by sequential addition of sulfobetaine 3-12 and Nonidet P-40, each at 3.7% final concentration (w/v). The mixture is incubated with stirring for 15 min at 30°, diluted with the same buffer to lower the detergent concentration to 1%, and centrifuged for 40 min at $215{,}000g_{max}$ at 4°.

Using a peristaltic pump operating at a flow rate of 1.5 ml/min the supernatant is applied to a MemSep 1000 cartridge (Millipore, Bedford, MA), previously equilibrated in the acetate/NaCl buffer containing 0.5% sulfobetaine 3-12 and 0.5% Nonidet P-40. The cartridge is washed with the same buffer, at 1 ml/min, until the radioactivity of the effluent is constant (about 20 ml). This radioactivity is mainly associated with phospholipids. The labeled protein is eluted by adding 200 m*M* NaCl to the same buffer. A 2.7-ml fraction is collected in a tube containing 10 μg/ml pepstatin (final concentration). PMSF is added to 1 m*M* final concentration, and the eluate is neutralized by addition of 2 *M* Tris buffer (pH 8.8).

Detergents are removed by adsorption at a flow rate of 0.4 ml/min on a 2-ml Extracti-Gel Column (Pierce, Rockford, IL), previously equilibrated in water. The labeled fractions (4 ml) are concentrated 20-fold by ultrafiltration on a Centricon 30 device (Amicon, Danvers, MA). The buffer in the concentrate is subsequently changed by a 10-fold dilution with 100 m*M* Tris-HCl (pH 6.8), and the sample is recentrifuged using the Centricon 30. More than 90% of the radioactivity is recovered in the concentrate (about 200 μl). The concentration of Tris-HCl (pH 6.8) in the sample is raised to 150 m*M*, and SDS and 2-mercaptoethanol are added at final concentrations of 2% (w/v) and 5% (v/v), respectively. After these steps, the protein is purified 10-fold with a 30% yield (Table I), as judged by SDS–PAGE, slicing of the gel, and summation of the radioactivity associated with the

TABLE I
PURIFICATION OF PHOTOLABELED VMAT[a]

Purification step	Protein (mg)	10^{-3} × Labeled transporter (cpm)	Yield (%)	Purification factor
Membranes	43.3	103	100	1
DEAE-cellulose eluate	1.47	35.9	35	10.2
First HPLC eluate	0.060	23.8	23	166
Second HPLC eluate	0.0041	2.7	2.6	274

[a] The results are for a purification starting with bovine chromaffin granule membranes,[8,9] obtained from 100 g of adrenal medulla. [Adapted from C. Sagné, M. F. Isambert, J. P. Henry, and B. Gasnier, *Biochem. J.* **316,** 825 (1996) with permission.]

73-kDa peak. At this step, protein concentration is measured by the technique of Schaffner and Weissman.[17]

The sample (300 μl) containing the proteins in SDS–2-mercaptoethanol is then heated at 100° for 10 min in a closed vial. Aggregates consisting of VMAT and only a limited number of proteins in the extract are formed. The aggregates are isolated by size-exclusion HPLC. The heated sample is filtered on a Durapore 0.45 μm (Millipore) and injected onto a 7.8 × 300 mm Protein-Pak 300 SW 10-μm column (Waters) using prefiltered 0.1% SDS, 200 m*M* sodium phosphate, pH 6.9, as eluent with a flow rate of 0.2 ml/min. The fractions collected (0.2 ml) are analyzed by γ counting and SDS–PAGE (Fig. 2). The photolabeled aggregates elute in the void volume (5.5 ml). This step is efficient, with a purification factor of 17 and a yield of 65% (Table I). From its specific activity, the protein can be considered as 50% pure. However, on SDS–PAGE it remains as an aggregate which does not enter the separating gel.

The next step is the dissociation of the aggregates, using the trifluoroacetic acid (TFA) treatment described by Hennessey and Scarborough.[18] Salts in the pooled fractions are diluted about 100-fold in three cycles of dilution with filtered (0.45 μm) water and centrifuged on a Centricon 30. The concentrate (about 200 μl) is freeze-dried in 1.5 ml polypropylene Microfuge tubes with a SpeedVac concentrator (Savant, Farmingdale, NY) and resuspended in 100 μl of anhydrous trifluoroacetic acid (TFA) under a chemical hood. Anhydrous TFA (sequencing grade) is purchased ampuled under nitrogen (Pierce). The TFA is then evaporated under a stream of argon. In order to achieve maximal removal of TFA, the tube is inclined and rotated manually during this operation. After evaporation of the TFA, the dried film on the walls of the tube is resuspended in 250 μl of prefiltered 150 m*M* sodium phosphate, 2% (w/v) SDS, 5% (w/v) 2-mercaptoethanol, and fractionated by size-exclusion HPLC under the conditions used for isolation of the aggregates. Two radioactive peaks are eluted at 5.5 and 6.3 ml, which correspond to residual aggregates and to the purified disaggregated transporter, respectively, as assessed by SDS–PAGE. The disaggregation yield, defined as the percentage of monomeric VMAT in the total specific radioactivity, is estimated from the areas of the major peaks. A mean value of 70% ± 12% (n = 14) has been calculated, with variations between 47 and 93%.

The various fractions are analyzed by SDS–PAGE. Careful observation of the stained proteins shows that the monomeric transporter, which appears as a homogeneous broad band, elutes from the size-exclusion column just

[17] W. Schaffner and C. Weissmann, *Anal. Biochem.* **56,** 502 (1973).
[18] J. P. Hennessey and G. A. Scarborough, *Anal. Biochem.* **176,** 284 (1989).

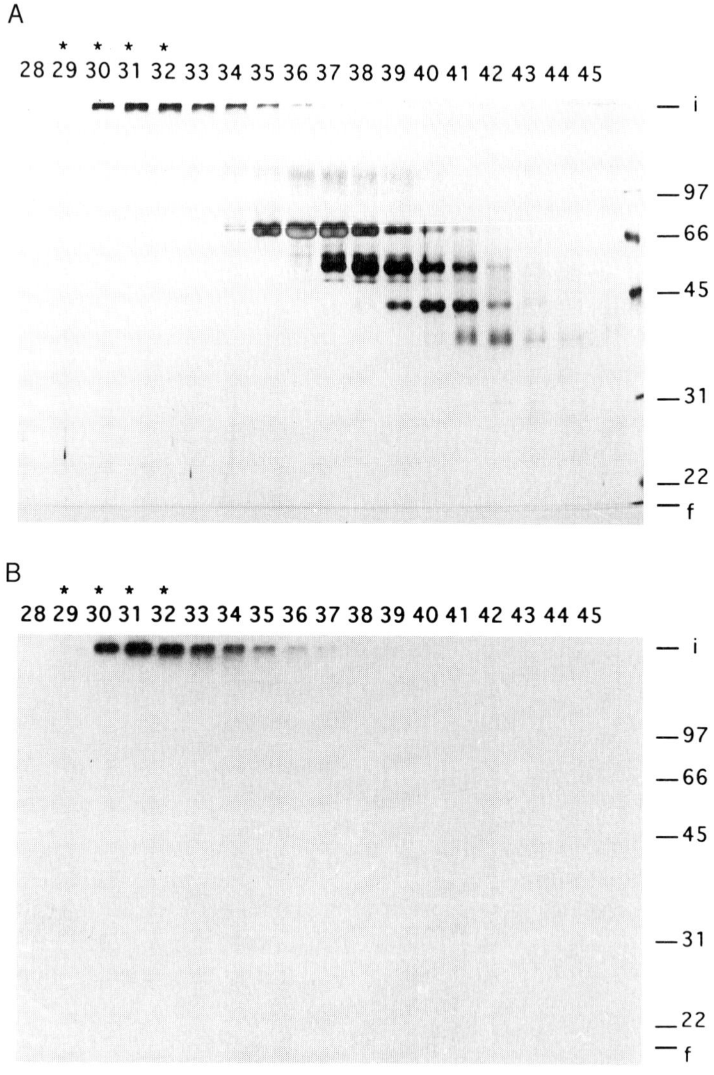

FIG. 2. SDS–PAGE analysis of the aggregated material fractionated by size exclusion HPLC. (A) Silver-stained gel; (B) autoradiogram. The partially purified [^{125}I]AZIK-labeled transporter, monitored by its radioactivity (B), is quantitatively aggregated by heating at 100° in SDS/2-mercaptoethanol. The aggregate is separated from the bulk of proteins by size-exclusion HPLC. [Reproduced from C. Sagné, M. F. Isambert, J. P. Henry, and B. Gasnier, *Biochem. J.* **316,** 825 (1996) with permission.]

after a 70-kDa doublet, which has been identified by immunoblotting as dopamine β-monooxygenase. Only fractions devoid of dopamine β-monooxygenase are pooled. The corresponding overall purification factor is 275, consistent with a purification of the transporter to homogeneity (Table I). However, the yield drops severely at the last step (about 10%), because of the variability of the disaggregation efficiency and of the overlap of VMAT and dopamine β-monooxygenase peaks.

Second Protocol

Purified aggregated VMAT can be useful for some purposes, such as protein sequencing[19] or antibody preparation. In this case, the protocol can be modified by adding another purification step, affinity chromatography on immobilized wheat germ agglutinin (WGA),[14] after DEAE chromatography and omitting the disaggregation step, which decreases the yield of the purification.

The neutralized eluate of the DEAE-MemSep cartridge is incubated with Sepharose 6MB-bound WGA (1 ml of wet resin prepared from a water suspension, Pharmacia, Piscataway, NJ), with gentle mechanical stirring for 2 hr in a 10-ml closed plastic chromatographic column. The column is centrifuged at low speed, and the supernatant is discarded. The resin is washed at flow rate of 1 ml/min by 12 volumes of 0.1% Nonidet P-40 and then 25 volumes of 200 m*M* NaCl/0.05% SDS/10 m*M* Tris-HCl (pH 6.8). The labeled protein is then eluted by incubation with 10 volumes of the last buffer containing 200 m*M* *N*-acetylglucosamine. The WGA eluate is concentrated about 50-fold by centrifugation on Centriprep 30 and Centricon 30 devices (Amicon). During this step, the buffer is exchanged for 100 m*M* Tris-HCl (pH 6.8) containing 0.05% SDS. The SDS and Tris-HCl (pH 6.8) in the concentrates are raised to 2% and 150 m*M*, respectively, and 2-mercaptoethanol is added to a 5% final concentration.

The sample (700 μl) is heated for 10 min at 100° in a closed vial and injected onto a size-exclusion HPLC column as before. The pooled radioactive fractions are concentrated, and their salt concentration is diluted 100-fold with water by centrifugation on Centricon 30 devices. The total yield of the three-step procedure is 20%. The purity of the preparation is checked by SDS–PAGE on an aliquot previously disaggregated by treatment with anhydrous TFA, as previously described. VMAT is the major component of the preparation as shown by the fact that N-terminal sequence analysis indicated only one component with the sequence of $bVMAT_2$.

[19] C. Sagné, M. F. Isambert, J. Vandekerckhove, J. P. Henry, and B. Gasnier, *Biochemistry* **36,** 3345 (1997).

[6] Functional Identification of Vesicular Monoamine and Acetylcholine Transporters

By HELENE VAROQUI and JEFFREY D. ERICKSON

Introduction

Neurotransmission depends on the regulated release of transmitter molecules such as the biogenic amines and acetylcholine (ACh). This requires the packaging of these molecules into the specialized secretory vesicles of neurons and neuroendocrine cells, a process mediated by specific vesicular transporters. Active transport of neurotransmitters into secretory organelles requires both the presence of a transmembrane H^+ electrochemical gradient, established and maintained by a vacuolar-type H^+-ATPase, and a vesicular transporter molecule which catalyzes the exchange of H^+ ions for neurotransmitter.[1-3] The endocrine-specific and neuronal isoforms of the vesicular monoamine transporter (VMAT1 and VMAT2, respectively) and the vesicular ACh transporter (VAChT) have now been cloned from several species. These transporters are structurally related and comprise a new gene family.[4,5] The development of active transport assays for the VMAT isoforms and for VAChT functionally distinguished these transporters and should enable the molecular basis for the differential pharmacology and functional parameters observed to be determined.

In this chapter, two vesicular transport assays are described which were developed to functionally identify and characterize the two VMAT isoforms and VAChT. Active transport of monoamines by VMAT1 and VMAT2 is performed in permeabilized fibroblasts transiently expressing these proteins by the recombinant T7 vaccinia virus system.[6] Active transport of ACh by human VAChT is performed in postnuclear homogenates containing secretory vesicles from stably transfected neuroendocrine PC-12 (rat adrenal pheochromocytoma) cells.[7]

[1] D. Njus, P. M. Kelley, and G. J. Harnadek, *Biochim. Biophys. Acta* **853,** 237 (1986).
[2] P. R. Maycox, J. W. Hell, and R. Jahn, *TINS* **13,** 83 (1990).
[3] S. M. Parsons, C. Prior, and I. G. Marshall, *Int. Rev. Neurobiol.* **35,** 279 (1993).
[4] T. B. Usdin, L. E. Eiden, T. I. Bonner, and J. D. Erickson, *TINS* **18,** 218 (1995).
[5] H. Varoqui and J. D. Erickson, *Mol. Neurobiol.* **15,** 165 (1997).
[6] J. D. Erickson, L. E. Eiden, and B. J. Hoffman, *Proc. Natl. Acad. Sci.* **89,** 10993 (1992).
[7] H. Varoqui and J. D. Erickson, *J. Biol. Chem.* **271,** 27229 (1996).

0076-6879/98 $25.00

Functional Identification of VMAT1 and VMAT2

In 1992, two groups independently identified VMAT1 and VMAT2 by expression cloning. In both cases the cDNA screening strategy relied on two processes of uptake by intact nonneuronal cells: diffusion of a quaternary amine (1-methyl-4-phenylpyridinium, MPP^+) or neutral amine (5-$[^3H]$hydroxytryptamine, $[^3H]$5HT) across the plasma membrane followed by sequestration of the charged species into an intracellular compartment which contains a vacuolar-type H^+-ATPase.[6,8]

The neuronal rVMAT2 isoform was cloned from a rat basophilic leukemia cell line (RBL-2H3) cDNA library by vesicular $[^3H]$5HT sequestration in intact fibroblasts.[6] Monkey kidney fibroblasts (CV-1 cells) were infected with a T7 polymerase-expressing vaccinia virus and then subsequently transfected with cDNA sublibraries driven by a T7 promoter. Clones conferring ability to accumulate $[^3H]$5HT in a reserpine-sensitive manner were identified by microscopy. The slow, pH-sensitive, and low-affinity accumulation of monoamines in intact CV-1 cells expressing rVMAT2 suggested that the amines cross the plasma membrane by passive diffusion of the neutral species. Once inside the cell, the monoamines become protonated and available to rVMAT2 located in the membranes of an intracellular compartment. To access this intracellular compartment directly, the plasma membrane was selectively permeabilized with digitonin. In permeabilized CV-1 cells expressing rVMAT2, high-affinity ATP-dependent uptake of ^3H-labeled monoamines occurred which was inhibited by nanomolar concentrations of reserpine or tetrabenazine and which was dependent on the presence of a transmembrane proton gradient (Fig. 1).

Permeabilized fibroblasts have been used for the functional identification of full-length cDNA clones for the human VMAT isoforms,[9,10] pharmacologic characterization of rat,[6,11,12] bovine,[13] and human VMATs,[10] and for structural–functional analysis of chimeric VMAT molecules[14] and site-specific VMAT mutants.[15,16] When VMAT1 and VMAT2 are transfected

[8] Y. Liu, D. Peter, A. Roghani, S. Schuldiner, G. G. Privé, D. Eisenberg, N. Brecha, and R. H. Edwards, *Cell* **70,** 539 (1992).

[9] J. D. Erickson and L. E. Eiden, *J. Neurochem.* **61,** 2314 (1993).

[10] J. D. Erickson, M. K.-H. Schäfer, T. I. Bonner, L. E. Eiden, and E. Weihe, *Proc. Natl. Acad. Sci.* **93,** 5166 (1996).

[11] E. Weihe, M. K.-H. Schäfer, J. D. Erickson, and L. E. Eiden, *J. Mol. Neurosci.* **5,** 149 (1994).

[12] J. D. Erickson, M. K.-H. Schäfer, L. E. Eiden, and E. Weihe, *J. Mol. Neurosci.* **6,** 277 (1995).

[13] M. Howell, A. Shirvan, Y. Stern-Bach, S. Steiner-Mordoch, J. E. Strasser, G. E. Dean, and S. Schuldiner, *FEBS Lett.* **338,** 16 (1994).

[14] J. D. Erickson, *Adv. Pharmacol.* **42,** 227 (1998).

[15] A. Shirvan, O. Laskar, S. Steiner-Mordoch, and S. Schuldiner, *FEBS Lett.* **356,** 145 (1994).

[16] S. Steiner-Mordoch, A. Shirvan, and S. Schuldner, *J. Biol. Chem.* **271,** 13048 (1996).

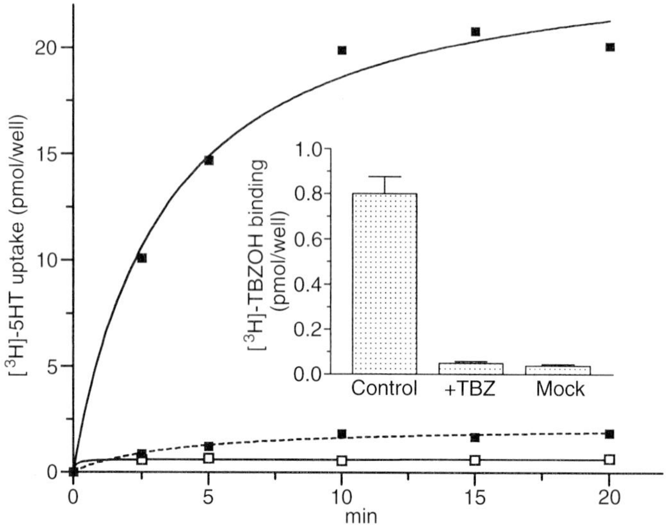

FIG. 1. Time course of [^{3}H]5HT (0.2 μM) accumulation at 37° by rat VMAT2 in digitonin-permeabilized CV-1 fibroblasts. (■), Control; (□) reserpine (0.5 μM); (dotted line), no ATP. *Inset:* [^{3}H]TBZOH (10 nM) binding to rat VMAT2 in the presence and absence of 2.5 μM tetrabenazine and in mock-transfected cells. Data are from a representative experiment performed in duplicate, or are expressed as the mean ± S.E. from three separate experiments.

into CV-1 fibroblasts, more than 80% of the cells in the culture express functional transporter. This is evidenced by immunohistochemical analysis with isoform-specific antisera and by autoradiographic visualization of [^{3}H]5HT uptake in permeabilized cells under bright-field microscopy.[9–11] In the presence of ATP, [^{3}H]5HT uptake mediated by either isoform reaches similar maximal levels, ≈50 times greater than nonspecific uptake into mock-transfected CV-1 cells by 15 min at 37°. Uptake is energy dependent (>90% ATP dependent) and is abolished by the proton-translocating ionophore carbonylcyanide *p*-trifluoromethoxyphenylhydrazone (5 μM; FCCP) or by the vacuolar H^+-ATPase inhibitor tri-(*n*-butyl)tin (50 μM; TBT). The transport process is saturable, with apparent affinity (K_m) for 5HT of 1.3 μM and 0.8 μM and V_{max} values of 37 and 42 pmol/min per 450,000 cells for hVMAT1 and hVMAT2, respectively.[10]

A characteristic feature of the VMATs is their broad range of substrate specificity. Biogenic amine substrates inhibit [^{3}H]5HT uptake by both VMAT isoforms with the following rank order of potency: 5HT > dopamine > epinephrine > norepinephrine > histamine. The high affinity displayed by VMAT2 for 5HT and the catecholamines (1–4 μM range) in permeabilized cells is similar to reported K_m values for monoamine uptake in synaptic vesicles isolated from the brain and heart and from cell lines.[17–22] The affinity of VMAT2 for histamine, as determined from the IC_{50} for inhibition of [^{3}H]5HT uptake or directly, as measured by reserpine- and tetrabenazine-sensitive [^{3}H]histamine uptake, is ≈200 μM.[9,10,12] In the RBL-2H3 cell line, VMAT2 also has a low affinity for histamine relative to the other monoamines.[22] Although VMAT1 and VMAT2 exhibit similar affinity for 5HT (≈2-fold difference) hVMAT2 shows ≈3- to 4-fold higher affinity for catecholamine substrates than does hVMAT1.[10]

The interaction of unsubstituted aromatic amines with the transporters is a key feature distinguishing both rat and human VMAT1 and VMAT2. HVMAT2 displays a ≈20-fold higher affinity than hVMAT1 for all unsubstituted aromatic amines tested so far, including phenylethylamine, amphetamine, and histamine, which lack strong electron-donating substituents on the aromatic ring, in permeabilized cells.[9,10] A chimeric hVMAT1/hVMAT2 molecule (2/1/2) has been constructed which dissociates the interaction of histamine from that of amphetamine.[14] HVMAT2 and the chimera exhibit an apparent affinity (IC_{50}) of ≈2 μM for amphetamine, whereas hVMAT1 displays a ≈20-fold lower affinity. On the other hand, hVMAT2 exhibits an apparent affinity (IC_{50}) of 200 μM for histamine, whereas both the chimera and hVMAT1 displays a ≈20-fold lower affinity. The 2/1/2 chimera exhibits hVMAT2-like sensitivity to amphetamine while it displays hVMAT1-like sensitivity to histamine. Thus, specific amino acid residues between Ile-265 and Tyr-418 of hVMAT2 may be important for the increased potency of histamine to compete with ^{3}H-5HT for uptake compared with hVMAT1.

The interaction of tetrabenazine with the transporters is a second key feature distinguishing VMAT1 from VMAT2. Tetrabenazine inhibits the uptake of [^{3}H]5HT by VMAT2 with a K_i of ≈ 80 nM, whereas concentrations of tetrabenazine as high as 20 μM do not affect transport mediated by hVMAT1.[9,10] Furthermore, only hVMAT2 shows significant binding of [^{3}H]TBZOH. Binding [^{3}H]TBZOH to VMAT2-transfected, permeabilized CV-1 cells occurs rapidly (saturable by 5 min at 37°) to levels more than

[17] F. J. Seidler, D. F. Kirksey, C. Lau, W. L. Whitmore, and T. A. Slotkin, *Life Sci.* **21,** 1075 (1977).

[18] L. Toll and B. D. Howard, *Biochemistry* **17,** 2517 (1978).

[19] R. Maron, Y. Stern, B. I. Kanner, and S. Schuldiner, *FEBS Lett.* **9,** 237 (1979).

[20] K. J. Angelides, *J. Neurochem.* **35,** 949 (1980).

[21] J. K. Disbrow and J. A. Ruth, *Life Sci.* **29,** 1989 (1981).

[22] B. I. Kanner and A. Bendahan, *Biochim. Biophys. Acta* **816,** 403 (1985).

20-fold higher than background and is completely displaceable by tetrabenazine at 2.5 μM (Fig. 1, inset). In contrast, [^{3}H]TBZOH binding to hVMAT1-expressing CV-1 cells under identical conditions is indistinguishable from binding to mock-transfected cells.

Development of Active Transport/Binding Assay in Permeabilized Cells

Several agents are known to selectively render the plasma membrane permeable while leaving intracellular membrane compartments intact and capable of maintaining electrochemical gradients.[23,24] Streptolysin O and staphylococcal α-toxin are channel-forming bacterial exotoxins which assemble in supramolecular amphiphilic polymers in the plasma membrane to form stable pores of various diameters. Proteins can pass through pores formed by streptolysin O, whereas α-toxin permits the exchange of small molecules only. Mild detergents such as digitonin and saponin have also been used for selective membrane permeabilization. Digitonin is a steroid glycoside that interacts specifically with 3β-hydroxysterols. Since cholesterol is enriched in the plasma membrane of most cells, this membrane is more vulnerable to digitonin. The conditions for detergent use must, however, be carefully controlled to restrict permeabilization to the plasma membrane and only to small molecules.

Permeabilization of the plasma membrane permits the exchange of the soluble contents of the cytoplasm with the buffer, allowing direct access of radiolabeled substrates or inhibitors to intracellular membrane compartments. We use an intracellular medium comprised of HEPES-buffered potassium tartrate. Polyacrylate- or tartrate-based intracellular media have been shown to prolong the functional viability of monoaminergic synaptic vesicles isolated from rat brain.[21,25] Media containing high concentrations of chloride ions should be avoided as they disrupt H^+ electrochemical gradients across secretory organelles due to the presence of vesicular chloride ion channels.[18,26–28] In addition, permeabilization of neuroendocrine cells must occur in a Ca^{2+}-free intracellular medium to avoid exocytosis of secretory organelles. Permeabilization methods have been used successfully

[23] G. Ahnert-Hilger, W. Mach, K. J. Fohr, and M. Gratzl, *Meth. Cell Biol.* **31,** 63 (1989).

[24] R. Diaz and P. D. Stahl, *Meth. Cell Biol.* **31,** 25 (1989).

[25] J. K. Disbrow, M. J. Gershten, and J. A. Ruth, *Experientia* **38,** 1323 (1982).

[26] R. P. Casey, D. Njus, G. K. Radda, and P. A. Sehr, *Biochem. J.* **158,** 583 (1976).

[27] X.-S. Xie, B. P. Crider, and D. K. Stone, *J. Biol. Chem.* **264,** 18870 (1989).

[28] J. D. Erickson, J. M. Masserano, E. M. Barnes, J. A. Ruth, and N. Weiner, *Brain Res.* **516,** 155 (1990).

in studies of chromaffin granule exocytosis,[29–32] mast cell exocytosis,[33] endosomal acidification,[34] Ca^{2+} transport across mitochondrial membranes,[35] and regulation of vesicular monoamine transporter function by protein kinases[36] and depolarization.[37]

An active transport or binding assay performed in permeabilized cells in culture is a convenient approach for cloning and functional expression, and for structure/function analysis of intracellular transport proteins. Several criteria must be met for functional cloning and characterization of intracellular transport proteins using permeabilized cells. Cell lines which abundantly express the transporter of interest should be available to examine the feasability of measuring active transport activity or binding of specific inhibitors under permeabilized conditions. The functional sensitivity, pharmacologic specificity, and bioenergetic requirements of the assay should be assessed and compared to results obtained with tissue preparations. Cultured cells should remain viable and adherent to the dish during the assay, which may require the coating of the dishes with collagen or another basement membrane component. Finally, a method of fixation must be developed to detect accumulated radiolabeled substrates by microscopy.

In this section, we describe a method for active vesicular transport of transport of monoamines using digitonin-permeabilized CV-1 fibroblasts expressing VMAT1 and VMAT2, PC12 cells which endogenously express VMAT1, and RBL-2H3 cells which express VMAT2. A comparison of the sensitivity of the [^{3}H]5HT uptake assay in these permeabilized cells is shown in Table I.

Cell Culture and Collagen Coating

Monkey kidney fibroblasts (CV-1 cells) are maintained at 37° in an atmosphere of 95% air, 5% CO_2 (v/v) in Dulbecco's modified Eagle's medium (DMEM), 10% fetal bovine serum (FBS), penicillin (100 U/ml), streptomycin (100 mg/ml), and glutamine (4 m*M*). PC12 cells are maintained as above with medium containing 7% fetal bovine serum and 7% heat

[29] S. P. Wilson and N. Kirshner, *J. Biol. Chem.* **258,** 4994 (1983).
[30] L. A. Dunn and R. W. Holz, *J. Biol. Chem.* **258,** 4989 (1983).
[31] T. Sarafian, D. Aunis, and M. F. Bader, *J. Biol. Chem.* **262,** 16671 (1987).
[32] P. E. Lelkes and H. B. Pollard, *J. Biol. Chem.* **262,** 15496 (1987).
[33] T. W. Howell, S. Cockcroft, and B. D. Gomperts, *J. Cell Biol.* **105,** 191 (1987).
[34] D. J. Yamashiro, S. R. Fluss, and F. R. Maxfield, *J. Cell Biol.* **97,** 929 (1983).
[35] W. P. Dubinsky and R. S. Cockrell, *FEBS Lett.* **59,** 39 (1975).
[36] N. Nakanishi, S. Onozawa, R. Matsumoto, R. Hasegawa, and N. Minami, *J. Biochem.* **118,** 291 (1995).
[37] C. Desnos, M.-P. Laran, K. Langley, D. Aunis, and J.-P. Henry, *J. Biol. Chem.* **270,** 16030 (1995).

TABLE I
VESICULAR TRANSPORT OF [^{3}H]5HT IN PERMEABILIZED CELLS[a]

Compound	hVMAT1	hVMAT2	PC-12	RBL-2H3
Control	336,666	372,009	449,432	502,025
Reserpine[b]	7253	7668	12,096	10,542
Tetrabenazine	337,951	9377	367,835	13,449

[a] Typical cpm values for [^{3}H]5HT uptake (0.2 μM) at 15 min by digitonin-permeabilized cells expressing the human VMAT cDNAs or cells endogenously expressing rat VMAT1 (PC-12) or rat VMAT2 (RBL-2H3).
[b] Reserpine and tetrabenazine are present at 2 μM concentration.

inactivated horse serum. RBL-2H3 cells are maintained in EMEM (Earle's modified Eagle's medium) with antibiotics/glutamine and supplemented with 15% fetal bovine serum.

Cells are plated on collagen-coated 6-well dishes (35 mm) for uptake characterization or Permanox 4-well chamber slides for microscopy. A thin coating of rat tail tendon collagen type I (Collaborative Biomedical Products, Bedford, MA) is used for CV-1 and RBL-2H3 cells; human placental collagen type VI (Sigma, St. Louis, MO) is used for PC-12 cells. Coated plates and slides may be used immediately or may be air dried and stored at 4° for up to 1 week under sterile conditions.

Rat tail type I collagen is purchased as a solution (5 mg/ml) in 0.02 *N* acetic acid. This material is diluted to 50 μg/ml using 0.02 *N* acetic acid. Two milliliters of this solution is added to each well and incubated at room temperature for 1 hr. The solution is aspirated and the wells are air dried. Three consecutive rinses with 2 ml of phosphate-buffered saline (PBS) are then performed.

Human placental type VI collagen is purchased dry and is stored as a stock solution (1 mg/ml) in 0.5 *M* acetic acid. This stock solution is diluted to 50 μg/ml with 20% (v/v) ethanol immediately before use. Apply 2 ml of this solution to each well and incubate at room temperature for 1 hr. Aspirate the solution and air dry the wells. The collagen is fixed with 20 m*M* NH_4OH and rinsed with three consecutive PBS washes.

Recombinant Vaccinia Virus Infection and Transfection of CV-1 Fibroblasts

CV-1 cells are transfected with T7 promoter-driven plasmids using the transient vaccinia virus/bacteriophage T7 hybrid system.[38] Cells are plated

[38] T. R. Fuerst, E. G. Niles, F. W. Studier, and B. Moss, *Proc. Natl. Acad. Sci.* **83,** 8122–8126 (1986).

at 2 × 10^6 per 10 cm^2 plate (for membrane preparations) or in collagen-coated 6-well dishes at 4 × 10^5 per well. The following day the cells are rinsed once with PBS. An aliquot of the vaccinia stock is mixed with an equal volume of filter-sterilized 0.25 mg/ml trypsin, incubated at 37° for 30 min, and sonicated for 30 sec in a bath sonicator. This is diluted with DMEM without serum to 10 plaque-forming units (pfu) per cell and 0.5 ml of virus/DMEM is added to each well or 1 ml is added to each plate. The plates are rocked gently at 37° for 30 min. A solution of DMEM (without serum) containing 1% LipofectACE (Life Technologies Inc., Gaithersburg, MD) is prepared in a sterile polystyrene tube. The DNA to be transfected is added to this solution 15 min prior to the addition to the cells. After 30 min of infection, 1 ml of the DNA/DMEM/LipofectACE mixture is added to the wells (or 2.5 ml/plate) and the cells are returned to the incubator without rocking. After 3 hr, medium containing serum can be added (optional). After 16 hr, uptake/binding is directly measured in permeabilized cells (6-well dishes), or cells in the plates are harvested and postnuclear supernatants arc prepared as described below.

Stocks of vaccinia virus expressing the T7 polymerase are available commercially from ATCC (Rockville, MD). Amplification and preparation of crude stocks are performed as previously described[39] using HeLa cells grown in 175-mm^2 flasks. In addition to titering the vaccinia stocks with a plaque assay, we titer them functionally by assaying the transport activity of VMAT2-expressing CV-1 cells prepared using serial dilutions of virus. Virus stocks can also be titered by expression of β-galactosidase in CV-1 cells.[40]

Permeabilized Cell Uptake/Binding Assay

Infected/transfected CV-1 cells, PC12 cells, and RBL-2H3 cells are rinsed with 1 ml PBS and then with 1 ml intracellular buffer containing 110 m*M* potassium tartrate/5 m*M* glucose/1 m*M* ascorbic acid/10 μM pargyline/20 m*M* HEPES (pH 7.4). The cells are then permeabilized for 10 min at 37° in uptake buffer containing 10 μM (12.5 μg/ml) digitonin. The medium is replaced with fresh buffer without digitonin containing 5 m*M* MgATP at room temperature. Fifty microliters of uptake buffer containing [^{5}]5HT (NEN Life Science Products, Boston, MA; 27.3 Ci/mmol) is then added and incubated at 37° for various intervals. Uptake is terminated with a 2.5 ml wash in buffer with 2 m*M* $MgSO_4$ on ice. The cells are then solubilized in 1 ml of 1% SDS and transferred to vials containing 10 ml of EcoScint scintillation fluid for quantitation by liquid scintillation

[39] F. M. Ausubel, R. Brent, R. E. Kingston, O. D. Moore, J. G. Seidman, J. A. Smith, and K. Struhl, "Current Protocols in Molecular Biology," Wiley Interscience, New York (1993).
[40] J. R. Sanes, J. L. R. Rubenstein, and J.-F. Nicolas, *EMBO J.* **5,** 3133 (1986).

counting. For determination of initial rates of uptake, various concentrations of radiolabeled and unlabeled 5HT are added and uptake is terminated after 2 min. For uptake inhibition studies (IC_{50}), the final concentration of [^{3}H]5HT is generally 100 nM and uptake is terminated after 2 min. Generally, three plates are processed simultaneously with wells containing control, mock, and seven different concentrations of 5HT or competitor/inhibitor, in duplicate. Inhibitors (from 100× stocks) are added immediately prior to addition of ^{3}H-5HT. Substrates are mixed with ^{3}H-5HT prior to addition to cells.

Stock solutions of digitonin [10 mg/ml in dimethyl sulfoxide (DMSO)] are stored at −80°. For permeabilization, 62.5 μl of the digitonin stock is added to 50 ml of the potassium tartrate medium. MgATP (Sigma) solutions are always prepared fresh and neutralized to pH 7.4 with KOH. Na_2ATP can also be used, but the uptake solution must then be supplemented with at least 2.5 mM $MgCl_2$. FCCP stock (100 mM) is prepared by addition of 393 μl DMSO to a 10-mg vial (Aldrich, Milwaukee, WI). One microliter is diluted 1:10 in absolute ethanol and then diluted again 1:10 in 50% (v/v) ethanol. Fifty microliters of this solution is then added to 10 ml of ATP-containing buffer, giving a final concentration of 5 μM. TBT (3.656 M solution; Aldrich) is diluted 1:10 in ethanol and then diluted again 1:26.5 in 50% ethanol. Fifty microliters of this solution is then added to 10 ml ATP-containing uptake buffer, giving a final concentration of 50 μM. Reserpine (Sigma) is dissolved in glacial acetic acid at a concentration of 50 mM, diluted 1:10 in 23% (v/v) ethanol, and diluted again 1:100 in buffer to obtain a 100× stock. A 10-mM solution of tetrabenazine (Fluka, Ronkonkoma, NY) is prepared in absolute ethanol and stored at −20°. A 100× stock solution is prepared by 1:100 dilution in buffer. When solvents are used, their final concentration is less than 0.25%, which has no effect on control [^{3}H]5HT uptake. Stock solutions (100×) of biogenic amine substrates, substituted and unsubstituted aromatic amines, and MPP^+ are dissolved in uptake buffer immediately prior to use.

The binding of α-[*O-methyl*-^{3}H]dihydrotetrabenazine ([^{3}H]TBZOH 2 nM; 152 Ci/mmol, Amersham Life Science, Inc., Cleveland, OH) is performed by addition of the radioligand to permeabilized fibroblasts exactly as described for [^{3}H]5HT. Binding is also performed in postnuclear homogenates from CV-1 cells expressing VMAT2 (see proceeding section).

Single-Cell Visualization of Functional Activity

The method of expression cloning that we use including mRNA isolation, plasmid cDNA library production and amplification, and transfection has been described.[41] Basically, the general strategy is to (1) construct a

[41] T. B. Usdin and M. C. Beinfeld, *Neuroprotocols* **5,** 144 (1994).

cDNA library in a plasmid expression vector from a cell line (or tissue) rich in mRNA encoding the protein of interest, (2) transfect the library into cells which lack this protein, (3) assay for functional activity, and (4) successively subdivide pools of the original library until a pool containing a single cDNA clone is obtained. To screen plasmid cDNA library subpools, the uptake/binding experiments described above are performed in collagen-coated chamber slides. The concentration of radioligand used for uptake/binding in a cloning strategy should approach the K_m/K_d of the protein of interest as determined by pilot experiments with the cell line from which a plasmid cDNA library has been constructed or from the literature. After an appropriate length of incubation (15–20 min) at 37°, the slides are placed on a metal rack on ice. The cells are rinsed three times with 1 ml ice-cold wash buffer, and then incubated with an appropriate fixative, the choice of which depends on the functional groups present on the radioligand. For ^{3}H-5HT uptake or ^{3}H-TBZOH binding experiments, we use 1 ml of wash buffer which contains 2.5% glutaraldehyde and 1% acrolein (Polysciences, Inc., Warrington, PA) at room temperature. The slides are kept in the fixative, on the ice tray for 1 hr. The cells are then incubated 3 × 5 min with 2 ml wash buffer on ice. The liquid is removed, the chambers are disassembled, and the slides are dipped in ice-cold water and allowed to air dry.

For subsequent autoradiography, the slides are dipped in nuclear emulsion (Kodak, Rochester, NY). In a darkroom, scoop out emulsion with a spatula into a Coplin jar and incubate at 40° for approximately 30 min before use. The emulsion may be mixed 1:1 with water. Place slides in plastic slide grips, dip into the emulsion, and hang dry for several hours. Place the slides into black slide boxes with Drierite capsules and store at 4° in a light safe. For development, dip the slides through D-19 (Kodak) for 2 min, water for 15 sec, and then Kodak Rapid Fix (without hardener) for 2 min. Once the slides are fixed, the lights may be turned on. Rinse the slides in tap water for 10 min; counterstain, if desired, for 30 sec in Nuclear Fast Red.

Using this method of single cell visualization of functional activity, hVMAT2-positive clones can be observed in permeabilized fibroblasts transfected with pools of ≈30,000 recombinants (for ^{3}H-5HT uptake) and of ≈5,000 recombinants (for ^{3}H-TBZOH binding) using human substantia nigra or pheochromocytoma cDNA libraries.[9]

Functional Identification of VAChT

In 1993, the *unc17* gene from the nematode *Caenorhabditis elegans* was cloned and found to encode a protein having ≈40% amino acid sequence identity with the VMATs and a similar predicted structure.[42] Based on this striking homology, its expression on synaptic vesicles in cholinergic nerve endings, and the observation of defects in cholinergic neurotransmission in *unc* mutants that were consistent with the lack of vesicular ACh storage, Rand and colleagues proposed that *unc17* was a VAChT. The model system for the study of VAChT has been synaptic vesicles purified from the electric organ of the marine ray *Torpedo*. The electric lobe, containing the cell bodies of neurons that innervate the electric organ, was used to prepare cDNA libraries and an *unc17* homolog was obtained by low-stringency screening.[43] The *Torpedo* protein, expressed in CV-1 fibroblast cells, possessed a high-affinity binding site for vesamicol, a drug which blocks *in vitro* and *in vivo* ACh accumulation in cholinergic vesicles. Using a probe derived from the *Torpedo* vesamicol-binding protein cDNA, mammalian homologs from rat neuroendocrine PC-12 and human neuroblastoma SK-N-SH cDNA libraries were obtained. Expression of the rat VAChT in fibroblasts enabled intact cells to sequester ACh in a vacuolar ATPase-containing compartment by a process which was inhibited by L-vesamicol.[44] Thus, a single cDNA encoding VAChT mediates vesamicol-sensitive ACh uptake.

The characteristics of active vesicular ACh transport have only been studied in highly purified synaptic vesicles from *Torpedo*.[3] It has not been possible to demonstrate ATP-dependent transport of ACh into permeabilized fibroblasts following VAChT transfection.[5,45] Furthermore, kinetic analysis of the ACh transport system was not possible using the intact fibroblast cell assay used above to identify the rat protein as VAChT. In mammalian synaptic vesicle preparations, ATP-dependent transport of ACh is very low and kinetic parameters of uptake have never been determined.[46,47] PC-12 cells synthesize, store, and secrete low levels of ACh as well as dopamine.[48–50] Subcellular fractionation and immunoelectron microscopy of PC-12 cells has revealed that the large dense cored vesicles possess a reserpine-

[42] A. Alfonso, K. Grundahl, J. S. Duerr, H. P. Han, and J. B. Rand, *Science* **261,** 617 (1993).

[43] H. Varoqui, M.-F. Diebler, F.-M. Meunier, J. B. Rand, T. B. Usdin, T. I. Bonner, L. E. Eiden, and J. D. Erickson, *FEBS Lett.* **342,** 97 (1994).

[44] J. D. Erickson, H. Varoqui, M. K.-H. Schäfer, W. Modi, M.-F. Diebler, E. Weihe, J. B. Rand, L. E. Eiden, T. I. Bonner, and T. B. Usdin, *J. Biol. Chem.* **269,** 21929 (1994).

[45] A. Roghani, J. Feldman, S. A. Kohan, A. Shirzadi, C. B. Gundersen, N. Brecha, and R. H. Edwards, *Proc. Natl. Acad. Sci.* **91,** 10620 (1994).

[46] R. Bauerfeind, A. Regnier-Vigouroux, T. Flatmark, and W. B. Huttner, *Neuron* **11,** 105 (1993).

[47] J. R. Haigh, K. Noremberg, and S. M. Parsons, *NeuroReport* **5,** 773 (1994).

[48] L. A. Green and G. Rein, *Nature* **268,** 349 (1977).

[49] L. A. Green and G. Rein, *Brain Res.* **129,** 247 (1977).

[50] D. Schubert and F. G. Klier, *Proc. Natl. Acad. Sci.* **74,** 5184 (1977).

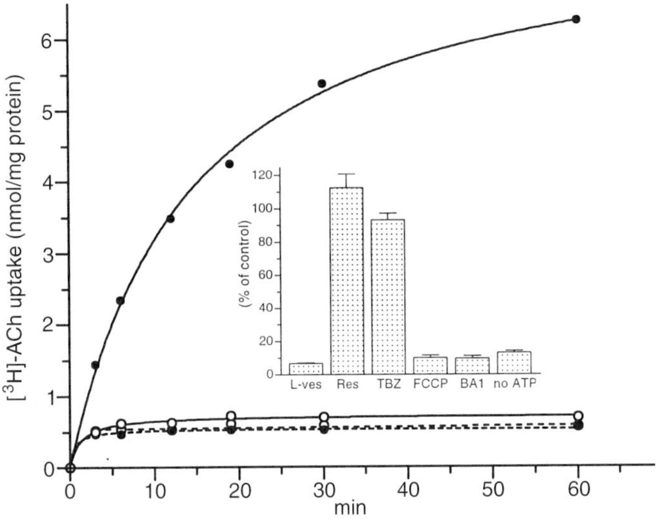

FIG. 2. Time course of [^{3}H]ACh (0.4 mM) accumulation in postnuclear supernatants from human VAChT-expressing PC-12 cells (●) and control PC-12 cells (○) in the presence (dotted line) and absence (solid line) of 2 μM L-vesamicol. *Inset:* Postnuclear supernatants from human VAChT-expressing PC-12 cells were incubated with 0.4 mM [^{3}H]ACh and 5 mM Mg^{2+}-ATP for 20 min at 37° in the presence and absence of various compounds: L-ves, 0.5 μM L-vesamicol; Res, 0.5 μM reserpine; TBZ, 0.5 μM tetrabenazine; FCCP, 2.5 μM; BA1, 1 μM bafilomycin; and without ATP. Uptake of ^{3}H-ACh observed at 4° was subtracted from the data and was less than 5% of the total uptake. Data are either from a representative experiment performed in duplicate or are expressed as the mean ± S.E. from three separate experiments.

sensitive transport system[46] and VMAT1,[51,52] whereas low levels of specific ACh transport,[46] vesamicol binding,[53] and VAChT[52] are found on the small synaptic-like microvesicles. The generation of species-specific antisera against VAChT has now enabled selection of rat PC-12 clones expressing high levels of the human VAChT protein. Using this system, the first kinetic parameters of mammalian VAChT have been obtained.[7]

The functional parameters of active transport of [^{3}H]ACh by hVAChT expressed in PC-12 cells are shown in Fig. 2. Specific vesamicol-sensitive uptake mediated by hVAChT cDNA is linear for ≈10 min with maximal steady-state levels attained by 30–60 min. Approximately 10 times more [^{3}H]ACh uptake is observed in the stable PC-12 cell line expressing hVAChT than in control PC-12 cells which only express the endogenous

[51] Y. Liu, E. S. Schweitzer, M. J. Nirenberg, V. M. Pickel, C. J. Evans, and R. H. Edwards, *J. Cell Biol.* **127,** 1419 (1994).

[52] E. Weihe, J.-H. Tao-Cheng, M. K.-H. Schäfer, J. D. Erickson, and L. E. Eiden, *Proc. Natl. Acad. Sci. USA* **93,** 3547 (1996).

[53] D. Blumberg and E. S. Schweitzer, *J. Neurochem.* **58,** 801 (1992).

rVAChT protein. Uptake mediated by the endogenous rVAChT protein was less than 2-fold greater than that observed in the presence of 2 μM L-vesamicol or at 4°. An analysis of the bioenergetics and specifity of [^{3}H]ACh accumulation by hVAChT indicates that uptake is dependent on exogenous ATP at 37° and is abolished by low temperature (4°), the proton ionophore FCCP (2.5 μm), and bafilomycin A_1 (1 μM), a specific inhibitor of the vesicular H^+-ATPase. Uptake is specifically inhibited by L-vesamicol, exhibiting an IC_{50} of ≈15 nM, while being unaffected by 0.5 μM reserpine or tetrabenazine. The initial rate of [^{3}H]ACh uptake, measured during the linear portion of the time course (6 or 10 min), is saturable, exhibiting an apparent K_m or ≈1 mM and V_{max} of ≈0.6 nmol/min per mg of protein. The functional parameters of ^{3}H-ACh transport by hVAChT measured in this system is similar to those obtained using highly purified synaptic vesicles from *Torpedo.*[3]

In this section, we describe the method for ATP-dependent ACh transport following transfection of neuroendocrine PC-12 cells with hVAChT. This method should be useful for the structure–function analysis of VAChT.

Generation of Stable Neuroendocrine Cell Lines

The development of species- and transporter-specific antibodies against the vesicular transporters has allowed the expression of the human proteins in neuroendocrine cells which contain secretory vesicles and express endogenous vesicular transporter molecules. Rabbit antisera were raised against C-terminal peptide fragments of rat and human VMAT1, VMAT2, and VAChT.[10,11,52,54] Their specificity was checked by their ability to stain transfected CV-1 cells expressing only the cognate protein, and neither of the other two, for a given mammalian species (rat or human). In addition, specific staining was abolished by preincubation of the antisera with a 25 μM concentration of the peptide immunogen. Human and rat VMAT1 antisera do not cross-react. Human and rat VMAT2 antisera are, however, completely cross-reactive. Despite the high degree of identity of the rat and human VAChT sequences, antisera show poor cross-reactivity.

cDNAs are subcloned into a mammalian expression vector such as pRc/CMV (Invitrogen, Carlsbad, CA), or pIRES (Clontech, Palo Alto, CA), which contain the cytomegalovirus promoter/enhancer. PC-12 cells are plated in 10 cm^2 dishes at a density of 2×10^6 cells. The following day, the medium is removed and the cells are rinsed twice with PBS. One-half hour prior to this rinse, 5 ml of serum-free medium containing 1% Lipofectin (Life Technologies, Inc.) and plasmid containing the cDNA (2 μg/ml) is prepared and incubated at room temperature. Cells are transfected with

[54] M. K.-H. Schäfer, E. Weihe, J. D. Erickson, and L. E. Eiden, *J. Mol. Neurosci.* **6,** 225 (1995).

the Lipofectin/plasmid solution for 6 hr at 37°. This medium is then removed and replaced with normal PC-12 medium containing 0.5 mg/ml geneticin (G418; Life Technologies, Inc.). The selection medium is changed every third day for approximately 2–3 weeks. At this time, individual G418-resistant colonies are clearly visible. Most of the medium is removed and colonies are picked using a sterile micropipette tip and placed in wells of a 24-well plate containing 1 ml of fresh medium. After 2 weeks of growth the cells are trypsinized and one-half of the culture is plated onto collagen-coated 4-well chamber slides. To the other half, in the same 24-well plate, 1 ml of fresh medium is added. Positive expressing clones are screened by immunocytochemistry. We then generally amplify several positive clones and prepare frozen stocks (10% DMSO in medium) before any further analysis of the stably transformed cells.

Immunohistochemistry

CV-1 cells, infected with the T7-expressing vaccinia virus and transfected with vesicular transporter cDNAs, and PC-12 cell colony picks, following lipofection and G418 selection, are grown for 18–24 hr on collagen-coated four-chamber Permanox slides. Primary antibody is visualized following incubation with biotinylated antirabbit secondary antibody, application of streptavidin-coupled peroxidase, and further incubation with peroxide and diaminobenzidine (DAB) as described (VectaStain ABC Elite Kit, Vector Laboratories, Inc., Burlingame, CA). All steps are performed at room temperature. Briefly, cells are fixed for 2 hr in 10% formalin/PBS and rinsed three times with PBS. Nonspecific binding is blocked by addition of diluent buffer containing 2% normal goat serum, 0.1% Triton X-100, in PBS for 10 min. Primary antibody (VMAT or VAChT antisera at a 1:2000 dilution in diluent buffer) is then added and incubated for 1 hr. For PC-12 cells, the antipeptide polyclonal antisera against hVAChT is preadsorbed with postnuclear homogenates from control PC-12 cells to eliminate the low background staining. Cells are rinsed once in PBS and treated with 3% H_2O_2 (v/v) in PBS for 10 min. Cells are then washed three times with PBS. A biotinylated anti-rabbit immunoglobulin G (IgG) is applied at 1:500 in diluent buffer for 1 hr. During the last 30 min, AB solution (supplied in the ABC Elite Kit) is made up by adding A to diluent at 1:125 followed by B at 1:125. Cells are rinsed three times with PBS and incubated in 1:125 AB solution for 30 min. After a further rinse in PBS, chambers are stripped off and incubations continued in Coplin jars. Cells are then rinsed twice in 0.1 *M* Tris (pH 7.5) and decanted. Fifty milliliters of DAB/peroxide solution are added and let stand for 5–10 min. Stock solutions of DAB (20 mg/500 μl in 0.1 *M* Tris, pH 7.5) are stored at −80°. To prepare DAB/peroxide solution, one aliquot of DAB is thawed and added to 50 ml 0.1

M Tris (pH 7.5). This solution is filtered and 120 μl of 0.3% H_2O_2 is added immediately prior to use. Slides are rinsed quickly in Tris buffer twice, followed by two quick water washes. Slides are then air dried and viewed.

Preparation of Postnuclear Homogenates

We generally prepare postnuclear homogenates containing the secretory organelles from confluent cultures of stably transformed PC-12 cells grown in T175 flasks. Postnuclear supernatants are also prepared from transfected CV-1 cells. The cells are first rinsed with PBS and then collected in PBS containing 10 mM EDTA (pH 7.4) following a sharp slap to the side of the flask or gentle scraping. The cell suspensions are centrifuged at 800g for 10 min and the cell pellets are homogenized (Dounce, type B pestle or ball-bearing homogenizer, H+Y Enterprise, Redwood City, CA) in ice-cold buffer containing 80 mM potassium tartrate, 20 mM HEPES, 1 mM EGTA, 1 mM phenylmethylsulfonyl fluoride (PMSF), 6 μg/ml leupeptin, 5 μg/ml aprotinin, and 10 μM echothiophate (Wyeth Ayerst Laboratories, Rouses Point, NY) adjusted to pH 7.0 with KOH. The homogenates are centrifuged at 800g 10 min, and the pellets are resuspended in same buffer, homogenized, and centrifuged again. The postnuclear supernatants are pooled. Aliquots of the homogenates are mixed with uptake buffer and used immediately for ^{3}H-ACh uptake assays. Aliquots are also frozen at −80° for subsequent AH5183, L-[*piperidinyl*-3,4-^{3}H] ([^{3}H]vesamicol 31 Ci/mmol; NEN Life Science Products), or [^{3}H]TBZOH binding assays and for Western blot analysis. Protein is measured by the Bradford assay using bovine serum albumin as the standard.[55]

ACh Uptake and Binding Assays

For [^{3}H]ACh uptake assay, aliquots (50 μl) of postnuclear homogenates containing 100–200 μg of protein are mixed with uptake/binding buffer (50 μl) containing 110 mM potassium tartrate, 20 mM HEPES (pH 7.4) and 50 μM echothiophate and incubated at 37° for 5 min in the presence and absence of various inhibitors. Following preincubation, a solution (100 μl) containing 10 mM Mg^{2+}-ATP (neutralized with KOH to pH 7.4) and [^{3}H]ACh (0.4 mM; 55.2 mCi/mmol, NEN Life Science Products) is added. (The final concentration of MgATP is 5 mM.) For kinetic analysis, various concentrations of radiolabeled and unlabeled ACh are added and uptake is terminated after 6 or 10 min. When concentrations of ACh superior to 0.4 mM are required, unlabeled ACh (Sigma preweighted vials) is used. The amount of ACh which is accumulated and retained on the glass fiber

[55] M. M. Bradford, *Anal. Biochem.* **73,** 248 (1976).

filters represents less than 1% of the total ACh added to the medium. Stocks (1000×) of L- or D-vesamicol (Research Biochemicals Inc., Natick, MA) are made up in water and stored at −80°. Bafilomycin A_1 (LC Laboratories, Woburn, MA) is dissolved in DMSO as a 1000× stock. Homogenates from control PC-12 cells and hVAChT-expressing PC-12 cells are always analyzed together to assess the specific ACh uptake derived from endogenous rat VAChT, which is less than 2-fold over that seen at 4° or in the presence of 2 μM vesamicol.

For [^{3}H]vesamicol or [^{3}H]TBZOH binding assay, aliquots (50 μl) of postnuclear homogenates containing 30 μg protein are mixed with uptake/binding buffer and various concentrations of [^{3}H]vesamicol (31 Ci/mmol, NEN Life Science Products) or ^{3}H-TBZOH are added. Nonspecific binding is determined by incubating parallel samples in the presence of a 300-fold excess of unlabeled L-vesamicol or tetrabenazine. The suspensions are warmed to 20° and incubated for 1 h.

Thc uptake and binding reactions are stopped by placing and tubes in ice-cold water, and vacuum filtering the samples through GF/B glass fiber filters and washing with 5 ml ice-cold uptake/binding buffer. Radioactivity bound to the filters is solubilized in 1 ml of 1% SDS followed liquid scintillation counting. K_m, IC_{50}, and K_d values are determined by nonlinear regression. Interassay variability is generally less than 10%.

[7] Photoaffinity Labeling of Vesicular Acetylcholine Transporter from Electric Organ of *Torpedo*

By STANLEY M. PARSONS, GARY A. ROGERS, and LAWRENCE M. GRACZ

Introduction

Direct biochemical study of acetylcholine (AcCh) storage in test tubes became possible after methods were developed for the purification of purely cholinergic synaptic vesicles from the electric organ of the marine ray *Torpedo*.[1-3] The purification has since been improved.[4] Vesicles of greater than 95% purity can be isolated in about 2½ days. When the purified vesicles are suspended in magnesium ion and ATP, they internalize exogenous

[1] A. Nagy, R. R. Baker, S. J. Morris, and V. P. Whittaker, *Brain Res.* **109,** 285 (1976).
[2] S. S. Carlson, J. A. Wagner, and R. B. Kelly, *Biochemistry* **17,** 1188 (1978).
[3] K. Ohsawa, G. H. C. Dowe, S. J. Morris, and V. P. Whittaker, *Brain Res.* **161,** 447 (1979).
[4] L. M. Gracz and S. M. Parsons, *Biochim. Biophys. Acta* **1292,** 293 (1996).

0076-6879/98 $25.00

AcCh via the vesicular AcCh transporter (VAcChT).[5] The availability of this assay combined with well-controlled work by Howard and colleagues in PC-12 (rat adrenal pheochromocytoma) cells led to characterization of the bioenergetics of AcCh storage. It has been shown that uptake is dependent on internal acidification of the vesicles by a V-type ATPase and that a separate VAcChT protein exchanges "luminal" protons for external (cytoplasmic) AcCh to effect storage of the neurotransmitter.[6–8]

Pharmacological characterization of the VAcChT began with the demonstration that it can be distinguished from other cholinergic proteins (such as the nicotinic and muscarinic receptors) by its relative insensitivity to known drugs.[9] The compound *trans*-2-(4-phenylpiperidino)cyclohexanol (then called AH5183 and now called vesamicol) was found to inhibit uptake noncompetitively[10,11] by binding to an enantioselective site in the VAcChT.[12] These findings confirmed the hypothesis by Marshall[13] about the action of vesamicol based on physiologic experiments. Extensive structure–activity studies of AcCh[14,15] and vesamicol.[16–20] have been carried out utilizing synaptic vesicles purified from *Torpedo*.

Synthesis of [^{3}H]Azidoacetylcholine (I) and [^{3}H]Azidoaminobenzovesamicol (II)

The VAcChT has been photoaffinity labeled with analogs of AcCh and vesamicol called [^{3}H]azidoAcCh **(I)** and [^{3}H]azidoABV **(II)**, respectively

[5] S. M. Parsons and R. Koenigsberger, *Proc. Natl. Acad. Sci. USA* **77,** 6234 (1980).
[6] L. Toll and B. D. Howard, *J. Biol. Chem.* **255,** 1787 (1980).
[7] D. C. Anderson, S. C. King, and S. M. Parsons, *Biochemistry* **21,** 3037 (1982).
[8] B. W. Hicks and S. M. Parsons, *J. Neurochem.* **58,** 1211 (1992).
[9] D. C. Anderson, S. C. King, and S. M. Parsons, *Molec. Pharmacol.* **24,** 48 (1983).
[10] B. A. Bahr and S. M. Parsons, *J. Neurochem.* **46,** 1214 (1986).
[11] B. A. Bahr, E. D. Clarkson, G. A. Rogers, K. Noremberg, and S. M. Parsons, *Biochemistry* **31,** 5752 (1992).
[12] B. A. Bahr and S. M. Parsons, *Proc. Natl. Acad. Sci. USA* **83,** 2267 (1986).
[13] I. G. Marshall, *Br. J. Pharmacol.* **40,** 68 (1970).
[14] G. A. Rogers and S. M. Parsons, *Mol. Pharmacol.* **36,** 333 (1989).
[15] E. D. Clarkson, G. A. Rogers, and S. M. Parsons, *J. Neurochem.* **59,** 695 (1992).
[16] G. A. Rogers, S. M. Parsons, D. C. Anderson, L. M. Nilsson, B. A. Bahr, W. D. Kornreich, R. Kaufman, R. S. Jacobs, and B. Kirtman, *J. Med. Chem.* **32,** 1217 (1989).
[17] G. A. Rogers, W. D. Kornreich, K. Hand, and S. M. Parsons, *Mol. Pharmacol.* **44,** 633 (1993).
[18] S. M. Efange, A. Khare, S. M. Parsons, R. Bau, and T. Metzenthin, *J. Med. Chem.* **36,** 985 (1993).
[19] S. M. Efange, A. B. Khare, C. Foulon, S. K. Akella, and S. M. Parsons, *J. Med. Chem.* **37,** 2574 (1994).
[20] S. M. Efange, R. H. Mach, C. R. Smith, A. B. Khare, C. Foulon, S. K. Akella, S. R. Childers, and S. M. Parsons, *Biochem. Pharmacol.* **49,** 791 (1995).

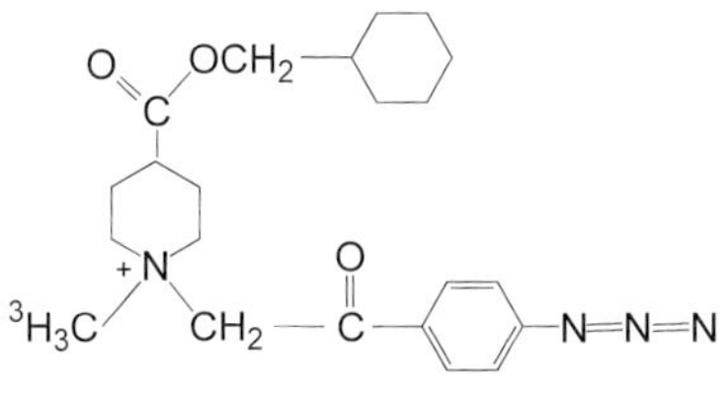

(I)

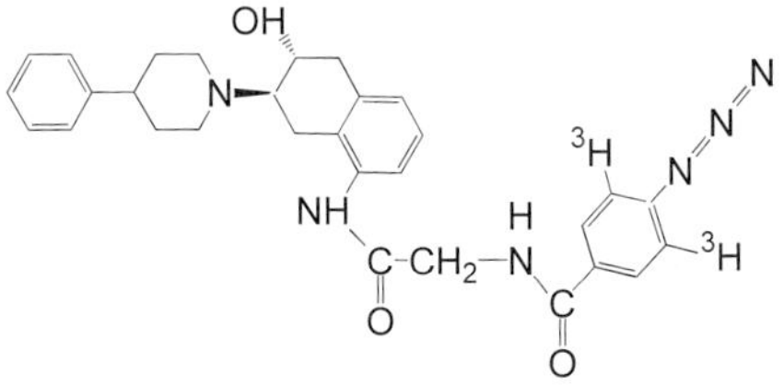

(II)

FIG. 1. Chemical structures of azidoAcCh [compound **(I)**] and azidoABV [compound **(II)**].

(Fig. 1). The structure–activity studies mentioned above yielded the synthetic routes to chemical intermediates required to make radioactive arylazido compounds predicted to have relatively high affinity for the AcCh and vesamicol binding sites. All reagents needed for the syntheses of these compounds are available from standard commercial sources except as noted below. The reader should consult the original references for interpretation of the spectroscopic data used in the structure proofs.

[^{3}H]Azidoacetylcholine **(I)**[21]

Cyclohexylmethyl Isonipecotate. Isonipecotic acid hydrochloride (2.1 g, 12.7 mmol) is suspended in 15 ml of thionyl chloride ($SOCl_2$) and heated to 40° for 12 hr. Excess $SOCl_2$ is removed on a rotary evaporator with the aid of dry CH_2Cl_2. The acid chloride is dissolved in dry CH_2Cl_2 and heated in a fume hood in order to remove the remaining $SOCl_2$. During several hours, aliquots of cyclohexylmethanol (a total of 3.5 equivalent) are added to the stirred solution, which is then refluxed overnight. The CH_2Cl_2 solution is washed with cold aqueous carbonate and dried over Na_2SO_4. Removal of the solvent results in an oil that is chromatographed on silica gel. Product ester is eluted with 5% (v/v) ethanol/$CHCl_3$ in 65% yield as a colorless oil. The hydrochloride salt, which is produced by precipitating the product

[21] G. A. Rogers and S. M. Parsons, *Biochemistry* **31,** 5770 (1992).

dissolved in ether by bubbling in anhydrous HCl gas, has melting point (mp) of 161.5–163°.

Cyclohexylmethyl-cis-N-(4-azidophenacyl)-N-methyl Isonipecotate Bromide. Cyclohexylmethyl isonipecotate (9 mg, 40 μmol) and $[^3H]CH_3I$ (0.65 μmol, 10.8 Ci/mmol, Amersham Corp., Arlington Heights, IL) are combined in 0.8 ml of toluene in a 1-ml screw-cap vial. After 30 hr in the dark, succinic anhydride (5 mg) plus 80 μl of CCl_4 are added. The mixture is vortexed for 5 min to effect solution. This step facilitates separation of the product because succinic anhydride reacts with the large excess of unreacted cyclohexylmethyl isonipecotate (which is a secondary amine) but not with the methylated product (which is a tertiary amine). A large excess of cyclohexylmethyl isonipecotate is used to suppress double methylation. After 20 hr in the dark, the reaction mixture is chromatographed on silica gel. The product elutes with 5% methanol/$CHCl_3$. Succinylated cyclohexylmethyl isonipecotate remains on the column. After the methanolic fractions are pooled and evaporated to dryness, the residue is dissolved in dry $(C_2H_5)_2O$ (200 μl). 4-Azidophenacyl bromide (5 mg, 21 μmol) plus one drop of *N,N*-diisopropylethylamine are dissolved in the ethereal solution, which is stored in the dark for 2 days. Chromatography of the reaction mixture on silica gel provides product that elutes with methanol in 35% radiochemical yield. Fractions containing most of the radioactive peak are pooled, solvent is removed under vacuum, and the residue is taken up in 1 ml ethanol [yielding about 228 *μM* (**I**)] for storage at −20°. The nonradioactive compound (FW 479) can be made similarly on a larger scale.

[³H]AzidoABV (**II**)[22]

N-(Trifluoroacetyl)-1-amino-5,8-dihydronaphthalene Oxide. A 1-liter, three-necked, round-bottom flask is equipped with a large cold-finger condenser filled with solid CO_2 and 2-propanol. To the flask 1-aminonaphthalene (79 g, 0.55 mol, carcinogenic!), $(C_2H_5)_2O$ (300 ml), *tert*-butanol (50 ml), and NH_3 (200–300 ml) are added with stirring. Sodium (30 g, 1.3 mol, dried with paper towels just before weighing and returned to mineral oil after weighing until just before use) is added in portions over a 4-hr period, followed by an additional 50 ml of *tert*-butanol. After 1 hr, absolute ethanol (100 ml) is added slowly. This mixture is allowed to stir overnight and then is quenched by careful addition of NH_4Cl (50 g) and H_2O (400 ml). The two liquid layers are separated, and the aqueous layer is extracted two times with ether. The ether extracts are combined with the organic layer, and this solution is extracted twice with water, once with a saturated NaCl

[22] G. A. Rogers and S. M. Parsons, *Biochemistry* **32,** 8596 (1993).

solution, dried over Na_2SO_4, and finally filtered through $MgSO_4$. Ether and most of the *tert*-butanol are removed at 65° on a rotary evaporator. This material may be used directly for the next step or colorless 1-amino-5,8-dihydronaphthalene (mp 37–39°) may be obtained in 97% yield after vacuum distillation.

1-Amino-5,8-dihydronaphthalene (60.1 g, 0.414 mol) is dissolved in 200 ml of stirred benzene and cooled to 0°. Trifluoroacetic anhydride (89 g, 0.42 mol) is added slowly because of the exothermicity of the reaction. The ammonium salt begins to precipitate almost immediately, but the reaction solution is homogeneous upon complete addition of the anhydride. This step is necessary to protect the amino group from oxidation during formation of the epoxide (below). The solution is maintained at 0° for 1 hr, after which benzene and trifluoroacetic acid are removed under reduced pressure. More benzene is added and again evaporated in order to promote the removal of trifluoroacetic acid.

The resulting amide is dissolved in 300 ml of stirred $(C_2H_5)_2O$ to which is added 3-chloroperoxybenzoic acid (85.0 g, 80–85% pure). The solution is maintained near 10° throughout the addition and then allowed to warm to 23° and stirred for 5 hr. The product epoxide containing some 3-chlorobenzoic acid is collected by filtration and washed with ether. The solid is resuspended in 400 ml of $(C_2H_5)_2O$, thoroughly washed by agitation, and recollected by filtration to yield 89.0 g of epoxide (mp 174–178.5°). After extraction of 3-chlorobenzoic acid from the ethereal mother liquor with aqueous carbonate, an additional 2.1 g of product can be isolated by evaporation of the ether for an overall yield of 85%. Purification of the epoxide by crystallization from $CHCl_3$ with a wash of cold $(C_2H_5)_2O$ raises the melting point to 179.5–181°.

(±)-trans-5-Amino-2-hydroxy-3-(4-phenylpiperidino)tetralin [(±)-ABV, FW 322]. *N*-(Trifluoroacetyl)-1-amino-5,8-dihydronaphthalene oxide (1.9 g, 7.4 mmol) is dissolved in 25 ml of absolute EtOH to which is added 4-phenylpiperidine (3.0 g, 19 mmol). The solution is maintained at 45° for 17 hr and then refluxed for 3 hr. After 7 hr at 23° a crystalline solid is collected by filtration and set aside. The solid is a positional isomer of (±)-ABV. The mother liquor is evaporated to an oil that is taken up in CCl_4. This process is repeated in order to remove most of the EtOH. When the oil is again dissolved in CCl_4, crystallization of 4-phenylpiperidinium trifluoroacetate commences. The remaining mother liquor is chromatographed on silica gel where *N*-(trifluoroacetyl)-4-phenylpiperidine elutes with CCl_4 and (±)-ABV with $CCl_4/CHCl_3$ (0.84 g, 35% yield). (±)-ABV is crystallized from $CHCl_3$/ethanol (mp 174–175°C).[16]

(−)-N-Glycyl-4-aminobenzovesamicol [(−)-GlyABV]. (±)-ABV and *N-tert*-butoxycarbonylglycine *p*-nitrophenyl ester (1 equivalent) are reacted

in acetic acid in a closed vial overnight at 23°, and the acetic acid is removed *in vacuo*. The condensation product is taken up in 50% (v/v) trifluoroacetic acid/CH_2Cl_2 and allowed to deblock in a sealed vial overnight at 45° to yield (±)-GlyABV trifluoroacetate after removal of solvent *in vacuo*. After dissolving the trifluoroacetate in CH_2Cl_2 and washing it with aqueous carbonate, neutral (±)-GlyABV is purified by chromatography on silica gel and then resolved into its enantiomers by chromatography on a Chiralpak AD semipreparative column (Daicel Chemical Industries, Exton, PA) in 20% 2-propanol/hexane.[17] The elution order of the enantiomers is extremely sensitive to the composition of the solvent and age of the column; thus, the optical rotations of the separated enantiomers should be taken to confirm their identities.

(−)-N-(4′-Azidobenzoylglycyl)-4-aminobenzovesamicol ([^{3}H]azidoABV). *N*-[*benzoate*-3,5-$^3H_2$]Succinimidyl 4-azidobenzoate (5.09 nmol, 49.1 Ci/mmol, NEN Division of DuPont, Inc., Boston, MA) is combined with 10.7 μg (28.2 nmol) of (−)-GlyABV in 100 μl of 2-propanol and placed in the dark. After 4 days the sample is diluted with 100 μl of ethanol. The product is purified by high-performance liquid chromatography on a semipreparative Chiralpak AD (Daicel Chemical Ind.) column (25% ethanol/hexane at 4 ml/min) with collection of fractions. It is important to turn the UV detector off as soon as elution of **(II)** begins at about 24 min. Unreacted (−)-GlyABV elutes at about 16 min.[22] Fractions containing most of the radioactive peak are pooled, solvent is removed under vacuum, and the residue is taken up in 1 ml ethanol [yielding about 3.26 μ*M* **(II)**] for storage at −20°. About 160 μCi of **(II)** in 64% radiochemical yield is obtained. The nonradioactive compound (FW 524) can be made similarly on a larger scale.

Isolation of Cholinergic Synaptic Vesicles from Electric Organ of *Torpedo*

Torpedo are obtained live from a commercial fisherman or a commercial supplier of marine animals. Different species of *Torpedo* are found in substantial abundance in most of the major oceans of the world, and they often are caught when feeding on commercially exploited fish stocks. *Torpedo californica* commonly weigh 20–40 pounds, and about 15% of the body weight is electric organ. The fish are rested in a filtered well-aerated tank of seawater for 3–4 days after delivery. They are stunned by a blow to the back of the head and pithed with a scalpel, and the nerve bundles innervating the electric organs are severed. Because the flat cartilaginous skull efficiently delivers the blow to the brain, the rays are immediately rendered unconscious and do not suffer as far as can be determined. The electric organs, which fill most of the wings, are readily recognized and removed by blunt dissection after the overlying skin is peeled back. They are frozen immediately in liquid nitrogen, after which they are fractured into chunks that are packaged into 800-g portions and stored at $-100°$ until use.[4]

The deep-frozen electric organ (800 g) is shaved very finely with a rapidly spinning blade to produce a "snow." The yield of vesicles decreases greatly if a very fine snow is not produced, and a very sharp blade seems to be the most important factor in obtaining favorable yields. A simple, commercially available machine used to make flavored "snow cones" at carnivals works well. The snow is allowed to warm slowly to $-2°$ with occasional mixing. The slush is suspended in 800 ml of ice-cold buffer composed of 0.76 *M* glycine containing 0.05 *M* *N*-(2-hydroxyethyl)piperazine-*N*′-2-ethanesulfonic acid (HEPES), 0.01 *M* ethylene glycol bis(β-aminoethyl ether)-*N,N,N*′,*N*′-tetraacetic acid (EGTA), 0.001 *M* ethylenediaminetetraacetic acid (EDTA), 0.02% NaN_3, and 2 μM 2,6-bis(1,1-dimethylethyl)-4-methylphenol (BHT) adjusted to pH 7.2 with 0.80 *M* NaOH. All subsequent steps in the vesicle purification are conducted at 4°. The slush is divided into portions that are slowly sheared with a blender for 20 sec, after which they are recombined. Approximately 300 ml is homogenized with five slow strokes using a tightly fitting, spinning Teflon pestle in a 350-ml glass mortar of the Potter–Elvehjem type. A drill press can be adapted to drive the pestle.

The tissue homogenate is centrifuged in six 500-ml polycarbonate bottles at 17,700*g* for 1 hr to remove membrane debris. Twenty 70-ml polycarbonate centrifuge tubes having O ring screw caps are filled with supernatant to within 1 cm of the neck. Using a cannulated syringe, 6 ml of isosmotic "cushion" composed of 650 m*M* sucrose, 7.5% (w/w) Ficoll 400 (Sigma, St. Louis, MO), 10 m*M* HEPES, 1 m*M* EDTA, and 1 m*M* EGTA (adjusted to pH 7.0 with 0.80 *M* KOH) having a density of 1.117 g/ml is injected under the supernatant onto the bottom of each tube. Ficoll 400 is used to obtain increased density beyond that provided by isosmotic sucrose because it contributes little to the osmotic pressure and its high viscosity stabilizes the interface between the supernatant and the cushion. After balancing them with additional supernatant, the tubes are centrifuged at 49,000*g* for 16–21 hr in two Beckman Type 21 rotors. The membranes focused on top of the cushions are removed with a cannulated syringe and pooled to yield about 160 ml of crude vesicles.

The vesicular pool is mixed with sufficient fresh cushion solution to increase the density of the mixture enough to layer under 0.40 *M* sucrose (see proceeding text). To accomplish this, 15 ml of fresh cushion solution is added to the pool, which then is mixed. A drop of the mixed pool is placed onto 0.40 *M* sucrose in a test tube to determine whether it falls to the bottom. Additional fresh cushion in 3 ml increments is added to the pool with mixing and the mixture is tested until it does fall. About 20–40 ml of cushion is required. Isosmotic density gradient solutions (0.5 liter each) composed of 0.01 *M* HEPES, 0.001 *M* EDTA, 0.001 *M* EGTA, and 0.02% (w/v) NaN_3, adjusted to pH 7.0 with 0.8 *M* KOH and containing either 0.80 *M* glycine or 0.65 *M* sucrose, are prepared. An isosmotic solution containing 0.40 *M* sucrose is prepared by mixing 50 ml of the 0 *M* sucrose solution with 80 ml of the 0.65 *M* sucrose solution. A typical density gradient is formed from 0 *M* sucrose (7 ml), 0.40 *M* sucrose (8 ml), pooled vesicles containing additional cushion (22 ml), and 0.65 *M* sucrose (4 ml). The solutions are applied to the bottom of a 1 × 3.5-inch dome-top polyallomer quick-seal tube (Beckman Coulter Instruments, Inc., Fullerton, CA) in the order stated with a cannulated syringe. The eight required tubes are heat sealed and centrifuged at 196,200*g* for 3.5 hr in a Beckman VTi 50 vertical rotor. They are opened by carefully slicing the tops off with a single-edge razor blade. In initial work, 20 fractions of 1.5 ml each are obtained from each gradient by bottom puncture. The fractions are assayed for ATP content using the luciferase assay.[23] Mature cholinergic synaptic vesicles isolated with this procedure (so-called VP_1 vesicles) band at 1.055 g/ml and contain ATP as well as AcCh. The densities of the fractions can be estimated from the refractive indices.[24] The vesicles are found in a faint milky layer near the top, and after experience is gained, they can be withdrawn rapidly from the opened centrifuge tube by use of a cannulated syringe and visual inspection.

Vesicles in the pool (50 ml) are further purified by size-exclusion chromatography on a Sephacryl S-1000 column (150 cm × 1.5 cm) equilibrated with buffer containing 820 m*M* glycine, 5 m*M* HEPES, 1 m*M* EDTA, 1 m*M* EGTA, and 0.02% NaN_3 (adjusted to pH 7.0 with 0.80 *M* KOH). The column effluent is monitored by apparent absorbance at 350 nm. The vesicles elute in the middle of the resolving volume as the second peak of apparent absorbance (due to light scattering). The fractions containing vesicles are pooled and concentrated with a Centricon centrifugal ultrafiltration device (Amicon, Danvers, MA, 10 kDa cutoff) to 1–2 mg protein/ml if desired. About 5 mg of vesicular protein is obtained as determined by the Bradford[25] dye assay. The VAcChT in these vesicles is stable for weeks when stored in a closed vessel at 4°.

[23] G. A. Kimmich, J. Randles, and J. S. Brand, *Anal. Biochem.* **69,** 187 (1975).

[24] P. Sheeler, "Centrifugation in Biology and Medical Science," pp. 246–247. Wiley, New York, 1981.

[25] M. M. Bradford, *Anal. Biochem.* **72,** 248–254 (1976).

Assay of Ligand Binding or Transport by Purified Synaptic Vesicles

Binding and transport assays are done similarly to each other. Purified vesicles at about 0.075 to 0.2 mg protein/ml are incubated with a radioactive compound in buffer containing 0.10 *M* HEPES, 0.70 *M* glycine, 0.001 *M* EDTA, and 0.001 *M* EGTA adjusted to pH 7.80 with 0.80 *M* KOH (TB). This pH is optimal for both the AcCh and vesamicol families of ligands in both transport and binding assays. The buffer also includes components such as paraoxon to inhibit AcCh esterase and MgATP to drive transport as needed. Samples are prepared in the absence and presence of an excess of nonradioactive ligand that blocks the specific binding or transport of the radioactive ligand. Samples are mixed well by use of a vortex device after addition of reagents. Glass-fiber filters (Whatman, Clifton, NJ, GF/F, 1.3 cm) that have been pretreated with polyethyleneimine are mounted on a 10-place filter manifold/collection box (Hoefer Scientific Instruments, San Francisco, CA, model FH 225V). To prepare them, 400 filters are soaked 3 hr in 400 ml of 0.5% (w/v) polyethyleneimine in water. They are poured gently into a Büchner funnel of 8 inch diameter and rinsed generously with water by filtration. The funnel with the filters is placed overnight in an oven at 37° for drying. Dry filters are removed carefully from the funnel with tweezers and placed in a plastic food baggy with zipper after which they can be stored at room temperature in the dark. After incubation of vesicles with a radioactive ligand, a portion (typically 50–100 μl) is applied with suction assistance onto a filter prewetted just before with ice-cold TB. The filter then is immediately and rapidly washed by 3–4 1-ml portions of ice-cold TB with suction assistance. The radioactivity trapped on the filter is determined by liquid scintillation spectroscopy. Filters must be extracted overnight by the scintillation cocktail (a type compatible with aqueous solutions) for accurate and full determination of bound radioactivity. AcCh becomes soluble in cocktail containing 10% water better than it does in dry scintillation cocktail. The decrease in bound radioactivity due to the presence of excess nonradioactive inhibitor is the specific binding or transport by the vesicles.

Reversible Binding of Compound (I)

Before irreversible photoaffinity labeling is attempted, it is important to demonstrate that a potential label binds in a well-behaved manner to the expected site. As compound **(I)** is stable to normal indoor lighting, its interaction with cholinergic synaptic vesicles can be studied under reversible conditions. Although the effect of sunlight on the compound has not been determined, it should be avoided by working in an inner room or in a room with shaded windows. These comments apply to compound **(II)** also (see below).

All TB contains 0.15 mM paraoxon. Purified synaptic vesicles at about 2 mg protein/ml in TB are incubated at 23° for 30 min to allow full inactivation of acetylcholinesterase. Two microliters of compound **(I)** in ethanol prepared as above is dried *in vacuo* and dissolved in 40 μl TB to yield about 10 μM of compound **(I)**. The exact concentration of compound **(I)** is determined by measuring the radioactivity in a 5 μl portion. An approximately isosmotic solution of AcCh in which 0.35 M AcCh chloride replaces the 0.70 M glycine in TB is prepared by dissolving 150 mg of preweighed AcCh chloride (Sigma) in 2.36 ml of TB lacking glycine. The pH of this solution should be checked and adjusted to 7.80 if necessary. Serial dilutions of this solution are made with TB to prepare additional isosmotic solutions containing 35, 3.5, and 0.35 mM AcCh. Also, 0.35 M AcCh chloride containing vesamicol is prepared by dissolving 150 mg of preweighed AcCh chloride in 2.36 ml of TB lacking glycine but containing 60 μM (−)-vesamicol hydrochloride (FW 295.5, Research Biochemicals International, Inc.). Vesamicol hydrochloride is soluble to at least 1 mM in aqueous solutions. Vesamicol is used to measure the extent of nonspecific binding of **(I)** because at equilibrium it is a competitive inhibitor of the binding of AcCh analogs.[11] Because they spontaneously degrade slowly, aqueous solutions of AcCh must be prepared shortly before use. Eighty-six microliters from each stock solution (including TB) is dispensed into individual tubes, and 4 μl compound **(I)** in TB is added and mixed in. Vesicular suspension (10 μl) is added and allowed to equilibrate for 15 min at 23°. Bound compound **(I)** is determined by filtration and washing of 90 μl of the suspension as described above.

The results of a typical binding experiment are shown in Fig. 2. Because the concentration of the binding site (about 40 nM) is comparable to the dissociation constant for **(I)**, the concentration of free **(I)** has to be calculated iteratively at each concentration of AcCh. Hence, for each total concentration of **(I)**, the free concentration in solution is calculated from Eqs. (1) and (2):

$$\mathbf{(I)}_{\mathrm{F}} = \mathbf{(I)}_{\mathrm{T}} - \mathbf{(I)}_{\mathrm{B}} \tag{1}$$

$$\mathbf{(I)}_{\mathrm{B}} = [B_{\max} \times \mathbf{(I)}_{\mathrm{F}} \times K_{\mathrm{IA}}]/(\mathbf{(I)}_{\mathrm{F}} \times K_{\mathrm{IA}} + \mathrm{AcCh} \times K_{(\mathrm{I})} + K_{(\mathrm{I})} \times K_{\mathrm{IA}}] \tag{2}$$

where $\mathbf{(I)}_{\mathrm{F}}$ is the free concentration of **(I)**, $\mathbf{(I)}_{\mathrm{T}}$ is the total concentration of **(I)**, $\mathbf{(I)}_{\mathrm{B}}$ is the bound concentration of **(I)**, $K_{(\mathrm{I})}$ is the dissociation constant for the specific VAcChT · **(I)** complex (240 nM[21]), K_{IA} is the dissociation constant for the specific VAcChT · AcCh complex, and AcCh is the concentration of AcCh. Regression analysis is done with the program Scientist

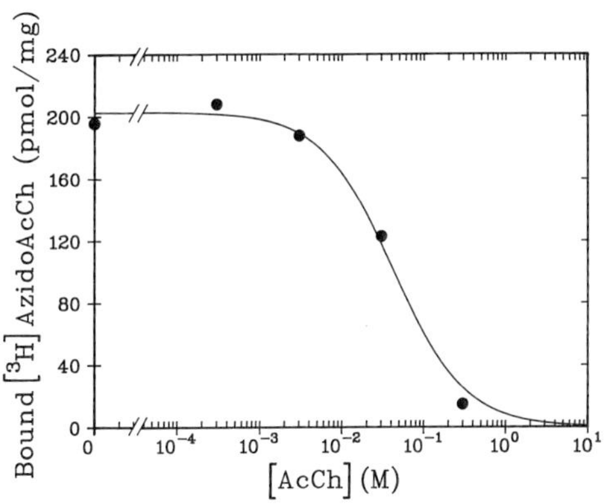

FIG. 2. Inhibition of the reversible binding of compound **(I)** by AcCh. Final concentrations of the vesicular protein and **(I)** were 0.2 mg/ml and 440 n*M*, respectively. Bound **(I)** was determined by filtration as described in the text. Nonspecifically bound **(I)** was estimated in the presence of 60 μM vesamicol (plus 300 m*M* AcCh) and was subtracted from the total amount bound. The estimated K_{IA} for AcCh from regression analysis is 17 ± 3 m*M* using Eqs. (1) and (2) in the text. (Reprinted with permission from G. A. Rogers and S. M. Parsons, *Biochemistry* **31,** 5770 (1992). Copyright 1992 American Chemical Society.)

(MicroMath Scientific Software, Salt Lake City, UT), which is available only for personal computers.

Nonradioactive AcCh inhibited the binding of compound **(I)** with a dissociation constant of 17 ± 3 m*M* in the experiment of Fig. 2. The highest concentration of AcCh investigated displaced essentially all **(I)**, as vesamicol displaced only a small additional amount. The equilibrium dissociation constant observed for AcCh is relatively high and thus possibly suspect as not arising from the correct AcCh binding site. However, the apparent dissociation constant for AcCh binding to this site decreases to about 0.3 m*M* under active transport conditions. The latter value is less than the estimated concentration of AcCh in cholinergic nerve terminals.[26] The apparent increase in affinity of AcCh under active transport conditions occurs because of a large kinetic component in the Michaelis constant.[11] Thus, analog **(I)** binds to the site for AcCh transport. This result demonstrates why it was necessary to carry out a structure–activity study before choosing an affinity label analog of AcCh for this system. AcCh and its close analogs bind to the VAcChT so weakly that they would be unlikely to provide specific labeling of the transporter sufficient to identify it.

[26] S. M. Parsons, C. Prior, and I. G. Marshall, *in* "International Review of Neurobiology," Vol. 35 (R. J. Bradley and R. A. Harris, Eds.), pp. 279–390. Academic Press, New York, 1993.

Reversible Binding of Compound (II)

Compound **(II)** has moderately high affinity for the VAcChT, and for this reason the high-affinity ligand ABV[17] is used in this experiment to block specific binding of compound **(II)**. Synaptic vesicles from the center of the vesicular peak (having about 30 μg protein/ml) in the size exclusion chromatography step of the purification are used without concentration. Vesicles are diluted 60-fold into TB to about 0.5 μg protein/ml. All aqueous solutions of compound **(II)** and ABV must be handled in glass containers coated with Sigmacote (Sigma Chemical Co.) or the compounds will be lost through adsorption to the vessel surface. Plastic containers are unsuitable. Compound **(II)** in ethanol (1.5 μl) prepared as above is dried *in vacuo* and taken up in 50 μl TB to yield a solution that is about 100 n*M* in **(II)**. Serial dilutions with TB to about 1 n*M* of **(II)** are prepared on a 50 μl scale with retention of all of the dilutions. ($\pm$)-ABV (0.322 mg) is dissolved in 10 ml ethanol to yield 100 μM, and 1 μl is added to each of 6 tubes. The ethanol is evaporated from the tubes. Vesicles (100 μl) are added to these tubes and to 6 additional ones. After incubation for 30 min, two 10-μl portions of each dilution of **(II)** are added to two samples of vesicles, one containing ($\pm$)-ABV and one not. After incubation for 11 hr in sealed tubes, the final total concentration of **(II)** in each sample is determined by measuring the radioactivity in a 10-μl portion with liquid scintillation spectroscopy. The bound radioactivity then is determined by filtration and washing of a 90-μl portion as described above.

Figure 3 shows the results of a typical experiment. Equations (3) and (4) are fit simultaneously to the data sets for total and nonspecific binding, respectively, using Scientist.

$$\mathbf{(II)}_T = B_{max} \times \mathbf{(II)}/[\mathbf{(II)} + K_{(II)}] + S \times \mathbf{(II)} + C \quad (3)$$

$$\mathbf{(II)}_N = S \times \mathbf{(II)} + C \quad (4)$$

where $\mathbf{(II)}_T$ is the total amount of bound **(II)**, B_{max} is the concentration of specific binding sites, **(II)** is the concentration of free **(II)** [assumed equal to all **(II)** in solution], $K_{(II)}$ is the dissociation constant for the specific VAcChT · **(II)** complex, $\mathbf{(II)}_N$ is the amount of nonspecifically bound **(II)**, S is the slope of the nonspecific binding of **(II)**, and C is the intercept of the nonspecific binding of **(II)**. The dissociation constant for specifically bound **(II)** was 2.1 $\pm$ 0.4 n*M* and the B_{max} was 0.29 $\pm$ 0.03 n*M* (corresponding to 580 $\pm$ 80 pmol/mg) in the experiment of Fig. 3.

Photoaffinity Labeling and Visualization of Acetylcholine-binding Protein

*Photolysis of Compound **(I)** Bound to Vesicles.* All TB contains 0.15 m*M* paraoxon. Synaptic vesicles (0.8 ml containing 0.8 mg protein) are

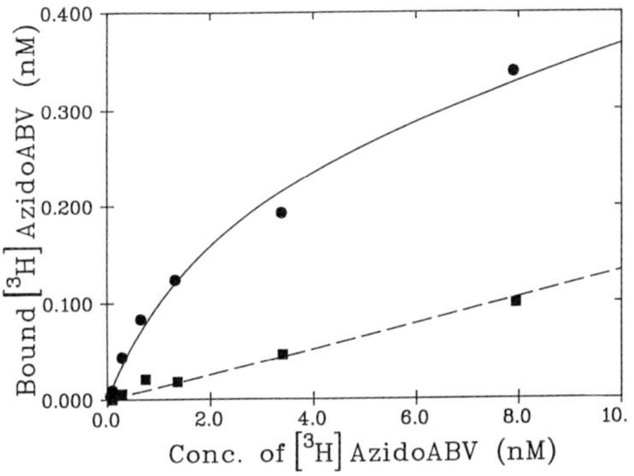

FIG. 3. Saturation of the reversible binding of compound **(II)**. Bound **(II)** was determined by filtration as described in the text in the absence (●) or presence (■) of 800 n*M* (±)-ABV to yield data sets representing the amount of total and nonspecific binding at the indicated concentrations of compound **(II)**, respectively. A simultaneous fit of Eqs. (3) and (4) in the text to the two sets of data yielded $K_{(II)}$ equal to 2.1 ± 0.4 n*M* and B_{max} equal to 0.29 ± 0.03 n*M* (corresponding to 580 ± 80 pmol/mg) for specific binding. (Reprinted with permission from G. A. Rogers and S. M. Parsons, *Biochemistry* **32,** 8596 (1993). Copyright 1993 American Chemical Society.)

incubated in TB for at least 30 min. Seven portions of 100 μl each are placed into separate tubes. Three microliters of compound **(I)** in ethanol prepared as above is dried *in vacuo* and dissolved in 60 μl TB to yield about 10 μ*M* of **(I)**. The exact concentration of compound **(I)** is determined by measuring the radioactivity in a 5-μl portion. Nonradioactive **(I)** (0.479 mg) is dissolved in 1 ml of TB [yielding 1 m*M* **(I)**]. (−)-Vesamicol hydrochloride is made up to 1 m*M* in TB by dissolving 0.296 mg in 1 ml. Concentrated MgATP in TB is adjusted to pH 7.80 with KOH and diluted to 100 m*M* with TB. AcCh is made up to 350 m*M* as above. Samples A and B (lanes 2 and 4 in Fig. 4) receive 145 μl of TB; sample C (lane 3) receives 25 μl of nonradioactive **(I)** and 120 μl of TB; sample D (lane 6) receives 25 μl of nonradioactive **(I)**, 25 μl of MgATP, and 95 μl of TB; sample E (lane 8) receives 71.4 μl AcCh and 73.6 μl of TB; sample F (lane 7) receives 2.14 μl AcCh and 142.9 μl of TB; and sample G (lane 9) receives 0.25 μl vesamicol and 144.75 μl of TB. All samples receive 5 μl radioactive **(I)**. Because of a high level of nonspecific binding, a concentration of **(I)** slightly below its dissociation constant value is used in this experiment. This yields an optimal compromise between good specific and excessive nonspecific labeling.

Samples destined for lanes 3–9 (Fig. 4) are irradiated simultaneously using a handheld 18-watt lamp (Blak-Ray, UVP, Inc., San Gabriel, CA) that emits light in a band of wavelengths centered at 366 nm. Irradiation is carried out in four periods of 2 min each with mixing after each period.

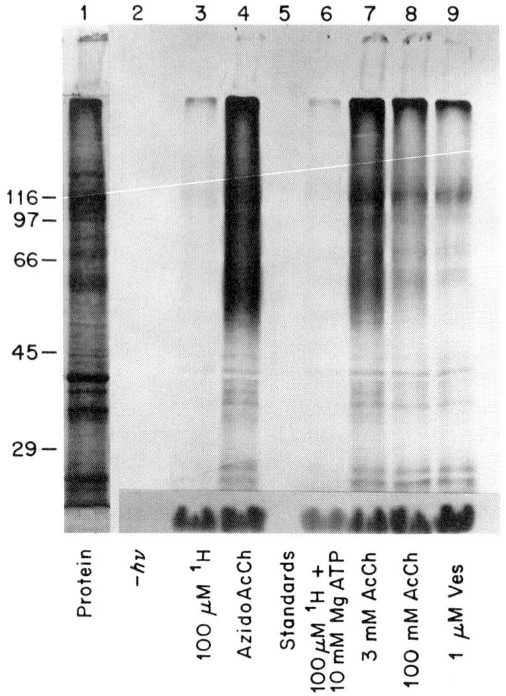

FIG. 4. Photoaffinity labeling and visualization of the AcCh-binding protein by compound **(I)**. Vesicles were labeled in the presence or absence of the indicated reagents and then subjected to SDS–PAGE and autofluorography as described in the text. The figure is a composite showing Coomassie Blue staining, a short-term film exposure to the dye front region of the gel, and a long-term exposure to the stacking and resolving regions of the gel. Lane 5 contained standard proteins with their masses (kDa) indicated to the left. Lanes 1 and 4 are the Coomassie staining and autofluorographic patterns, respectively, of the same sample of vesicles. AzidoAcCh refers to compound **(I)**. Lanes 2 and 3 were controls that received no irradiation or 100 μM of nonradioactive **(I)**, respectively. Lanes 7 and 8 contained 3 or 100 mM AcCh, respectively, and lane 9 contained 1 μM vesamicol. Lane 6 contained 100 μM of nonradioactive **(I)** and 10 mM of MgATP. (Reprinted with permission from G. A. Rogers and S. M. Parsons, *Biochemistry* **31,** 5770 (1992). Copyright 1992 American Chemical Society.)

Under similar conditions, the half-life for photolysis of the azido group is about 1 min as determined by ultraviolet spectroscopy. After photolysis, an aliquot of 20 μl is taken from each sample and filtered. The filtered samples are washed repeatedly with TB at 23°, and bound tritium is determined. Vesamicol (1 μM) blocks 45 pmol of label incorporation/mg of protein. This corresponds to about 9% of the specific binding assessed separately with 250 μM [^{3}H]vesamicol (NEN division of Du Pont, Inc.).

The remaining volume of each sample is centrifuged at 160,000*g* and 23° in a Beckman Airfuge for 60 min in order to pellet the vesicles, which are immediately analyzed by SDS–PAGE.

*SDS–PAGE and Autofluorography of Vesicles Photolabeled with Compound **(I)**.* Photolabeled and pelleted synaptic vesicles are subjected to sodium dodecyl sulfate–polyacrylamide gel electrophoresis (SDS–PAGE) using a 10% resolving gel, a 3% stacking gel, and the discontinuous buffer system of Laemmli.[27] For this purpose, after removal of the supernatant the pelleted vesicles are dissociated in 80 μl of a treatment buffer made by dissolving 4.8 g urea in 4 ml of 10% SDS and diluting with 5 ml of 0.75 *M* Tris-HCl (pH 8.8). The samples are vortexed repeatedly during a 50-min period with the addition of 12 μl of 2-mercaptoethanol to each after 20 min. Finally, all samples are heated to 90° for 2 min. One-half of each sample is stored at −20° for possible reanalysis and the other half is subjected to electrophoresis. After electrophoresis, the gel is stained with Coomassie Brilliant Blue R in order to visualize proteins and then treated with Intensify (Du Pont, Inc.) as instructed by the manufacturer before being dried onto filter paper under vacuum. The dried gel is autofluorographed with XAR-5 film (Kodak, Rochester, NY) at −80°.

A continuous region from about 50 kDa to the top of the resolving gel is heavily labeled by photolyzed **(I)** (Fig. 4, lane 4). The protein stain (lane 1) of the same gel lane demonstrates that the diffuseness of the fluorogram does not arise from poorly focused proteins. No protein corresponding to the diffuse pattern of radiolabeling is stained clearly. An absence of extensive nonspecific aggregation is indicated by the lack of protein staining or radioactivity in the stacking gel. Lane 2 (Fig. 4) demonstrates that photolysis is required for labeling, and lanes 3 and 6 show that excess nonradioactive **(I)** blocks labeling of virtually all proteins, whether or not vesicles are energized by ATP. Lanes 7 and 8 (Fig. 4) show increasing levels of protection from labeling by 3 and 100 m*M* ACh, respectively. The rather modest amount of protection obtained with 3 m*M* AcCh is consistent with the K_{IA} value of 17 m*M*. Lane 9 (Fig. 4) shows that vesamicol at only 1 μM nearly completely blocks the diffuse labeling seen in lane 4. In summary, the protection pattern seen in Fig. 4 is fully consistent with specific labeling of the VAcChT at the AcCh binding site.

Photoaffinity Labeling and Visualization of Vesamicol-Binding Protein

*Photolysis of Compound **(II)** Bound to Vesicles.* Four glass test tubes treated with Sigmacote each receive 4.6 μl of the ethanolic solution of **(II)**

[27] U. K. Laemmli, *Nature* **227,** 680 (1970).

prepared above. Five microliters of a 1 mM solution of (−)-vesamicol hydrochloride in ethanol is added to the second tube (final concentration to be 50 μM (−)-vesamicol), and 1 μl of a 100 μM ethanolic solution of (±)-ABV is added to the third tube [final concentration to be 1 μM (±)-ABV]. A fifth treated tube receives 1.1 μl of the ethanolic (±)-ABV. The ethanol in all tubes is evaporated. Next, 110 μl of synaptic vesicles that are 0.65 mg of protein/ml TB is added to the fifth tube and incubated 1 hr to prebind ABV. Volumes of 100 μl of the fresh vesicular suspension are added to the first three test tubes, and 100 μl of the vesicular suspension prebound to (±)-ABV is added to the fourth test tube. All four suspensions are incubated for 10 min with occasional mixing and then illuminated with an 18-W UV lamp (Ultra-Violet Products Inc., San Gabriel, CA) that emits light in a band of wavelengths centered at 254 nm. For 10 min the tubes are alternately illuminated and vortexed to yield a total illumination time of about 5 min each. Following the photolysis, 5 μl of the solution is removed from each tube and the radioactivity measured.

SDS–PAGE and Autofluorography of Vesicles Photolabeled with Compound ***(II)***. Fifty microliters is removed from each sample of photolabeled vesicles from above and diluted with 100 μl TB. The remaining portion of each sample is stored at −20° for possible reanalysis. The diluted vesicles are pelleted in an Airfuge (Beckman Instrument Co.) at 73,000 rpm for 70 min. The supernatant is carefully removed, and the pellets are covered with 20 μl of treatment buffer that contains 10% sodium dodecyl sulfate (SDS) and 20% glycerol in 0.125 M Tris-HCl (pH 6.8). The samples are vortexed occasionally over a 6-hr period at 23°. Finally, each is diluted with 20 μl of a 10% (w/v) solution of dithiothreitol (DTT) and heated in a 90° bath for 3 min. Molecular weight standards are treated with the same solutions but are given only a brief incubation prior to addition of the DTT. The dissociated samples are subjected to SDS–PAGE using a 5–15% linear gradient resolving gel, a 3% stacking gel, and the discontinuous buffer system of Laemmli.[27] Electrophoresis is halted when the tracking dye is about 1 cm from the bottom of the gel. After electrophoresis, the gel is stained with Coomassie Brilliant Blue R in order to visualize proteins and then treated with Intensify (DuPont) before being dried onto filter paper under vacuum. The dried gel is autofluorographed with XAR-5 film (Kodak) at −80° for 3 days.

The amount of radioactivity in each sample after photolysis should be similar and correspond to about 150 nM compound **(II)**. Because the concentration of binding sites (about 450 nM as determined by dilution of a small volume of the untreated vesicles into 250 nM [^{3}H]vesamicol) is in excess of both this concentration and the dissociation constant for compound **(II)** (see above), almost all of **(II)** present in the solution will be specifically bound in the absence of competing ligand. Protein staining of the SDS–PAGE gel should reveal well-focused bands throughout the molecular mass range of 12–200 kDa. Autofluorography of the unprotected sample (Fig. 5, sample 1) demonstrates heavy labeling of a continuous region from about 50 to 200 kDa. Quantitation of the absorbance indicates that 94% of the total labeling of proteins (exclusive of the labeling at the dye front) is in this region. In addition, four polypeptides with M_r values of about 22,600, 33,000, 35,100, and 37,500 are specifically labeled. The 33-kDa band is somewhat diffuse, and correlation with the Coomassie image

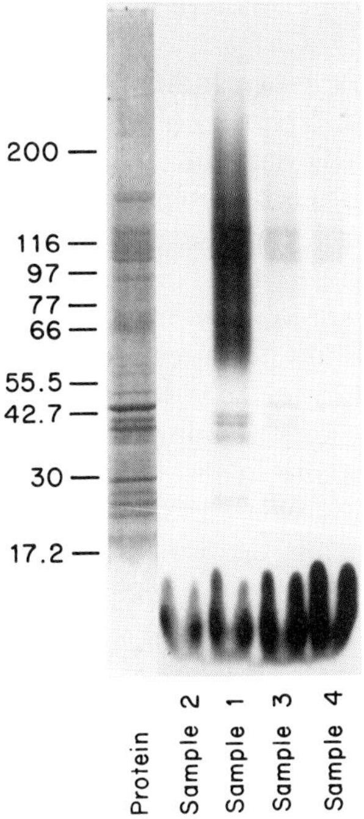

FIG. 5. Photoaffinity labeling and visualization of the vesamicol-binding protein with compound **(II)**. Vesicles were labeled in the absence or presence of competing ligands and then subjected to SDS–PAGE and autofluorography as described in the text. The figure is a composite showing Coomassie Blue staining (the lane called Protein) of Sample 1 and the autofluorograph of all four gel lanes. Sample 1 is the autofluorograph for nonblocked vesicles. Samples 2 and 3 show labeling when vesamicol or 1 μM (±)-ABV was added simultaneously with the addition of compound **(II)**, respectively, and sample 4 shows labeling when 1 μM (±)-ABV was added 1 hr prior to the addition of **(II)**. The masses (kDa) and positions of standard proteins are indicated to the left. Specifically labeled bands at 22.6, 33, 35.1, and 37.5 kDa in Sample 1 were visible in the original film, but may not be visible in the reproduction. (Reprinted with permission from G. A. Rogers and S. M. Parsons, *Biochemistry* **32,** 8596 (1993). Copyright 1993 American Chemical Society.)

suggests that the diffuse appearance arises as the result of a doublet of closely spaced bands at 32.4 and 33.8 kDa. Addition of 50 μM vesamicol coincident with compound **(II)** completely blocks incorporation of radioactivity into components with M_r greater than 12 kDa (sample 2). At the lower concentration of 1 μM, the higher affinity (±)-ABV protected nearly all sites above 12 kDa from being labeled, whether it was applied coincidentally (sample 3) or prior (sample 4) to exposure of the vesicles to compound **(II)**. Neither vesamicol nor (±)-ABV produced any visible change in the protein staining pattern (data not shown). Whether the specifically labeled species of M_r 22–38 kDa are proteolytic fragments of the VAcChT is not known. Overall, the protection pattern seen in Fig. 5 is fully consistent with specific labeling of the VAcChT at the vesamicol binding site.

The different electrophoretic patterns for specific labeling in Figs. 4 and 5 are due to the use of different types of resolving gels. Vesicles labeled with compounds **(I)** and **(II)** yield similar patterns of specific labeling after SDS–PAGE utilizing the same gel conditions. The extreme streaking of the specifically labeled species is not due to inadequate sample preparation, as far as can be determined. Changes in the sample dissociation conditions, such as use of higher concentrations of SDS, urea, low temperature, and different reducing agents, do not affect the results. Thus, the AcCh and vesamicol binding sites reside in the same protein as judged by SDS–PAGE, and this protein exhibits poor staining and migration behavior in SDS–PAGE. Possibly the VAcChT in synaptic vesicles of the electric organ is extraordinarily heavily glycosylated, or even high concentrations of SDS do not succeed in dissociating the protein fully.

[8] Bioenergetic Characterization of γ-Aminobutyric Acid Transporter of Synaptic Vesicles

By JOHANNES W. HELL and REINHARD JAHN

Introduction

Neurotransmitters are stored in synaptic vesicles and, on depolarization and subsequent calcium influx, released from nerve terminals by exocytosis. The main inhibitory neurotransmitter in the mammalian brain is γ-aminobutyric acid (GABA), whereas glycine is the prevailing inhibitory neurotransmitter in the spinal cord. The GABA uptake into synaptic vesicles is mediated by the vesicular GABA transporter, which probably works as an electrogenic GABA proton antiporter in contrast to the sodium-driven

0076-6879/98 $25.00

GABA transporter in the plasma membrane.[1-3] The proton gradient is created by an ATP-dependent proton pump, the vesicular or V-type ATPase.[2] In the absence of anions for which the synaptic vesicle membrane is permeable (e.g., chloride), the vesicular ATPase creates a large membrane potential ($\Delta\Psi$) but a negligible proton gradient (ΔpH).[2,4] Synaptic vesicles possess a conductivity for chloride which can provide a charge balance for the proton transport.[5] In the presence of 150 m*M* chloride, $\Delta\Psi$ as created by limited proton uptake is largely, but not completely, dissipated by concomitant chloride influx. As a result, the ATPase can transport more protons. The effect of adding 150 m*M* chloride to the incubation medium is, therefore, a strong reduction of $\Delta\Psi$ and a corresponding increase of ΔpH without a change in the total electrochemical potential $\Delta\mu_{H^+}$.[4,6]

In this article we will describe the technical details for measuring GABA uptake in synaptic vesicles, for the solubilization and functional reconstitution of the vesicular GABA transport system, which includes the GABA transporter as well as the vesicular proton ATPase, into artificial proteoliposomes, and for the creation of $\Delta\Psi$ and ΔpH by salt gradients and ionophores to energize the reconstituted GABA transporter independently from the vesicular ATPase.

Materials

Reagents

[2,3-^{3}H]GABA (1-1.5 TBq/mmol) and [2-^{3}H]glycine (1.1-2.2 TBq/mmol) is available from Du Pont–New England Nuclear (Boston, MA), sodium cholate (Ultrol grade) from Calbiochem (San Diego, CA), carbonyl cyanide *p*-(trifluoromethoxy)phenylhydrazone (FCCP) and nigericin from Sigma (St. Louis, MO), and valinomycin from Boehringer (Mannheim, Germany). All other reagents can be obtained from standard commercial sources.

Synaptic Vesicles

Synaptic vesicles are usually isolated by a combination of differential centrifugation, sucrose gradient centrifugation, and size-exclusion chroma-

[1] E. M. Fykse and F. Fonnum, *J. Neurochem.* **50,** 1237 (1988).
[2] J. W. Hell, P. R. Maycox, H. Stadler, and R. Jahn, *EMBO J.* **7,** 3023 (1988).
[3] J. W. Hell, L. Edelmann, J. Hartinger, and R. Jahn, *Biochemistry* **30,** 11795 (1991).
[4] P. R. Maycox, T. Deckwerth, J. W. Hell, and R. Jahn, *J. Biol. Chem.* **263,** 15423 (1988).
[5] R. Jahn, J. W. Hell, and P. R. Maycox, *J. Physiol.* (*Paris*) **84,** 128 (1989).
[6] J. W. Hell, P. R. Maycox, and R. Jahn, *J. Biol. Chem.* **265,** 2111 (1990).

tography. The classic strategy starts with the preparation and isolation of nerve terminals which reseal when pinched off the axons during a gentle homogenization step, forming so-called synaptosomes. Subsequently, synaptosomes are lysed by hypoosmotic shock to release synaptic vesicles which are further purified by sucrose gradient centrifugation and size-exclusion chromatography.[7,8] This method allows the preparation of very pure synaptic vesicle fractions with less than 5% contamination. The enrichment factor is in the range of 23, which is in good agreement with calculations indicating that synaptic vesicle proteins constitute about 5% of the total brain protein. However, this method is not only very tedious, it also takes about 24 hr. The loss of GABA uptake activity of synaptic vesicles during such a long isolation procedure is significant. As a result, GABA uptake is difficult to detect in a preparation such as that.

An alternative method for the isolation of synaptic vesicles was introduced by Hell *et al.*[2] Rat brains are frozen in liquid N_2, pulverized under liquid N_2 in a mortar, and homogenized in 320 m*M* sucrose; synaptic vesicles are then isolated by differential centrifugation and finally by molecular sieve chromatography. This method is not only more efficient with a threefold higher yield, but also significantly faster than the classic procedure, allowing the isolation of synaptic vesicles within 8 hr. The enrichment factor is about 20, just slightly lower than for the classic protocol. It was not until developing this method that we were able to detect and characterize GABA uptake in a highly enriched and well-defined synaptic vesicle preparation.[2] For these reasons, this method, which is described in detail elsewhere,[9] is recommended for studies of GABA uptake by synaptic vesicles. Following this protocol, synaptic vesicles are about fivefold enriched in the fraction obtained after differential centrifugation with respect to synaptic vesicle marker proteins. Proton gradient-dependent GABA uptake in this fraction shows the same characteristics as in the highly enriched synaptic vesicle fraction isolated by subsequent gel-filtration chromatography. It is, therefore, possible to study the GABA uptake in the crude synaptic vesicle fraction prepared by differential centrifugation, if the characteristics of this uptake are carefully monitored, especially with respect to the low affinity of the vesicular GABA transporter and its proton gradient dependency (see below).

[7] A. Nagy, R. R. Baker, S. J. Morris, and V. P. Whittaker, *Brain Res.* **109,** 285 (1976).

[8] W. B. Huttner, W. Schiebler, P. Greengard, and P. De Camilli, *J. Cell Biol.* **96,** 1374 (1983).

[9] J. W. Hell and R. Jahn, *in* "Cell Biology: A Laboratory Handbook" (J. E. Celis, Ed.), 2d ed., Vol. 2, pp. 102–110. Academic Press, London (1998).

Phospholipids

The source and quality of phospholipids is often crucial for reconstituting a transport system. Systematic experiments using commercially available phospholipids were unsuccessful in our hands. Phospholipids isolated from bovine brain following a procedure described by Nelson *et al.*,[10] however, are suitable for the functional reconstitution of the GABA transport system of synaptic vesicles. We have found that this procedure can significantly be simplified as follows.[3] A volume of 1.1 liter of absolute ethanol is preflushed with N_2; then, 26 g of bovine brain cortex is frozen and pulverized in liquid N_2 using a mortar. The powder is resuspended in 60 ml ethanol with 10 strokes at 2000 rpm in a glass–Teflon homogenizer. One liter of ethanol is added and the suspension stirred for 1 hr under N_2 before filtration through Whatman (Clifton, NJ) 3MM paper and reduction of the solvent by evaporation under vacuum at 30°. The phospholipid suspension can be kept at −20° under N_2 for several weeks before use.[3]

GABA Uptake in Synaptic Vesicles

Synaptic vesicles should be prepared according to Ref. 2 as described in detail in Ref. 9 and resuspended after collection by ultracentrifugation in assay buffer by passing through a 23- and subsequently a 27-gauge hypodermic needle. Foaming has to be avoided, and the final protein concentration should be in the range of 1 mg/ml. Standard assay buffer consists of 320 m*M* sucrose, 4 m*M* KCl, 4 m*M* $MgSO_4$, and 10 m*M* HEPES–KOH, pH 7.3.

Preincubate 100 μl of the synaptic vesicle suspension at 32° for 5 min. Add 50 μl of assay buffer containing 74 kBq (2 μCi) [2,3-^{3}H]GABA (final concentration at 450 μM) and various additions (see below) and incubate for another 5 min at 32°. Add 3–5 ml ice-cold assay buffer and filter onto nitrocellulose (prewet in assay buffer; pore diameter 0.45 μm). Rinse the test tube once and wash the filter three to four times with 3–5 ml ice-cold assay buffer. Incubate filters for at least 1 hr in 5–10 ml scintillation cocktail and quantify in a scintillation counter.

Since the GABA transporter of synaptic vesicle is energized by $\Delta\mu_{H^+}$ which is created by the vesicular ATP-dependent proton pump, it can be detected as the difference in radioactive uptake or binding between samples incubated in the presence and in the absence of ATP. ATP-dependent uptake is induced by application of 2 m*M* ATP (final concentration),

[10] N. Nelson, S. Cidon, and Y. Moriyama, *Methods Enzymol.* **157,** 619 (1988).

whereas ATP-independent binding is determined by omitting ATP from the assay buffer. In addition, it is crucial to confirm that the ATP-dependent GABA uptake is blocked by dissipating the proton gradient ΔpH with a protonophore (e.g., FCCP; 46 μM final concentration).

To test if the sodium GABA cotransporter of the plasma membrane contributes in a given synaptic vesicle preparation to GABA uptake, the incubation is started with 50 μl assay buffer in which 240 mM sucrose have been replaced with 120 mM NaCl, resulting in a final concentration of 40 mM NaCl. In this way an inward-directed sodium gradient is created which can drive the sodium GABA cotransporter of the plasma membrane in the absence of ATP. Control incubations are performed in the absence of any sodium. Even at relatively low GABA concentrations (40 μM) which is orders of magnitude below the K_D value of the synaptic vesicle GABA transporter but in a range optimal for the sodium-dependent GABA transporter of the plasma membrane (K_D: 2.5 μM[11]), we never detected sodium-dependent GABA uptake in synaptic vesicle fractions obtained by differential centrifugation with or without subsequent gel filtration chromatography. However, ouabain-sensitive sodium potassium ATPase activity, a specific marker enzyme of the plasma membrane, is detectable in synaptic vesicle fractions before and even after molecular sieve chromatography.[2] These observations can be best explained by the possibility that plasma membrane fragments containing the sodium-driven GABA transporter cannot form sealed vesicle structures as necessary to detect uptake or the plasma membrane GABA transporter is not functional after freeze-thawing.

Reconstitution of the GABA Uptake System

In this section we describe two different procedures for solubilization and functional coreconstitution of the GABA transporter of synaptic vesicles and the vesicular proton ATPase. Synaptic vesicles and phospholipids are in both cases solubilized with the ionic detergent cholate. Cholate has a high critical micellar concentration (cmc), i.e., the formation of colloidal micelles from monomeric cholate occurs at relatively high concentrations. This property allows effective removal of cholate by methods such as dialysis or molecular sieve chromatography. The first protocol[6] involves a simple dilution step which reduces the cholate concentration from 1% to 0.01%, permitting the formation of proteoliposomes. This method is fast and may be used to study the GABA transporter in conjunction with the proton ATPase. However, it results in a large incubation volume (about 1 ml) and cannot easily be used to create ion gradients appropriate for driving the

[11] B. I. Kanner, *Biochemistry* **17,** 1207 (1978).

GABA transporter independently from the proton ATPase. For that purpose we adapted a second reconstitution method[3] which is based on reducing the cholate concentration by gel filtration chromatography.

Liposomes formed as described in the following paragraphs in the absence of protein accumulate GABA and its three-carbon homolog β-alanine when an artificial ΔpH (inside positive) is created.[3] Liposomes consisting of lipids from commercial sources (asolectin or defined mixtures of synthetic lipids obtained from Avanti Polar Lipids, Alabaster, AL) showed the same accumulation of GABA (Hell and Jahn, unpublished observations). Obviously, the liposome membrane is permeable for GABA and β-alanine which may adopt a six- or five-ring structure, respectively, to form internal salt bridges between their amino and the carboxyl groups. Inside the liposomes these compounds are protonated because of the lower pH and trapped because protonated GABA or β-alanine is presumably not membrane permeable. The same liposomes are, however, not permeable for glycine, which is a two-carbon homolog of GABA and a substrate for the GABA transporter of synaptic vesicles.[12] Intact biological membranes are obviously not permeable for GABA, and we did not test whether proteoliposomes formed in the presence of protein show nonsaturable and, therefore, nonspecific GABA uptake when artificial ion gradients are used to create ΔpH. However, we recommend substituting [2-^{3}H]glycine for [2,3-^{3}H]GABA when measuring uptake activity of the GABA transporter after solubilization and reconstitution into proteoliposomes.

A sample of the phospholipid suspension (see "Materials") corresponding to a dry weight of about 62 mg is dried under a stream of N_2 and then under vacuum for 3–4 hr. Phospholipids are solubilized with 0.5 ml 1% sodium cholate in 10 m*M* HEPES–KOH, pH 7.3, by ultrasonication with a microtip (e.g., Branson cell disrupter B15, power at position 4). The solution should become clear after 0.5–2 min. Larger fragments are removed by a short centrifugation step (1 min, $9000g_{max}$). To solubilize synaptic vesicles, 25 μl of 5% sodium cholate, 500 m*M* KCl, 10 m*M* HEPES–KOH, pH 7.3, are added to 100 μl of a synaptic vesicle fraction containing 10 mg/ml protein. After careful mixing, the solution is incubated for 5 min on ice and nonsolubilized material sedimented by ultracentrifugation (e.g., $250{,}000g_{max}$ for 15 min in a Beckman TLA 100.2 rotor at 4°). The supernatant is collected and 25 μl of the phospholipid solution is added. After 5 min on ice the sample is quickly frozen in liquid N_2, thawed at room temperature, and stored on ice for up to 2 hr.

To reconstitute GABA uptake by dilution, 10 μl of solubilized synaptic vesicles supplemented with phospholipids and treated as described in the

[12] P. M. Burger, J. Hell, E. Mehl, C. Krasel, F. Lottspeich, and R. Jahn, *Neuron* **7,** 287 (1991).

previous paragraph are pipetted into 0.9 ml of assay buffer prewarmed at 32°. After 2 min, uptake is started by adding 100 μl of assay buffer containing 250 kBq (7 μCi) [2-^{3}H]glycine (final concentration at 450 μM), ATP (final concentration 2 mM), and additional compounds as desired (e.g., FCCP to test for dependence on ΔpH). Samples are incubated for 10 min at 32°. Uptake is stopped with 3 ml ice-cold assay buffer, and proteoliposomes are filtered onto nitrocellulose, washed, and counted as described above.

To reconstitute GABA uptake by gel filtration chromatography, 125 μl of solubilized synaptic vesicles and phospholipids are poured over a Sephadex G-50 superfine column (0.7 cm × 15 cm). Chromatography can be performed in sucrose buffer (320 mM sucrose, 10 mM MOPS–tetramethylammonium hydroxide, pH 7.3 or 6.5) or KCl buffer (160 mM KCl, 10 mM MOPS–KOH, pH 7.3 or 6.5). The fraction containing the proteoliposomes elutes between 4 and 5.6 ml and can easily be recognized and collected without the use of a spectrophotometer because of its turbidity. Proteoliposomes can be kept on ice for up to 2 hr. ATP-dependent uptake can be initiated by adding 100 μl assay buffer containing 74 kBq (2 μCi) [2-^{3}H]glycine (final concentration at 450 μM), ATP (final concentration at 2 mM), and various additions (see previous text) to 200 μl proteoliposomes formed in sucrose buffer at pH 7.3. After 10 min at 32°, uptake is stopped with 3 ml ice-cold assay buffer, and proteoliposomes are filtered onto nitrocellulose, washed, and counted as described above.

GABA Uptake Driven by $\Delta\Psi$ and ΔpH as Created by Artificial Salt Gradients

The GABA transporter of synaptic vesicles can be energized by both components of $\Delta\mu_{H^+}$, $\Delta\Psi$, and ΔpH (see preceeding section). To study the GABA transporter independent of the vesicular proton ATPase, salt gradients can be generated between the interior and exterior of proteoliposomes and converted into either $\Delta\Psi$ or ΔpH by the use of appropriate ionophores. To establish an inside acidic ΔpH, proteoliposomes are formed by molecular sieve chromatography in KCl buffer (pH 7.3), collected by ultracentrifugation (e.g., 250,000g_{max} for 15 min in a Beckman TLA 100.2 rotor at 4°), and resuspended by three passages through a hypodermic needle (27-gauge) in 1.5 ml sucrose buffer (pH 7.3). Proteoliposomes should be used immediately for the uptake assay. To create ΔpH and start the uptake, 200 μl of this fraction are added to 100 μl sucrose buffer containing 74 kBq (2 μCi) [2-^{3}H]glycine (final concentration at 450 μM), and nigericin (final concentration 250 nM). The uptake is terminated after 20 min at 32° by filtration onto nitrocellulose as described above.

To create an inside positive $\Delta\Psi$, proteoliposomes are generated by size exclusion chromatography in sucrose buffer at pH 6.5. They can be stored on ice for up to 2 hr. Uptake is initiated by pipetting 200 μl of 110 mM K_2SO_4, 74 kBq (2 μCi) [2-^{3}H]glycine (final concentration at 450 μM), valinomycin (final concentration at 125 nM), MOPS–KOH, pH 6.5, to 200 μl of the proteoliposome suspension prewarmed for 5 min at 32°. After 2 min at 32°, uptake is terminated with 3 ml of an ice-cold solution containing 110 mM K_2SO_4 and MOPS–KOH, pH 6.5. Proteoliposomes are filtered onto nitrocellulose filters and washed three times with this solution. Uptake is determined by scintillation counting. Although the GABA transporter of synaptic vesicles can be driven by $\Delta\Psi$ in the absence of ΔpH,[3,6] it requires a pH of 6.5 or less inside (although not necessarily outside) the vesicles or proteoliposomes.[6] This restriction can be met by performing the whole assay at pH 6.5. A lower pH seems unfavorable (Hell and Jahn, unpublished results, 1990).

Outward-directed salt gradients are more stable than inward-directed gradients. Transferring K^+-filled proteoliposomes into sucrose medium to establish ΔpH with nigericin is, therefore, presumably more efficient for measuring the GABA transporter than creating $\Delta\Psi$ with valinomycin in the presence of an inward-directed K^+ gradient. However, a small, inside negative $\Delta\Psi$ could be detected in proteoliposomes with an outward-directed K^+ gradient even in the absence of nigericin (which did not cause further changes in $\Delta\Psi$).[3] This observation can be explained by the assumption that the proteoliposome membrane is somewhat permeable for the monovalent potassium cation but not for the divalent sulfate anion. A slight inside negative $\Delta\Psi$ would be expected to counteract the electrogenic GABA transporter, thereby limiting its efficacy to energize this transporter. In our hands ΔpH based on an outward-directed K^+ gradient was stable for more than 1 hr, although specific uptake was highest after 20 min. In contrast, $\Delta\Psi$ based on an inward-directed K^+ gradient declines within minutes.[3] Although uptake is very fast under the latter conditions (significant accumulation can be detected after a 30 sec incubation), it is strongly reduced after 5–10 min.[3]

Properties of GABA Uptake System of Synaptic Vesicles

Intact synaptic vesicles can accumulate 300 pmol/mg protein within 5 min at a concentration of 450 μM GABA. Studies of the vesicular GABA transporter face the problem that this transporter has a very low affinity. Its K_D for GABA is in the range of 10 mM,[1,2,12] limiting the ratio which can be achieved for specific GABA uptake versus nonspecific GABA binding. It should be pointed out, however, that such a high K_D value is not unexpected for a transporter which functions inside nerve terminals where the cytosolic concentration of GABA is thought to be in the range of 5 mM or even higher.[13] The high K_D value makes it difficult to test the substrate specificity, although specific uptake of [2,3-^{3}H]GABA or [2-^{3}H]glycine is reduced by 50% or more in the presence of 20 mM unlabeled GABA.[12] Interestingly, the structural analogs of GABA, β-alanine and glycine, compete for GABA transport activity with a similar affinity (i.e., 20 mM of either amino acid diminishes the GABA uptake into synaptic vesicles to the same degree as 20 mM unlabeled GABA).[12] Additional evidence indicates that [2-^{3}H]glycine is transported by the same carrier activity as GABA,[12] although it is presently unclear whether this activity comprises a single transporter or two or more transporters with slightly different properties. However, the transport activity as measured under the experimental conditions described above mediates the uptake of GABA and glycine with similar efficacy and affinity. [2-^{3}H]Glycine can, therefore, substitute for [2,3-^{3}H]GABA as is crucial for experiments involving reconstitution of the GABA transporter into proteoliposomes especially when a ΔpH component is present (see above).

The GABA uptake of synaptic vesicles depends on an intravesicular pH not higher than 6.5.[6] This property has to be taken into account whenΔpH is completely dissipated as can be achieved by adding 10 mM $(NH_4)_2SO_4$.[6] NH_3 can permeate the vesicle membrane and neutralizes intravesicular protons. When GABA uptake is measured in the absence of ΔpH as in the presence of $(NH_4)_2SO_4$ or when ΔpH is created by valinomycin in combination with an inward-directed potassium gradient, extravesicular pH has to be in the range of 6.5 to match the intravesicular pH requirement.[6]

A legitimate concern of the electrochemical characterization of the GABA uptake using a preparation of synaptic vesicles in which GABAergic vesicles constitute only a fraction of the total pool of synaptic vesicles is that the GABAergic subpopulation may behave differently from the majority of synaptic vesicles. After solubilization and randomized reconstitution, however, the properties of the GABA/glycine uptake, of the glutamate uptake system of synaptic vesicles,[4] and of the balance between $\Delta\Psi$ and ΔpH are virtually identical with those of intact synaptic vesicles with respect to the effect of increasing chloride concentration or $(NH_4)_2SO_4$ on $\Delta\Psi$ and ΔpH,[4,6] corroborating the validity of the described approach of a thermodynamic characterization of the vesicular GABA uptake. It is our hope that the strategies and observations outlined in this article stimulate further characterization of this transport system.

[13] O. P. Ottersen and J. Storm-Mathisen, *Hdbk. Chem. Neuroanat.* **3,** 141–246 (1984).

[9] Solubilization and Reconstitution of Synaptic Vesicle Glutamate Transport System

By SCOTT M. LEWIS and TETSUFUMI UEDA

Introduction

Amassed evidence indicates that glutamate is the major excitatory amino acid neurotransmitter in the vertebrate central nervous system (CNS).[1–5] Of particular interest is its involvement in cognitive disorders (such as Alzheimer's disease and Huntington's disease), psychiatric disorders,[6,7] and epilepsy.[8] The proteins involved in glutamatergic neurotransmission have been extensively characterized postsynaptically and presynaptically, including all the types of glutamate receptors,[2,4,5,9,10] the plasma membrane glutamate transporter,[11–14] and the synaptic vesicle glutamate transport system.[15–19] Of these proteins, only the synaptic vesicle glutamate transport system (consisting of all necessary proteins and their activities) remains to be structurally characterized.

The synaptic vesicle glutamate transport system performs a key function in neurotransmission by loading the synaptic vesicle with glutamate. Given

[1] F. Fonnum, *J. Neurochem.* **42,** 1 (1984).
[2] D. T. Monaghan, R. J. Bridges, and C. W. Cotman, *Ann. Rev. Pharmacol. Toxicol.* **29,** 365 (1989).
[3] D. Nicholls and D. Attwell, *Trends in Pharm. Sciences* **11,** 462 (1990).
[4] J. C. Watkins, P. Krogsgaard-Larsen, and T. Honore, *Trends Pharmacol. Sci.* **11,** 25 (1990).
[5] S. Nakanishi, *Science* **258,** 597 (1992).
[6] S. Sahai, *Eur. Arch. Psychiatry Clin. Neurosci.* **240,** 1211 (1990).
[7] P. J. Shaw, *Curr. Opin. Neurol. Neurosurg.* **6,** 414 (1993).
[8] J. O. McNamara, *Curr. Opin. Neurol. Neurosurg.* **6,** 583 (1993).
[9] M. Masu, Y. Tanabe, K. Tsuchida, R. Shigemoto, and S. Nakanishi, *Nature* **349,** 760 (1991).
[10] M. Hollman and S. Heinemann, *Ann. Rev. Neurosci.* **17,** 31 (1994).
[11] Y. Kanai and M. Hediger, *Nature* **360,** 467 (1992).
[12] G. Pines, N. C. Danbolt, M. Bjoras, Y. Zhang, A. Bendahan, L. Eide, H. Koepsell, J. Storm-Mathisen, E. Seeberg, and B. I. Kanner, *Nature* **360,** 464 (1992).
[13] T. Storck, S. Schutle, K. Hofmann, and W. H. Stoffel, *Proc. Natl. Acad. Sci. USA* **89,** 10955 (1992).
[14] J. Arriza, W. A. Fairman, J. I. Wadiche, G. H. Murdoch, M. P. Kavanaugh, and S. G. Amara, *J. Neurosci.* **14,** 5559 (1994).
[15] S. Naito and T. Ueda, *J. Biol. Chem.* **258,** 696 (1983).
[16] S. Naito and T. Ueda, *J. Neurochem.* **44,** 99 (1985).
[17] P. R. Maycox, T. Deckwerth, J. W. Hell, and R. Jahn, *J. Biol. Chem.* **263,** 15423 (1988).
[18] E. M. Fykse, H. Christensen, and F. Fonnum, *J. Neurochem.* **52,** 946 (1989).
[19] S. Cidon and T. S. Sihra, *J. Biol. Chem.* **264,** 8281 (1989).

METHODS IN ENZYMOLOGY, VOL. 296
0076-6879/98 $25.00

its vital role in glutamate neurotransmission, directing glutamate to the neurotransmitter pathway away from the metabolic pathway, it potentially represents an important site of regulation of central nervous system pathogenesis. Synaptic vesicle glutamate transport requires a proton ATPase of the vacuolar type and chloride.[16–21] These function to provide the appropriate intravesicular pH and electrochemical gradients that are necessary to drive glutamate transport. It is not known whether chloride-stimulated, ATP-dependent glutamate transport is accomplished by several proteins, each functioning independently, or if more than one function is performed by one protein. Previous studies on vesicular glutamatergic transporter solubilization have not demonstrated structural separation of the activities. Carlson *et al.*[22] and Maycox *et al.*[17] demonstrated solubilization and functional reconstitution of the vesicular glutamate transport system; however, the separation of proton-pump ATPase and glutamate transport activities was not achieved. Maycox *et al.*[23] showed that glutamate transport could be driven by an alternative proton pump, bacteriorhodopsin; however, the study did not structurally separate the proton-pump ATPase from the putative glutamate transporter and, in fact, demonstrated that the ATPase was still present in the reconstituted proteoliposome. In order to answer questions regarding the structural composition of the vesicular glutamate transport system, it will be necessary to isolate the protein functions, if possible, and then characterize the proteins responsible. This is complicated by the fact that a purified glutamate transporter may have no activity of its own and, hence, that it can only be identified once it is reconstituted into a vesicle containing the other necessary protein components.

Solubilization of the synaptic vesicle proteins followed by separation and functional reconstitution would allow these questions to be addressed. In this chapter we review the solubilization and reconstitution of the vesicular glutamate transport system and report an extension of these techniques, resulting in a partial purification of the vesicular glutamate transport system and a partial separation of the glutamate transporter and proton-pump ATPase. As these techniques require a minimum of 10–20 mg synaptic vesicle protein, we also include an updated synaptic vesicle purification protocol for large-scale preparation of vesicles similar to that described by Kish and Ueda.[24]

[20] J. S. Tabb, P. E. Kish, R. Van Dyke, and T. Ueda, *J. Biol. Chem.* **267,** 15412 (1992).
[21] J. Hartinger and R. Jahn, *J. Biol. Chem.* **268,** 23122 (1993).
[22] M. D. Carlson, P. E. Kish, and T. Ueda, *J. Biol. Chem.* **264,** 7369 (1989).
[23] P. R. Maycox, T. Deckwerth, and R. Jahn, *EMBO J.* **9,** 1465 (1990).
[24] P. E. Kish and T. Ueda, *Methods Enzymol.* **174,** 9 (1989).

Large-Scale Preparation of Glutamatergic Synaptic Vesicles

Reagents for Vesicle Preparation

Solution A: 0.15 *M* NaCl

Solution B: 0.32 *M* sucrose, 1 m*M* magnesium acetate, 0.5 m*M* calcium acetate, 0.2 m*M* phenylmethylsulfonyl fluoride (PMSF, added just prior to use from a 200 m*M* stock solution in ethanol), 1 m*M* $NaHCO_3$, pH 7.4

Solution C: 1.28 *M* sucrose

Solution D: 6 m*M* Tris–maleate, 0.2 m*M* PMSF (added just prior to use), pH 8.3

Solution E: 0.32 *M* sucrose, 1 m*M* $NaHCO_3$, and 1 m*M* dithiothreitol (DTT), pH 7.2.

Procedure. Glutamatergic synaptic vesicles are prepared from two bovine cerebra obtained fresh the morning of the preparation from a local slaughterhouse and transported on ice to the laboratory. All procedures are performed either on ice or at 4°. The fresh brains are rinsed in 2 liters of Solution A following removal of the brain stem, cerebellum, and excess white matter. This yields approximately 600–700 g of wet cerebral tissue. The brains (150 g at a time) are then blended (in a Waring blender at high speed) in 1.5 liters of Solution B with three 7-sec bursts separated by 5-sec intervals, giving a final volume of 6 liters. This suspension is filtered through a plastic mesh strainer and then diluted to 10 liters by addition of 4 liters of Solution B. The brain suspension is then homogenized in a large cylindrical (14 × 30 cm), continuous-flow Plexiglas homogenizer with a circular disc Teflon pestle at 990 rpm. The pestle is driven by a Craftsman Free Standing Variable Speed Drill Press, Model 113.213130. The pestle is weighted by 2.2 foot-lbs torque. The homogenate is fed through the floor of the homogenizer between the pestle and the Plexiglas floor by a peristaltic pump. The homogenized material drains through a port above the pestle. The resulting homogenate is centrifuged in a Sorvall GSA rotor at 2500 rpm ($1000g_{max}$) for 10 min. The supernatants are collected by decanting, pooled, and further centrifuged in a Sorvall GSA rotor at 13,000 rpm ($27{,}000g_{max}$) for 15 min. The resulting supernatants are decanted and discarded. The pellets are resuspended in Solution B, pooled to give a final volume of 1.5 liters, and homogenized to smoothness (3 strokes in a 500-ml glass homogenizer with a Teflon pestle at 990 rpm; all subsequent homogenizations are performed with a glass homogenizer–Teflon pestle at 990 rpm). This is diluted to 3.0 liters by slow addition (with stirring) of an equal volume of Solution C. The resulting 0.8 *M* sucrose suspension is then centrifuged in a Sorvall GSA rotor at 13,000 rpm ($27{,}000g_{max}$) for 45 min.

The supernatant and floating myelin bands are carefully removed by aspiration, and myelin adhering to the sides of the tubes is removed by wiping with a disposable tissue. The pellets (comprising a crude synaptosomal fraction) are resuspended in 2 liters Solution D and homogenized with 1–2 strokes in a 500 ml glass homogenizer. The homogenate is brought to a final volume of 6 liters in Solution D and allowed to complete lysing by incubation for 45 min at 4° with gentle stirring. The lysed synaptosomes are then centrifuged in a Sorvall SS-34 rotor at 19,000 rpm ($43,500g_{max}$) for 15 min. The resulting supernatant is collected, pooled, and concentrated to 400 ml, using an Amicon (Danvers, MA) Spiral Ultraconcentrator (30,000 molecular weight cutoff). The concentrate is then brought to a maximum of 750 ml by flushing with 300–350 ml of Solution D to collect the lysate occupying the internal volume of the ultraconcentrator.* The concentrated synaptosomal lysate is then centrifuged in a Beckman Ti45 rotor at 43,000 rpm ($200,000g_{max}$) for 70 min. The pellet containing crude synaptic vesicles is resuspended (by homogenization) in 180 ml of Solution D and 15-ml aliquots are layered on 12 discontinuous sucrose gradients (20 ml 0.4 *M* and 25 ml 0.6 *M*). Following centrifugation for 2 hr at 35,000 rpm ($140,000g_{max}$) in a Beckman Ti45 rotor at 4°, the 0.4 *M* layer and the 0.4 *M*/0.6 *M* interface layer are collected separately. Each layer is diluted with Solution D to less than 0.18 *M* sucrose and centrifuged for 70 min at 43,000 rpm ($200,000g_{max}$) in a Beckman Ti45 rotor. The pellet is resuspended (by homogenization) in Solution E and stored in liquid nitrogen until use. This typically yields approximately 100 mg (for the combined fractions of the 0.4 *M* and 0.4 *M*/0.6 *M* interface layers) of synaptic vesicle protein as intact synaptic vesicles.

Solubilization of Synaptic Vesicle Transport System

Reagents for Synaptic Vesicle Glutamate Transporter Solubilization

Solubilization Solution A: 6 m*M* Tris–maleate, pH 8.1

Solubilization solution B: 0.132 *M* KCl, 8.8 m*M* 4-(2-hydroxyethyl)-1-piperazineethanesulfonic acid (HEPES), 1.2% (w/v) sodium cholate, and 2% (w/v) crude soybean phospholipids (40% phosphatidylcholine), pH 7.4. The buffer is prepared by sonication (Branson sonicator with a microtip, model W185, 20 W) under nitrogen for 5–10 min.

Procedure. The synaptic vesicle suspension obtained above is thawed. In a minor modification of Carlson *et al.*[22], approximately 10–12 mg of

* Alternatively, the concentration by the Amicon Spiral Ultraconcentrator can be replaced by centrifugation in a Beckman Ti45 rotor at 43,000 rpm ($200,000g_{max}$) for 70 min. However, given the volume of the solution, the use of the Ultraconcentrator is more efficient.

vesicles are placed in a Beckman Ti50 centrifuge tube and diluted with Solubilization Solution A, resulting in a final volume of 10 ml. This is centrifuged at 47,000 rpm (200,000g_{max}) for 90 min at 4°. The pellet is used immediately or stored overnight at −80°. If stored overnight, the pellet is thawed on ice before use. The pellet is then resuspended by hand homogenization in a 1 ml glass homogenizer with a Teflon pestle in Solubilization Solution B to a final concentration of 10 mg/ml. Following a 30-min incubation on ice, the mixture is centrifuged at 47,000 rpm (200,000g_{max}) in a Beckman Ti50 rotor for 30 min at 4°. The supernatant, containing the solubilized proteins, is removed by hand pipetting with care to avoid the pellet. The gel filtration chromatography profile of the solubilized material versus intact vesicles is shown in Fig. 1, along with the glutamate uptake activity observed upon reconstitution of the solubilized fractions.[22] The solubilized glutamate transport system has a Stokes radius of 69–84 Å and a molecular weight of 490,000 to 640,000, assuming the protein(s) have a globular shape.[22] The best results for solubilization are achieved by using 5–20 mg of vesicles per tube, as judged by the ATP-dependent glutamate uptake following reconstitution. This method can be used on rat or bovine synaptic vesicles. The solubilized proteins can be successfully reconstituted after 24 hr on ice, but with a 30% loss in glutamate uptake activity.

Of the detergents tested, cholate extracted the maximal ATP-dependent glutamate uptake activity, as determined upon functional reconstitution.[22] The optimal concentration is 1.2% (see Fig. 2).[22] CHAPS (1.2%) and urea (2.5 *M*) also solubilized the glutamate transporter, but resulted in considerably less glutamate uptake activity on reconstitution (2-fold higher glutamate uptake activity in the presence of ATP than in its absence, compared to 10- to 15-fold for cholate extraction).[22] Extraction with 1.2% Brij 35, $C_{12}E_9$, Triton X-100, Triton CF-32, and Lubrol PX did not yield proteoliposomes with significant ATP-dependent glutamate uptake activity. Sodium cholate (1–1.1%) has also been successfully used to solubilize the monoamine[25] and GABA[26] vesicular transport systems. Changing the solubilizing buffer by removing the KCl from the solubilizing buffer and replacing it with 0.3 *M* sucrose reduced the recovered ATP-dependent activity by over 40%.[22]

Comparison to Another Solubilization Procedure. Maycox *et al.*[17,23] described an alternative solubilizing buffer of 500 m*M* KCl, 15 m*M* $MgCl_2$, and 5% (w/v) sodium cholate. This is added in a ratio of 1:4 (v/v) to a suspension of rat brain synaptic vesicles (10 mg/ml). Subsequent centrifuga-

[25] R. Maron, H. Fishkes, B. I. Kanner and S. Schuldiner, *Biochemistry* **18,** 4781 (1979).
[26] J. W. Hell, P. R. Maycox, and R. Jahn, *J. Biol. Chem.* **265,** 2111 (1990).

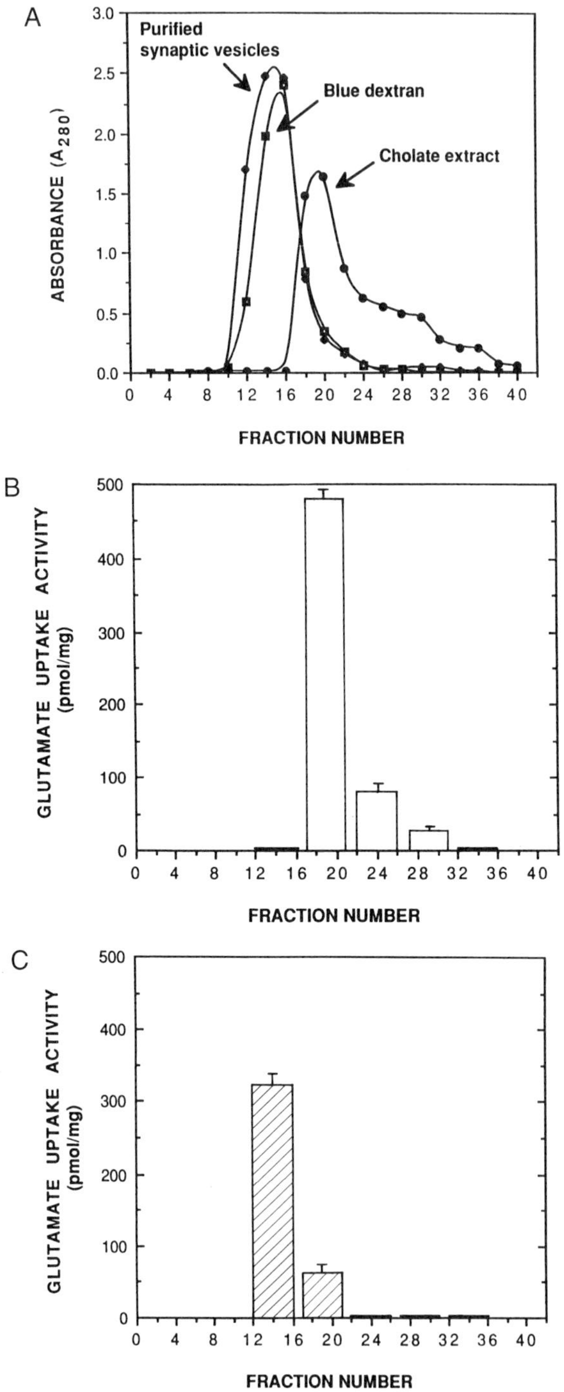
A
Purified
synaptic vesicles
Blue dextran
Cholate extract
ABSORBANCE (A_{280})
FRACTION NUMBER
B
GLUTAMATE UPTAKE ACTIVITY
(pmol/mg)
FRACTION NUMBER
C
GLUTAMATE UPTAKE ACTIVITY
(pmol/mg)
FRACTION NUMBER

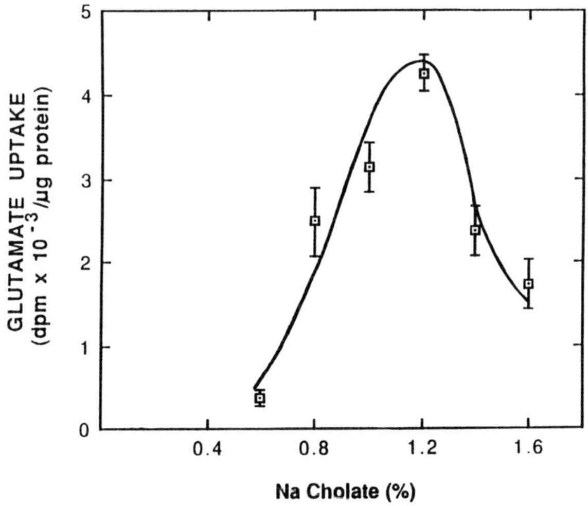

FIG. 2. Effect of various concentrations of cholate on the solubilization of the ATP-dependent glutamate uptake system. Assay conditions are described in Carlson *et al.*[22]

tion at 250,000g_{max} for 15 min in a Beckman TLA 100.2 rotor did not affect the results on reconstitution. A revised procedure by Maycox *et al.*[23] included the centrifugation step, but for 5 min at 250,000g_{max} in a Beckman TLA 100.3 rotor instead of the centrifugation described above. Also, the

---

FIG. 1. Gel-filtration chromatography of the synaptic vesicles and the cholate extract. The purified synaptic vesicles, the cholate extract, and blue dextran were applied separately (10 mg of each in 1.3 ml) to a Bio-Gel A-1.5 m gel filtration column at 4° (Bio-Rad agarose beads; 0.9 × 58 cm; equilibrated with 0.15 *M* KCl, 10 m*M* HEPES (pH 7.4), and 5 m*M* monothioglycerol; exclusion limit 1.5 million). For chromatography of the solubilized vesicular glutamate uptake system, the column buffer was supplemented with 0.3 (w/v) sodium cholate and 0.3% (w/v) soybean phospholipid. (A) Protein eluted as determined by absorbance at 280 nm. (B) ATP-dependent glutamate uptake activity of the reconstituted proteoliposomes derived from various fractions of the cholate extract. Five groups (5 ml each) of the indicated pooled fractions were reconstituted into liposomes by gel filtration on Bio-Gel P6-DG (Bio-Rad, 2.5 × 10 cm, equilibrated with 0.25 *M* sucrose and 6 m*M* Tris–maleate, pH 8.1), followed by a 2-fold dilution in 6 m*M* Tris–maleate (pH 8.1) and centrifugation for 70 min at 225,000g_{max} (43,000 rpm, Beckman Ti50.2 rotor, 4°). The pellets were resuspended in Solution E and assayed for ATP-dependent glutamate uptake activity using 50 μM glutamate (0.8 Ci/mmol).[22] C, ATP-dependent glutamate uptake activity of various unsolubilized synaptic vesicle fractions. Five groups (5 ml each) of the indicated pooled fractions were treated as in B and assayed for ATP-dependent glutamate uptake activity.[22] ATP-dependent activity is the activity in the presence of ATP minus the activity in the absence of ATP.

solubilized proteins are incubated on ice for 10 min prior to centrifugation. Further comparisons will be made below in the section on reconstitution.

Reconstitution of Synaptic Vesicle Glutamate Transport System

Reagents for Reconstitution of Synaptic Vesicle Transport System

Reconstitution Solution A: 0.25 *M* sucrose, 6 m*M* Tris–maleate, pH 8.1
Reconstitution Solution B: 6 m*M* Tris–maleate, pH 8.1
Reconstitution Solution C: 0.32 *M* sucrose, 1 m*M* $NaHCO_3$, 1 m*M* dithiothreitol, pH 7.2
Solubilization Solution B: 0.132 *M* KCl, 8.8 m*M* HEPES, 1.2% (w/v) sodium cholate, and 2% (w/v) crude soybean phospholipids (40% phosphatidylcholine), pH 7.4. The buffer is prepared by sonication (Branson sonicator with a microtip, model W185, 20 W) under nitrogen for 5 –10 min.

Procedure. In Carlson *et al.*,[22] the solubilized material (from rat synaptic vesicles) is supplemented with additional lipid in Solubilization Solution B. The optimal phospholipid concentration in the supplementing solution is 2% (w/v), as shown in Fig. 3.[22] The supplemental lipid solution is added to the cholate extract containing solubilized proteins at a ratio of 2:3

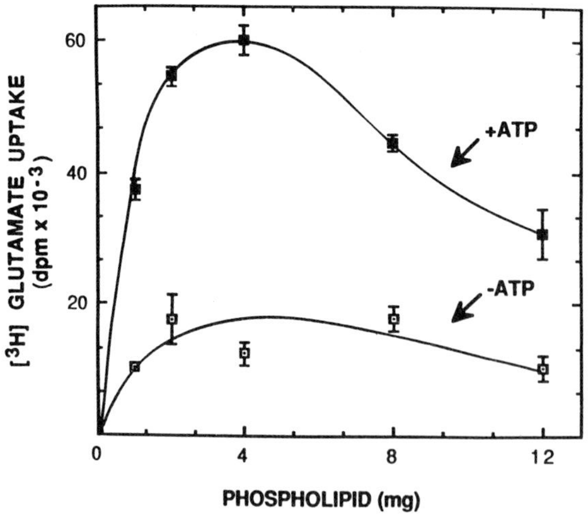

FIG. 3. Effect of various amounts of phospholipid on the reconstitution of the ATP-dependent glutamate uptake system. The amount of phospholipid in the solubilizing buffer was constant at 2%. The amount of crude soybean phospholipid added (mg/0.2 ml Solubilization Solution B) to the cholate extract prior to reconstitution was varied.

(v/v). The mixture is applied to a Sephadex G-50 column (grade fine, 0.9 × 18 cm) previously equilibrated with Reconstitution Solution A at room temperature. The turbid fractions eluting at the void volume (10–15 300-μl fractions) are collected, pooled, diluted 3-fold with Reconstitution Solution B, and centrifuged at 200,000g_{max} for 30 min at 4°. The pellet is resuspended in Reconstitution Solution C. Alternatively, this procedure can be performed substituting a 50-fold dilution step (with Reconstitution Solution B) for the Sephadex G-50 column step.[22] The ATP dependence is slightly lower (5- to 8-fold ATP stimulation) compared to that observed with the use of the Sephadex G-50 column (10- to 15-fold ATP stimulation).[22]

This procedure results in a protein yield of 25% with an increase in the specific activity of approximately 50% (1083 pmol glutamate/mg protein to 1560 pmol glutamate/mg protein).[22] The reconstituted transport system is essentially identical to the native transporter in the time course of glutamate uptake, K_m for glutamate, chloride effect, response to electrochemical proton gradient inhibitors, and response to glutamate analogs and glutamate receptor ligands.[22]

Crude soybean phospholipid is the most effective of the phospholipids examined in the solubilization and reconstitution.[22] Pure preparations of phosphatidylcholine (soybean or bovine), soybean phosphatidylethanolamine, soybean phosphatidylinositol, or bovine phosphatidylserine are not effective, resulting in less than 20% of the ATP-dependent glutamate uptake activity of that obtained with the crude soybean phospholipid. The crude soybean phospholipid contains approximately 40% phosphatidylcholine, 30% phosphatidylethanolamine, and less than 10% amounts of phosphatidic acid, triolein, phosphatidylinositol, cholesterol, and phosphatidylserine. Soybean phospholipid was also used during extraction and reconstitution of the monoamine transport system.[25]

Omitting the phospholipid from Solubilization Solution B during the solubilization step did not abolish the glutamate uptake activity following reconstitution, but it did significantly lower the ATP stimulation to 3- to 4-fold.[22] Omitting the additional phospholipid from the reconstitution step resulted in no recovered activity, indicating that additional lipid is an absolute requirement for reconstitution.[17,22] The activity following reconstitution is also dependent on the buffer conditions of the column. Substitution of 0.15 *M* KCl for the 0.25 *M* sucrose or changing the pH to 7.4 resulted in a 50% or greater reduction in the ATP stimulation.[22]

An alternative procedure for reconstitution will be described below, for use following the fractionation of the soluble proteins on a glycerol gradient.

Comparison to Another Reconstitution Procedure. Maycox *et al.*[17,23] reconstituted the rat brain vesicular glutamate transport system with purified brain phospholipids. The protocol uses different reconstitution solutions, but is similar in the required lipid concentration. It also differs in the methods of reconstitution: theirs employs freezing and thawing followed by dilution. The phospholipid suspension (125 mg/ml in 10 m*M* Tris–MES, pH 7.0, 1 m*M* dithiothreitol, and 1% sodium cholate, w/v) is added in a ratio of 1 : 5 (v/v) to the solubilized extract. The mixture is shock-frozen in methanol/dry ice, thawed slowly on ice, and shock-frozen again. Following the second thaw, the mixture is diluted 100-fold in sucrose buffer (0.32 *M* sucrose and 10 m*M* HEPES, pH 7.4). However, the resulting ATP-dependent glutamate uptake activity (specific activity) is less than 50% of the starting uptake activity.

Maycox *et al.*[23] later reported a modified procedure that eliminated the freeze–thaw step as the glutamate carrier was damaged by this method. Brain phospholipid (resuspended by tip sonication instead of bath sonication) is added as described and the mixture is incubated on ice for 5 min and then incubated at 32° for 5 min. Reconstitution is performed by dilution of 150 μl of this mixture into 9 ml of their assay buffer. They report a 2.5-fold increase in the time-dependent specific activity, compared to the starting vesicles, with a 30% recovery of protein. Direct comparisons to the procedure described by Carlson *et al.*[22] are difficult, as Maycox *et al.*[17] use different assay conditions (4-fold higher glutamate concentration) and a time-dependent specific activity (as opposed to a steady-state specific activity); however, the two procedures are comparable in that both result in an improvement in specific activity in comparison to the starting material.

Further Purification of Synaptic Vesicle Glutamate Transport System

The vesicular glutamate transport system can be further purified prior to reconstitution by applying the cholate solubilized material to a continuous glycerol gradient. In addition to the purification, this gradient can partially separate proteins of the glutamate transporter system, as described in the section on coreconstitution. All procedure are performed on ice or at 4°.

Reagents for Glycerol Gradient Purification and Reconstitution of the Solubilized Glutamate Transport System

Gradient Solution A: glycerol (variable concentration, v/v), 10 m*M* Tris–maleate, 10 m*M* dithiothreitol, 0.7% (w/v) sodium cholate, pH 7.4

Gradient Reconstitution Solution A: 10 m*M* HEPES, 1 m*M* dithiothreitol, 1% (w/v) sodium cholate, 5% (w/v) crude soybean phospholipid (sonicated 5–10 min under nitrogen), pH 7.4

Gradient Reconstitution Solution B: 6 m*M* Tris–maleate, pH 8.1

Gradient Reconstitution Solution C: 10 m*M* HEPES, 250 m*M* sucrose, pH 7.4

Purification of Synaptic Vesicle Proteins by Glycerol Gradient Centrifugation. Continuous glycerol gradients are formed using 25 and 38% (v/v) glycerol in Gradient Solution A. The gradients are formed in 5.1-ml Beckman quick seal polyallomar tubes. Approximately 2.5 ml of each solution are necessary to fill each tube to 4.5–4.6 ml (the remaining solution being lost in transfer). The extracts from cholate-solubilized bovine vesicles (0.5–0.6 ml/5.1 ml gradient) are layered onto the gradient using a 25-gauge needle on a syringe. The tube is balanced, sealed, and centrifuged for 3 hr at 70,000 rpm (390,000g_{max}) at 4° using a vertical rotor (Beckman VTi80). The glycerol gradient is fractionated (17–18 fractions with a fraction volume of 300 μl).

Reconstitution of Solubilized Proteins Fractionated on Glycerol Gradient. The solubilized synaptic vesicle proteins are reconstituted with additional phospholipid. Gradient Reconstitution Solution A is added at a ratio of 1:5 (v/v) to the gradient fraction; the solution is mixed and diluted at least 10-fold with Gradient Reconstitution Solution B, followed by centrifugation at 47,000 rpm (200,000g_{max}) for 90 min at 4° using a Beckman Ti50 rotor. The pellet (proteoliposomes) is resuspended, by hand homogenization in a 1-ml glass–Teflon homogenizer, in Gradient Reconstitution Solution C. The yield is increased if Gradient Reconstitution Solution A is freeze–thawed prior to addition to the gradient fractions. Proteoliposomes are stored at −80° prior to assay. The synaptic vesicle glutamate transport system can be successfully reconstituted from longer gradient centrifugation times of 6 hr (Beckman VTi80 rotor at 70,000 rpm, 390,000g_{max}) to 14 hr (Beckman SW41 rotor at 41,000 rpm, 288,000g_{max}).

Glutamate Uptake and ATPase Activity Profiles of the Glycerol Gradient. The ATP-dependent glutamate uptake activity peaks in fractions 8 and 9 (see Fig. 4), with the activity declining rapidly on either side. The ATP-independent glutamate uptake activity remains low throughout the gradient fractions except for a slight peak detected in fraction 3. A similar profile is obtained for ATP hydrolytic activity; however, the activity is distributed more broadly and peaks in fractions 9 through 13. All proteins necessary for glutamate transport are present in fractions 8 and 9; however, separation of these proteins may be occurring on the sides of this profile, as suggested by high ATP hydrolytic activity in fractions 10 through 13 while ATP-dependent glutamate uptake activity is declining. The ATP hydrolytic profile is *N*-ethylmaleimide (NEM) sensitive. To ensure that the ATP hydrolytic profile reflects the protein-pump ATPase, the gradient fractions were assayed for the ability to generate a pH gradient in the presence of a high concentration of chloride, using methylamine uptake as a probe

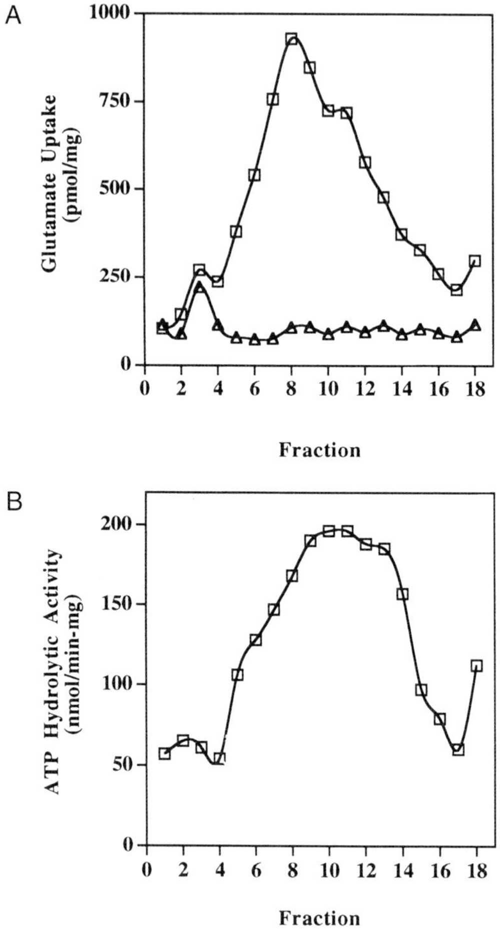

FIG. 4. Gradient profile for glutamate uptake and proton-pump ATPase activity. Gradient fractions 1 to 18 represent 38–25% glycerol, respectively. (A) Glutamate uptake activity (pmol/mg protein) in the presence (squares) or absence (triangles) of ATP. Glutamate uptake was measured using tritium-labeled L-glutamate by a modification of the procedure described by Naito and Ueda.[16] The assay medium contained 10 mM HEPES (pH 7.4), 2 mM potassium aspartate, 4 mM magnesium sulfate, 250 mM sucrose, 4 mM potassium chloride, and 5–30 μg of protein (in the form of proteoliposomes) in a final volume of 120 μl. The uptake was initiated by adding potassium L-glutamate (with 2 μCi tritiated glutamate) with or without ATP to a final concentration of 50 μM and 2 mM, respectively. The incubation temperature was 37° and the incubation time was 30 min. The sample was filtered (using glass fiber filters, grade C, obtained from Cambridge Technology, Inc.) and washed with 0.15M KCl (kept at 0°). Radioactivity on the filter was determined in a liquid scintillation spectrophotometer (Beckman LS 9000) using Cytoscint ES scintillation cocktail (ICN, Irvine, CA). A filter blank was subtracted from all samples. (B) ATP hydrolytic activity (nmol/min-mg protein). ATP

(with and without nigericin to dissipate the proton gradient). As shown in Fig. 5, the profile of methylamine uptake parallels the ATP hydrolytic activity. The activity declines rapidly on either side of the peak fractions. The uptake is sensitive to the addition of nigericin, indicating the uptake is due to a pH gradient and therefore the presence of a proton-pump ATPase. The protein concentration profile showed the highest concentrations in gradient fractions 11–14 (low glycerol concentrations), with progressively less protein in the lower gradient fractions (high glycerol concentrations).

Using this procedure with the vesicles obtained from the 0.4 *M* layer (see preceding text), the specific activity can be increased 69 ± 6% (1014–1507 pmol glutamate/mg protein to 1660–2635 pmol glutamate/mg protein) with an increase in ATP stimulation from 14.2 ± 0.7 to 31 ± 6-fold. The protein yield is 1–2% for the peak fraction. If the vesicles obtained from the 0.4 *M*/0.6 *M* interface layer are used, purification is also achieved, but to a less significant extent. The sedimentation coefficient in 0.7% sodium cholate for the proteins in the peak fraction is 3.8 S, using myoglobin (2.1 S), bovine serum albumin (4.3 S), and catalase (11.3 S) as the standards.

Coreconstitution of Vesicular Glutamate Transporter and Endogenous/Exogenous Proton Pumps

Coreconstitution of two separate solubilized protein preparations allows questions to be asked about the number and nature of the proteins required for synaptic vesicle glutamate transport. Specifically, if the glutamate transporter is structurally separate from the proton-pump ATPase, it would allow detection of the separated transporter by providing the proton pump necessary to generate electrochemical proton gradients. The proton pump could be endogenous or exogenous. If endogenous, an extract from glutamatergic synaptic vesicles would be required to be high in proton-pump ATPase activity, but to contain little or no ATP-dependent glutamate

---

hydrolysis was measured by a colormetric assay of released phosphate based on a malachite green/phosphomolybdate complex as described by P. A. Lanzetta, L. J. Alvarez, P. S. Reinsch, and O. A. Candia, *Anal. Biochem.* **100,** 95 (1979). Briefly, the reaction medium contained 20 m*M* HEPES (pH 7.4), 250 m*M* sucrose, 4 m*M* potassium chloride, 7 m*M* magnesium sulfate, 20–50 μg protein (in the form of proteoliposomes) and 5 m*M* ATP. The reaction was initiated by the addition of ATP. The reaction volume was 250–500 μl and the incubation was carried out at 37°. At several time points, a 50-μl aliquot was removed and assayed for phosphate. This was compared to a standard phosphate solution to determine the nanomoles of phosphate released. Protein was determined by the Lowry assay using bovine serum albumin as a standard (Reprinted with permission from O. H. Lowry, N. J. Rosebrough, A. L. Farr, and R. J. Randall, *J. Biol. Chem.* **193,** 265 [1951]).

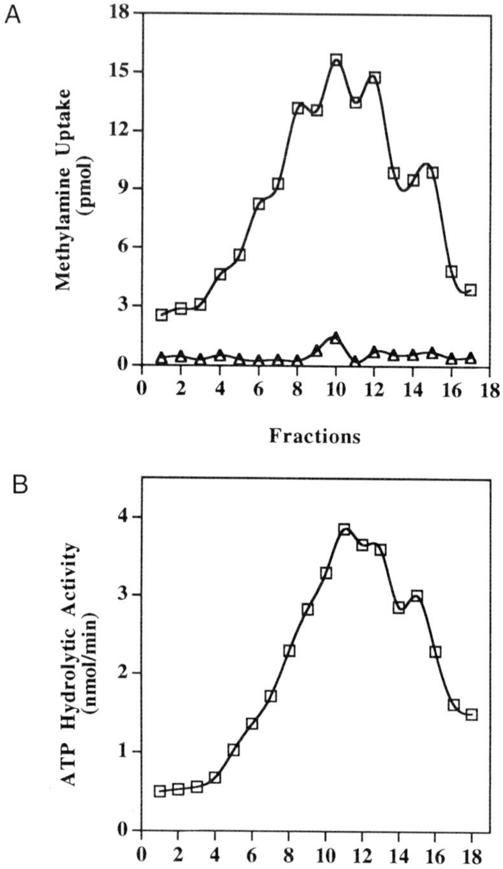

FIG. 5. Comparison of ATP hydrolytic activity with methylamine uptake with and without nigericin. Gradient fractions 1–18 represent 38–25% glycerol, respectively. (A) Methylamine uptake activity (pmol) normalized to fraction volume with (triangles) and without (squares) nigericin. Methylamine uptake was assayed in 10 m*M* HEPES (pH 7.4), 4 m*M* magnesium sulfate, 140 m*M* potassium chloride, 50 μM methylamine (containing 1 μCi ^{14}C-methylamine), 20–50 μg protein (in the form of proteoliposomes), and 2 m*M* ATP. The reaction was initiated by the addition of ATP and methylamine; the reaction volume was 120 μl. After incubation for 10 min at 37°, the reaction mixture was filtered as for the glutamate uptake mixture described in Fig. 4. A filter blank was subtracted from all values. The uptake in the presence of nigericin was performed using a final concentration of 0.15 μM. (B) ATP hydrolytic activity (nmol/min) normalized to fraction volume. ATP hydrolytic activity was assayed as in Fig. 4.

uptake activity. This section will detail techniques for coreconstitution of the vesicular glutamate transporter and a proton pump along with an example of its application to the detection of a putative separated glutamate transporter obtained from the glycerol gradient.

Reagents for Coreconstitution of Vesicular Glutamate Transporter and Endogenous/Exogenous Proton Pumps

Solubilization Solution C: 0.132 *M* KCl, 8.8 m*M* HEPES, 1.2% (w/v) CHAPS (3-[(3-cholamidopropyl)dimethylammonio]-1-propanesulfonic acid), 2% (w/v) crude soybean phospholipids (40% phosphatidylcholine), pH 7.4. The buffer is prepared by sonication (Branson sonicator with a microtip, model W185, 20 W) under nitrogen for 5–10 min.

Preparation of Endogenous Proton-Pump ATPase-Rich Extract. The preparation is obtained by solubilizing bovine synaptic vesicles with Solubilization Solution C according to the method described above. We found that CHAPS, while effective in extracting ATPase from synaptic vesicles, was less effective in extracting glutamate uptake activity (specific activity for ATP-dependent glutamate uptake, 218 ± 68 pmol/mg), extending the results of Carlson *et al.*[22] The protein yield is 11%. Extraction by alternate detergents (1.2%, w/v, sodium cholate, Triton X-100, $C_{12}E_9$, Brij 35, Triton CF-32, and Lubrol PX; 2.5 *M* urea) resulted in lower ratios of ATPase to glutamate transport activity. No detergent was found that extracted a glutamate transport-free ATPase from the vesicle. The following variations did not improve the purity of the CHAPS-solubilized ATPase-rich extract: sodium cholate–glycerol gradient centrifugation of the CHAPS extract, solubilization by 1.2% sodium cholate of the reconstituted CHAPS extract followed by sodium cholate–glycerol gradient centrifugation, or solubilization using 0.9% CHAPS in Solubilization Solution C. Fractionation, by sodium cholate–glycerol gradient centrifugation, of the sodium cholate-solubilized vesicular glutamate transport system did not yield a better ATPase-rich extract.

Recombination of Glycerol Gradient Fractions with Endogenous Proton-Pump ATPase-Rich Extract. The nonpeak (glutamate uptake and ATP hydrolytic activity) glycerol gradient fractions were screened for a separated glutamate transporter by coreconstitution with the proton-pump ATPase-rich CHAPS extract. Glycerol gradient fractions and CHAPs extract were combined by mixing the glycerol gradient fraction with the CHAPS extract (2 : 1, v/v) prior to the addition of Gradient Reconstitution Solution A as described above. Alternative ratios of 1 : 2, 1 : 1, and 5 : 1 were least effective, whereas a ratio of 3 : 1 was intermediate. The fractions were reconstituted as described and assayed for ATP-dependent glutamate uptake and ATP hydrolytic activity.

Application: Evidence for Partial Separation of Proton-Pump ATPase and Putative Glutamate Transporter. If a gradient fraction contains a separated putative glutamate transporter, reconstitution of that fraction with the ATPase-rich CHAPS extract would enhance the ATP-dependent glutamate uptake. Such fractions were found, corresponding to glycerol gradient fractions 3–5, representing 34–35% glycerol. The peak enhancement was most often found in fraction 4. This area was distinct from the peaks of ATP-dependent glutamate uptake and ATP hydrolysis activity. The glutamate uptake activity in the presence and absence of ATP, the ATP-dependent glutamate uptake, and the ATP hydrolytic activity are shown in Fig. 6 for fraction 4 and the CHAPS extract, alone or in combination. The ATP-dependent glutamate uptake is higher in the combination of fraction 4 and the CHAPS extract than in either alone, representing a 2.3 ± 0.3 (SEM)-fold enhancement in activity; the fold enhancement was defined as the ratio of the ATP-dependent glutamate uptake specific activity for the combination to the sum of the specific activity for fraction 4 and the CHAPS extract. As two specific activities cannot be added, the sum of activity for fraction 4 and the CHAPS extract was calculated by dividing the total ATP-dependent glutamate uptake (pmol) by the total protein (mg) for the two fractions forming an average specific activity. The enhancement is not seen in ATP-independent uptake, as shown in Fig. 6. This enhancement correlates with the addition of ATP hydrolytic activity. Resolution of fraction 4 is best if only two glycerol gradients are pooled, rather than three or more. Reconstitution of fractions 3–5 with a cholate (1.2%) extract of coated vesicles (as an alternative source of proton-pump ATPase) yielded similar results, but reconstitution with a cholate (1.2%) extract of chromaffin granules was not effective; the coated vesicles were prepared by a modification[27] of Nandi *et al.*,[28] and the chromaffin granules were obtained from th elaboratory of David Njus.

The enhanced activity is sensitive to the removal of chloride, consistent with the known properties of the native synaptic vesicle glutamate transport system[16,18,21,29] (see Table I). This indicates that the chloride stimulation of the native transporter is retained in the coreconstituted system. The chloride stimulation was observed in both fraction 4 and the CHAPS extract, but to a smaller extent. The removal of aspartate from the standard assay mixture had no effect, indicating that the activity observed is not due to the plasma membrane glutamate transporter. As shown in Table I, the

[27] X. Xie, D. K. Stone, and E. Racker, *J. Biol. Chem.* **259,** 11676 (1984).

[28] P. K. Nandi, G. Irace, P. P. Van Jaarsveld, R. E. Lippoldt, and H. Edelhoch, *Proc. Natl. Acad. Sci. USA* **79,** 5881 (1982).

[29] J. S. Tabb and T. Ueda, *J. Neurosci.* **11,** 1822 (1991).

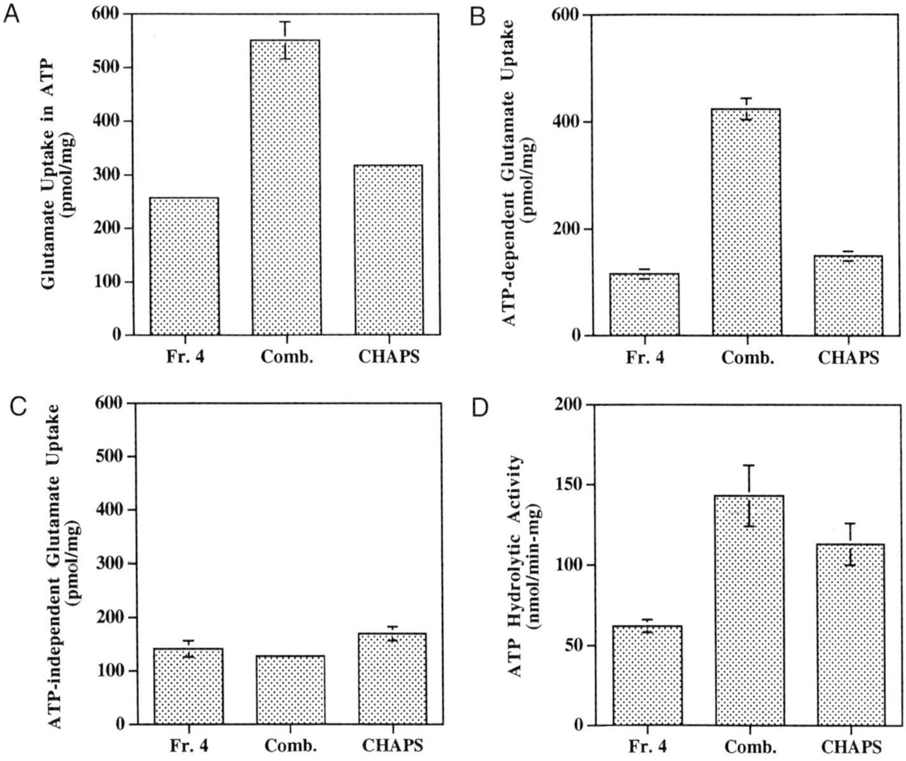

FIG. 6. Coreconstitution of a putative vesicular glutamate transporter fraction with an ATPase-rich extract. Glycerol gradient centrifugation fraction 4 and the CHAPS extract were assayed for glutamate uptake activity and ATP hydrolytic activity upon reconstitution, alone or in combination (see Fig. 4). (A) Glutamate uptake activity in the presence of ATP. (B) ATP-dependent glutamate uptake activity (activity in the presence of ATP minus the activity in the absence of ATP). (C) ATP-independent glutamate uptake activity. (D) ATP hydrolytic activity. (A)–(C) Representative experiment and the error is the range of duplicate results. The error in (D) is the standard error of the mean. Columns without error bars have errors below the plotting resolution.

ligand effects on the enhanced activity are identical to those known for the native synaptic vesicle glutamate transport system.[16,22,29,30] In addition, the glutamate uptake activity is sensitive to bafilomycin, a v-type ATPase inhibitor,[31] indicating a successful reconstitution with the proton-pump ATPase.

These data support the notion that the synaptic vesicle glutamate transporter and the associated proton-pump ATPase represent distinct proteins.

[30] H. C. Winter and T. Ueda, *Neurochemical Research* **18,** 79 (1992).

[31] E. J. Bowman, A. Siebers, and K. Altendorf, *Proc. Natl. Acad. Sci. USA* **85,** 7972 (1988).

TABLE I
EFFECTS OF CHLORIDE AND LIGANDS ON GLUTAMATE UPTAKE ENHANCEMENT

Sample	Concentration	Enhancement (percent of control)[a]
Control		
Omission		
Chloride	4 m*M*	2
Aspartate	2 m*M*	94
Additions		
trans-ACPD	1 m*M*	0
cis-ACPD	1 m*M*	75
γ-Methylglutamate	5 m*M*	19
Glutamine	5 m*M*	92
α-Ketoglutarate	5 m*M*	85
L-α-Aminoadipate	5 m*M*	75
Glutaric acid	5 m*M*	110
Bafilomycin	1 μM	0
Oligomycin	1 μM	97
Vanadate	5 m*M*	76

[a] The listed agents were omitted from or added to the standard assay mixture described in Fig. 4. The control enhancement was 1.52 ± 0.08 (defined as the ATP-dependent glutamate uptake (pmol) of the combination divided by the sum of the ATP-dependent glutamate uptake (pmol) of fraction 4 and the CHAPS extract normalized to fraction volume). The addition of 5 m*M* *N*-methyl-D-aspartate, kainate, or quisqualate had no effect. ACPD, 1-aminocyclopentane-1,3-dicarboxylate.

The fraction identified as containing the putative glutamate transporter occurs at a higher glycerol density in the gradient than the unseparated system. It is conceivable that this fraction represents an aggregated form of the transporter, to which a small amount of proton-pump ATPase is attached as a contaminant, and that all or part of this aggregate is available for functional coreconstitution with a proton-pump ATPase. For a given experiment, this fraction coincides with a small peak of ATP-independent glutamate uptake. Analysis of fractions 3–5 by sodium dodecyl sulfate–polyacrylamide gel electrophoresis (SDS–PAGE), revealed many proteins, two of which were enriched. These had apparent molecular weights of 44,000 and 60,000.

As seen in Fig. 6, the separation is not complete. No fraction, completely devoid of ATP-dependent glutamate uptake, was found that demonstrated ATP-dependent uptake upon combination with a proton-pump ATPase. Attempts to produce such fractions, by varying the gradient detergent (cholate vs Triton X-100), the gradient detergent concentration (0.7% vs 1.2% sodium cholate), the gradient glycerol concentration (10–45%, v/v), the gradient centrifugation time (2–14 hr), the solubilizing buffer protein concentration (1 mg/ml, 10 mg/ml, and 25 mg/ml), and the nature of the vesicles prior to solubilization (cold inactivation[32] of the ATPase), did not result in improved functional reconstitution or improved enhancement. This may reflect the labile nature of the transporter, in a form dissociated from the proton-pump ATpase or from its native lipid environment, due to its interaction with certain detergents.

Our observations are in accord with other structurally characterized vesicular neurotransmitter transport systems,[33] such as the monoamine[34–36] and acetylcholine[37–39] transport systems. In each case the transporter is separate from the proton-pump ATPase.[40–42] The observations are also compatible with the clathrin-coated vesicle chloride transport system.[43] Figure 6 represents evidence for a structural separation of the protein components of the synaptic vesicle glutamate transport system.

Coreconstitution of Vesicular Glutamate Transport System and Exogenous Proton Pump. Maycox *et al.*[23] reported coreconstitution of the rat synaptic vesicle glutamate transport system with monomeric bacteriorhodopsin or bacteriorhodopsin liposomes. For monomeric bacteriorhodopsin, 30 μl of the monomer solution is added to 150 μl of solubilized glutamate transporter/brain phospholipid mixture. This is reconstituted by incubation at 32° for 5 min followed by dilution into 9 ml of the assay buffer. For bacteriorhodopsin, 40 μl of a bacteriorhodopsin liposome suspension are added to 150 μl of the solubilized glutamate transporter/brain phospholipid mixture. This is frozen in liquid nitrogen, stored on dry ice 5 min, and thawed slowly on ice. The cycle is repeated once, followed by dilution into 9 ml of assay buffer. The 32° incubation is omitted. Details concerning the preparation of bacteriorhodopsin can be found in Maycos *et al.*[23] Comparisons to the technique with the endogenous proton-pump ATPase are difficult as both approaches yielded functionally active glutamate transport, but are designed to ask two different questions. Maycox *et al.*[23] examined

[32] Y. Moriyama and N. Nelson, *J. Biol. Chem.* **264,** 3577 (1989).
[33] S. Schuldiner, A. Shirvan, and M. Linial, *Physiological Reviews* **75,** 369 (1995).
[34] Y. Stern-Bach, J. N. Keen, M. Bejerano, S. Steinermordoch, M. Wallach, J. B. C. Findley, and S. Schuldiner, *Proc. Natl. Acad. Sci. USA* **89,** 9730 (1992).
[35] Y. Liu, D. Peter, A. Roghani, S. Schuldiner, G. G. Prive, D. Eisenberg, N. Brecha, and R. H. Edwards, *Cell* **70,** 539 (1992).
[36] J. D. Erickson, L. E. Eiden, and B. J. Hoffman, *Proc. Natl. Acad. Sci. USA* **89,** 10993 (1992).
[37] D. C. Anderson, S. C. King, and S. M. Parsons, *Biochemistry* **21,** 3037 (1982).
[38] B. A. Bahr and S. M. Parsons, *Biochemistry* **31,** 5763 (1992).
[39] A. Alfonso, K. Grundahl, J. S. Duerr, H. Han, and J. B. Rand, *Science* **261,** 617 (1993).
[40] S. Cidon and N. Nelson, *J. Biol. Chem.* **261,** 9222 (1986).
[41] Y. Moriyama and N. Nelson, *J. Biol. Chem.* **262,** 1 (1987).
[42] E. Floor, P. S. Leventhal, and S. F. Schaeffer, *J. Neurochem.* **55,** 1663 (1990).
[43] X. Xie, B. P. Crider, and D. K. Stone, *J. Biol. Chem.* **264,** 18870 (1989).

whether the glutamate transporter can be driven by an alternative proton pump. The study reported in this chapter asked whether the glutamate transporter can be structurally separated from the proton-pump ATPase.

Acknowledgments

This work was supported by NIH Grant NS 26884, Javits Neuroscience Investigator Award (TU), and NIMH Training Grant 1 F32 NS09737-01 and Postdoctoral Fellowship NS09737-02 (SML).

[10] Analysis of Neurotransmitter Transport into Secretory Vesicles

By J. PATRICK FINN III, ANDREW MERICKEL, and ROBERT H. EDWARDS

Introduction

Two distinct classes of neurotransmitter transport activities participate in synaptic transmission. Transport activities located at the plasma membrane function to remove transmitter from the synapse and so terminate signaling. They use the inwardly directed Na^+ gradient to drive cotransport of the transmitter.[1,2] In the case of monoamines, drugs that block plasma membrane transport include cocaine and antidepressants such as fluoxetine. In contrast, vesicular transport packages transmitter so that it can be released by regulated exocytosis and relies on a proton electrochemical gradient generated by a vacuolar H^+-ATPase.[3,4] In particular, the vesicular transport of monoamines involves the exchange of two lumenal protons for one molecule of cytoplasmic transmitter.[5,6] Vesicular monoamine transport also shows inhibition by reserpine and tetrabenazine rather than cocaine or antidepressants. In addition to these bioenergetic and pharmacologic differences, the two classes of transport activity reside on different membrane compartments.

[1] S. G. Amara and M. J. Kuhar, *Ann. Rev. Neurosci.* **16,** 73 (1993).
[2] B. I. Kanner, *J. Exp. Biol.* **196,** 237 (1994).
[3] R. H. Edwards, *Curr. Opin. Neurobiol.* **2,** 586 (1992).
[4] S. Schuldiner, A. Shirvan, and M. Linial, *Physiol. Ref.* **75,** 369 (1995).
[5] J. Knoth, M. Zallakian, and D. Njus, *Biochemistry* **20,** 6625 (1981).
[6] R. G. Johnson, *Physiol. Ref.* **68,** 232 (1988).

0076-6879/98 $25.00

The subcellular localization of neurotransmitter transport activities influences the methods involved in their functional characterization. Since the plasma membrane transport activities remove transmitter from the extracellular space, they can be measured using whole cells, synaptosomes, or plasma membrane vesicles. In contrast, vesicular transport occurs into intracellular membranes and hence cannot be studied using intact cells or synaptosomes. Rather, it must be studied using membrane vesicles prepared from cells or neural tissue. Although artificial ionic gradients can be used to drive transport, the presence of a vacuolar H^+-ATPase in the membrane vesicles can also provide the driving force.

Classical studies of vesicular transport have relied on the availability of tissues expressing robust activity. The bioenergetic analysis of vesicular amine transport has largely involved chromaffin granules purified from the bovine adrenal medulla.[6–8] However, brain synaptic vesicles[9,10] and platelets[11,12] also exhibit vesicular monoamine transport activity. The analysis of vesicular acetylcholine (ACh) transport has used synaptic vesicles prepared from the electric organ of *Torpedo*.[13,14] The study of amino acid transport has largely involved synaptic vesicles from the rat brain.[15–21] However, the use of native tissues may complicate the analysis of transport activity. In the case of monoamines, bovine chromaffin granules contain extremely high concentrations of transmitter (>1 *M*), which complicates the interpretation of transport assays. To circumvent this problem, investigators have subjected the granules to repeated osmotic lysis, producing chromaffin ghosts. Heterologous expression of the molecularly cloned transport proteins in a range of cell types now provides an additional way to circumvent this as well as other limitations of transport analysis using native tissues.

[7] A. Carlsson, N. A. Hillarp, and B. Waldeck, *Acta Physiol. Scand.* **264,** 7369 (1963).
[8] N. Kirschner, *J. Biol. Chem.* **237,** 2311 (1962).
[9] D. Scherman, *J. Neurochem.* **47,** 331 (1986).
[10] Y. Moriyama, K. Amakatsu, and M. Futai, *Arch. Biochem. and Biophys.* **305,** 271 (1993).
[11] J. M. Rabey, A. Lerner, M. Sigal, E. Graff, and Z. Oberman, *Life Sciences* **50,** 65 (1992).
[12] D. Chatterjee and G. M. Anderson, *Arch. Biochem. and Biophys.* **302,** 439 (1993).
[13] D. C. Anderson, S. C. King, and S. M. Parsons, *Biochemistry* **21,** 3037 (1982).
[14] S. M. Parsons, C. Prior, and I. G. Marshall, *Int. Rev. Neurobiol.* **35,** 279 (1993).
[15] S. Naito and T. Ueda, *J. Biol. Chem.* **258,** 696 (1983).
[16] P. R. Maycox, T. Deckwerth, J. W. Hell, and R. Jahn, *J. Biol. Chem.* **263,** 15423 (1988).
[17] E. M. Fykse and F. Fonnum. *J. Neurochem.* **50,** 1237 (1988).
[18] J. W. Hell, P. R. Maycox, H. Stadler, and R. Jahn, *EMBO J.* **7,** 3023 (1988).
[19] M. D. Carlson, P. E. Kish, and T. Ueda, *J. Neurochem.* **53,** 1889 (1989).
[20] P. E. Kish, C. Fischer-Bovenkerk, and T. Ueda, *Proc. Natl. Acad. Sci. USA* **86,** 3877 (1989).
[21] P. M. Burger, J. Hell, E. Mehl, C. Krasel, F. Lottspeich, and R. Jahn, *Neuron* **7,** 287 (1991).

Molecular cloning of vesicular transporters for monoamines (VMAT1 and 2)[22–24] and ACh (VAChT)[25–27] has enabled their expression in heterologous cell systems. However, the measurement of transport activity requires expression on intracellular vesicles capable of generating a proton electrochemical gradient, such as synaptic vesicles. Although the standard heterologous expression systems involve nonneural cell lines such as COS cells that do not contain synaptic vesicles, both the VMATs and VAChT sort to endosomes when expressed in these cells.[22,28] Like other synaptic vesicle proteins, the transporters contain signals for internalization that are recognized by a wide range of cell types. Endosomes also contain the vacuolar H^+-ATPase and so provide the appropriate driving force for vesicular neurotransmitter transport. Thus, heterologous expression of the VMATs in nonneural cells confers transport activity. This circumvents any problems associated with endogenous transmitter and facilitates the analysis of mutant transport proteins. However, VAChT shows essentially no activity when expressed in nonneural cells.[26] Rather, overexpression in a neuroendocrine cell line such as rat pheochromocytoma PC-12 cells appears required for VAChT function,[29] possibly because of higher level expression, localization to a more appropriate vesicular compartment such as synaptic vesicles, or association with a neural-specific protein. For simplicity, we describe the methods used by our group to characterize the function of VMATs in nonneural cells. Nonetheless, these can be adapted with little difficulty to the analysis of vesicular transport of other neurotransmitters using membranes prepared from other cells or tissues.

Glyoxylic Acid-Induced Fluorescence

In a qualitative assay for vesicular transport in intact cells, we have taken advantage of the intrinsic fluorescence of dopamine. Classical staining for monoamines used formaldehyde to induce their fluorescence, but glyox-

[22] Y. Liu, A. Roghani, and R. H. Edwards, *Proc. Natl. Acad. Sci. USA* **89,** 9074 (1992).

[23] Y. Liu, D. Peter, A. Roghani, S. Schuldiner, G. G. Prive, D. Eisenberg, N. Brecha, and R. H. Edwards, *Cell* **70,** 539 (1992).

[24] J. D. Erickson, L. E. Eiden, and B. J. Hoffman, *Proc. Natl. Acad. Sci. USA* **89,** 10993 (1992).

[25] A. Roghani, J. Feldman, S. A. Kohan, A. Shirzadi, C. B. Gundersen, N. Brecha, and R. H. Edwards, *Proc. Natl. Acad. Sci. USA* **91,** 10620 (1994).

[26] J. D. Erickson, H. Varoqui, M. D. Schafer, W. Modi, M. F. Diebler, E. Weihe, J. Rand, L. E. Eiden, T. I. Bonner, and T. B. Usdin, *J. Biol. Chem.* **269,** 21929 (1994).

[27] S. Bejanin, R. Cervini, J. Mallet, and S. Berrard, *J. Biol. Chem.* **269,** 21944 (1994).

[28] Y. Liu, E. S. Schweitzer, M. J. Nirenberg, V. M. Pickel, C. J. Evans, and R. H. Edwards, *J. Cell Biol.* **127,** 1419 (1994).

[29] H. Varoqui and J. D. Erickson, *J. Biol. Chem.* **271,** 27229 (1996).

ylic acid provides a much simpler alternative.[30,31] In cells that contain monoamines, the staining usually indicates sites of transmitter storage. However, the presence of endogenous transporters in cells that synthesize monoamines complicates the analysis of introduced sequences. Thus, we have used the method in transfected fibroblast cells. Since these cells do not synthesize dopamine, however, they must be incubated in dopamine. Further, high concentrations of dopamine (1 m*M*) are required since the cells also lack high-affinity monoamine reuptake systems. At these high concentrations, dopamine enters the cells by nonspecific, low-affinity uptake. In the cytoplasm, vesicular transporters expressed in the cells can then package the dopamine into intracellular (probably endocytic) vesicles. Thus, cells that express the VMATs show strong particulate fluorescence for catecholamine using glyoxylic acid-induced fluorescence, whereas cells that do not express the VMATs show diffuse cytoplasmic staining.[22,23]

Procedure

1. Grow cells on poly(L-lysine)-coated coverslips in the presence of 1 m*M* dopamine for 6–18 hr. This concentration of dopamine is toxic for many cultured cells and particularly neural cells, but CHO cells tolerate it reasonably well for 8–12 hr in standard medium with serum.
2. Wash each coverslip three times with 0.1 *M* sodium phosphate, pH 7.4.
3. Incubate each coverslip at 4° in 2% (w/v) glyoxylic acid and 0.0005% (w/v) $MgCl_2$, pH 4.9–5.0 for 3 min.
4. After draining thoroughly, dry each coverslip at 45° with a hair dryer, heat rapidly to 80° in an oven for 5 min, and examine under epifluorescence.

Comments

Brilliant blue fluorescence with an excitation peak at 410 nm and a primary emission peak at 480 nm is characteristic of catecholamines treated with glyoxylic acid.[30] Serotonin also fluoresces after glyoxylic acid treatment, but exhibits excitation and emission wavelengths distinct from catecholamines. However, the availability of specific antibodies to serotonin enables simple immunostaining of fixed tissue.

Vesicular Transport Activity

Vesicular monoamine transport uses a proton electrochemical gradient to concentrate transmitter at levels vastly exceeding those in the cytoplasm

[30] K. M. Knigge, G. Hoffman, D. E. Scott, and J. R. Sladek, *Brain Research* **120,** 393 (1977).
[31] J. C. de la Torre, *J. Neurosci. Meth.* **3,** 1 (1980).

(10^4–10^5 : 1 vesicle : cytoplasm). However, the mechanism by which proton translocation drives substrate recognition on the cytoplasmic face of the vesicle, translocation across the membrane, release into the lumen, and reorientation back to the cytoplasm remains unknown. In one model of the transport cycle, either the energy of binding or perhaps the outward translocation of one proton induces the movement of substrate across the vesicle membrane. A reduction in affinity perhaps related to the acidic environment then releases the transmitter into the lumen. Outward translocation of a second proton then reorients the substrate recognition site back to the cytoplasm so that it can bind another molecule of transmitter. This model derives from previous studies using assays of transport and drug binding by chromaffin ghosts. We have now adapted these assays to heterologous expression systems, enabling us to determine the structural basis for vesicular transport in molecular detail using site-directed mutants. The availability of two closely related VMAT isoforms that differ in their functional characteristics also facilitates the molecular analysis of transporter function through the production of first chimeras[32,33] and then point mutants.[34] To examine transport activity or drug binding, membrane vesicles are prepared from an appropriate heterologous expression system—in the case of the VMATs, we use COS1 cells transiently expressing the appropriate cDNA. Alternatively, other investigators have used viral expression systems, most notably a vaccinia virus system requiring transfection of the cDNA under the control of a T7 promoter and simultaneous infection with vaccinia virus encoding the T7 RNA polymerase.[24]

Transient Expression

1. Grow COS1 cells to confluence in Dulbecco's modified Eagle's medium (DMEM) containing 10% calf serum at 37° in 5% CO_2. Feed the cells with fresh DMEM containing serum 24 hours before electroporation.
2. Rinse the cells with calcium/magnesium-free phosphate-buffered saline (CMF-PBS), detach them with trypsin, pellet them in DMEM containing serum by centrifugation, and resuspend each 15-cm plate in 800 μl of prewarmed PBS containing calcium and magnesium.
3. Transfer the cell suspension to a 0.4-cm gap cuvette (Bio-Rad, Richmond, CA) containing 15 μg of DNA, electroporate immediately at 0.4 kV, 960 microfarads (μF), and replate the cells from a single 15-cm plate onto another 15 cm-plate in fresh DMEM-containing serum. (Even with the death of up to 50% of the cells during this procedure, the cells do become

[32] D. Peter, J. Jimenez, Y. Liu, J. Kim, and R. H. Edwards, *J. Biol. Chem.* **269,** 7231 (1994).
[33] D. Peter, T. Vu, and R. H. Edwards, *J. Biol. Chem.* **271,** 2979 (1996).
[34] J. P. Finn III and R. H. Edwards, *J. Biol. Chem.* **272,** 16301 (1997).

overgrown over the next 2–4 days, but this does not interfere with the expression of transport activity.)

4. Two days later, refeed the cells with fresh DMEM containing serum since the large number of cells depletes nutrients rapidly.

5. On the third day after electroporation, wash the transfected COS1 cells in CMF-PBS, detach them from the plate with trypsin, and collect the cells by centrifugation. Although expression in COS cells peaks any time between 2 and 4 days after transfection, we usually prepare membranes on the third day. Changing the medium the day before appears important for maximal activity.

Membrane Preparation

1. Resuspend each pellet of transfected COS1 cells from a 15-cm plate in 360 μl cold 0.32 *M* sucrose, 10 m*M* HEPES–KOH, pH 7.4 (SH buffer) containing 2 μg/ml aprotinin, 2 μg/ml leupeptin, 1 μg/ml E-64 [*trans*-epoxysuccinyl-L-leucylamido(4-guanido)butane], 1 μg/ml pepstatin A, and 0.2 m*M* diisopropyl fluorophosphate.

2. Sonicate the cell suspension in an ice-cooled water bath sonicator at medium intensity for 20 1-sec pulses. Alternatively, homogenize the cells with a 0.01-mm clearance cell homogenizer by passing the cell suspension through the clearance 10 times. However, we prefer sonication over homogenization because sonication enables the cells to be resuspended in a smaller volume, resulting in a higher concentration of membranes. In addition, sonication allows many more samples to be prepared in a shorter time. However, cell biological studies of localization by density gradient centrifugation should use the cell homogenizer rather than the sonicator since sonication disrupts membrane vesicles.

3. Remove the cell debris by centrifugation at 16,000*g* for 5 min at 4°.

4. Collect the resulting supernatant and divide it into aliquots for storage at −80°. Membranes prepared in this manner can be maintained for up to 6 months with no loss of transport activity.

Comments

The use of semipermeabilized cells provides an alternative to membrane preparation in the analysis of vesicular neurotransmitter transport. Appropriate treatment with a detergent such as digitonin (10 μM for 10 min) can selectively permeabilize the plasma membrane of cultured cells while not significantly disturbing the integrity of intracellular membranes.[24] Cells treated in this way can therefore exhibit uptake while still adherent. However, the assay depends on the presence of equal numbers of cells in different wells, their continued adherence throughout the assay (requiring pretreatment of the wells with an appropriate matrix such as collagen), and efficient recovery of the retained radioactivity. Although the affinities reported using this preparation appear severalfold lower than those measured using standard membranes, the ability to study multiple mutants simultaneously in a multiwell plate can provide an advantage for large-scale screening.

Uptake Assay

1. Either use fresh membranes or thaw a frozen aliquot of membranes on ice. The standard reaction solution contains SH buffer with 4 m*M* KCl, 2 m*M* $MgSO_4$, 2.5 m*M* ATP, and 20 n*M* radiolabeled substrate, such as $[^3H]$serotonin, $[^3H]$dopamine, $[^3H]$epinephrine, $[^3H]$norepinephrine, or $[^3H]$histamine (NEN, Boston, MA). For each time point, add 10 μl membranes (50–100 μg protein) to 200 μl of reaction solution that has been prewarmed to 29° in 12 × 75-mm polypropylene culture tubes.
2. Incubate the reaction solution plus membranes for varying intervals at 29°. We use this temperature rather than 37° because it generally shows a higher signal to background, possibly as a result of increased membrane leakiness at higher temperatures.
3. Terminate the transport reaction by adding 1.5 ml cold SH buffer to the reaction mixture and filter rapidly through 0.2 μm Supor 200 membrane filters (Gelman, Ann Arbor, MI) presoaked in water. Wash the reaction tube with an additional 1.5 ml cold SH buffer and rapidly filter this material as well. The filtration step takes a total of 20 sec. If a weak vacuum significantly prolongs the filtration step, this may increase variability in the results due to the increased efflux of transmitter.
4. Dry the filters and measure the bound radioactivity by scintillation counting in 2.5 ml scintillation fluid designed for small-volume aqueous solutions (e.g., CytoScint, ICN, Costa Mesa, CA).

Comments

Membranes prepared from untransfected cells indicate the true background of the transport assay. However, reserpine and tetrabenazine can also be used to inhibit transport and so indicate the background. Reserpine (2–10 μM, made from a stock prepared in 1 *M* acetic acid) potently inhibits both VMAT isoforms ($IC_{50} \sim 0.1\ \mu M$). Although it can bind essentially irreversibly to the VMATs, binding takes several minutes and the acute effect of reserpine probably involves reversible, competitive inhibition. This is very important because we do not usually preincubate the membranes with the inhibitors. Determination of the true background thus depends on rapid inhibition at the start of the assay. In addition, tetrabenazine

(2–10 μM, prepared from a stock made in 4 mM HCl) potently inhibits the VMAT2 isoform ($IC_{50} \sim 0.5\ \mu M$) but not the VMAT1 isoform ($IC_{50} \sim 4.2\ \mu M$).

To determine the time course for uptake, we commonly use 0-, 0.5-, 1-, 2-, 5-, and 10-min incubations with the zero time point performed at 0°, rather than 29°. For all experiments, each time point is performed in duplicate and the uptake value for zero time at 0° subtracted as background. Plotting picomoles of radioactive substrate/mg protein versus time in minutes indicates the time interval during which transport remains linear. To determine K_m and V_{max}, nonradioactive substrate is added to the reaction solution containing 20 nM radioactive substrate at concentrations ranging from 100 nM to several micromolar. For accuracy, K_m and V_{max} values are determined using an incubation time within the linear range of transport. The protein concentration of each membrane sample is determined by protein assay and the K_m and V_{max} values calculated from analysis of a Lineweaver–Burke plot. We find that measurements made in duplicate on at least three separate occasions using at least two different membrane preparations produce highly consistent results.

Bioenergetic Considerations

Ionophores such as carbonylcyanide *m*-chlorophenylhydrazone (CCCP), nigericin, and valinomycin inhibit vesicular neurotransmitter transport. CCCP is a proton ionophore that inhibits vesicular transport by dissipating the proton electrochemical gradient ($\Delta\mu_{H^+}$). Nigericin is an ionophore that exchanges Na^+ and H^+ and so dissipates the pH component (ΔpH) of $\Delta\mu_{H^+}$ to a greater extent than the electrical component ($\Delta\Psi$). In contrast, valinomycin is a potassium ionophore that dissipates $\Delta\Psi$ more than ΔpH. Thus, stronger inhibition by nigericin than valinomycin suggests an activity more dependent on ΔpH than $\Delta\Psi$ and stronger inhibition by valinomycin suggests a $\Delta\Psi$-driven process. Importantly, this and other methods have been used to demonstrate the dependence of vesicular monoamine and ACh transport principally on ΔpH and of vesicular glutamate and to a lesser extent GABA transport on $\Delta\Psi$. In terms of the concentrations, we use 5 μM CCCP, 5 μM nigericin, and 20 μM valinomycin, dissolving each in ethanol. However, dissipation of the proton electrochemical gradient will also reduce nonspecific trapping of monoamines in acidic compartments, making them unsuited to assess the uptake actually due to expression of the transport proteins.

The conditions of the reaction can also help to reveal the bioenergetics of transport. Permeant anions such as thiocyanate and, in many cases, chloride neutralize the charge of protons entering the vesicle and so enable development of a greater ΔpH at the expense of ΔΨ. Thus, the use of these anions or 150 m*M* KCl enhances reactions dependent on ΔpH and abolishes those dependent exclusively on ΔΨ. Conversely, buffers containing sucrose[23] or potassium tartrate,[24] an impermeant anion, will maximize ΔΨ. In practice, however, vesicular amine transport occurs at similar, high rates in both 150 m*M* KCl and HEPES-buffered sucrose, presumably because this activity can use either component of the proton electrochemical gradient. Importantly, vesicular glutamate transport requires low levels of Cl^- but whether this relates to an allosteric requirement or a small dependence on ΔpH remains unclear.[15,16,19,35,36]

Since the active transport of neurotransmitter couples to the proton electrochemical driving force, changes in $\Delta\mu_{H^+}$ can also be used to measure activity. Although an indirect measure of transport function, Jahn and colleagues have used these elegant fluorimetric methods to characterize the bioenergetics of amino acid transport into synaptic vesicles. An increase in ΔpH quenches the fluorescence of acridine orange by trapping it within the vesicles. Thus, the accumulation of glutamate within synaptic vesicles, which consumes predominantly ΔΨ and thus increases ΔpH, quenches acridine orange fluorescence and so provides another index of transport activity. Similarly, dyes such as oxonol V can be used to monitor changes in ΔΨ. Since these methods rely on the consumption of one component of the proton electrochemical gradient and an increase in the other, they are probably limited to the measurement of those activities such as vesicular glutamate transport that are driven largely by one or the other but not both components. In addition, the methods rely on a high turnover number because unlike the uptake of radiolabeled tracers, they measure total changes in vesicle content. Perhaps most important, the methods rely on a relatively pure population of vesicles since the presence of contaminating vesicles will obscure changes in ΔpH and ΔΨ.

Drug Binding

Reserpine

In addition to their use as specific inhibitors of transport activity, drugs such as reserpine provide powerful tools to measure transporter number and hence turnover, characterize the substrate recognition site, assess protein folding, and determine coupling to the proton electrochemical gradient. [^{3}H]Reserpine binds essentially irreversibly to both VMATs and has even

[35] J. Hartinger and R. Jahn, *J. Biol. Chem.* **268,** 23122 (1993).
[36] J. S. Tabb, P. E. Kish, R. Van Dyke, and T. Ueda, *J. Biol. Chem.* **267,** 15412 (1992).

been used to monitor purification of the solubilized protein.[37] Importantly, the presence of a proton electrochemical gradient accelerates reserpine binding, providing a measure of coupling to the driving force even in the absence of transport activity.[38,39] Presumably, the translocation of one proton out of the vesicle reorients the substrate recognition site to the cytoplasmic face of the membrane where it can bind to another molecule of transmitter. Indeed, we have shown stimulation of [^{3}H]reserpine binding by $\Delta\mu_{H^+}$ in mutants incapable of active transport, providing evidence of intact bioenergetic coupling as well as normal expression levels.[40] Monoamines inhibit ^{3}H-reserpine binding to the transporters at concentrations close to their K_m, further suggesting that reserpine binds at the site of substrate recognition.[38,41] Interestingly, histamine inhibits reserpine binding very poorly, suggesting that histamine interacts at a site distinct from other monoamines and reserpine.[42] Similarly, a series of mutations that eliminate transport activity show normal levels of reserpine binding but no inhibition of reserpine binding by serotonin, indicating a selective defect in substrate recognition.[40] In the absence of transport activity, a normal level of ^{3}H-reserpine binding also suggests proper protein folding and expression. The irreversible binding of reserpine also enables the determination of transporter number and hence the measurement of substrate turnover. Since values obtained with both reserpine and tetrabenazine are similar, the drugs presumably bind with a similar stoichiometry.[32] Last, the preservation of reserpine binding but absence of acceleration by $\Delta\mu_{H^+}$ indicates a lack of coupling to the driving force.

Procedure

1. For each time point, add 10 μl membranes (50–100 μg protein) to 200 μl SH buffer at 29° containing 4 mM KCl, 4 mM $MgSO_4$, 5 mM ATP, and 2 nM [^{3}H]reserpine. We commonly use a 5-min incubation for studies of competition, a 30-min incubation to determine transporter protein number, and 0-, 2-, 5-, 15-, and 30-min incubations to assess the acceleration of binding by $\Delta\mu_{H^+}$.

2. After incubation at 29°, separate the free reserpine from the reserpine bound to vesicular membranes by centrifugation (4000g) for 2 min at room

[37] Y. Stern-Bach, N. Greenberg-Ofrath, I. Flechner, and S. Schuldiner, *J. Biol. Chem.* **265,** 3961 (1990).

[38] J. H. Weaver and J. D. Deupree, *Eur. J. Pharmacol.* **80,** 437 (1982).

[39] D. Scherman and J. P. Henry, *Mol. Pharmacol.* **25,** 113 (1984).

[40] A. Merickel, P. Rosandich, D. Peter, and R. H. Edwards, *J. Biol. Chem.* **270,** 25798 (1995).

[41] S. Schuldiner, Y. Liu, and R. H. Edwards, *J. Biol. Chem.* **268,** 29 (1993).

[42] A. Merickel and R. H. Edwards, *Neuropharmacol.* **34,** 1543 (1995).

temperature through a prepacked 1.5-ml column of Sephadex LH-20 (Pharmacia LKB Biotechnology Inc., Piscataway, NJ). The extremely hydrophobic nature of reserpine and the consequent nonspecific adherence to many materials mandate this particular form of separation (rather than simple filtration). Equilibrate the Sephadex LH-20 in SH buffer for 12 to 18 hr and pack by centrifugation (4000*g*) for 1.5 min immediately before use.

3. Measure the radioactivity in the excluded volume by scintillation counting in 5 ml of scintillation fluid designated for moderate to large volume aqueous solutions (e.g., EcoLume, ICN).

Comments

Nonspecific binding is measured in the presence of 2 μM unlabeled reserpine and subtracted as background. Since reserpine binding is accelerated by the presence of an electrochemical gradient, the addition of 5 μM of the proton ionophore CCCP to a solution containing the ATP necessary to activate the vacuolar H^+-ATPase dramatically slows reserpine binding. Further, we recommend that the 1.5-ml columns be prepared using a 3 ml syringe (Monoject) with a 8.5-mm diameter disc cut from 18 × 18 × 1/16 PE POR SH70 filter sheets (Bel-Art Products, Pequannock, NJ) since these materials show less nonspecific reserpine binding.

To calculate turnover number, we measure V_{max} and divide this by the number of transporters in the preparation as measured by drug binding. However, it is very important to measure both the activity and the number of transporters with small amounts of protein so that both assays remain in the linear range. The turnover number for both VMATs is ~400/min.

Additional Drugs

Monoamine Transport: Tetrabenazine. Like reserpine, tetrabenazine provides a powerful tool to assess protein folding and the level of expression. However, the recognition site for tetrabenazine differs from that for reserpine. First, $\Delta\mu_{H^+}$ does not stimulate the binding of [^{3}H]dihydrotetrabenazine, indicating that the binding of this drug does not reflect bioenergetic coupling. Second, monoamine substrates inhibit the binding of [^{3}H]dihydrotetrabenazine at only very high concentrations, suggesting that this drug does not interact at the site of substrate recognition.[43] In addition, tetrabenazine does not inhibit VMAT1 transport.[32,44] Interestingly, histamine inhibits tetrabenazine but not reserpine binding, suggesting that tetrabenazine may interact at the site for recognition of histamine but not other monoamines.[42]

[43] F. Darchen, E. Scherman, and J. P. Henry, *Biochemistry* **28,** 1692 (1989).

[44] J. D. Erickson, M. K. H. Schafer, T. I. Bonner, L. E. Eiden, and E. Weihe, *Proc. Natl. Acad. Sci. USA* **93,** 5166 (1996).

ACh Transport: Vesamicol. Vesamicol potently inhibits the transport of ACh into synaptic vesicles. Although previous studies had suggested that [^{3}H]vesamicol might bind to a protein distinct from the transporter, molecular cloning has demonstrated that the drug binds directly to the transport protein.[26,29] Thus, vesamicol can be used to estimate the number of ACh transporters. However, it may not bind to the site of substrate recognition and has not been shown to be sensitive to bioenergetic coupling.[14]

Amino Acid Transport. Unfortunately, specific and potent inhibitors of the vesicular GABA and glutamate transporters have not been identified. The stilbene inhibitor DIDS can block glutamate transport, apparently by interacting at the site of Cl^- recognition, but clearly interacts with a wide range of other proteins as well.[35]

Procedure

1. Prepare the standard reaction solutions described above containing 2 n*M* [^{3}H]dihydrotetrabenazine (Amersham) in place of the radiolabeled substrate.
2. To assess the inhibition of tetrabenazine binding by substrates, we first determine the time course of binding. Add 50–100 μg membrane protein to 200 μl of the reaction solution and incubate the reactions at 29° for 1, 2, 5, 10 and 20 min. After performing this time course, use an interval during which binding remains linear for competition studies. To measure total binding, incubate for a longer time such as 30 min.
3. Terminate each reaction by rapid dilution with 4 ml cold SH buffer containing 125 μM unlabeled tetrabenazine and filter through 0.2-μm Supor 200 membrane filters presoaked in SH buffer also containing 125 μM unlabeled tetrabenazine. This unlabeled tetrabenazine helps reduce the sticky, nonspecific binding of [^{3}H]dihydrotetrabenazine. Rinse the tube with an additional 4 ml cold SH buffer and filter this rapidly as well.

Comments

Nonspecific binding is measured by inhibition in the presence of 100 μM unlabeled tetrabenazine and subtracted as background. As control, we also use membranes prepared from cells not expressing the transporters. Typically, COS1 cells transiently transfected with VMAT2 bind 20-fold more [^{3}H]dihydrotetrabenazine than untransfected cells.

Flux Reversal

In addition to forward progress through the transport cycle, vesicular transport proteins can also catalyze flux reversal. If this involves full progress through the transport cycle but in reverse, it will result in the net efflux of transmitter. However, transport proteins can also mediate the exchange of lumenal substrate for cytoplasmic substrate (no net flux). In the case of vesicular monoamine transport, exchange may have a physiological role since the biosynthesis of epinephrine involves import of dopamine into the vesicle, conversion there to norepinephrine, export to the cytoplasm, conversion there to epinephrine, and finally, uptake of epinephrine. In particular, the export of norepinephrine may well involve exchange for cytosolic monoamine. Interestingly, net efflux would be expected to occur after the dilution of preloaded vesicles into medium without transmitter, simply due to the reequilibration of a previously established concentration gradient. However, bovine chromaffin ghosts show no evidence of net efflux after such dilution.[45] Dissipation of the proton electrochemical gradient does induce rapid efflux under these conditions, suggesting a kinetic barrier to reversal of the transport cycle. Indeed, amphetamines may increase synaptic concentrations of monoamines by first inducing efflux from vesicular stores through a weak base effect that dissipates $\Delta\mu_{H^+}$.[46] Vesicular efflux may also contribute to the toxicity of these psychostimulants.[47] Interestingly, VMAT inhibitors such as reserpine and tetrabenazine inhibit exchange but not efflux using the standard assays. This has suggested that net efflux is transporter independent. Alternatively, the drugs may simply not be able to inhibit this reaction and more recent observations strongly support a role for the transport proteins in efflux as well as exchange. We will now focus on the measurement of flux reversal catalyzed by the VMATs, although the same principles can be used to study other vesicular neurotransmitter transport activities. In particular, the amino acid transporters appear to mediate a high rate of spontaneous net efflux that can be blocked by stilbene inhibitors such as DIDS.[35]

Procedure

1. Add [^{3}H]serotonin to a final concentration of 5 μM to fresh or thawed membranes prepared as above in SH buffer. Incubate at 4° for ~8 hr to load the vesicles nonspecifically (i.e., even in the absence of vesicular transport activity).

[45] R. Maron, Y. Stern, B. I. Kanner, and S. Schuldiner, *J. Biol. Chem.* **258,** 11476 (1983).
[46] D. Sulzer and S. Rayport, *Neuron* **5,** 797 (1990).
[47] J. F. Cubells, S. Rayport, G. Rajendran, and D. Sulzer, *J. Neurosci.* **14,** 2260 (1994).

2. To measure net efflux, add 5 μl preloaded membranes to 1.5 ml SH buffer containing 2.5 mM $MgSO_4$ and 4 mM KCl at 27°. In the absence of proton ionophores, we have observed very little spontaneous net efflux, confirming previous results with bovine chromaffin ghosts.[45] We use either 5 μM nigericin or 20 μM valinomycin to stimulate efflux by dissipating ΔpH and $\Delta\Psi$, respectively. To study exchange, include substrate in the dilution buffer.

3. After incubation at 27° for varying intervals, terminate the reaction by rapid dilution with 1.5 ml cold SH buffer and filter through 0.2-μm Supor 200 membrane filters presoaked in water. Rinse the reaction tube with an additional 1.5 ml cold SH buffer and filter this rapidly as well.

4. Dry the filters and measure the bound radioactivity by scintillation counting in 2.5 ml scintillation fluid designed for small volume aqueous solutions (e.g., CytoScint, ICN).

Comments

Since this assay does not require active transport to load the membrane vesicles, it can be used to study flux reversal catalyzed by mutants incapable of active transport. Indeed, comparison with membranes not containing the VMATs shows that only membranes containing the VMATs catalyze efflux even though drugs have not previously inhibited this reaction. Similar to the uptake assay, the analysis of vesicles loaded with different concentrations of serotonin can also indicate the K_m for flux reversal and hence important features of the transport cycle such as the affinity of the lumenally oriented substrate recognition site.

Purification and Functional Reconstitution

Although heterologous expression systems provide a powerful way to study transporter function, a number of studies require relatively large amounts of purified transport protein. In the case of the VMATs, labeling with [^{3}H]reserpine has enabled the purification of a candidate transport protein from the bovine adrenal medulla,[37,48] but heterologous expression systems now provide additional sources. This section describes the methods used in our laboratory to purify VMAT2 protein that retains function. Similar procedures have been used to reconstitute vesicular GABA and glutamate transport activities solubilized from synaptic vesicles.[16,49–51]

[48] Stern-Bach, J. N. Keen, M. Bejerano, S. Steiner-Mordoch, M. Wallach, J. B. Findlay, and S. Schuldiner, *Proc. Natl. Acad. Sci. USA* **89,** 9730 (1992).

[49] M. D. Carlson, P. E. Kish, and T. Ueda, *J. Biol. Chem.* **264,** 7369 (1989).

[50] J. W. Hell, P. R. Maycox, and R. Jahn, *J. Biol. Chem.* **265,** 2111 (1990).

[51] J. W. Hell, L. Edelmann, J. Hartinger, and R. Jahn, *Biochemistry* **30,** 11795 (1991).

Source

The expression of molecularly cloned transport proteins in bacteria, yeast, or baculovirus systems may yield large quantities of recombinant protein. However, in our experience, the protein often lacks function, presumably because of aggregation, abnormal folding, or incorrect posttranslational modification. On the other hand, mammalian expression systems usually produce functional protein but in relatively small amounts. To increase expression in these systems, we have taken advantage of the ability of VMATs to protect against the neurotoxin *N*-methyl-4-phenylpyridinium (MPP^+)[22]. *N*-Methyl-1,2,3,6-tetrahydropyridine (MPTP) produces a form of parkinsonism through the specific accumulation of its active metabolite MPP^+ in monoamine cells.[52] The toxin acts by inhibiting respiration.[53–55] VMATs protect against MPP^+ toxicity by sequestering the toxin inside vesicles, away from its primary site of action in mitochondria.[23,55] Thus, selection of stable CHO cell transformants in 1 m*M* MPP^+ results in high levels of functional VMAT2 protein (~10μg/confluent 15-cm plate).

To purify VMAT2, we have used metal chelate-affinity chromatography. To enable recognition by the column, we have added a six-histidine tag at the C terminus of the protein. In addition, we inserted a Factor Xa cleavage site (Ile-Glu-Gly-Arg) between VMAT2 and the His_6 tag to facilitate the purification. Importantly, heterologous expression shows that this modified VMAT2 is functionally indistinguishable from the wild-type protein. The following procedures were optimized for solubilization and purification of this Hig-tagged VMAT2 protein stably expressed in Chinese hamster ovary (CHO) cells but can probably be adapted to other expression systems with relatively few changes.

Solubilization

1. Wash 20 15-cm plates of CHO cells stably expressing His-tagged VMAT2 with CMF-PBS, detach the cells from the plate with trypsin, and collect them by centrifugation.

2. Resuspend the resulting pellet in 9 ml cold SH buffer containing 2 μg/ml aprotonin, 2 μg/ml leupeptin, 1 μg/ml E-64, 1 μg/ml pepstatin A, and 0.2 m*M* diisopropyl fluorophosphate.

3. Homogenize the cell suspension at 0.01-mm clearance and remove the cellular debris by centrifugation at 5000*g* for 5 min at 4°.

[52] J. W. Langston, P. Ballard, J. W. Tetrud, and I. Irwin, *Science* **219,** 979 (1983).

[53] W. J. Nicklas, I. Vyas, and R. E. Heikkila, *Life Sci.* **36,** 2503 (1985).

[54] R. R. Ramsay and T. P. Singer, *J. Biol. Chem.* **261,** 7585 (1986).

[55] R. R. Ramsay, M. J. Krueger, S. K. Youngster, M. R. Gluck, J. E. Casida, and T. P. Singer, *J. Neurochem.* **56,** 1184 (1991).

4. Collect the supernatant, adjust the volume to 15 ml with SH buffer containing the same protease inhibitors, and collect the membranes by centrifugation at 160,000*g* (SW41, Beckman) for 45 min at 4°. This large volume helps to remove soluble proteins.

5. Resuspend the resulting membrane pellet in 1 ml 10 m*M* HEPES–KOH (pH 7.4), again containing protease inhibitors, adjust the volume to 15 ml with the same hypotonic buffer containing protease inhibitors, and collect the membranes by centrifugation at 160,000*g* for 45 min at 4°. Hypotonic lysis helps to eliminate soluble proteins within the secretory compartment and other organelles.

6. Solubilize the resulting membrane pellet in 3 ml 500 m*M* NaCl, 20 m*M* Tris, pH 7.8, 5 m*M* imidazole, and 1.5% cholate and incubate on ice for 15 min. Cholate is preferred over other detergents since it has a high critical micelle concentration (cmc) that facilitates its removal by dialysis and hence promotes the formation of proteoliposomes. In addition, cholate efficiently solubilizes VMAT2 protein and VMAT2 protein solubilized with cholate remains active on reconstitution.

7. Remove the insoluble material by centrifugation at 121,000*g* for 30 min at 4° and collect the supernatant.

Comments

To reconstitute the transport protein directly from this crude preparation, solubilize the membrane pellet in 1 ml SH buffer containing protease inhibitors and 1.5% cholate. For crude protein preparations, the solubilized protein can be frozen at −80° and maintained for up to 6 months with no detectable reduction in transport activity on reconstitution into artificial membranes.

Purification

1. Add 1 ml iminodiacetic acid conjugated to Sepharose 6B Fast Flow resin (Sigma) to a convenient small column, rinse the resin with 50 ml of water, charge the resin with 5 ml 100 m*M* $NiCl_2$, and equilibrate the resin with 50 ml 500 m*M* NaCl, 20 m*M* Tris, pH 7.9, 5 m*M* imidazole, and 1% cholate (binding buffer). While charging, the resin changes from white to blue. These columns can be reused by stripping the nickel with 5 ml 500 m*M* EDTA, followed by a rinse with 50 ml water.

2. Apply the solubilized protein to the column. We normally apply the solubilized protein 3–4 times (over ~1 hr) to maximize binding to the metal column.

3. Rinse the column with 5 ml binding buffer followed by 50 ml binding buffer containing 20 m*M* imidazole. Although this low concentration of

imidazole does not usually suffice to elute the His_6-tagged protein, it may be necessary to confirm this in a particular system. Conversely, it may be possible to increase the imidazole concentration of this wash step without eluting the protein and so improve the purification.

4. Elute the protein with 2.5 ml binding buffer containing 200 m*M* imidazole. Again, it may be necessary or even desirable to adjust the imidazole concentration for maximal yield and purification. However, metal chromatography rarely results in a complete purification after this single step. Collect the eluate in a Centricon-10 concentrator (Amicon) and spin at 5000*g* until the volume is reduced to 150 μl.

5. Reduce the concentration of NaCl to 100 m*M* by dilution, maintaining 20 m*M* Tris, 2 m*M* $CaCl_2$, and 1% cholate and a small volume.

6. Remove the His_6 tag from VMAT2 by incubation with 10 μg Factor Xa protease (NEB) at 23° for 18 hr. Depending on the surrounding sequence and exact reaction conditions, Factor Xa often cleaves inefficiently, requiring adjustment of either the Factor Xa concentration or incubation time.

7. Adjust the reaction solution to 4.5 ml 500 m*M* NaCl, 20 m*M* Tris, pH 7.9, 5 m*M* imidazole, and 1% cholate.

8. Prepare a second column as described above and apply the partially purified, digested protein solution to the column, again over ~1 hr (3–4 passes).

9. Collect the flow-through containing the cleaved, purified VMAT2 protein and concentrate the solution from 4.5 ml to 200 μl using the Centricon column as before. This second column essentially eliminates contaminating proteins that bind to the metal column—Factor Xa affects the binding of recombinant but not these other proteins, separating them at this stage. The purified protein can also be frozen at −80° and maintained for up to 6 months with no loss in transport activity on reconstitution.

Comments

Western analysis can be used to follow the yield of the protein and silver or even Coomassie staining to monitor purification. Activity is monitored by reconstituting the protein into proteoliposomes.

Reconstitution into Proteoliposomes

1. Mix 5 mg total bovine brain lipids (Avanti Polar Lipids, Inc., Birmingham, AL) and 0.5 mg asolectin (Avanti Polar Lipids, Inc.) in a cold, completely dry 50-ml pear-shaped flask. The exact source and status of these lipids can be crucial for functional reconstitution. Deliver a stream of argon and rotate the flask to dry a thin layer of the lipid.

2. Add 2.5 ml 150 mM NH_4Cl, 10 mM HEPES–KOH (pH 7.4), 1 mM $MgSO_4$, and 1.5% cholate to the flask. To solubilize the dried lipids, sonicate the mixture at high intensity for 12 1-sec pulses three times in an ice-cooled bath sonicator (Branson) and then vortex thoroughly.

3. Mix the cholate-solubilized transport protein with 1 ml solubilized lipid in a 12,000–14,000 molecular weight cutoff Spectra/Por dialysis tube (Spectrum Medical Industries, Inc., Houston, TX).

4. Dialyze for 12–18 hr against 1 liter 150 mM NH_4Cl, 10 mM HEPES–KOH (pH 7.4), and 1 mM $MgSO_4$ at 4° and 2 hr against 1 liter 150 mM NH_4Cl, 10 mM HEPES–KOH (pH 8.5), and 1 mM $MgSO_4$ at 4°. As dialysis removes the cholate, proteoliposomes form.

5. Concentrate the resultng proteoliposomes by centrifugation at 121,000g for 90 min at 4°. The preparation of the membranes in NH_4Cl and final dilution into a solution without NH_4Cl results in the diffusion of ammonia (but not H^+) out of the vesicle to generate an outwardly directed pH gradient that can drive active transport.[37] Alternatively, the neurotransmitter transporter can be coreconstituted with another protein such as the H^+-ATPase present in crude membrane preparation or bacteriorhodopsin[56] that can generate the driving force for active transport by the addition of ATP or by simple illumination.

Transport using Proteoliposomes

1. For each time point, add 10 μl proteoliposomes to 200 μl 150 mM sodium isethionate, 10 mM HEPES–KOH (pH 8.5), 2.5 mM $MgSO_4$, 4 mM KCl, and 20 nM [^{3}H]serotonin at 29°.

2. After incubation at 29°, terminate the transport reaction by rapid dilution with 1.5 ml cold 150 mM sodium isethionate, 10 mM HEPES–KOH (pH 8.5), 2.5 mM $MgSO_4$, and 4 mM KCl and filter through 0.2-μm Supor 200 membrane filters. Wash the tube with an additional 1.5 ml cold isethionate buffer and filter this rapidly as well.

3. Dry the filters and measure the bound radioactivity by scintillation counting in 2.5 ml scintillation fluid designed for small volume aqueous solutions.

Comments

Controls include the reconstitution of protein isolated from cells not expressing VMATs, liposomes made without protein, and the addition of 5 μM reserpine. Typically, proteoliposomes containing VMATs accumulate 300% more [^{3}H]serotonin than these controls. Importantly, this activity

[56] P. R. Maycox, T. Deckwerth, and R. Jahn, *EMBO J.* **9,** 1465 (1990).

is invariably lower than using intact membranes prepared directly from heterologous expression systems, presumably because of incompletely or improperly reconstituted transport protein, increased leakiness of the artificial membranes, and the inherently transient nature of the diffusion potential used to drive transport.

Conclusions

The assays described here address the basic functional characteristics of vesicular neurotransmitter transport. They have already helped to identify the proteins responsible and characterize their general properties. More recently, they have helped to identify the individual amino acid residues responsible for features of transport function. However, they can also help us to understand changes in the function of these transport proteins that occur normally during information processing and abnormally in such diseases as Parkinson's disease.

Acknowledgments

We thank E. Fon, Y. Liu, D. Peter, and R. Reimer for helpful discussions. This work was supported by a predoctoral fellowship from NINDS (to J.P.F.), a fellowship from the UCLA Biotechnology Training Program (to A.M.), and grants from NIDA and NIMH (to R.H.E.).

Section II

Pharmacological Approaches and Binding Studies

Articles 11 through 15

[11] Inhibitors of γ-Aminobutyric Acid Transport as Experimental Tools and Therapeutic Agents

By POVL KROGSGAARD-LARSEN, BENTE F. FRØLUND, and ERIK FALCH

Introduction

The neutral amino acid γ-aminobutyric acid (GABA) is an inhibitory neurotransmitter in the central nervous system (CNS), and GABA is also involved as a transmitter and/or a paracrine effector in the regulation of a variety of physiological mechanisms in the periphery.[1–4] So far, the role of GABA in the CNS has been most extensively studied, and a number of processes and mechanisms associated with GABA-operated synapses have been shown to be effective targets for therapeutic intervention in certain disease conditions. The most flexible approaches to stimulation of GABA neurotransmission processes appear to be facilitation of the effect of synaptically released GABA via GABA receptor modulatory sites[4] or by interfering with the inactivation of GABA.[5,6]

In light of the limited knowledge of the processes underlying the termination of the synaptic action of GABA, it is at present not possible to single out with certainty the transport mechanism(s) most susceptible to pharmacological intervention. Based on extensive molecular and cellular pharmacological studies, it has been demonstrated that neuronal and glial GABA transport mechanisms have dissimilar substrate specificities, which in turn are distinctly different from those of pre- and postsynaptic GABA receptors.[2,5,7] Thus, focusing on GABA transport mechanisms as pharmacological targets, the most logical strategies for such pharmacological interventions with the purpose of stimulating GABA neurotransmission seem to be (1) effective blockade of neuronal as well as glial GABA uptake in

[1] N. G. Bowery and G. Nistico (Eds.) "GABA: Basic Research and Clinical Applications." Pythagora Press, Rome, 1989.

[2] P. Krogsgaard-Larsen, B. Frølund, F. S. Jørgensen, and A. Schousboe, *J. Med. Chem.* **37,** 2489 (1994).

[3] C. Tanaka and N. G. Bowery (Eds.) "GABA: Receptors, Transporters and Metabolism." Birkhäuser Verlag, Basel, 1996.

[4] S. J. Enna and N. G. Bowery (Eds.) "The GABA Receptors." The Humana Press, New Jersey, 1996.

[5] P. Krogsgaard-Larsen, *Med. Res. Rev.* **8,** 27 (1988).

[6] P. Krogsgaard-Larsen, E. Falch, O. M. Larsson, and A. Schousboe, *Epilepsy Res.* **1,** 77 (1987).

[7] P. Krogsgaard-Larsen, H. Hjeds, E. Falch, F. S. Jørgensen, and L. Nielsen, *Adv. Drug Res.* **17,** 381 (1988).

0076-6879/98 $25.00

order to enhance the inhibitory effects of synaptically released GABA, or (2) selective blockade of glial GABA uptake in order to increase the amount of GABA taken up into the nerve terminals. It was predicted that selective neuronal GABA transport inhibitors might conceivably produce convulsions as the result of depletion of GABA from nerve terminals, and earlier animal behavioral studies actually have demonstrated that neuronal GABA transport inhibitors/substrates produce convulsions, whereas selective inhibitors of glial GABA transport show anticonvulsant effects.[6]

Heterogeneity of GABA Transport Mechanisms

Molecular cloning studies have disclosed that, like virtually all other biomechanisms, GABA transporters are heterogeneous.[8] Although the individual GABA transporters have been given different acronyms by the research groups which have been involved in the molecular cloning studies, rat GABA transporters are now most commonly named rGAT-1, rGAT-2, rGAT-3, and rBGT-1, the last being capable of transporting GABA as well as the osmolyte betaine.[9–13] Species differences have been observed, and since homologs transporters identified in different animals have sometimes been given different acronyms, the literature should be read cautiously. Thus, whereas human GABA transporters are named hGAT-1, hGAT-2, hGAT-3, and hBGT-1, respectively,[14] mouse GAT1 is the homolog of rGAT-1, mouse GAT2 is the homolog of rBGT-1, mouse GAT3 is the homolog of rGAT-2, and mouse GAT4 is the homolog of rGAT-3.[15] A taurine transporter (TAUT) has also been cloned from rat brain.[16] This transporter, which is sensitive to inhibition by β-alanine, transports GABA

[8] M. E. A. Reith (Ed.) "Neurotransmitter Transporters: Structure, Function, and Regulation." Humana Press, New Jersey, 1997.

[9] J. Guastella, N. Nelson, H. Nelson, L. Czyzyk, S. Keynan, M. C. Miedel, N. Davidson, H. A. Lester, and B. I. Kanner, *Science* **249,** 1303 (1990).

[10] L. A. Borden, K. E. Smith, P. R. Hartig, T. A. Branchek, and R. L. Weinshank. *J. Biol. Chem.* **267,** 21098 (1992).

[11] J. A. Clark, A. Y. Deutch, P. Z. Gallipoli, and S. G. Amara, *Neuron* **9,** 337 (1992).

[12] A. Yamauchi, S. Uchida, H. M. Kwon, A. S. Preston, R. B. Robey, A. Garcia-Perez, M. B. Burg, and J. S. Handler, *J. Biol. Chem.* **267,** 649 (1992).

[13] Q.-R. Liu, B. Lopez-Corcuera, S. Mandiyan, H. Nelson, and N. Nelson, *J. Biol. Chem.* **268,** 2106 (1993).

[14] T. G. M. Dhar, L. A. Borden, S. Tyagarajan, K. E. Smith, T. A. Branchek, R. L. Weinshank, and C. Gluchowski, *J. Med. Chem.* **37,** 2334 (1994).

[15] K. E. Smith, E. L. Gustafson, L. A. Borden, T. G. M. Dhar, M. M. Durkin, P. J.-J. Vaysse, T. A. Branchek, C. Gluchowski, and R. L. Weinshank, *in* "GABA: Receptors, Transporters and Metabolism" (C. Tanaka and N. G. Bowery, Eds.), p. 63. Birkhäuser, Basel, 1996.

[16] K. E. Smith, L. A. Borden, C.-H. D. Wang, P. R. Hartig, T. A. Branchek, and R. L. Weinshank, *Mol. Pharmacol.* **42,** 563 (1992).

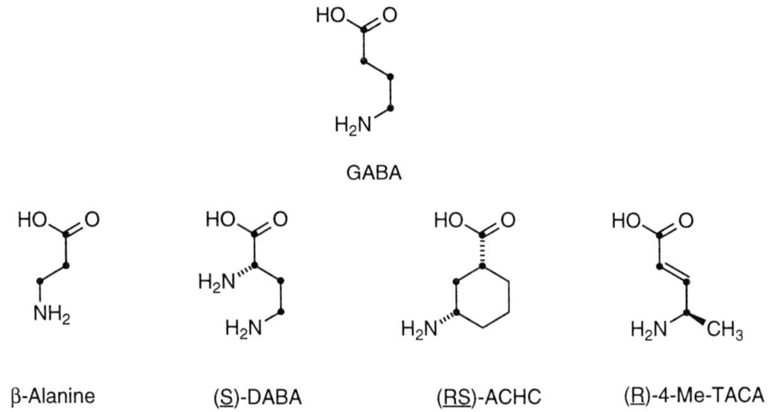

FIG. 1. Structures of GABA and some GABA analogs with inhibitory effects on GABA transport systems.

with low affinity and thus should be considered a potential functional GABA transporter in the CNS.

In light of the previously described dissimilar pharmacological characteristics of neuronal and glial GABA transport systems,[2,5,7] the neuronal versus glial localizations of GABA transporter mRNAs have been studied.[15] Whereas GAT-1, GAT-3, and BGT-1 mRNAs were detected in both neurons and glial cells, GAT-2 is the only GABA transporter being selectively expressed in nonneuronal cells, and BGT-1 was shown to be the only high-affinity GABA transporter mRNA present in Type 1 astrocytes. It has been proposed that the dissimilar pharmacological data obtained using cultured neurons and glial cells may reflect the presence of various combinations of the cloned GABA and taurine transporters.[15]

GABA Analogs as GABA Transport Inhibitors

A very large number of GABA analogs have been studied as inhibitors of neuronal and glial GABA transport.[6] (*S*)-2,4-Diaminobutyric acid [(*S*)-DABA] and *β*-alanine (Fig. 1) have been used as markers for neuronal and glial GABA transport mechanisms, respectively.[17,18] The lack of specificity of (*S*)-DABA as a substrate for neuronal GABA transport,[19,20] and

[17] L. L. Iversen and J. S. Kelly, *Biochem. Pharmacol.* **24,** 933 (1975).

[18] D. L. Martin, *in* "GABA in Nervous System Function" (E. Roberts, T. N. Chase, and D. B. Tower, Eds.), p. 347. Raven Press, New York, 1976.

[19] A. J. Kennedy and M. J. Neal, *J. Neurochem.* **30,** 459 (1978).

[20] A. Schousboe, O. M. Larsson, J. D. Wood, and P. Krogsgaard-Larsen, *Epilepsia* **24,** 531 (1983).

the observation that β-alanine appears to be transported with equal efficiency by neuronal and glial uptake mechanisms, in both cases without using GABA transport carriers,[21] have markedly reduced the value of these amino acids as marker substrates. Like (*S*)-DABA, *cis*-3-aminocyclohexanecarboxylic acid [(*RS*)-ACHC] (Fig. 1), considered to be a selective substrate for neuronal GABA transport mechanism(s),[22,23] actually interacts with glial as well as neuronal GABA transport.[20,24] Whereas the (*S*)-form of *trans*-4-aminopent-2-enoic acid (4-Me-TACA) is a specific $GABA_A$ agonist, (*R*)-4-Me-TACA (Fig. 1) is a GABA transport inhibitor showing no detectable GABA receptor affinity, but (*R*)-4-Me-TACA is only slightly more potent as an inhibitor of neuronal GABA transport than as an inhibitor of GABA uptake into glial cells.[25]

Nipecotic Acid, Guvacine, and Related GABA Transport Inhibitors

Muscimol, a psychoactive constituent of *Amanita muscaria*, has been used as a lead structure for the design and development of compounds interacting specifically with GABA synaptic recognition sites.[2,26,27] Muscimol is a high-affinity and potent $GABA_A$ agonist, which also interacts with neuronal and glial GABA transport mechanisms.[25,27] The muscimol analog (*RS*)-dihydromuscimol (DHM) shows a similar pharmacological profile,[27] but whereas the GABA transport affinities of DHM reside exclusively in (*R*)-DHM, (*S*)-DHM is a specific $GABA_A$ agonist (Fig. 2).[28] The bicyclic analog of muscimol, 4,5,6,7-tetrahydroisoxazolo[5,4-*c*]pyridin-3-ol (THIP), and the structurally related monocyclic amino acids, isoguvacine and isonipecotic acid, are specific $GABA_A$ agonists.[29,30] On the other hand, the isomeric compounds 4,5,6,7-tetrahydroisoxazolo[4,5-*c*]pyridin-3-ol (THPO), guvacine, and nipecotic acid are GABA transport inhibitors showing no

[21] O. M. Larsson, R. Griffiths, I. C. Allen, and A. Schousboe, *J. Neurochem.* **47,** 426 (1986).
[22] M. J. Neal and N. G. Bowery, *Brain Res.* **138,** 169 (1977).
[23] M. J. Neal, J. R. Cunningham, and J. Marshall, *Brain Res.* **176,** 285 (1979).
[24] O. M. Larsson, G. A. R. Johnston, and A. Schousboe, *Brain Res.* **260,** 279 (1983).
[25] A. Schousboe, P. Thorbek, L. Hertz, and P. Krogsgaard-Larsen, *J. Neurochem.* **33,** 181 (1979).
[26] P. Krogsgaard-Larsen, E. Falch, and H. Hjeds, *Prog. Med. Chem.* **22,** 67 (1985).
[27] P. Krogsgaard-Larsen, G. A. R. Johnston, D. R. Curtis, C. J. A. Game, and R. M. McCulloch, *J. Neurochem.* **25,** 803 (1975).
[28] P. Krogsgaard-Larsen, L. Nielsen, E. Falch, and D. R. Curtis, *J. Med. Chem.* **28,** 1612 (1985).
[29] P. Krogsgaard-Larsen, G. A. R. Johnston, D. Lodge, and D. R. Curtis, *Nature* **268,** 53 (1977).
[30] P. Krogsgaard-Larsen, H. Hjeds, D. R. Curtis, D. Lodge, and G. A. R. Johnston, *J. Neurochem.* **32,** 1717 (1979).

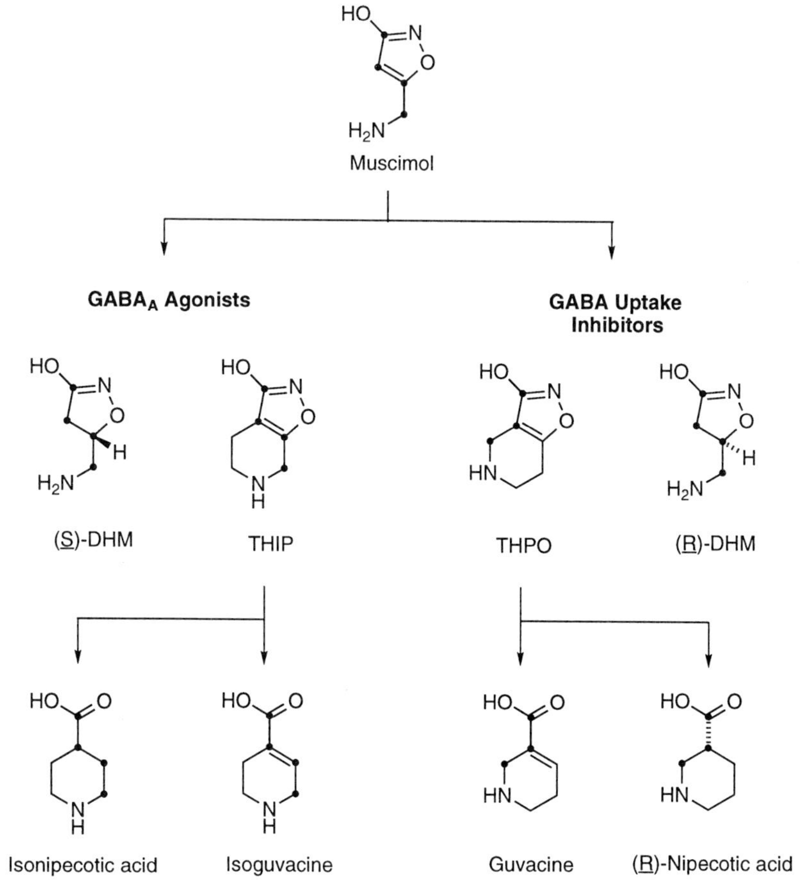

FIG. 2. A schematic illustration of the "separation" of the $GABA_A$ agonist and GABA uptake (transport) effects of muscimol, leading to the development of specific $GABA_A$ agonists and GABA transport inhibitors.

affinity for $GABA_A$ receptors.[6,31–33] The compounds shown in Fig. 2 have become the standard $GABA_A$ agonists and GABA transport inhibitors, and guvacine and nipecotic acid have been further developed into clinically active inhibitors of GABA transport[2,6] (see later section).

[31] P. Krogsgaard-Larsen and G. A. R. Johnston, *J. Neurochem.* **25,** 797 (1975).

[32] G. A. R. Johnston, P. Krogsgaard-Larsen, and A. Stephanson, *Nature* **258,** 627 (1975).

[33] G. A. R. Johnston, P. Krogsgaard-Larsen, A. L. Stephanson, and B. Twitchin, *J. Neurochem.* **26,** 1029 (1976).

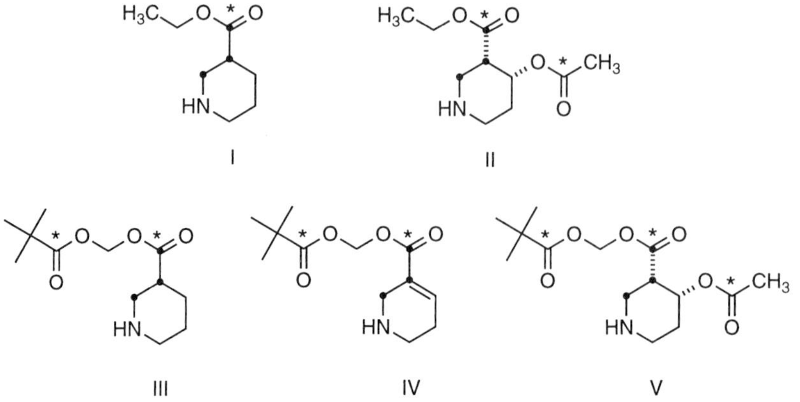

FIG. 3. Structures of a number of ester prodrugs of the GABA transport inhibitors nipecotic acid, guvacine, and *cis*-4-hydroxynipecotic acid. The asterisks indicate the ester bonds susceptible to spontaneous and/or enzyme-catalyzed hydrolysis.

Prodrugs of GABA Transport Inhibitors

The amino acids guvacine and nipecotic acid (Fig. 2) show very weak pharmacological effects after systemic administration to animals, reflecting that the blood–brain barrier (BBB) essentially prevents these compounds from entering the brain from the bloodstream.[6] The prodrug approach has therefore been used to study the pharmacological effects of these and related GABA uptake inhibitors in animals.[34–36] An ideal prodrug is a derivative of a biologically active compound without pharmacological effects per se, which is converted into the parent compound in the brain tissue without formation of by-products showing undesired effects. The compounds shown in Fig. 3 are bioreversible derivatives of nipecotic acid, guvacine, or *cis*-4-hydroxynipecotic acid, which are potential prodrugs for these GABA transport inhibitors. Although nipecotic acid ethyl ester **(I)** shows anticonvulsant properties in mice,[35] muscarinic cholinergic effects of this compound[34,35] and of the double esters of *cis*-4-hydroxynipecotic acid, compounds **II** and **V**,[36] prevent the use of these ester derivatives as prodrugs.[6] The anticonvulsant effects of the pivaloyloxymethyl esters of nipecotic acid and guvacine, compounds **III** and **IV**, respectively, were not accompanied by cholinergic side effects,[36] and these compounds are potentially useful as anticonvulsant drugs.[6] The relatively high doses of these

[34] M. J. Croucher, B. S. Meldrum, and P. Krogsgaard-Larsen, *Eur. J. Pharmacol.* **89,** 217 (1983).
[35] H.-H. Frey, C. Popp, and W. Löscher, *Neuropharmacology* **18,** 581 (1979).
[36] E. Falch, B. S. Meldrum, and P. Krogsgaard-Larsen, *Drug. Des. Delivery* **2,** 9 (1987).

two prodrugs required for effective suppression of seizures in animals may, however, prevent their use in the human clinic.

GABA Transport Inhibitors Containing 4,4-Diphenyl-3-butenyl Group and Related Compounds

Introduction of small substituents on the secondary amino groups of nipecotic acid or guvacine leads to compounds with decreased affinity for the GABA transport carriers.[6] Quite surprisingly, substitution of the very bulky 4,4-diphenyl-3-butenyl (DPB) group for the hydrogen atom of the amino group of nipecotic acid to give *N*-DPB-nipecotic acid (SKF-89976-A) (Fig. 4) provided a GABA transport inhibitor markedly more potent than nipecotic acid itself.[37] Similarly, DPB-guvacine[37] and DPB-THPO[38,39] were shown to be at least an order of magnitude more potent than the parent GABA transport inhibitors, guvacine and THPO. The DPB analogs of GABA, compound **VI**, and of ACHC, compound **VIII**, were, however, shown to be weaker than the parent amino acids as GABA transport inhibitors.[37] These unexpected structure–activity relationships may reflect that the two DPB analogs **VI** and **VIII** contain secondary amino groups (Fig. 4). This proposal is supported by the observation that *N*-methylation of **VI** to give **VII**, which contains a tertiary amino group, provided a substantially more potent GABA transport inhibitor.[40] Whereas saturation of the double bond of DPB-substituted GABA transport inhibitors such as SKF-89976-A to give IX (Fig. 4) results in virtual loss of activity,[37] replacement of this double bond in DPB analogs by an ether function, as exemplified by CI-966,[41] has limited influence on the effects of the compounds on GABA transport systems.[42] A variety of analogs of these DPB-containing GABA transport inhibitors have been developed, including the thiophene analog Tiagabine,[43] which is used therapeutically as an antiepileptic agent.

[37] F. E. Ali, W. E. Bondinell, P. A. Dandridge, J. S. Frazee, E. Garvey, G. R. Girard, C. Kaiser, T. W. Ku, J. J. Lafferty, G. I. Moonsammy, H. J. Oh, J. A. Rush, P. E. Setler, O. D. Stringer, J. W. Venslavsky, B. W. Volpe, L. M. Yunger, and C. L. Zirkle, *J. Med. Chem.* **28,** 653 (1985).

[38] E. Falch, O. M. Larsson, A. Schousboe, and P. Krogsgaard-Larsen, *Drug. Dev. Res.* **21,** 169 (1990).

[39] H. S. White, J. Hunt, H. H. Wolf, E. A. Swinyard, E. Falch, P. Krogsgaard-Larsen, and A. Schousboe, *Eur. J. Pharmacol.* **236,** 147 (1993).

[40] E. Falch and P. Krogsgaard-Larsen, *Eur. J. Med. Chem.* **26,** 69 (1991).

[41] C. P. Taylor, M. G. Vartanian, R. D. Schwarz, D. M. Rock, M. J. Callahan, and M. D. Davis, *Drug Dev. Res.* **21,** 195 (1990).

[42] E. Falch and P. Krogsgaard-Larsen, *Drug Des. Delivery* **4,** 205 (1989).

[43] K. E. Andersen, C. Braestrup, F. C. Grønwald, A. S. Jørgensen, E. B. Nielsen, U. Sonnewald, P. O. Sørensen, P. D. Suzdak, and L. J. S. Knutsen, *J. Med. Chem.* **36,** 1716 (1993).

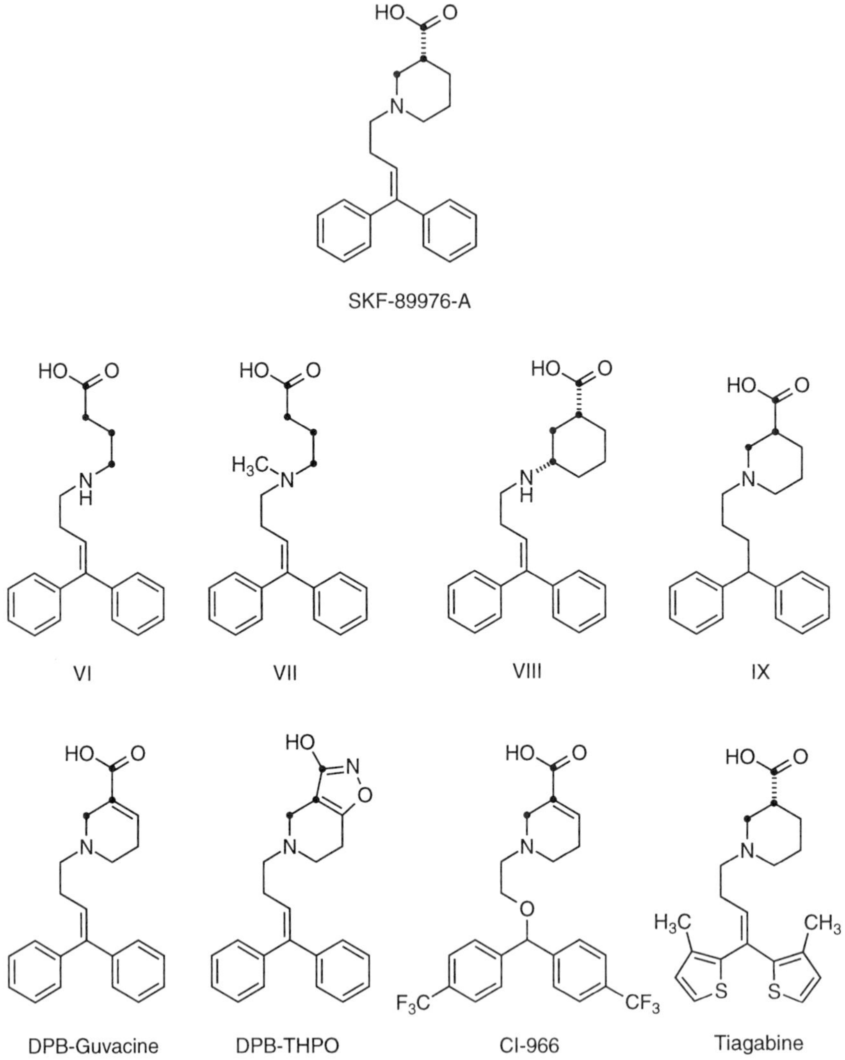

FIG. 4. Structures of *N*-4,4-diphenyl-3-butenylnipecotic acid (*N*-DPB-nipecotic acid, SKF-89976-A) and a number of other GABA transport inhibitors containing DPB or structurally related *N*-substituents.

The mechanisms underlying the effective interaction of these DPB analogs of the classical GABA transport inhibitors with the GABA carrier systems are essentially unknown, although extensive structure–activity studies have shown that the amino acid moieties of these compounds are

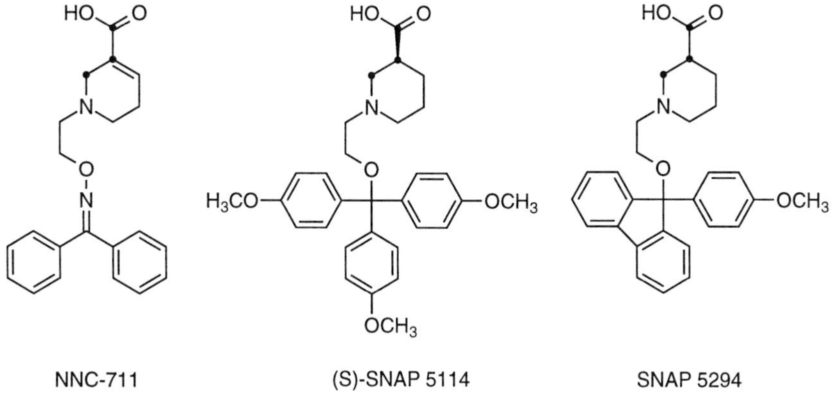

FIG. 5. Structures of some GABA transport inhibitors containing different lipophilic *N*-substitutents.

recognized and bound by the carrier systems.[6] However, whereas these underivatized monocyclic amino acids, including nipecotic acid,[6,44] are acting as substrates/inhibitors, the DPB analog, SKF-89976-A, has been shown not to be a substrate for the GABA transport carrier.[45]

Whereas GABA shows high or moderately high affinity for hGAT-1, rGAT-1, rGAT-2, hGAT-3, rGAT-3, as well as hBGT-1, nipecotic acid and guvacine show high affinity for hGAT-1, rGAT-1, and rGAT-2, rather low affinity for hGAT-3 and rGAT-3, and very low affinity for hBGT-1.[46] Interestingly, all of the DPB analogs and related lipophilic compounds so far studied, including CI-966, SKF-89976-A, and tiagabine (Fig. 4), interact selectively with hGAT-1 and rGAT-1, showing the highest affinity for hGAT-1.[46] Similar relative GABA transport affinities were demonstrated for the structurally related compound NNC-711, containing an oxime group (Fig. 5).[46]

Very different affinity profiles were observed for the more recently synthesized analogs, SNAP 5114 and SNAP 5294[14,15] (Fig. 5). SNAP 5114 binds with some selectivity to hGAT-3 (IC_{50} 5 μM) and shows decreasing potency at rGAT-2 (IC_{50} 21 μM), hBGT-1 (140 μM), and hGAT-1 (388 μM).[14] SNAP 5294, on the other hand, preferentially binds to rGAT-2

[44] G. A. R. Johnston, A. L. Stephanson, and B. Twitchin, *J. Neurochem.* **26,** 83 (1976).

[45] O. M. Larsson, E. Falch, P. Krogsgaard-Larsen, and A. Schousboe, *J. Neurochem.* **50,** 818 (1988).

[46] L. A. Borden, T. G. M. Dhar, K. E. Smith, R. L. Weinshank, T. A. Branchek, and C. Gluchowski, *Eur. J. Pharmacol. Mol. Pharmacol. Sect.* **269,** 219 (1994).

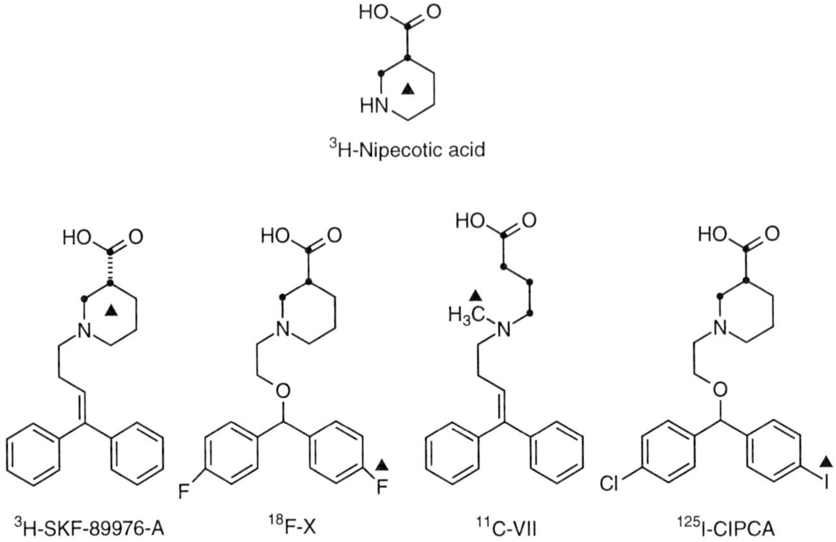

FIG. 6. Structures of a number of radiolabeled GABA transport inhibitors. The locations of the radioactive isotopes in the molecules are indicated by triangles.

(IC_{50} 51 μM) and shows lower affinity for hGAT-1 (IC_{50} 133 μM), hGAT-3 (IC_{50} 142 μM), and hBGT-1 ($IC_{50} \gg 100\ \mu M$).[14]

Radiolabeled GABA Transport Inhibitors as Research Tools

Radiolabeled forms of a number of GABA transport inhibitors have been synthesized and used as research tools. Using [^{3}H]nipecotic acid it has been demonstrated that this classical GABA transport inhibitor is a substrate for neuronal[44] as well as glial[47] GABA transport carriers. Labeled SKF-89976-A, which also contains tritium atoms in the piperidine ring (Fig. 6), was shown to penetrate easily into cultured neurons and astrocytes but not to be a substrate for the GABA transport carriers in these cell systems.[45] [^{125}I]CIPCA has been synthesized for brain distribution studies and may be a useful radiotracer for studies of the GABA transport system.[48] GABA transport inhibitors may be useful tools for *in vivo* studies of central GABA neurotransmission in animals and man using positron emission

[47] O. M. Larsson, P. Krogsgaard-Larsen, and A. Schousboe, *J. Neurochem.* **34,** 970 (1980).
[48] M. E. Van Dort, D. L. Gildersleeve, and D. M. Wieland, *J. Labelled Compd. Radiopharm.* **36,** 961 (1995).

tomography (PET) techniques. These prospects have prompted the synthesis of ^{11}C-labeled **VII**[49] and the GABA transport inhibitor X labeled with ^{18}F($[^{18}F]X$),[50] as potential PET ligands (Fig. 6). The results of studies using these tools have not yet been reported.

[49] D. Le Bars, E. Falch, and P. Krogsgaard-Larsen, unpublished.

[50] M. R. Kilbourn, M. R. Pavia, and V. E. Gregor, *Appl. Radiat. Isot.* **41,** 823 (1990).

[12] Design and Synthesis of Conformationally Constrained Inhibitors of High-Affinity, Sodium-Dependent Glutamate Transporters

By A. RICHARD CHAMBERLIN, HANS P. KOCH, and RICHARD J. BRIDGES

Introduction

The value of conformationally constrained analogs rests with the hypothesis that restricting the number of attainable conformations of a flexible, acyclic molecule such as L-glutamate should limit the number of glutamate binding sites with which the analog can interact, thereby increasing its selectivity as a pharmacological probe. As the spatial positions occupied by both the functional groups (e.g., COO^-, NH_3^+) and the backbone of the active analogs become increasingly well defined, it becomes possible to begin delineating a binding site pharmacophore; that is, the optimal chemical characteristics required for binding to the transporter. These pharmacophore models then become tools in the development of the "next generation" analogs that will be used to refine the models and identify more selectively acting agents, such as those that can distinguish among closely related transporter subtypes. Rather than a linear progression, this approach relies on a cyclic strategy, where each round of modeling, synthesis, and assays yields a more detailed picture of the binding site pharmacophore and a more selective library of compounds (Fig. 1).

The utility of this approach has been amply demonstrated in the identification and study of the neurotransmitter receptors within the EAA (excitatory amino acid) system.[1] Although progress in delineating the pharmacology of the glutamate transporters has not yet reached a comparable point, numerous advances have been made, particularly in the instance of the high-

[1] A. R. Chamberlin and R. J. Bridges, "Conformationally Constrained Acidic Amino Acids as Probes of Glutamate Receptors and Transporters." Raven, New York, 1993.

0076-6879/98 $25.00

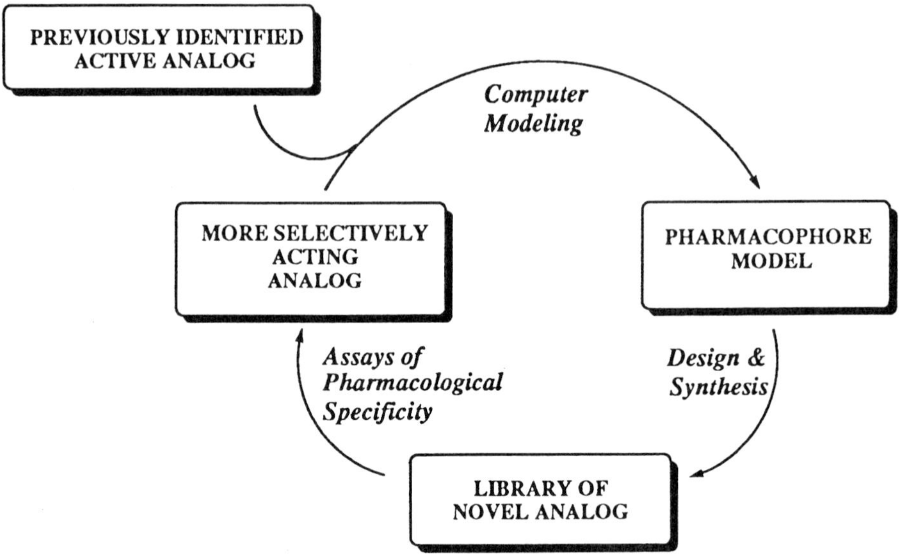

FIG. 1. Iterative strategy for discovering new glutamate transport inhibitors.

affinity, sodium-dependent systems. Building on early structure–function studies that relied on systematic substitutions along the length of the acyclic backbones of glutamate and aspartate,[2,3] a number of novel conformationally constrained transport inhibitors have been identified (Fig. 2), including dihydrokainate (DHK), L-*trans*-2,4-pyrrolidine dicarboxylate[4] (L-*trans*-2,4-PDC), *cis*-1-aminocyclobutane 1,3-dicarboxylate[5] (*cis*-ACBD), and (2*S*,3*S*,4*R*)-2-(carboxycyclopropyl)glycine[6] (CCG-III). The goal of this chapter is to provide an outline of an approach to analyzing these conformationally constrained inhibitors and using this information to develop new compounds with novel pharmacological characteristics.

[2] A. Schousboe, P. Larsson, P. Krogsgaard-Larsen, J. Drejer, and L. Hertz, *in* "The Biochemical Pathology of Astrocytes," pp. 381–394. Alan R. Liss, New York, 1987.

[3] P. J. Roberts and J. C. Watkins, *Brain Res.* **85,** 120 (1975).

[4] R. J. Bridges, M. S. Stanley, M. W. Anderson, C. W. Cotman, and A. R. Chamberlin, *J. Med. Chem.* **34,** 717 (1991).

[5] E. J. Fletcher, K. N. Mewett, C. A. Drew, R. D. Allan, and G. A. R. Johnston, *Neurosci. Lett.* **121,** 133 (1991).

[6] Y. Nakamura, K. Kataoka, M. Ishida, and H. Shinozaki, *Neuropharmacol.* **32,** 833 (1993).

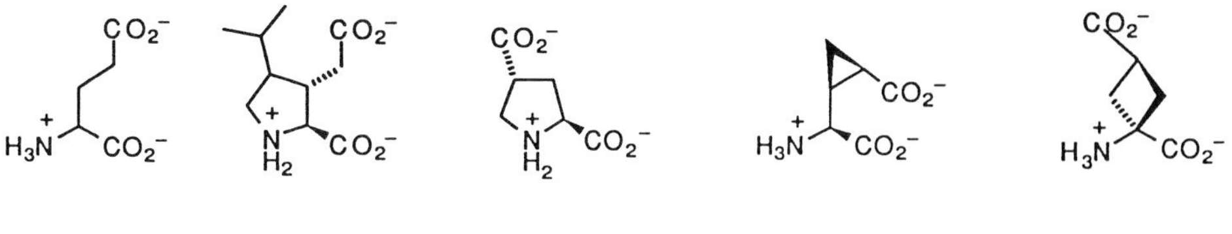

FIG. 2. Conformationally constrained glutamate transport inhibitors.

Conformational Analysis of Glutamate: A Starting Point

As an acyclic molecule, glutamate can adopt nine staggered conformations, resulting from rotation about the $\alpha\beta$ and $\beta\gamma$ sp^3 bonds (Fig. 3), that place the functional groups in a variety of orientations. The fact that conformationally restricted analogs mimic relatively few of these conformations is the basis for the increased selectivity that is often observed for rigid glutamate analogs. This selectivity implies that glutamate itself may bind to each receptor in a distinct conformation, and one might assume that any bound form of L-glutamate would tend to resemble one of the staggered solution conformers,[7] since binding of conformations other than local minima (e.g., eclipsed conformers) would incur an energy penalty. However, particularly favorable binding interactions might compensate for the unfavorable torsional interactions in non-ground-state structures, so that the possibility of glutamate binding in some intermediate form cannot be ignored.

Unstrained cyclic analogs would be expected to approximate staggered conformations quite closely with respect to the relative positions of the polar functional groups. On the other hand, other analogs might resemble intermediate rotamers because of bond angle and length deformations due to ring strain, as well as differences in preferred torsional angles between cyclic and acyclic hydrocarbon chains. In this context, it should also be kept in mind that a number of structural features defined by the relative and absolute stereochemistry of each compound can affect binding affinities, including the relative positions of the charged heteroatoms, the orientation (i.e., directionality) of the dipoles, and the position of the hydrocarbon backbone. Thus, it is appropriate to explore both enantiomers of glutamate analogs, as well as of diacids with other chain lengths (e.g., aspartate) that can assume conformations in which the polar functional groups attain a glutamate-like relationship.

[7] N. S. Ham, *in* "Molecular and Quantum Pharmacology" (E. Bergmann and B. Pullman, Eds.), pp. 261–268. D. Riedel, Dordrecht-Holland, 1974.

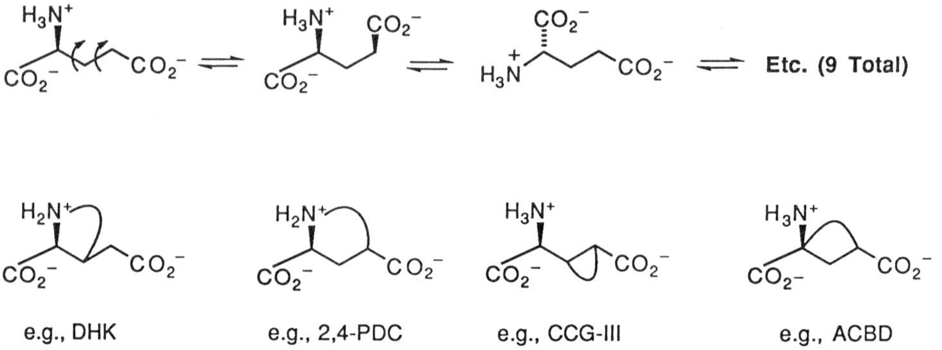

FIG. 3. Glutamate conformer interconversion and analog constraints.

Bearing all of these factors in mind, glutamate may be analyzed to determine how rotation about the two central C–C bonds can be restricted without unduly perturbing the basic skeleton. In addition to the relatively obvious strategy of mimicking glutamate conformers by embedding the diacid chain in carbocyclic or azacyclic ring systems, it is also possible to reduce the number of available conformations by simply shortening the chain. In this regard, it is appropriate to include aspartate analogs (e.g., DL-β-*threo*-hydroxyaspartate, L-aspartate-β-hydroxymate)[2,3] in this analysis, since they lack one of the conformationally mobile σ bonds of glutamate while the functional groups common to both overlap well with "folded" glutamate conformers.

Calculation of Low Energy Conformations of Acidic Amino Acids: Creating a Conformer Library

In order to compare the structures of the active transport inhibitors, computational libraries of the energetically accessible conformations of each must be generated. From a practical point of view, this task must be carried out using molecular mechanics rather than higher levels of theory (for a good general introduction to molecular modeling, see the review by Cohen[8]). Energy minimizations and conformational searches are carried out using a modified AMBER force field,[9,10] although other force fields

[8] N. C. Cohen, J. M. Blaney, C. Humblet, P. Gund, and D. C. Barry, *J. Med. Chem.* **33,** 883 (1990).

[9] S. J. Weiner, P. A. Kollman, D. A. Case, U. C. Singh, C. Ghio, G. Alagona, S. Profeta, Jr., and P. Weiner, *J. Am. Chem. Soc.* **106,** 765 (1984).

[10] W. D. Cornell, P. Cieplak, C. I. Bayly, I. R. Gould, K. M. Merz, Jr., D. M. Ferguson, D. C. Spellmeyer, T. Fox, J. W. Caldwell, and P. A. Kollman, *J. Am. Chem. Soc.* **117,** 5179 (1995).

can also be employed to generate the required libraries. For example, MM2 and AMBER give similar libraries for these acidic amino acids, although the calculated relative energies differ somewhat. This difference is of little consequence, however, since relative energies are ignored in the subsequent structural comparisons because of the notoriously unreliable predictions (at any level of theory) for multiply charged, conformationally mobile molecules. In particular, medium effects are likely to be a critical factor in determining relative energies of conformers, and there is no satisfactory way of simulating the medium in such calculations. Even if an aqueous environment could be perfectly simulated, the microenvironment of the binding site is probably much more hydrophobic and certainly less homogeneous than bulk solution, properties which would skew energies relative to those in solution. More importantly, even if relative energies could be calculated with precision, it cannot be assumed that any ligand binds in its lowest energy form. For these reasons, all conformers with calculated energies within 3–4 kcal/mol of the global minimum are considered in the analysis, thus ensuring inclusion of all structures present in amounts greater than approximately one-tenth of a percent. Within this group, however, relative energies are ignored.

Molecular modeling is performed on a Silicon Graphics Indigo 2 workstation using BIOSYM or MACROMODEL molecular modeling software. Conformational searches are conducted as follows: First, a local minimum is found by application of 1000 steps of a conjugate gradient minimization until the RMS (root mean square) deviation is less than 0.001 Å. Conformational space is then explored by use of a dynamics simulation where the molecule is effectively subjected to 600 K for 500,000 iterations at 1-fsec intervals. The molecular geometry of the compound is recorded after every 5000 fsec, yielding a total of 100 high-energy geometries. These high-energy geometries are then minimized using the conjugate gradient approach described above. Unique conformations with energies within 3 kcal of the lowest energy conformer are deemed significant and are saved. The molecules are modeled as having a positively charged N atom and two negatively charged carboxylates, as would exist at approximately neutral pH, with the dielectric constant set at $\delta = 80$.

Comparisons of Libraries and Pharmacophore Delineation: Overlaying Active Analogs

The next objective is to assess the results of binding data in terms of conformational and/or configurational information obtained by the molecular modeling. To do so requires correlating structure and function using a variant of the "active analog" approach,[11] which entails a systematic search of available conformations for all analogs as described above, followed by a comparison of the resultant structures to find conformers whose functional groups show similar spatial arrangements. In general, comparisons of active inhibitors are conducted using an overlay method in which three points of comparison (the amine functionality, the proximal carboxylate, and the distal carboxylate/anionic center) are specified for superimposition. The lowest RMS deviation of atoms among the pairwise comparisons is taken as the best fit, which is generally within a few tenths of an angstrom. Positional averages are then calculated for each of the three comparison atoms; these averages are then used as comparison points for all subsequent overlays. This procedure results in structures that have a common arrangement of functional groups and that define the pharmacophore of the transporter.

Such comparisons are best made among groups of conformationally restricted analogs that exhibit the greatest selectivity of action. For example, the fact that L-*trans*-2,4-PDC was found to be a highly selective and potent transport inhibitor[4] suggested that a comparison of available pyrrolidine envelopes with glutamate conformations might reveal a unique conformation of glutamate preferred for optimal binding at the transporter. Thus, minimized conformations of the nine staggered conformers of L-glutamate and several envelope forms of L-*trans*-2,4-PDC were compared, yielding a best fit pair (Fig. 4a) that defines the preferred conformations for the bound forms of both molecules, with a carboxyl–carboxyl distance of 4.58 ± 0.08 Å and a distal carboxyl–amino distance of 2.45 ± 0.02 Å. The L-*trans*-2,4-PDC conformer is one in which the distal carboxyl group occupies an axial-like position at the flap of the pyrrolidine envelope, and the glutamate conformer corresponds to a relatively abundant solution conformation. This compound is a considerably more potent transport inhibitor than DHK, the well-characterized weak competitive inhibitor. Nonetheless, there should be conformations of PDC and DHK that overlay tightly, and indeed this analysis does yield a pair that correspond well (Fig. 4b).[4] Further, the structures of two other potent conformationally restricted transport inhibitors, *cis*-ACBD[5] and CCG-III[6] (Fig. 2), are consistent with this model, that is, both yield a good fit with L-*trans*-2,4-PDC and the corresponding glutamate conformer in regard to overlap of the charged groups. These results are consistent with the hypothesis that the partially folded form of glutamate shown is the active conformation at the Na^+-dependent trans-

[11] G. R. Marshall and I. Motoc, *in* "Molecular Graphics and Drug Design, Topics in Molecular Pharmacology" (A. S. V. Burgen, G. C. K. Roberts, and M. S. Tute, Eds.), p. 115. Elsevier, Amsterdam, 1986.

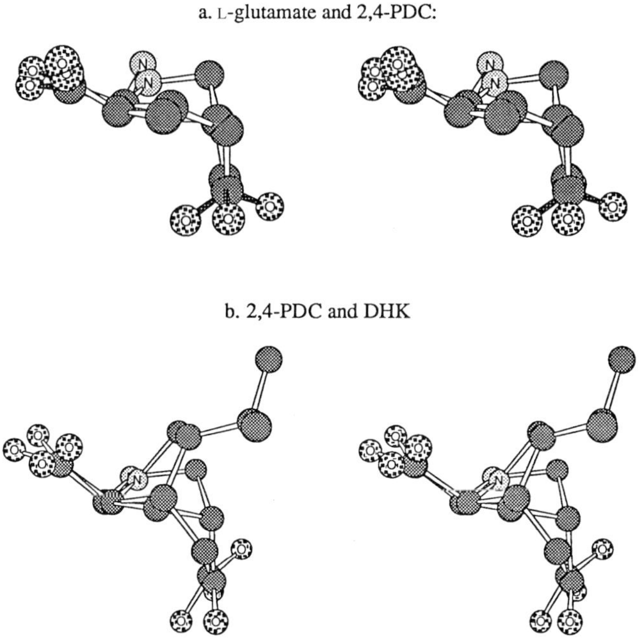

FIG. 4. Overlays of L-glutamate and transport inhibitors.

porter and provide a firm basis for the design and testing of new inhibitors. This latter process of checking the fit of known inhibitors to the proposed pharmacophore is obviously an important part of the process, providing necessary support for the validity of the newly defined pharmacophore.

Design and Synthesis of Next Generation Compounds

Once the pharmacophore has been proposed and tested, it becomes the key to the development of new inhibitors. The primary objective of this endeavor should be to identify synthetically accessible conformer mimics that can be prepared in reasonable quantities in pure form. The issue of whether it is desirable to pursue enantiomerically pure products, rather than racemic mixtures that might be easier to prepare, must be given careful thought in studies such as these. It is generally preferable to develop enantioselective routes for a number of reasons, especially considering that enantiomers of a given compound often have different selectivities and potencies that could not have been anticipated based on the enantiomeric series to which they belong. In addition, it is rarely the case that enantioselective synthesis is significantly more laborious. On the other hand, in instances where an enantioselective route would clearly be much more time-consuming than a nonselective synthesis, one should not hesitate to first prepare racemic material for preliminary biological characterization.

Although the stated objective of this project is the synthesis of conformationally defined glutamate analogs, it must be remembered that one of the most important factors in binding is the relative positions of the charged functional groups; therefore, alternative ways of "connecting" these functional groups should be considered, as should varying the steric bulk in the intervening hydrocarbon chain. This being the case, it would be judicious to prepare a number of conformationally defined acidic amino acids with aspartate, glutamate, and homoglutamate embedded in molecular frameworks with limited and predictable conformational preferences. As discussed earlier, a few such analogs already exist. However, having a defined pharmacophore allows one to significantly expand the list of target compounds as well as to better understand the chemical basis of binding to the transporter.

Although there are databases and software packages that can be used to search for molecular similarity to a given compound, the glutamate pharmacophore is simple enough that the search can easily be conducted "manually." In considering appropriate candidates, a primary concern beyond good overlap with the pharmacophore model must be whether the compound can be prepared readily. A reliable strategy for design is to choose a few structural motifs with precedented synthetic accessibility, and build upon those. Some examples in the case of the glutamate transporter would include 2,4-PDCs derived from 4-hydroxyprolines[4] or from kainate,[12] 2,3-PDCs derived from aspartate,[13] 3,4-methanoprolines derived from hydroxyproline,[14] and 2,4-methanoprolines via intramolecular photochemical cycloaddition.[15] Designing inhibitors based on these ring systems simply requires modeling appropriately substituted derivatives of each, followed by comparison with the transporter pharmacophore; those fitting best are chosen as synthetic targets. New analogs should be patterned after individual conformations of known inhibitors. In this way, for example, *meso*-2,4-methano-2,4-PDC (MPDC) and L-*anti-endo*-3,4-methano-pyrrolidine dicarboxylate (L-*anti-endo*-MPDC) were identified as candidates based on their

[12] F. Lovering, Ph.D Dissertation, University of California, Irvine, 1993.

[13] J. M. Humphrey, R. J. Bridges, J. A. Hart, and A. R. Chamberlin, *J. Org. Chem.* **59,** 2467 (1993).

[14] R. J. Bridges, F. E. Lovering, H. Koch, C. W. Cotman, and A. R. Chamberlin, *Neurosci. Lett.* **174,** 193 (1994).

[15] C. S. Esslinger, H. P. Koch, M. P. Kavanaugh, D. P. Philips, A. R. Chamberlin, C. M. Thompson, and R. J. Bridges, *Bioorg. Med. Chem. Lett.*, submitted (1998).

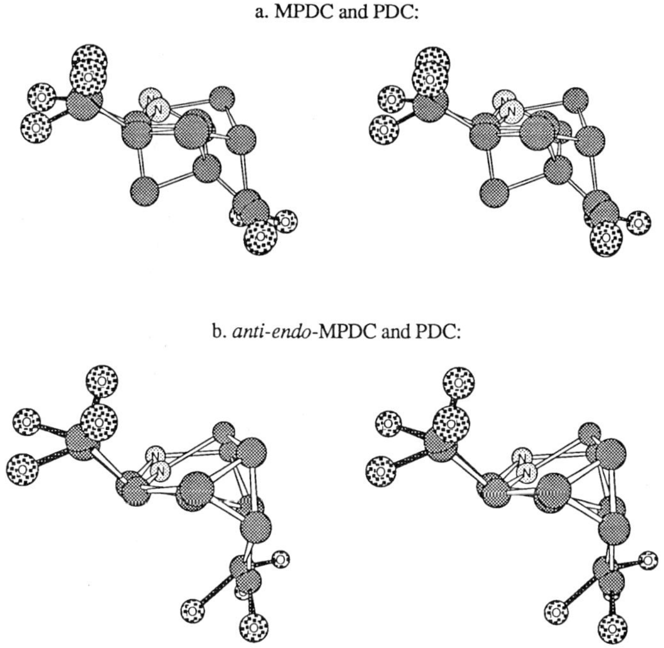

FIG. 5. Overlays of MPDC and *anti-endo*-MPDC with 2,4-PDC.

structural similarity to 2,4-PDC (Figs. 5a and 5b, respectively), and both of these very rigid analogs proved to be active transport inhibitors.[14] In retrospect, L-*anti-endo*-MPDC can also be thought of as a DHK derivative in which the β-side chain is connected to the 4-position normally occupied by the isopropenyl group, or a rigidified CCG analog in which the cyclopropyl methylene group is connected to the amino group via a methylene bridge.

As this example illustrates, human reasoning—and even chemical intuition—can still play an important role in the design process despite readily available computing power. When analogs are designed based on such known structural templates, preparation of the targets then requires relatively straightforward modification of existing synthetic routes to provide the desired substitution pattern. Of course, the synthetically more adventurous, or adept, have the option of venturing further from established chemistry. Either way, structures of all targets must be carefully confirmed by complete spectral analysis, including H NMR NOE experiments and often X-ray crystallography.

The initial design and synthesis phases focus on positioning and orientation of the charged groups, implicitly ignoring their connectivity. The analogs can be "fine-tuned," however, by modifying the hydrocarbon backbone of various active analogs. For example, attaching alkyl substituents to the PDC ring might be anticipated to perturb the normal conformer populations of the pyrrolidine ring, perhaps in favor of the proposed pharmacophore-like conformation. In addition, more subtle steric interactions between the substituent and the boundaries of the binding sites might be expected to result in different pharmacological properties compared with the parent analog. The preparation of one such derivative was easily achieved by recognizing that kainate itself (which is commercially available) could be used as starting material for a 3-substituted 2,4-PDC, since the isopropenyl group can serve as a latent carboxyl and the β-substituent (in reduced form) as the appendage of interest.[12] Likewise, the published routes to 2,3-PDCs can readily be modified to produce both 3- and 5-substituted 2,3-PDCs; these relatively minor structural modifications can, indeed, affect the activity dramatically.[16]

Finally, replacing one or both of the carboxylates with other negatively charged groups should be considered, since there is ample precedent (even among acid amino acid receptor ligands) that doing so can markedly affect the activity. For example, the distal carboxyl group of PDC can be changed simply by modifying the synthetic route to *trans*-2,4-PDC, so that the cyanide nucleophile that gives rise to the 4-carboxy substituent in PDC is replaced with a phosphonate anion or a thiolate nucleophile to give the 4-phosphonate and 4-sulfonate analogs of PDC (dubbed PPC and PSC, respectively).[17]

Assessing Inhibitory Activity

The development of conformationally constrained inhibitors of the glutamate transporters, as well as the testing of proposed pharmacophore models, is ultimately dependent upon assessing the ability of the analogs to bind to the substrate site on the transport proteins. This has typically been accomplished by quantifying the capacity of analogs to competitively inhibit the uptake of a radiolabeled substrate (e.g., L-[^{3}H]glutamate or D-[^{3}H]aspartate) into any one of a number of CNS preparations (e.g., brain slices,[18] primary cultures of neurons or glia,[19] or synaptosomes[14]). As synaptosomes can be rapidly prepared and easily used to screen large numbers of analogs, this preparation has been widely employed in structure–activity studies.[4–6]

[16] C. L. Willis, D. L. Dauenhauer, J. M. Humphrey, A. R. Chamberlin, A. L. Buller, D. T. Monaghan, and R. J. Bridges, *Toxicol. App. Pharmacol.* **144,** 45 (1997).

[17] H. P. Koch, S. Esslinger, A. R. Chamberlin, and R. J. Bridges, *Society for Neuroscience Abstracts* **22,** Abs. #619.6.

[18] V. Balcar and G. Johnston, *J. Neurochem.* **19,** 2657 (1972).

[19] A. Garlin, A. Sinor, S. Lee, J. Grinspan, and M. Robinson, *J. Neurochem.* **64,** 2572 (1995).

Synaptosomal Transport Assay

Reagents

Homogenization Buffer (HB): 10 m*M* Tris–acetate, 0.32 *M* sucrose, pH 7.4 (glacial acetic acid)

Ficoll: (Type 400 DL, Sigma) 7.5%, 12.0%, 0.32 *M* sucrose, 40 μM EDTA (potassium salt), pH 7.4 (KOH)

Assay Buffer (AB): 10 m*M* Tris–base, 128 m*M* NaCl, 10 m*M* D-glucose, 5 m*M* KCl, 1.5 m*M* NaH_2PO_4, 1 m*M* $MgSO_4$, 1 m*M* $CaCl_2$, pH 7.4 (glacial acetic acid)

Synaptosomal Preparation. Numerous procedures are available for the preparation of CNS synaptosomes (see Robinson, this volume).[19a] The following protocol, which employs a discontinuous Ficoll gradient, is based on the method of Booth and Clarke.[20] Two male Sprague-Dawley rats are sacrificed by rapid decapitation. The brains are dissected free from the skull, the cerebellums are removed, and the tissue is placed into ≈20 ml of ice-cold HB. The HB is decanted and the tissue resuspended in fresh, ice-cold HB (10 ml/brain). While submerged in the HB, the forebrains are diced into small pieces with scissors and then transferred to a chilled Wheaton (Millville, NJ) glass/glass homogenizer (15 ml, 0.001- to 0.003-inch clearance). The tissue is homogenized by hand (12 strokes) and then centrifuged (1100*g*, 3 min, 4°, Sorvall SS-34 rotor). The supernatant is collected and recentrifuged (14,600*g*, 10 min, 4°). The resulting pellet is resuspended in 10 ml 12% Ficoll using a motorized Potter-type glass/glass (0.045–0.065 mm clearance, 5 strokes) homogenizer (Braun, Allentown, PA) at 925 RPM. The suspension is then used as the bottom layer of a discontinuous Ficoll gradient on which is sequentially layered 10 ml of 7.5% Ficoll and 10 ml of HB. The gradient is centrifuged (111,000*g*, 30 min, 4°, swinging bucket SW-28 rotor). The synaptosomal layer that forms at the interface of the 12% and 7.5% Ficoll layers is collected with a Pasteur pipette and homogenized (motorized Potter-type, glass/glass, 0.045- to 0.065-mm clearance, 5 strokes, 925 RPM). The suspension is gently mixed with ≈25 ml of HB and centrifuged (28,150*g*, 20 min, 4°, SS-34 rotor). The resulting pellet is collected and suspended in AB (0.2 ml/mg wet wt). This results in a synaptosomal suspension of approximately 0.2 mg protein/ml. The synaptosomes, which are stored on ice, are used for transport assays within a few hours of preparation.

Synaptosomal Transporter Activity. The following protocol is based on

[19a] M. B. Robinson, *Methods Enzymol.* **296,** [13], (1998) (this volume).

[20] R. F. G. Booth and J. B. Clark, *Biochem. J.* **176,** 365 (1978).

the method of Kuhar and Zarbin.[21] A 100 μl aliquot of the synaptosomal suspension is added to an assay tube and preincubated at 25° for 5 min. The assay can be conveniently carried out in duplicate. Uptake is initiated by the addition of a 100 μl aliquot of AB containing D-[^{3}H]aspartate (1-20 μM, final concentration). In those assays evaluating inhibitor activity, the 100 μl aliquot of AB contains both the D-[^{3}H]aspartate and potential inhibitors to ensure simultaneous addition. Following a 2-min incubation at 25°, the assay tubes are rapidly transferred to ice and quenched by the addition of 4 ml of ice-cold AB. The synaptosomal suspension is quickly vacuum ($\approx$20 mm Hg) filtered through GF/F (Whatman, Clifton, NJ) glass microfiber filters. The assay tubes and filters are sequentially rinsed with an additional 4 ml of ice-cold AB. The termination of the assay requires less than 20 sec to complete. The filters containing the synaptosomes are transferred to vials containing liquid scintillation fluid and allowed to dissolve for 24 hr. Retained radioactivity is quantified by liquid scintillation counting and corrections for nonspecific uptake (e.g., leakage, binding) are made by subtracting the amount of D-[^{3}H]aspartate accumulated at 4°. Initial assays should confirm that uptake is both sodium-dependent, by utilizing sodium-free (choline-substituted) buffer, and linear with respect to time and protein content. Kinetic constants of substrates and inhibitors can be determined using this same procedure and appropriately varying the concentrations of D-[^{3}H]aspartate and inhibitor with respect to K_m and K_i values (for review, see Ref. 22).

Assessing Substrate Activity

Although the various preparations of CNS tissue may provide a more relevant pharmacological overview of transport as it occurs *in vivo*, the systems do suffer from potential complications associated with heterogeneity (see Robinson, this volume).[19a] As an alternative, the recent cloning and expression of distinct glutamate transporters has allowed each of the subtypes to be selectively characterized (see Chan *et al.*,[22a] this volume). Advantageously, expression of the individual transporters in oocytes permits uptake to be quantified both biochemically and electrophysiologically.[23] Measurement of uptake-induced currents provides an approach to assess the suitability of glutamate analogs as transporter substrates that is

[21] M. J. Kuhar and M. Zarbin, *J. Neurochem.* **31,** 251 (1978).

[22] H. Christensen, "Biological Transport." W. A. Benjamin, Reading, Massachusetts, 1975.

[22a] A. Cha, N. Zerangue, M. Kavanaugh, and F. Bezanilla, *Methods Enzymol.* **296,** [38], (1998) (this volume).

[23] J. L. Arriza, W. A. Fairman, J. I. Wadiche, G. H. Murdoch, M. P. Kavanaugh, and S. G. Amara, *J. Neurosci.* **14,** 5559 (1994).

not dependent on the availability of radiolabeled analogs. In contrast, it must be remembered that competitive inhibition studies with radiolabeled substrates (see above) address only whether or not an analog binds to the transporter, not if it is also translocated. This limitation can be in part overcome in a more traditional neurochemical preparation by testing for the process of heteroexchange. Early studies of glutamate uptake demonstrated that the addition of alternative substrates could stimulate the efflux of another substrate that had previously been transported ("loaded") into the synaptosomes.[24] Thus, if an analog is able to undergo heteroexchange, it can be further classified as a substrate and differentiated from nontransportable inhibitors that bind to the substrate site on the carrier protein, but are not transported. Caution must be exercised in these experiments to ensure that the initial internal content of radiolabeled substrate is similar when comparisons of efflux rates are made and that the observed efflux occurs specifically via the transporter and not by a secondary mechanism (e.g., nonspecific leakage, receptor-activated event). The specificity of the leakage can be confirmed with a previously identified nontransportable inhibitor (e.g., DHK[23]) and the potential contribution of EAA receptors can be excluded through the use of EAA agonists and antagonists. The process of exchange has been used to demonstrate that the conformationally constrained analogs L-*trans*-2,4-PDC[4] and *cis*-ACBD[5] are transportable inhibitors of high-affinity glutamate uptake in cerebellar granule cells and cortical astrocytes[25] and that L-*trans*-2,3-PDC is a nontransportable inhibitor of the high-affinity glutamate transporter in forebrain synaptosomes.[26]

Synaptosomal Exchange Activity. Synaptosomes are prepared as described above, except that the final suspension is made in a smaller volume of AB (0.04 ml/mg wet wt). A 9.9-ml aliquot of this suspension is transferred to an assay tube containing 100 μl of D-[^{3}H]aspartate (2.5 μM, final concentration) and incubated at 25° for 15 min. The synaptosomal suspension is centrifuged (28,150g, 20 min at 4°) to reisolate the D-[^{3}H]aspartate-containing synaptosomes, which are then suspended in 5.0 ml of ice-cold AB and maintained in an ice bath. The assay can be conveniently carried out in duplicate. The exchange assays are initiated by the addition of 100 μl of the suspension to 2.9 ml of AB preequilibrated to 37° and containing potential substrates. The assay is terminated 1–10 min later by rapidly quenching with 4 ml of ice-cold AB followed by vacuum filtration ($\approx$20

[24] M. Erecinska and M. Troeger, *FEBS Lett.* **199,** 95 (1986).

[25] R. Griffiths, J. Dunlop, A. Gorman, J. Senior, and A. Grieve, *Biochem. Pharmacol.* **47,** 267 (1994).

[26] C. L. Willis, J. M. Humphrey, H. P. Koch, J. A. Hart, T. Blakely, L. Ralston, C. A. Baker, S. Shim, M. Kadri, A. R. Chamberlin, and R. J. Bridges, *Neuropharmacol.* **35,** 5531 (1996).

mmHg) through GF/F glass fiber filters. The assay tubes and filters are sequentially rinsed with an additional 4 ml of ice-cold AB. The filters containing the synaptosomes are transferred to vials containing liquid scintillation fluid and allowed to dissolve for 24 hr. Retained radioactivity is quantified by liquid scintillation counting and used to calculate changes in rate of efflux of D-[^{3}H]aspartate produced by the presence of the various analogs. The initial content of the synaptosomes ("0" time) is established by adding the 100-μl aliquot of synaptosomes to 2.9 ml of ice-cold AB followed immediately by filtering and rinsing. This value is determined at the beginning, middle, and end of each experiment to ensure that the amount of D-[^{3}H]aspartate in the synaptosomes does not significantly change while they are being maintained in the ice bath. To ensure that appropriate comparisons are made between experiments, the synaptosomes must initially contain similar amounts of D-[^{3}H]aspartate, as internal content could effect rates of efflux. Substrates can be identified by their ability to increase the rate of efflux observed in AB alone. In contrast, nontrans-

TABLE I

SYNAPTOSOMAL EFFLUX OF D-[^{3}H] ASPARTATE: COMPARISON OF TRANSPORTABLE AND NONTRANSPORTABLE INHIBITORS

Compound (μM)[a]	Inhibition of synaptosomal D-[^{3}H] aspartate uptake[b] K_i (μM)	Synaptosomal efflux of D-[^{3}H] aspartate[c] (pmol/min/mg protein)
L-Glutamate (50)	5 ± 1	167 ± 9
L-Glutamate (10)	—	147 ± 8
DHK (300)	28 ± 2	6 ± 0.3
L-Glutamate (10) and DHK (300)	—	30 ± 6
L-*trans*-2,4-PDC (15)	1.5 ± 0.5	104 ± 14
L-*anti-endo*-3,4-MPDC (50)	5 ± 2	60 ± 9
L-*trans*-2,3,-PDC (300)	33 ± 6	9 ± 3
L-Glutamate (10) and L-*trans*-2,3-PDC (300)	—	38 ± 8

[a] To ensure comparable levels of transporter binding, compounds were included in the exchange assays at a concentration of approximately 10× K_i.

[b] Lineweaver–Burk plots and associated kinetic analysis of the transport inhibitors were carried out using k_{cal} kinetic program (BioMetallics Inc., Princeton, NJ) with weighting based on constant relative error. K_i values were estimated on the basis of a replot of K_m, approximate values.

[c] Initial content of D-[^{3}H] aspartate in the synaptsomes was 1475 ± 200 pmol/mg protein. Rates were determined over a 2 min period and have been corrected for efflux of D-[^{3}H] aspartate at 37° in the absence of added compound (114 ± 17 pmol/min/mg). Values are reported as mean ± SEM ($n \geq 8$ duplicate determination).

portable inhibitors can be identified by both their inability to increase the rate of efflux and their ability to competitively inhibit the heteroexchange of other substrates (see Table I). In this respect, well-characterized non-transportable inhibitors (e.g., dihydrokainate) can be used to inhibit substrate-induced D-[^{3}H]aspartate efflux and verify that heteroexchange is occurring via the sodium-dependent transporters.

Conclusion

An iterative strategy employing cycles of modeling, synthesis, and assays has been very successful in the quest to develop new glutamate transport inhibitors and to define the structural requirements of the transporter binding site. These basic principles can be applied to the study of any small molecule–protein interaction for which there is an assay available, including other transporters, receptors, and enzymes. The approach does suffer from some limitations. In particular, it requires choosing specific atoms to be compared and it ignores the likelihood that complementary functional groups in a binding site can interact with a given functional group in different ways. For example, a carboxylate group can interact with complementary groups at a binding site in several of different orientations (e.g., "straight on" to one or both oxygen atoms via the inside electron lone pairs, or to only one of the oxygens via its outside lone pair). Nonetheless, it has provided a reasonable picture of the transporter binding site requirements and has successfully predicted the activities of several new transport inhibitors. As the specificity of the assay methods increase, whether in regard to transporter subtypes or substrate/inhibitor activity, the resulting differences in activities can be incorporated into the computational modeling studies. In turn, this should lead to the development of more detailed transporter pharmacophores and the development of even more selective analogs.

[13] Examination of Glutamate Transporter Heterogeneity Using Synaptosomal Preparations

By MICHAEL B. ROBINSON

Introduction

For almost three decades, synaptosomes have been used to study Na^+-dependent high affinity transport of several neurotransmitters, including dopamine, choline, norepinephrine, glycine, γ-aminobutyric acid (GABA),

0076-6879/98 $25.00

and glutamate. As the names imply, the activity of these transporters is dependent on the Na^+ electrochemical gradient to actively accumulate substrate and maintain a transmembrane gradient. The identification of transport activities for each of these molecules was used as evidence to support their neurotransmitter status by providing a route for inactivation of neurotransmitter after release.

Synaptosomes are a membrane fraction enriched in synaptic nerve terminals that have pinched off from the cell bodies and resealed. Based on electron microscopic studies,[1,2] the synaptosomes contain mitochondria and are presumably capable of generating ATP, which maintains the electrochemical gradients required for transporter activity. Although this membrane fraction is enriched in nerve terminals, these preparations are generally thought to be contaminated with glial elements.[3] These glial elements can reseal to form gliosomes. Therefore, it is possible that the transport activities in these preparations are due to uptake into neuronal and/or glial elements. For glutamate transporters, this may be a significant issue since there are different subtypes of transporters expressed by glia and by neurons.[4,5] At present, the relative contribution of glial and neuronal glutamate transporters to the activity observed in synaptosomes has not been determined.

Using these synaptosomes or slightly less purified preparations, frequently called crude synaptosomes or P2, several groups have studied the properties of Na^+-dependent L-glutamate transport. These studies have included the examination of the pharmacological properties of this activity in preparations derived from different brain regions (for reviews, see[6–8]). The results from these studies provided strong evidence for the existence of subtypes of Na^+-dependent glutamate transporters. The cloning of several different cDNAs that can be used to express Na^+-dependent glutamate transport in heterologous systems proved that there are subtypes or isozymes of this system.[9–12] It has been somewhat difficult to demonstrate that

[1] C. W. Cotman and D. Taylor, *J. Cell Biol.* **55,** 696 (1972).

[2] V. P. Whittaker, *in* "The Subcellular Fractionation of Brain Tissue with Special Reference to the Preparation of Synaptosomes and Their Component Organelles," pp. 1–52. Marcel Dekker, New York, 1972.

[3] F. A. Henn, D. J. Anderson, and D. G. Rustad, *Brain Res.* **101,** 341 (1976).

[4] J. D. Rothstein, L. Martin, A. I. Levey, M. Dykes-Hoberg, L. Jin, D. Wu, N. Nash, and R. W. Kuncl, *Neuron* **13,** 713 (1994).

[5] K. P. Lehre, L. M. Levy, O. P. Ottersen, J. Storm-Mathisen, and N. C. Danbolt, *J. Neurosci.* **15,** 1835 (1995).

[6] V. J. Balcar and Y. Li, *Life Sciences,* **51,** 1467 (1992).

[7] N. C. Danbolt, *Prog. Neurobiol.* **44,** 377 (1994).

[8] M. B. Robinson and L. A. Dowd, *Adv. Pharm.* **37,** 69 (1997).

[9] Y. Kanai and M. A. Hediger, *Nature* **360,** 467 (1992).

the pharmacological properties and regional distribution of the cloned transporters recapitulate those predicted with synaptosomes prepared from different brain regions. In this chapter, the methods used to study the pharmacological properties of L-[^{3}H]glutamate transport in crude synaptosomes and some limitations of this approach will be discussed.

Methods

The following reagents will be required to measure transport activity. Normally the buffers and the sucrose are prepared the night before an experiment and stored at 4°.

0.32 M sucrose.

Na^+-Containing Transport Buffer. (About 1 liter is needed.) 5 m*M* Tris base, 10 m*M* HEPES, 140 m*M* NaCl, 2.5 m*M* KCl, 1.2 m*M* $CaCl_2$, 1.2 m*M* $MgCl_2$, 1.2 m*M* K_2HPO_4, 10 m*M* dextrose. This buffer is usually about pH 7.2 after the salts are added and should not need to be adjusted. If the pH is not close to pH 7.2, this usually indicates that something has not been added correctly.

Na^+-Free or Choline-Containing Buffer. (At least 4 liters are needed; most will be used for the rinses with the filtration machine.) The recipe for this buffer is the same as for the sodium-containing buffer except that 140 m*M* choline chloride is substituted for the NaCl.

Stock Solutions of Glutamate and Inhibitors. Normally, 10 m*M* stocks are made in Na^+-containing buffer and the solutions are diluted as needed to provide 10× solutions for assays as described in Table I. With many of the excitatory amino acids, neutralizing the solutions with NaOH facilitates solubilization and is necessary to avoid changing the pH of the buffer when high concentrations of inhibitor are used. Use KOH or another base to neutralize stock solutions made up in choline-containing buffers.

Cocktail. Two 10 μM cocktails are made: one with Na^+-containing buffer and one with choline-containing buffer. Nonradioactive material makes up the bulk of the Glu in the cocktail (9.9 μM) and the rest is radioactive. We routinely purchase L-[^{3}H]Glu from DuPont (NEN/Dupont, Boston,

[10] G. Pines, N. C. Danbolt, M. Bjørås, Y. Zhang, A. Bendahan, L. Eide, H. Koepsell, J. Storm-Mathisen, E. Seeberg, and B. I. Kanner, *Nature* **360,** 464 (1992).

[11] T. Storck, S. Schulte, K. Hofmann, and W. Stoffel, *Proc. Natl. Acad. Sci. USA* **89,** 10955 (1992).

[12] W. A. Fairman, R. J. Vandenberg, J. L. Arriza, M. P. Kavanaugh, and S. G. Amara, *Nature* (*London*) **375,** 599 (1995).

TABLE I
GUIDE FOR PIPETTING GLUTAMATE TRANSPORT ASSAYS

Tube	Membranes (μl)	Cocktail (μl)	Buffer (μl)	Inhibitor
Pipetting pharmacological studies[a]				
Total	50	25	425	None
Blank	50	25	425	None
1 μ*M* inhibitor	50	25	375	50 μl of 10 μ*M* stock
3 μ*M* inhibitor	50	25	375	50 μl of 30 μ*M* stock
etc.				
Pipetting for analysis of kinetic constants[b]				
				Additional non-radioactive Glu
0.5 μ*M* total	50	25	425	None
0.5 μ*M* blank	50	25	425	None
1 μ*M* total	50	50	400	None
1 μ*M* blank	50	50	400	None
3 μ*M* total	50	50	300	100 μl of 10 μ*M* stock
3 μ*M* blank	50	50	300	100 μl of 10 μ*M* stock
etc.				

[a] When performing many of these assays, it is convenient to dilute the cocktail 1 part with 1 part buffer. With this diluted cocktail 50 μl can be added and it is easier to use a repeater pipette.

[b] Individuals in the laboratory are strongly encouraged to make up a table like one of these for each set of 24 tubes.

MA) which is shipped as 1 mCi/ml and has a specific activity of between 40 and 60 Ci/mmol. If one uses the specific activity in microliters to make 10 ml of cocktail, the concentration is 0.1 μ*M*. For example, our current lot is 49.4 Ci/mmol, so 49.4 μl is put in 10 ml. To make cocktails, 99 μl of 1 m*M* Glu in either Na^+- or choline-containing buffer and the appropriate volume of radioactive L-Glu are combined. The volume is brought to 10 ml with either Na^+- or choline-containing buffer. With the example of the specific activity described above, there should be 548,000 disintegrations per minute (dpm) per 50 μl of cocktail.

Preparation of Crude Synaptosomes

The procedure routinely used has been adapted from previously published protocols.[2,13] All steps in this procedure are done while maintaining

[13] M. B. Robinson, M. Hunter-Ensor, and J. Sinor, *Brain Res.* **544,** 196 (1991).

the tissue at 4°. The volumes are based on starting wet weight and rounded to the nearest 0.5 ml. The volumes for the dilutions sometimes exceed the capacity of the homogenizer; when this is the case we homogenize in a constant volume of 5 or 10 ml and then dilute to the larger volume. The procedure generally takes about 1.5 hr from start to finish. We routinely use the tissue within 1–2 hr of preparation.

1. After euthanasia by decapitation, the brain tissue is rapidly harvested and dissected on a metal plate that is cooled on ice. Although finer dissection of the brain tissue is possible, six brain regions, including cortex, striatum, hippocampus, midbrain, brain stem, and cerebellum, are relatively easy to dissect from rat brain with the smallest brain region weighing approximately 60 mg.
2. The tissues are homogenized in 20 volumes of ice-cold 0.32 *M* sucrose at 400 rpm for 7 strokes using a Teflon/glass homogenizer (Wheaton, Millville, NJ) and centrifuged at 800*g* for 10 min in a floor model centrifuge with refrigeration. We use a Beckman J2-21 centrifuge with a JA20.1 rotor. With this rotor the speed is 2500 rpm.
3. The supernatant from this centrifugation step is poured into another tube and centrifuged in the same rotor at 12,500 rpm (20,000*g*) for 20 min.
4. After removal of the supernatant, the resultant pellet is resuspended in 40 volumes of ice-cold sucrose by vortexing and is centrifuged at 12,500 rpm (20,000*g*) for 20 min.
5. This washed P2 pellet (crude synaptosomes) is resuspended in 50 volumes of sucrose by homogenizing at 400 rpm for 2 strokes. This solution is then kept on ice until use.
6. Proteins are measured using the procedure originally described by Lowry.[14] With this procedure, the final protein concentration is between 25 and 35 μg/50 μl after dilution into 50 volumes of sucrose. We usually put the crude synaptosomes into plastic 20-ml scintillation vials. We do not routinely measure the protein prior to measuring the properties of transport; this is done at some later time.

Notes: It is possible to further enrich the synaptosomes using discontinuous sucrose density gradients. This step reduces the contamination with myelin and mitochondria.[1,2] We do not routinely use this additional step, but we have examined some of the pharmacological properties of transport

[14] O. H. Lowry, N. J. Rosenberg, A. L. Farr, and R. J. Randall, *J. Biol. Chem.* **193,** 265 (1951).

in cortex and cerebellum after this additional step and found that they are comparable to that observed in the crude synaptosomal preparation.[15]

Measurement of Na^+-Dependent L-[^{3}H]Glutamate Transport Activity

In the original characterization of the transport activities for these transporters, a centrifugation step was used to stop the reaction and separate accumulated substrate from free substrate. With the availability of filtration machines, it has become much less time consuming. It is relatively easy for a single person to do more than 250 assays in a day.

The night before or prior to starting the membrane preparation, it is advisable to make up all of the stock solutions and label the test tubes. Twenty-four test tubes are put in each test tube rack. This number of test tubes works well for the timing of the assays. The racks must have the specific spacing required by the cell harvester. Assays are initiated by the addition of 50 μl of crude synaptosomes. Therefore, we generally pipette the other components of the assay into test tubes during the centrifugation steps used to prepare the crude synaptosomes.

Table I serves as a guide for pipetting the assays. After all reagents are added, place the test tube racks in a 37° water bath to equilibrate them to the assay temperature. Place an empty test tube rack into a water–ice slurry. On completion of the assay, each duplicate set of tubes is moved from the warm water bath to the ice bath for rapid cooling which stops all transport activity. Place a beaker with ice-cold choline-containing buffer in the ice bath or a separate ice bucket and set up a repeater pipette with a tip set at 2 ml.

Each assay is incubated for 3 min. Crude synaptosomes are added to duplicate tubes at 15-sec intervals. The tubes are vortexed and placed back in the 37° bath. With this timing, synaptosomes can be added to all 24 tubes in 2 min, 45 sec. Starting at 3 min, 2 ml of ice-cold choline-containing buffer are added to each set of duplicate tubes, and after vortexing the tubes are placed in the test tube rack that is in the water/ice slurry. After all reactions are stopped, the tubes are filtered using a Brandell cell harvester. Prewet a #32 glass fiber filter with ice-cold choline-containing buffer; filter the assays and wash three times with ice-cold choline-containing buffer (2 ml). At the end of the experiment, the filters are removed with forceps and placed in scintillation vials with 5 ml of scintillation cocktail (cytoscint ES from ICN Biochemicals Inc., Irvine, CA).

[15] M. B. Robinson, J. D. Sinor, L. A. Dowd, and J. F. Kerwin, Jr., *J. Neurochem.* **60,** 167 (1993).

Notes: Although metabolism of L-Glu by these preparations during the time of assay has not been observed,[16] it is at least a theoretical concern. Therefore, many groups use D-[^{3}H]aspartate as a substrate for this transporter. Another advantage of D-[^{3}H]aspartate is that there is less Na^+-independent accumulation than with L-[^{3}H]Glu. In spite of these advantages, D-Asp is not the endogenous substrate and there is some evidence that the relative turnover numbers for the subtypes of transporters may be different.[17] Some groups have compared the pharmacology of L-[^{3}H]Glu and D-[^{3}H]Asp transport in various brain regions and found evidence for modest differences.[6,18]

Data Calculations and Properties of Assay

Depending on the brain region, accumulation in the presence of Na^+ is usually between 15,000 and 30,000 dpm at low concentrations of Glu (0.5 μM). In the absence of Na^+, the accumulation is generally between 500 and 1000 dpm. Less than 5% of the total accumulation of radiolabel is Na^+-independent when measured at 0.5 μM. The intraassay variability is quite low; duplicate measures are within 5%. With this high signal-to-noise ratio, it is not necessary to examine the sensitivity of Na^+-independent accumulation to compounds being tested as inhibitors of Na^+-dependent transport activity. At higher concentrations of Glu, the ratio of Na^+-dependent to Na^+-independent accumulation decreases. If the sensitivity of transport to compounds is going to be examined at higher concentrations of L-Glu, it may be necessary to confirm that the compound does not inhibit Na^+-independent accumulation, which may represent Cl^--dependent transport.[19]

Using the assay conditions described above, Na^+-dependent transport of L-[^{3}H]Glu is linear for at least 5 min, and less than 10% of the exogenous substrate is accumulated. The Na^+-dependent accumulation is also linear with protein. The accumulation of radiolabel in choline-containing buffer is comparable to that observed with incubations performed at 4°.[13]

The data are calculated in Eq. (1):

[16] J. Ferkany and J. T. Coyle, *J. Neurosci. Res.* **16,** 491 (1986).

[17] J. L. Arriza, W. A. Fairman, J. I. Wadiche, G. H. Murdoch, M. P. Kavanaugh, and S. G. Amara, *J. Neurosci.* **14,** 5559 (1994).

[18] A. D. Mitrovic and G. A. R. Johnston, *Neurochem. Int.* **24,** 583 (1994).

[19] R. Zaczek, M. Balm, S. Arlis, H. Drucker, and J. T. Coyle, *J. Neurosci. Res.* **18,** 425 (1987).

Velocity

$$= \frac{(\text{dpm in presence of Na}^+ - \text{dpm in absence of Na}^+)}{(2.22 \times 10^{12}\ \text{dpm/Ci} \times \text{Specific activity of Glu in Ci/mol} \times 3\ \text{min} \times \text{protein in mg})}. \quad (1)$$

If the specific activity is in Ci/mol then the velocity calculated using the equation described above is in mol/mg protein per min. With the example described in the directions for making cocktail, the specific activity would be 494 Ci/mol. If the concentration of glutamate is changed, it is important to remember to change the specific activity based on the amount of glutamate added.

The kinetic properties of L-[^{3}H]Glu transport have been examined by several groups. Although some observe two sites, a high-affinity site and a low-affinity site, others observe only a single site and a nonsaturable or a Na^+-independent component (for a recent summary, see Ref. 8). Although we do not observe a low affinity site, we do not routinely use high concentrations of Glu. In our laboratory, the K_m values for L-[^{3}H]Glu obtained using crude synaptosomes prepared from different brain regions are between 1 and 5 μM. In fact, cerebellum and brainstem consistently have 2- to 3-fold lower K_m values than forebrain regions, but it is uncertain if this statistically significant difference is biologically relevant.[13] The V_{max} values for transport are quite different for different brain regions. In cerebellum and brain stem the V_{max} values are approximately 0.5 nmol/mg protein per min and in forebrain/midbrain regions the V_{max} values are approximately 1 to 3 nmol/mg protein per min.

We have examined the dependence of transport activity on Na^+ by varying the concentration of Na^+ using crude synaptosomes prepared from cerebellum or cortex.[15] The apparent K_m values and Hill slopes for Na^+ were examined at both low (0.1) and high concentrations of L-Glu (50 and 100 μM). At the higher concentrations of L-Glu, the K_m values for Na^+ are approximately 7 to 10 mM and the Hill slopes are approximately 2. At the lower concentrations of L-Glu, the K_m values for Na^+ are significantly higher (12 to 20 mM) and the Hill slopes are different in cerebellum and cortex. In cerebellum, the Hill slope is approximately 3 and in cortex the Hill slope is approximately 2. Since it is likely that the bulk of transport activity in cerebellum is mediated by a different transporter(s) than the activity in cortex, it is possible that there are differences in the number of Na^+ ions transported per molecule of L-Glu. A difference in stoichiometry was also proposed based on electrophysiological measurements using cloned transporters expressed in heterologous systems.[20]

In both cortex and cerebellum, the K_m and the V_{max} for L-Glu transport are affected by the concentration of Na^+. As the concentration of Na^+ is increased, the K_m values decrease and the V_{max} values increase.[15] The observation that Glu affects the apparent K_m for Na^+ and that Na^+ affects the apparent K_m for Glu is consistent with either a random order of association of Na^+ and Glu with the transporter or an ordered reaction with binding of Na^+ : Glu : Na^+.[21]

Pharmacological Analyses

Initially, each compound is tested at a high concentration (100 μM to 3 mM), depending on the availability of compound. If one assumes simple competitive inhibition of a homogeneous population of transporters, these preliminary studies can be used to estimate the IC_{50} value using Eq. (2):

$$F = \frac{1}{(1 + I/IC_{50})}. \tag{2}$$

In Eq. (2), F is the fraction of transport that is not inhibited; this is obtained by calculating the velocity of Na^+-dependent transport in the presence of inhibitor and dividing this value by the velocity in the absence of inhibitor. I is the concentration of inhibitor and IC_{50} is the concentration of inhibitor that inhibits 50% of the activity.

After estimating the IC_{50} value in initial experiments, we usually examine the concentration dependence for inhibition of glutamate transport using three concentrations in every log unit of concentration (e.g., 1, 3, 6, 10, and 30 μM). These concentrations are chosen to center on the IC_{50} value and should span from 10% inhibition to 90% inhibition.

When fitting the data, our goals are to identify the IC_{50} value and to determine if the data are consistent with a single population of sites. The data are first fit by linear regression analysis to the Hill equation. Most compounds are consistent with a single site with Hill slopes near 1. It is important to examine each individual experiment because averaging data from experiments with dramatically different IC_{50} values can give a Hill slope less than 1. If the Hill slope is near 1, a weighted IC_{50} value is calculated using Eq. (2). The weights assigned are based on the slope of a

[20] N. Zerangue and M. P. Kavanaugh, *Nature* **383,** 634 (1996).

[21] W. D. Stein, "Transport and Diffusion across Cell Membranes." Academic Press, New York, 1986.

theoretical curve with data near the IC_{50} receiving the highest weighting. If the Hill coefficient for the inhibition is less than 0.7, the data are fit to one or two sites using nonlinear regression analysis minimizing the sum of the square of the residuals. The NIH sponsored PROPHET system works well for this comparison providing a statistical comparison of the one- and two-site fits of the data. An Excel spreadsheet that performs the same function can also be used.

As is true for most pharmacological studies, many of the compounds tested in these systems also interact with excitatory amino acid receptors and may also interact with other proteins or nonexcitatory amino acid receptors. It is important to determine if the data for inhibition of transport are consistent with a competitive mechanism of action. This is done by examining the concentration dependence for glutamate transport in the absence and presence of increasing concentrations of inhibitor. Although it is possible to use this data for any one of a number of different plots to determine the K_i for the inhibitor, we use this data to determine if the inhibitor changes the K_m or the V_{max} for glutamate transport and if the changes in K_m are consistent with a simple competitive mechanism of action. For example, the concentration of inhibitor that inhibits transport by 50% (using low concentrations of glutamate—0.5 μM) should double the K_m.[22] It is important to remember that even if a compound increases the K_m, it is theoretically possible that the compound is acting by a noncompetitive mechanism.

Comparison of the pharmacological properties of Na^+-dependent L-Glu transport in different brain regions reveals dramatically different sensitivity to inhibition by some, but not all compounds (see Table II). For example, L-α-aminoadipate inhibits transport activity in cerebellum with an IC_{50} value of approximately 40 μM and inhibits transport in hippocampus, striatum, cortex, and midbrain with an IC_{50} value of approximately 600 μM. The opposite pattern is observed with dihydrokainate, which inhibits transport in the forebrain/midbrain regions with an IC_{50} value of approximately 150 μM, but does not inhibit transport in cerebellum (approximately 10% inhibition at 3 mM). In brain stem, the data for inhibition of transport by dihydrokainate and L-α-aminoadipate are best fit to two sites and consistent with approximately a 50 : 50 mixture of the two sites. These pharmacological comparisons have been extended to several other excitatory amino acid analogs. Now there are at least six compounds that show greater than 10-fold selectivity for inhibition of transport in either cortex or cerebellum. This differentiation of cerebellar and forebrain transport activity suggests

[22] C. Cheng and W. H. Prusoff, *Biochem. Pharmacol.* **22,** 3099 (1973).

TABLE II
POTENCIES OF EXCITATORY AMINO ACID ANALOGS FOR INHIBITION OF Na^+-DEPENDENT GLUTAMATE TRANSPORT IN SYNAPTOSOMES

	IC_{50} values (μM) for brain region					
Compound	Cortex	Striatum	Hippocampus	Midbrain	Brain stem	Cerebellum
L-Aspartate	3.5[a]	3.3[b]	N.D.	N.D.	N.D.	1.7[a]
D-Aspartate	3.1[a]	4.9[b]	N.D.	N.D.	N.D.	1.7[a]
L-*trans*-Pyrrolidine 2,4-dicarboxylate	2.3[a]	N.D.	N.D.	N.D.	N.D.	1.6[a]
DL-*threo*-β-Hydroxy-aspartate	2.3[c]	1.6[c]	2.1[c]	1.5[c]	2.3[c]	3.4[c]
Dihydrokainate	120[c]	80[c]	170[c]	160[c]	100 (65%)[c] 16,000 (35%)	>3000[c]
L-α-Aminoadipate	800[c]	650[c]	650[c]	590[c]	40 (50%)[c] 700 (50%)	40[c]
Kainate	72 (71%)[a] 3500 (21%)	N.D.	N.D.	N.D.	N.D.	53 (83%)[a] Insensitive (23%)

[a] Data from[13].
[b] Data from[15].
[c] Data from[16].

that cerebellar transport in crude synaptosomes is mediated by a different transporter(s) than the activity observed in forebrain crude synaptosomes.

Although the data for inhibition of transport activity by many of the compounds are consistent with a single homogeneous population of sites, data for inhibition by some of the compounds are consistent with two sites. Hill slopes for inhibition of transport activity by several compounds are less than 0.7, and the data for inhibition are best fit to two sites (see Table II). In our studies with kainate, every inhibition curve is best fit to two sites. Approximately, 80% of the sites have high affinity (50–70 μM), and 20% of the sites either have low affinity or are insensitive.

Comments

Although this approach has provided valuable information about the properties of transporter activity, it is important to recognize the potential limitations of this assay.

Potential Indirect Effects

As is true for most experimental preparations that are not purified to homogeneity, it is important to recognize that inhibitors may not block transport by directly interacting with the transporter. For example, compounds that block the Na^+,K^+-ATPase can inhibit transport activity. Examination of the concentration dependence in the absence and presence of increasing concentrations of inhibitor can be helpful. An increase in the apparent K_m is consistent with a competitive interaction, but this kinetic approach cannot be used to unambiguously identify the mechanism of action. It is at least theoretically possible that inhibitors can increase the K_m through an indirect mechanism. For example, excitatory amino acid analogs could stimulate release of Glu or Asp, which would inhibit transport activity and appear competitive.

Differentiation of Pure Blockers from Substrate Inhibitors

With this assay, it is not possible to differentiate compounds that are pure blockers from those that are substrates. It is critical to differentiate these two different types of inhibitors because they are frequently used to determine if transport activity limits the duration of rapid synaptic signaling and to determine if transport activity limits the toxicity of Glu.[7,8] Because most of the inhibitors of transport activity available to date are either substrates or are nonselective, it is likely that compounds identified in the future as selective inhibitors of transport activity will be used in other functional assays to examine the role of transport activity in limiting physiological and pathological effects of Glu.

Two different strategies have been developed to differentiate pure blockers from substrates. These strategies are described elsewhere in this volume and will be described briefly here. The first strategy depends on the measurement of current.[23] Transport of substrates, such as L-Glu and D-Asp, by these transporters is accompanied by a net inward flux of positive charge. Therefore, compounds that block the current generated during the transport of a substrate but generate little or no current when applied alone are likely to be non-substrate-inhibitors. This electrogenic property can be used to study transporter activity in heterologous expression systems or intact cellular systems that endogenously express the transporters. The second strategy depends on the measurement of release of radioactivity from synaptosomes preloaded with a substrate.[24] Compounds that are substrates stimulate the efflux of radioactivity and compounds that are blockers

[23] A. Cha, N. Zerangue, M. Kavanaugh, and F. Bezanilla, *Methods Enzymol.* **296,** [38], (1998) (this volume).

[24] A. R. Chamberlain, H. P. Koch, and R. J. Bridges, *Methods Enzymol.* **296,** [12], (1998) (this volume).

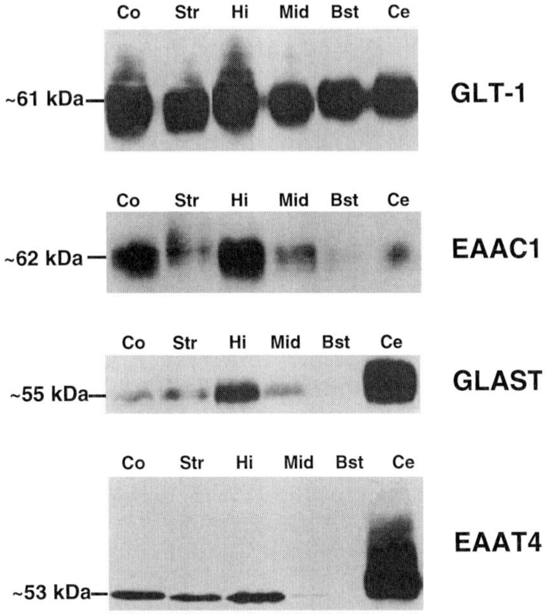

FIG. 1. Immunoblots of subtypes of glutamate transporters in crude synaptosomal membranes prepared from cortex (Co), striatum (Str), hippocampus (Hi), midbrain (Mid), brain stem (Bst), and cerebellum (Ce). These immunoblots were prepared as previously described[25] using antibodies provided by Dr. Jeff Rothstein. Except for the GLT-1 immunoblots, 50 μg of protein was loaded in each of the lanes. For the GLT-1 immunoblots, 5 μg of protein was loaded in each lane. The immunoblot presented is representative of two independent experiments.

block this efflux. The limitations of each of these strategies are discussed in these articles.

Multiple Subtypes Present in Synaptosomal Preparation

With the cloning of multiple transporters, several groups have developed antibodies that specifically interact with subtypes of these transporters. We have used antibodies from the laboratory of J. Rothstein to determine which transporters are present in the crude synaptosomal preparations derived from different brain regions (see Fig. 1). Using immunoblots to qualitatively examine the distribution, GLT-1 (also known as EAAT2) has a relatively uniform distribution in the six different brain regions with

[25] L. A. Dowd, A. J. Coyle, J. D. Rothstein, D. B. Pritchett, and M. B. Robinson, *Mol. Pharm.* **49,** 465 (1996).

modest enrichment in hippocampus and cortex. EAAC1 (also known as EAAT3) is enriched in cortex and hippocampus with relatively low levels in brain stem and cerebellum. In contrast, GLAST (also known as EAAT1) and EAAT4 are selectively enriched in cerebellum compared to forebrain and brainstem regions. This indicates that in each brain region transport activity may be mediated by a heterogeneous population of transporters. In considering the impact of the expression of multiple subtypes of transporters, it is important to remember that both the V_{max} and the K_m of each subtype may affect the relative contribution of each subtype to the pharmacology at any concentration of Glu. The K_m values for the subtypes of transporters may be different. Therefore, at the low concentrations of Glu that we use routinely, transporters with lower apparent affinities but with comparable capacities may not contribute significantly to the activity measured. Unfortunately, the relative abundance of each of the transporters has not been clearly defined in many brain regions, but antisense oligonucleotides have been used to examine the contribution of each of the transporters to activity in some brain regions. This strategy is described elsewhere in this volume.[26]

Since there are multiple transporters in crude synaptosomes prepared from the different brain regions, one might expect that subtype selective inhibitors of transport activity would provide evidence for heterogeneity of transport within a particular brain region. This heterogeneity may, in part, explain why it has been difficult to reconcile the endogenous pharmacology of the transporter activity with that observed for the cloned transporters expressed in heterologous expression systems.[8]

Acknowledgments

The author is supported by NIH grants NS29868 and NS36465. The author would also like to thank Joanna Vondrasek and Brian Schlag for helpful comments for the revisions of this manuscript, Brian Schlag for the preparation of the immunoblots that are presented in Fig. 1, and Anjali Gupta for excellent editorial assistance.

[26] L. A. Bristol and J. Rothstein, *Methods Enzymol.* **296,** [35], (1998) (this volume).

[14] Specificity and Ion Dependence of Binding of GBR Analogs

By AMADOU T. CORERA, JEAN-CLAUDE DO RÉGO, and JEAN-JACQUES BONNET

Several of the GBR[1] compounds described in 1980[2] were the first potent inhibitors of the neuronal dopamine uptake which displayed a nanomolar affinity for the dopamine transporter (DAT) and a lower affinity for transporters of norepinephrine and serotonin.[2-4] Among the most potent and selective compounds, GBR 12783 and GBR 12935 have been used as radioligands for DAT[5,6] and some of their derivatives are known as photoaffinity probes for DAT.[7-9] Studies with [^{3}H]GBR 12783 have revealed that concentrations of sodium in the 10–30 m*M* range are sufficient for a maximal stimulation of specific binding of this ligand to DAT. [^{3}H]GBR 12783 has also been instrumental in the demonstration of the inhibitory effects of other cations on binding. These data have been verified with other DAT ligands. However, GBR derivatives also display some affinity for sites distinct from the DAT, including σ sites and two other sites called the piperazine acceptor site[10] and "site 2."[11]

In the present chapter, we will consider methodological aspects which play an important role in the binding of GBR derivatives to DAT, starting from a description of standard experimental procedures used for a GBR binding study. Special attention will be paid to binding specificity and to

[1] GBR, Gist BRocades.
[2] P. van der Zee, H. S. Koger, J. Gootjes, and W. Hespe, *Eur. J. Med. Chem.* **15,** 363 (1980).
[3] J.-J. Bonnet and J. Costentin, *Eur. J. Pharmacol.* **121,** 199 (1986).
[4] R. E. Heikkila and L. Manzino, *Eur. J. Pharmacol.* **103,** 241 (1984).
[5] J.-J. Bonnet, P. Protais, A. Chagraoui, and J. Costentin, *Eur. J. Pharmacol.* **126,** 211 (1986).
[6] A. Janowsky, P. Berger, F. Vocci, R. Labarca, P. Skolnick, and S. M. Paul, *J. Neurochem.* **46,** 1272 (1986).
[7] D. E. Grigoriadis, A. A. Wilson, R. Lew, J. S. Sharkey, and M. J. Kuhar, *J. Neurosci.* **9,** 2664 (1989).
[8] F. R. Sallee, E. L. Fogel, E. Schwartz, S.-M. Choi, D. P. Curran, and H. B. Niznik, *FEBS Lett.* **256,** 219 (1989).
[9] S. P. Berger, R. E. Martenson, P. Laing, A. Thurkauf, B. de Costa, K. C. Rice, and S. M. Paul, *Molec. Pharmacol.* **39,** 429 (1991).
[10] P. H. Andersen, *J. Neurochem.* **48,** 1887 (1987).
[11] H. C. Akunne, C. M. Dersch, J. L. Cadet, M. H. Baumann, G. U. Char, J. S. Partilla, B. R. de Costa, K. C. Rice, F. I. Carroll, and R. B. Rothman, *J. Pharmacol. Exp. Ther.* **268,** 1462 (1994).

0076-6879/98 $25.00

changes in Na^+ dependence resulting from modifications of the composition of incubation medium.

Standard Experimental Procedures for Binding Assays

Membrane Preparation

All procedures for membrane preparation are performed at 0–2°. Male Sprague-Dawley rats (150–300 g) purchased from Charles River are killed by decapitation and the striata are dissected out and homogenized in 10 volumes (w/v) of ice-cold 0.32 *M* sucrose, with 10 up-and-down strokes of a Teflon–glass homogenizer (800 rpm). The nuclear material is removed by centrifugation at 1000*g* for 10 min and the supernatant (S1) is stored. The P1 pellet is resuspended in an equal volume of sucrose and recentrifuged in order to improve the yield of the preparation. The two supernatants are combined and the mixture is centrifuged at 17,500*g* for 30 min. The resultant pellet P2 is resuspended in 15 volumes of a 130 m*M* Na^+ medium by sonication for 5 sec (3-mm microprobe diameter; Sonics & Materials, Danbury, CT). The 130 m*M* Na^+ medium contains (m*M*): NaH_2PO_4, 0.3, $NaHCO_3$, 9.7, NaCl, 120, pH 7.5 ± 0.1. The membranes are centrifuged at 50,000*g* for 10 min, and resuspended in the same medium by sonication. The protein concentration is determined and the membrane preparation is diluted to obtain a concentration of 50–100 μg of membrane protein/ml of a 10 m*M* Na^+ medium containing (m*M*): NaH_2PO_4, 0.3, $NaHCO_3$, 9.7, pH 7.5 ± 0.1. A slight modification of the ratio $NaH_2PO_4/NaHCO_3$ can be necessary to obtain pH 7.5 at incubation temperature.

Binding Experiments

The binding is started by addition of 400-μl aliquots of [^{3}H]GBR 12783 to silicone-coated tubes containing 1 ml of membrane preparation and drug solutions or their vehicle in a total volume of 3.6 ml (10 m*M* Na^+ incubation medium). The incubation period (10 min at 37°, 1 hr at 25°, 4 hr at 0°) is stopped by rapid vacuum filtration through GF/B filters previously soaked for at least 1 hr in 0.5% polyethyleneimine. Each tube is rinsed once and filters are washed three times with 5 ml of ice-cold 10 m*M* Na^+ medium. The radioactivity is counted by liquid scintillation spectrometry. The specific binding is calculated by subtracting the nonspecific binding determined in the presence of mazindol (Sigma) from the total binding.

Purity of the radioligand can decline quite rapidly and/or specific activity can be different from that stated in the data sheets.[11–13] Consequently, repurification of the radioligand should be achieved by high-performance liquid chromatography (HPLC)[11] and determination of the specific activity should be performed by homologous competition binding assays.[14] An exact determination of the specific activity is complicated by adsorption of the radioligand to containers. For this reason, in binding assays, we routinely use silicone-coated tubes from a commercial source; others have used chromic acid-cleaned glass containers and precoated pipette tips.[15] In any case, saturation curves should be constructed as a function of effective free ligand concentrations.[16] Addition of bovine serum albumin to the incubation medium to reduce adsorption has been either suspected or demonstrated to actually lower the free concentration of some of the compounds present in the medium, including that of the radioligand.[15,17]

Equilibrium dissociation constants (K_d), density of binding sites (B_{max}) and IC_{50} are usually estimated with the nonlinear curve-fitting program Ligand (Biosoft, Cambridge, UK). K_i values are calculated from IC_{50} values as described by Cheng and Prusoff[18] for competitive inhibitors. Data can also be analyzed by the method of binding surface analysis[19–21] in which binding surfaces are generated by competition between two concentrations of the radioligand. At each radioligand concentration, 8–10 different concentrations of the cold ligand are assayed in the absence and presence of binding inhibitors. Curve-fitting procedures are achieved using the program MLAB-PC (Civilized Software, Bethesda, MD).

Reliable experimental conditions for GBR binding to slices can be found in reports from Richfield.[15,22] Other procedures for preparation and incubation of membranes are described in the literature. These include the preparation of membranes by homogenization in 30 m*M* (sodium phosphate–0.32 *M* sucrose buffer and centrifugation (40,000*g* for 20 min) fol-

[12] M. E. A. Reith and G. Selmeci, *Naunyn-Schmiedeberg's Arch. Pharmacol.* **345,** 309 (1992).
[13] H. C. Akunne, C. Dersch, G. U. Char, J. S. Partilla, B. R. de Costa, K. C. Rice, and R. B. Rothman, *Soc. Neurosci. Abstr.* **18,** 1433 (1992).
[14] H. L. Wiener and M. E. A. Reith, *Anal. Biochem.* **207,** 58 (1992).
[15] E. K. Richfield, *Brain Res.* **540,** 1 (1991).
[16] P. Seeman, C. Ulpian, K. A. Wreggett, and J. W. Wells, *J. Neurochem.* **43,** 221 (1984).
[17] J.-J. Bonnet, S. Benmansour, N. Amejdki-Chab, and J. Costentin, *Eur. J. Pharmacol.* **266,** 87 (1994).
[18] Y. C. Cheng and W. H. Prusoff, *Biochem. Pharmacol.* **22,** 3099 (1973).
[19] R. B. Rothman, *Alcohol Drug Res.* **6,** 309 (1986).
[20] P. McGonigle, K. A. Neve, and P. B. Molinoff, *Molec. Pharmacol.* **30,** 329 (1986).
[21] G. E. Rovati, D. Rodbard, and P. J. Munson, *Anal. Biochem.* **184,** 172 (1990).
[22] E. K. Richfield, *Molec. Pharmacol.* **43,** 100 (1993).

lowed by resuspension in the same medium and identical centrifugation.[23,24] Yields of membrane preparations are subject to subtle changes depending on the exact procedures used to obtain them. However, more important sources of variation in membrane preparations are (1) tissue and species used in the preparation, (2) freezing and thawing, postmortem intervals, (3) composition of preparation–incubation medium, (4) temperature of incubation, and (5) determination of nonspecific binding.

1. *Origin of DAT.* Binding of GBR is not influenced by membrane polarization[25] or by endogenous dopamine concentration as it has been shown that treatments of animals with reserpine and/or α-methyl-p-tyrosine do not alter binding.[5,26] Binding of GBR derivatives to DAT can be carried out on preparations from different animal species.[27,28] The specificity of these preparations will be discussed later, but as a general rule rat and mouse preparations give GBR binding exhibiting the higher specificity, followed by guinea pig, dog, monkey and human preparations.[27,28] Preparations from bovine striatum may be useful because they possess a high density of binding sites, but their specificity needs to be fully characterized.[29,30] Finally, the use of GBR derivatives as radioligands for DAT cDNAs expressed in cells has been problematic thus far because it exhibits a low ratio of specific/total binding.[31]

2. *Freezing and Thawing, Postmortem Intervals.* The use of frozen tissues for membrane preparation can lead to contradictory results. In fact, procedures which appear rather similar result in either a 20% decrease in [^{3}H]GBR 12935 binding[32] or no change at all.[33] Freezing membranes prepared from frozen tissues induces a substantial labeling of the so-called piperazine acceptor site in canine caudate.[26] Also consistent with this, a

[23] A. L. Kirifides, J. A. Harvey and V. J. Aloyo, *Life Sci.* **50,** PL 139 (1992).
[24] M. E. A. Reith and L. L. Coffey, *J. Neurosci. Meth.* **51,** 31 (1994).
[25] G. Billaud, J. Costenin, and J.-J. Bonnet, *Eur. J. Pharmacol.* **247,** 333 (1993).
[26] H. C. Akunne, J. N. Johannessen, B. R. de Costa, K. C. Rice, and R. B. Rothman, *Neurochem. Res.* **17,** 261 (1992).
[27] M. Hirai, N. Kitamura, T. Hashimoto, T. Nakai, T. Mita, O. Shirakawa, T. Yamadori, T. Amano, S. A. Noguchi-Kuno, and C. Tanaka, *Jap. J. Pharmacol.* **47,** 237 (1988).
[28] N. A. Sharif, J. L. Nunes, A. D. Michel, and R. L. Whiting, *Neurochem. Int.* **15,** 325 (1989).
[29] C. J. Cao, M. M. Young, J. B. Wong, L. G. Mahran, and M. E. Eldefrawi, *Membr. Biochem.* **8,** 207 (1989).
[30] A. J. Eshleman, D. O. Calligaro, and M. E. Eldefawi, *Membr. Biochem.* **10,** 129 (1993).
[31] Z. B. Pristupa, J. M. Wilson, B. J. Hoffman, S. J. Kish, and H. B. Niznik, *Molec. Pharmacol.* **45,** 125 (1994).
[32] A. Janowsky, F. Vocci, P. Berger, I. Angel, N. Zelnik, J. E. Kleinman, P. Skolnick, and S. M. Paul, *J. Neurochem.* **49,** 617 (1987).
[33] J.-M. Maloteaux, M.-A. Vanisberg, C. Laterre, F. Javoy-Agid, Y. Agid. and P. M. Laduron, *Eur. J. Pharmacol.* **156,** 331 (1988).

methodological study performed with [^{3}H]Win 35428[24] demonstrates that freezing and thawing rat striatum can have no effect on binding or can decrease the affinity and/or binding density, depending on the composition of medium used for preparation–incubation, or on the type of membrane preparation (crude membrane suspension vs P_2 pellet). On the other hand, freezing and thawing are not responsible for appearance of a second population of [^{3}H]Win 35428 binding sites.[24] Thus, it is possible to use cryopreserved tissues for binding experiments, or for uptake studies,[34–36] but it is necessary to determine ahead of time that freezing and thawing has no effect.

The reduced importance of the postmortem interval seems rather well established. Different studies performed with human caudate and putamen demonstrate that postmortem intervals of 10–80 hr at 4° do not result in a time-dependent decrease in [^{3}H]GBR 12935 binding.[37–39] Experiments performed with rat preparations confirm this.[32,40] Nevertheless, an increase in temperature markedly enhances the decline in [^{3}H]GBR 12783 binding.[40]

3. *Composition of Preparation–Incubation Medium.* No cation different from Na^+ should be present in the preparation–incubation medium. Under these conditions, low Na^+ concentrations stimulate binding of GBR derivatives and a maximal binding is observed for 10–30 m*M* Na^+.

Monovalent cations such as Li^+, K^+, Rb^+, and Cs^+ cannot be used as a substitute for Na^+, and furthermore, they block binding of GBR derivatives and dopamine uptake at low millimolar concentrations.[6,15,25,41–44] It is noteworthy that the inhibitory potency of K^+ increases when Na^+ decreases.[25,41,44] Every organic buffer which has been assayed actually inhibits [^{3}H]GBR 12783 and/or [^{3}H]GBR 12935 binding.[12,22,41,42] In particular, $Tris^+$ impairs GBR binding, inhibits dopamine uptake, and also inhibits the binding of [^{3}H]cocaine, [^{3}H]Win 35428, and [^{3}H]mazindol to DAT.[12,22,41–48]

[34] J. A. Hardy, P. R. Dodd, A. E. Oakley, R. H. Perry, J. A. Edwardson, and A. M. Kidd, *J. Neurochem.* **40,** 608 (1983).
[35] N. Haberland and L. Hetey, *J. Neural Trans.* **68,** 289 (1987).
[36] J. D. Elsworth, J. R. Taylor, P. Berger, and R. H. Roth, *Neurochem. Int.* **23,** 61 (1993).
[37] N. Zelnik, I. Angel, S. M. Paul, and J. E. Kleinman, *Eur. J. Pharmacol.* **126,** 175 (1986).
[38] P. Allard and J. O. Marcusson, *Neurobiol. Aging* **10,** 661 (1989).
[39] C. Czudek and G. P. Reynolds, *J. Neural Transm.* **77,** 227 (1989).
[40] F. Thibaut, J.-J. Bonnet, and J. Costentin, *Neurosci. Lett.* **96,** 335 (1989).
[41] J.-J. Bonnet, S. Benmansour, J.-M. Vaugeois, and J. Costentin, *J. Neurochem.* **50,** 759 (1988).
[42] N. Amejdki-Chab, S. Benmansour, J. Costentin, and J.-J. Bonnet, *J. Neurochem.* **59,** 1795 (1992).
[43] H. E. Milner and S. M. Jarvis, *Biochem. Soc. Transac.* **20,** 243S (1992).
[44] C. Héron, G. Billaud, J. Costentin, and J.-J. Bonnet, *Eur. J. Pharmacol.* **301,** 195 (1996).
[45] L. T. Kennedy and I. Hanbauer, *J. Neurochem.* **41,** 172 (1983).
[46] D. O. Calligaro and M. E. Eldefrawi, *Membr. Biochem.* **7,** 87 (1988).

TABLE I

EFFECT OF RINSES ON KINETIC CONSTANTS OF SPECIFIC BINDING OF [^{3}H]Win 35428[a]

Experimental conditions	K_{d1} (nM)	B_{max1} (pmol/mg protein)	K_{d2} (μM)	B_{max2} (pmol/mg protein)
Set 1				
120 mM KCl	0.79 ± 0.03[b]	1.20 ± 0.05	1.12 ± 0.18[b]	23.8 ± 4.8[b]
120 mM NaCl	0.47 ± 0.10	1.35 ± 0.13	0.12 ± 0.04	9.5 ± 2.8
Set 2				
120 mM KCl	0.49 ± 0.12	0.76 ± 0.09	0.23 ± 0.06	13.0 ± 6.0
120 mM NaCl and rinses	0.48 ± 0.13	0.77 ± 0.14	0.10 ± 0.03	7.9 ± 1.1

[a] Provoked by an initial exposure to a 120 mM KCl containing medium. Rat striatal membranes are exposed for 20 min at 20° to a 10 mM Na^+ containing either 120 mM KCl or 120 mM NaCl and then centrifuged (50,000g, 10 min) and resuspended in a 10 mM Na^+ medium (set 1). In set 2, the exposure is followed by two rinses for 20 min at 20° by a 130 mM Na^+ medium. Saturations are performed at 0° for 2 hr in a 10 mM Na^+ medium (final volume, 500 μl); they are obtained by isotopic dilution of a 0.3 nM [^{3}H]Win 35428 concentration (range 0.3–1000 nM). Nonspecific binding is estimated with 30 μM cocaine. Mean ± S.E.M. values of six experiments performed in duplicate. Kinetic constants are estimated with the nonlinear curve-fitting program Ligand.

[b] Significantly different from 120 mM NaCl with $P < 0.05$, 0.01, and 0.001, respectively (Student's t test). Similar results are obtained when nonspecific binding is determined with 10 μM mazindol.

The presence of an inhibitory cation during the membrane preparation can alter the results in subsequent binding experiments. Exposure of rat striatal membranes to a medium containing a 120 mM K^+ concentration decreases the [^{3}H]Win 35428 specific binding as a consequence of decreases in B_{max} and/or binding affinity. This effect disappears when exposure to K^+ is followed by two rinses with a 130 mM Na^+ medium (Table I). Similarly, $Tris^+$ behaves as a competitive inhibitor for various DAT radioligands, including GBR derivatives, when membranes are prepared and incubated in a Na^+ medium,[12,41,47] but if $Tris^+$ is present only in the preparation medium it produces a decrease in B_{max} with or without changes in K_d.[30,45]

Some agents such as choline chloride and N-methyl-D-glucamine chloride which could be used to substitute for Na^+ without altering the chloride concentration should be avoided as they inhibit binding of GBR derivatives at millimolar concentrations.[42–44]

[47] I. Zimanyi, A. Lajtha, and M. E. A. Reith, *Naunyn-Schmiedeberg's Arch. Pharmacol.* **340,** 626 (1989).

[48] L. L. Coffey and M. E. A. Reith, *J. Neurosci. Meth.* **51,** 23 (1994).

TABLE II
EFFECT OF KCl, NaCl, AND $CaCl_2$ ON DISSOCIATION OF [^{3}H]GBR 12935 SPECIFIC BINDING[a]

Dissociation medium	$t_{1/2}$ (min)
10 m*M* Na^+	17.3 ± 0.5
10 m*M* Na^+ + 30 m*M* KCl	9.8 ± 0.8[b]
10 m*M* Na^+ + 30 m*M* NaCl	13.3 ± 1.2[b]
10 m*M* Na^+ + 3 m*M* $CaCl_2$	13.2 ± 0.5[b]

[a] 20°. Binding equilibrium is achieved by incubation at 20° in a silicone-coated beaker of rat striatal membranes (1.3 mg protein) and 0.8 n*M* [^{3}H]GBR 12935 in a total volume of 50 ml of a 10 m*M* Na^+ medium containing (m*M* concentration): NaH_2PO_4 0.3, $NaHCO_3$ 9.7, pH 7.5 ± 0.1. At the end of the incubation period, 1-ml aliquots were diluted in 50 ml of a 10 m*M* Na^+ medium containing the chloride form of the tested ions. Dissociation is stopped by rapid vacuum filtration through GF/B filters previously soaked for at least 1 hr in 0.5% polyethyleneimine. Zero time dissociations are made by dilution and immediate filtration. The constancy of the binding is checked and demonstrated by repeating this procedure at regular intervals during the dissociation period. Mean ± S.E.M. values of five experiments performed in duplicate.

[b] Significantly different from 10 m*M* Na^+ medium with $P < 0.01$ (Dunnett's *t* test).

Divalent cations such as Ca^{2+} and Mg^{2+} inhibit the binding of GBR derivatives; depending on the experimental conditions, this inhibition can occur at submillimolar concentrations in the presence of a low Na^+ concentration, and it decreases when the Na^+ concentration increases.[6,15,25,41,44] Binding of [^{3}H]cocaine, [^{3}H]Win 35428, and [^{3}H]mazindol are also blocked by these divalent cations.[45–48]

K^+, Ca^{2+}, and Mg^{2+} enhance the dissociation of [^{3}H]GBR 12783 and/or [^{3}H]GBR 12935 binding[44] (Table II), and increases in Na^+ block the dissociation of [^{3}H]GBR 12783 elicited by these inhibitory cations.[44] Other dissociation experiments reveal that cation concentrations present in extracellular fluids (128 m*M* Na^+, 4 m*M* K^+, 1 m*M* Mg^{2+}, and 1 m*M* Ca^{2+} [49,50]) do not modify the dissociation of [^{3}H]GBR 12783 binding, whereas cytosolic

[49] B. Alberts, D. Bray, J. Lewis, M. Raff, K. Roberts, and J. D. Watson, *in* "Molecular Biology of the Cell," p. 301. Garland Publishing, New York, 1989.

[50] J. Darnell, H. Lodish, and D. Baltimore, *in* "Molecular Cell Biology," p. 537. W. H. Freeman and Company, New York, 1990.

K^+ concentrations (150 mM) produce a dramatic binding dissociation.[44] This could explain the important contribution of the transmembrane K^+ gradient to the extracellular binding of inhibitors to DAT, and in the preferential inward transport of dopamine.

The importance of ions present in physiological fluids can also be verified by determining the anionic requirements for binding. The stimulatory effects of Na^+ on binding are amplified by the presence of HCO_3^- and/or phosphate anions[30,51] which buffer physiological fluids. These anions are likely to be the most stimulatory ones at low concentrations ($\leq$20–30 mM), whereas I^- is very potent at 50–200 mM.[51] Cl^-, Br^-, F^-, NO_3^-, SO_4^{2-}, and the acetate ion are quite equipotent but less efficient.[6,41,45,51]

Thus, it appears that the best media for preparation–incubation should simply contain 10–30 mM Na^+ concentrations in a phosphate or phosphate/HCO_3^- form. Consistent with this, some results suggest that tonicity and Na^+ concentration of the preparation medium are devoid of any effect on binding results. So, the shape of Na^+ dependence of binding of [^{3}H]GBR 12783 or [^{3}H]mazindol is not dependent on the concentration of Na^+ in the preparation medium.[41,47] However, the situation is not so clear and different data support the use of a preparation medium of higher tonicity. Use of a Krebs medium (containing some inhibitory cations and a 130 mM Na^+ concentration) for membrane preparation produces [^{3}H]GBR 12783 binding values which are lower, but also more stable, than those obtained using a 10 mM Na^+ medium.[5,41] In the same way, the preparation of membranes of synaptosomes in a 10 mM Na^+ medium rather than in a 135 mM Na^+ medium increases the K_d value for [^{3}H]GBR 12783 binding studied in a 10 mM Na^+ medium, an effect which disappears when incubations are performed in a 135 mM Na^+ medium.[25] Also consistent with this is the significant and reproducible loss of [^{3}H]GBR 12783 binding which is observed when membranes are washed with a 10 mM Na^+ medium rather than with a 130 mM Na^+ medium.[17,52] Similarly, when striatal membranes are washed with a 10 mM Na^+ medium and not with a 130 mM Na^+ medium, a decrease in [^{3}H]Win 35428 binding is observed as a consequence of reductions in affinity and B_{max} for the two populations of binding sites. In fact, the best values of [^{3}H]Win 35428 binding are obtained with preparation and incubation in a 30 mM sodium phosphate medium containing 0.32 M sucrose,[24] an agent which is devoid of any inhibitory effect on binding.[24,42] However, the [^{3}H]GBR 12783 binding is not improved by using a medium containing sucrose for incubation. Thus, it appears that a 10–30 mM sodium

[51] M. E. A. Reith and L. L. Coffey, *J. Neurochem.* **61,** 167 (1993).

[52] F. Refahi-Lyamani, S. Saadouni, J. Costentin, and J.-J. Bonnet, *Naunyn-Schmiedeberg's Arch. Pharmacol.* **351,** 136 (1995).

phosphate or phosphate HCO_3^- buffered medium containing 0.32 *M* sucrose or 0.12 *M* NaCl is a suitable medium for membrane preparation, whereas a 10–30 m*M* sodium phosphate or phosphate HCO_3^- medium is preferable as an incubation medium.

4. *Temperature of Incubation.* GBR derivatives display lower affinity for DAT when the incubation temperature is increased. K_d values for [^{3}H]GBR 12783 binding increase with temperature, from 0.3–0.5 n*M* at 0° to 0.7–0.9 n*M* at 20–25° and 2–4 n*M* at 37° without any modification of the B_{max}.[53,54] In similar experimental conditions, increasing the temperature from 0 to 25° increases the K_d value for [^{3}H]GBR 12935 binding from 0.3 to 0.6 n*M*.[12] An increase in incubation temperature also decreases affinity for all DAT inhibitors (but not for substrates),[53,54] and this has been verified for the binding of two cocaine derivatives, [^{3}H]Win 35428 and [^{125}I]βCIT, to DAT.[12,55]

It is noteworthy that a temperature of incubation of 0–2° may improve specificity of GBR binding. This is strongly suggested by results demonstrating that methylphenidate concentrations which perfectly discriminate between DAT and the piperazine acceptor site at 0° give monophasic competition curves at 25–30°.[10] Incubations performed at 0–2° lead to association and dissociation rates so low that binding equilibrium is achieved after a 2 to 4-hr incubation period with membrane preparations,[12,41,53] and an even longer period with slices.[15] Despite the long time necessary to achieve equilibrium, a 30-hr incubation period at 0–2° is preferred to the thermal instability observed at 25°.[15] Dissociation rates of [^{3}H]GBR 12783 binding measured in medium devoid of inhibitory cations reach such low values at 0–2° ($t_{1/2}$: 120–140 min)[5,44] that extensive washings of tubes and filters can be envisaged, resulting in lower binding to non-DAT sites.

5. *Determination of Nonspecific Binding.* The nonspecific binding of tritiated GBR derivatives is generally quantified with mazindol, a competitive inhibitor of the binding to DAT in rat, mouse, and human tissues.[12,56–58] A mazindol concentration of 3 μM was used in studies of [^{3}H]GBR 12783

[53] J.-J. Bonnet, S. Benmansour, J. Costentin, E. M. Parker, and L. X. Cubeddu, *J. Pharmacol. Exp. Ther.* **253,** 1206 (1990).

[54] G. Billaud, J.-F. Menard N. Marcellin, J.-M. Kamenka, J. Costentin, and J.-J. Bonnet, *Eur. J. Pharmacol.* **268,** 357 (1994).

[55] M. Laruelle, S. S. Giddings, Y. Zea-Ponce, D. S. Charney, J. L. Neumeyer, R. M. Baldwin, and R. B. Innis, *J. Neurochem.* **62,** 978 (1994).

[56] J. Marcusson and K. Eriksson, *Brain Res.* **457,** 122 (1988).

[57] P. Berger, J. D. Elsworth, M. E. A. Reith, D. Tanen, and R. H. Roth, *Eur. J. Pharmacol.* **176,** 251 (1990).

[58] P. O. Allard, K. Eriksson, S. B. Ross, and J. O. Marcusson, *Neuropsychobiol.* **23,** 177 (1990–91).

binding to DAT from rat striatum performed in Na^+-containing medium at 0–2°, whereas 10–30 μM mazindol was preferred at higher temperatures. Major involvement of a second population of binding sites could be excluded because linear Scatchard plots and inhibitor- or substrate-induced competition curves with Hill coefficients very close to unity are observed under these conditions.[5,25,41,53,54,59] However, considering the high proportion of non-DAT sites present in some preparations and the reduction in radioligand affinity resulting from the too-common use of media containing inhibitory cations, the nonspecific binding should be quantified with mazindol concentrations not exceeding 1 μM, a concentration which generates a specific binding similar to that defined with 1 mM dopamine.[56,60,61] A 0.3–0.5 μM mazindol concentration actually blocks more than 90% of the binding of [^{3}H]GBR 12935 to DAT[6,10,11,62,63] without affecting the binding of the radioligand to other sites such as the piperazine acceptor site[10,63] and the so-called "site 2."[11] Competition curves of mazindol with [^{3}H]GBR 12935 are either biphasic or with a Hill coefficient markedly lower than unity, suggesting that mazindol discriminates between DAT and other binding sites.[11,27,28,63,64] This is consistent with the observation of linear Scatchard plots of [^{3}H]mazindol binding with preparations from different animal species.[47,53,65] In contrast, GBR derivatives, benztropine and nomifensine are likely not to constitute reliable discriminative agents.[10,11,27,28,63] Replacement of the 5 μM mazindol concentration used for the determination of nonspecific binding of [^{3}H]GBR 12935 to human putamen membranes by 20 μM benztropine induces a disappearance of the characteristics of the specific binding to DAT, (i.e., linear Scatchard plot, Na^+ dependence, and a decrease in patients with Parkinson's disease).[27] Cocaine derivatives (Win 35428, Win 35065-2) could be interesting for determining nonspecific binding because they display a 300- to 1000-fold higher affinity for DAT than for "site 2,"[11] and no affinity for the piperazine acceptor site.[63] However, cocaine itself does not discriminate between these binding sites,[11,63] consistent with its ability to inhibit canine cytochrome P450IID1 and to recognize

[59] S. Benmansour, J.-J. Bonnet, P. Protais, and J. Costentin, *Neurosci. Lett.* **77,** 97 (1987).

[60] P. Seeman and H. B. Niznik, *FASEB J.* **4,** 2737 (1990).

[61] H. B. Niznik, E. F. Fogel, F. F. Fassos, and P. Seeman, *J. Neurochem.* **56,** 192 (1991).

[62] P. Berger, A. Janowsky, F. Vocci, P. Skolnick, M. M. Schweri, and S. M. Paul, *Eur. J. Pharmacol.* **107,** 289 (1985).

[63] H. B. Niznik, R. F. Tyndale, F. R. Sallee, F. J. Gonzalez, J. P. Hardwick, T. Inaba, and W. Kalow, *Arch. Biochem. Biophys.* **276,** 424 (1990).

[64] P. Allard, M. Danielsson, K. Papworth, and J. O. Marcusson, *J. Neurochem.* **62,** 338 (1994).

[65] C. M. Dersch, H. C. Akunne, J. S. Partilla, G. U. Char, B. R. de Costa, K. C. Rice, F. I. Carroll, and R. B. Rothman, *Neurochem. Res.* **19,** 201 (1994).

σ sites.[66,67] Finally, using other agents to determine nonspecific binding is questionable because they may recognize site(s) different from DAT (BTCP and methylphenidate:[10,11,63,67]), they may produce variable results, and/or they may not recognize DAT sites or states bound by other DAT inhibitors (amfonelic acid: [11,28,31,68]).

Binding Specificity

GBR derivatives with a nanomolar affinity for DAT also display a 10–100 n*M* affinity for several other sites[69] such as the vesicular amine transporter,[3,70] $\sigma 1$ and $\sigma 2$ sites,[71–73] and a piperazine acceptor site.[10] The use of wheat germ agglutinin lectin chromatography provided evidence that the piperazine acceptor site could be a protein which binds [^{3}H]GBR 12935 with a nanomolar affinity, in a Na^+-independent manner, and with pharmacological characteristics which are similar to those of cytochrome P450IID1 (debrisoquine/sparteine monooxygenase).[63] The presence and activity of P450IID1 in mammalian brain have been demonstrated.[67] In fact, $\sigma 1$ and $\sigma 2$ sites could constitute subpopulations of the cytochrome P450 superfamily.[72,73] However, the population of mazindol-insensitive binding sites for [^{3}H]GBR 12935 in brain and liver exceeding by far those for classical $\sigma 2$ and/or $\sigma 1$ radioligands, the piperazine acceptor site is likely to comprise several different populations of sites.[72] Flupentixol recognizes the piperazine acceptor site nonstereospecifically[10] and with a 20–600 m*M* affinity[10,15,63,74] (i.e., an affinity which can be close to that of flupentixol for DAT).[11,63,75] Consequently, the use of flupentixol to inhibit binding of GBR to non-DAT sites remains controversial.[11,15] Other properties of this acceptor site include sensitivity to several piperazines, including GBR 12935 and GBR 12909 (in the 10 n*M* range), sensitivity to micromolar concentrations of mazindol and nomifensine and to submillimolar concentrations of dopamine,[15,63,76] insensitivity to or blockade by high Na^+ concen-

[66] J. Sharkey, K. A. Glen, S. Wolfe, and M. J. Kuhar, *Eur. J. Phamacol.* **149,** 171 (1988).
[67] R. F. Tyndale, R. Sunahara, T. Inaba, W. Kalow, F. J. Gonzalez, and H. B. Niznik, *Molec. Pharmacol.* **40,** 63 (1991).
[68] S. Izenwasser, L. L. Werling, J. G. Rosenberger, and B. M. Cox, *Neuropharmacol.* **29,** 1017 (1990).
[69] P. H. Andersen, *Eur. J. Pharmacol.* **166,** 493 (1989).
[70] M. E. A. Reith, L. L. Coffey, C. Xu, and N.-H. Chen, *Eur. J. Pharmacol.* **253,** 175 (1994).
[71] P. C. Contreras, M. E. Bremer, and T. S. Rao, *Life Sci.* **47,** PL 133 (1990).
[72] S. B. Ross, *Pharmacol. & Toxicol.* **68,** 293 (1991).
[73] M. Klein, P. D. Canoll, and J. M. Musacchio, *Life Sci.* **48,** 543 (1991).
[74] I. Gordon, R. Weizman, and M. Rehavi, *Brain Res.* **674,** 205 (1995).
[75] M.-H. Lemasson, J.-J. Bonnet, and J. Costentin, *Biochem. Pharmacol.* **33,** 2137 (1984).
[76] I. Gordon, R. Weizman, and M. Rehavi, *Life Sci.* **55,** 189 (1994).

trations,[63,76] and lack of sensitivity to lesions of dopaminergic neurons by 6-OHDA.[10]

Another population of mazindol- and nomifensine-insensitive sites for [^{3}H]GBR 12935 is described as "site 2."[11] From a pharmacological point of view, this "site 2" is quite similar to the piperazine acceptor site and the rather low affinity of SKF 525A and dextromethorphan for this site is consistent with their affinity for the piperazine acceptor site[63,67] and for $\sigma 2$ sites which constitute a subpopulation of this.[77,78] On the contrary, its dramatic sensitivity to lesions by 6-OHDA differentiates the "site 2" from the piperazine acceptor site.[11] These points require additional studies, but now it appears possible that "site 2" could be a GBR binding site with moderate to high affinity for selected DAT ligands and is likely to be different from the piperazine acceptor site.[11] It is worth noting that the interspecies variability of properties of cytochrome P450 and σ sites may explain the difficulties in comparing binding data from different species.[11]

Binding of GBR derivatives can be obtained in preparations from rat or mouse brain area containing dopaminergic cells and projections such as (in decreasing density order) striatum, olfactory tubercules (olf.tub.) nucleus (n.) accumbens, ventral tegmental area, and substantia nigra.[5,15,68,79–82] Pharmacological characteristics of [^{3}H]GBR 12935 binding in olf.tub., n. accubens, and striatum are rather similar.[57,68] A [^{3}H]GBR 12935 binding is also detected in other rat brain area such as (pre)frontal cortex,[74,79,81] which contains a very low dopamine concentration and displays a reduced but typical neuronal dopamine uptake with functional and pharmacological properties similar to those found in striatum.[36,83,84] However, the meaning of this [^{3}H]GBR 12935 binding remains questionable since it is likely to concern quite exclusively non-DAT sites. As a matter of fact, the percentage of non-DAT sites increases when the DAT density decreases, and the 50 μM mazindol-sensitive binding of [^{3}H]GBR 12935 studied in rat frontal cortex displays properties of a binding to a Na^+-inhibited, flupentixol-sensitive, and dopamine-insensitive piperazine acceptor site.[74]

[77] B. L. Largent, A. L. Gunglach, and S. H. Snyder, *Proc. Natl. Acad. Sci. USA* **81,** 4983 (1984).

[78] J. M. Walker, W. D. Bowen, F. O. Walker, R. R. Matsumoto, B. R. de Costa, and K. C. Rice, *Pharmacol. Rev.* **42,** 355 (1990).

[79] T. M. Dawson, D. R. Gehlert, and J. K. Wamsley, *Eur. J. Pharmacol.* **126,** 171 (1986).

[80] I. Leroux-Nicollet and J. Costentin, *Neurosci. Lett.* **95,** 7 (1988).

[81] F. Mennicken, M. Savasta, P. Peretti-Renucci, and C. Feuerstein, *J. Neural Transm.* **87,** 1 (1992).

[82] J. M. Wilson, J. N. Nobrega, M. E. Carroll, H. B. Niznik, K. Shannak, S. T. Lac, Z. B. Pristupa, L. M. Dixon, and S. J. Kish, *J. Neurosci.* **14,** 2966, (1994).

[83] D. F. Kirksey and T. A. Slotkin, *Br. J. Pharmacol.* **67,** 387 (1979).

[84] J. A. Hardy, P. Webster, I. Backstrom, J. Gottfries, L. Oreland, A. Stenstrom, and B. Winblad, *Neurochem. Int.* **10,** 445 (1987).

Another factor affecting the decrease in DAT density is the change of animal species from rat or mouse to canine and monkey or human.[27,28,85] So, [^{3}H]GBR 12935 and [^{125}I]FAPP recognize DAT from canine striatum with a nanomolar affinity[8,28], but the density of binding sites is markedly lower than in striatum from rodents[28]. Consequently, the choice of experimental conditions plays a critical role in the specificity of GBR binding. Studies of [^{3}H]GBR 12935 binding carried out in a sodium phosphate medium can effectively involve DAT quite exclusively since the shape of Na^+ dependence of binding to canine striatal membranes[43] is similar to those displayed by tritiated DAT ligands in rodent preparations. This is strengthened by the 90% decrease in [^{3}H]GBR 12935 specific binding resulting from a subtotal destruction of dopaminergic terminals by MPTP[26], although a pharmacologically distinct class of [^{3}H]GBR 12935 binding sites on dopaminergic terminals has also been described[11]. A decrease in [^{3}H]GBR 12935 affinity caused by differences in preparation–incubation media and/or DAT solubilization results in a 20–40% involvement of non-DAT sites in the "specific" binding of [^{3}H]GBR 12935.[28,63,86] The pharmacological characterization of the [^{3}H]GBR 12935 binding studied in these conditions may result from either an incomplete competition between drugs, or a biphasic or shallow [n(Hill) $\leq$0.5] competition curve.[28,63,86]

The binding of [^{3}H]GBR 12935 to non-DAT sites is more significant in human putamen than in rat striatum,[27] in agreement with the fact that [^{3}H]GBR 12935 and [^{125}I]FAPP bind to P450IID6 and to several proteins different from DAT in human preparations.[61] Consequently, the high-affinity binding of [^{3}H]GBR 12935 to DAT corresponds to 20–80% of its binding to human caudate and putamen preparations.[27,31–33,56,87,88] [^{3}H]GBR 12935 is commonly used in human caudate and putamen preparations for quantitation of DAT changes associated with aging[37,38,89] and different pathological states such as Parkinson's disease, progressive supranuclear palsy,[33] schizophrenia,[39,61] Huntington's chorea,[61] and dementia of the Alzheimer's type associated with extrapyramidal symptoms.[88,90] Results obtained from patients with Parkinson's disease constitute an example of the pivotal importance of determining nonspecific binding When nonspecific binding is defined with 0.3 μM mazindol, the specific binding of [^{3}H]GBR 12935 in

[85] M. A. Fahey, D. R. Canfield, R. D. Spealman, and B. K. Madras, *Soc. Neurosci. Abstr.* **15,** 1090 (1989).

[86] R. Lew, D. E. Grigoriadis, J. Sharkey, and M. J. Kuhar, *Synapse* **3,** 372 (1989).

[87] J. De Keyser, J.-P. DeBacker, G. Ebinger, and G. Vauquelin, *J. Neurochem.* **53,** 1400 (1989).

[88] P. O. Allard, J. Rinne, and J. O. Marcusson, *Brain Res.* **637,** 262 (1994).

[89] J. De Keyser, G. Ebinger, and G. Vauquelin, *Ann. Neurol.* **27,** 157 (1990).

[90] P. Allard, I. Alafuzoff, A. Carlsson, K. Eriksson, E. Ericson, C.-G. Gottfries, and J. O. Marcusson, *Eur. Neurol.* **30,** 181 (1990).

putamen membranes from patients with Parkinson's disease represents 3% of the control values,[88] in agreement with a similar fall in dopamine concentration[91,92] and with the disappearance of a protein of apparent M_r 62,000 labeled with [^{125}I]FAPP and corresponding to DAT.[61] In contrast, 10-fold higher mazindol concentrations result in limited decreases in [^{3}H]GBR 12935 binding in these patients, reaching about 60% in putamen and 50% in caudate.[27,32,33]

Studies of GBR binding in human brain structures other than caudate and putamen are complicated by the large proportion of non-DAT sites.[64,93] Nevertheless, the density of specific binding sites for [^{3}H]GBR 12935 agrees rather well with the density of DAT which could be expected from dopamine concentrations and dopamine uptake in these structures,[84,91] with higher B_{max} values in caudate and putamen followed by olf.tub., n. accumbens, and substantia nigra.[56,87] The K_d values for [^{3}H]GBR 12935 binding are 1–2 nM in all of these structures.[56,87] However, media containing inhibitory cations (Tris$^+$, K^+, Mg^{2+}), high temperature of incubation, and/or inappropriate determination of nonspecific binding can result in questionable binding data. So, no significant reduction in [^{3}H]GBR 12935 binding is observed in n.accumbens from patients with Parkinson's disease[32] which display an important loss in dopamine and metabolites.[92,94] Similarly, a population of [^{3}H]GBR 12935 binding sites in human frontal cortex[95] has been reported that displays properties (density, Na^+ and K^+ dependences) which are not expected for a binding to DAT.[41,47,51,84,91,96] The unusual pharmacological profile of this binding, with submicromolar affinity for GBR 12909 and benztropine, lack of high affinity for mazindol,[64] and no affinity for cocaine, amphetamine, or dopamine even in millimolar concentrations is not characteristic of DAT and more likely corresponds to those of the piperazine acceptor site and/or the "site 2." Finally, the [^{3}H]GBR 12935 binding to frontal cortex is reduced by 6-OHDA, but this neurotoxin also decreases GBR binding to "site 2."[11] Taken together, these data strongly suggest that this binding mainly involves sites different from DAT.[64,95]

[91] B. Scatton, F. Javoy-Agid, L. Rouquier, B. Dubois, and Y. Agid, *Brain Res.* **275,** 321 (1983).

[92] B. Bokobza, M. Ruberg, B. Scatton, F. Javoy-Agid, and Y. Agid, *Eur. J. Pharmacol.* **99,** 167 (1984).

[93] P. Allard, J. O. Marcusson, and S. B. Ross, *J. Neurochem.* **62,** 342 (1994).

[94] K. Price, I. Farley, and O. Hornykiewicz, *Adv. Biochem. Physchopharmacol.* **19,** 293 (1978).

[95] A. Hitri, D. Venable, H. Q. Nguyen, M. F. Casanova, J. E. Kleinman, and R. J. Wyatt, *J. Neurochem.* **56,** 1663 (1991).

[96] M. E. A. Reith, B. E. Meisler, H. Sershen, and A. Lajtha, *Biochem. Pharmacol.* **35,** 1123 (1986).

Na^+ Dependence

Na^+ dependence of the binding is generally considered as a major criterion for binding to DAT. A Na^+-dependent binding is reported for classical ligands ([^{3}H]GBR 12783 and [^{3}H]GBR 12935) in different rat cerebral structures,[5,32] as well as for irreversible derivatives such as 3-azido[^{3}H]GBR 12935[9] and [^{125}I]FAPP[61] in membranes from rat striatum and human caudate. In interpreting curves of Na^+ dependence it is important to distinguish between those obtained in physiological buffered media from those observed in media containing inhibitory cations.

GBR compounds, like other dopamine uptake blockers, need low Na^+ concentrations to reach a maximal level of binding to DAT. The very low specific binding of [^{3}H]GBR 12783 obtained with rat striatal membranes in a 1–2 m*M* $NaHCO_3$ medium is greatly improved by low concentrations of $NaHCO_3/NaH_2PO_4$ buffer so that a maximal binding is reached for 10–30 m*M* Na^+ concentrations[5] (Fig. 1). This dependence on low Na^+ concentrations is related neither to procedures of membranes preparation nor to animal species. A quite identical Na^+-dependence shape is reported for [^{3}H]GBR 12935 binding on canine striatal membranes,[43] confirming preliminary data indicating that [^{3}H]GBR 12935 binding on solubilized

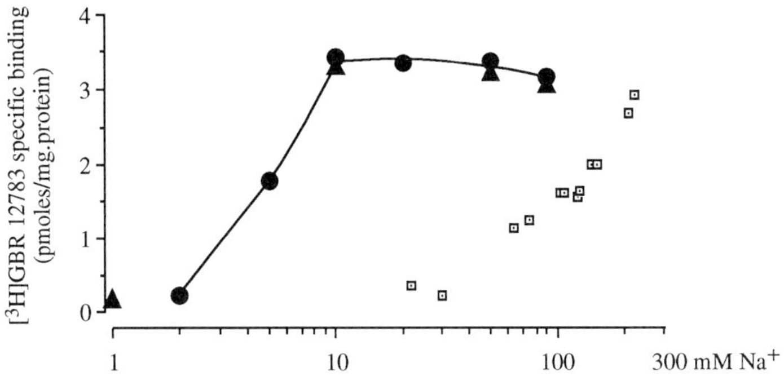

FIG. 1. Effects of experimental conditions on the Na^+ dependence of [^{3}H]GBR 12783 binding. Rat striatal membranes are incubated at 0° for 18 hr (nonspecific binding in the presence of 3 μM mazindol). In a set of experiment, membranes (□) are prepared and incubated (1 n*M* [^{3}H]GBR 12783) in a Krebs–Ringer-like medium containing various Na^+ concentrations, a 225 m*M* Cl^- concentration being maintained with choline chloride. Data are from two experiments performed in duplicate. In another set of experiments, membranes are either prepared and incubated (0.4 n*M* [^{3}H]GBR 12783) in four different Na^+ media (▲), or prepared in a 10 m*M* Na^+ medium and incubated (0.4 n*M* [^{3}H]GBR 12783) in six different Na^+ media (●). Data are means of five to seven experiments performed in duplicate. [Adapted with permission from J.-J. Bonnet, S. Benmansour, J.-M. Vaugeois, and J. Costentin, *J. Neurochem.* **50,** 759 (1988).]

canine DAT is maximal in a 10 mM Na_2HPO_4/0.32 M sucrose buffer.[86] In any case, Na^+ stimulates binding by increasing the affinity of the radioligand with no effect on B_{max}.[41,43] Similar increases in binding affinity are observed for [^{3}H]mazindol, [^{3}H]cocaine, and [^{3}H]Win 35428.[47,51,96,97] These results establish that 10–20 mM Na^+ concentrations are sufficient for a maximal binding to DAT, and this is confirmed by the fact that an increase in Na^+ from 10 to 130 mM does not improve the affinity of substrates and of most of the pure uptake blockers.[42,59] A further increase in Na^+ up to 100–300 mM results in a slight but significant decrease in [^{3}H]GBR 12783 binding to striatal membranes,[41,59] but does not alter the specific binding of [^{3}H]GBR 12935 to rat brain slices.[15] This low sensitivity to the inhibitory effects of high Na^+ concentrations is found again for [^{3}H]mazindol binding,[46,47] but it differs from the marked reduction in binding observed for [^{3}H]cocaine and [^{3}H]Win 35428.[30,46,51,96] Nevertheless, in any case, high Na^+ induces binding decreases as a consequence of reductions in binding affinity.[30,41,47,51]

The presence of inhibitory cations in the incubation medium modifies the Na^+ dependence of GBR binding so that high Na^+ concentrations are required to obtain better binding values. So, a maximal binding of [^{3}H]GBR 12783 is observed for a 225 mM Na^+ concentration and a null concentration of choline when this latter cation is used as substitute for Na^+ (Fig. 1).[41] A similar example is described for a Li^+/Na^+ substitution and binding of [^{125}I]βCIT to DAT.[98] Thus, the presence of increasing concentrations of inhibitory substitutes markedly modifies the shape of Na^+ dependence. In contrast substitutions with $Tris^+$, a common buffer used at fixed concentrations, generally results in a more conventional Na^+-dependence curve. In these conditions, Na^+ stimulates [^{3}H]GBR 12935 binding so that it reaches a plateau at Na^+ concentrations up to 70–200 mM; this has been observed with rat striatal membranes, as well as with solubilized DAT from canine striatum.[6,63,74,76] In the same way, the [^{3}H]cocaine binding in Tris-buffered media is stimulated by 50–100 mM Na^+ concentrations[45,46,99]; in one study, the Na^+-induced binding stimulation is followed by an inhibition at higher Na^+ concentrations,[45] in agreement with results already described in physiological medium.

A report concerning binding of [^{3}H]GBR 12935 to rat DAT is at variance with the experimental data discussed here.[10]

[97] S. Saadouni, F. Refahi-Lyamani, J. Costentin, and J.-J. Bonnet, *Eur. J. Pharmacol.* **268,** 187 (1994).

[98] S. C. Wall, R. B. Innis, and G. Rudnick, *Molec. Pharmacol.* **43,** 264 (1993).

[99] B. K. Madras, M. A. Fahey, J. Bergman, D. R. Canfield, and R. D. Spealman, *J. Pharmacol. Exp. Ther.* **251,** 131 (1989).

Conclusion

GBR compounds have played an important role in the characterization of structural and functional properties of DAT. A better definition of the nature of the nonspecific sites and the use of rational experimental conditions of binding, i.e., in keeping with our knowledge of the sensitivity of binding to ions, temperature, and other dopamine uptake inhibitors, should allow the study of DAT with radiolabeled GBR compounds to be expanded for better comparison of the properties of their binding with those of other DAT radioligands.

[15] Cocaine and GBR Photoaffinity Labels as Probes of Dopamine Transporter Structure

By ROXANNE A. VAUGHAN

Introduction

The dopamine transporter (DAT) is a neuronal plasma membrane protein which terminates dopaminergic neurotransmission by the rapid reuptake of dopamine from the synapse into the presynaptic neuron. DAT belongs to a family of structurally homologous Na^+ and Cl^--dependent transporters for many neurotransmitters and amino acids. Strutural modeling of DAT and its homologs based on hydrophobicity analysis and other theoretical considerations predicts 12 transmembrane spanning domains with N and C termini oriented intracellularly. Three or four consensus glycosylation sites are found in the interhelical loop between putative TMDs 3 and 4, and numerous consensus phosphorylation sites are found in presumed intracellular aspects.[1]

Many compounds of social and physiological significance such as tricyclic antidepressants and drugs of abuse bind to neurotransmitter transporters and exert their effects by inhibiting substrate transport. Ligands which bind to DAT with high affinity and inhibit transport of dopamine include the structurally diverse molecules cocaine, amphetamine, benztropine, mazindol, nomifensine, and GBR ([2-(diphenylmethoxy)ethyl]-4-(3-phenylpropyl)piperazine) compounds.[2] The molecular mechanisms underlying substrate transport and action of uptake blockers are unknown, and currently

[1] S. Amara and M. Kuhar, *Ann. Rev. Neurosci.* **16,** 73 (1993).

[2] F. I. Carroll, A. H. Lewin, and M. J. Kuhar, *in* "Neurotransmitter Transporters: Structure, Function, and Regulation" (Maarten Reith, Ed.), pp. 263–296. Humana Press, 1997.

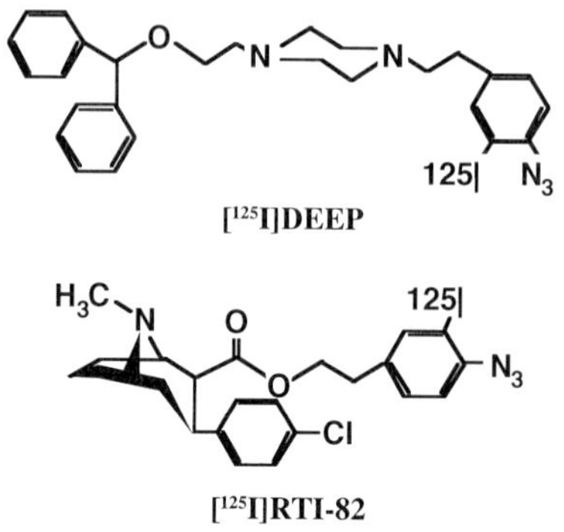

FIG. 1. Chemical structures of [^{125}I]DEEP and [^{125}I]RTI 82. [Reprinted with permission from R. A. Vaughan, *Mol. Pharmacol.* **47,** 956 (1995).]

there is little experimental evidence pertaining to the tertiary structure of the protein, its topological orientation, or sites of action of substrates or antagonists. Knowledge of DAT structure–function properties is requisite for determining the molecular basis of transporter function, and may make possible the development of therapeutic substitution treatments for psychostimulant drugs.

Several structural properties of DAT have been elucidated using photoaffinity labels based on two different classes of uptake blockers: [^{125}I]DEEP ([^{125}I]-1-[2-(diphenylmethoxy)ethyl]-4-[2-azido-3-iodophenyl)ethyl]piperazine), a compound from the GBR series,[3] and [^{125}I]RTI 82 ([^{125}I]-3β-(*p*-chlorophenyl)tropane-2β-carboxylic acid, 4′-azido-3-iodophenylethyl ester), a cocaine analog.[4] These compounds label dopamine transporters with nanomolar affinity, and the high energy and specific activity of the ligands make them superior reagents for DAT visualization. Proteolytic mapping and epitope-specific immunoprecipitation of photolabeled DATs have been used to investigate the binding sites of these compounds and have elucidated several structural properties of the protein. These techniques may be adaptable for use in identifying additional functional or structural domains of DAT and its homologs.

Preparation of Photoaffinity Labeled DATs

The chemical structures of [^{125}I]DEEP and [^{125}I]RTI 82 are shown in Fig. 1. The synthesis and characterization of these ligands have been pre-

[3] D. E. Grigoriadis, A. A. Wilson, R. Lew, J. S. Sharkey, and M. J. Kuhar, *J. Neurosci.* **9,** 2664 (1989).

[4] A. Patel, J. W. Boja, J. Lever, R. Lew, R. Simantov, F. I. Carroll, A. H. Lewin, A. Philip, Y. Gao, and M. J. Kuhar, *Brain Res.* **576,** 173 (1992).

viously described.[3,4] For photoaffinity labeling, 10–200 mg tissue from striatum or other brain region of interest are homogenized in 10 ml ice-cold 10 mM Na_2HPO_4 buffer, pH 7.4, containing 0.32 M sucrose, using a Brinkman Polytron homogenizer at setting 6 for 10 sec. The homogenate is centrifuged at 20,000g for 12 min at 4°, the supernatant is discarded, and the homogenization and centrifugation are repeated. The membranes are suspended in sucrose–phosphate buffer at 5–10 mg/ml original wet weight, and radioligand is added to a final concentration of 5 nM. Pharmacological specificity of photolabeling can be demonstrated by inclusion of a dopamine uptake blocker such as 30 μM (−)-cocaine, which inhibits photoincorporation of both radioligands into DAT. Binding is carried out for 1 hr at 0°, then the sample is transferred to a shallow vessel such as a petri plate, and is irradiated with UV light at a distance of 1–3 cm for 45 sec. Ambient light does not photoactivate these compounds, allowing manipulations to be performed in ordinary room illumination.

After UV irradiation, membranes are washed twice by centrifugation and prepared for subsequent analysis. For direct electrophoresis or gel purification of DAT, membranes are solubilized with SDS–PAGE sample buffer [62.5 mM Tris, pH 6.8, 2% SDS, 10% (v/v) glycerol, 10 mM dithiothreitol (DTT)] at 20 mg/ml original wet weight. For immunoprecipitations, membranes are similarly solubilized using an SDS concentration of 0.5% rather than 2%. For *in situ* proteolysis, unsolubilized membranes are resuspended in 50 mM Tris-HCl, pH 8.0, at 50 mg/ml original wet weight. Solubilized or unsolubilized samples can be stored at −20°.

DATs from caudate nucleus, nucleus accumbens, and olfactory tubercle show the most prominent labeling due to the high expression level of the protein in these regions,[3] but DATs expressed in other brain regions such as prefrontal cortex, substantia nigra, and ventral tegmental area can also be labeled at lower levels.[5] [^{125}I]DEEP and [^{125}I]RTI 82 label DATs from many species, including rat, mouse, dog, monkey, and human, and from DATs expressed heterologously in COS 7 (monkey kidney) and LLC-PK_1 (pig kidney) cells.[5–7] A limitation of the ligands is that their specificity for DAT is not absolute, and several other proteins in brain and cell membranes become labeled.[3–5] Although DAT is the only protein in the 80-kDa range which becomes labeled, and the other labeled proteins show no DAT pharmacological or immunological specificity, this cross-reactivity nevertheless reduces the usefulness of the photolabeled samples for certain applications.

[5] R. A. Vaughan, V. L. Brown, M. McCoy, and M. J. Kuhar, *J. Neurochem.* **66,** 2146 (1996).
[6] A. Patel, G. Uhl, and M. J. Kuhar, *J. Neurochem.* **61,** 496 (1993).
[7] R. A. Huff, R. A. Vaughan, M. J. Kuhar, and G. R. Uhl, *J. Neurochem.* **68,** 225 (1997).

Immunoprecipitation

Analysis of DAT functional and structural domains has been made possible by the availability of several epitope-specific antipeptide antibodies capable of immunoprecipitating DAT. These rabbit polyclonal antisera are directed against five distinct regions of the DAT primary sequence in the N and C termini and two internal regions.[8,9] For immunoprecipitations these sera are used at dilutions of 1:30–1:400. A limiting factor for using low-titer antisera in immunoprecipitations can be the large amount of immunoglobulin G (IgG) which is eventually loaded onto the gel. To circumvent this problem, IgG can be covalently attached to the protein A Sepharose resin using dimethylpimelimidate prior to use.[10]

Solubilized samples are diluted in an immunoprecipitation buffer (IPB) consisting of phosphate-buffered saline (PBS), pH 7.4, containing 0.05% SDS and 0.1% Triton X-100. Other neutral buffers such as 50 m*M* Tris-HCl, pH 8.0, may be used in place of PBS. To prevent denaturation of antibodies during immunoprecipitations, SDS concentrations should be 0.1% or less. Thus, membranes solubilized with 0.5% SDS require five-fold dilution, whereas samples prepared by electroelution (described in proceeding text) can be precipitated with little to no dilution. It is also possible to avoid SDS by solubilizing membranes with nondenaturing detergents such as Triton X-100, CHAPS, or digitonin.

Solubilized, diluted DATs are incubated at 4° for 1–3 hr with protein A Sepharose beads precoupled with antiserum. If precoupled beads are not used, DAT samples are diluted into IPB containing appropriate concentrations of antisera and incubated for 1–3 hr at 4°, followed by addition of protein A beads for an additional hour. Immunological specificity of precipitation can be demonstrated by adding a molar excess of the immunizing peptide to the sample during this step to block the recognition of the protein or peptide fragments by the antibodies. Sample size for immunoprecipitations can be adjusted as needed, depending on signal intensity and dilution requirements, but when performing the assay in microfuge tubes, final volumes should be 1250 μl or less to allow room for sample and beads to mix. After formation of immune complexes, beads are pelleted at low speed, supernatants are removed, and beads are washed 1–2 times with IPB. Samples are then eluted from the beads with SDS–PAGE sample buffer and subjected to electrophoresis. Unproteolyzed DAT is electrophoresed on 8–10% polyacrylamide gels, whereas proteolyzed samples are analyzed on 15–17% gels, which will resolve peptides as small as 2–4 kDa.

[8] R. A. Vaughan, *Mol. Pharmacol.* **47,** 956 (1995).

[9] R. A. Vaughan and M. J. Kuhar, *J. Biol. Chem.* **271,** 21672 (1996).

[10] C. Schneider, R. A. Newman, D. R. Sutherland, U. Asser, and M. F. Greaves, *J. Biol. Chem.* **257,** 10766 (1982).

DAT Proteolysis

Two different procedures have been used for peptide mapping of DAT: proteolysis after gel purification and electroelution, and *in situ* proteolysis of DATs in membrane suspensions. Each has scientific and technical advantages and disadvantages.

Gel Purification and Electroelution

The gel purification strategy is used in order to separate DAT from the nonspecifically photolabeled proteins present in the crude membrane samples.[8] In this procedure, solubilized photolabeled membranes are electrophoresed on 8–10% gels, DAT is identified by autoradiography, and the gel region containing DAT is excised. Radioactive ink or phosphorescent orientation markers are placed on the gel backing paper before film exposure to provide precise alignment of the gel and film. Excised gel pieces are briefly hydrated in SDS–PAGE electrophoresis buffer (25 m*M* Tris, 192 m*M* glycine, 0.1% SDS, pH 6.8), and the paper backing is removed. Up to five to six gel pieces are placed in the bottom of a Bio-Rad (Richmond, CA) electroelution chamber, upper and lower reservoirs are filled with electrophoresis buffer, and samples are subjected to a current of 10 mA per chamber for 4–6 hr. Either 3500 or 15,000 molecular weight cutoff dialysis caps can be used to collect samples, which have a final volume of about 500 μl per chamber. Based on radioactivity, recovery of DAT from the gel pieces is 50–70%. Electroeluted samples can be stored at $-20°$ without loss of DAT solubility, and if desired, electroeluted samples can be concentrated several-fold using a Centricon 30 concentration device (Amicon, Danvers, MA). Gel purification and electroelution produce radioactively pure DAT, as no other photoaffinity-labeled protein comigrates with DAT on the first gel. The electroeluted sample can then be subjected to proteolysis, and because DAT is the only labeled protein in the sample, all radioactive fragments produced are DAT products.

Staphylococcus aureus V8 protease and trypsin proteolyze electroeluted DAT at concentrations of 0.02–1 mg/ml and 0.4–3 mg/ml, respectively. Suitable conditions are 25–50 μl sample plus an equal volume of enzyme, incubated at 22° for 1–4 hr. After proteolysis, samples can be prepared directly for SDS–PAGE by the addition of sample buffer and subjected to electrophoresis on 15–17% gels. This allows the visualization of the complete spectrum of photolabeled fragments (Fig. 2). The ability to detect all fragments, regardless of their immunoprecipitability, is one of the greatest advantages of this technique. Protease-treated samples can also be analyzed by epitope-specific immunoprecipitation as described above. Recognition of a fragment with site-specific antibodies demonstrates the presence of

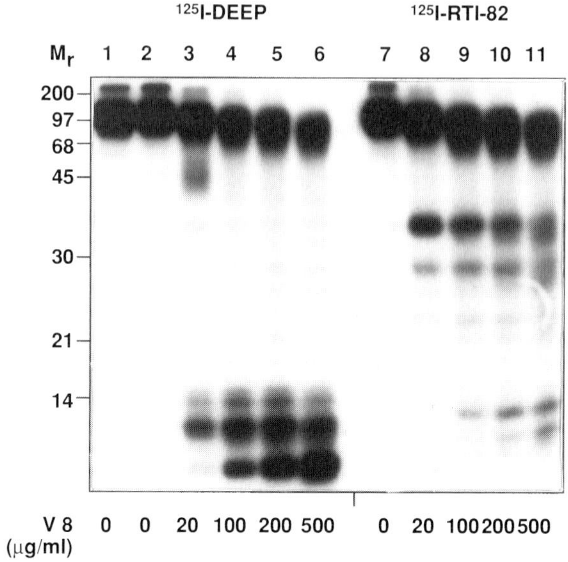

FIG. 2. V8 proteolysis of gel purified DATs. Dopamine transporters were photoaffinity labeled with [^{125}I]DEEP or [^{125}I]RTI 82, and were gel purified and electroeluted. Samples were then incubated with the indicated concentrations of V8 protease and electrophoresed on a 15% polyacrylamide gel followed by autoradiography. Molecular mass standards for all gels are indicated in kilodaltons. [Reprinted with permission from R. A. Vaughan, *Mol. Pharmacol.* **47,** 956 (1995).]

the epitope in the fragment and provides an indication of the fragment's position in the protein primary sequence.

A major concern of proteolysis work is the ongoing protein digestion which occurs after completion of desired protease treatment times. Denaturation of the sample with SDS does not completely inhibit protease activity, and boiling or extensive heat treatment results in transporter aggregation, eliminating this as a potential mechanism for protease inhibition. It is therefore essential to add protease inhibitors and up to 1 mg/ml carrier protein to serve as competitor substrate for the protease before proceeding with sample manipulation. This should be done even if samples will be directly electrophoresed, as proteolysis can occur within the gel matrix. Protease activity can also bleed into adjacent gel lanes during electrophoresis, producing digestion of untreated control samples, so protease inhibitors and carrier protein should also be added to these samples, or they should be placed in lanes separated from protease-containing lanes. It is also helpful to keep samples cold during precipitation and electrophoresis, and to minimize electrophoresis time. One advantage of trypsin compared to

V8 protease is that trypsin inhibitor at equivalent w/w ratio with the enzyme substantially reduces subsequent trypsin activity.

Proteolysis of electroeluted DATs labeled with [^{125}I]DEEP and [^{125}I]RTI 82 is shown in Fig. 2. V8 treatment of [^{125}I]DEEP-labeled DATs produces fragments of 10, 7, and 4 kDa, whereas 34 and 29 kDa fragments are generated from [^{125}I]RTI 82-labeled DATs. The production of different proteolysis patterns indicates that the two ligands are incorporated in different regions of the protein, an interpretation verified by epitope-specific immunoprecipitation.[8] The 10- and 7-kDa [^{125}I]DEEP fragments can be immunoprecipitated with antiserum 16 directed against amino acids 42–59 near the N terminus of the protein, demonstrating that [^{125}I]DEEP is incorporated near TMDs 1 and 2. Conversely, the 34 kDa [^{125}I]RTI 82-labeled fragment is not precipitated by serum 16, but is recognized by three antibodies directed against residues in the C-terminal half of the protein, indicating a more C-terminal incorporation of this ligand. Localization of the [^{125}I]RTI 82 binding site could not be determined more precisely with the electroelution method, as [^{125}I]RTI 82-labeled fragments smaller than 34 kDa were not consistently produced.

In Situ Proteolysis

DATs can also be analyzed proteolytically using an *in situ* proteolysis strategy.[9] In this procedure, photolabeled membrane suspensions are treated with enzyme, followed by separation of membranes and supernatants, and immunoprecipitation of fragments. Because of the nonspecifically photolabeled proteins present in the membrane samples, DAT proteolytic fragments generated with this protocol cannot be unequivocally identified without immunoprecipitation, and fragments may be produced which escape detection due to loss of antibody epitopes. The advantages of this strategy are that the starting DAT sample is in its native rather than a denatured form, less enzyme is required to digest the protein, and the presence or absence of integral membrane structure in the fragments can be determined. Retention of a fragment in the membranes denotes the presence of integral membrane structure, whereas proteolytic fragments not containing membrane structure will be released as soluble forms and can be recovered in the supernatant.[11]

For this procedure, membranes are photoaffinity labeled and suspended in 50 mM Tris-HCl, pH 8.0, at 50 mg/ml original wet weight. The membrane suspensions are treated with 1–200 μg/ml trypsin at 22°. Convenient volumes are 25 μl membranes plus 25 μl enzyme, although other volume

[11] D. Hereld, R. Vaughan, J.-Y. Kim, J. Borleis, and P. Devreotes, *J. Biol. Chem.* **269,** 7036 (1994).

combinations can be used. Dose–response and time course studies showed that enzyme dose was the more important factor, and at these tissue and enzyme concentrations, proteolysis is complete in 5 min. Following proteolysis, supernatants and membranes are separated by centrifugation and analyzed by SDS–PAGE either directly or after immunoprecipitation (Fig. 3). The previously mentioned concerns relative to ongoing proteolysis remain relevant to this procedure, although continued sample proteolysis is reduced in the membrane fractions because the enzyme is removed with the supernatant.

The spectrum of photolabeled DAT fragments generated by *in situ* proteolysis displays both similarities and differences to that of the fragments produced by proteolysis of solubilized DATs, and additional DAT structural properties were elucidated by characterization of the unique fragments. *In situ* trypsinization of [^{125}I]DEEP-labeled DATs generates 45- and 14-kDa fragments which immunoprecipitate with N-terminal antiserum 16 (Fig. 3). While this result is consistent with the results from the electroelution method with respect to the incorporation of [^{125}I]DEEP near TMDs 1 and 2, the generation of these larger fragments with the *in situ*

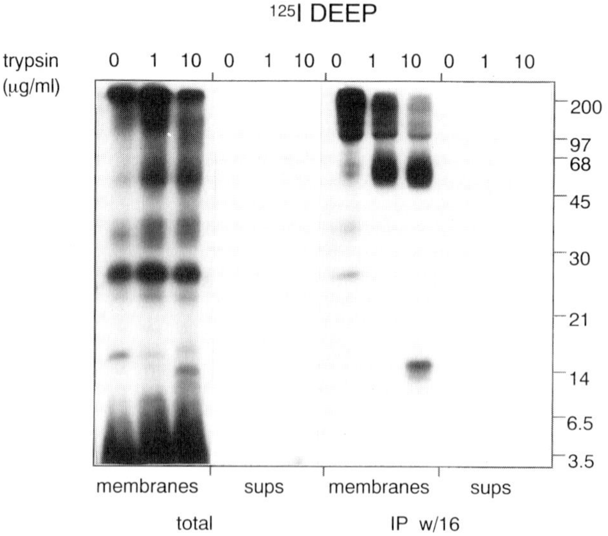

FIG. 3. *In situ* proteolysis of [^{125}I]DEEP-labeled DATs. Rat striatal membranes labeled with [^{125}I]DEEP were treated for 5 min with 0, 1, or 10 μg/ml trypsin as indicated. Samples were centrifuged and membranes and supernatants were separated. Aliquots were either analyzed directly (totals) or were immunoprecipitated with antiserum 16 (IP w/16) directed against amino acids 42–59. Samples were then electrophoresed on a 17% polyacrylamide gel followed by autoradiography. [Reprinted with permission from R. A. Vaughan and M. J. Kuhar, *J. Biol. Chem.* **271,** 21672 (1996).]

technique made it possible to identify and localize the sites of N-linked glycosylation in the DAT primary sequence (described below). Two fragments of 32 and 16 kDa are generated by *in situ* trypsinization of [^{125}I]RTI 82-labeled DATs.[9] The 32 kDa fragment is immunologically identical to the [^{125}I]RTI 82-labeled 34 kDa fragment produced from solubilized DATs and represents most of the C-terminal half of the protein. The 16-kDa fragment is immunoprecipitated only by serum 5 generated against amino acids 225–236, a region just N-terminal to putative TMD 4. This positions the 16-kDa fragment in the central region of the protein near putative TMDs 4–7, thus localizing the [^{125}I]RTI 82 incorporation site much more precisely than was possible with the gel purification strategy.

All four of the photolabeled fragments produced by *in situ* proteolysis were retained in membranes, demonstrating that they contain at least one transmembrane domain or other integral membrane structure. This indicates that the 14- and 16-kDa fragments precipitated by sera 16 and 5 contain one or more transmembrane domain within 125–145 residues of the antibody epitopes, and suggests that incorporation of the radioligands occurs close to or within transmembrane domains.

Carbohydrate Structure of Dopamine Transporter

The photoaffinity labels have also proved to be valuable tools for examining DAT glycosylation properties. Deglycosylation of DAT with *N*-glycanase or neuraminidase removes about 25- or 5-kDa in mass, respectively, demonstrating the presence of N-linked oligosaccharides and terminal sialic acids on the protein. While these treatments can be done using photolabeled membrane suspensions,[6,12,13] gel purification or immunoprecipitation of the protein prior to deglycosylation makes small changes in molecular mass easier to detect and requires less enzyme.[5,8,9]

For treatment of electroeluted DATs, 10 μl of sample are adjusted to contain 0.7% Nonidet P-40 and 5 m*M* EDTA, and treated at 22° with 0.1 U neuraminidase or 1.5 U *N*-glycanase for 1 hr or overnight, respectively. SDS–PAGE sample buffer is added and samples are electrophoresed on 10% gels. For treatment of immunoprecipitated samples, DATs are carried through the immunoprecipitation protocol to the final washing step before elution. The beads containing the immune complexes are suspended in 30 μl gel running buffer (25 m*M* Tris, 192 m*M* glycine, 0.1% SDS, pH 6.8) containing 0.7% Nonidet P-40 and 5 m*M* EDTA. Neuraminidase (0.01 U)

[12] R. Lew, D. Grigoriadis, A. Wilson, J. W. Boja, R. Simantov, and M. J. Kuhar, *Brain Res.* **539,** 239 (1991).

[13] R. Lew, R. Vaughan, R. Simantov, A. Wilson, and M. J. Kuhar, *Synapse* **8,** 152 (1991).

and/or *N*-glycanase (0.3 U) treatments are performed at 22° for 1 hr or overnight, respectively.[5,9] At the end of the treatment, the supernatant is removed, the beads are washed once, and samples are eluted with SDS–PAGE sample buffer followed by electrophoresis.

The region in the DAT primary sequence containing glycosylation sites was identified by a combination of enzymatic deglycosylation, immunoprecipitation, and peptide mapping.[9] [^{125}I]DEEP or [^{125}I]RTI 82-labeled DATs were subjected to *in situ* proteolysis and site-specific immunoprecipitation as described above, followed by treatment with *N*-glycanase or neuraminidase (Fig. 4). Of the four photolabeled proteolytic fragments produced by this technique, the only one affected by deglycosylation was the [^{125}I]DEEP-labeled 45-kDa fragment, which showed reduced molecular mass after treatment with both enzymes in a manner identical to full-length DAT. The lack of effect on the 14-kDa [^{125}I]DEEP-labeled fragment demonstrates that the proteolytic cut which reduces the 45-kDa fragment to the 14-kDa form removes the glycosylated region. Based on size, the 45-kDa fragment

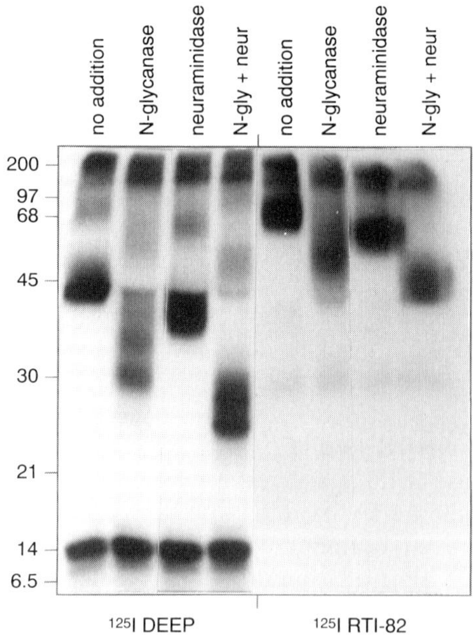

FIG. 4. Carbohydrate analysis of DAT tryptic fragments. Membranes photoaffinity labeled with the indicated ligand were treated with trypsin followed by immunoprecipitation of the fragments with serum 16 for [^{125}I]DEEP samples or serum 5 directed against amino acids 225–236 for [^{125}I]RTI 82 samples. Aliquots were given the indicated enzymatic deglycosylation treatments, followed by electrophoresis and autoradiography. [Reprinted with permission from R. A. Vaughan and M. J. Kuhar, *J. Biol. Chem.* **271,** 21672 (1996).]

would be expected to contain the consensus glycosylation sites present at positions 182, 188, 196, and 204 between putative TMDs 3 and 4. This result indicates that native DATs are glycosylated at one or more of these sites, shows that N-linked carbohydrates and sialic acids are present on the same peptide fragment, and indicates that the loop between TMDs 3 and 4 is oriented extracellularly as predicted.

Deglycosylation after immunoprecipitation was also used to examine the immunological and carbohydrate characteristics of DATs expressed in several species and brain regions.[5] The sensitivity provided by photolabeling in combination with the use of immunoprecipitation to concentrate DAT samples from large amounts of tissue made it possible to examine these properties in DATs expressed in midbrain cell bodies and frontal cortex where DAT expression is exceedingly low. DATs from all species and brain regions examined contain comparable amounts of N-linked carbohydrates and sialic acids, indicating substantial phylogenetic and regional conservation of carbohydrate structures. These results also demonstrate that DATs expressed in midbrain cell body regions possess mature carbohydrate structure and are functional for binding transport blockers. No DATs containing immature carbohydrate structure were observed, indicating that there are few incompletely glycosylated DATs in the cell body regions and/or that immature DAT forms are not active for ligand binding.

Summary

Several aspects of DAT structure and function have been elucidated using a combination of photoaffinity labeling, proteolysis, enzymatic deglycosylation, and epitope-specific immunoprecipitation.[8,9] The two photolabels are incorporated in different regions of the protein, suggesting that the binding sites for the ligands are distinct or partially nonoverlapping, consistent with results produced by site-directed mutagenesis and analysis of chimeras.[14–17] These studies have also verified several aspects of DAT structure previously hypothesized based only on theoretical considerations, including the presence of at least one transmembrane helix or other membrane-anchoring structure in two different regions of the protein, identification of the glycosylated domain, and some topological properties. It should be possible to extend and adapt these techniques to further delineate DAT

[14] S. Kitayama, S. Shimada, H. Xu, L. Markham, D. M. Donovan, and G. R. Uhl, *Proc. Natl. Acad. Sci. USA* **89,** 7782 (1992).

[15] J. B. Wang, S. Davis, and G. Uhl, *Soc. Neurosci. Abst.* **19,** 745 (1993).

[16] B. Giros, Y.-M. Wang, S. Suter, S. B. McLeskey, C. Pifl, and M. G. Caron, *J. Biol. Chem.* **269,** 15985 (1994).

[17] K. J. Buck, and S. G. Amara, *Proc. Natl. Acad. Sci. USA* **91,** 12584 (1995).

structural properties and to identify other functional domains such as phosphorylation sites or active sulfhydryl moieties.

Acknowledgments

The author thanks Mr. Mike McCoy for outstanding technical assistance.

Section III

Transport Assays and Kinetic Analyses

Articles 16 through 19

[16] Ion-Coupled Neurotransmitter Transport: Thermodynamic vs. Kinetic Determinations of Stoichiometry

By GARY RUDNICK

Introduction

Neurotransmitter transporters are molecular machines and, like other machines, they require an input of energy in order to perform useful work. The work of these machines is to move a neurotransmitter across a lipid bilayer from a low concentration in the extracellular medium to a higher concentration in the cytoplasm. The energy source that powers this work is the unequal distribution of Na^+, K^+, and Cl^- and the electrical potential across the membrane (Fig. 1). Ion channels use the same ion gradients and potential to drive the ionic currents that are used for signaling by the cell. Because the Na^+, K^+-ATPase generates these ion gradients, we can think of ATP hydrolysis as the ultimate source of energy for neurotransmitter transport. However, each transporter has its own way of coupling neurotransmitter accumulation to the ion gradients and potential generated by the ATPase.

All plasma membrane neurotransmitter transporters couple the symport (movement together across the membrane, cotransport) of their cognate neurotransmitter substrate with Na^+ ions, but the number of Na^+ ions moving with each neurotransmitter molecule (the Na^+ stoichiometry) varies from one to three depending on the transporter. Many, but not all, transporters also catalyze symport of Cl^- with their substrate. Likewise, some transporters couple the movement of K^+ out of the cell to the entry of neurotransmitter (antiport or countertransport). In synaptic vesicle transport systems, antiport of H^+ ions drives neurotransmitter accumulation. Depending on the number of ions moving, their direction of movement, and the charge on the substrate neurotransmitter, there may be a net movement of charge across the membrane with each neurotransmitter molecule transported.

If ions and net charge move across the membrane coupled to neurotransmitter flux, then ion gradients and membrane potential will influence the steady-state distribution of neurotransmitter. The ultimate concentration of extracellular neurotransmitter will be determined by these driving forces and how they are coupled to transport. This coupling, or stoichiometry, becomes important because the ability of reuptake systems to decrease

0076-6879/98 $25.00

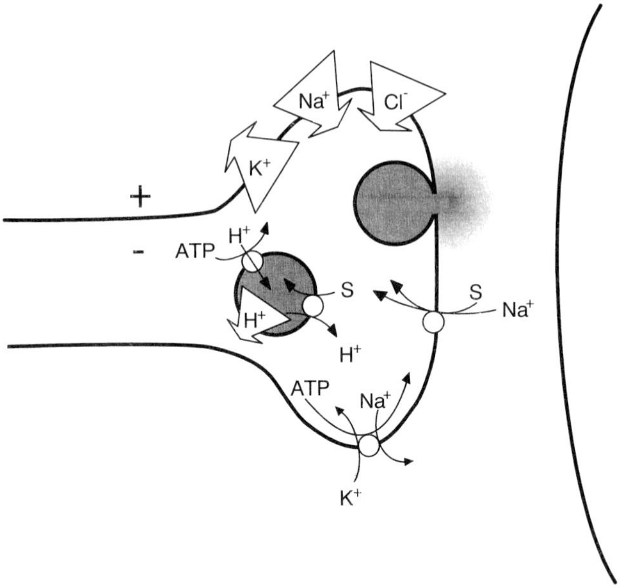

FIG. 1. Overall scheme of neurotransmitter reuptake and storage. The Na^+, K^+-ATPase generates gradients of Na^+ (out > in), and K^+ (in > out), and a transmembrane electrical potential, $\Delta\psi$ (interior negative), that leads to a Cl^- concentration difference (out > in). Each of these components can serve as a driving force for the accumulation of extracellular neurotransmitter, S, in the cytoplasm. Once inside the cell, the neurotransmitter is taken up by synaptic vesicles. The vacuolar ATPase in the synaptic vesicle generates an electrochemical H^+ potential (interior positive and acidic) that drives neurotransmitter uptake by H^+ antiport.

synaptic levels of neurotransmitter is likely to be responsible for the level of receptor activation during and between periods of synaptic activity. For example, a reevaluation of the Na^+ stoichiometry for glutamate transport indicates that three Na^+ ions are cotransported.[1] This finding has altered our estimate of glutamate excitotoxicity during ischemia in the brain, as in stroke. If glutamate were cotransported with only two Na^+ ions, as previously thought,[2] there would not be enough driving force during ischemia to lower glutamate levels to below the excitotoxic threshold.

Kinetics vs Thermodynamics

The number of ions co- or countertransported with a neurotransmitter affects the driving force, and therefore the maximum neurotransmitter

[1] N. Zerangue and M. P. Kavanaugh, *Nature* **383,** 634 (1996).

[2] Y. Kanai, S. Nussberger, M. Romero, W. Boron, S. Hebert, and M. Hediger, *J. Biol. Chem.* **270,** 16561 (1995).

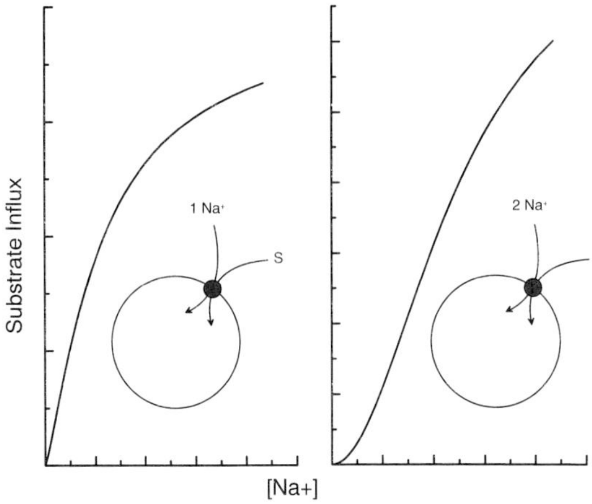

FIG. 2. Predicted Na^+ dependence for substrate accumulation by symport with one or two Na^+ ions. *Left:* Symport of substrate with a single Na^+ ion leads to a hyperbolic dependence of influx on the Na^+ concentration. *Right:* Symport of two Na^+ ions with substrate, under special circumstances, leads to a hyperbolic Na^+ dependence with a Hill coefficient of 2. In practice, the Hill coefficient may be considerably less than 2 if Na^+ binding is not highly cooperative (see text).

gradient, imposed by a given solute gradient. Likewise, the kinetic behavior of a transporter is affected by the number of ions participating in the reaction. For example, 10-fold transmembrane gradient of Na^+ will drive some accumulation of a transport substrate if the Na^+ : substrate stoichiometry is 1 : 1 and more accumulation if the coupling is 2 : 1. If two cotransported Na^+ ions bind with appropriate affinity and kinetics, the Na^+ dependence of transport rate will be sigmoidal (Fig. 2). Analysis of this sigmoidicity will reveal a second-order dependence on Na^+, or a Hill coefficient approaching 2. By comparison, if a single Na^+ is required for transport, the Na^+ dependence will be hyperbolic (Hill coefficient of 1). Thus, the number of ions cotransported with a substrate frequently is reflected in the ion dependence of transport rate. However, it would be incorrect to assume that ion coupling stoichiometry is always given by the kinetic behavior of transport.[3]

Kinetics of Ion Dependence

For a reaction in which a neurotransmitter substrate, S, is cotransported with one or more Na^+ ions, the transport equation may be written as

$$S_o + n\mathrm{Na}_o^+ \rightarrow S_i + n\mathrm{Na}_i^+ \tag{1}$$

[3] R. J. Turner, *Methods Enzymol.* **191,** 479 (1990).

where the subscripts o and i refer to the external and internal solutes, respectively, and n is the Na^+ stoichiometry. Under certain conditions, when all of the cotransported Na^+ ions bind with high positive cooperativity so that essentially no transporters are partially filled with Na^+, then

$$V = \frac{[Na^+]^n \times V_{max}}{[Na^+]^n + (K_{Na})^n} \tag{2}$$

where V is the rate of transport, V_{max} is the rate at saturating Na^+, and K_{Na} is a constant related to the Na^+ dependence under the given conditions. The rate will vary with Na^+ in a manner that depends on n, and fitting V vs Na^+ data to Eq. (2) will provide a value for n.

In all other cases, however, the value of n derived from kinetic analysis is not the same as the cotransport stoichiometry. One possibility is that multiple ions must bind to the transporter for translocation to occur, but only one of the ions is actually translocated. In this case, kinetics might indicate $n > 1$ even though one Na^+ ion was cotransported. Alternatively, if the Na^+ ions bind independently, there will be significant buildup of transporters with partially filled Na^+ sites. Under these conditions, the kinetically determined n will be less than the total number of Na^+ ion binding sites. Thus, a kinetically determined $n = 1$ may mask a case where multiple Na^+ ions are cotransported. The discrepancy between sigmoidicity and stoichiometry has been directly observed with the Na^+ pump of pig kidney.[4]

Thermodynamic Determinations

The ion coupling of a transporter can be compared to the gearing on a bicycle (Fig. 3). In a low gear, the pedals must turn many times for each turn of the wheel, and the rider can pedal up a steep hill. This is analogous to a transporter in which many Na^+ ions are cotransported per molecule of substrate. The "mechanical" advantage means that a given Na^+ gradient can drive substrate up a steep concentration "hill." If fewer Na^+ ions are cotransported, the process is more efficient, like the bicycle in a higher gear, but a smaller gradient of substrate can be generated from the same Na^+ gradient. As long as the Na^+, K^+-ATPase can keep up, a high Na^+ stoichiometry will not dissipate the Na^+ gradient. The extracellular Na^+ concentration is likely to always be in vast excess over the neurotransmitter concentration.

[4] S. J. Karlish and W. D. Stein, *J. Physiol.* (*London*) **359,** 119 (1985).

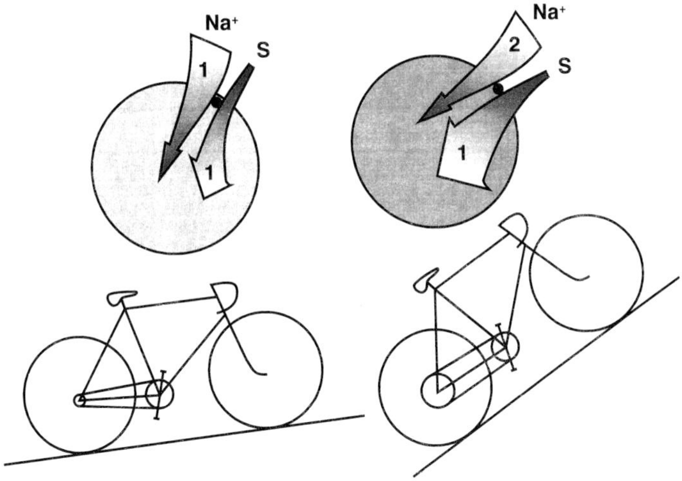

FIG. 3. Dependence of the accumulated substrate gradient on the Na^+ symport stoichiometry. *Left:* A 1:1 symport stoichiometry (above) is shown as being similar to a bicycle in a high gear which is able to climb only a mild hill (below). *Right:* A 2:1 symport stoichiometry leads to much more substrate accumulation with the same Na^+ gradient (above), analogous to a bicycle in a low gear climbing a steeper hill (below).

From the magnitude of the neurotransmitter gradient formed in response to a given ion gradient, the coupling theoretically can be deduced.[5] For each solute A that is distributed across the membrane, there is a chemical potential

$$\Delta\mu_A = RT \ln \frac{[A]_i}{[A]_o} \tag{3}$$

where $[A]_i$ and $[A]_o$ represent the concentrations of internal and external solute, respectively, R is the gas constant ($1.9872\ \mathrm{cal \cdot deg^{-1} \cdot mol^{-1}}$), and T is the temperature in kelvins. If the solute is charged, the total electrochemical potential $\Delta\tilde{\mu}_A$ is determined by the transmembrane concentration difference and electrical potential as follows:

$$\Delta\tilde{\mu}_A = RT \ln \frac{[A]_i}{[A]_o} + z_A F \Delta\psi \tag{4}$$

where z_A is the charge on solute A, F is the faraday constant ($23.061\ \mathrm{cal \cdot mV^{-1} \cdot mol^{-1}}$), and $\Delta\psi$ is the electrical potential (cytosolic–external) in

[5] P. S. Aronson, *in* "Electrogenic transport: Fundamental Principles and Physiological Implications" (M. P. Blaustein and M. Lieberman, Eds.), p. 49. Raven Press, New York, 1984.

millivolts. If the transport of n_A molecules of solute A is coupled to n_B molecules of solute B in the same direction, then at equilibrium

$$n_A \, \Delta\tilde{\mu}_A = -n_B \, \Delta\tilde{\mu}_B \tag{5}$$

Expanding Eq. (5)

$$n_A \frac{RT}{F} \ln \frac{[A]_i}{[A]_o} = n_B \frac{RT}{F} \ln \frac{[B]_o}{[B]_i} - \Delta\psi(n_A z_A + n_B z_B) \tag{6}$$

When A is a neurotransmitter and B is Na^+ ion, the coupling stoichiometry (n_B/n_A) determines how the neurotransmitter gradient varies with changes in the Na^+ gradient. Equation (6) can be modified if other co- or countertransported ions participate in the reaction. For example, for the glutamate transporter EAAT3, which cotransports 3 Na^+ ions and 1 H^+ ion with glutamate, and countertransports 1 K^+ ion,

$$\begin{aligned} \frac{RT}{F} \ln \frac{[glu^-]_i}{glu^-]_o} = 3\frac{RT}{F} \ln \frac{[Na^+]_o}{Na^+]_i} + \frac{RT}{F} \ln \frac{[H^+]_o}{[H^+]_i} \\ + \frac{RT}{F} \ln \frac{[K^+]_i}{[K^+]_o} - \Delta\psi(-1 + 3 + 1 - 1) \end{aligned} \tag{7}$$

Methods for Determining Ion Coupling Stoichiometry

Several methods have been developed to determine the coupling of transmembrane ion gradients and electrical potentials to substrate accumulation. Some are relatively simple in both concept and in execution but are amenable only to a few transporters. Others are suitable for a wider variety of transporters but require particular experimental systems such as membrane vesicles or electrophysiological techniques. The methods described below have been used for a variety of transport systems.

Flux Ratios

The most direct method is to follow the movement of both substrate and coupled ion with time. If the amount of substrate-dependent ion flux is equal to the ion-dependent substrate flux, for example, one could conclude that the coupling stoichiometry is 1:1. If ion flux is twice substrate flux, then 2:1 coupling is indicated, etc. (Fig. 4). This approach requires that a measurable part of the ion flux be substrate-dependent. For many transporters, this is not the case. Basal Na^+, Cl^-, and K^+ fluxes are frequently many orders of magnitude greater than the V_{max} for transport. When this is so, any increase in ion flux due to substrate transport is undetectable.

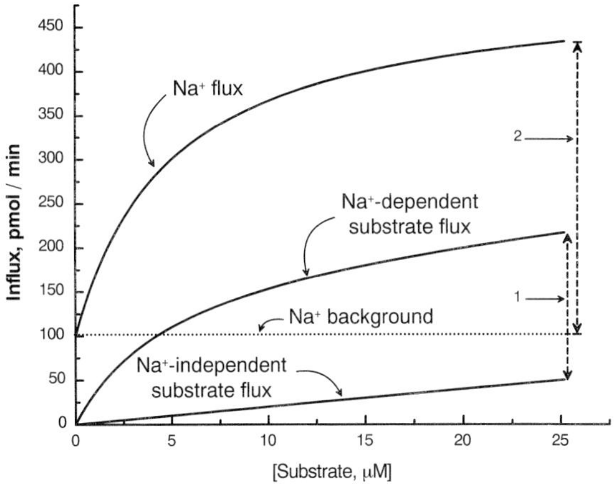

FIG. 4. Stoichiometry by comparing Na^+ and substrate influx. A hypothetical example of 2:1 Na^+ symport stoichiometry is shown. Addition of increasing concentrations of substrate increases both substrate influx and Na^+ influx. the substrate-dependent Na^+ influx and Na^+-dependent substrate saturate with the same K_m for substrate, but the Na^+ influx is twice that for substrate. This method is feasible only when the Na^+ background is low enough that the increase in Na^+ due to substrate can be measured accurately.

Additionally, any uncoupled ion flux stimulated by neurotransmitter transport (see below) will complicate measurements. A further complication is that intact cells may have significant endogenous concentrations of the substrate. Exchange between the internal, unlabeled substrate and radiolabeled external substrate will lead to apparent substrate entry but not Na^+ entry, thereby decreasing the apparent Na^+ stoichiometry. There are, however, exceptional cases where the technique has proven useful.

In the ε^t cell line characterized by Bulloch *et al.*,[6] glutamate increases Na^+ influx measurably, with a dependence on glutamate concentration that mirrors the saturation of glutamate influx itself.[7] The initial rate of $^{22}Na^+$ influx was roughly twice the rate of glutamate influx, suggesting a 2:1 coupling stoichiometry. In the same study, Stallcup *et al.*[7] measured the Na^+ dependence of glutamate influx and observed no evidence for sigmiodal behavior as would be expected for multiple, independent Na^+ ions participating in the reaction. This demonstrates the unreliability of kinetic determinations for stoichiometry, since conclusions from the kinetics alone would erroneously indicate a coupling of one Na^+ ion per glutamate. In other systems, measurements of the Na^+ dependence of glutamate influx showed

[6] K. Bulloch, W. B. Stallcup, and M. Cohn, *Life Sci.* **22,** 495 (1978).
[7] W. B. Stallcup, K. Bulloch, and E. E. Baetge, *J. Neurochem.* **32,** 57 (1979).

a sigmoidal response consistent with a coupling stoichiometry of two or more.[8]

In cases where high basal ion fluxes obscure substrate-dependent coupled ion fluxes, the transporter can be moved to an environment where it represents most of the ion permeability of the membrane. The rat brain γ-aminobutyric acid (GABA) transporter, GAT-1, was partially purified by Kanner and co-workers and reconstituted with lipids to form proteoliposomes.[9] In this relatively pure preparation, many other proteins, originally present in the neuronal membrane, had been removed by purification. The ion permeabilities corresponding to these other proteins were much lower in the reconstituted system, and GABA-dependent Na^+ and Cl^- fluxes were observed. From the ratio of Na^+ and Cl^- fluxes, measured with ^{22}Na and ^{36}Cl, to the GABA fluxes, Keynan and Kanner[9] calculated a stoichiometry of two Na^+ ions and one Cl^- ion per molecule of GABA. Although this method gave a very satisfying result for the GABA transporter, it is applicable only for those transporter proteins that are functional after solubilization, purification, and incorporation into proteoliposomes. Some transporters, for example, the 5-HT (5-hydroxytryptamine) transporter, so far have defied efforts to functionally reconstitute the purified protein. For these transporters, indirect methods are required to determine the coupling stoichiometry.

Steady-State Gradients in Vesicles

If the fluxes of two solutes are coupled, one can predict from Eq. (6) how the transmembrane concentration gradient of solute A will be affected by changes in the gradient of solute B. Membrane vesicles—isolated from plasma membrane or intracellular organelles and free of cytoplasmic components—have proven to be very useful in transport studies.[10] In membrane vesicle systems, the internal and external media can be independently manipulated, and there is no energetic contribution from cellular metabolism. It should be possible, therefore, to impose a defined gradient of B, introduce solute A in a radiolabeled form, and wait for an equilibrium to establish. From the distribution of A between the vesicle lumen and the medium, the transmembrane gradient of A can be determined, compared with the gradient of B, and the coupling stoichiometry, (n_B/n_A), can be calculated. In practice, the procedure is somewhat more involved, but it basically uses the response in the accumulation of a substrate, such as A, to the imposed gradient of a coupled ion, B.

[8] U. Klockner, T. Storck, M. Conradt, and W. Stoffel, *J. Biol. Chem.* **268,** 14594 (1993).
[9] S. Keynan and B. I. Kanner, *Biochem.* **27,** 12 (1988).
[10] H. R. Kaback, *Science* **186,** 882 (1974).

In some membrane vesicles, the ion gradients can be generated *in situ* by ion pumps. For example, in secretory organelles such as synaptic vesicles, a vacuolar ATPase generates a transmembrane electrochemical H^+ potential using cytoplasmic ATP (Fig. 1). Since the cytoplasmic face of the isolated vesicle faces the medium, addition of ATP can drive transport *in vitro*. Addition of permeant anions such as Cl^- or SCN^- dissipate the electrical portion, $\Delta\psi$, of the $\Delta\tilde{\mu}_{H^+}$, permeant weak bases, such as methylamine, dissipate the ΔpH, and H^+ ionophores, such as FCCP, dissipate both components of $\Delta\tilde{\mu}_{H^+}$. Using these agents to manipulate $\Delta\psi$ and ΔpH, and measuring the effect with radiolabeled methylamine and SCN^-, it has been possible to use the endogenous ATPase to generate a range of electrochemical H^+ potentials. From the response of transport systems that use these potentials, the H^+ stoichiometry of vesicular neurotransmitter transporters has been extensively studied.[11–15] This approach is not possible for mammalian plasma membrane vesicles since the active site of the Na^+, K^+-ATPase also is on the cytoplasmic side of the membrane and therefore inaccessible to medium ATP in right-side-out vesicles. In this case, a Na^+ ion gradient, for example, can be imposed by equilibrating vesicles in Na^+-free medium and then diluting them into Na^+-containing medium.

Most membrane vesicles containing a Na^+ cotransport system will accumulate the substrate (for example, a neurotransmitter) to an internal concentration higher than that of the medium if internal Na^+ is lower than medium Na^+. The imposed Na^+ gradient is not permanent, however, since other pathways for Na^+ flux exist in many cells, and as the Na^+ gradient decays, so does the substrate gradient. In extreme cases, the time course for substrate accumulation becomes a sharp peak, with a rising phase as substrate enters in response to the Na^+ gradient and a falling phase as the Na^+ gradient decays (Fig. 5). In such cases, when the Na^+ gradient had dissipated significantly before the substrate had achieved equilibrium, there was no chance to use the equilibrium method to determine stoichiometry. Other methods are available when the vesicles are leaky to Na^+, but in vesicles with relatively slow Na^+ leaks, a variation of the equilibrium method is useful.

In suitable membrane vesicles, substrate accumulation in response to a Na^+ gradient reaches a plateau value after an initial uptake phase. If the plateau is maintained for long enough, it is reasonable to assume that

[11] J. Knoth, M. Zallakian, and D. Njus, *Biochem.* **20,** 6625 (1981).
[12] R. G. Johnson, S. E. Carty, and A. Scarpa, *J. Biol. Chem.* **256,** 5773 (1981).
[13] P. R. Maycox, T. Deckwerth, J. W. Hell, and R. Jahn, *J. Biol. Chem.* **263,** 15423 (1988).
[14] J. W. Hell, P. R. Maycox, and R. Jahn, *J. Biol. Chem.* **265,** 2111 (1990).
[15] J. Tabb, P. Kish, R. VanDyke, and T. Ueda, *J. Biol. Chem.* **267,** 15412 (1992).

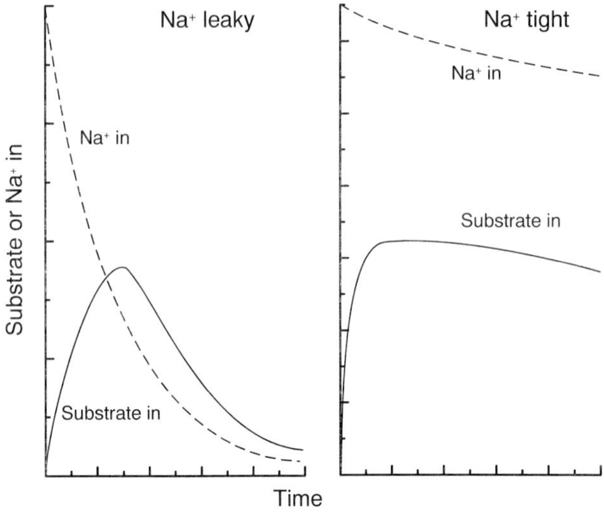

FIG. 5. Effect of membrane vesicle Na^+ permeability on the time course of substrate accumulation. *Left:* The Na^+ gradient imposed by dilution (dashed line) decays rapidly relative to substrate influx. As a consequence, substrate accumulation peaks at a time when substantial dissipation of the Na^+ gradient has occurred and then rapidly decays with the Na^+ gradient. *Right:* Slow Na^+ efflux relative to Na^+ gradient decay. The lack of significant Na^+ efflux before the peak of substrate accumulation leads to a sustained plateau during which the substrate and Na^+ gradients are in equilibrium.

a transient equilibrium was established between the Na^+ and substrate gradients. It is difficult to say how long is long enough for any given system, but as a rule of thumb, substrate accumulation should remain at 90% of the peak uptake level for 4–5 times longer than it took to reach that 90% level. To determine Na^+ stoichiometry, a series of transport measurements should be made with different imposed Na^+ gradients. The results can be viewed as a logarithmic plot of the accumulated substrate gradient against the imposed Na^+ gradient. If $\ln([Na^+]_o/[Na^+]_i)$ is plotted on the abscissa and $\ln([\text{substrate}]_i/[\text{substrate}]_o)$ on the ordinate, then the slope of the line should approximate the Na : substrate stoichiometry. To accurately determine the internal substrate concentration, it is necessary to know the internal volume of the vesicle population. However, the more relevant parameter for stoichiometry determinations is the relative internal concentration and how that changes with the imposed Na^+ gradient. Assuming that the intravesicular volume does not change when the Na^+ gradient is altered, and most of the substrate remains extravesicular, the amount of internal substrate can be substituted for the transmembrane substrate concentration gradient.

Likewise, if external Na^+ is held constant, and internal Na^+ is used to vary the Na^+ gradient, the relationship may be simplified even further to $\ln[\text{substrate}]_i$ vs $-\ln[Na^+]_i$. There is reason to vary internal Na^+ rather than external Na^+. Frequently, the rate of accumulation depends on the external Na^+ concentration. Decreasing external Na^+ to lower the Na^+ gradient may have the unwanted effect of slowing down the time course of accumulation so that equilibrium of the substrate gradient with the Na^+ gradient takes longer than at optimal external Na^+. No matter which method is used to vary the Na^+ gradient (or other gradients, for that matter), a time course for accumulation should be determined at the extremes of internal or external Na^+ to ensure that the time of measurement is close to equilibrium in all cases.

Many transport systems are electrogenic. For these systems, the coupled movement of substrate with ions across the membrane involves the concomitant translocation of net charge. Electrogenic transporters are sensitive to changes in transmembrane potential. Since varying the Na^+ gradient could alter the potential, manipulations of the Na^+ gradient also might affect accumulation indirectly through these potential changes. To avoid this complication in membrane vesicles, the ionophore valinomycin has been used with equal internal and external K^+ to keep the membrane potential near zero. Valinomycin is a specific K^+ carrier, and addition of up to 1 μM of this agent to a suspension of membrane vesicles allows K^+ to permeate the membranes extremely rapidly. Any shift in potential due to alteration in the Na^+ gradient will be rapidly compensated for by valinomycin-mediated K^+ movements. A high K^+ concentration is used to ensure that the K^+ gradient does not significantly change.

The same approach described here for Na^+ can be used for other ions as well. Many neurotransmitter transporters are stimulated by external Cl^-.[16] In some cases, this stimulation has been attributed to Cl^- cotransport.[17–19] One method to determine if Cl^- is cotransported with substrate is to vary the Cl^- gradient and measure the effect on the substrate gradient. This method also has the advantage that it can determine the coupling stoichiometry between Cl^- and substrate. If both Na^+ and Cl^- are cotransported, both must be present in the medium for transport to occur. If either the Na^+ or the Cl^- gradients is varied, the other gradient should be kept constant, although it need not be kept at zero. The Na^+ gradient could even contribute to substrate accumulation when measuring the effect of

[16] M. J. Kuhar and M. A. Zarbin, *J. Neurochem.* **31,** 251 (1978).

[17] P. J. Nelson and G. Rudnick, *J. Biol. Chem.* **257,** 6151 (1982).

[18] R. Radian and B. I. Kanner, *Biochem.* **22,** 1236 (1983).

[19] H. H. Gu, S. Wall, and G. Rudnick, *J. Biol. Chem.* **271,** 6911 (1996).

the Cl^- gradient without affecting the measurement of Cl^- stoichiometry. If the effect of the Cl^- gradient on substrate accumulation is measured, and not the absolute substrate concentration gradient, then the presence of other invariant gradients that contribute to substrate accumulation will not affect the response to Cl^-, but only the absolute level of accumulation.

Some neurotransmitter transporters are coupled to exchange with K^+. Both glutamate[20] and 5-HT transporters[21] are stimulated by a transmembrane K^+ gradient (in > out). Testing for K^+ antiport is more complicated than for Na^+ or Cl^- symport because valinomycin cannot be used to clamp the membrane potential. Moreover, the basal permeability of some cell membranes is dominated by a K^+ conductance. Thus, changing the K^+ gradient may influence transport indirectly through the potential rather than by antiport. To test if K^+ is involved directly, the effect of K^+ can be tested in the absence of a K^+ gradient. If K^+ is transported as part the reaction, then the rate of transport will be very sensitive, or absolutely dependent, on the presence of K^+.[22] Once the dependence on K^+ is established, under conditions where K^+ cannot contribute to the potential ($K^+_{in} = K^+_{out}$), the K^+ stoichiometry can be determined by varying the K^+ gradient. To clamp the membrane potential, a H^+ ionophore such as 2,4-dinitrophenol or carbonyl cyanide *p*-trifluoromethoxyphenylhydrazone (FCCP) can be added with strong pH buffering in both internal and external medium.

Static Head Method

In some cases, it may be difficult or impossible to obtain membrane vesicles containing the transporter of interest that also maintain stable ion gradients. In this case, a modification of the typical membrane vesicle approach has been useful.[23] In the static head method, there is no delay between the time when vesicles are diluted into external medium to generate ion gradients and the time when transport equilibrium is measured. As with the vesicle systems described above, membrane vesicles equilibrated in one solution are rapidly diluted into a different solution. With the static head method, however, internal and external solutions contain different concentrations of radiolabeled substrate so that a substrate gradient (in > out) also is generated along with the ion gradients.

If the ion gradients are just sufficient to maintain that substrate gradient, the intravesicular substrate concentration will remain stable, at least for a

[20] B. I. Kanner and E. Marva, *Biochem.* **21,** 3143 (1982).
[21] G. Rudnick and P. J. Nelson, *Biochem.* **17,** 4739 (1978).
[22] P. J. Nelson and G. Rudnick, *J. Biol. Chem.* **254,** 10084 (1979).
[23] Y. Fukuhara and R. J. Turner, *Biochem. Biophys. Acta* **770,** 73 (1984).

short time until the ion gradients decay significantly. If the ion gradients are initially larger than required to maintain the substrate gradient, then net external substrate will enter the vesicle, and if the ion gradients are insufficient, the internal substrate will exit the vesicle. By performing different dilutions to create different Na^+ or Cl^- gradients, for example, one can experimentally determine the ion gradient required to just balance the substrate gradient. The process can be repeated with different substrate gradients so that a series of ion gradients are obtained, each corresponding to a substrate gradient. A plot of $\ln([Na^+]_o/[Na^+]_i)$ vs $\ln([\text{substrate}]_i/[\text{substrate}]_o)$, for example, should give a slope equal to the Na^+ substrate stoichiometry for the Na^+ gradients just sufficient to balance each substrate gradient.

The same approach can be applied to whole cells. In this case, cells are allowed to accumulate substrate for a specific period of time, and then the external medium is diluted to different extents to lower the substrate concentration. At each external Na^+ concentration, for example, there will be an external substrate concentration that will just maintain the accumulated internal substrate. At higher external substrate concentrations, accumulation will continue; at lower concentrations, efflux will occur. From the relationship between the external Na^+ and substrate concentrations, a coupling ratio may be deduced. This calculation depends on the assumption that during the short time after dilution, when substrate influx or efflux occurs, the internal Na^+ concentration does not change. By definition, the internal substrate concentration does not change, so the relationship reduces to $\ln[Na^+]_o$ vs $-\ln[\text{substrate}]_o$. The slope of this relationship should be equal to the Na^+ substrate coupling stoichiometry. Likewise, external Cl^- H^+ and K^+ could be varied to obtain the coupling stoichiometry for these ions. The requirement for any of these determinations is that the manipulations in external ion concentrations do not rapidly change internal ion concentrations or membrane potential.

Electrical Measurements

An electrogenic transporter is also rheogenic. That is, ion-coupled substrate transport generates an electrical current across the membrane. The magnitude of this current relative to substrate flux is determined by the number of co- and countertransported ions and their charges. In principle, the direction and magnitude of transport-associated currents can directly confirm or contradict a particular stoichiometry deduced from flux studies. In practice, however, uncoupled currents associated with transport[24] can confound these determinations.

[24] M. S. Sonders and S. G. Amara, *Current Opinion in Neurobiology* **6,** 294 (1996).

Electrophysiological measurements have given new insight into the function of neurotransmitter transporters. These studies have measured currents that correspond to partial reactions, such as Na^+ binding,[25,26] and have demonstrated currents due to coupled and uncoupled ion fluxes.[27–32] Attempts have been made to use these current measurements to determine the ion coupling stoichiometry of electrogenic transporters. This procedure has met with some success in the case of the glutamate transporters[1,2] for which uncoupled fluxes require an anion such as Cl^-. However, for many other transporters, the uncoupled currents are carried by the same ions that are required for transport. The persistence of these uncoupled currents confounds attempts to use current measurements to determine stoichiometry.

In the case of the glutamate transporter EAAC1, Kanai *et al.*[2] measured glutamate flux, Na^+ flux, and current in the absence of Cl^- using *Xenopus* oocytes injected with EAAC1 cRNA. They concluded that glutamate transport occurred in symport with two Na^+ ions. Subsequently, Zerangue and Kavanaugh[1] used a more sensitive approach with the EAAT3 glutamate transporter and concluded that three Na^+ ions were cotransported. This study was also performed in oocytes, but the approach differed in that the two-electrode voltage clamp was used to determine the potential at which glutamate flux was exactly balanced by the ion gradients. This approach is similar to the static head technique described above, except that the membrane potential, rather than the Na^+ gradient, was varied to balance glutamate flux.

In practice, the direction of glutamate flux was determined over a range of potentials by adding an inhibitor of the transporter to see if it blocked an inward current (glutamate influx) or an outward current (glutamate efflux). The reversal potential, at which the glutamate transport current was zero, was determined as a function of external glutamate, Na^+, K^+, and H^+. The reversal potential shifted approximately 30 mV for each 10-

[25] S. Mager, N. Kleinberger-Doron, G. I. Keshet, N. Davidson, B. I. Kanner, and H. A. Lester, *J. Neurosci.* **16,** 5405 (1996).

[26] S. Mager, J. Naeve, M. Quick, C. Labarca, N. Davidson, and H. A. Lester, *Neuron* **10,** 177 (1993).

[27] J. N. Cammack, S. V. Rakhilin, and E. A. Schwartz, *Neuron* **13,** 949 (1994).

[28] S. Mager, C. Min, D. J. Henry, C. Chavkin, B. J. Hoffman, N. Davidson, and H. A. Lester, *Neuron* **12,** 845 (1994).

[29] W. A. Fairman, R. J. Vandenberg, J. L. Arriza, M. P. Kavanaugh, and S. G. Amara, *Nature* **375,** 599 (1995).

[30] J. I. Wadiche, S. G. Amara, and M. P. Kavanaugh, *Neuron* **15,** 721 (1995).

[31] F. Lin, H. A. Lester, and S. Mager, *Biophys. J.* **71,** 3126 (1996).

[32] M. S. Sonders, S.-J. Zhu, N. R. Zahniser, M. P. Kavanaugh, and S. G. Amara, *J. Neurosci.* **17,** 960 (1997).

fold increase in the external concentration of H^+ and glutamate, −30 mV for K^+, and about 100 mV for Na^+. These results are most consistent with symport of three Na^+ ions and one H^+ ion with each glutamate ion, and antiport of one K^+ ion in the same cycle.

For transporters that conduct uncoupled currents under any conditions compatible with transport, this approach will not work, although measurement of the influence of ion gradients on substrate accumulation still will be valid. In addition, the response of transport to an imposed potential could be used to assess the charge movements accompanying transport. Thus, 5-HT transport was unaffected by altering the potential at which oocytes expressing SERT cRNA were voltage clamped,[28] whereas dopamine uptake by oocytes expressing DAT cRNA was increased by plasma membrane hyperpolarization.[32] These results are consistent with electroneutral 5-HT transport and electrogenic DA transport, but an important assumption inherent in these studies must be stated.

In both cases, the effect of membrane potential on the rate, rather than the extent, of transport was measured. As discussed previously, the effect of driving forces on the steady-state or equilibrium accumulation of substrate should reflect the charge stoichiometry of the overall reaction. Although it might be expected that the transport rate would respond to potential in a similar manner, the interpretation of a rate effect is different from that of an equilibrium effect. Measurements of the potential dependence of transport rates are additionally affected by the fact that the effect reflects the potential dependence of a rate-determining step in the reaction. The potential dependence of this step does not necessarily reflect all the charge movements in the overall reaction. It is possible to imagine a situation where charge moves in more than one step in the cycle. Each of those steps could have its own response to potential changes and each response could be different from the potential dependence of the overall reaction. It is, therefore, preferable to measure the effect of potential on the steady-state accumulation of substrate.

Obviously, the same considerations apply no matter how the potential is manipulated. In membrane vesicles, for example, the effect of potential on transport is frequently measured by imposing a K^+ diffusion potential with valinomycin. If net positive charge enters the vesicle with substrate, imposition of a K^+ gradient (in $>$ out) to generate an interior negative potential will stimulate the equilibrium accumulation of substrate. The K^+ diffusion potential also may stimulate the rate of substrate influx, but the effect on rate is not as good a measure of charge movement in the full transport cycle.

[17] Inhibition of [^{3}H]Dopamine Translocation and [^{3}H]Cocaine Analog Binding: A Potential Screening Device for Cocaine Antagonists

By MAARTEN E. A. REITH, CEN XU, F. IVY CARROLL, and NIAN-HANG CHEN

Introduction

There is considerable interest in developing a cocaine antagonist for use as an adjunct in the treatment of cocaine addiction or as a counteractive medication to combat acute central overstimulation. Because the dopamine transporter in the plasma membrane is one of the targets considered to be crucial for its centrally stimulatory activity,[1] it would be logical to conceive of a compound that interferes with the action of cocaine at the transporter level. Antagonism of cocaine action could be accommodated in a model depicting separate sites of action for uptake blockers and substrate recognition. In the case of the serotonin transporter, early work with radiolabeled imipramine led to proposals involving separate sites[2,3] perhaps with an endogenous ligand[4] acting only on the blocker site to explain how upregulated imipramine binding sites could cause reduced neuronal uptake of serotonin.[5] The possibility of an endogenous ligand has been also entertained for the dopamine transporter,[6,7] but such compounds have remained elusive.[8–10] In a model assuming separate blocker and substrate recognition

[1] M. W. Fischman and C.-E. Johanson, *in* "Pharmacological Aspects of Drug Dependence: Towards an Integrated Neurobehavioral Approach" (C. R. Schuster and M. J. Kuhar, Eds.), p. 159 (1996).

[2] M. S. Briley, S. Z. Langer, R. Raisman, D. Sechter, and E. Zarifian, *Science* **209,** 303 (1980).

[3] N. Brunello, D. M. Chuang, and E. Costa, *Science* **215,** 1112 (1982).

[4] E. Costa, M. L. Barbaccia, O. Gandolfi, and D. M. Chuang, *in* "New Vistas in Depression" (S. Z. Langer, R. Takahashi, T. Segawa, and M. Briley, Eds.), p. 31. Pergamon Press, Oxford, 1985.

[5] M. L. Barbaccia, O. Gandolfi, D. M. Chuang, and E. Costa, *Proc. Natl. Acad. Sci. USA* **80,** 5134 (1983).

[6] I. Hanbauer, L. T. Kennedy, M. C. Missale, and E. C. Bruckwick, *in* "New Vistas in Depression" (S. Z. Langer, R. Takahashi, T. Segawa, and M. Briley, Eds.), p. 41. Pergamon Press, Oxford, 1985.

[7] M. E. A. Reith, *NIDA Res. Monogr.* **88,** 23 (1988).

[8] M. E. A. Reith, H. Sershen, and A. Lajtha, *Neurochem. Res.* **5,** 1291 (1980).

[9] C. R. Lee, A. M. Galzin, M. A. Taranger, and S. Z. Langer, *Biochem. Pharmacol.* **36,** 945 (1987).

METHODS IN ENZYMOLOGY, VOL. 296
0076-6879/98 $25.00

sites, it is possible to envision an antagonist that binds to the blocker binding site but does not induce the required conformational change to prevent substrate binding and translocation; this antagonist then could interfere with the binding of a blocker such as cocaine.[7] Although equilibrium binding experiments with radioligands for the dopamine transporter and uptake inhibition studies have generally supported the involvement of common binding domains for cocaine and dopamine,[11–14] there is also evidence for the additional involvement of separate domains from studies on kinetic modeling of uptake data collected by rotating disk voltammetry,[15] thermodynamic analysis of radioligand binding,[16] protection against action of sulfhydryl reagents at blocker binding sites,[17–19] and construction of transporter chimeras[20,21] or mutants with site-directed alterations.[22] All data taken together support the existence of a common binding domain in the recognition of cocaine and dopamine, as well as separate domains. It is possible that a cocaine antagonist could be developed that only interferes with a cocaine binding domain that is not involved in the recognition of dopamine, thereby preventing the action of cocaine without disturbing dopamine uptake itself.[18,22,23] Although the conformational model as advanced by Saadouni *et al.*[14] could also accommodate the existence of a cocaine antagonist at the dopamine transporter level, no direct evidence for this model beyond thermodynamic analysis is available. The group of Rothman has

[10] F. Artigas, E. Martinez, and A. Adell, *Eur. J. Pharmacol.* **181,** 9 (1990).

[11] E. Richelson and M. Pfenning, *Eur. J. Pharmacol.* **104,** 277 (1984).

[12] B. K. Krueger, *J. Neurochem.* **55,** 260 (1990).

[13] M. E. A. Reith, B. De Costa, K. C. Rice, and A. E. Jacobson, *Eur. J. Pharmacol.* **227,** 417 (1992).

[14] S. Saadouni, F. Refahi-Lyamani, J. Costentin, and J. J. Bonnet, *Eur. J. Pharmacol.* **268,** 187 (1994).

[15] J. S. McElvain and J. O. Schenk, *Biochem. Pharmacol.* **43,** 2189 (1992).

[16] J. J. Bonnet, S. Benmansour, J. Costentin, E. M. Parker, and L. X. Cubeddu, *J. Pharmacol. Exp. Ther.* **253,** 1206 (1990).

[17] K. M. Johnson, J. S. Bergmann, and A. P. Kozikowski, *Eur. J. Pharmacol.* **227,** 411 (1992).

[18] M. E. A. Reith, C. Xu, and L. L. Coffey, *Biochem. Pharmacol.* **52,** 1435 (1996).

[19] C. Xu, L. L. Coffey, and M. E. A. Reith, *Naunyn-Schmiedeberg's Arch. Pharmacol.* **355,** 64 (1997).

[20] K. J. Buck and S. G. Amara, *Proc. Natl. Acad. Sci. USA* **91,** 12584 (1994).

[21] S. Povlock and S. G. Amara, *in* "Neurotransmitter Transporters: Structure, Function, and Regulation" (M. E. A. Reith, Ed.), p. 1. Humana Press, Totowa, New Jersey, 1996.

[22] S. Kitayama, S. Shimada, H. Xu, L. Markham, D. M. Donovan, and G. R. Uhl, *Proc. Natl. Acad. Sci. USA* **89,** 7782 (1992).

[23] F. I. Carroll, A. H. Lewin, and M. J. Kuhar, *in* "Neurotransmitter Transporters: Structure, Function, and Regulation" (M. E. A. Reith, Ed.), p. 263. Humana Press, Totowa, New Jersey, 1996.

reported the possibility of "partial agonism" in the context of effects on extracellular dopamine by GBR 12909 in awake animals[24]; *in vivo* interrelating systems may be required for this intriguing phenomenon as *in vitro* approaches have not been able to show partial agonism.[25] There is an *in vitro* observation demonstrating a moderate antagonistic effect of (7α-methoxy)cocaine against cocaine-induced dopamine uptake blockade.[26]

Impact of Assay Conditions

The general approach in screening for a potential cocaine antagonist consists of assessing the inhibitory effect of a test compound on the translocation of dopamine by the dopamine transporter, usually in rat striatal synaptosomal preparations, as well as its inhibitory effect on the binding of a radiolabeled cocaine analog. The aim is to find a compound that potently inhibits binding without affecting dopamine uptake. Routinely, such assays are performed under conditions optimized for the intended measure, uptake, or binding. However, it is known that experimental conditions can greatly affect the observed characteristics of uptake or binding phenomena as well as the potency of inhibitors. For instance, dopamine uptake rates depend on the assay conditions,[27,28] and inhibitors will be more or less active as a function of ionic content of the incubation mixture.[29–31] Binding of cocaine analogs has been shown to vary depending on the conditions used to generate the brain preparation and on the assay buffer,[32,33] and conditions also affect observed potencies of inhibitors.[31,34,35] One important condition, not usually taken into account, is the pH of the assay buffer. An example of this, observed in our laboratory, is the impact of pH on the dependency of the shape of the Na^+-dependency curve of the binding of ^{3}H-labeled 2β-carbomethoxy-3β-(4-fluorophenyl)tropane (WIN

[24] M. H. Baumann, G. U. Char, B. R. De Costa, K. C. Rice, and R. B. Rothman, *J. Pharmacol. Exp. Ther.* **271,** 1216 (1994).

[25] A. N. Gifford, J. S. Bergmann, and K. M. Johnson, *Drug Alcohol Depend.* **32,** 65 (1993).

[26] D. Simoni, J. Stoelwinder, A. P. Kozikowski, K. M. Johnson, J. S. Bergmann, and R. G. Ball, *J. Med. Chem.* **36,** 3975 (1993).

[27] R. P. Shank, C. R. Schneider, and J. J. Tighe, *J. Neurochem.* **49,** 381 (1987).

[28] I. Zimanyi, A. Lajtha, and M. E. A. Reith, *Naunyn Schmiedebergs Arch. Pharmacol.* **340,** 626 (1989).

[29] S. C. Wall, R. B. Innis, and G. Rudnick, *Mol. Pharmacol.* **43,** 264 (1993).

[30] N. Amejdki-Chab, J. Costentin, and J. J. Bonnet, *J. Neurochem.* **58,** 793 (1992).

[31] R. B. Rothman, K. M. Becketts, L. R. Radesca, B. R. De Costa, K. C. Rice, F. I. Carroll, and C. M. Dersch, *Life Sci.* **53,** PL267 (1993).

[32] M. E. A. Reith and L. L. Coffey, *J. Neurochem.* **61,** 167 (1993).

[33] L. L. Coffey and M. E. A. Reith, *J. Neurosci. Methods* **51,** 23 (1994).

[34] D. O. Calligaro and M. E. Eldefrawi, *Membr. Biochem.* **7,** 87 (1987).

[35] C. Xu and M. E. A. Reith, *J. Pharmacol. Exp. Ther.* **282,** 920 (1997).

35,428), a phenyltropane analog of cocaine, to rat striatal membrane (Fig. 1). At pH 7.4 in the presence of 0.32 *M* sucrose, used in many dopamine transporter studies to enhance binding, a peak of [^{3}H]WIN 35,428 binding was observed at [Na^+] ~30–50 m*M* along with a reduction in binding at higher [Na^+] (Fig. 1A). When sucrose was not present, binding was much lower with a plateau at 30 m*M* continuing up to 200 m*M* Na^+. The stimulatory effect of sucrose was more pronounced at low rather than high [Na^+]

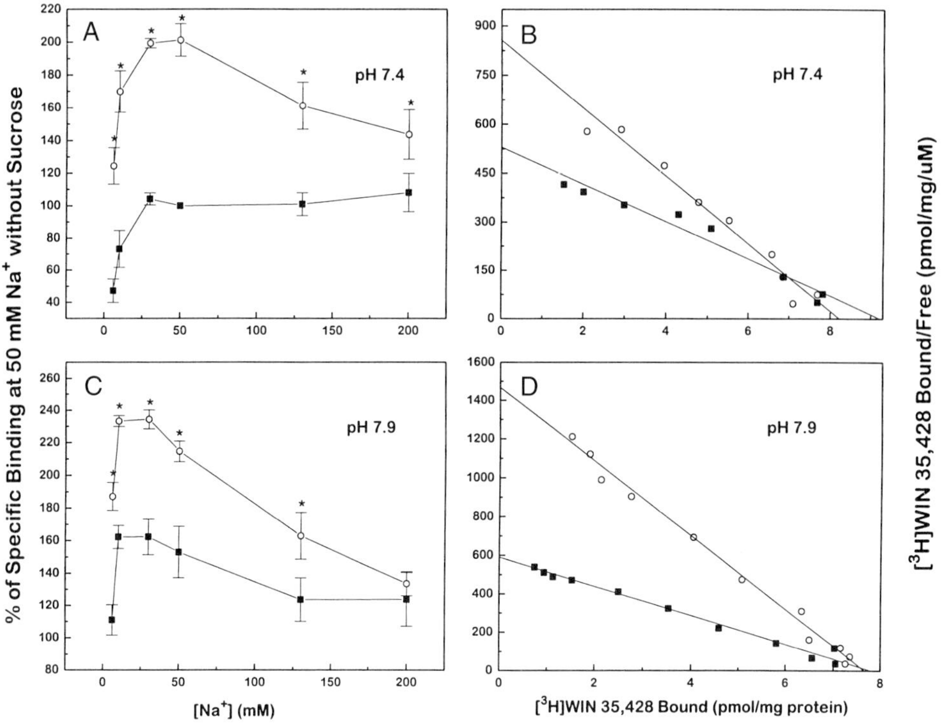

FIG. 1. Na^+ dependency (A and C) and saturation analysis (B and D) of [^{3}H]WIN 35,428 binding at pH 7.4 and 7.9 in the presence (○) and absence (■) of sucrose. (A, C) [^{3}H]WIN 35,428 binding to the rat striatal membranes was measured with varying concentrations of Na^+ from 0 to 200 m*M*. Binding is expressed as % of the binding at 50 m*M* Na^+ without sucrose measured with the same membrane preparation on the same day. Values are mean ± SEM of three independent experiments carried out in triplicate. The average binding at 50 m*M* Na^+ without sucrose was 880 fmol/mg of protein. (B, D) The radioligand was present at 3.5 n*M* at pH 7.4 and 1.5 n*M* at pH 7.9, and increasing concentrations of unlabeled WIN 35,428 were added up to 500 n*M*; nonspecific binding was defined with 30 μM cocaine. The straight line represents the best fit chosen by the LIGAND program. Both panels show a typical experiment that was assayed in triplicate. The experiment was carried out three times with independent striatal membrane preparations (for averages see text). $P < 0.05$ compared with absence of sucrose at same [Na^+] (Student's *t*-test).

resulting in a changed shape of the Na^+ curve. A similar effect was also observed at pH 7.9 (Fig. 1C), although the binding peak appeared at a lower $[Na^+]$ ~10–30 m*M* both with and without sucrose. In addition, a reduction in binding at higher $[Na^+]$ was evident both in the absence and, more strongly, in the presence of sucrose. For pH 7.4, the effect of 0.32 *M* sucrose was tested on the binding constants at 18 m*M* Na^+, on the ascending portion of the Na^+ curve (Fig. 1B). The presence of sucrose in the assay buffer reduced the K_d value from 18 ± 2 to 11 ± 1 n*M* ($P < 0.05$, Student's *t*-test) (average ± SEM for 3 independent experiments) without affecting the B_{max} value (9 ± 1.0 compared with 8 ± 0.3 pmol/mg of protein) (Fig. 1B). The application of the one-site binding model to [^{3}H]WIN 35,428 binding data under the present conditions is discussed in detail in our previous studies.[21,36] At pH 7.9 and 30 m*M* Na^+ (peak binding in the Na^+ curve), again the presence of 0.32 *M* sucrose reduced the K_d from 12 ± 1 to 6 ± 1 n*M* without affecting the B_{max} (6 ± 0.9 compared with 7 ± 0.7 pmol/mg of protein) ($P < 0.05$) (Fig. 1D). Thus, both pH and sucrose are important factors determining the shape of the Na^+ curve. At a relatively higher pH (7.9) the binding peak tends to occur at 10–30 m*M* Na^+ and inhibition by higher $[Na^+]$ is observed with or without sucrose, whereas at a relatively lower pH (7.4) the binding peak is shifted rightwardly to ~30–50 m*M* Na^+ and no inhibition by higher $[Na^+]$ is seen in the absence of sucrose. This puts a different light on our previously proposed distinction between cocaine-/methyl phenidate-like ligands and GBR-/mazindol-like ligands, inhibited and not inhibited, respectively, by high $[Na^+]$.[32] In most studies on [^{3}H]cocaine and [^{3}H]WIN 35,428 binding reviewed by us previously[32] either the pH was ≥7.7 or sucrose was present, favoring the occurrence of high $[Na^+]$ inhibition, whereas in most studies on [^{3}H]GBR 12935/12783 and [^{3}H]mazindol binding the pH was ≤7.6 and sucrose was absent, favoring the occurrence of a Na^+ plateau.

In searching for a cocaine antagonist, it is important to reduce the chance of finding false positives, inhibiting binding more than uptake, and minimize the risk of missing potential lead compounds, inhibiting uptake more than binding. Given the impact of conditions described above, it is therefore important to carry out the two assays under comparable experimental conditions, as discussed later.

Kinetic Considerations in Comparing Uptake and Binding

There are special kinetic features that need to be considered in comparing uptake and binding assays. For example, uptake needs to be studied

[36] C. Xu, L. L. Coffey, and M. E. A. Reith, *Biochem. Pharmacol.* **49,** 339 (1995).

in the *initial velocity* phase (i.e., duing a short time period) whereas binding is ideally assessed on *equilibration* (i.e., after a relatively longer time period). Because some inhibitors need time to equilibrate, a short uptake assay can therefore yield a higher IC_{50} for uptake inhibition than the IC_{50} for binding inhibition observed in a longer binding assay, erroneously leading to the conclusion that the compound is a "partial agonist." The method we have adopted for rat striatal synaptosomes circumvents these problems, making use of a rapidly equilibrating ligand for translocation. [^{3}H]dopamine itself, and a rapidly equilibrating radioligand for binding, [^{3}H]WIN 35,428. In the uptake assay, the inhibitor is added at time 0, [^{3}H]dopamine is added at exactly 7 min, and termination occurs at exactly 8 min; in the binding assay, inhibitor and [^{3}H]WIN 35,428 are present from time 0, and termination occurs at 8 min. Consideration of the rate constants involved of radioligands and inhibitor demonstrates that both assays "catch" the inhibitor at the same time point of its approach toward equilibrium (7–8 min) so that uptake and binding potencies are comparable in that respect even if the inhibitor has not fully equilibrated (for full details, see our previous paper[36]). Furthermore, both [^{3}H]dopamine and [^{3}H]WIN 35,428 can be considered to be at equilibrium with their binding sites on the transporter under the allotted times in the assays.[36]

An important study in this context is that of the group of Rothman *et al.*[31] pioneering the approach of carrying out uptake and binding assays under identical conditions. In their study, uptake of [^{3}H]dopamine was compared with binding of ^{125}I-labeled 2β-carbomethoxy-3β-(4-iodophenyl)tropane (RTI-55), another phenyltropane analog of cocaine. [^{125}I]RTI-55 equilibrates appreciably slower than [^{3}H]WIN 35,428, and the timing approach is therefore necessarily different. Rothman *et al.*[31] preincubated the inhibitor for 1 hr, and then conducted either the uptake or binding assay for an extra 20 min. Although in this design [^{125}I]RTI-55 will not equilibrate in the time allowed, the inhibitory potencies of test compounds will be comparable in the uptake and binding assays as long as the concentration of [^{125}I]RTI-55 is smaller than its K_d so that the inhibitor equilibrium is not shifted appreciably.

Results Obtained under Identical Conditions for Uptake and Binding

Uptake ([^{3}H]dopamine) and binding ([^{3}H]WIN 35,428) inhibitory potencies were obtained pairwise with rat striatal synaptosomal preparations on the same experimental day under the conditions described above for cocaine, WIN 35,428, benztropine, nomifensine, mazindol, BTCP, and GBR 12909 (Fig. 2). The data for GBR 12909 were subject to appreciable variation between days, with uptake and binding IC_{50} values changing in tandem, presumably because of variable adsorption onto the walls of tubes of the

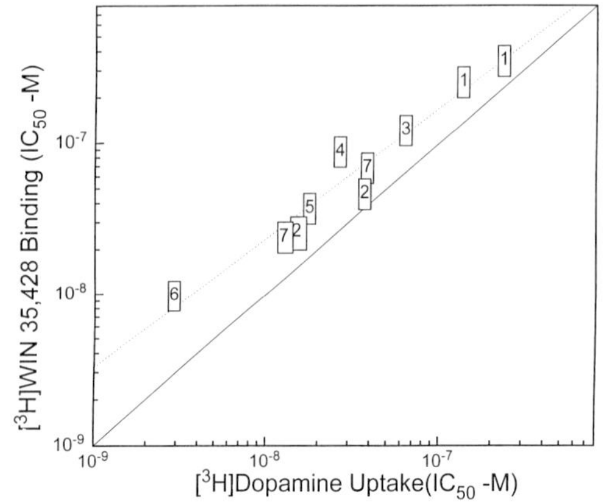

FIG. 2. IC_{50} values of compounds in inhibiting [^{3}H]WIN 35,428 binding and [^{3}H]dopamine uptake under identical conditions. Each point represents a paired observation of both the binding and uptake IC_{50} obtained with the same membrane preparation and the same stocks of drugs. The data were obtained with 8 min for [^{3}H]WIN 35,428 binding and 1 min for [^{3}H]dopamine uptake. The solid straight line represents a theoretical line describing a perfect one-to-one relationship between IC_{50} values for binding and uptake. The dashed line represents the results of least squares linear regression analysis of the data ($r = 0.97$, $n = 10$, $P < 0.001$). (Reprinted in adapted form by permission of the publisher from "Translocation of dopamine and binding of WIN 35,428 [2β-carbomethoxy-3β-(4-fluorophenyl)tropane] measured under identical conditions in rat striatal synaptosomal preparations: inhibition by various blockers," M. E. A. Reith, L. L. Coffey, and C. Xu, *Biochemical Pharmacology* **49,** No. 3, pp. 339–350. Copyright 1995 by Elsevier Science Inc.) 1, Cocaine; 2, WIN 35,428; 3, benztropine; 4, nomifensine; 5, mazindol; 6, BTCP; and 7, GBR 12909.

stock solutions (for discussion see Ref. 36). The correlation line composed of all points (Fig. 2, broken line, $r = 0.97$, $n = 10$, $P < 0.001$) was displaced to the left compared with the theoretical line depicting a perfect one-to-one ratio (solid straight line). Binding over uptake IC_{50} ratios were on the average 2.3, indicating a more potent uptake inhibitory activity. Several reasons for this are discussed in our previous report,[36] including the possibility of a spare receptor reserve speculated to derive from transporter dimers that carry two binding sites for WIN 35,428, of which only one needs to be occupied for preventing uptake by that dimer.

The same experimental approach was taken to reassess the potencies of three phenyltropane analogs of cocaine that had been demonstrated previously to be stronger in inhibiting [^{3}H]WIN 35,428 binding than [^{3}H]dopamine uptake under assay conditions optimized for the respective measure (Table I). When assayed under identical binding/uptake conditions, the compounds did not differ from WIN 35,428 itself in their binding over uptake IC_{50} ratio, again being approximately 1.4–2.

TABLE I
INHIBITION OF [^{3}H]WIN 35,428 BINDING AND [^{3}H]DOPAMINE UPTAKE UNDER NONIDENTICAL AND IDENTICAL CONDITIONS

Compound	Nonidentical conditions, binding over uptake IC_{50} (ratio)[a]	Identical conditions: Binding IC_{50}[b,c] (nM)	Binding Hill	Uptake IC_{50}[b,c] (nM)	Uptake Hill	Binding over uptake IC_{50} (ratio)[d]
WIN 35,428	1.0 ± 0.2	21.1 ± 3.2	1.1 ± 0.1	14.2 ± 1.2	1.0 ± 0.1	1.5 ± 0.1
RTI-32	0.24 ± 0.04	5.8 ± 0.2	1.1 ± 0.1	2.9 ± 0.3	1.0 ± 0.1	2.0 ± 0.2
RTI-114	0.23 ± 0.02	3.7	1.2	2.6	1.0	1.4
RTI-121	0.15 ± 0.02	2.8 ± 0.2	1.2 ± 0.1	2.2 ± 1.0	1.1 ± 0.0	1.7 ± 0.8

[a] Computed from mean ± SEM of binding and uptake data of Carroll and colleagues.[39]

[b] Average for 2 independent experiments each carried out in triplicate (±SEM = range × 0.5 because n = 2), except for RTI-114, which was examined in one experiment.

[c] There are significant different differences (two-way ANOVA) between drugs [factor A, F(3,13) = 38.47, $P < 0.0005$] and assays [factor B, binding/uptake, F(1,13) = 6.41, $P < 0.05$], but there was no interaction [F(3,13) = 1.54, $P = 0.30$].

[d] Average of ratios obtained in 2 independent experiments (±SEM = range × 0.5 because n = 2); each ratio was calculated per experiment with the parallel binding and uptake assays; RTI-114 was examined in one experiment.

Experimental Protocols

General Considerations

Ideally, the binding and uptake assays are performed in parallel on the same P_2 preparation with the same stocks of test compounds. This ensures a valid comparison between binding and uptake results. It does require the participation of at least two or three people (see proceeding text) in the experiment, because the striatal suspensions, once prepared, need to be processed for binding and uptake in a timely fashion. When uptake experiments are carried out by single filtration, incubations are staggered in time; we routinely start each incubation 1 min apart, and 60 assays will therefore result in the striatal suspension sitting on ice between 5 and 10 min for the first tube, and 65–70 min for the last tube. Over that time frame, uptake activity only decreases marginally, for which one can control by repeating total and nonspecific conditions in the second half of the set of incubations. This approach will require two people handling the uptake assays (one for starting preincubation/incubation and one for stopping and filtering), or if a harvester is used, one individual. An additional person needs to participate to run the binding assays which are routinely harvested in most labs.

Protocols

1. Set up tubes for binding and uptake assays with every condition in triplicate. The following description is for binding assays carried out in a total volume of 0.2 ml in 1-ml ministrip tubes (Skatron, Sterling, VA), for harvesting with a Skatron miniharvester, and uptake assays in a total volume of 0.4 ml in borsosilicate culture tubes (12 × 75 mm), for workup by the single filtration assay. Adjust details for other assay designs. For instance, one might want to assay for binding in 96-well plates, and terminate with a harvester that has the 96-well format. Or one might terminate the uptake assays with a harvester with pins in a format of 24 or 48 positions as the standard racks that hold the 12 × 75-mm tubes. We have measured [^{3}H]dopamine uptake (in parallel with binding), in sets of 24 assay mixtures by harvesting in the Brandel (Gaithersburg, MD) cell harvester with Whatman (Clifton, NJ) GF/C filters. In that case, incubations with drug and [^{3}H]dopamine were started with less than 5 sec between samples. Immediately after addition of ice-cold stop solution on the same time schedule, all 24 samples were filtered collectively. This procedure results in a variable waiting time (0–2 min) between adding a stop solution and filtering the mixture, but in control experiments (data not shown) it was found that there was no loss of accumulated [^{3}H]dopamine taken up during waiting times up to 6 min, the longest interval tested. With GF/C filters the filtration by the Brandel harvester was more rapid than with GF/F filters, and the uptake results were the same.

The final concentrations of all components, in both binding and uptake assays, are 73 m*M* NaCl, 3 m*M* KCl, 0.7 m*M* $MgSO_4$, 6 m*M* glucose, 0.6 m*M* $CaCl_2$, 0.006 m*M* nialamide, 0.08 *M* sucrose, 18 m*M* Na^+, and 10 m*M* phosphate from a mixture of primary and secondary phosphate buffer (giving a pH of 7.4 at room temperature), approximately 0.25 mg (for binding) or 0.12 mg (for uptake) of P_2 protein per ml assay medium, and test drug (a total of 8 concentrations evenly spaced around the IC_{50}). The radioactive [^{3}H]WIN 35,428 stock for the binding assays can be prepared at this point. The stock should be prepared such that addition of 20-μl aliquots to the binding assays results in a final concentration of 4 n*M* radioactive [^{3}H]WIN 35,428 (approximately 80 Ci/mmol, DuPont—New England Nuclear, Boston, MA). For binding (uptake) measurements, the tubes are filled with 20 (0) μl [^{3}H]WIN 35,428 stock, 120 (240) μl buffer stock, and 10 (20) μl test drug and kept on ice.

2. Dissect striatal tissue from rats (choose a strain, age, and gender and use same in followup experiments). Homogenize in 15 volumes of ice-cold 0.32 *M* sucrose in a glass homogenizer with a motor-driven Teflon pestle (7 strokes up and down at 800 rpm). Rinse the homogenizer and pestle with 30 volumes of 0.32 *M* sucrose. Combine this fluid and the homogenate,

and centrifugate at 1000 *g* for 10 min at 0–4°. Centrifuge the supernatant fraction subsequently at 17,000 *g* for 20 min at 0–4°.

3. Prepare the dopamine radioisotope stocks while the striatal preparation is taken through the final centrifuge spin. The stock should be prepared such that addition of 40-μl aliquots to the uptake assays results in 4 n*M* [^{3}H]dopamine (from supplier, specific activity 30–35 Ci/mmol) along with 46 n*M* unlabeled dopamine. Isotope solutions that come from the supplier with a gas other than air in the vial are accessed through the cap with a Hamilton syringe in a manner that does not introduce air in the container. Keep the prepared stocks on ice, and shield the dopamine-containing stock from light by wrapping the tube with aluminum foil.

4. Take the tube(s) with the striatal preparation(s) from the centrifuge, and *homogenize* the resulting pellet (P_2) in approximately 0.04 ml of 0.32 *M* sucrose per mg of initial tissue weight *with the glass–Teflon homogenizer* (not with a Polytron or similarly forceful device); dilute part of this homogenate 2-fold for the uptake experiments. At this stage of the experiment, initiation of the actual binding and uptake assays is approximately 10 min away. Binding assays are handled in multiples of 12 (harvester), and uptake tubes individually. Transfer the first set of assay tubes (#1–12 for binding and #1 for uptake) to a waterbath at 25° so that tubes are allowed to become equilibrated at that temperature; set the shaker at a moderate setting (for example, 50 rpm). The goal is to warm up each tube for approximately 10 min prior to adding brain membranes (see proceeding text; the smaller-sized binding tubes may require less time to warm up). Transfer next sets of binding tubes to the waterbath at 1-min intervals allowing completion of subsequent steps (see proceeding text); transfer uptake tubes sequentially at 1-min intervals.

For the binding assays, after prewarming in the bath, add 50 μl P_2 suspension followed by gentle vortexing. Because a multiple of 12 tubes need to be harvested at the end of the incubation, make additions sequentially as fast as possible, which can be done within half a minute for all additions to all tubes; the use of a multichannel pipette will enhance the process. Incubate the assay mixture at 25° for 8 min. Terminate the reaction by the addition of an excess of ice-cold wash buffer (73 m*M* NaCl, 3 m*M* KCl, 0.7 m*M* $MgSO_4$, 6 m*M* glucose, 0.6 m*M* $CaCl_2$, 30 m*M* Na^+, and 17 m*M* phosphate from a mixture of primary and secondary phosphate buffer, giving a pH of 7.4 at room temperature) and filtration over Skatron receptor binding filter mats (glass fiber filter, 1-μm retention, No. 11734, equivalent to Whatman GF/B) with the Skatron miniharvester. Because this binding assay is of the equilibrium type, a difference of 0.5 min in total incubation time between the first and last tube of each set of 12 has little or no impact on the final results.

For the uptake assays, approximately 10 min after prewarming in the bath, add 100 μl P_2 suspension followed by gentle vortexing. Incubate the assay mixture at 25° for 7 min and add, exactly at 7 min, 40 μl radioligand stock. At exactly 8 min, terminate the reaction by the addition of an excess of ice-cold wash buffer (see previous text) and filtration over Whatman GF/F glass fiber filters with a single-manifold Millipore filtration apparatus.

All filters are pretreated with 0.05% (w/v) poly(L-lysine) (molecular weight 15,000–30,000) (other groups have successfully used polyethyleneimine instead of polylysine for these radioligands). After the first filtration, filters are washed three times with 1 (binding) or 4 (uptake) ml of ice-cold wash buffer, and asssayed for radioactivity by liquid scintillation counting. Filters are cut individually and counted in separate scintillation vials. These counting procedures can be simplified if a scintillation counter is available that accepts cassettes containing filter mats with assayed material in designated spots.

Nonspecific binding or uptake is defined with 100 μM cocaine. Nonspecific binding or uptake is usually no more than 4% of the total value determined in the absence of inhibitor.

5. Data analysis can be done in many ways with a number of software packages available. We routinely compute IC_{50} values and pseudo-Hill numbers with the equation of the ALLFIT program of De Lean *et al.*[37] entered into the Microsoft ORIGIN curve-fitting and plotting software. We usually run this nonlinear regression program with total and nonspecific binding (uptake) centered as constants; if there are enough data points of low levels of test compound showing little or no inhibition, total binding values are better estimated by letting that parameter float. When WIN 35,428 or dopamine is the inhibiting compound in the binding or uptake assays, respectively, the data can be analyzed with the nonlinear computer fitting program LIGAND.[38] In this case, we routinely use data files in which nonspecific uptake (binding) (N_1) has not been subtracted; the results obtained by entering N_1 as a constant in the fitting procedures are usually close to estimates obtained by having N_1 float, and generally we choose the former approach, relying on a pharmacological definition of nonspecific binding for the target site.

[37] A. DeLean, P. J. Munson, and D. Rodbard, *Am. J. Physiol.* **235,** E97 (1978).

[38] P. J. Munson and D. Rodbard, *Anal. Biochem.* **107,** 220 (1980).

[39] F. I. Carroll, P. Kotian, A. Dehghani, J. L. Gray, M. A. Kuzemko, K. A. Parham, P. Abraham, A. H. Lewin, J. W. Boja, and M. J. Kuhar, *J. Med. Chem.* **38,** 379 (1995).

Acknowledgment

We would like to thank the National Institute on Drug Abuse (DA 08379 to M.E.A.R.) and National Natural Sciences Foundation of China (39570812 to N.-H.C.) for support, and Lori L. Coffey for dedicated participation in developing the described protocols in our laboratory.

[18] Transport and Drug Binding Kinetics in Membrane Vesicle Preparation

By HEINZ BÖNISCH

Introduction

Molecular cloning has shown that neurotransmitter transporters of the plasma membrane form at least two families.[1-3] One family consists of the sodium- and chloride-dependent transporters, and members of this family include the transporters for the monoamines dopamine, norepinephrine (NE), and serotonin and those for a series of amino acids such as γ-aminobutyric acid (GABA) or glycine. Members of the second family are the sodium- and potassium-dependent excitatory amino acid transporters such as glutamate or aspartate. Transport of neurotransmitters is driven by the transmembrane Na^+ gradient, established by the Na^+,K^+-ATPase, and the process is analogous to other secondary active transport systems such as the intestinal Na^+/glucose transporter. However, unlike Na^+/glucose transport, additional ions are required for transport of most of the neurotransmitters, such as extracellular Cl^- and intracellular K^+. Even before the cloning of neurotransmitter transporters, studies on plasma membrane vesicles had contributed significantly to our present knowledge on energetic and pharmacological properties of many neurotransmitter transport systems. For example, plasma membrane vesicles have been used to study transport and/or binding characteristics of the transporters for GABA,[4,5]

[1] S. G. Amara and M. J. Kuhar, *Annu. Rev. Neurosci.* **16,** 73 (1993).
[2] B. Borowsky and B. J. Hoffman, *Int. Rev. Neurobiol.* **38,** 139 (1995).
[3] M. E. A. Reith, "Neurotransmitter Transporters." Humana Press, Totowa, New Jersey, 1997.
[4] B. I. Kanner, *Biochemistry* **17,** 1207 (1978).
[5] R. Roskoski, *J. Neurochem.* **36,** 544 (1981).

0076-6879/98 $25.00

NE,[6,7,8] serotonin,[9–18] and glutamate.[19,20] Membrane vesicles isolated from stably transfected cells expressing either the serotonin, dopamine, or NE transporter were used to define the ion coupling stoichiometry of these transporters.[21,22]

Neurotransmitter transporters are the primary targets of a series of psychoactive compounds. For example, cocaine is an inhibitor of the monoamine transporters and amphetamine is a substrate of the dopamine and the NE transporter; both substances cause an increase in the synaptic concentration of dopamine which ultimately is responsible for their abuse potential. However, neurotransmitter transporters are also targets for useful therapeutic drugs. Substances such as tiagabine which bind to the GABA transporter and thereby inhibit GABA reuptake are used as antiepileptic drugs. For several classes of antidepressant drugs, the primary mode of action is to inhibit neurotransmitter transport by the NE transporter and/or the serotonin transporter. Binding studies using membrane vesicles isolated from tissues or cells expressing one of these transporters have contributed to our understanding of their interaction with these transporters.[10,17,23,24]

This article will focus on the isolation of plasma membrane vesicles from PC12 rat pheochromocytoma cells which express the NE transporter[23,25–27] and will show how plasma membrane vesicles can be used to study kinetics of substrate transport and drug binding to the transporter.

[6] R. Harder and H. Bönisch, *Biochim. Biophys. Acta* **775,** 95 (1984).
[7] R. Harder and H. Bönisch, *J. Neurochem.* **45,** 1154 (1985).
[8] S. Ramamoorthy, P. D. Prasad, P. Kulanthaivel, F. H. Leibach, R. D. Blakely, and V. Ganapathy, *Biochemistry* **32,** 1346 (1993).
[9] D. R. Cool, F. H. Leibach, and V. Ganapathy, *Biochemistry* **29,** 1818 (1990).
[10] D. R. Cool, F. H. Leibach, and V. Ganapthy, *Am. J. Physiol.* **259,** C196 (1990).
[11] G. Rudnick, *J. Biol. Chem.* **252,** 2170 (1977).
[12] G. Rudnick and P. J. Nelson, *Biochemistry* **17,** 4739 (1978).
[13] J. Talvenheimo, P. J. Nelson, and G. Rudnick, *J. Biol. Chem.* **254,** 4631 (1979).
[14] P. J. Nelson and G. Rudnick, *J. Biol. Chem.* **254,** 10084 (1979).
[15] P. J. Nelson and G. Rudnick, *J. Biol. Chem.* **257,** 6151 (1982).
[16] S. R. Keyes and G. Rudnick, *J. Biol. Chem.* **257,** 1172 (1982).
[17] J. Talvenheimo, H. Fishkes, P. J. Nelson, and G. Rudnick, *J. Biol. Chem.* **258,** 6115 (1983).
[18] G. Rudnick, R. Bencuya, P. J. Nelson, and R. A. Zito, *Mol. Pharmacol.* **20,** 118 (1981).
[19] B. I. Kanner and I. Sharon, *Biochemistry* **17,** 3949 (1978).
[20] E. G. Schneider, M. R. Hammerman, and B. Sacktor, *J. Biol. Chem.* **255,** 7650 (1980).
[21] H. H. Gu, S. C. Wall, and G. Rudnick, *J. Biol. Chem.* **269,** 7124 (1994).
[22] G. Rudnick, *Methods Enzymol,* **296,** [16], (1998) (this volume).
[23] H. Bönisch and R. Harder, *Naunyn-Schmiedeberg's Arch. Pharmacol.* **334,** 404 (1986).
[24] G. Rudnick, J. Talvenheimo, H. Fishkes, and P. J. Nelson, *Psychopharmacol. Bull.* **19,** 545 (1983).
[25] L. A. Greene and G. Rein, *Brain Res.* **128,** 247 (1977).
[26] U. Friedrich and H. Bönisch, *Naunyn-Schmiedberg's Arch. Pharmacol.* **333,** 246 (1986).
[27] M. Brüss, P. Pörzgen, L. J. Bryan-Lluka, and H. Bönisch, *Mol. Brain Res.* **52,** 257 (1997).

Plasma Membrane Vesicles

Plasma membrane vesicles isolated from a tissue or preferably from a clonal cell population expressing large amounts of a specific transporter have provided important insights into the driving forces and chemical interconversions associated with transport of molecules across cell membranes.[28,29] Investigating a vectorial transport process in a cell-free system offers considerable additional information not obtainable from studies using isolated tissues or intact cells. The transport system can be selectively assayed in a functional state free from many of the problems arising from internal compartmentalization and intracellular metabolism. The magnitude and polarity of a chemical or electrochemical driving force can be imposed or dissipated across the membrane, enabling an estimation of their contribution to the transport process. Plasma membrane vesicles can also be used to investigate the first step of a transport cycle (i.e., the formation of the ligand–carrier complex by means of binding studies using labeled competitive inhibitors of the transport system under study). In addition, purified plasma membrane vesicles may be used to solubilize the transporter or as starting material for reconstitution (as proteoliposomes) of the transport system and for purification of the transport protein, as was achieved with the GABA transporter.[30]

Isolation of Plasma Membrane Vesicles from PC12 Cells

Isolation procedures for membrane vesicles derived from a variety of tissues or cells expressing a transporter for serotonin (e.g., platelets, human placenta), norepinephrine (e.g., PC-12 cells, human placenta), dopamine (brain), GABA (brain), or glutamate (brain, kidney) are described in the literature cited in the introduction. As a practical example, the isolation of plasma membrane vesicles from PC-12 cells will be described. These membrane vesicles were the first cell-free system used to study in some detail the ion requirements and kinetics of the NE transport system, and to solubilize the transporter in a functional state.[6,7,23,31]

Preparation of plasma membrane vesicles for transport studies requires a large initial cell number, a high membrane yield, and the production of sealed, osmotically sensitive vesicles. To obtain a sufficient number of cells, PC-12 cells should first be adapted to grow in suspension (spinner) culture as described previously (Harder and Bönisch[6]). About 24 hr before cell

[28] J. E. Lever, *CRC Crit. Rev. Biochem.* **7,** 187 (1980).

[29] G. Sachs, R. J. Jackson, and E. C. Rabon, *Am. J. Physiol.* **238,** G151 (1980).

[30] R. Radian, A. Bendahan, and B. J. Kanner, *J. Biol. Chem.* **261,** 15437 (1986).

[31] E. Schömig and H. Bönisch, *Naunyn-Schmiedeberg's Arch. Pharmacol.* **334,** 412 (1986).

harvesting, 10 μM reserpine (dissolved in 0.1% dimethyl sulfoxide) is added to the culture medium to deplete endogenous stores of catecholamines.

Two methods (A, B) are described for the isolation of plasma membrane vesicles from PC-12 cells; one (method A) has previously been described in some detail.[6] An outline of Method A is provided in Fig. 1.

Method A

1. Cells from 1–3 liters of spinner culture with a cell density below 3×10^7 cells/ml are sedimented (45 min at 1*g*).
2. Sedimented cells are washed twice with phosphate-buffered saline (PBS) and centrifuged at 50*g* (for 5 min).
3. The cell pellet is weighed and all volumes in subsequent steps are based on this wet weight. The remaining steps in the protocol are carried out at 4°.
4. 5 ml per gram wet weight of ice-cold "lysis buffer" (10 mM Tris-HCl, pH 7.2, 2 mM $MgCl_2$, 1 mM dithiothreitol (DTT), 1 mM PMSF [phenylmethylsulfonyl fluoride]) are added and cells are incubated for 10 min on ice.
5. Thereafter, 7 ml aliquots of the cell suspension are homogenized with 10 strokes of a tight-fitting Dounce homogenizer (Wheaton).
6. The homogenate is diluted 1:8 in cold "dilution buffer" (140 mM

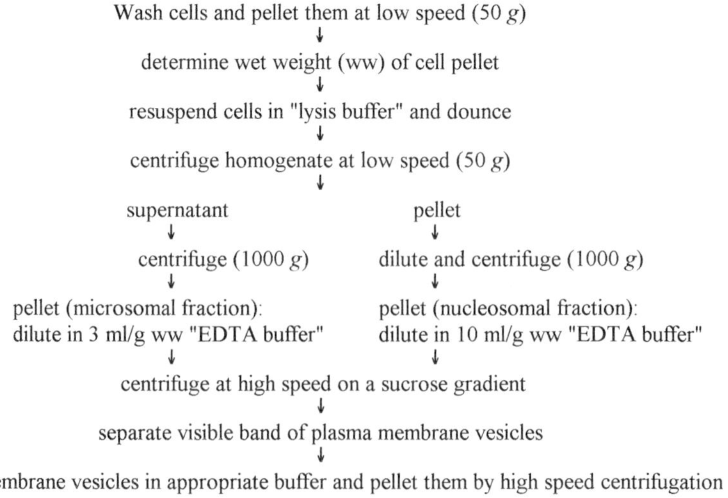

FIG. 1. Scheme for preparation of plasma membrane vesicles from PC-12 cells; for details, see text.

NaCl, 10 m*M* Tris-HCl, pH 7.4, 2 m*M* $MgCl_2$, 0.4 m*M* PMSF, 0.1% mercaptoethanol) and centrifuged at 50*g* for 15 min.

7. Both the supernatant (S) and the pellet (P), which is resuspended in 10 ml "dilution buffer" (per gram original wet weight), are centrifuged at 1000*g* for 5 min.

8. The pellet from the supernatant ("microsomal" fraction) is resuspended in 3 ml and that of the resuspended pellet ("nucleosomal" fraction) in 10 ml (per gram wet weight each) "EDTA buffer" (140 m*M* NaCl, 10 m*M* Tris-HCl, pH 7.4, 2 m*M* Na_2EDTA, 1 m*M* $CaCl_2$, 0.1% mercaptoethanol) and layered on top of a linear sucrose gradient (20–56% w/v) containing 135 m*M* potassium phosphate (pH 7.4), 1 m*M* Na_2EDTA, 0.3 m*M* $CaCl_2$ and 0.1% mercaptoethanol.

9. Either 3 ml membrane suspension/gradient are spun for 2 hr at 110,000*g* in a fixed-angle rotor (38-ml buckets, Kontron TST 28.38) or 10-ml suspension are spun in a high-capacity fixed angle rotor for 1.5 hr at 145,000*g* (94-ml tubes, Kontron TFT 45.94). We usually use the high-capacity fixed angle rotor with 94-ml polycarbonate centrifuge tubes. To prevent sticking of biological material to the tube wall, tubes are rinsed before use with a silicon solution (SIGMA, St. Louis, MO) and thereafter dried at ambient temperature.

10. As demonstrated in Fig. 2, centrifugation of the nucleosomal membranes results in a single visible band of plasma membrane vesicles (at a density of 1.120 g/cm^3), whereas microsomal fractions yield a further visible band at higher density (1.135 g/cm^3). These bands are collected by a Pasteur pipette or (more efficiently) by underlaying a 65% sucrose solution to the gradient and fractionating the gradient by means of an UV monitor (e.g., ISCO UA-5); see Fig. 3.

11. The isolated plasma membrane fraction is diluted 1 : 8 in "potassium phosphate buffer" (135 m*M* potassium phosphate, pH 7.4, 1 m*M* $MgCl_2$, 1 m*M* ascorbic acid, 1 m*M* dithiothreitol) and pelleted (2 hr, 145,000*g*, TFT 45.95 rotor). The pellet is resuspended in 2.5 ml potassium phosphate buffer per gram original wet weight and stored in liquid nitrogen until use.

The purity of plasma membrane vesicles isolated according to the following, simpler procedure (method B) has not been studied in detail. In any case, the membranes are suitable for binding studies.

Method B

PC-12 cells are sedimented and washed with PBS as described earlier and centrifuged at 50*g* for 15 min. The cell pellet (6–8 g wet weight) is resuspended in buffer A (140 m*M* KCl, 20 m*M* HEPES, pH 7.0, 2 m*M* $MgCl_2$, 1 m*M* dithiothreitol). The cell suspension is transferred to a Parr

FIG. 2. Separation of "nucleosomal" (left) and "microsomal" (right) fractions by high-speed centrifugation on a linear sucrose gradient in a high-capacity fixed angle rotor. The upper visible bands represent plasma membrane vesicles, the lower band (right tube) are mitochondrial membranes.

cell disruption bomb and the cells are disrupted by nitrogen cavitation: the cell suspension is stirred with a magnetic stirrer; cells are equilibrated for 30 min with 25 atm N_2 and then released dropwise. Immediately after release, EDTA and mercaptoethanol are added to the homogenate to obtain a final concentration of 2 m*M* EDTA and 0.1% mercaptoethanol. To disrupt aggregates the homogenate is pressed through hypodermic needles of decreasing size (21-gauge × 1.5, 25-gauge × 1, 25-gauge × 0.75) and then diluted 1:1 in buffer B (140 m*M* KCl, 20 m*M* HEPES, pH 7.0, 2 m*M* $MgCl_2$, 0.1% mercaptoethanol). The diluted homogenate is layered on top of a 32% sucrose cushion (containing 135 m*M* potassium phosphate, pH 7.4, 2 m*M* EDTA, 0.3 m*M* $CaCl_2$, and 0.1% mercaptoethanol) and centrifuged (at 4°) for 1 hr at 110,000*g* in a Kontron TST 28.35 swinging

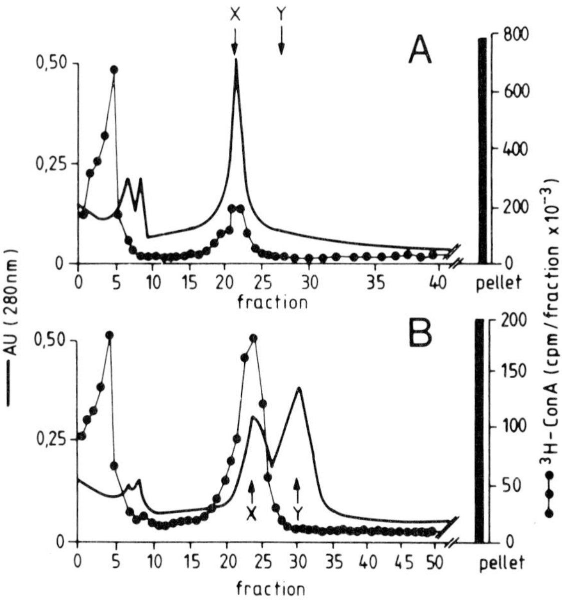

FIG. 3. Distribution of UV absorbance (280 nm; arbitrary units, AU) and [^{3}H]concanavalin A ([^{3}H]ConA) within a linear sucrose gradient of the "nucleosomal" (A) and "microsomal" (B) fraction. For further details, see text. [Reprinted with permission from R. Harder and H. Bönisch, *Biochim. Biophys. Acta* **775,** 95 (1984).]

bucket rotor. The plasma membranes which appear as a band at the interface between sucrose and buffer are collected using a Pasteur pipette. The collected fractions are diluted 6- to 10-fold in either "potassium phosphate buffer" (for transport studies) or, for example, for binding studies, in "sodium buffer" (135 m*M* NaCl, 10 m*M* Tris-HCl, pH 7.4, 5 m*M* KCl, 1 m*M* $MgSO_4$, 1 m*M* dithiothreitol) and centrifuged for 1 hr at 200,000*g*. The final membrane pellet is suspended in 1 ml of the same buffer per gram cell wet weight.

As shown in Fig. 3 and Table I, the vesicle fraction (designated as *X* in Fig. 3 and Table I) isolated by method A is characterized by 24-fold and 8-fold enrichments (compared to the cell homogenate) in the activities of the plasma membrane markers Na^+,K^+-ATPase and phosphodiesterase, respectively. In addition, when intact PC-12 cells are exposed to tritiated concanavalin A to label sugar residues of membrane proteins, the isolated vesicle fraction contains nearly all of the recovered radioactivity of the fractionated gradient (Fig. 3). The band at higher density of the microsomal fraction consists of mitochondrial and endoplasmic reticulum membranes

TABLE I
MARKER ENZYME ACTIVITIES WITHIN FRACTIONS OBTAINED DURING ISOLATION OF PLASMA MEMBRANE VESICLES FROM PC-12 CELLS[a]

Fraction	(Na^+,K^+)-ATPase	PDE	SDH	NADH dehydrogenase
Homogenate	1.0	1.0	1.0	1.0
Nucleosomal membranes	23.7	8.4	0.6	0.9
Microsomal membranes (*X*)	23.9	5.3	1.7	1.1
Microsomal membranes (*Y*)	1.6	0.3	18.5	10.1

[a] Mean activities from 3–6 preparations (relative to those of the homogenate) of the marker enzymes for plasma membranes [phosphodiesterase (PDE) and (Na^+,K^+)-ATPase], mitochondria [succinate dehydrogenase (SDH)], and endoplasmic reticulum [NADH dehydrogenase]. The letters *X* and *Y* refer to fractions shown in Fig. 3. Results taken from: R. Harder and H. Bönisch, *Biochim. Biophys. Acta* **775,** 95 (1984).

as indicated by the relative activities of corresponding marker enzymes (see Table I).

Testing Some Characteristics of Isolated Plasma Membrane Vesicles

Before determining the kinetics of transport for a labeled substrate, it is important to prove that the measured retention of radioactivity is really due to accumulation in the aqueous intravesicular space and not the result of binding to the vesicular membrane. This may be accomplished by measuring uptake in an incubation medium of varying osmolarity as described by Harder and Bönisch.[6] A plot of uptake versus the reciprocal of osmolarity (i.e., volume) should result in a zero intercept at infinite osmolarity if uptake but not binding is occurring.

For calculation of the degree of intravesicular accumulation of a transported substrate, it is necessary to determine the intravesicular water space. This can be accomplished by resuspension of pelleted membrane vesicles in a buffer (e.g., the incubation buffer used for uptake experiments) containing tritiated water and incubation (at 4°) overnight. Next, vesicles are centrifuged in a small tube through a cushion of silicon oil, the tip of the tube is frozen (at −80°) and cut, and the radioactivity (and protein content) of the pellet is determined. Using this method, we determined a vesicular space of 1.85 μl/mg protein for membrane vesicles isolated according to method A.[6]

It is also useful to determine the time course of uptake of a labeled substrate at different temperatures to assess whether the transported substrate is accumulated, whether accumulation is dependent on the membrane protein concentration, and whether uptake increases linearly with time during the entire assay. Measurement of initial rates is a prerequisite for determination of saturation kinetics (see proceeding text), and inward transport is unidirectional (initial) as long as accumulation increases linearly with time. Depending on the temperature, this may be for only a few seconds. Thereafter, the uptake curve becomes flatter, reaches a plateau, and may decline thereafter, especially at higher temperatures, which can cause collapse of the artificially imposed ion gradients, leading to reversal of transport (a property of most neurotransmitter transporters). Uptake should be inhibited by standard inhibitors of the transport system, and at zero degrees there should be no net uptake. Figure 4 shows an example of a time course of radiolabel uptake. Figure 4 demonstrates that 37° is too high a temperature for measuring initial rates or for determining the maximum degree of accumulation at steady state (when net uptake is zero). These data illustrate why temperatures between 20 and 30° commonly are used. In the example shown in Fig. 4 maximum intravesicular accumulation of tritiated NE (at 25°) was shown to be 550-fold,[6] a value similar to that obtained for the accumulation of serotonin in platelet plasma membrane vesicles.[12] In addition, the rate of uptake was linearly dependent on the membrane protein concentration up to 1 mg protein/ml.[6]

A time course may also be used to demonstrate ion dependence of transport and accumulation. Because the kinetic constants determined in saturation kinetics are influenced by the imposed ion gradients, care should be taken to ensure optimal conditions. Details of such studies are provided in the references given in the introduction and in another article in this volume.[22] Specific protocols for uptake and binding assays are given below.

Transport and Binding

General Background

Transport by carriers exhibits a number of identifying characteristics. In addition to the temperature dependence of transport with temperature coefficients above that for simple diffusion, the general characteristics of carrier-mediated transport include (1) saturability of transport; (2) structural specificity and/or stereoselectivity of transport; (3) competition for transport by structurally related other substrates; and (4) sensitivity to specific (competitive or noncompetitive) inhibitors which differ chemically from the solute under study. These properties are determined by the transmembrane transporter protein (carrier) which enables a vectorial reaction, i.e., substrate movement from one side of the membrane to the other side. According to the most accepted model, translocation is through a "single

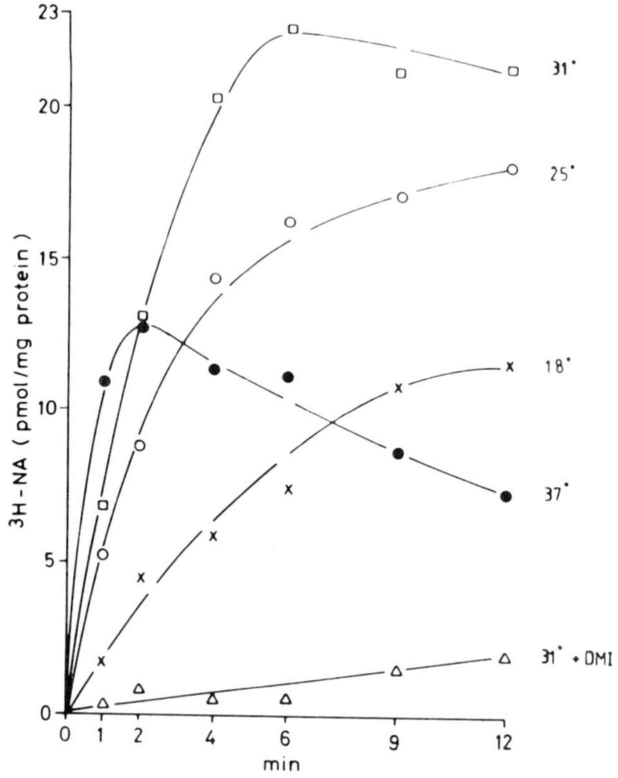

FIG. 4. Time course of temperature-dependent accumulation of tritiated norepinephrine ([^{3}H]NA) in plasma membrane vesicles from PC-12 cells and its inhibition by desipramine (DMI). [Reprinted with permission from R. Harder and H. Bönisch, *Biochim. Biophys. Acta* **775,** 95 (1984).]

center gated pore mechanism" with alternating access of the solute (from the external or internal membrane site) to the binding site in the pore.[32] The overall reaction includes the following steps: (1) Binding of the substrate (i.e., the transported ligand) to a specific binding site(s) on one side of the membrane, (2) translocation, (3) dissociation and release of substrate to the other side of the membrane, and (4) return reactions, i.e., either the carrier returns in its unloaded state and catalyzes net transport or it returns in a substrate-loaded state catalyzing exchange. It is generally accepted that the translocation step, which is the essential switch in the transport

[32] R. Krämer, *Biochim. Biophys. Acta* **1185,** 1 (1994).

reaction, involves a substrate-induced conformational change of the carrier protein or its binding site.

A carrier may also be characterized by means of binding of a ligand to the binding site; in this case only the first step in the transport process is examined. Basically, the characteristics mentioned above to identify a transport system also hold true for ligand binding to the carrier. Although the nucleotide sequence and the derived amino acid sequence are used to classify a transport system, functional properties such as ion dependence and especially affinities for various ligands finally determine its typical character (and often its name). These characteristics are derived from transport and ligand binding kinetics.

Kinetics of Transport or Binding

Theoretical Background

The principles of transport kinetics are basically identical with enzyme kinetics[33] and they have been described in detail in numerous reviews[32,34–42]; this also holds true for the principles of binding kinetics.[43–45] Thus, the following discussion will provide only a general overview of basic transport kinetics.

Transport Kinetics

In mathematical models various assumptions have been applied to describe initial rates of unidirectional transport. In these models the substrate–carrier complex (CS) is generally considered a more or less constant quan-

[33] I. H. Segel, "Enzyme Kinetics: Behavior and Analysis of Rapid Equilibrium and Steady-State Enzyme Systems." Wiley, New York, 1975.
[34] W. Wilbrandt and T. Rosenberg, *Pharmacol. Rev.* **13,** 109 (1961).
[35] W. D. Stein, "The Movement of Molecules across Cell Membranes." Academic Press, New York, 1967.
[36] S. G. Schultz and P. F. Curran, *Physiol. Rev.* **50,** 637 (1970).
[37] R. K. Crane, *Rev. Physiol. Biochem. Pharmacol.* **78,** 99 (1977).
[38] A. Kotyk and K. Janacek, "Cell Membrane Transport." Plenum Press, New York, 1975.
[39] A. Kotyk and K. Janacek, "Membrane Transport." Plenum Press, New York, 1977.
[40] H. N. Christensen, "Biological Transport." W. A. Benjamin, Massachusetts, 1975.
[41] M. Höfer, "Transport Across Biological Membranes." Pitman Publishing Ltd., London, 1981.
[42] W. D. Stein, *Am. J. Physiol.* **250,** C523 (1986).
[43] L. T. Williams and R. J. Lefkowitz, "Receptor Binding Studies in Adrenergic Pharmacology." Raven Press, New York, 1978.
[44] H. I. Yamamura, S. J. Enna, and M. J. Kuhar, "Neurotransmitter Receptor Binding." Raven Press, New York, 1978.
[45] P. D. Hrdina, *in* "Neuromethods 4: Receptor Binding" (A. A. Boulton, G. B. Baker, and P. D. Hrdina, Eds.), p. 455. Humana Press, Clifton, New Jersey, 1986.

tity. According to the "equilibrium assumption" of Michaelis and Menten[46] and in analogy to enzyme kinetics, the catalysis of transport is formulated by Eq. (1):

$$S + C \underset{k_{-1}}{\overset{k_1}{\rightleftarrows}} CS \tag{1}$$

The "equilibrium assumption" should be avoided when possible (Christensen[40]). More generally accepted models are based on the "steady state approximation" of Briggs and Haldane.[47] According to Christensen,[40] the catalysis of transport can be described by Eq. (2):

$$S_o + C \underset{k_{-1}}{\overset{k_1}{\rightleftarrows}} CS \overset{k_2}{\longrightarrow} C + S_i \tag{2}$$

where S_i and S_o indicate the localization of the substrate (inside or outside). The rate of transport is given by the rate constant k_2 (reflecting translocation) times the concentration of the carrier–substrate complex (CS). Equation (3) shows that the model is more complex if reversibility of a translocation step and additionally a dissociation step are incorporated:

$$S_o + C \underset{k_{-1}}{\overset{k_1}{\rightleftarrows}} (CS)_o \underset{k_{-2}}{\overset{k_2}{\rightleftarrows}} (CS)_i \underset{k_{-3}}{\overset{k_3}{\rightleftarrows}} C + S_i \tag{3}$$

For secondary active transport systems such as Na^+-coupled neurotransmitter transport, formulation of the catalysis of transport might still be more complex because the coupling ion is assumed to affect substrate binding. An interesting model for such a system has been proposed by Yamato.[48]

However, each of the different models results in the same simple Eq. (4) in which the rate of transport is defined by the substrate concentration [S], K_m, and V_{max}:

$$v = \frac{V_{max}[S]}{K_m + [S]} \tag{4}$$

In Eq. (4) K_m is the substrate concentration at which the maximum transport rate (V_{max}) is half-maximal.

Binding Kinetics

The reversible interaction of a ligand (L) with a binding site on the transporter protein ("receptor," R) to form the receptor–ligand complex

[46] L. Michaelis and M. L. Menten, *Biochem. Z.* **49,** 333 (1913).

[47] G. E. Briggs and J. B. S. Haldane, *Biochem. J.* **19,** 338 (1925).

[48] I. Yamato, *FEBS Lett.* **298,** 1 (1992).

(RL) as given in Eq. (5) underlies the law of mass action and follows the principles of Michaelis–Menten kinetics:

$$\mathrm{L} + \mathrm{R} \underset{k_{-1}}{\overset{k_1}{\rightleftarrows}} \mathrm{LR} \tag{5}$$

In Eq. (5), which is analogous to Eq. (1), k_1 and k_{-1} represent the association and dissociation rate constants. At equilibrium of binding, the equilibrium dissociation constant (K_D) is given by Eq. (6):

$$K_D = \frac{k_{-1}}{k_1} = \frac{[\mathrm{R}][\mathrm{L}]}{[\mathrm{RL}]} \tag{6}$$

where [R], [L], and [RL] represent the concentrations of free receptor, free ligand, and bound ligand (or occupied receptor), respectively.

The dependence of [RL] on [L] is described by a rectangular hyperbola defined by Eq. (7):

$$[\mathrm{LR}] = \frac{[\mathrm{RL}]_{max}[\mathrm{L}]}{K_D + [\mathrm{L}]} \quad \text{or} \quad B = \frac{B_{max}[\mathrm{L}]}{K_D + [\mathrm{L}]} \tag{7}$$

where $[\mathrm{LR}]_{max}$ (or B_{max}) is the total number of binding sites (i.e., receptors). When [L] equals K_D half of the receptor are occupied by the ligand.

Meaning of Kinetic Constants

Only if the equilibrium assumption holds true does K_m represent the ratio k_{-1}/k_1 (i.e., the dissociation constant of the carrier–substrate complex or the substrate affinity constant). Otherwise, K_m is a complex value in which the several rate constants all play a role in determining its dimension, and K_m only mirrors an "apparent" affinity constant. From Eq. (2), K_m represents the ratio $(k_2 + k_{-1})/k_1$.

Affinity values for a series of ligands (substrates or inhibitors, including stereoisomers) provide an important pharmacological fingerprint for the transporter of interest. For radiolabeled ligands, such values are obtained in transport studies from the half-saturation constant, K_m, or in binding studies from the equilibrium dissociation constant, K_D. For unlabeled compounds, the IC_{50}, i.e., the concentration which causes half-maximum inhibition of either uptake or binding of a labeled ligand, is a rough estimate of its affinity for the transporter. If the K_m (or K_D) of the radiolabeled ligand is known, and if independent evidence for the type of inhibition (competitive or noncompetitive) is available, an IC_{50} value can be converted to a K_i value according to equations described by Cheng and Prusoff.[49] If a K_i

[49] Y. Cheng and W. H. Prusoff, *Biochem. Pharmacol.* **22,** 3099 (1973).

value was obtained in uptake studies, it reflects a K_m value (in case of a substrate) or a K_D value (in case of a nontransported inhibitor). However, a K_i value obtained from inhibition of binding always reflects the K_D. Because a K_m may not reflect the dissociation constant (see above), K_i values for substrates determined from inhibition of transport may well be different from K_i values for the same substrates obtained from inhibition of binding (reflecting dissociation constants), and differences between these two types of K_i values do not necessarily indicate that the binding site measured in binding studies differs from the site of interaction in transport.[23,50]

K_i as well as K_D values do not allow one to distinguish between a transported ligand (substrate) or a nontransported ligand (inhibitor). Often it is difficult to decide whether a compound is a transported substrate. A good example is amphetamine, which causes inhibition of dopamine or NE uptake by the corresponding transporters and which, because of its high lipophilicity, may not need a special transporter to enter dopaminergic or noradrenergic cells. Thus, amphetamine is often regarded as a competitive inhibitor of these transporters, although such a property cannot explain the phenomenon of countertransport, i.e., acceleration of outward transport (of dopamine and NE) induced by amphetamine (for discussion see, e.g., Rudnick[51]). However, tritiated amphetamine has been shown to be taken up into reserpine pretreated PC-12 cells through the cocaine- and desipramine-sensitive NE transporter.[52] Thus, the most direct way to identify a substrate is to demonstrate that it is transported. However, since only substrates can induce countertransport, the ability of a carrier ligand to induce countertransport is a strong hint that it is a substrate.[53,54]

The maximum transport rate, V_{max}, is expected to be about the same for all substrates under the equilibrium assumption, since under this assumption it is a direct measure of the carrier concentration. However, if it is derived from Eq. (2), V_{max} equals k_2 times the carrier concentration [C]; for example V_{max} is proportional to the total number of carriers times a rate constant reflecting the translocation step ($V_{max} = k_2[C]$). V_{max} values of substrates for a given transport system mirror their "transport efficacy" (like an agonist efficacy for a certain receptor), and thus can also be used to define a transport system. For example, the NE transporter and the

[50] E. Schömig, M. Körber, and H. Bönisch, *Naunyn-Schmiedeberg's Arch. Pharmacol.* **337,** 626 (1988).

[51] G. Rudnick, *in* "Neurotransmitter Transporters" (M. E. A. Reith, Ed.), p. 73. Humana Press, Totowa, New Jersey, 1997.

[52] H. Bönisch, *Naunyn-Schmiedeberg's Arch. Pharmacol.* **327,** 267 (1984).

[53] R. Wölfel and K. H. Graefe, *Naunyn-Schmiedeberg's Arch Pharmacol.* **345,** 129 (1992).

[54] H. Bönisch and U. Trendelenburg, *Handb. Exp. Pharmacol.* **90/I,** 247 (1988).

dopamine transporter exhibit about the same apparent affinity (K_m) for NE and dopamine; however, V_{max} values for their natural substrates are much higher than for the "false" substrate.[55] At substrate concentrations far below the K_m, the substrate concentration in the denominator of the Michaelis–Menten equation [Eq. (4)] can be neglected, and Eq. (4) simplifies to $v = S(V_{max}/K_m)$. The ratio V_{max}/K_m then describes the rate constant at very low substrate concentration, and the rank order of this ratio for various substrates also characterizes a transport system.[56]

Assuming a 1:1 interaction between a labeled ligand and its binding site at a carrier, a B_{max} value reflects the density (number) of carrier sites; this value should be the same for different ligands. If a B_{max} value and a V_{max} value for a substrate are measured in the same experimental system (vesicles or intact cells), the turnover number (or the duration for one transport cycle) for the substrate can be calculated from the ratio V_{max} to B_{max}.[23]

Carrying Out Transport or Binding Assays in Membrane Vesicles

Uptake of substrates into plasma membrane vesicles by secondary active transport systems such as the Na^+- and Cl^--dependent neurotransmitter transporters requires artificially imposed transmembrane gradients of Na^+ and Cl^- (out > in) and K^+ (in > out). To achieve these gradients, vesicles are loaded with potassium phosphate during the isolation procedure (last centrifugation). Transport is started by severalfold (e.g., 10-fold) dilution of the final vesicle fraction within an incubation buffer containing a physiological concentration of NaCl. To determine K_m and V_{max} for a given substrate, membrane vesicles are incubated in various concentrations (below and above the expected K_m) of the radiolabeled substrate for a very short time period (to measure initial rates) and at a defined temperature. Because K_m values are mostly in the high nanomolar to micromolar range, the labeled substrate is usually diluted with unlabeled substrate to reveal high substrate concentrations. The time period at which initial rates are measurable should first be determined at the highest substrate concentration. For determination of nonspecific uptake, the uptake assay is carried out either in the presence of a high concentration (>100 times K_i) of a competitive and specific uptake inhibitor or at 4° or, in the case of sodium-dependent transport, in the absence of the driving ion gradient (e.g., external Na^+ replaced by Li^+). Finally, the vesicles are separated from the incubation buffer by filtration.

[55] K. J. Buck and S. G. Amara, *Proc. Natl. Acad. Sci. USA* **91,** 12584 (1994).

[56] K.-H. Graefe and H. Bönisch, *Handb. Exp. Pharmacol.* **90/I,** 193 (1988).

Binding studies are usually performed with broken membranes. If membrane vesicles are used, they must first be loaded with the corresponding incubation buffer to exclude electrochemical gradients over the vesicle membrane. The radiolabeled ligand (containing either ^3H or ^{125}I and no impurities) should bind with high affinity (K_D in the low nanomolar range or lower). As a rule, only inhibitors fulfill this criterion since a low K_D implies that the ligand slowly dissociates from its binding site, a characteristic not compatible with the properties of a substrate. The membrane concentration should be low to exclude a pronounced (more than 10%) reduction of free ligand at equilibrium of binding. Since the time to reach equilibrium of binding increases with decreasing ligand concentration, a time course of binding should first be determined at the lowest ligand concentration. Nonspecific binding, which by definition is neither of high affinity nor saturable, can be determined using a high concentration of a competing unlabeled ligand which would block only the specific binding but not the nonspecific binding. The technique most widely used to rapidly separate bound and free ligand has been filtration followed by rapid washing of the filters with cold buffer. Total elapsed time for filtration and washing should be less than 20 sec if the K_D is in the low nanomolar range, it can be longer for ligands with K_D values in the subnanomolar range, and it is extremely short (less than 1 s and thus not practicable) if the K_D is in the high nanomolar range or higher.[57] If glass fiber filters are used to separate the membranes from the incubation medium, it is often helpful to presoak the filters with 0.1% polyethyleneimine to reduce nonspecific binding to the filters. The radioligand should not be diluted with unlabeled ligand if K_D and B_{max} are determined by saturation equilibrium binding. If a K_D is determined kinetically (from the ratio of the rate constant for dissociation, k_{-1}, over the rate constant of association, k_1), dissociation of the bound ligand can either be induced by extreme (at least 100-fold) dilution of the radioligand or by the addition (at equilibrium of radioligand binding) of an excess (1000-fold) of a competing unlabeled ligand.

To determine IC_{50} values for inhibition of uptake or binding by a competing ligand, uptake or binding experiments are carried out at a low substrate or radioligand concentration relative to its affinity constant and in the presence of various concentrations of the competitor (below and above the expected IC_{50}). However, the type of inhibition can only be derived from change(s) in the kinetic constants (K_m, K_D, and/or V_{max}, B_{max}) obtained in saturation kinetics of transport or binding in the presence of the competitor. Competitive (reversible) inhibitors cause an increase in

[57] J. P. Bennet, *in* "Neurotransmitter Receptor Binding" (H. I. Yamamura, S. J. Enna, and M. J. Kuhar, Eds.), p. 57. Raven Press, New York, 1978.

the affinity constant without to change V_{max} or B_{max}, whereas noncompetitive (irreversible) inhibitors also reduce maximum transport or binding.

Analysis of Transport or Binding Data

A plot of the rate of specific uptake (v) versus substrate concentration (S) results in a rectangular hyperbolic curve. This curve mirrors saturation of the uptake rate with increasing substrate concentrations, and the rate of uptake obeys Michaelis–Menten type kinetics characterized by the equation $v = (V_{max}S)/(K_m + S)$. To obtain the half-saturation constant K_m (apparent affinity constant) and V_{max} (the maximum transport rate), the curve can be analyzed by several graphical methods in which the data are converted to result a linear plot. According to Henderson,[58] good estimates result from a plot S/v against S; from the slope ($=1/V_{max}$) of the resulting regression, the intercept with the ordinate (K_m/V_{max}), and from the intercept with the ordinate ($-K_m$), the kinetic constants (K_m, V_{max}) can be derived. K_D and B_{max} are estimated from a Scatchard plot, in which the ratio of bound ligand (B) over free ligand (F) is plotted against the amount of bound ligand (B). This should result in a straight line which can be fitted by linear regression. K_D is estimated from the negative reciprocal of the slope ($-1/K_D$) and B_{max} is given by the x-intercept. An example of such a plot obtained from binding data with [^{3}H]desipramine to the NE transporter of the plasma membrane of PC-12 cells is shown in Fig. 5A. A curvilinear Scatchard plot indicates either two or more binding sites or a cooperative interaction. The latter can be determined from a Hill plot in which the logarithm of the ratio B over B_{max} minus B (log $[B/(B_{max} - B)]$ is plotted against the logarithm of the free radioligand concentration (log F), where B_{max} was estimated from a Scatchard plot. The slope of the line is denoted as Hill coefficient or Hill number (n_H). Since $K_D = k_{-1}/k_1$ [see Eq. 6)], the K_D can also be estimated from the rate constants for association and dissociation (as mentioned above). From a time course of radioligand binding, the amount of ligand bound at a given time (B_t) and the amount of ligand bound at equilibrium (B_{eq}) is obtained. A plot of ln $[B_{eq}/(B_{eq} - B_t)]$ against the time of incubation results in a straight line in which the slope represents the "observed" association rate constant (k_{ob}) as shown in an example in Fig. 5B (left, inset). The association rate constant (k_1) can be calculated from the equation $k_1 = (k_{obs} - k_{-1})/F$, where F is the free ligand concentration and k_{-1} is the dissociation rate constant. As shown in Fig. 5B (right insert), the dissociation rate constant is obtained from the negative

[58] P. J. F. Henderson, *in* "Enzyme Assays, a Practical Approach" (R. Eisenthal and M. J. Danson, Eds.), p. 277. Oxford University Press, Oxford, 1992.

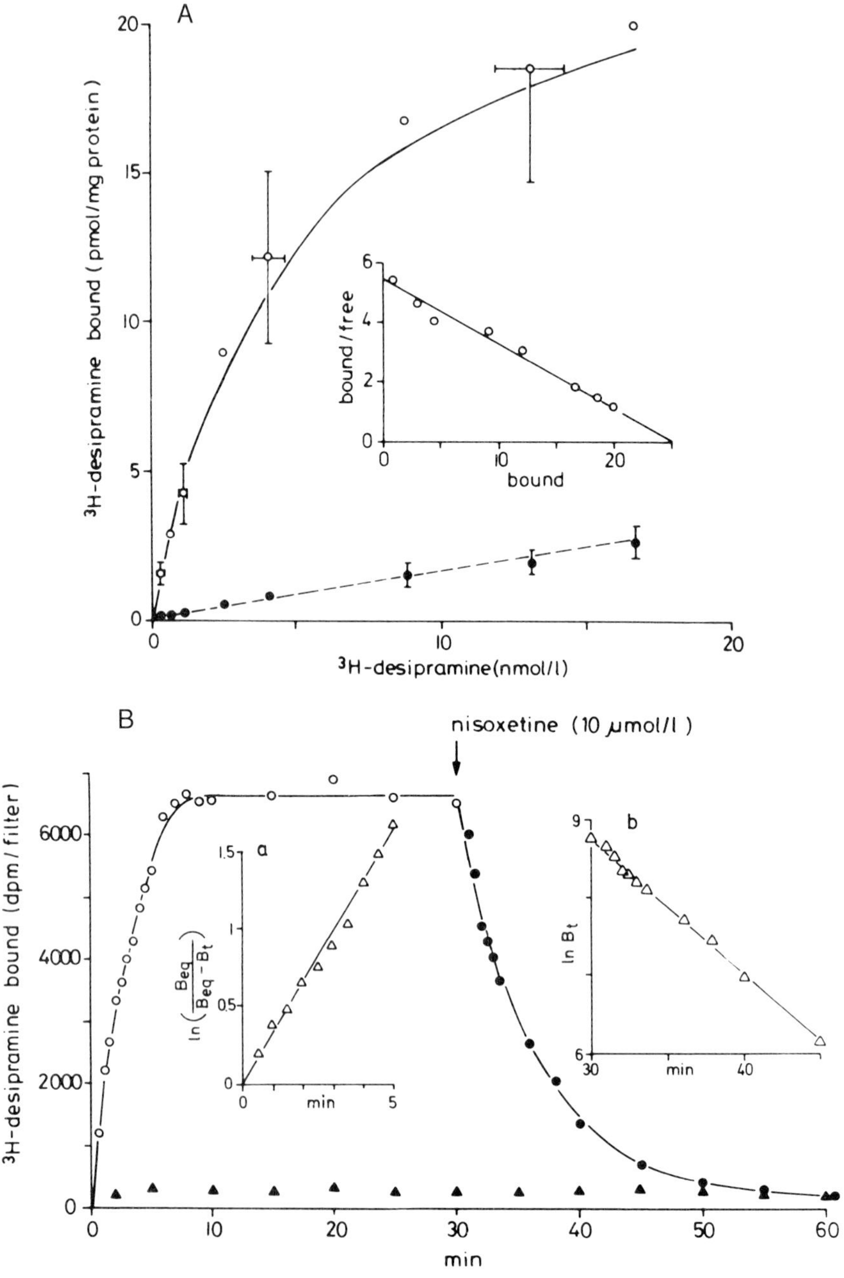
A
20
15
10
5
0
³H-desipramine bound (pmol/mg protein)
0
10
20
³H-desipramine(nmol/l)
6
4
2
0
bound / free
0
10
20
bound
B
nisoxetine (10 μmol/l)
6000
4000
2000
0
3H-desipramine bound (dpm/filter)
0
10
20
30
40
50
60
min
a
1.5
1
0.5
0
ln (Beq / (Beq - Bt))
0
min
5
b
9
6
ln Bt
30
min
40

slope of a plot of ln B_t (i.e., ligand remaining bound) versus the time after onset of dissociation.

Kinetic data obtained in transport or binding experiments are much more accurately analyzed with the help of computers using published computer programs[59] or commercially available computer software (e.g., GraphPad PRISM).

Specific Protocols for Uptake and Binding Assays

Next we present two protocols that have been used in our laboratory to carry out uptake and binding studies at the NE transporter of plasma membrane vesicles isolated from PC12 cells.[7,23]

Transport Assay

To determine initial rates of uptake of a labeled substrate, 18 μl of thawed membrane vesicles (7–14 μg protein) are mixed (on ice) with 2 μl of potassium phosphate buffer containing, if desired, desipramine (to determine nonspecific uptake) or the drug under study. After reaching ambient temperature (after about 10 min), uptake of the labeled substrate (e.g., [^{3}H]NE) is started by the addition of 200 μl prewarmed (usually 31°) incubation buffer containing the labeled substrate. As incubation buffer we use a Na^+/Tris buffer consisting of 140 m*M* NaCl, 10 m*M* Tris-HCl (pH 7.4), 1 m*M* $MgCl_2$, 1 m*M* dithiothreitol, and 1 m*M* ascorbic acid (to protect catecholamines against oxidation). Uptake is terminated by withdrawal of 200 μl of the incubation mixture, rapid mixing with 2 ml ice-cold 150 m*M* NaCl solution, and immediate filtration of the mixture through a small Whatman GF/F filter. After two rapid washes with 2 ml of the ice-cold NaCl solution, the radioactivity is extracted overnight from the filters by shaking in scintillation cocktail.

Binding Assay

Plasma membrane vesicles are first equilibrated in buffer C (135 m*M* NaCl, 10 m*M* Tris-HCl, pH 7.4, 5 m*M* KCl, 1 m*M* $MgSO_4$) containing 1 m*M*

[59] P. J. Munson and D. Rodbard, *Anal. Biochem.* **107,** 220 (1980).

FIG. 5. (A) Saturation equilibrium binding and Scatchard plot (inset) of binding of [^{3}H]desipramine to the NE transporter of the plasma membrane of PC-12 cells. (B) Time course of association [and its analysis (Inset, left)] and dissociation [and its analysis (Inset, right)] of binding of [^{3}H]desipramine to the NE transporter of the plasma membrane of PC-12 cells. [Reprinted with permission from H. Bönisch and R. Harder, *Naunyn-Schmiedeberg's Arch. Pharmacol.* **334,** 404 (1986).]

dithiothreitol and then pelleted (as described earlier). For determination of equilibrium binding of [^{3}H]desipramine to the NE transporter, 20 μl of membrane suspension (about 12 μg protein) is mixed with 10 μl buffer C with or without 10 μM nisoxetine (to determine nonspecific binding) and warmed to 25°. To start the binding reaction, 200 μl of buffer C containing [^{3}H]desipramine (and eventually a competing drug) is added to the membrane mixture and incubation is continued for the appropriate time (usually 30 min). At the end of incubation, 200 μl incubation mixture is withdrawn, rapidly mixed with 2 ml ice-cold buffer C, and rapidly filtered through Whatman GF/F glassfiber filters. The filters are rapidly washed with 4 × 2 ml ice-cold buffer and the radioactivity is extracted by shaking the filters overnight in scintillation cocktail.

[19] Use of Human Placenta in Studies of Monoamine Transporters

By VADIVEL GANAPATHY, PUTTUR D. PRASAD, and FREDERICK H. LEIBACH

Introduction

Term placentas represent one of the most readily available human tissues for biomedical research. As long as the mother and the baby associated with the placental source are not identified, the use of the placental tissue in research has an "exempt" status (exemption no. 45 CFR 46 101(b)5 as indicated in the Department of Health and Human Services guidelines). The protocols are generally exempted from a full review by Institutional Human Assurance Committees and are approved with ease. Human term placentas express high levels of the serotonin transporter[1] and the norepinephrine transporter.[2,3] Dopamine transporter is not expressed in this tissue.[2,3] Platelets from human peripheral blood have been, and still are, widely used for studies involving the serotonin transporter. Human placental tissue offers an attractive alternative with several notable advantages for the study

[1] D. F. Balkovetz, C. Tiruppathi, F. H. Leibach, V. B. Mahesh, and V. Ganapathy, *J. Biol. Chem.* **264,** 2195 (1989).

[2] S. Ramamoorthy, F. H. Leibach, V. B. Mahesh, and V. Ganapathy, *Am. J. Physiol.* **262,** C1189 (1992).

[3] S. Ramamoorthy, P. D. Prasad, P. Kulanthaivel, F. H. Leibach, R. D. Blakely, and V. Ganapathy, *Biochemistry* **32,** 1346 (1993).

0076-6879/98 $25.00

of this monoamine transporter. For investigations involving the norepinephrine transporter, there is no human tissue that is available as readily as the placenta. Both of these monoamine transporters are expressed in the maternal-facing brush border membrane of the placenta.[1-3] These membranes are easy to prepare from the placental tissue using reproducible techniques. Since the membrane preparations obtained by these techniques exist predominantly in the form of right-side-out vesicles,[4,5] these preparations are suitable to study the transport function of these transporters[1-3] as well as the binding of specific ligands to these transporters.[6,7] Choriocarcinoma cells which are of human placental origin also constitutively express the serotonin transporter,[8] and these cells (JAR and BeWo) have proved to be extremely useful to investigate the transcriptional and posttranscriptional events involved in the regulation of the serotonin transporter expression and function[8-13] and also in the analysis of the function of the promoter region in the human serotonin transporter gene.[14] Curiously, these cells do not express the norepinephrine transporter. Human choriocarcinoma cells are the most convenient cell culture model systems currently available to study the transcriptional regulation of the human serotonin transporter gene. Human platelets are not suitable for this purpose because of their nonnucleated nature. Human lymphocytes, which constitutively express the serotonin transporter,[15] may be used for this purpose but are not suitable for continuous passages. Continuous cell lines of human lymphocytes expressing the serotonin transporter are not readily available, though they

[4] J. D. Glazier, C. J. P. Jones, and C. P. Sibley, *Biochim. Biophys. Acta* **945,** 127 (1988).

[5] N. P. Illsley, Z. Q. Wang, A. Gray, M. C. Sellers, and M. M. Jacobs, *Biochim. Biophys. Acta* **1029,** 218 (1990).

[6] D. R. Cool, F. H. Leibach, and V. Ganapathy, *Am. J. Physiol.* **259,** C196 (1990).

[7] L. D. Jayanthi, P. D. Prasad, S. Ramamoorthy, V. B. Mahesh, F. H. Leibach, and V. Ganapathy, *Biochemistry* **32,** 12178 (1993).

[8] D. R. Cool, F. H. Leibach, V. K. Bhalla, V. B. Mahesh, and V. Ganapathy, *J. Biol. Chem.* **266,** 15750 (1991).

[9] S. Ramamoorthy, D. R. Cool, V. B. Mahesh, F. H. Leibach, H. E. Melikian, R. D. Blakely, and V. Ganapathy, *J. Biol. Chem.* **268,** 21626 (1993).

[10] L. D. Jayanthi, S. Ramamoorthy, V. B. Mahesh, F. H. Leibach, and V. Ganapathy, *J Biol. Chem.* **269,** 14424 (1994).

[11] J. D. Ramamoorthy, S. Ramamoorthy, A. Papapetropoulos, J. D. Catravas, F. H. Leibach, and V. Ganapathy, *J. Biol. Chem.* **270,** 17189 (1995).

[12] S. Ramamoorthy, J. D. Ramamoorthy, P. D. Prasad, G. K. Bhat, V. B. Mahesh, F. H. Leibach, and V. Ganapathy, *Biochem. Biophys. Res. Commun.* **216,** 560 (1995).

[13] R. Kekuda, V. Torres-Zamorano, F. H. Leibach, and V. Ganapathy, *J. Neurochem.* **68,** 1443 (1997).

[14] A. Heils, A. Teufel, S. Petri, G. Stober, P. Riederer, D. Bengel, and K. P. Lesch, *J. Neurochem.* **66,** 2621 (1996).

[15] B. A. Faraj, Z. L. Olkowski, and R. T. Jackson, *Int. J. Immunopharmacol.* **16,** 561 (1994).

can be prepared by viral transformation.[16] Thus, human placental brush border membrane vesicles and human choriocarcinoma cells offer valuable tools for the study of the serotonin transporter and/or the norepinephrine transporter.

Preparation of Brush Border Membrane Vesicles from Human Term Placenta

Tissue Collection

Placentas arising from vaginal deliveries or cesarian deliveries can be used for the preparation of membrane vesicles. Fresh placentas obtained immediately after delivery are preferable. Placentas obtained within 1 hr after delivery can be used if the tissue was stored at 4° during the interval.

Gross Placental Structure

Placenta is discoidal in shape, one side representing the maternal surface or decidual surface through which the placenta was attached to the uterine wall prior to delivery and the other side representing the fetal surface which contains the amniochorion, chorionic plate, and umbilical cord. On the maternal surface a thin layer (0.25 cm thick) of the tissue is actually of maternal origin arising from the uterine wall. The remaining portion of the placental tissue is of fetal origin. The spongy tissue lying between the decidual layer and the chorionic plate is called the villous tissue, which forms the starting material for the isolation of brush border membrane vesicles. *In vivo,* the villous tissue is perfused with maternal blood flowing through the intervillous spaces without capillary network on the mother side and with fetal blood flowing through the capillary network on the fetal side. In term placentas, these two blood circulations are separated essentially by a single cell layer of fetal origin consisting of the syncytiotrophoblast. This layer is a true syncytium, composed of a single, multinuclear syncytiotrophoblast. Lateral cell membranes are absent in this layer. The syncytiotrophoblast is a polarized epithelium, with one surface facing the maternal side and in direct contact with maternal blood and the other surface facing the fetal side. The maternal-facing surface consists of the brush border membrane, also called the apical or microvillous membrane, and the fetal-facing surface consists of the basal membrane. These two membranes are structurally, biochemically, and functionally distinct. The

[16] K. P. Lesch, D. Bengel, A. Heils, S. Z. Sabol, B. D. Greenberg, S. Petri, J. Benjamin, C. R. Muller, D. H. Hamer, and D. L. Murphy, *Science* **274,** 1527 (1996).

procedure described below is for the preparation of the brush border membranes which express the serotonin and norepinephrine transporters. Whether these transporters are expressed in the basal membrane is not known, but preparative procedures are available to isolate the basal membranes,[5,17] if needed for any purpose.

Procedure for the Isolation of Brush Border Membrane Vesicles

The various steps involved in the procedure[18] are outlined in Fig. 1. Two buffered solutions are needed. The wash buffer is isotonic and consists of 10 m*M* HEPES/Tris, 300 m*M* mannitol, pH 7.5. The preloading buffer is of varied composition, depending upon the nature of the experiments planned with the membrane vesicles, because this buffer is used to suspend the final membrane preparations. For transport assays involving the serotonin and norepinephrine transporters, the preloading buffer normally consists of 20 m*M* HEPES/Tris, 100 m*M* potassium gluconate, 100 m*M* mannitol, pH 7.5. For ligand binding assays, the composition of the preloading buffer can be changed as desired. All procedures are generally performed at 4°.

Step 1. Remove the umbilical cord and the amniochorion, and shave off the decidual layer (~0.25 cm thick) from the maternal surface. Separate the villous tissue from the chorionic plate and tease out the spongy tissue in fine pieces (1–2 mm thick) from the fetal blood vessels. Collect the tissue pieces (~250 g) in a beaker, add 300–400 ml of the wash buffer, swirl gently with the aid of a blunt glass rod, and filter through a single layer of nylon membrane (30- to 100-μm pore size, Nitex). Discard the flowthrough. Wash the tissue pieces two more times by a similar procedure. This washing step removes most of the blood from the tissue. The use of the isotonic wash buffer prevents hemolysis. At this stage, set aside a small amount of the washed tissue to prepare the homogenate for measurement of marker enzymes.

Step 2. Place the washed tissue pieces in a 1 liter beaker and add 300 ml of the wash buffer to make a slurry. Stir the tissue slurry on a magnetic stirrer at maximum speed for 30 min. This step separates the membrane components of the syncytiotrophoblast. Vigorous stirring is enough for this purpose because of the absence of lateral membranes connecting the brush border membrane to the basal membrane. Homogenization of the tissue is not needed. Following the 30-min stirring, sieve the slurry through two layers of cheese cloth and collect the flowthrough.

[17] L. K. Kelley, C. H. Smith, and B. F. King, *Biochim. Biophys. Acta* **734,** 91 (1983).

[18] D. F. Balkovetz, F. H. Leibach, V. B. Mahesh, L. D. Devoe, E. J. Cragoe, Jr., and V. Ganapathy, *Am. J. Physiol.* **251,** C852 (1986).

Placenta

Shave off the thin layer of maternal tissue from the decidual surface
Remove chunks of the villous tissue (400-500 g) and cut the tissue into fine pieces (1-2 mm)

Suspend the tissue pieces (~250 g) in wash buffer (300-400 ml), swirl gently with a glass rod and filter through one layer of Nitex nylon membrane. Discard the flowthrough. Repeat twice

Suspend the washed tissue in ~300 ml wash buffer in a one liter beaker and stir on a magnetic stirrer at maximum speed for 30 min

Sieve the slurry through two layers of cheese cloth.
Collect the flowthrough and centrifuge at 7,500xg for 15 min

Pellet (discard)

Supernatant
Centrifuge at 70,000xg for 30 min

Pellet
Suspend in 50 ml wash buffer and homogenize using 10 strokes in a glass/glass Dounce homogenizer and transfer to a beaker.
Add 600 μl of 1 M $MgCl_2$, stir for one minute and let stand for 15 min

Supernatant (discard)

Centrifuge at 2,000xg for 15 min

Pellet (discard)

Supernatant
Dilute to 100 ml with wash buffer and centrifuge at 70,000xg for 30 min

Pellet
Suspend in one ml of preloading buffer using a glass rod and homogenize the suspension by passing through a 25 gauge needle several times. Dilute the membrane suspension with 25 ml of the preloading buffer and centrifuge at 70,000xg for 30 min

Supernatant (discard)

Pellet
Suspend in preloading buffer and make a homogeneous suspension as in the previous step. Estimate the protein content in the homogenate and dilute it to the desired concentration. Freeze aliquots of the membrane preparation in liquid nitrogen.

Supernatant (discard)

FIG. 1. Flow chart for the preparation of brush border membrane vesicles from human placenta.

Step 3. Centrifuge the flowthrough in 250-ml transparent polycarbonate bottles at 7500*g* for 15 min. (The *g* values used here and in the following steps represent g_{max}, calculated for the bottom of the centrifuge bottle or tube.) A refrigerated centrifuge, either Beckman model JA-14 or IEC with rotor no. 872, can be used at this step. In both centrifuge models, the 7500g translates to 7000 rpm. At this step, discard the pellet and save the supernatant. The pellet contains sedimented red blood cells at the bottom covered with a layer of cells and cell debris arising primarily from the loose connective tissue which supports the stromal core of the villus in intact placenta. The nuclei of the syncytiotrophoblast are also most likely sedimented in the pellet.

Step 4. Centrifuge the supernatant in a Beckman ultracentrifuge (type 30 rotor, 30-ml transparent polycarbonate tubes) at 70,000g (25,000 rpm) for 30 min to sediment the membranes.

Step 5. Suspend the pellets in a toal volume of 50 ml of the wash buffer with the aid of a blunt glass rod and transfer the suspension into a glass–glass Dounce homogenizer (Kontes Scientific glassware/instruments) and homogenize with 10 strokes. Transfer the homogenate into a 100-ml beaker. Add 0.6 ml of 1 M $MgCl_2$ to the homogenate (final concentration of $MgCl_2$, 12 mM), stir for 1 min, and incubate for 15 min. In this step, non-brush-border membranes are aggregated by the divalent Mg^{2+}. It is believed that the relatively sparse anionic charges on the nonbrush-border membranes leads to the formation of intermembranous chelates with Mg^{2+} which results in the aggregates.[19] In contrast, brush border membranes possess relatively high density of anionic charges which leads to the formation of intramembranous chelates with Mg^{2+}, thus resisting aggregation. Ca^{2+} can be substituted for Mg^{2+} for this purpose, but is not generally used because exposure of the brush border membranes to high concentrations of Ca^{2+} during the preparative procedure is known to affect the function of certain transport systems.[20]

Following the 15-min incubation, centrifuge the suspension at 2000g (3500 rpm) for 15 min in a 250-ml polycarbonate bottle using a refrigerated centrifuge as described in Step 3. The Mg^{2+}-aggregated non-brush-border membranes are sedimented in this step. Discard the pellet. Save the supernatant which now contains brush border membranes.

Step 6. Centrifuge the supernatant at 70,000g (25,000 rpm) for 30 min in 30-ml polycarbonate tubes as described in Step 4. Brush border membranes are sedimented in this step. Suspend the pellet in each tube in 0.8 ml of the preloading buffer with the aid of a blunt glass rod and pass the suspension through a 25-gauge needle several times to make the suspension homogeneous. Dilute the suspension with 25 ml of the preloading buffer in the same centrifuge tubes and centrifuge again at 70,000g for 30 min.

[19] A. G. Booth and A. J. Kenny, *Biochem. J.* **142,** 575 (1974).

[20] P. Kulanthaivel, Y. Miyamoto, V. B. Mahesh, F. H. Leibach, and V. Ganapathy, *Placenta* **12,** 327 (1991).

Suspend the resulting pellets in a small volume (~0.8 ml/pellet) of the preloading buffer and make a homogeneous suspension using a 25-gauge needle as described previously.

Step 7. Estimate the protein content of the membrane suspension. Dilute the suspension with the preloading buffer appropriately to give the desired protein concentration (5 or 10 mg/ml). Freeze the membrane suspension in small aliquots in liquid nitrogen and keep a small sample for assay of marker enzymes.

Determination of Purity of Brush Border Membrane Preparations

The purity of the brush border membrane preparations can be assessed by determining the enrichment of the brush border membrane marker enzymes alkaline phosphatase and 5′-nucleotidase compared to the tissue homogenate. The homogenate is prepared from the washed tissue set aside in Step 1. Assay of alkaline phosphatase is much easier than that of 5′-nucleotidase. We use a procedure which is a slight modification of the one described by Forstner *et al.*[21] for the assay of alkaline phosphatase. The procedure involves determination of *p*-nitrophenol liberated from the substrate *p*-nitrophenylphosphate (18 m*M*) in 50 m*M* glycine/NaOH buffer (pH 9.2) in the presence of 0.5 m*M* $MgCl_2$ and 1 m*M* $ZnCl_2$. Appropriately diluted homogenate or membrane preparation is incubated with the substrate for 15 min at 37° in a final volume of 0.6 ml. The reaction is terminated by the addition of 2.5 ml of 0.02 *N* NaOH. Free *p*-nitrophenol is determined spectrophotometrically at 400 nm. Alternatively, commercially available assay kits (e.g., Sigma, St. Louis, MO) can be used for the purpose. The enrichment of alkaline phosphatase activity in brush border membrane preparations over the tissue homogenate usually ranges between 20- and 30-fold.

Uptake Measurements in Membrane Vesicles

Placental brush border membrane vesicles are highly suitable for studying transport function of the serotonin and norepinephrine transporters. This experimental system allows us to alter transmembrane gradients for Na^+, K^+, Cl^-, and H^+ and membrane potential as desired and thus makes it possible to analyze systematically the driving forces involved in the function of these transporters.[1–3,22]

[21] G. G. Forstner, S. M. Sabesin, and K. J. Isselbacher, *Biochem. J.* **106,** 381 (1968).
[22] V. Ganapathy, S. Ramamoorthy, and F. H. Leibach, *Trophoblast Res.* **7,** 35 (1993).

Uptake measurements in membrane vesicles involve a rapid filtration technique. Uptake is initiated by rapidly mixing 40 μl of the membrane suspension with 160 μl of uptake buffer containing radiolabeled substrate. Uptake is terminated after a desired time by the addition of 3 ml of ice-cold stop buffer. The mixture is immediately filtered through a Millipore (Bedford, MA) filter (DAWP type, 25-mm diameter, 0.65-μm pore size) under vacuum. The filtration apparatus from Micro Filtration Systems (model no. KG25) or any other comparable setup can be used for this purpose. The membrane vesicles stay on top of the filter during filtration. The filter is washed three times with 3 ml of the ice-cold stop buffer. The filter is then transferred to a scintillation vial. Scintillation cocktail (7.5 ml) is added to the vial and the radioactivity associated with the filter is determined. Measurements can be made at room temperature for the purpose of convenience. However, higher temperatures such as 30° or 37° can be used to obtain appreciably increased uptake rates. Kinetic analysis of the transport function should be done by measuring initial uptake rates with the use of appropriate incubation times.

The stop buffer usually consists of 10 m*M* HEPES/Tris, 150 m*M* KCl, pH 7.5. If desired, inhibitors of the serotonin and norepinephrine transporters such as imipramine or desipramine (10 μM) can be added to the stop buffer to prevent any efflux of the transported substrates out of the vesicles. However, since the filtration is rapid and the stop buffer is ice-cold, we find it unnecessary to include these inhibitors. The composition of the uptake buffer can be modified as desired depending on the purpose of the experiment. In general, the uptake buffer consists of 10 m*M* HEPES/Tris, 150 m*M* NaCl, pH 7.5. The role of membrane potential in the transport function can be investigated using several different approaches. Inside-negative or inside-positive membrane potentials can be generated with the aid of either the K^+-ionophore valinomycin or the H^+-ionophore carbonylcyanide *p*-trifluoromethoxyphenylhydrazone in the presence of appropriate transmembrane K^+ or H^+ gradients. The membrane vesicles can also be voltage-clamped by keeping an equal concentration of K^+ in the intravesicular medium as well as in the uptake medium and by adding valinomycin.

Measurements of Ligand Binding

The characteristics of the binding of specific ligands to the serotonin and norepinephrine transporters in the placental brush border membrane vesicles can be studied by a technique[6,7] similar to that described for uptake measurements. The free ligand can be separated from the ligands bound to the membrane vesicles after equilibrium binding by the same rapid filtration technique. We usually use glass fiber filters (GF/F, 0.7-μm diameter) soaked for 1 hr in 0.3%polyethyleneimine for this purpose. The ligands specific to the monoamine transporters are generally cationic, and nonspecific binding of these ligands to the filter is effectively reduced by soaking the filter in a polyethyleneimine solution prior to use. Because the human placental brush border membranes express the serotonin transporter as well as the norepinephrine transporter, only those ligands which are relatively specific to either of these transporters can be used in binding experiments in order to be able to distinguish the ligand binding to one transporter from that to the other. Paroxetine is suitable for the studies involving the serotonin transporter[6] and nisoxetine is suitable for studies involving the norepinephrine transporter.[7]

Use of Human Choriocarcinoma Cells in Studies of Monoamine Transporters

Human choriocarcinoma cells (JAR and BeWo) express a serotonin transporter which is identical to the transporter present in normal tissues.[8,13] However, even though the norepinephrine transporter is expressed in normal placenta, these choriocarcinoma cells do not express this monoamine transporter.[22] Interestingly, neither the choriocarcinoma cells nor the normal placenta express the vesicular monoamine transporter.[23] The placental trophoblast cells are unique in this respect. The JAR choriocarcinoma cells have been widely used for studies involving the regulatory aspects of the serotonin transporter,[8–14] but these cells do not polarize in culture as normal syncytiotrophoblast does.[24] The BeWo choriocarcinoma cells are known to form a polarized monolayer.[25,26] Studies have been conducted to investigate the differential sorting of the monoamine transporters by stably transfecting the kidney cell lines LLC-PK_1 and MDCK, which form a polarized monolayer in culture, with the transporter cDNAs.[27] However, since neither the normal kidney tubular cells nor the kidney cell lines constitutively express the monoamine transporters, the BeWo cells may offer a more suitable experimental model system for this purpose.

[23] P. D. Prasad, B. J. Hoffman, A. J. Moe, C. H. Smith, F. H. Leibach, and V. Ganapathy, *Placenta* **17,** 201 (1996).

[24] A. M. Mitchell, A. S. Yap, E. J. Payne, S. W. Manley, and R. H. Mortimer, *Placenta* **16,** 31 (1995).

[25] B. Wice, D. Menton, H. Geuze, and A. L. Schwartz, *Exp. Cell Res.* **186,** 306 (1990).

[26] T. C. Furesz, C. H. Smith, and A. J. Moe, *Am. J. Physiol.* **265,** C212 (1993).

[27] H. H. Gu, J. Ahn, M. J. Caplan, R. D. Blakely, A. I. Levey, and G. Rudnick, *J. Biol. Chem.* **271,** 18100 (1996).

Cell Culture

JAR cells are cultured at 37° in a humidified atmosphere of 95% air and 5% CO_2 (v/v) in RPMI 1640 medium and BeWo cells are cultured in 1:1 mixture of Dulbecco's modified Eagle's medium (DMEM) and F-12 medium. Each medium is supplemented with 10% fetal bovine serum, penicillin (100 U/ml), and streptomycin (100 μg/ml). Cells are cultured in 75-cm^2 culture flasks for maintenance. For uptake measurements, cells in 75-cm^2 flasks are released by treatment with 0.1% trypsin–1 m*M* EDTA for 5–10 min and washed free of trypsin and EDTA by washing with regular culture medium. The cells are then seeded in 35-mm petri dishes or 24-well culture plates at a cell density of 150,000 cells/cm^2 culture area. Twenty-four hours after subculturing, the medium is replaced with fresh culture medium. The cells are used for uptake measurements on day 3 (the day of subculturing is taken as day 1). When the role of hormones or second messenger systems in the regulation of the serotonin transporter is studied, the replacing medium used on the second day is devoid of fetal bovine serum and is substituted with a hormonally defined medium. The defined medium consists of RPMI 1640 (JAR) or DMEM/F-12 (1:1 v/v) (BeWo), supplemented with insulin (5 μg/ml), apotransferrin (5 μg/ml), prostaglandin E_1 (2.5×10^{-5} mg/ml), hydrocortisone (5×10^{-8} *M*), and thyroxine (5×10^{-12} *M*). The hormonal conposition of the defined medium can be modified as desired for the purpose of individual experiments.

Uptake Measurement in Cultured Cells

The culture dishes containing confluent cells are taken out of the incubator and kept at room temperature for 2 hr. In our experience, the activity of the serotonin transporter measured immediately after the cells are taken out of the incubator is significantly lower than the activity measured after 2 hr of incubation of the cells at room temperature. We do not know the exact reasons for this change in the transporter activity. However, the activity stabilizes after 2 hr. Following the 2-hr stabilization period, the culture medium is aspirated and the cells are washed with the uptake buffer. Disruption of the cell layer is carefully avoided during this step. Aspiration can be done with convenience and ease using a setup that operates under vacuum suction. The suction tube can be hooked to a Pasteur pipette and the medium in the culture dish can be removed fast by placing the pipette tip at the inside edge of the dish and slightly tilting the dish as the medium is aspirated. Washing is done two times. The uptake buffer (1 ml for each 35-mm dish and 0.5 ml for each well in a 24-well culture plate) containing radiolabeled serotonin is added to the washed cells and incubated for a desired time, usually for 3 min. Uptake is terminated by aspirating the buffer and subsequently washing the cells three times with fresh uptake buffer. The cells are lysed with 0.2 *N* NaOH/1% sodium dodecyl sulfate (SDS) (1 ml for each 35-mm dish and 0.5 ml for each well in a 24-well culture plate), and the lysate is transferred to scintillation vials for determination of radioactivity. The composition of the uptake buffer is, in general, 25 m*M* HEPES/Tris, pH 7.5, 140 m*M* NaCl, 5.4 m*M* KCl, 1.8 m*M* $CaCl_2$, 0.8 m*M* $MgSO_4$, 5 m*M* glucose, and 0.1 m*M* iproniazid, an inhibitor of monoamine oxidases. Seritonin uptake that occurs independent of the serotonin transporter (i.e., diffusion) is determined by measuring the uptake in the presence of 0.1 m*M* imipramine. This component is usually less than 10% of total uptake measured in the absence of imipramine.

Preparation of Plasma Membrane Vesicles from Cultured Choriocarcinoma Cells

Plasma membrane vesicle preparations from choriocarcinoma cells are useful for several purposes. If a hormone or a second messenger modulates the serotonin transporter activity in intact cells, it may be desirable to see if the change in the transport activity is detectable in plasma membrane vesicles prepared from treated cells. The membrane preparations are also useful for the determination of the serotonin transporter density by ligand binding measurements. Even though the serotonin transporter activity is easily measurable in JAR cells as well as in BeWo cells, the transporter density is low and is not easily measurable with [^{3}H]paroxetine or [^{3}H]imipramine because of the low specific radioactivity of these ligands from commercial sources. We routinely use [^{125}I]RTI-55 for this purpose. RTI-55 is a cocaine analog which is a high affinity ligand to the serotonin transporter and also to the dopamine and norepinephrine transporters. However, since the choriocarcinoma cells do not express the dopamine and norepinephrine transporters, RTI-55 can be considered as a selective ligand for the serotonin transporter in these cells. The high specific radioactivity of [^{125}I]RTI-55 available from commercial sources makes it feasible to determine the transporter density by Scatchard analysis even though the transport density in these cells is low. Again, specific binding of [^{125}I]RTI-55 is low when intact cells or the cell homogenates are used because of the low transporter density. Therefore, purified plasma membrane preparations are desirable for Scatchard analysis.

Cells are cultured in large flasks to provide enough starting material for the preparation of plasma membrane vesicles. We usually use 225-cm^2 culture flasks for this purpose. All steps involved in the membrane isolation procedure are performed at 0–4°. The culture medium is aspirated from the flasks and the cells are washed with ice-cold phosphate-buffered saline

(10 ml/flask). Following this, 25 ml of the lysis buffer (10 mM Tris/HCl, pH 7.5) is added to the flask. The cells lyse because of the hypotonicity of the buffer. The lysed cells are scraped with the aid of a cell scraper and the contents are transferred to a glass beaker. The lysate is homogenized for 30 sec in an Ultra-Turrax (Tekmar Company, Cincinnati, OH) homogenizer. Other kinds of motor-driven homogenizers can also be used. The final volume of the homogenate is brought to 50 ml with the lysis buffer. A stock solution of 1 M $MgCl_2$ is added to the homogenate to a final $MgCl_2$ concentration of 12 mM. The mixture is stirred for 1 min and let stand for 15 min. The divalent cation Mg^{2+} leads to preferential aggregation of nonplasma membrane fragments. The aggregated material is sedimented by centrifugation at 2,000g for 15 min as described earlier for the preparation of placental brush border membranes. The plasma membranes present in the resulting supernatant are then collected by centrifugation at 70,000g for 30 min. The final membrane pellets are suspended by passing through a 25-gauge needle in a small volume of either a 10 mM HEPES/Tris buffer, pH 7.5, containing 100 mM potassium gluconate and 100 mM mannitol (for serotonin uptake measurements) or a 10 mM Tris-HCl buffer, pH 7.5 (for RTI-55 binding measurements). Protein concentration in the final membrane preparations is usually adjusted to 1 mg/ml for RTI-55 binding or 2.5 mg/ml for serotonin uptake. Measurements of uptake and binding can be made as described earlier for placental brush border membrane vesicles.

Concluding Remarks

It is clear that the human placental tissue and the human choriocarcinoma cells offer excellent experimental tools for studies involving the serotonin transporter and the norepinephrine transporter. There is another aspect that is equally or even more important in the use of human placenta and choriocarcinoma cells for the study of these two monoamine transporters. Placenta is a noninnervated tissue and the fact that this tissue expresses high levels of the serotonin and norepinephrine transporters is interesting and intriguing. The physiological functions of these transporters in the placenta have not been identified. We speculate that these transporters are involved in the clearance of the vasoactive monoamines serotonin and norepinephrine from the intervillous space.[28] This process may be crucial to the normal placental function and for optimal development of the fetus. What is readily apparent, however, is that the functions of these transporters in the placenta are heavily compromised when the mother uses drugs such

[28] V. Ganapathy and F. H. Leibach, *in* "Placental Toxicology" (B. V. Ramasastry, Ed.) p. 161. CRC Press, Boca Raton, Florida, 1995.

as cocaine and amphetamines during pregnancy and that these effects are directly relevant to the pathogenesis of maternal and fetal complications induced by these drugs.[28–30] It is hoped that the ease with which the human placental tissue and human choriocarcinoma cells can be used to investigate various aspects of the serotonin and norepinephrine transporters would facilitate future studies and result in the unraveling of the physiological role of these monoamine transporters in the placenta and hence in the growth and development of the fetus.

Acknowledgments

The authors thank Ida O. Walker for excellent secretarial assistance. This work was supported by National Institutes of Health Grant DA 10045.

[29] V. Ganapathy and F. H. Leibach, *Placenta* **15,** 785 (1994).

[30] J. D. Ramamoorthy, S. Ramamoorthy, F. H. Leibach, and V. Ganapathy, *Am. J. Obstet. Gynecol.* **173,** 1782 (1995).

Section IV

Biochemical Approaches for Structure–Function Analyses

Articles 20 through 27

[20] Analysis of Transporter Topology Using Deletion and Epitope Tagging

By JANET A. CLARK

Introduction

Several strategies have been employed in an effort to elucidate the structure and transmembrane topology of transport proteins. A number of these strategies are outlined in this volume, and as a result of their use, modifications are being made to predicted hydropathy models. An understanding of transporter structure is crucial for determining the correlation between transporter structure and function. In addition, structural knowledge will aid in the design of more specific transporter reagents which may prove therapeutically useful. The transmembrane topology of the Na^+- and Cl^--dependent transporter family was studied using the GAT-1 GABA transporter, deletion analysis, and epitope tagging in *Xenopus laevis* oocytes.[1] The methods employed in study of GAT-1 transmembrane topology are the focus of the chapter to follow.

Current studies of the transmembrane topology of integral membrane proteins using deletion analysis and epitope tagging derive from gene fusion methods designed to study secretion[2] and topology[3] of bacterial proteins. In these original experiments, a modified form of the *Escherichia coli phoA* gene, which encodes the secreted periplasmic enzyme alkaline phosphatase, was fused to the C-terminal portion of a protein of interest. Alkaline phosphatate is active in the periplasm of bacteria, where it normally resides, but is inactive when located in the cytoplasm. Therefore, the level of alkaline phosphatase activity is a direct indication of the location of the enzyme in bacteria and a useful tool for discerning the secretion or topogenic properties of bacterial proteins. The *phoA* gene fusion method has been used to determine the transmembrane topology of the C-terminal half of the bacterial glutamate transporter in *E. coli.*[4] Another bacterial enzyme, β-galactosidase (*lacZ*), has been used in fusion studies done in both *E. coli* and *Saccharomyces cerevisiae* to examine protein localization and transmem-

[1] J. A. Clark, *J. Biol. Chem.* **272,** 14695 (1997).
[2] C. S. Hoffman and A. Wright, *Proc. Natl. Acad. Sci. USA* **82,** 5107 (1985).
[3] C. Manoil and J. Beckwith, *Science* **233,** 1493 (1986).
[4] D. J. Slotboom, J. S. Lolkema, and W. N. Konings, *J. Biol. Chem.* **271,** 31317 (1996).

0076-6879/98 $25.00

brane topology, respectively.[5,6] β-Galactosidase has properties complementary to those of *phoA*. That is, β-galactosidase is active in the cytoplasm but inactive in the periplasm. In a very recent report, the *phoA* and *lacZ* gene fusion techniques were extended for the first time to a eukaryotic system where the transmembrane topology of the human vasopressin V2 receptor was studied in COS cells.[7] Use of these methods has greatly contributed to our understanding of protein localization and transmembrane topology in virtually all cell types.

A second approach has been developed to study nascent chain translocation and transmembrane topology of integral membrane proteins. This approach is a natural extension of the gene fusion methods and involves the truncation of membrane proteins to defined locations and fusion of an epitope tag. The location of the epitope tag, determined by various experimental means, indicates the membrane sidedness of the residues to which the tag is fused. Alternatively, the epitope tag can be fused to isolated sequences from a protein, such as hydrophobic domains, in order to determine the topogenic properties of those sequences. Detection of N-linked glycosylation of the epitope or protection of an epitope from protease digestion are the assays used to determine membrane sidedness of the epitope tag. The most commonly used epitope for such studies is the C-terminal 144 amino acids (codons 56–199) of the secretory protein prolactin, although several epitopes have been used for such studies, including β-lactamase[8] and chloramphenicol acetyltransferase.[9] This C-terminal prolactin fragment lacks intrinsic translocation activity and has previously been shown to serve as a faithful reporter for translocation when following topogenic sequences in chimeric proteins.[10] Protease protection assays are used to determine the transmembrane spanning abilities of hydrophobic domains and the cellular location of regions connecting these domains. Nascent chains in the endoplasmic reticulum (ER) are oriented such that extracellular domains are located within the ER lumen and cytoplasmic domains are on the cytosolic face of the ER membrane. Proteinase K exposure of nascent chains in vesicles prepared from ER membranes results in cleavage of cytoplasmically exposed domains while domains in the ER

[5] C. Manoil, *J. Bacteriol.* **172,** 1035 (1990).

[6] B. Fiedler and G. Scheiner-Bobis, *J. Biol. Chem.* **271,** 29312 (1996).

[7] R. Schülein, C. Rutz, and W. Rosenthal, *J. Biol. Chem.* **271,** 28844 (1996).

[8] C. P. Cartwright and D. J. Tipper, *Mol. Cell. Biol.* **11,** 2620 (1991).

[9] Y. Xie, S. A. Langhans-Rajasekaran, D. Bellovino, and T. Morimoto, *J. Biol. Chem.* **271,** 2563 (1996).

[10] R. E. Rothman, D. W. Andrews, M. C. Calayag, and V. R. Lingapappa, *J. Biol. Chem.* **263,** 10470 (1988).

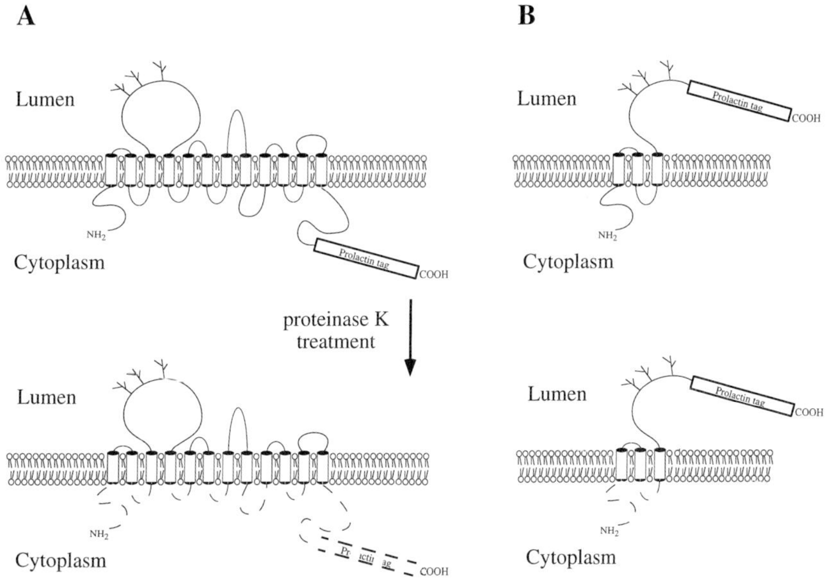

FIG. 1. Schematic of proteinase K treatment of transporter–prolactin chimeras in membrane vesicles. (A) Epitopes on the cytosolic face of the vesicle membrane are digested and no protected fragments will be detected by SDS–PAGE analysis of samples. (B) Epitopes in the vesicle lumen are protected from protease and protected fragments detected by analysis on SDS–PAGE. [Adapted with permission from J. A. Clark, *J. Biol. Chem.* **272,** 14695 (1997).]

lumen are protected (Fig. 1A and B). Since proteinase K cleaves nonspecifically, all regions of the protein exposed to the cytoplasm will be subject to digestion (Fig. 1). This method has been used to examine the transmembrane topology of several proteins, including the human P-glycoprotein (MDR1),[11–13] the GluR3 glutamate receptor[14] and nicotinic acetylcholine receptor subunits.[15] Studies of MDR1[11–13] and GluR3[14] have shown that models based on hydropathy analyses can be misleading and have resulted in the generation of novel topological profiles for these proteins.

Protease protection assay of GAT-1-prolactin chimeras in vesicles prepared from *Xenopus laevis* oocyte membranes has shown that loops connecting HD3 and HD4, and HD7 and HD8 are accessible to protease.[1]

[11] W. R. Skach and V. R. Lingappa, *J. Biol. Chem.* **268,** 23552 (1993).

[12] W. R. Skach, L.-B. Shi, C. Calayag, A. Frigeri, V. R. Lingappa, and A. S. Verkman, *J. Cell Biol.* **125,** 803 (1994).

[13] W. R. Skach and V. R. Lingappa, *Cancer Res.* **54,** 3202 (1994).

[14] J. A. Bennett and R. Dingledine, *Neuron* **14,** 373 (1995).

[15] R. A. Chavez and Z. W. Hall, *J. Cell Biol.* **116,** 385 (1992).

These data suggest the presence of pore loop-type structures which may be involved in formation of a substrate binding pocket. Exposure of these regions of GAT-1 to cytosol is not predicted by hydropathy analysis. The transmembrane topology of GAT-1 hydrophobic domains as determined with these methods agrees with the predicted hydropathy model for Na^+- and Cl^--dependent transporters.

In addition to elucidating the transmembrane topology of GAT-1, these methods reveal processes governing the membrane assembly of this transporter. It has been proposed that proteins which span the membrane several times acquire their unique topology by a series of alternating signal and stop-transfer sequences.[16,17] Although assembly of the carboxy-terminal half of GAT-1 appears to require such a pattern of sequences, amino-terminal GAT-1 assembly seems more complex, requiring the cooperative actions of several topogenic sequences.[1] Data obtained from the use of gene fusion methods in the study of the transmembrane topology of Na^+,K^+-ATPase[9] and H^+,K^+-ATPase[18] reveal that membrane assembly of these proteins is similar in complexity to that reported for GAT-1.[1] Interpretation of data from gene fusion studies of bacterial protein topology can be equally problematic because of the presence of topogenic sequences governing membrane assembly.[19] Without the use of additional methods to aid in the interpretation of protease protection data, determination of GAT-1 transmembrane topology would have been difficult. Methods used in analyzing the transmembrane topology of GAT-1 are described below and outlined in Fig. 2. Studies of the amino terminal membrane assembly and transmembrane topology of GAT-1 will be discussed to illustrate the challenges in interpreting data obtained using transporter deletion and epitope tagging.

Experimental Procedures

Plasmid Construction

Construction of gene fusion plasmids can be expedited by constructing a plasmid containing the epitope tag and flanked by convenient restriction sites. The cDNA of interest is placed downstream of a bacteriophage promoter such as T7, T3, or SP6 which is needed for the synthesis of transcripts. Addition of 5′ untranslated regions of alfalfa mosaic virus or *Shaker* and

[16] V. R. Lingappa, J. R. Linappa, and G. Blobel, *Nature* **281,** 117 (1979).
[17] G. Blobel, *Proc. Natl. Acad. Sci. USA* **77,** 1496 (1980).
[18] K. Bamberg and G. Sachs, *J. Biol. Chem.* **269,** 16909 (1994).
[19] E. S. Hennessey and J. K. Broome-Smith, *Curr. Opin. Struct. Biol.* **3,** 524 (1993).

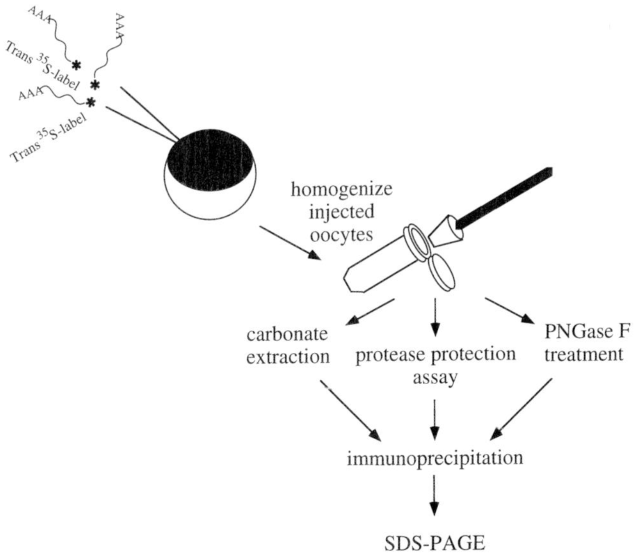

FIG. 2. Outline of methods used to study GAT-1 transmembrane topology. Oocytes are injected with transcripts of truncated and epitope-tagged transporters and Tran ^{35}S-label. Oocytes are incubated for 4–6 hr, then homogenized. Oocyte homogenates are used to assay for (1) chimera membrane integration status by extraction in sodium carbonate, pH 11.5; (2) membrane sidedness of epitope using protease protection assay; (3) N-linked glycosylation state of chimeras by PNGase F treatment. All chimeras and fragments are solubilized and immunoprecipitated with rabbit anti-ovine prolactin polyclonal serum. Immunoprecipitated proteins are analyzed by running on SDS–PAGE followed by autoradiography. [Adapted with permission from J. A. Clark, *J. Biol. Chem.* **272,** 14695 (1997).]

poly(A) added to the 3′ region have proven useful in enhancing transporter expression in oocytes.[20] However, expression of GAT-1 and GAT-1-prolactin chimeras was robust and did not require further enhancement.[1] A prolactin epitope containing plasmid, JGPRO, was created by ligating an *Eco*RI/*Sph*I fragment from pSPSp^{+1}L.ST.gG.p^{T}, obtained from W. R. Skach and V. R. Lingappa, into digested JG3.6.[21] JGPRO has a *Bst*EII site prior to the sequence encoding the epitope tag. Truncated transporter constructs are generated using polymerase chain reaction (PCR) with an oligo containing a convenient restriction site and directed against the 5′ noncoding region paired with oligos directed against defined regions in the transporter se-

[20] S. Mager, J. Naeve, M. Quick, C. Labarca, N. Davidson, and H. A. Lester, *Neuron* **10,** 177 (1993).

[21] G. F. Graminski, C. K. Jayawickreme, M. N. Potenza, and M. R. Lerner, *J. Biol. Chem.* **268,** 5957 (1993).

quence and containing a *Bst*EII site. Standard methods for PCR[22] are used to amplify the desired fragments. Fragments should be sequenced to ensure that mutations have not been introduced during amplification. Specific sites for truncation in the transporter were chosen such that a maximal number of residues were located between the previous hydrophobic domain and the epitope tag. This placement was an attempt to ensure that protected digested fragments were maximal in size. PCR fragments are digested with *Kpn*I and *Bst*EII and ligated into digested JGPRO. PRO, a construct with full-length prolactin, was generated by ligation of a *Pst*I and *Hind*III fragment from a prolactin construct (BPI) obtained from W. R. Skach and V. R. Lingappa into pBSSKII (−). The full-length prolactin serves as an important control in the protease protection assays to determine the integrity of the membrane vesicles and the conditions for protease treatment.

RNA Transcription

Transporter fusion protein constructs are linearized with the appropriate restriction enzyme. Following digestion the linearized construct is treated with proteinase K (0.2 mg/ml) at 37° for 30 min and extracted once with an equal volume of phenol–chloroform and once with an equal volume of chloroform. The extracted sample is precipitated with an equal volume of 5 *M* ammonium acetate and 2.5 volumes of ethanol. Following centrifugation the pellet is resuspended in RNase-free water treated with diethyl pyrocarbonate (DEPC). Several kits for the transcription of mRNA are now on the market and yield 25–50 μg of transcript starting with 1 μg of linearized construct. Methods employed by these kits follow established protocols for the preparation of synthetic RNA for injection into *Xenopus* oocytes.[23,22] mRNA for the study of GAT-1 was transcribed with T7 RNA polymerase at 37° using the mMessage mMachine protocol (Ambion, Inc., Austin, TX) which includes the addition of 5′ cap analog to assure the protection of the transcript from nucleolytic degradation once injected into the oocyte.[23] Following the transcription reaction, DNA template is removed with DNase (20 μg/ml final concentration) for 10 min at 37°. The transcription reaction is extracted once with an equal volume of phenol–chloroform, once with an equal volume of chloroform, and precipitated as described above three times. The transcripts are dissolved in DEPC-treated water for injection and stored at −80°.

[22] J. Sambrook, E. F. Fritsch, and T. Maniatis, *in* "Molecular Cloning" (C. Nolan, Ed.). Cold Spring Harbor Laboratory Press, Cold Spring Harbor, 1989.

[23] D. A. Melton, *Methods Enzymol.* **152,** 288 (1987).

Xenopus laevis Oocyte Expression

Several different heterologous expression systems have been used to study the transmembrane topology of gene fusions. Choice of the appropriate system depends on the protein being studied. It is important to ensure that the protein of interest expresses well, is processed correctly, and, if possible, is functional in the system of choice. Gene fusion studies of the topology of mammalian integral membrane proteins have typically used either cell-free translation in rabbit reticulocyte lysate supplemented with canine microsomal membranes or *Xenopus laevis* oocytes for expression. Although cell-free translation has proven adequate for the study of many proteins, it is possible that components needed for proper translocation and integration of membrane proteins may be missing in a cell-free lysate. *Xenopus* oocytes possess the capacity to faithfully translate and process nonamphibian protein with exceptional efficiency.[24,25] In addition, many Na^+/Cl^--dependent transporters, especially GAT-1, have been studied in oocytes and shown to have transport kinetics indistinguishable from those observed in mammalian cells.[26,27]

Ovarian follicles from *Xenopus laevis* (Xenopus; Ann Arbor, MI) are surgically removed[25] and placed in calcium-free oocyte medium (96 m*M* NaCl, 2 m*M* KCl, 5 m*M* MgCl, and 5 m*M* HEPES, pH 7.5). Follicles are dissected into small pieces and incubated in 2 mg/ml collagenase (Sigma, St. Louis, MO) in calcium-free oocyte medium for 0.5–2 hr at 25°. Each lot of collagenase varies in activity; therefore, the length of incubation must be determined for each lot of enzyme. Enzyme activity is quenched with calcium-supplemented (0.6 m*M*) oocyte medium by washing several times with excess volume. Oocytes are stored at 18° in calcium-supplemented oocyte medium until injection.

Approximately 25 μCi of Tran^{35}S-label (ICN, Costa Mesa, CA) (0.25 μl of 10X concentrated solution) is added to 1 μl of 250 ng/ml transcript and 50 nl of this solution injected per oocyte, 16–24 hr following collagenase treatment. After incubation at 18° for 4–6 hr, oocytes are pooled and homogenized on ice in homogenization buffer (0.2 *M* sucrose, 50 m*M* potassium acetate, 5 m*M* magnesium acetate, 1.0 m*M* dithiothreitol (DTT), 50 m*M* Tris, pH 7.5) using a miniature Teflon pestle in 1.5-ml microfuge tubes. Specifically, 9 oocytes are pooled per transporter–prolactin chimera and are homogenized in 3 μl of homogenization buffer per oocyte. Follow-

[24] T. P. Snutch, *Trends Neurosci.* **11,** 250 (1988).

[25] J. B. Gurdon and M. P. Wickens, *Methods Enzymol.* **101,** 370 (1983).

[26] J. Guastella, N. Nelson, H. Nelson, L. Czyzyk, S. Keynan, M. C. Miedel, N. Davidson, H. Lester, and B. Kanner, *Science* **249,** 1303 (1990).

[27] R. D. Blakely, M. B. Robinson, and S. G. Amara, *Proc. Natl. Acad. Sci. USA* **85,** 9846 (1988).

ing homogenization, $CaCl_2$ is added to 10 m*M* final concentration to stabilize the vesicles.

Protease Protection Assay

Protease protection assays are performed with methods derived from Skach and Lingappa[11] and Chavez and Hall.[15] The nonspecific protease proteinase K (Boehringer Mannheim; 0.2 mg/ml final) is added to aliquots of oocyte homogenates (10 μl) in the presence or absence of 1% Triton X-100 and incubated on ice for 1 hr. The addition of detergent is used to test the protease resistance of protected fragments. Inactivation of protease is a crucial step in this assay. If inactivation is not done efficiently, protease that is not inactivated will continue to digest fusion proteins when vesicles are disrupted. Inactivation of protease is efficiently accomplished by addition of 10 m*M* phenylmethylsulfonyl fluoride (PMSF) in dimethyl sulfoxide (DMSO), and immediate dilution with 10 volumes 1% SDS, 0.1 *M* Tris, pH 8.0, that has been preheated to 95–100°. The choice of DMSO as solvent for PMSF is not trivial since the use of 2-propanol disrupts membrane vesicles prior to inactivation of the protease. AEBSF [4-(2-aminoethyl)benzenesulfonyl fluoride] in DMSO (10 m*M*) is as effective as PMSF at inactivation of protease and may be substituted for PMSF. The diluted sample is then boiled for 5–10 min to complete inactivation and denature the transporter–prolactin chimeras for immunoprecipitation. Samples are diluted with 4 volumes of 1.25× lysis buffer (187.5 m*M* NaCl, 1.25% Triton X-100, 1.25% DOC, 62.5 m*M* Tris, pH 8.0, 2.5 m*M* EDTA, 0.125% SDS) and set at 4° with rocking ~12–16 hr. The concentrations and types of detergents in the lysis buffer needed to successfully solubilize transporter may vary for different proteins. The lysis buffer components used to immunoprecipitate GAT-1–prolactin chimeras were based on conditions used previously for immunoprecipitation of GAT-1 transporter using antibody against a partially purified transporter.[28] Following solubilization, samples are spun at 13,000*g* for 15 min in a refrigerated microcentrifuge (4°) to remove particulate matter and proteins are immunoprecipitated from the supernatant as described later.

Immunoprecipitation

Fusion proteins are immunoprecipitated with rabbit anti-ovine prolactin polyclonal serum (ICN Biomedicals, Inc.) at 1:500 for 4 hr with rocking at 4°. Protein A Affi-Gel (Bio-Rad, Richmond, CA) is prepared by washing three times in excess sodium phosphate buffer (50 m*M* sodium phosphate,

[28] S. Keynan, Y.-J. Suh, B. I. Kanner, and G. Rudnick, *Biochem.* **31,** 1974 (1992).

pH 7.5, 0.05% sodium azide, w/v) to remove free protein A. To pellet, protein A Affi-Gel is centrifuged at 10,000 rpm in a refrigerated microcentrifuge (4°) for 15 sec. Washed Affi-Gel is resuspended in a volume of sodium phosphate buffer equal to the volume of resin. Protein A Affi-Gel (50 μl) is added and samples are set at 4° with rocking for an additional 3 hr. Incubation with protein A Affi-Gel is followed by three washes with cold 1× lysis buffer to remove all proteins not precipitated by antibody and two washes with cold 0.1 *M* NaCl, 0.1 *M* Tris, pH 8.0, to remove detergents from the samples. Proteins are eluted from Protein A Affigel in 2× Laemmli buffer[29] (0.25 *M* Tris, pH 6.8, 4% SDS, 0.4% glycerol, 1.43 *M* 2-mercaptoethanol, 0.01% bromphenol blue) at 55° for 15 min. Alternatively proteins may be eluted from protein A Affi-Gel by addition of 100 m*M* glycine, pH 2.5, followed by addition of 2× Laemmli buffer and adjustment of sample pH. Boiling in Laemmli buffer is not recommended, as this can lead to aggregation of hydrophobic proteins[30] which may be problematic with the longer fusion proteins. Eluted samples are analyzed by sodium dodecyl sulfate–polyacrylamide gel electrophoresis (SDS–PAGE). In order to visualize the full range of fragments and the full-length fusion protein on the same gel it is recommended that samples be run on an appropriate gradient gel system. Both 4–20% and 8–16% gradient gels were used to run the GAT-1–prolactin chimeras and fragments. Gels are dried and either apposed to BioMax film (Kodak, Rochester, NY) for 4–6 days or apposed to a phosphorimaging plate (Fuji, Stamford, CT) for 2–4 days. Sizes of chimeras and fragments are estimated by determining their relative mobility.

Carbonate Extraction

Alkaline extraction, based on the methods of Fujiki *et al.*,[31] is used to determine the integration status of an integral membrane protein as it is assembled. Membrane vesicles exposed to alkaline pH are converted to open sheets of membranes. Soluble proteins and proteins peripherally associated with the membrane are recovered in the supernatant of carbonate treated vesicles. Proteins that are integrated into the membrane are pelleted. Oocyte homogenates (15 μl) are diluted 400-fold in either 0.1 *M* sodium carbonate, pH 11.5, or 0.1 *M* Tris, pH 7.5, and set on ice for 30 min. Membranes are pelleted by centrifugation at 230,000*g* for 30 min at 4°. Proteins in the supernatants were precipitated with 15% trichloroacetic acid (TCA) and pelleted in a microcentrifuge at 13,000 rpm for 10 min at

[29] U. K. Laemmli, *Nature* **227,** 680 (1970).

[30] R. J. Gregory, S. H. Cheng, D. P. Rich, J. Marshall, S. Paul, K. Hehir, L. Ostedgaard, K. W. Klinger, M. J. Welshe, and A. E. Smith, *Nature* **347,** 382 (1990).

[31] Y. Fujiki, A. L. Hubbard, S. Fowler, and P. B. Lazarow, *J. Cell Biol.* **93,** 97 (1982).

4°. Following a wash with ice-cold acetone and a second centrifugation, the pellets are air dried. TCA pellets and membrane pellets are dissolved in 1% SDS, 0.1 *M* Tris, pH 8.0, 100 μM AEBSF (ICN Biomedicals, Inc.) and boiled for 5–10 min. Following addition of 4 volumes of 1.25× lysis buffer, transporter–prolactin chimeras are precipitated as described above.

PNGase Treatment

Peptide: *N*-Glycosidase F (PNGase) treatment was used to determine the glycosylation state of transporter–prolactin chimeras. PNGase F cleaves between the innermost sugar and asparagine residues of high mannose, hybrid, and complex oligosaccharides on N-linked glycoproteins. Oocyte homogenates (10 μl) were denatured in 0.5% SDS, 1% 2-mercaptoethanol at 100° for 10 min. Following denaturation sodium phosphate, pH 7.5, and Nonidet P-40 (NP-40) were added to final concentrations of 50 m*M* and 1%, respectively. Samples were incubated without or with PNGase F (1000 U per sample) at 37° for 1 hr. Samples were solubilized in 1× lysis buffer and subsequently immunoprecipitated, as described earlier.

Discussion

Full-length prolactin was used in the study of GAT-1 transmembrane topology to determine the protease assay conditions and serve as an internal control for each experiment. Proteinase K treatment of homogenates from oocytes expressing full-length prolactin shows that the majority of membrane vesicles in the homogenate are intact as prolactin is fully protected from the protease (Fig. 3A). The intensity of the protected prolactin band is nearly identical to the intensity of the prolactin control band (Fig. 3A) indicating that the majority of vesicles in the *Xenopus* oocyte membrane homogenate are in the correct orientation. Prolactin is digested in the presence of proteinase K and a nonionic detergent (Fig. 3A) indicating that full-length prolactin is not protease insensitive. A protease resistant band of 14–15 kDa was detected for the full-length prolactin as well as each of the GAT-1–prolactin chimeras tested. This fragment corresponds to the predicted size of the prolactin tag and is not indicative of a particular membrane orientation.

Data obtained using transporter deletion and epitope tagging may be interpreted in two ways. First, data may be interpreted qualitatively. That is, the protection status of a transporter chimera is used to infer the membrane sidedness of the epitope. Detection of a protected fragment is interpreted to mean that the point at which the epitope is fused to the protein of interest is located in the lumen of the vesicles and therefore on the extracellular face

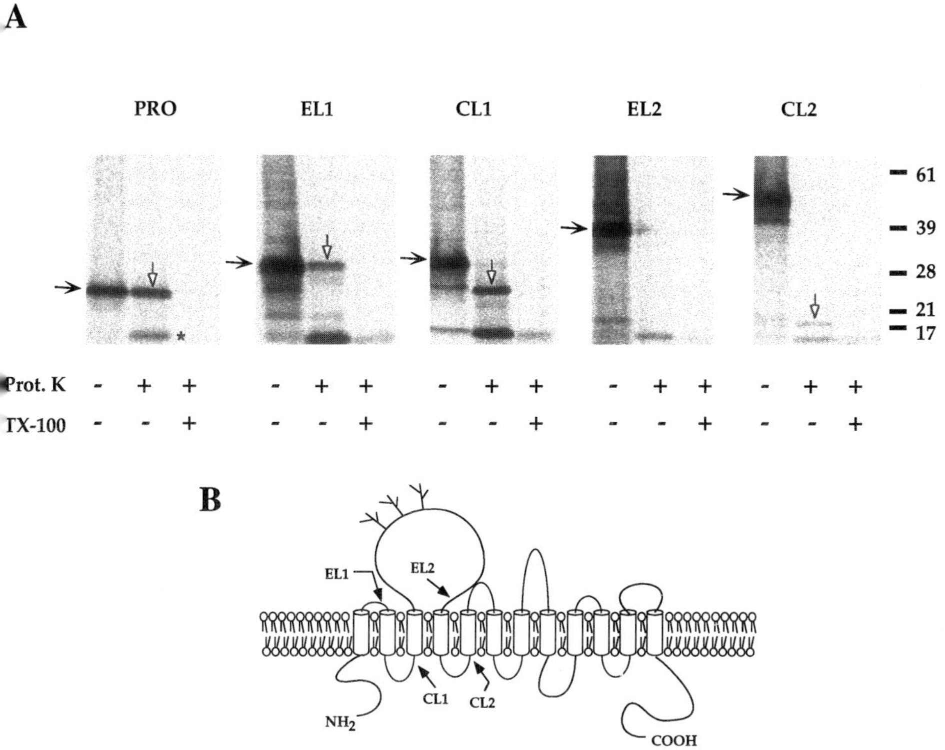

FIG. 3. Proteolysis and immunoprecipitation of GAT-1–prolactin chimeras EL1, CL1, EL2, and CL2. (A) Homogenates of *Xenopus* oocytes expressing chimeras were treated without protease, with proteinase K (Prot. K), or with protease and Triton X-100 (TX-100), as indicated. Exposure to protease was terminated after 60 min, digested chimeras were immunoprecipitated with prolactin antisera, and immunoprecipitated proteins were analyzed by SDS–PAGE using 4–20% acrylamide gels. Closed arrows indicate untreated fusion proteins, open arrows indicate protected fragments, and the asterisk denotes the prolactin epitope. Units for molecular mass standards are kilodaltons. Three to five independent experiments were performed with each construct yielding highly reproducible results. Full-length prolactin and EL1 are protected from protease digestion, but digested in the presence of protease and detergent. The prolactin epitope is protected from protease in chimeras CL1 and CL2, and susceptible to protease in chimera EL2 under nondetergent conditions. (B) Predicted transmembrane topology of GAT-1 based on hydropathy analysis. Sites of truncation and epitope fusion for EL1, CL1, EL2, and CL2 have been denoted. [Adapted with permission from J. A. Clark, *J. Biol. Chem.* **272,** 14695 (1997).]

of the plasma membrane. However, in extremely hydrophobic regions of a protein it would be difficult to distinguish between extracellular localization of the tag and its entrapment in the membrane. Conversely, inability to detect a protected fragment is interpreted to mean that the epitope is located in the cytoplasm and digested by protease. Figure 3B shows the predicted hydropathy model for GAT-1 with the sites of truncation for GAT-1–prolactin chimeras EL1, CL1, EL2, and CL2. These chimeras were designed to determine the topology of the first four hydrophobic domains of GAT-1. Protected fragments are detected following protease treatment of vesicles with EL1, CL1, or CL2 but no protected fragment was detected for EL2 (Fig. 3A). Although the cellular location of epitope for these chimeras was not expected based on the hydropathy model of GAT-1, it is believed that the hydrophobic domains (HD), once integrated into the membrane, do confirm the predicted topology (see membrane assembly discussion below). Second, data may be interpreted somewhat quantitatively, that is, the size of the fragment protected can yield information concerning the site of protease digestion. Since a protected fragment must retain the epitope, one can determine the general region of protease digestion by determining the molecular weight of the fragment, subtracting from that value the molecular weight of the epitope, and using the subtracted value to estimate the location of the site of protease digestion. Protease treatment of EL6 generated several protected fragments (Fig. 4A). Using the fragment size, without the epitope, it was determined that two of the digestion sites corresponded to the fourth and fifth intracellular loops (Fig. 4A and B). This is consistent with placement of these loops in the cytosol. Two additional sites of digestion were localized to the fourth extracellular loop (Fig. 4A and B). Although this data was not predicted with respect to the hydropathy model for GAT-1, it is consistent with data obtained for two other GAT-1–prolactin chimeras in this study.[1] Most of the GAT-1–prolactin chimeras show a single protected fragment following protease treatment (Fig. 3A). However, the longer constructs fused at sites in the fifth and sixth extracellular loops show several protected fragments (Fig. 4A). The reason for detection of multiple fragments is not understood.

Although the transmembrane topology of an integral membrane protein can be inferred from data derived from deletion analysis and epitope tagging, it should be remembered that these methods were originally developed to study topogenic properties and assembly of proteins. The transmembrane topology of proteins which assemble in a predictable manner, such as acetylcholine receptor subunits,[15] is readily determined using the methods described above. However, when the assembly of an integral membrane protein is complex, as it is for Na^+,K^+-APase,[9] H^+,K^+-ATPase,[18] and GAT-1,[1] data derived from such studies must be interpreted with caution.

Transmembrane assembly of GAT-1 appears to require the cooperative actions of several topogenic sequences. The presence of topogenic sequences throughout GAT-1 made it impossible to determine transmembrane topology based on data from individual chimeras. Therefore, it is

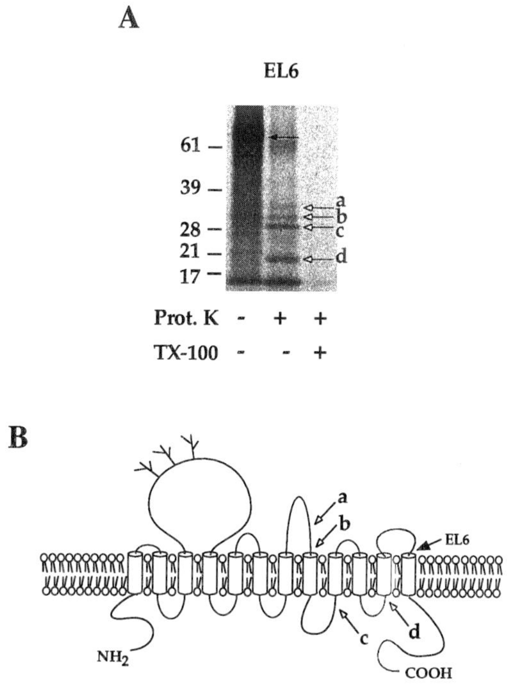

FIG. 4. Proteolysis and immunoprecipitation of GAT-1–prolactin chimera EL6. (A) Homogenates of *Xenopus* oocytes expressing EL6 were treated as described in the legend for Fig. 3 and samples were analyzed on an 8–16% acrylamide gel. Closed arrow indicates untreated fusion protein and open arrows indicate protected fragments. Three independent experiments were performed yielding similar results. The prolactin epitope is protected from protease in EL6 and digested in the presence of detergent. (B) Predicted hydropathy model for GAT-1 with site of truncation and epitope fusion for EL6. Approximate sites of protease cleavage for EL6 protected fragments shown in A) (a, b, c, and d) have been denoted with open arrows. [Adapted with permission from J. A. Clark, *J. Biol. Chem.* **272,** 14695 (1997).]

necessary to consider data from all of the constructs in order to develop a two-dimensional picture of this protein. Cooperative actions are necessary for assembly of the first four hydrophobic domains of GAT-1, as membrane integration for each of these hydrophobic domains did not occur until the next downstream hydrophobic domain was present. Protease protection data for EL1 (Fig. 3A) shows that EL1 has not been digested, suggesting that this chimera has been translocated into the ER lumen but not integrated into the membrane. Alkaline extraction of EL1-containing vesicles (Fig. 5A) shows a profile that resembles that for the secreted protein prolactin, in contrast to the profile for CL1 which is membrane integrated. The first HD is not integrated into the membrane in EL1 until HD2 is added, as demonstrated by digestion of the amino terminus upon protease treatment

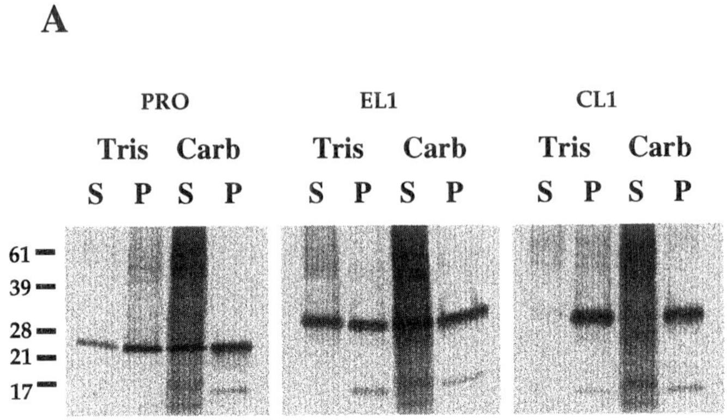

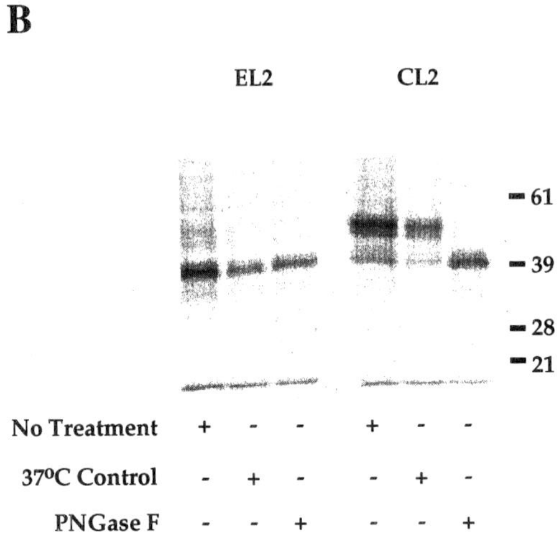

FIG. 5. Membrane integration status and glycosylation state of amino-terminal GAT-1–prolactin chimeras. (A) *Xenopus* oocyte homogenates expressing PRO, EL1, or CL1 were incubated either at pH 7.5 (Tris) or at pH 11.5 (Carb). Membranes were pelleted by centrifugation and proteins immunoprecipitated from both the supernatant (S) and the pellet (P) with the prolactin antisera as described. The distribution of EL1 protein resembles that for the secretory protein prolactin, whereas the chimera CL1 was found associated with the pellet only. (B) *Xenopus* oocyte homogenates expressing GAT-1–prolactin chimeras were incubated in the absence or presence of *N*-glycosidase F (PNGase F) for 60 min and proteins immunoprecipitated with the prolactin antisera as described. PNGase F treatment did not change the migration of EL2 as compared to untreated chimera. Mobility of CL2 was significantly increased following PNGase F treatment. [Adapted with permission from J. A. Clark, *J. Biol. Chem.* **272,** 14695 (1997).]

of CL1 (Fig. 3A). No protected fragments were detected following treatment of EL2 with protease (Fig. 3A), suggesting that the large extracellular loop connecting hydrophobic domains three and four is not extracellular. However, as revealed by the difference in the glycosylation status of EL2 and CL2 (Fig. 5B), HD3 does become integrated in the membrane when HD4 is added in CL2. Thus, the large loop connecting hydrophobic domains three and four must be extracellular and glycosylated in full-length GAT-1. Finally, while glycosylation of CL2 confirms the integration of HD3, the size of the protected fragment suggests that HD4 in CL2 is not yet integrated into the membrane. Therefore, data from GAT-1 prolactin chimeras EL1, CL1, EL2, and CL2 must be analyzed together to determine the transmembrane topology of the first three hydrophobic domains in the mature protein.

The transmembrane topologies of GAT-1 and many other integral membrane proteins,[1,11,15] have been determined using deletion and epitope tagging despite potential limitations to the methodology. A combination of methods may be used to confirm and define conclusions derived from study of transporter transmembrane topology using these methods, as was done for GAT-1.[1] Alternative methods for the study of transmembrane topology are presented elsewhere in this volume.[32,33] Convergent data from several methods will be important for confirming and extending each study of transporter topology.

[32] R. P. Seal, B. H. Leighton, and S. Amara, *Methods Enzymol.* **296,** [22], (1998) (this volume).
[33] R. Vaughn, *Methods Enzymol.* **296,** [15], (1998) (this volume).

[21] Selective Labeling of Neurotransmitter Transporters at the Cell Surface

By GWYNN M. DANIELS and SUSAN G. AMARA

Introduction

Neurotransmitter transporters, like all complex integral membrane proteins, undergo an extensive maturation process before becoming fully functional. During this maturation process, transporter molecules are found on the surface of the endoplasmic reticulum (ER), Golgi apparatus,

0076-6879/98 $25.00

and associated vesicles, as well as on the plasma membrane. Although the transporter protein half-life appears to be quite long (estimated at up to 6 days), under normal circumstances a significant amount of carrier protein will be found in subcellular compartments, rather than at the cell surface. This is especially true in transient transfection systems, where gene expression is maximized, and high concentrations of transporter protein can overwhelm the cellular maturation machinery. In this case, large amounts of immature protein may remain in the ER and Golgi, rather than being delivered to the plasma membrane. Therefore, selective labeling of only those molecules found at the surface of the cell may be desirable when examining a number of aspects related to transporter function.

Cell surface biotinylation serves as an important tool in several experimental approaches used to study neurotransmitter transport proteins, by identifying those molecules present only at the cell surface. This method is particularly useful for experiments that clarify changes in the functional properties of transporters, such as in the analysis of mechanisms of regulation of transport activity, or in the interpretation of results obtained from mutagenesis. Differences in the transport rate observed in these studies can have several plausible explanations, including alterations in biosynthesis, stability of the protein, catalytic activity, or the surface expression of the carrier. In structure–function studies, mutations become much more informative if it can be established that the carriers are efficiently targeted to and maintained at the cell surface, where they have the potential to be functional. Similarly, in experiments examining the modulation of transporter activity, the technique of cell-surface labeling offers a means to help distinguish whether changes in transport activity are due to recruitment or removal of carriers from the cell surface, or whether they reflect actual changes in the catalytic activity and/or substrate affinity of the transporter itself. Finally, the approach also offers the possibility of considering some of the biochemical properties of the transporters, such as the nature and extent of glycosylation or other posttranslational modifications, that are present on the plasma membrane.

The protocol presented below describes the use of membrane impermeant modified biotin molecules for the selective labeling of only those proteins present at the plasma membrane. With this approach, cell surface proteins are biotinylated with a hydrophilic, membrane-impermeant biotin derivative and affinity purified using an avidin resin; the carrier proteins are then identified using specific antibodies on immunoblots of biotinylated proteins.

Cell Surface Biotinylation

Introduction

Cell surface biotinylation is a powerful tool that is used to selectively label proteins found on the plasma membrane. It utilizes the high affinity of avidin for biotin ($K_d \sim 10^{-15}$) to separate integral membrane proteins from those residing in the cytoplasm or on the membranes of subcellular organelles. A protocol developed by Sargiacomo *et al.*[1] has been used extensively to examine the differential distribution of membrane proteins in polarized epithelial cells. Adaptations of this method were used to determine the localization of neurotransmitter transporters in MDCK cells,[2-4] as well as to assess the efficiency of delivery of the carriers to the cell surface.[5]

This technique utilizes a number of modifications of biotin, all of which are linked to an *N*-hydroxysuccinimide (NHS) ester. This highly reactive ester group forms stable amide bonds via nucleophilic attack of free amines. In intact cells, these membrane-impermeant molecules react with unblocked NH_2-terminal amino acid residues, or more commonly with the epsilon amine groups of lysine residues exposed at the cell surface. This procedure is therefore useful for studying membrane proteins, such as the neurotransmitter transporters, which are predicted to have extracellular loops containing multiple lysine residues. Sulfonation of the NHS group, which is soluble only in organic solvents, allows the biotinylation reagent to be dissolved in aqueous solutions. The addition of a long spacer arm between the NHS and biotin moieties increases the efficiency of avidin binding by reducing steric hindrance, resulting in greater access of the biotin to the biotin binding site on the avidin molecule. The sulfosuccinimidyl-2-(biotinamido)ethyl-1,3-dithiopropionate (NHS-SS-biotin), used in the protocol described below, is also synthesized with a cleavable disulfide bond within the spacer arm. The biotin–avidin complex is readily released from proteins labeled with this reagent using concentrations of reducing agents normally associated with protein sample buffers.

[1] M. Sargiacomo, M. Lisanti, L. Graeve, A. Le Bivic, and E. Rodriguez-Boulan, *J. Membr. Biol.* **107,** 277 (1989).

[2] G. Pietrini, Y. J. Suh, L. Edelmann, G. Rudnick, and M. J. Caplan, *J. Biol. Chem.* **269,** 4668 (1994).

[3] J. Ahn, O. Mundigl, T. R. Muth, G. Rudnick, and M. J. Caplan, *J. Biol. Chem.* **271,** 6917 (1996).

[4] H. H. Gu, J. Ahn, M. J. Caplan, R. D. Blakely, A. I. Levey, and G. Rudnick, *J. Biol. Chem.* **271,** 18100 (1996).

[5] T. T. Nguyen and S. G. Amara, *J. Neurochem.* **67,** 645 (1996).

Cell Culture

Cell surface biotinylation has been used successfully to label plasma membrane proteins in a variety of cell types. They include polarized epithelial cells and cells grown in suspension, as well as many types of adherent cells. Cells are generally grown under standard culture conditions for the cell type being used. A variety of tissue culture dishes are suitable for this purpose, although it is recommended that they not be polylysine coated, as the free amine groups of the lysine will compete with the membrane proteins in the biotinylation reaction. The size of the plates or wells used is determined by the expression level of the protein of interest and the sensitivity of the detection assay. Therefore, the optimum surface area will need to be assessed for each protein examined.

Polarizing epithelial cells are used specifically to examine the targeting of plasma membrane proteins to the apical or basolateral surface of the cell (for a discussion of this, see Ref. 6). Madin–Darby canine kidney (MDCK), LLC-PK_1 (porcine kidney), and Caco-2 (human colon adenocarcinoma) cells are the cell lines most commonly used. Stably transfected cells are plated at high density on polycarbonate filter supports (Costar Transwell, Cambridge, MA) and grown for several days until they form a confluent monolayer. Cells grown in this manner will develop tight junctions, which separate the plasma membrane into distinct apical and basolateral domains. The tight junctions also create an effective barrier between the medium in the upper (apical) chamber of the well and that in the lower (basolateral) chamber, allowing the membrane proteins to be labeled exclusively on one surface or the other. The integrity of the tight junctions in the filter-grown cells is tested by one of two methods. Leakage across the monolayer is assessed by adding [^{3}H]inulin or [^{3}H]ouabain to the apical chamber. Following a 2-hr incubation, an aliquot of the medium from the lower chamber is counted by liquid scintillation spectrometry. Leakage across the monolayer is expressed as the percent of the total counts added to the apical chamber that leaked into the basolateral chamber. If leakage across the monolayer is greater than 1% of the total counts, the monolayer is not considered tight, and the cells should be discarded. Alternatively, the transepithelial electrical resistance can be measured using a Millicell-ERS apparatus (Millipore, Bedford, MA). The electrical resistance, which will vary from cell type to cell type, is measured in ohms (Ω) per cm^2 and increases dramatically when the cells form a tight monolayer.

Nonpolarizing cell types should be plated at moderate density and given adequate time in culture to recover from trypsinization and express the

[6] K. Matter and I. Mellman, *Curr. Op. Cell Biol.* **6,** 545 (1994).

membrane protein of interest. The nonpolarized delivery of membrane proteins to the cell surface of epithelial cells can also be assessed in cells that are grown on standard tissue culture plates rather than on filter supports.

Biotinylation Reaction

Reagents

PBS^{++}: Phosphate-buffered saline containing 0.1 m*M* $CaCl_2$ and 1 m*M* $MgCl_2$

NHS-SS-Biotin (Pierce, Rockford, IL)

Biotinylation Buffer: 2 m*M* $CaCl_2$, 150 m*M* NaCl, 10 m*M* triethanolamine, pH 7.5–9.0

Quench Buffer: 100 m*M* glycine in PBS^{++}

Procedure. All steps are carried out at 0–4° to reduce internalization of the biotin label. Growth medium is aspirated, and the cells washed three times for 10 min with ice-cold PBS^{++} to remove extraneous amines. Freshly prepared biotin solution (0.5–2.0 mg/ml) is added in a quantity sufficient to cover the cell layer. When labeling polarized epithelia, the surface of the cells not receiving NHS-SS-biotin is incubated in biotinylation buffer alone. Incubation in biotin solution is carried out for 20–25 min at 4° with gentle agitation. As hydrolysis of the NHS ester in aqueous solutions is the major competing reaction, repeating the incubation with fresh biotin solution may improve incorporation of the label.

Following the biotinylation reaction, unreacted NHS-SS-biotin is quenched by the addition of 100 m*M* glycine. The cells are rinsed twice with ice-cold quench buffer, followed by incubation in additional quench buffer at 4° for 20 min with gentle agitation. After quenching the cells are rinsed twice with ice-cold PBS^{++}.

Comments. The polarized distribution of membrane proteins can only be assessed accurately if the biotinylation reagent is confined to either the apical or the basolateral surface of the filter-grown epithelial cells. Therefore, when working with polarized epithelial cells, particular care should be taken not to disrupt the integrity of the monolayer during the aspiration and pipetting steps. In addition, buffers containing calcium are required to maintain epithelial tight junctions which disintegrate rapidly in the absence of free calcium. However, the presence of calcium is not necessary when working with nonpolarized cells.

As any free amino group will compete with surface proteins in the amide bond formation, it is important to avoid buffers containing free amines (e.g., Tris, glycine, azide). Interestingly, the replacement of phosphate buffer with 10 m*M* triethanolamine has been shown to increase the efficiency of NHS-biotin incorporation.[7] Since *N*-hydroxysuccinimide reacts preferentially with amines in their unprotonated state, the efficiency of the biotinylation reaction may be improved by increasing the pH of the biotinylation buffer. Therefore, it is advisable to compare the results of experiments done over a range of pH values, from neutral to basic, to determine the pH for optimal labeling of a particular protein.

Both time and concentration dependence of the biotinylation reaction must also be experimentally determined. Concentrations of NHS-SS-biotin ranging from 0.5 to 2.0 mg/ml are standard. At higher concentrations it may be necessary to first dissolve the NHS-SS-biotin in room-temperature buffer and then immediately transfer the solution to ice.

Extensive washing with quench buffer ensures that any remaining NHS-SS-biotin is fully reacted before cell lysis, to avoid the possibility of labeling intracellular proteins.

Nonspecific binding of unlabeled proteins in the subsequent recovery procedure can be accounted for by including a control for each reaction condition in which the biotinylation buffer does not contain biotin.

Cell Lysis

Reagents

Lysis Buffer: 1% Triton X-100, 150 m*M* NaCl, 5 m*M* EDTA, 50 m*M* Tris, pH 7.5

Procedure. Cell monolayers grown on permeable filters are carefully excised from the support cups with a sharp scalpel. The cells are then scraped from the wells or filters into 1 ml lysis buffer. The cell lysate is triturated gently and transferred to a 1.6-ml microfuge tube. The membrane proteins are extracted on ice for 1 hr with occasional brief vortexing.

Following detergent extraction, the lysates are cleared by centrifuging at 14,000*g* for 10 min at 4°. Then, 900 μl of the supernatant is transferred to a new microfuge tube. Care is taken to avoid transferring any of the pelleted material.

Comments. In addition to forming polarized monolayers on filter membranes, epithelial cells grow in a monolayer on the side of the filter cup, often extending as high as the meniscus of the apical medium.[8] Because they are not growing on the filter, these cells do not form the strong tight junctions which exclude the biotinylation reagent from the basolateral surface when it is applied to the apical medium. Likewise, biotinylation solution which is applied to the basolateral surface does not reach this

[7] C. J. Gottardi and M. J. Caplan, *Science* **260,** 552 (1993).

[8] C. J. Gottardi, L. A. Dunbar, and M. J. Caplan, *Am. J. Physiol.* **268,** F285 (1995).

population of cells. Inclusion of these nonpolarized cells in the data analysis can lead to misinterpretation of the results; therefore, it is important to carefully separate them from the cells growing directly on the filter.

When comparing the surface distribution of proteins expressed in the same cell line under different conditions, it can be assumed that the number of cells per tissue culture dish will remain constant when the cells are plated at the same density. However, when comparing surface expression across cell lines, it will be impossible to maintain a consistent cell number. Therefore, it is necessary to normalize the data by assessing the total protein concentration. An aliquot of the cleared lysate should be assayed for protein concentration by BCA (Pierce), or another acceptable method, and cleared lysate containing an equivalent amount of protein carried through to the next step.

Recovery of Biotinylated Proteins

Reagents

UltraLink Immobilized NeutrAvidin (Pierce)
Lysis Buffer: 1% Triton X-100, 150 m*M* NaCl, 5 m*M* EDTA, 50 m*M* Tris, pH 7.5
High-Salt Wash Buffer: 0.1% Triton X-100, 500 m*M* NaCl, 5 m*M* EDTA, 50 m*M* Tris, pH 7.5
No-Salt Wash Buffer: 50 m*M* Tris, pH 7.5

Procedure. Biotin-labeled proteins are separated from unlabeled proteins by adding 50–200 μl NeutrAvidin resin (50% slurry) to the cleared lysate. The biotinylated proteins are allowed to bind to the NeutrAvidin resin by incubating at 4° with end-over-end mixing for 1–16 hr. The NeutrAvidin resin is pelleted by centrifuging at 5000*g* for 15 min at 4°. The supernatant is transferred to another microfuge tube and saved for further analysis. The pellet is then washed three times by vortexing briefly in 1 ml lysis buffer, after which the NeutrAvidin resin is repelleted by centrifuging at 5000*g* for 2 min at 4°. The pellet is subsequently washed twice in 1 ml high-salt wash buffer, and once in 1 ml no-salt wash buffer. After the no-salt wash buffer has been aspirated, the pellets are again centrifuged at 5000*g* for 2 min at 4° and as much residual buffer as possible is removed.

Comments. The NeutrAvidin pellet is quite loose, so care must be taken when aspirating to avoid losing any of the pelleted resin. It is better to leave behind a portion of the buffer than to risk losing part of the sample.

The volume of NeutrAvidin slurry used must be in excess of the number of binding sites present on the biotinylated proteins in the cleared lysate.

Failure to include sufficient NeutrAvidin resin will lead to incomplete biotin binding and a subsequent reduction in protein recovery.

The biotin–avidin binding reaction is quite rapid, and binding of the labeled proteins to the NeutrAvidin beads should be complete within 1 hr. However the binding step may proceed overnight to ensure that binding is complete.

Recovery of biotinylated proteins with UltraLink immobilized NeutrAvidin has a number of distinct advantages over the use of native avidin, or even streptavidin. Nonspecific protein binding is reduced significantly, improving the signal-to-noise ratio. The modification of charged amino acid residues found at the surface of the avidin molecule provides a more neutral isoelectric point, and the binding capacity has been increased more than 2-fold over that of NeutrAvidin-agarose by coupling the NeutrAvidin moiety to 3M Emphaze biosupport medium AB1.

The use of both high and low ionic strength washes should ensure that all nonbiotinylated proteins are removed from the NeutrAvidin resin, resulting in a preparation free from contaminating proteins.

Alternatively, biotin-labeled surface proteins can be immunoprecipitated with a specific antibody, separated by SDS–PAGE, transferred to a membrane support, and visualized using an antibiotin antibody. This method has the advantage of reducing contaminating proteins which may be nonspecifically labeled on a Western blot, but requires either the use of a nonhydrolyzable NHS-biotin conjugate or nonreducing conditions.

A two-step recovery method can also be utilized. With this technique the protein of interest is first purified by immunoprecipitation using a specific antiserum, and the biotinylated portion of the immunoprecipitated protein recovered by binding to the NeutrAvidin resin as described.

Electrophoresis and Immunoblotting

Reagents

- 2× SDS Sample Buffer containing 143 m*M* 2-mercaptoethanol
- Immobilon-P transfer membrane (Millipore)
- Blocking Buffer: 3% (w/v) bovine serum albumin (BSA), 2% (w/v) nonfat dry milk in PBS^{++}
- Renaissance Western blot chemiluminescence reagent (NEN, Boston, MA)

Procedure. The biotinylated proteins are released from the pelleted NeutrAvidin resin by the addition of 40 μl of 2× SDS sample buffer. An aliquot of 20 μl of the supernatant (unbound fraction) from the NeutrAvidin pellet (see Recovery of Biotinylated Proteins) is added to 20 μl of 2× SDS sample buffer. After being vortexed briefly, the proteins are incubated at room temperature for 10 min, followed by incubation at 37° for 30 min. The samples are separated by electrophoresis through an 8% SDS–polyacrylamide gel. Separated proteins are subsequently electroblotted to Immobilon-P membrane. After thorough drying, the membrane is wetted in methanol and rinsed well with distilled water in preparation for Western blotting.

The rehydrated membrane is incubated at room temperature in blocking buffer for 1 hr. Primary antiserum is added to the blocking buffer at dilutions ranging from 1:500 to 1:2000 (v/v). The blots are incubated in diluted primary antibody at room temperature for 1–3 hr. Following incubation in primary antibody, the buffer is aspirated and the blots are washed twice for 10 min in blocking buffer. The blots are then incubated for 1 hr at room temperature in horseradish peroxidase-conjugated secondary antibody diluted according to the manufacturer's instructions in blocking buffer. After incubation in secondary antibody the blots are washed extensively in blocking buffer diluted 1:10 in phosphate-buffered saline (PBS), and the bands are visualized using chemiluminescence reagent.

Comments. We have found that heating transporter proteins to 100° in SDS sample buffer tends to promote aggregation. Therefore, in order to retain the majority of the protein in the monomeric form, labeled proteins are heated to no more than 37° prior to electrophoresis.

An aliquot of the unbound fraction from the NeutrAvidin resin may be included in the electrophoresis and immunoblotting steps as a convenient control for internalization of the biotinylation reagent as well as to assess the efficiency of incorporation of the biotin label. Depending on the sensitivity of the antibody used to probe the Western blot, it may be necessary to concentrate the sample on a Centricon (Amicon, Beverly, MA) column or by acetone precipitation in order to visualize the protein bands. If the membrane glycoproteins were efficiently labeled by the NHS-SS-biotin, the glycosylated form of the protein of interest should be effectively depleted from the unbound fraction. If the biotinylation reagent was not internalized, the immature unglycosylated membrane protein will be absent from the fraction bound to the NeutrAvidin resin and will be found only in the unbound fraction.

The blocking buffer is used to coat the Immobilon-P membrane with proteins which will not be recognized by the antibodies. A number of protein sources used at a variety of concentrations can be used for this purpose. The blocking buffer described above is just one example of a variety of buffers that have been used successfully in the immunoblotting of neurotransmitter transporters.

Discussion

Cell surface biotinylation has been useful in assessing the delivery of the norepinephrine transporter and its glycosylation mutants to the plasma membrane in both transiently and stably transfected cells.[5] Biotinylated transporters are strongly labeled in the biotin treated lanes, whereas no transporter protein is visible in cells which have been treated in an identical manner, but without the addition of NHS-SS-biotin (Fig. 1, A and D). In contrast, nonbiotinylated proteins (Fig. 1, B)[5] show roughly equivalent labeling in both the treated and untreated cells. The majority of the norepi-

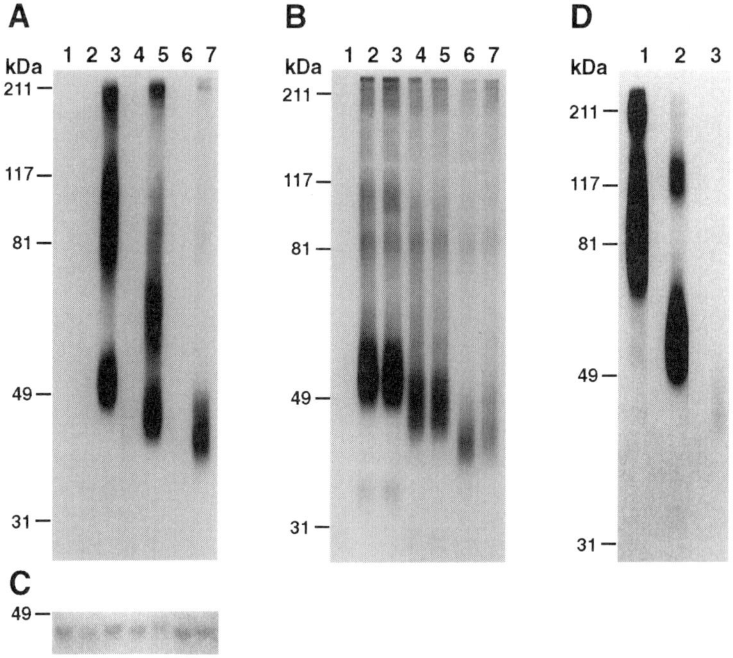

FIG. 1. Cell surface expression of the norepinephrine transporter (NET) and its glycosylation mutant proteins (NQQ and QQQ) in transiently transfected HeLa and stably transfected MDCK cells (A, B, and C) Control pBSK- (lane 1), NET (lanes 2 and 3), NQQ (lanes 4 and 5), or QQQ (lanes 6 and 7) transfected HeLa cells incubated in the presence (lanes 1, 3, 5, and 7) or absence (lanes 2, 4, and 6) of NHS-SS-biotin. (D) MDCK cells stably expressing NET (lane 1), NQQ (lane 2), or QQQ (lane 3). Biotinylated membrane proteins were isolated from 200 μg total protein with NeutrAvidin resin. Protein bound to the resin (A, C, and D) and 20 μg unbound protein (B) were separated by SDS–PAGE and transferred to membrane support. Western blots (A, B, and D) were probed with affinity purified α-NET antibody, followed by incubation with horseradish peroxidase-conjugated secondary antibody, and visualized by enhanced chemiluminescence. The blot from (A) was stripped and reprobed with α-actin monoclonal antibody (C). [Reproduced with permission from T. T. Nguyen and S. G. Amara, *J. Neurochem.* **67,** 645 (1996).]

nephrine transporter protein in the unbiotinylated fraction is found in the unglycosylated form, suggesting an intracellular location, whereas the fully glycosylated mature form of the protein is enriched at the cell surface.

Transfection systems can vary widely in the amount of protein produced, which can in turn result in marked differences in the interpretation of the data. Transient transfection systems, in general, produce large amounts of protein over a short period of time, while stable transfections yield lower protein levels at a steady state. Therefore, stable cell lines reflect expression levels more comparable to those found in the native state, and may provide a more accurate picture of what is happening *in situ*. The vaccinia virus: T7 expression system, in which cells are transfected with a gene under the control of the T7 promoter and infected with a recombinant vaccinia virus expressing the T7 polymerase, results in particularly high protein levels.[9] When the vaccinia virus: T7 expression system is used, large amounts of unglycosylated protein reach the cell surface (Fig. 1 A, lower band lanes 3, 5, and 7). Presumably this is because the secretory pathway of these cells is overwhelmed with transporter protein, allowing unglycosylated carriers, which are not normally delivered to the cell surface, to be sent to the plasma membrane. In contrast, cells which stably express the transporter have only the glycosylated form at the cell surface. The unbound fraction from NeutrAvidin isolation of surface proteins from these cells shows transporter primarily in the unglycosylated form (data not shown), further demonstrating that only the glycosylated form is directed to the cell surface.

To rule out the possibility that intracellular proteins were being biotinylated nonspecifically during the course of the biotinylation reaction, the blot from panel A was stripped and reprobed with an antibody directed against actin, a highly abundant cytoplasmic protein (Fig. 1, C). Only negligible amounts of actin were detected in the proteins bound to the NeutrAvidin beads, and the amount of actin observed was equivalent whether or not the cells were exposed to the biotinylation reagent. This result supports the idea that the actin detected reflects the nonspecific adsorption of this protein to the NeutrAvidin resin, and argues against the possibility that the biotinylation reagent gained access to intracellular proteins through cell lysis or leakage during the biotinylation reaction.

As the results of cell surface biotinylation experiments are qualitative, it is necessary to consider the efficiency of the labeling and recovery reactions when utilizing this procedure. Many conditions, such as the pH of the labeling buffer and the number of biotin binding sites available during NeutrAvidin recovery, can affect the outcome of the experiment. Incomplete labeling, nonsaturating NeutrAvidin binding, and loss of sample during wash steps can all result in low signal and misleading results. In addition,

[9] S. L. Povlock and S. G. Amara, *Methods Enzymol.* **296,** [29], 1998 (this volume).

conditions which allow access of the biotin label to intracellular proteins can produce results which may be misinterpreted. However, cell surface biotinylation, when used judiciously, can provide important information on the localization of transporter protein.

[22] Transmembrane Topology Mapping using Biotin-Containing Sulfhydryl Reagents

By REBECCA P. SEAL, BARBARA H. LEIGHTON, and SUSAN G. AMARA

Introduction

The highest resolution image of protein structure is currently provided by the powerful technique of X-ray crystallography. Although this technique has been extremely successful when applied to a variety of soluble proteins, it has been less successful for integral membrane proteins, because well-ordered, three-dimensional crystals of proteins that normally reside within lipid environments have been tremendously difficult to grow.[1] Advances in the methodologies for membrane protein crystallization and for generating high resolution diffraction patterns, however, indicate that X-ray crystallography will soon become a viable tool for producing highly detailed structural information of these proteins, although further refinements are still needed.[2]

In light of these difficulties, insight into the structure of membrane proteins has come through the development of a number of lower resolution approaches designed to reveal the topological arrangement of membrane spanning segments in protein subunits. The strategies are quite varied and include the detection of glycosylation sites inserted into hydrophilic domains.[3–5] determination of the orientation of a reporter domain linked to a series of C-terminal truncations using a protease protection assay,[6–8] the development of site-directed antibodies for use in protease protection

[1] R. M. Garavito, D. Picot, and P. J. Loll, *J. Bioenerg. Biomembr.* **28**(1), 13 (1996).

[2] E. Pebay-Peyroula, G. Rummel, J. P. Rosenbusch, and E. M. Landau, *Science* **277,** 1676 (1997).

[3] R. A. Chavez and Z. W. Hall, *J. Biol. Chem.* **266,** 15532 (1991).

[4] J.-T. Zang and V. Ling, *J. Biol. Chem.* **266,** 18224 (1991).

[5] K. Bamberg and G. Sachs, *J. Biol. Chem.* **269,** 16909 (1994).

[6] M. L. Jennings, *Annu. Rev. Biochem.* **58,** 999 (1989).

[7] R. A. Chavez and Z. W. Hall, *J. Cell. Biol.* **116,** 385 (1992).

[8] J. A. Clark, *Methods Enzymol.* **296,** [20], 1998 (this volume).

0076-6879/98 $25.00

and indirect immunofluorescence assays,[9–12] and the chemical modification of specific amino acid residues using membrane-permeant and -impermeant reagents.[13,14] Because each strategy has its own inherent strengths and limitations, a topological model that is based on the results of several approaches is likely to be much more informative. Many of the strategies used to examine transmembrane topology require modification of the protein being studied, which often results in changes in the functional properties of the protein. Interpretations that can be made regarding the topology are limited in this case, because perturbations in the structure of the protein may underlie the changes observed in the functional properties. Although valuable structural information has been obtained from such studies, greater confidence is achieved when protein function is maintained. Neurotransmitter transporters, like many multispanning membrane proteins, contain domains that are highly conserved and are not tolerant of primary sequence alterations, such as the insertion of a reporter epitope or the mutagenesis of several consecutive amino acid residues. These regions do, however, appear to allow minor modifications, such as the substitution of single amino acid residues. Because substitution with cysteine appears to be very well tolerated, one attractive approach to evaluate the topology of these proteins is to determine the membrane orientation of individually placed cysteine residues using sulfhydryl-reactive reagents.[13] In this way, topology can be assessed under conditions in which protein function, and therefore presumably protein structure, remains unaltered.

In this chapter, we describe two methods that are used to evaluate the intracellular or extracellular orientation of single cysteine residues within membrane proteins. These methods, which are variations on the same approach, utilize compounds that react specifically with the sulfhydryl group of cysteine. Both approaches require two types of sulfhydryl-modifying reagents: One must be membrane impermeant, and the other must be membrane permeant and have a covalently linked biotin moiety. Maleimide compounds, which form thioether bonds with cysteine residues, are used in the first method described in this chapter; the second method is based on the use of methanethiosulfonate (MTS) derivatives, which form mixed disulfide bonds with cysteine residues. A series of single cysteine mutants are created by substituting individual amino acid residues that are thought

[9] S. W. Bahouth, H. Y. Wang, and C. C. Malbon, *Trends Pharmacol. Sci.* **12,** 338 (1991).
[10] J. Borjigin and J. Nathans, *J. Biol. Chem.* **269,** 14715 (1994).
[11] B. M. Conti-Fine, S. Lei, and K. E. Mclane, *Annu. Rev. Biophys. Biomol. Struct.* **25,** 197 (1996).
[12] T. M. Shih and A. L. Goldin, *J. Cell. Biol.* **136,** 1037 (1997).
[13] T. W. Loo and M. C. Clark, *J. Biol. Chem.* **270,** 843 (1995).
[14] T. Kimura, M. Ohnuma, T. Sawai, and A. Yamaguchi, *J. Cell. Biol.* **272,** 580 (1997).

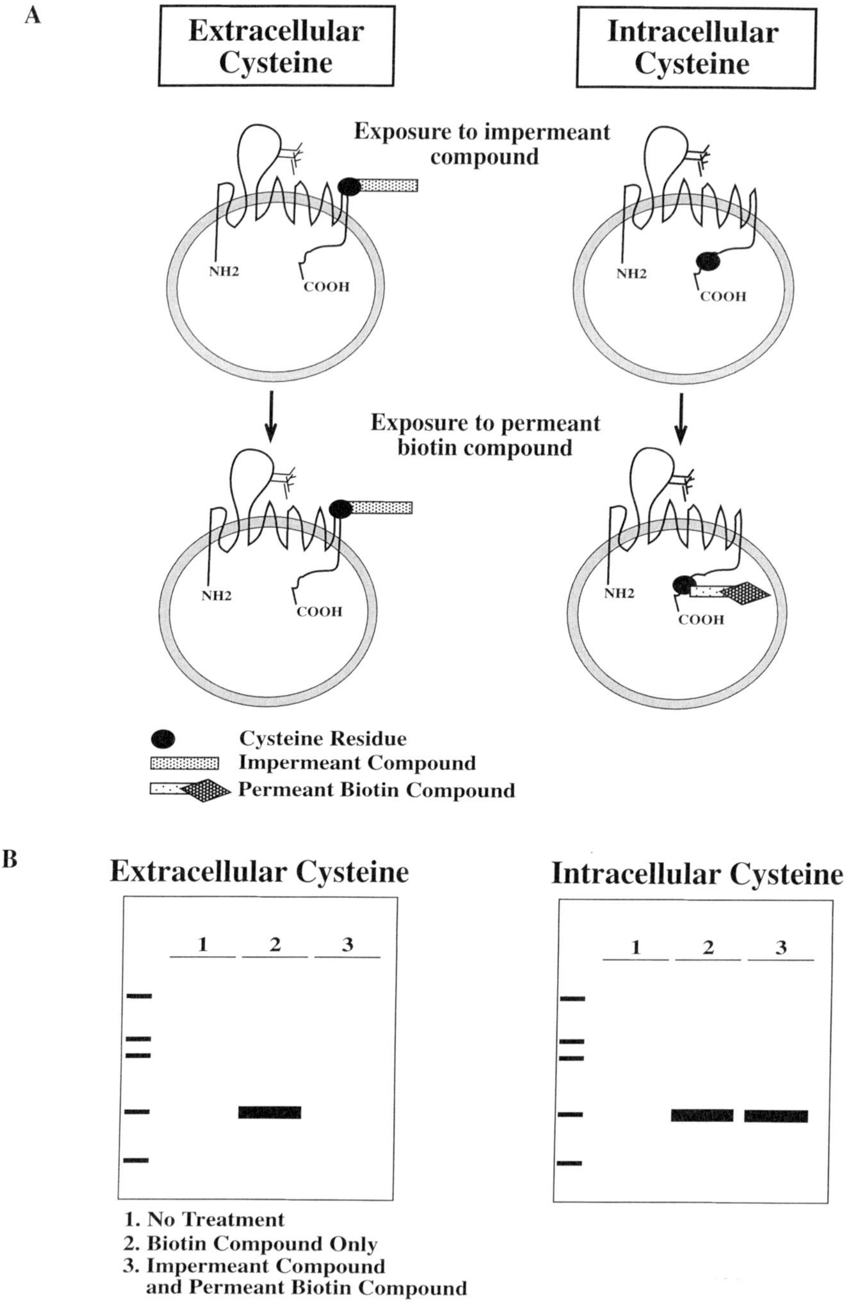
A
Extracellular Cysteine
Intracellular Cysteine
Exposure to impermeant compound
NH2
COOH
NH2
COOH
Exposure to permeant biotin compound
NH2
COOH
NH2
COOH
Cysteine Residue
Impermeant Compound
Permeant Biotin Compound
B
Extracellular Cysteine
1
2
3
Intracellular Cysteine
1
2
3
1. No Treatment
2. Biotin Compound Only
3. Impermeant Compound
and Permeant Biotin Compound

to reside in the putative intracellular or extracellular loops of the protein with cysteine. The important consideration for selecting a particular residue for cysteine substitution is whether it is likely to reside in an aqueous environment as might be expected for an extracellular or cytoplasmic loop. Although the process of choosing residues is often empirical, the rationale for deciding which residues to alter can be based on any of several factors, including predicted secondary structures, hydropathy analyses, the presence of charged residues, and the likelihood that a particular substitution will be well tolerated. The mutant proteins are expressed in an exogenous cell culture system, and only those exhibiting normal functional properties are evaluated. Intact cultured cells expressing a single-cysteine mutant are first incubated with the membrane-impermeant reagent, followed by incubation with the membrane-permeant, biotin-containing reagent (Fig. 1A). If the cysteine residue resides on the cytoplasmic face of the membrane, then only the membrane-permeant, biotin-containing reagent will gain access to and react with the residue (Fig. 1B). If, on the other hand, the cysteine residue resides on the extracellular surface of the cell, then the membrane-impermeant reagent will react first, blocking the subsequent reaction of the biotin-containing reagent, and the residue will not be detected (Fig. 1B). As a control, cells expressing the single cysteine mutant are incubated without the biotin-containing reagent, and in this case, the residue should not be detected. In another important control, cells are incubated with the biotin-containing reagent alone (Fig. 1B). If the residue is not detected, then it can be assumed that the biotinylated reagent was unable to react with it and the position of the cysteine in relation to the membrane cannot be evaluated by this method.

To begin these studies, a version of the protein should be created that has had as many of the endogenous cysteine residues as possible replaced, with an effort to minimize the impact of these cysteine replacements on

FIG. 1. Strategy for revealing the membrane orientation of individually placed cysteine residues. (A) Cells that express a single-cysteine mutant are incubated with the membrane-impermeant sulfhydryl-reactive reagent, and then excess reagent is washed away. The cells are then exposed to the membrane-permeant biotin-containing sulfhydryl reagent. (B) Illustration of Western blots showing the hypothetical results for both an extracellular and an intracellular cysteine residue. Lane 1 illustrates an important experimental control. Protein should not be detected from cells that are not exposed to the biotin-containing reagent. Lane 2 illustrates another critical control, that when cells are exposed to only the reagent containing the biotin group, the protein should be detected. Lane 3 illustrates the results obtained when the cells are incubated as described in (A). The cysteine residue is not detected if it is located extracellularly, but is detected when located intracellularly.

the overall function of the protein. Ideally, the protein should be devoid of endogenous cysteine residues, because these residues may also react with the sulfhydryl-modifying reagents or they may form disulfide bonds with the engineered cysteines and prevent their reaction with the reagents. Although it might be possible to apply these approaches to a protein with endogenous cysteine residues, provided that they remain unreactive, the possibility that these additional reactions might occur would be an important consideration in interpreting the results.

Method I: Maleimide Compounds

Introduction

Membrane-permeant biocytin maleimide and membrane-impermeant stilbene disulfonate maleimide are the two sulfhydryl-specific reagents used in this method. The double bond of the maleimide group reacts specifically with the sulfydryl group of cysteine residues to form a thioether bond which is not reversible with reducing agents.

A series of single cysteine mutants are created by substituting individual amino acid residues in the putative intracellular or extracellular loops of the protein with cysteine. The membrane orientation of the engineered cysteine residues are determined by expressing the resulting functional mutants in an exogenous cell culture system. Cells expressing a mutant are first incubated with stilbene disulfonate maleimide and then with biocytin maleimide. In parallel, cells expressing the same protein are incubated with biocytin maleimide alone. To detect the presence of the biotin labeling, the protein of interest is immunoprecipitated, resolved by SDS–PAGE and then subjected to Western blotting with streptavidin conjugated to horseradish peroxidase (HRP). When the cells are incubated with stilbene disulfonate maleimide and then biocytin maleimide, detection of biotin labeling indicates that the cysteine residue is intracellular, whereas no observed labeling implies an extracellular location. Biotin labeling should always be detected when only biocytin maleimide is used. One advantage of this method is that by reprobing the Western blot with an antibody specific to the protein being studied, one can be assured that the protein is present even if biotin labeling is not detected. If an antibody specific to the protein is not available, then an epitope tag with an available antibody should be incorporated into the protein sequence. It is important that protein function not be compromised by the presence of the epitope tag.

Maleimide Binding

Reagents

PBS Phosphate-buffered saline
Biocytin maleimide [3-(*N*-maleimidylpropionyl) biocytin] (Molecular Probes, Eugene, OR)
Stilbene disulfonate maleimide (Molecular Probes)

Procedure. All of the experiments are carried out in 6-well plates using cultured cells that express the target protein either 48 hours after transient transfection or, in the case of stably transfected cell lines, when the cells are confluent. Cells are washed three times with 1 ml of PBS and then incubated with membrane-permeant biocytin maleimide in 1 ml of PBS at 25°. Experimental concentrations for labeling extracellular or intracellular cysteines with this compound must be determined empirically. Typically, extracellular cysteines can bind at concentrations as low as 1 μM, whereas intracellular cysteines may require 200 μM and a longer incubation period.[13] Incubation times range from 5–30 min.[13] If dilutions are made from biocytin maleimide dissolved in dimethyl sulfoxide (DMSO), then the DMSO concentration in PBS should not exceed 1%. To block extracellular biocytin maleimide labeling, cells are preincubated with 200 μM membrane-impermeant stilbene disulfonate maleimide in PBS for 30 minutes, washed three times with PBS and then incubated with biocytin maleimide.[13] Next, cells are washed three times with PBS, 2% 2-mercaptoethanol in PBS, and twice with PBS to quench excess maleimide reactivity. Plates should then be placed on ice for cell lysis.

Cell Lysis

Reagents

Triple Detergent Lysis Buffer: 150 mM NaCl, 0.1% sodium dodecyl sulfate (SDS), 1% Nonidet P-40 (NP-40), 0.5% sodium deoxycholate, 0.02% NaN_3, 100 μg/ml phenylmethylsulfonyl fluoride (PMSF), 1 μg/ml aprotinin, 50 mM Tris-Cl, pH 8.0
Single Detergent Lysis Buffer: 150 mM NaCl, 1% NP-40, 0.02% NaN_3, 100 μg/ml PMSF, 1 μg/ml aprotinin, 50 mM Tris-Cl, pH 8.0
No-Salt Lysis Buffer: 1% NP-40, 100 μg/ml PMSF, 1 μg/ml aprotinin, 50 mM HEPES, pH 7.0

Procedure. Cell lysis is carried out on ice for 30 min on a rotating shaker. A volume of 250 μl of lysis buffer per 35-mm dish is appropriate, although the volume of lysis buffer will be determined by the size of the well. The correct lysis buffer must be determined for each protein. For example, glutamate transporters are effectively solublized with triple detergent lysis buffer.[15] Two other commonly used lysis buffers, which differ in the types of detergent and the salt concentration, are listed below.[16] After incubation, the cells are scraped from the plates with a rubber policeman and transferred to 1.6-ml microfuge tubes. The tubes are spun at 15,000*g* at 4° and the supernatant transferred to a new tube. The pellets may be saved for Western blotting to confirm that the proteins are in the soluble fraction.

Immunoprecipitation

Reagents

Immunoprecipitation (IP) Buffer: 150 m*M* NaCl, 0.1% NP-40, 1 m*M* EDTA, 0.25% gelatin, 0.02% NaN_3, 50 m*M* Tris-HCl, pH 7.5
No-Salt Wash Buffer: 0.1% NP-40, 10 m*M* Tris-HCl, pH 7.5
2× SDS Sample Buffer containing 200 m*M* dithiothreitol (DTT)
Antibody to the protein of interest (Primary Antibody)
Protein A–Sepharose CL-4B (Sigma, St. Louis, MO)

Procedure. The volume of the supernatant from the cell lysis step is increased to 1 ml with IP buffer and an antibody specific to the protein of interest is added. The amount of antibody added must be determined by the concentration of the antigen and the antibody titer, as too much antibody may increase the nonspecific background. A useful range for polyclonal serum may be 0.5–5 μl per ml of IP buffer.[16] The lysate and antibody are mixed at 4° for 1 hr and then transferred to a new tube containing protein A–Sepharose. A 50% (v/v) slurry of protein A–Sepharose is spun at 12,000*g* at 4° for 1 min and the buffer removed prior to adding the lysate. The lysate/Sepharose mixture is vortexed and then incubated for 1 hr at 4° with mixing. The tubes are spun at 12,000*g* at 4° for 1 min and the supernatant aspirated away using a 23-gauge needle that has been bent twice.[17] The needle should be inserted into the Sepharose and all traces of the buffer aspirated away. One milliliter of IP buffer is used to wash the sepharose for 20 min on a nutator at 4°. After washing, the sample is spun at 12,000*g*, 4° for 1 min and all traces of wash buffer aspirated away. The wash is repeated with IP buffer and then with no-salt wash buffer. The protein is recovered from the Sepharose by adding 60 μl of 2× SDS sample buffer and heating to 100° for 3–5 min. Samples are spun at 12,000*g* at 4° for 1 min and supernatants transferred to a new tube. Samples may be frozen at −20°.

[15] B. H. Leighton, R. P. Seal, and S. G. Amara, unpublished observations, 1997.

[16] J. Sambrook, E. F. Fritch, and T. Maniatis, "Molecular Cloning," 2nd ed., Cold Spring Harbor Laboratory, Cold Spring Harbor, New York, 1989.

[17] E. Harlow and D. Lane, "Antibodies: A Laboratory Manual," Cold Spring Harbor Laboratory, Cold Spring Harbor, New York, 1988.

Streptavidin Western Blotting

Reagents

TBS: 150 m*M* NaCl, 10 m*M* Tris–Base, pH 7.4
TST: 150 m*M* NaCl, 0.1% Tween 20, 10 m*M* Tris–Base, pH 7.4
Bovine Serum Albumin: Crystalline BSA (Sigma)
Normal Horse Serum (Gibco-BRL, Gaithersburg, MD)
Horseradish Peroxidase (HRP)-conjugated Streptavidin (Vector, Burlingame, CA)
Renaissance Western Blot Chemiluminescence Reagent (NEN, Boston, MA)

Procedure. Proteins are resolved by SDS–PAGE and transferred to a membrane to detect the presence of biotin labeling. Methods for both SDS–PAGE and transferring of proteins to membranes are provided in the chapter on labeling neurotransmitter transporters at the cell surface.[18] Membranes with the transferred proteins are blocked with either 2% (w/v) crystalline BSA or 3–5% (v/v) normal horse serum in TBS for 1 hr at 25°. Blots are then incubated with HRP–streptavidin diluted in blocking buffer for 1 hr at 25°. The blots are washed four times with TST for 10 min per wash and the biotin-labeled protein bands visualized by chemiluminescence. The membranes can be reprobed with an antibody specific to the protein of interest to assess the level of protein present in each lane.

Method II: Methanethiosulfonate Derivatives

Introduction

Several methanethiosulfonate (MTS) derivatives are used in this method.[19] *N*-Biotinylaminoethylmethanethiosulfonate (MTSEA-biotin) and *N*-biotinylcaproylaminoethylmethanethiosulfonate (MTSEA-biotincap) each contain a biotin moiety that is linked to methanethiosulfonate, although MTSEA-biotincap contains a longer linker group. Reaction of these compounds with cysteine results in attachment of the biotin moiety to the thiol of cysteine through a disulfide linkage. The biotin-containing MTS derivatives lack the charged amine group found in the parent compound, MTSEA, and thus appear to be more membrane permeant. In support of this, both MTSEA-biotin and MTSEA-biotincap react readily with cysteines substituted at the EAAT1 (glutamate transporter) C terminus, a domain which has been established to be cytoplasmic by other

[18] G. M. Daniels and S. G. Amara, *Methods Enzymol.* **296,** [21], 1998 (this volume).
[19] M. H. Akabas, D. A. Stauffer, M. Xu, and A. Karlin, *Science* **258,** 307 (1992).

approaches. In addition, when intact cells are incubated with either of the two compounds, both appear to react with cysteine residues of a well-established intracellular protein, α-actin, further indicating that they are indeed membrane permeant. Membrane impermeant MTS derivatives such as 2-sulfonatoethylmethanethiosulfonate (MTSES) and [2-(trimethylammonium)ethyl]methanethiosulfonate (MTSET) can be used in conjunction with MTSEA-biotin or MTSEA-biotincap. These reagents introduce either an $S{-}CH_2CH_2SO_3^-$ group (MTSES) or an $S{-}CH_2CH_2N(CH_3)_3^+$ group (MTSET) onto the thiol through a disulfide bond. Larger membrane-impermeant reagents such as stilbene disulfonate maleimide can also be used.

Although the strategy for determining the membrane orientation of single cysteine residues placed into the putative hydrophilic loops of a protein is the same as described in method I, the procedure for identifying the labeled cysteine residue is different in this method because MTS derivatives react with the sulfhydryl group of the cysteine residues to form mixed disulfide bonds, and thus, reducing agents cannot be used until after the labeled proteins have been isolated.

Cells expressing a single cysteine mutant are incubated with the membrane-impermeant sulfhydryl-reactive reagents MTSES or MTSET, and then incubated with the membrane-permeant reagents, MTSEA-biotin or MTSEA-biotincap. Cells are lysed and all of the biotinylated proteins are isolated by incubating the supernatants with NeutrAvidin beads, which bind biotin with high affinity.[20] The proteins are then resolved on denaturing polyacrylamide gels and transferred to membranes for Western blotting with an antibody specific to the protein or specific to an exogenous epitope incorporated into the protein, as was discussed in method I. If the protein is not detected, then it is assumed that the biotin labeling was blocked by the membrane-impermeant compound and thus, the cysteine residue is accessible from the extracellular side. As a control, cells expressing the same mutant protein are incubated in parallel with only the membrane-permeant MTSEA-biotin or MTSEA-biotincap reagent. The cysteine residue must be detected with at least one of these reagents; otherwise, they cannot be used to determine the membrane orientation of this residue, and other membrane-permeant biotinylated reagents will need to be tested. Another consideration is that MTSES or MTSET may permeate the membrane when used at higher concentrations or in certain cell types, and thus concentrations and incubation times may have to be adjusted to minimize their access to the cytoplasmic side of the membrane.[21] This can be tested

[20] Sargiacomo, M. Lisanti, L. Graeve, A. Le Bivic, and E. Rodriguez-Boulan, *J. Membr. Biol.* **107,** 277 (1989).

[21] M. Holmgren, M. E. Jurman, and G. Yellen, *Neuropharmacol.* **35,** 797 (1996).

by observing whether or not these reagents react with cysteine residues placed into regions of the protein which have been clearly established as intracellular (for example, in the N or C termini for the plasma membrane neurotransmitter transporter families).

Application of Methanethiosulfonate Derivatives

Reagents

Phosphate-Buffered Saline (PBS)
MTS-biotin (Toronto Research Chemicals, Toronto, Canada)
MTS-biotincap (Toronto Research Chemicals)
MTSES (Toronto Research Chemicals)
MTSET (Toronto Research Chemicals)
Stilbene disulfonate maleimide (Molecular Probes)

Procedure. Cultured cells either stably or transiently expressing a single-cysteine mutant are grown in 6-well plates. The cells are incubated first with the membrane impermeant MTS derivatives, MTSES or MTSET, in 1 ml PBS under empirically determined conditions. In the case of most neurotransmitter transporters or receptors, cells are incubated with concentrations of 10 m*M* MTSES or 1 m*M* MTSET for 2–5 min.[22,23] Stilbene disulfonate maleimide can be incubated at concentrations ranging from 200 μM to 2 m*M* for 20 min; however, these parameters should be adjusted as necessary. Next, the cells are washed three times with 1 ml of PBS and then incubated with 1 ml PBS containing either MTSEA-biotin or MTS-biotincap, also under empirically determined conditions. The biotin-containing compounds must be dissolved in DMSO first and then diluted into PBS, with the final concentration of DMSO not to exceed 1%. Concentrations ranging from 0.5–2 m*M* for 10–30 min are suggested for these compounds.[23] Residues that are accessible from the extracellular side of the membrane may label efficiently at lower concentrations (0.5 m*M*) and in a shorter amount of time (10 min) than those which are located intracellularly. The unbound reagent is again removed by washing the cells three times with 1 ml of PBS.

Cell Lysis

Reagents

Lysis Buffer: 1% Triton X-100, 150 m*M* NaCl, 5 m*M* EDTA, 50 m*M* Tris, pH 7.5

[22] J. Javitch, D. Fu, J. Chen and A. Karlin, *Neuron* **14,** 825 (1995).
[23] R. P. Seal and S. G. Amara, unpublished observations, 1997.

Procedure. One milliliter of lysis buffer is added to each well and the cells scraped off with a rubber policeman. The contents of each well are then transferred to a 1.6-ml microfuge tube and incubated on ice for 1 hour to extract the proteins from the membranes. The insoluble material is removed by spinning the tubes at 15,000*g* for 15 min at 4°, and the supernatants are transferred to a new microfuge tube.

Recovery of Biotinylated Proteins

Reagents

UltraLink Immobilized NeutrAvidin (Pierce, Rockford, IL)
Lysis Buffer: 1% Triton X-100, 150 m*M* NaCl, 5 m*M* EDTA, 50 m*M* Tris, pH 7.5
High-Salt Wash Buffer: 0.1% Triton X-100, 500 m*M* NaCl, 5 m*M* EDTA, 50 m*M* Tris, pH 7.5
No-Salt Wash Buffer: 50 m*M* Tris, pH 7.5

Procedure. To recover the biotinylated proteins, 150 μl of 50% slurry of NeutrAvidin beads are added to the supernatants and incubated overnight at 4° with mixing.[20] The next day, the beads are washed three times with lysis buffer, twice with high-salt wash buffer, and once with no-salt wash buffer.[20] The protein is recovered from the beads by adding 80 μl of 2× sample buffer and vortexing. At this point, samples can either be frozen at −20° or incubated at room temperature for 30 min before they are loaded onto denaturing polyacrylamide gels. Heating to temperatures greater than 37° should be avoided for some membrane proteins, as it is thought to promote aggregation. A more in-depth discussion of the biotin–NeutrAvidin interaction can be found in the chapter on labeling neurotransmitter transporters at the cell surface, in this volume.[18]

SDS–PAGE and Western Blotting

Reagents

TST: 150 m*M* NaCl, 0.1% Tween 20, 10 m*M* Tris–base, pH 7.4
TBS: 150 m*M* NaCl, 10 m*M* Tris–Base, pH 7.4
Blocking Buffer: TBS, 5% nonfat powdered milk, 1% BSA
Antibody to the protein of interest (Primary Antibody)
Antibody to the primary antibody and conjugated to HRP (Secondary Antibody)
Renaissance Western Blot Chemiluminescence Reagent (NEN)

Procedure. Proteins are resolved on denaturing polyacrylamide gels and transferred to membranes for Western blotting, as described.[18] Each of the following steps is carried out for 1 hr at room temperature. After blocking, the primary antibody is incubated with the membranes in TST buffer. The membranes are then washed four times in TST and the secondary antibody diluted in TST buffer is added. Membranes are again washed four times with TST and the proteins are visualized using the chemiluminescence reagents.

Discussion

The two methods described in this chapter offer a means to examine the transmembrane topology of a protein by determining the orientation of single cysteine residues placed into putative hydrophilic loops. Method I is designed for compounds that form bonds with cysteines which are not reversible with reducing reagents. Method II is effective for compounds that form disulfide bonds with cysteine residues and are, therefore, reversed by reducing reagents. In the first method, thioether bonds are formed between the maleimide compounds and the cysteine residues, and because these bonds cannot be reversed by reducing agents, it is possible to immunoprecipitate the target protein first and then detect the biotin-labeled protein by Western blotting with streptavidin conjugated to horseradish peroxidase. This method has the advantage that the presence of the protein on the Western blot can be verified when biotin labeling is not detected. In the second method, methanethiosulfonate derivatives form mixed disulfide bonds with the thiol group of cysteine residues and thus prohibit the use of reducing agents until after the selection of the biotin-labeled proteins with NeutrAvidin beads is complete. This method has the advantage that MTS reagents are typically more reactive with cysteine residues than are the maleimide compounds.

As is true of most approaches, there are some limitations to using sulfhydryl-specific reagents to determine the topological orientation of cysteine residues. First, one must use caution in assigning an intracellular localization to a cysteine residue when the membrane-impermeant reagent fails to block binding of the biotin-containing compound, because the same result would be observed if the residue is located extracellularly, but is not accessible to the membrane-impermeant compound. For this reason, several membrane-impermeant compounds should be tried, including ones that are as small or smaller than the biotin-containing compounds, as well as compounds that differ in their electrostatic properties. A second limitation may be that only a small number of cysteine residues react with the biotin-containing compounds. Cysteines that reside close to or within the putative transmembrane domains appear to be much less likely to react with the biotin-containing compounds than those located towards the middle of extended hydrophilic domains. Also, the secondary structure of a hydrophilic domain will influence whether or not the cysteine residue is accessible. Therefore, it may be necessary to try cysteine substitution at several sites within the putative hydrophilic domains of the protein. It also may be necessary to use several different biotin-containing reagents to find one that reacts with several of the engineered cysteine residues.

An additional caveat to the use of these techniques is that the approach does not necessarily distinguish between residues that are extracellular and those that residue within accessible hydrophilic domains within the membrane. Ion channels and transporters can have aqueous pores that hydrophilic compounds may enter from the outside.[21] Unless a region has been previously defined as a hydrophilic porelike domain within the membrane, the observation that membrane-impermeant compounds block the biotin labeling of cysteine residues introduced into these domains would lead to the incorrect conclusion that these residues were extracellular. A thorough evaluation of transmembrane topology by this approach is, therefore, likely to involve the use of several sulfhydryl-reactive compounds and the confirmation of the results by multiple methods that are described in this and other chapters.

In addition to evaluating the transmembrane topology, further structural analysis of a protein can be carried out using the functional cysteine substitution mutants created with this approach. For example, interactions between membrane-spanning segments of a protein can be evaluated using sulfhydryl-specific cross-linking reagents.[24] Conformational changes that occur with certain functional properties of membrane proteins, such as channel gating or protein–protein associations, can also be monitored using sulfhydryl-specific spin labels.[25–28] Additionally, amino acid residues that are thought to underlie protein function can be evaluated using a method called the substituted-cysteine accessibility method (SCAM).[18,29–32] In this method, individual amino acid residues are substituted with cysteine and the effects of cysteine modification on channel, receptor, or transporter function (i.e., ion conductance, ligand binding, or substrate transport) are evaluated using hydrophilic sulfhydryl-specific reagents that differ with

[24] C. Altenbach, T. Martini, H. G. Khorana, and W. L. Hubbell, *Science* **248,** 1088 (1990).
[25] W. L. Hubbell and C. Altenbach, *Curr. Opin. Struct. Biol.* **4,** 566 (1994).
[26] N. Yang and R. Horn, *Neuron* **15,** 213 (1995).
[27] D. L. Farrens, C. Altenbach, K. Yang, W. L. Hubbell, and H. G. Khorana, *Science* **274,** 768 (1996).
[28] J. Wu and H. R. Kaback, *Proc. Nat. Acad. Sci. USA* **93**(25), 14498 (1996).
[29] M. Xu and M. H. Akabas, *J. Biol. Chem.* **268,** 21505 (1993).
[30] J. M. Pascual, C.-C. Shieh, G. E. Kirsch, and A. M. Brown, *Neuron* **14,** 766 (1995).
[31] J.-G. Chen, S. Liu-Chen, and G. Rudnick, *Biochemistry* **36,** 1479 (1997).
[32] J. Javitch, *Methods Enzymol.* **296,** [23], 1998 (this volume).

respect to size and charge. Thus, the functional cysteine substitution mutants created to analyze transmembrane topology by the approach described in this chapter are also useful in the analysis of additional aspects of protein structure.

[23] Probing Structure of Neurotransmitter Transporters by Substituted-Cysteine Accessibility Method

By JONATHAN A. JAVITCH

Cysteine substitution and covalent modification have been used to study structure–function relationships and the dynamics of protein function in a variety of membrane proteins.[1–6] Moreover, charged, hydrophilic, lipophobic, sulfhydryl reagents have been used to probe systematically the accessibility of substituted cysteines in putative membrane-spanning segments of a number of proteins. This approach, the substituted-cysteine accessibility method (SCAM), has been used to map channel-lining residues in several proteins, including the nicotinic acetylcholine receptor,[7–9] the $GABA_A$ (γ-aminobutyric acid) receptor,[10,11] the cystic fibrosis transmembrane conductance regulator,[12] the UhpT (uptake of hexose phosphate) transporter,[13] and potassium channels.[14] We have adapted this approach to map the surface of the binding-site crevice in the dopamine D_2 receptor, a member of the G-protein-coupled receptor superfamily.[15–18]

[1] A. P. Todd, J. Cong, F. Levinthal, C. Levinthal, and W. L. Hubbell, *Proteins* **6,** 294 (1989).
[2] K. S. Jakes, C. K. Abrams, A. Finkelstein, and S. L. Slatin, *J. Biol. Chem.* **265,** 6984 (1990).
[3] C. Altenbach, T. Marti, H. G. Khorana, and W. L. Hubbell, *Science* **248,** 1088 (1990).
[4] C. L. Careaga and J. J. Falke, *Biophys. J.* **62,** 209 (1992).
[5] A. A. Pakula and M. I. Simon, *Proc. Natl. Acad. Sci. USA* **89,** 4144 (1992).
[6] K. Jung, H. Jung, J. Wu, G. G. Prive, and H. R. Kaback, *Biochemistry* **32,** 12273 (1993).
[7] M. H. Akabas, D. A. Stauffer, M. Xu, and A. Karlin, *Science* **258,** 307 (1992).
[8] M. H. Akabas, C. Kaufmann, P. Archdeacon, and A. Karlin, *Neuron* **13,** 919 (1994).
[9] M. H. Akabas and A. Karlin, *Biochemistry* **34,** 12496 (1995).
[10] M. Xu, D. F. Covey, and M. H. Akabas, *Biophys. J.* **69,** 1858 (1995).
[11] M. Xu and M. H. Akabas, *J. Biol. Chem.* **268,** 21505 (1993).
[12] M. H. Akabas, C. Kaufmann, T. A. Cook, and P. Archdeacon, *J. Biol. Chem.* **269,** 14865 (1994).
[13] R. T. Yan and P. C. Maloney, *Proc. Natl. Acad. Sci. USA* **92,** 5973 (1995).
[14] J. M. Pascual, C. C. Shieh, G. E. Kirsch, and A. M. Brown, *Biophys. J.* **69,** 428 (1995).
[15] D. Fu, J. A. Ballesteros, H. Weinstein, J. Chen, and J. A. Javitch, *Biochemistry* **35,** 11278 (1996).
[16] J. A. Javitch, D. Fu, and J. Chen, *Biochemistry* **34,** 16433 (1995).

0076-6879/98 $25.00

Neurotransmitter reuptake by transport proteins is a major mechanism for terminating synaptic transmission. These transporters operate by coupling the net electrochemical gradient for sodium, chloride, and neurotransmitter to the transmembrane translocation of these substrates.[19] Such a transporter must contain a water-accessible transport pathway, lined by residues from among the membrane-spanning segments. Neurotransmitter, sodium, and chloride must bind to specific sites within this pathway and cause a conformational change that alters the exposure of the bound substrates from the extracellular milieu to the intracellular milieu and reduces their binding affinity to facilitate intracellular release. Although the transport pathway bears some functional resemblance to an ion channel, the transporter gating mechanism must be considerably more complex. Likewise, the presence of binding sites for neurotransmitter and ions in membrane-spanning segments is reminiscent of the G-protein-coupled receptors, but substrate translocation requires a mechanism by which structural changes in the transporter alter the exposure of the binding sites and their affinities. We have now begun to use the substituted-cysteine accessibility method to determine the residues that line the transport pathway, as well as those that form the surface of the binding sites for substrates and inhibitors in the dopamine transporter (DAT).

Substituted-Cysteine Accessibility Method

SCAM provides an approach to map systematically the residues on the water-accessible surface of a protein. These residues are identified by substituting them with cysteine and assessing the reaction of charged, hydrophilic, sulfhydryl reagents with the engineered cysteines. Consecutive residues in putative membrane-spanning segments are mutated to cysteine, one at a time, and the mutant proteins are expressed in heterologous cells. In the case of a transporter, if ligand binding to and transport by a cysteine-substitution mutant is near-normal, we assume that the structure of the mutant transporter is similar to that of the wild type and that the substituted cysteine lies in a similar orientation to that of the wild-type residue. In the membrane-spanning segments, the sulfhydryl of a cysteine can face into the transport pathway, into the interior of the protein, or into the lipid bilayer; sulfhydryls facing into the water-accessible transport pathway

[17] J. A. Javitch, D. Fu, J. Chen, and A. Karlin, *Neuron* **14,** 825 (1995).
[18] J. A. Javitch, X. Li, J. Kaback, and A. Karlin, *Proc. Natl. Acad. Sci. USA* **91,** 10355 (1994).
[19] B. Giros and M. G. Caron, *Trends Pharmacol. Sci.* **14,** 43 (1993).

should react much faster with charged, hydrophilic, lipophobic, sulfhydryl-specific reagents.

Conformational changes in a protein may result in changes in the accessibility of substituted cysteines as assessed by their rates of reaction with polar sulfhydryl-specific reagents. For example, residues lining the channel of the nicotinic acetylcholine receptor change in accessibility upon activation of the receptor and opening of the channel.[7,8] Similarly, it should be possible to determine changes in the accessibility of residues in the transporter in different functional states. Removal of sodium and/or chloride combined with the presence or absence of substrate will likely yield information regarding conformational changes concomitant with substrate and ion binding and transport.

For such polar sulfhydryl-specific reagents, we use derivatives of methanethiosulfonate (MTS): positively charged MTS-ethylammonium (MTSEA) and MTS-ethyltrimethylammonium (MTSET), and negatively charged MTS-ethyl sulfonate (MTSES).[20] These reagents differ somewhat in size, with MTSET > MTSES > MTSEA. The largest, MTSET, fits into a cylinder 6 Å in diameter and 10 Å long; thus, the reagents are approximately the same size as dopamine. The MTS reagents form mixed disulfides with the cysteine sulfhydryl, covalently linking $-SCH_2CH_2X$, where X is NH_3^+, $N(CH_3)_3^+$, or SO_3^-. The MTS reagents are specific for cysteine sulfhydryls and do not react with disulfide-bonded cysteines or with other residues. The reagents are charged and quite hydrophilic. Moreover, they reportedly react with the ionized thiolate (RS^-) more than a billion times faster than with the un-ionized thiol (RSH),[21] and only cysteines accessible to water are likely to ionize. The hydrophilic, negatively charged, organomercurial *p*-chloromercuribenzene sulfonate (pCMBS) also has been used to probe the accessibility of substituted cysteines in membrane-spanning segments of a number of membrane proteins.[13,22,23]

Detection of Reaction

Reaction can be detected either directly or indirectly by measuring the effect of reaction on a functional property of the protein. Because of the very small quantities of protein produced in most heterologous expression

[20] D. A. Stauffer and A. Karlin, *Biochemistry* **33,** 6840 (1994).

[21] D. D. Roberts, S. D. Lewis, D. P. Ballou, S. T. Olson, and J. A. Shafer, *Biochemistry* **25,** 5595 (1986).

[22] R. T. Yan and P. C. Maloney, *Cell* **75,** 37 (1993).

[23] Y. Olami, A. Rimon, Y. Gerchman, A. Rothman, and E. Padan, *J. Biol. Chem.* **272,** 1761 (1997).

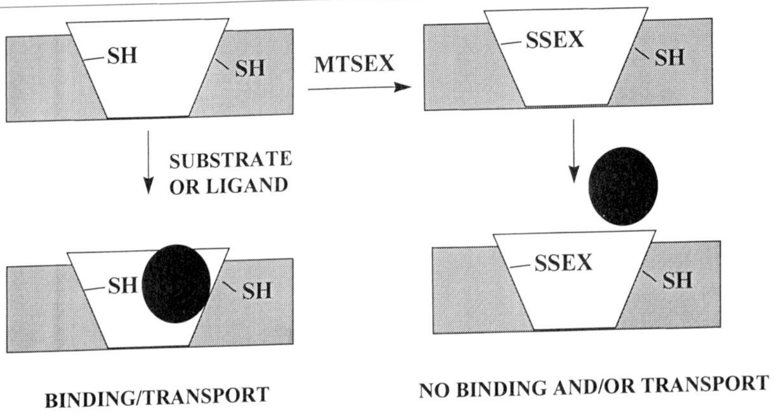

FIG. 1. Schematic representation of the reaction of the MTS reagents with a cysteine exposed in the binding-site crevice/transport pathway. The membrane is represented by a stippled rectangle; the binding-site crevice/transport pathway is indicated within the plane of the membrane, and the solid oval represents substrate/ligand. SEX represents SCH_2CH_2X [where X is NH_3^+, $N(CH_3)_3^+$ or SO_3^-] which is covalently linked to the water-accessible cysteine sulfhydryl. In the bound state, represented in the lower left-hand panel, substrate/ligand is reversibly bound at the binding site within the binding-site crevice/transport pathway. In the unbound state, represented in the upper left-hand panel, the binding site is unoccupied. After irreversible reaction with MTSEX, represented in the upper right-hand panel, substrate/ligand can no longer bind. The cysteine sulfhydryl facing lipid or the interior of the protein does not react with MTSEX. MTSEX can only react with a sulfhydryl in the binding-site crevice/transport pathway of unbound transporter and not of bound transporter. Thus, substrate/ligand retards the rate of reaction of transporter with MTSEX, thereby protecting subsequent ligand binding and transport.

systems, we cannot rely upon the direct detection of reaction.[24] Instead, we use the irreversible modification of function to assay the reaction. In a transporter, the reaction of an MTS reagent with an engineered cysteine in the transport pathway should alter binding and/or transport irreversibly (Fig. 1). Additionally, reaction with a cysteine near a binding site should be retarded by the presence of inhibitor and/or substrate.

[24] It is possible to react the protein in intact cells with (and without) an MTS reagent, solubilize the protein, label exhaustively with ^{3}H-labeled *N*-ethylmaleimide (NEM), and then purify the protein by immunoprecipitation. By assessing the ability of the MTS reagent to inhibit subsequent radiolabeled NEM incorporation, one could determine directly the extent of reaction.[23] The success of this approach would depend on the number of endogenous cysteines that might also react with NEM to increase the background and on the ease and efficiency of purification.

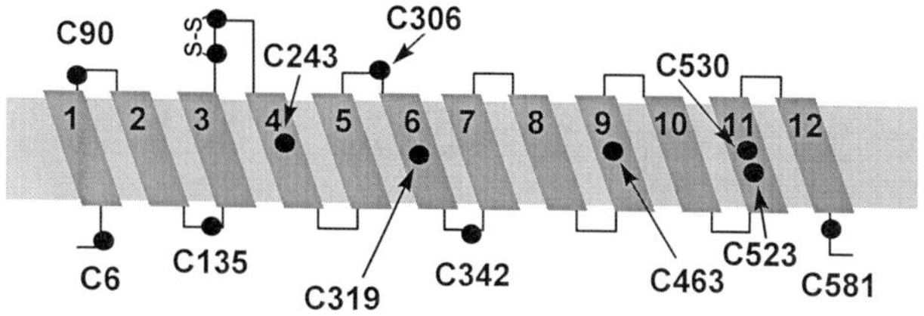

FIG. 2. Predicted topology of the dopamine transporter showing 12 putative membrane-spanning segments. The predicted positions of the 13 endogenous cysteines are shown. The cysteines in the second extracellular loop between M3 and M4 are thought to be disulfide bonded in the dopamine transporter and the serotonin transporter (see text). (Adapted from J. V. Ferrer and J. A. Javitch, *Proc. Natl. Acad. Sci. USA*, in press.)

Cysteines in Wild-Type Dopamine Transporter

The function of the protein used as the background for SCAM must not be affected by the sulfhydryl reagents. In some cases endogenous cysteines are not accessible to reaction with the MTS reagents, while in other cases, such as the dopamine D_2 receptor and DAT, endogenous cysteines are accessible and must first be identified and mutated to other residues. DAT contains 13 cysteines; based on topology predictions, four are extracellular, four are intracellular, and five are in membrane-spanning segments (Fig. 2). Of the extracellular residues, two are very highly conserved and are likely to be disulfide-bonded,[25,26] and thus unavailable for reaction with the sulfhydryl reagents.

Sulfhydryl reagents, including *N*-ethylmaleimide (NEM) and mercurial compounds, inhibited both dopamine transport and the binding of radiolabeled inhibitors.[27–33] Although some studies have suggested that cocaine protected more effectively than dopamine against reaction with NEM,[33] others have found that under appropriate conditions, both substrate and cocaine protected against this reaction.[27,32] Furthermore, we have found that when wild-type DAT is expressed in HEK 293 (human embryonal

[25] J. G. Chen, S. Liu-Chen, and G. Rudnick, *Biochemistry* **36,** 1479 (1997).
[26] J. B. Wang, A. Moriwaki, and G. R. Uhl, *J. Neurochem.* **64,** 1416 (1995).
[27] M. E. Reith and G. Selmeci, *Naunyn-Schmiedebergs Arch. Pharmacol.* **345,** 309 (1992).
[28] M. M. Schweri, *Neuropharmacol.* **29,** 901 (1990).
[29] M. M. Schweri, *Synapse* **16,** 188 (1994).
[30] C. J. Cao, M. M. Young, J. B. Wong, L. G. Mahran, and M. E. Eldefrawi, *Membrane Biochem.* **8,** 207 (1989).
[31] S. M. Meiergerd and J. O. Schenk, *J. Neurochem.* **62,** 998 (1994).
[32] S. Saadouni, F. Refahi-Lyamani, J. Costentin, and J. J. Bonnet, *Eur. J. Pharmacol.* **268,** 187 (1994).
[33] K. M. Johnson, J. S. Bergmann, and A. P. Kozikowski, *Eur. J. Pharmacol.* **227,** 411 (1992).

kidney) cells, binding and transport are blocked irreversibly by sulfhydryl reagents, and these reactions are retarded in the presence of ligand.[34] This suggests that at least one sulfhydryl is present in the binding site and transport pathway.

It is important, however, to consider alternative mechanisms for the inhibition of transport before assuming that the reagents are reacting with an endogenous cysteine in DAT. For example, the reagents could inhibit transport indirectly by reacting with an endogenous protein to cause a rapid rundown of the sodium gradient. Alternatively, the reagents might react with another protein that interacts directly with the transporter. If the critical reaction is in a protein other than DAT, then all of the endogenous cysteines could be removed from DAT without rendering the function of the transporter insensitive to the reagents. Such is the case with the minK potassium channel, in which all of the endogenous cysteines can be removed without eliminating the effect of MTSEA or MTSET on conductance.[35] Such a finding indicates the existence of another protein that interacts with minK to form the channel, and indeed the second subunit has been identified.[36,37]

Nonetheless, the ability of cocaine and dopamine to slow the rate of reaction with the MTS reagents implies a direct effect on DAT and not an indirect effect. Moreover, the parallel effects on [^{3}H]2-β-carbomethoxy-3-β-(4-fluorophenyl)tropane (CFT) binding and dopamine and *N*-methyl-4-phenylpyridinium (MPP^+) transport are consistent with a direct reaction with endogenous cysteines in DAT.[38] Therefore, before applying the substituted-cysteine accessibility method, it was necessary to identify the sensitive cysteine(s) in DAT, just as we did in the dopamine D_2 receptor.[18] This process was relatively straightforward in the case of the dopamine D_2 receptor where replacement of a single endogenous cysteine (Cys-118) with serine resulted in a 100-fold decrease in the reactivity of the receptor with MTSEA and MTSET. C118S expresses normally and has unaltered binding properties; this mutant was used as the background for further cysteine substitutions.[15–17]

[34] J. V. Ferrer and J. A. Javitch, *Proc. Natl. Acad. Sci. USA,* in press.

[35] K.-K. Tai, K.-W. Wang, and S. A. N. Goldstein, *J. Biol. Chem.* **272,** 1654 (1997).

[36] M. C. Sanguinetti, M. E. Curran, A. Zou, J. Shen, P. S. Spector, D. L. Atkinson, and M. T. Keating, *Nature* **384,** 80 (1996).

[37] J. Barhanin, F. Lesage, E. Guillemare, M. Fink, M. Lazdunski, and G. Romey, *Nature* **384,** 78 (1996).

[38] These findings, however, do not rule out the possibility that some of the effects of modification are on another protein that directly interacts with the transporter and is necessary for both transport and binding.

We typically substitute serine or alanine for endogenous cysteine.[39] The mutant transporter must transport substrate and bind ligand with near-normal properties and must be relatively insensitive to the MTS reagents. The ideal starting point would be to create a cysteineless protein with normal expression and function. Such a construct has been possible with the lactose permease,[6] the NhaA-Na^+/H^+ antiporter,[23] and a glutamate transporter,[40] but has not yet been achieved with the dopamine D_2 receptor or DAT.[41] We have succeeded, nonetheless, in identifying four endogenous cysteine residues in wild-type DAT that react with the MTS reagents.[34] The resulting mutant with these four endogenous cysteines replaced by other residues is expressed at reduced but workable levels and binds [^{3}H]mazindol and [^{3}H]CFT and transports dopamine in near-normal fashion. Most significantly, these functions are significantly less sensitive to inhibition by the MTS reagents.

Application of Substituted-Cysteine Accessibility Method to Dopamine Transporter

Using the mutant DAT insensitive to the MTS reagents as background, the substituted-cysteine accessibility method now can be used to determine residues that line the surface of the transport pathway.[42] Cysteines are substituted, one at a time, for residues in a putative membrane-spanning segment (see site-directed mutagenesis). The mutant transporter is expressed in heterologous cells (see transient and stable transfection), and ligand binding and transport assays are performed (see binding and uptake). If the transporter binds ligand and transports substrate, then the effect of reaction with the MTS reagents is determined (see use of the MTS reagents), always comparing the effects with those on the transporter used as the background for the cysteine substitutions.

[39] If the cysteine is not conserved in all related transporters, the choice of residue can be influenced by examining a sequence alignment of the related transporters and considering other residues present in the aligned position. For example, Cys-319 in the putative sixth membrane-spanning segment of DAT is aligned with phenylalanine in every other related transporter. Our initial attempt to substitute this residue with serine led to a transporter that expressed poorly; when we substituted phenylalanine instead, expression was substantially improved.[34]

[40] R. P. Seal and S. G. Amara, *Soc. Neurosci. Abstr.* **22,** 1575 (1996).

[41] If a water-accessible cysteine sulfhydryl is critical to the expression or function of the transporter and if reaction of such a cysteine with the MTS reagents affects binding or transport, then it may be impossible to produce the appropriate MTS-insensitive mutant. Thus, it is possible that a given protein will not be amenable to study with this method, and unfortunately this cannot be predicted in advance.

[42] This approach has recently been applied to the serotonin transporter: J. G. Chen, A. Sachpatzidis, and G. Rudnick, *J. Biol. Chem.* **272,** 28321 (1997).

Methods

Site-Directed Mutagenesis. The cDNA encoding wild-type human DAT was generously provided by Mark Sonders and Susan Amara. It was subcloned into pciNEO (Promega, Madison, WI) for transient expression in mammalian cells. Cysteine mutations are generated using the Chameleon mutagenesis kit (Stratagene, LaJolla, CA) with oligonucleotides incorporating a change in a restriction site as well as the desired mutation. Mutants are screened by restriction mapping, and the mutations are confirmed by DNA sequencing.

Transient Transfection. HEK 293 cells are grown in Dulbecco's modified Eagle's medium (DMEM)/F12 (1/1) containing 3.15 g/liter glucose in 10% bovine calf serum at 37° and 5% CO_2. Thirty-five-millimeter dishes of HEK 293 cells at 70–80% confluence are cotransfected with 1.6 μg of wild-type or mutant DAT cDNA in pciNEO and 0.4 μg pRSVTag using 9 μl of lipofectamine (GIBCO, Grand Island, NY) and 1 ml of OPTIMEM (Gibco). Five hours after transfection and again at 24 hr after transfection, the medium is changed. Forty-eight hours after transfection, cells are washed with phosphate-buffered saline (PBS: 8.1 m*M* NaH_2PO_4, 1.5 m*M* KH_2PO_4, 138 m*M* NaCl, 2.7 m*M* KCl, pH 7.2), briefly treated with PBS containing 5 m*M* EDTA, and then dissociated in PBS. Cells are pelleted at 1000*g* for 5 min at 4°, and resuspended for binding, uptake, or treatment with MTS reagents (see below).

Stable Transfection. Wild-type or mutant DAT cDNA has been subcloned into the bicistronic vector pcin4[43] (kindly provided by S. Rees). This vector is very useful for the rapid creation of pools of HEK 293 cells which stably express large amounts of DAT. Transfection is performed as described earlier with 2 μg pcin4-DAT (without pRSVTag). Twenty-four hours after transfection, the cells are split into a 100-mm plate and grown in medium containing 0.7 mg/ml G418. After selection, cells are maintained in medium containing 0.3 mg/ml G418.

[^{3}H]Mazindol and [^{3}H]CFT Binding. Binding is determined to DAT expressed in intact, dissociated cells. For transient transfections, cells from a 35-mm plate are suspended in 1 ml of uptake buffer (5 m*M* Tris, 7.5 m*M* HEPES, 120 m*M* NaCl, 5.4 m*M* KCl, 1.2 m*M* $CaCl_2$, 1.2 m*M* $MgSO_4$, pH 7.4). [^{3}H]CFT (~10 n*M*) or [^{3}H]mazindol (~4 n*M*) (Du Pont/NEN, Boston, MA) are incubated with 75μl of cell suspension in a final volume of 100 μl in polystyrene tubes in a 96-well format. The mixture is incubated at 4° for

[43] S. Rees, J. Coote, J. Stables, S. Goodson, S. Harris, and M. G. Lee, *BioTechniques* **20,** 1002 (1996).

2 hr and then filtered using a Brandel cell harvester through Whatman (Clifton, NJ) 934AH glass fiber filters. The filter is washed twice with 5 ml of 10 m*M* Tris-HCl, 120 m*M* NaCl, pH 7.4, at 4°. Specific binding is defined as total binding less nonspecific binding in the presence of 100 μM cocaine (Research Biochemicals, Natick, MA).

Uptake. Since filtration through glass-fiber filters gives inconsistent results (presumably due to variable cell breakage during filtration), we have chosen to assay uptake with attached cells. HEK 293 cells, however, are not robustly adherent, and care is required to wash cells without detaching them from tissue culture plates. Furthermore, we have found that incubation with the MTS reagents dissociates the cells from standard tissue culture plates. Coating the plates with polylysine, polyornithine, collagen, fibronectin, or Cell-Tac failed to prevent cell dissociation. We have found, however, that dissociated cells adhere tightly to lectin-coated 96-well plates, and, even after treatment with the MTS reagents, the wells can be washed without significant cell loss.

Fifty microliters of a 5 μg/ml solution of *Ulex europaeus* agglutinin I (Vector, Burlingame, CA) in lectin binding buffer (10 m*M* HEPES, 140 m*M* NaCl, 0.1 m*M* $CaCl_2$, pH 8.4) is added to each well of an Immulon 4 96-well plate (Dynatech, Chantilly, VA). After overnight incubation at 4°, the solution is removed and the plates are washed once with 200 μl per well of lectin binding buffer. Stably transfected HEK 293 cells (see above) are treated with PBS/EDTA as described above, dissociated in OPTI-MEM (Gibco), and transferred to the Immulon 4 plates. A nearly confluent 100-mm plate is used for a 96-well plate (approximately 50,000–100,000 cells/well). The plates are incubated at 37° for 1 hr to allow the cells to settle and adhere to the lectin-coated plates. The solution is removed and the wells are washed once with 200 μl of uptake buffer. Fifty microliters of uptake buffer or of freshly prepared MTS reagent (see below) diluted into uptake buffer is added per well and the plate is incubated at room temperature for 2 min. The solution is gently removed and the wells are washed once with 200 μl of uptake buffer. Twenty-five microliters of uptake buffer (with or without 100 μM cocaine to define background) and 25 μl of [^{3}H]dopamine (50 n*M*) or [^{3}H]MPP^+ (50 n*M*) are added, and the plate is incubated at room temperature for 5 min. To stop the reaction the wells are diluted with 200 μl of ice-cold uptake buffer and the solution is removed. The wells are washed once with 200 μl ice-cold uptake buffer. The cells are solubilized with 100 μl of 1% Triton X-100 and transferred for scintillation counting.

Use of Methanethiosulfonate Reagents. At pH 7 and 22°, MTSEA, MTSET, and MTSES (Toronto Research Chemicals) rapidly hydrolyze with a half-time of 5–20 min.[44] At lower pH and lower temperature, hydrolysis is appreciably slower. Stock reagent should be stored desiccated at 4°. A frequently used stock can be kept desiccated at room temperature, and it can be replenished from the 4° stock after appropriate warming to room temperature. The reagents should be weighed but kept dry until immediately before use. The reagents are relatively stable unbuffered in water at 4°; if it is necessary to dissolve the reagents or perform an intermediate dilution in buffer at a physiological pH, this should be done immediately before starting the reaction.

Cells from a 35 mm plate are suspended in 100 μl uptake buffer. Aliquots (22.5 μl) of cell suspension are incubated with 2.5 μl of freshly prepared MTS reagents at room temperature for 2 min. Alternatively, for uptake, the incubation is done with cells attached to lectin-coated 96-well plates (see earlier discussion). For screening, we typically use MTSET at 1 m*M*, MTSES at 10 m*M*, and MTSEA at 2.5 m*M* to normalize for the intrinsic reactivities of the reagents with sulfhydryls in solution.[20] Cell suspensions are then diluted 10-fold with uptake buffer and 75-μl aliquots in triplicate are used to assay for binding as described above. The fractional inhibition is calculated as 1 − [(specific binding after MTS reagent)/(specific binding without reagent)].

Determination of Rates of Reaction. The second-order rate constant (k) for the reaction of an MTS reagent with each susceptible mutant is estimated by determining the extent of reaction (as determined by extent of inhibition of binding or uptake) after a fixed time, typically 2 min, with six concentrations of reagent. The fraction of initial binding, Y, is fit to e^{-kct}, where k is the second-order rate constant, c is the concentration of MTS reagent, and t is the time (120 sec).

Protection. To assess the ability of substrate or inhibitor to retard the rate of reaction of the MTS reagents, cells are incubated in uptake buffer at room temperature for 20 min in the presence or absence of cocaine or dopamine and then MTS reagents are added, in the continued presence or absence of cocaine or dopamine, for 2 min. It is important to note that this protocol examines the ability of a reversible reaction to slow an irreversible reaction. Therefore, the concentration of MTS reagent and the time of reaction should result in less than complete inhibition, and the concentration of protecting ligand should be approximately 100-fold greater than its K_I or K_T in each particular mutant. For uptake assays, this protocol is followed with adherent cells as described above. For binding, the dissociated cells are incubated in 96-well multiscreen plates containing GFB filters (Millipore, Bedford, MA) and the solution is removed by filtration. The

[44] A. Karlin and M. H. Akabas, *Methods Enzymol.* **293,** 123 (1998).

wells are washed with uptake buffer to remove residual cocaine and dopamine. Binding to the washed cells is performed as described above. Protection is calculated as 1-[(inhibition in the presence of drug)/(inhibition in the absence of drug)].

Interpretation of Results: Assumptions of SCAM

In order to interpret the results of SCAM we make the following assumptions: (1) the highly polar MTS reagents react only at the water-accessible surface of the protein; (2) in membrane-spanning segments, access of highly polar reagents to side chains is only through the binding-site crevice and transport pathway; (3) the addition of $-SCH_2CH_2X$ to a cysteine at the surface of the binding site or transport pathway alters binding and/or transport irreversibly, and, reciprocally, (4) that for substituted cysteines that line the binding site, inhibitors and substrates should retard the reaction with the MTS reagents. It is difficult to predict the effects of substrates and inhibitors on the reaction of the MTS reagents within the transport pathway but outside of their binding sites: the binding of an inhibitor might still retard reaction by blocking access of the MTS reagents to positions deeper in the transport pathway. Dopamine might also block access to particular residues, but it could also increase the reaction with residues that become accessible during transport.

The effects of the addition of $-SCH_2CH_2X$ to the engineered cysteine could be a result of steric block, electrostatic interaction, or indirect structural changes. Thus, reaction could inhibit or potentiate binding and transport, or even have opposite effects on binding and transport. Regardless, an irreversible effect is evidence of reaction, and, therefore, of the accessibility of the engineered cysteine. This can be illustrated in the dopamine D_2 receptor by the mutation of Asp-108.[17] Mutation to cysteine of this residue at the extracellular end of the third membrane-spanning segment reduced the receptor's affinity for antagonist binding about threefold. Reaction of the positively charged MTSEA or MTSET at this position significantly inhibited binding. In contrast, reaction of the negatively charged MTSES restored the negative charge at this position and shifted the affinity toward that of wild-type receptor, thereby increasing occupancy and potentiating binding.

Similarly, in DAT we have observed that reaction of MTSEA or MTSET with two endogenous cysteines potentiated [^{3}H]mazindol and [^{3}H]CFT binding.[34] The fact that reaction can potentiate function necessitates care in experimental design; a potentiation of binding could be missed by measuring binding at too high a ligand concentration relative to the K_D (Fig. 3A). For example, if assayed at a ligand concentration equal to the K_D, a fivefold

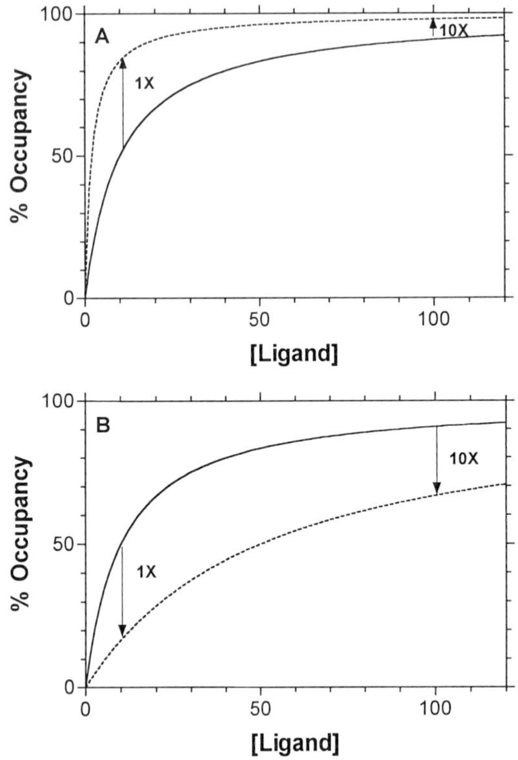

FIG. 3. Theoretical binding isotherms demonstrating the effects of a change in binding affinity on binding measured at different ligand concentrations. In (A), the dotted line represents a fivefold increase in affinity as compared with the solid line. At 1× the K_D, binding is increased by 68%, while at 10× the K_D, binding is increased by only 8%. In (B), the dotted line represents a fivefold decrease in affinity as compared with the solid line. At 1× the K_D, binding is decreased by 67%, while at 10× the K_D, binding is decreased by 27%.

increase in affinity would result in approximately a 68% increase in binding. In contrast, if assayed at a ligand concentration 10 times the K_D, the potentiation would be only about 8%, a change which might not be statistically significant, especially if the background were not completely unreactive. In contrast, a 5-fold decrease in affinity can be readily detected even at a ligand concentration 10 times the K_D (Fig. 3B).

In the dopamine D_2 receptor, we have observed that reaction of MTSEA at certain positions has a much greater effect on the binding of particular ligands: for example, reaction of MTSEA with the highly reactive endogenous cysteine, Cys-118, causes only a negligible decrease in the affinity for the antagonist haloperidol, but causes a 3000-fold decrease in the affinity for the antagonist sulpiride.[45] If reaction produces a negligible effect on the affinity of binding or the function of the receptor or transporter, a substituted cysteine which reacts with the reagents might be inferred to be inaccessible, resulting in a false negative determination. Thus, while it seems unlikely that a residue forming the surface of the transport pathway could be covalently modified by the addition of the charged $-SCH_2CH_2X$ without interfering with binding or transport, such a result is nonetheless possible. Moreover, a residue which is water-accessible might not react with the reagents because of steric factors. These potential complications demonstrate the importance of systematically mutating to cysteine consecutive residues along an entire membrane-spanning segment: although mutation of any individual residue might be subject to misinterpretation because of silent reaction or steric factors, this is unlikely to be a systematic problem affecting the overall pattern of accessibility of multiple residues in a membrane-spanning segment.

Cysteine Substitution

The ability to substitute cysteine residues for other residues and still obtain functional transporter is central to this approach. In dopamine D_2 receptor, 91 of 96 cysteine-substitution mutants tested to date bound antagonists with near-normal affinity.[15–17] These tolerated substitutions were for hydrophobic residues (alanine, leucine, isoleucine, methionine, and valine), polar residues (asparagine, serine, threonine), neutral residues (proline), acidic residues (aspartate), aromatic residues (phenylalanine, tryptophan, tyrosine), and glycine.

There are several reasons why cysteine substitution may be so well tolerated. Cysteine is a relatively small amino acid with a volume of 108 $Å^3$; only glycine, alanine, and serine are smaller.[46] In globular proteins, roughly half of non-disulfide-linked cysteines are buried in the protein interior and half are on the water-accessible surface of the protein.[47] Furthermore, cysteine has little preference for a particular secondary structure.[48,49]

A cysteine-substitution mutant which does not function cannot be studied by the substituted-cysteine accessibility method (or by traditional site-directed mutagenesis). Either the residues that cannot be mutated to cysteine are accessible in the binding-site crevice or the transport pathway and

[45] J. A. Javitch, D. Fu, and J. Chen, *Molec. Pharmacol.* **49,** 692 (1996).
[46] T. E. Creighton, "Proteins: Structures and Molecular Properties," W. H. Freeman and Co., New York, 1993.
[47] C. Chothia, *J. Molec. Biol.* **105,** 1 (1976).
[48] M. Levitt, *Biochemistry* **17,** 4277 (1978).
[49] P. Y. Chou and G. D. Fasman, *J. Molec. Biol.* **115,** 135 (1977).

make a crucial contribution to binding or transport, or they make a crucial contribution to maintaining the structure of the site and/or to the folding and processing of the transporter. The determination by the substituted-cysteine accessibility method of the accessibility of the neighbors of a crucial residue may allow us to infer the secondary structure of the segment containing this residue, and, thus, whether it is likely to be accessible as well. If it is not accessible, then the functional effect of its mutation is likely due to an indirect effect on structure.

Secondary Structure

To infer a secondary structure, we must assume that, if binding to or transport by a mutant is not affected by the MTS reagents, then no reaction has occurred, and that the side chain at this position is not accessible in the binding site or transport pathway. In an α-helical structure one would expect the accessible residues to form a continuous stripe when the residues are represented on a helical net. For example, in the third membrane-spanning segment of the dopamine D_2 receptor, the pattern of accessibility is consistent with this membrane-spanning segment forming an α helix with a stripe of about 140° facing the binding-site crevice[17] (Fig. 4). In contrast,

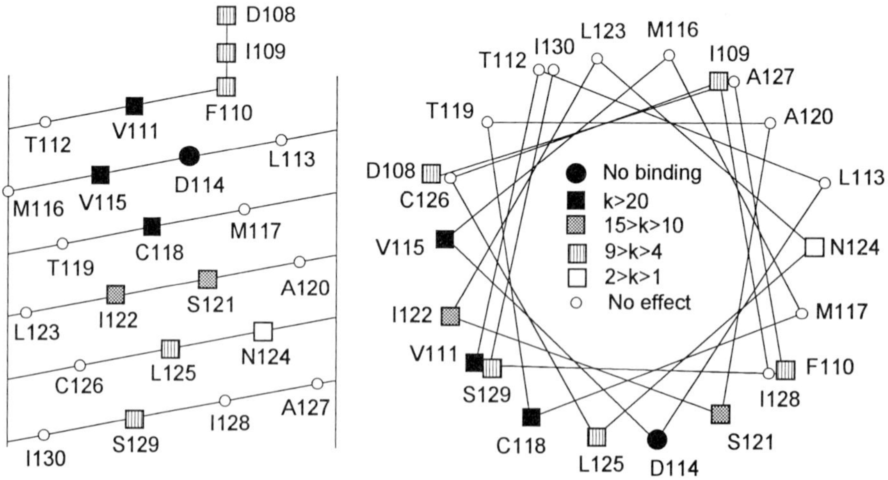

FIG. 4. Helical net (A) and helical wheel (B) representations of the residues in and flanking the M3 segment of the dopamine D_2 receptor, summarizing the effects of MTSEA on the binding of the antagonist, [^{3}H]YM-09151-2. Reactive residues are represented by squares, where the fill indicates the range of the second-order rate constants in $M^{-1} \cdot s^{-1}$ for reaction with MTSEA. Small open circles indicate that MTSEA had no effect on binding. Closed circle indicates no binding after cysteine substitution. D108 and I109 are represented outside of the α helix in the loop from the second membrane-spanning segment. [Adapted with permission from J. A. Javitch, D. Fu, J. Chen, and A. Karlin, *Neuron* **14,** 825 (1995).]

in a β strand, one would expect every other residue to be accessible to the reagents.

More complex or irregular patterns of accessibility can be more difficult to interpret, but these findings can also be rather informative. For example, we have found that cysteines substituted, one at a time, for 10 consecutive residues in the fifth membrane-spanning segment (M5) of the dopamine D_2 receptor are accessible to MTSEA and MTSET.[16] This pattern of accessibility is not consistent with M5 being a fixed α helix with one side facing the binding site crevice. We also observed an unusual pattern of accessibility in the seventh membrane-spanning segment (M7) of the dopamine D_2 receptor, but in this case our data were in agreement with a molecular model which predicted, based on previous data, that M7 is an α helix with a significant kink resulting from the presence of the highly conserved Asn-Pro in this membrane-spanning segment.[15] Thus, "irregular" patterns of accessibility can lead to new insights or questions for further experimental pursuit.

Topology

If the MTS reagents are membrane-impermeant, then when applied from the extracellular side of the membrane in the absence of substrate, they might only react with residues on the extracellular side of a constriction or gate in the transport pathway. MTSET is a quaternary ammonium with a fixed charge and MTSES is fully ionized at neutral pH. Both are membrane-impermeant.[23,50] MTSEA, however, is a weak base and has been reported to enter membrane vesicles readily[23,50]; it presumably crosses the membrane in the unprotonated state. Thus, although MTSEA should react vastly faster with ionized thiolates and should, therefore, only react with water-accessible cysteines, the reagent may gain access to these residues by passing through the membrane and then entering the transport pathway from the cytoplasmic side. Therefore, a residue which is water-accessible from the cytoplasmic side but not from the extracellular side of the membrane should not react at an appreciable rate with extracellularly applied MTSET or MTSES but might react with MTSEA.[51]

[50] M. Holmgren, Y. Liu, Y. Xu, and G. Yellen, *Neuropharmacol.* **35,** 797 (1996).

[51] In our mapping of three entire membrane-spanning segments of the dopamine D2 receptor expressed in HEK 293 (human embryonal kidney) cells, we have observed an essentially identical pattern of reaction for the positively charged reagents MTSEA and MTSET.[15–18] MTSEA is slightly more reactive than is MTSET, with accessible cysteines in the dopamine D2 receptor; however, MTSEA is smaller than MTSET (and the negatively charged MTSES), and it might gain more rapid access to certain residues in the water-accessible binding-site crevice simple because of reduced steric hindrance. Moreover, MTSEA shares with dopamine the ethylamine moiety, and its increased reactivity relative to MTSET may also result from an affinity reaction with the dopamine binding site.

Distinctions between SCAM and Typical Mutagenesis

A distinction between our approach and that of typical mutagenesis experiments is that we do not rely on the functional effects of a given mutation. The interpretation of the effects of typical mutagenesis experiments requires one to assume that functional changes caused by a mutation, such as changes in binding affinity or transport, are due to local effects at the site of the mutation and not due to indirect effects of the mutation on protein structure. The validity of this assumption is rarely proven for individual mutations, but the structure of the λ phage receptor, maltoporin, showed that 50% of the residues which, on the basis of mutagenesis experiments, had been implicated in λ phage recognition are actually buried.[52] Thus, mutation of these buried residues alters λ phage recognition indirectly. Likewise, the crystal structures of several dihydrofolate reductase mutants have demonstrated that a mutation approximately 15 Å from the substrate binding pocket exerts an effect on catalytic activity through an extended structural perturbation.[53]

In contrast to mutagenesis approaches which only detect perturbations in protein function, the substituted-cysteine accessibility method allows one to determine whether a residue is on the water-accessible surface of the binding site and/or transport pathway when the mutant has near-normal function. Other advantages of the approach include the ability to probe binding sites by assessing the ability of ligand to retard the reaction of the MTS reagents with particular substituted cysteines and the ability to probe the steric constraints and electrostatic potential of sites by comparing the rates of reaction of reagents of varying size and charge.

Acknowledgments

I would like to thank Myles Akabas and Arthur Karlin for helpful discussion and for comments on this manuscript and Thomas Caughey, Jiayun Chen, and Jamine Ferrer for assistance with experiments on the dopamine transporter. This work was supported in part by an American Heart Association Grant-in-Aid and by NIH grants MH57324, MH54137, and DA11495.

[52] T. Schirmer, T. A. Keller, Y. F. Wang, and J. P. Rosenbusch, *Science* **267,** 512 (1995).

[53] K. A. Brown, E. E. Howell, and J. Kraut, *Proc. Natl. Acad. Sci. USA* **90,** 11753 (1993).

[24] Biosynthesis, N-Glycosylation, and Surface Trafficking of Biogenic Amine Transporter Proteins

By SAMMANDA RAMAMOORTHY, HALEY E. MELIKIAN, YAN QIAN, and RANDY D. BLAKELY

Introduction

Except for acetylcholine, whose actions are limited by extracellular enzymatic hydrolysis, neurotransmitters, including L-glutamate, γ-aminobutyric acid (GABA), dopamine (DA), nonrepinephrine (NE), and serotonin (5HT), are inactivated by synaptic transport systems that rapidly clear neurotransmitter after release.[1–4] Molecular cloning and heterologous expression studies[5–12] have provided new opportunities to evaluate structural determinants of transporter function and regulation. For example, we now know that transport activities responsible for clearance of individual neurotransmitters are the products of single genes, though the more abundant neurotransmitters (L-glutamate, GABA, glycine) possess transporter genetic variants whose sites of expression and pharmacologic characteristics can be discriminated *in vivo* (with some overlap in specificity evident).[13–21]

[1] L. L. Iversen, *Br. J. Pharmacol.* **41,** 571 (1971).
[2] B. I. Kanner and S. Schuldiner, *CRC Crit. Rev. Biochem.* **22,** 1 (1987).
[3] U. Trendelenburg, *Trends Pharmacol. Sci.* **32,** 334 (1991).
[4] G. Rudnick and J. Clark, *Biochim. Biophys. Acta* **1144,** 249 (1993).
[5] J. Guastella, N. Nelson, H. Nelson, L. Czyzyk, S. Keynan, M. C. Miedel, N. Davidson, H. A. Lester, and B. I. Kanner, *Science* **249,** 1303 (1990).
[6] T. Pacholczyk, R. D. Blakely, and S. G. Amara, *Nature* **350,** 350 (1991).
[7] R. D. Blakely, H. E. Berson, R. T. Fremeau, Jr., M. G. Caron, M. M. Peek, H. K. Prince, and C. C. Bradley, *Nature* **354,** 66 (1991).
[8] B. J. Hoffman, E. Mezey, and M. J. Brownstein, *Science* **254,** 579 (1991).
[9] S. Ramamoorthy, A. L. Bauman, K. R. Moore, H. Han, T. Yang-Feng, A. S. Chang, V. Ganapathy, and R. D. Blakely, *Proc. Natl. Acad. Sci. USA* **90,** 2542 (1993).
[10] J. E. Kilty, D. Lorang, and S. G. Amara, *Science* **254,** 578 (1991).
[11] S. Shimada, S. Kitayama, C. Lin, A. Patel, E. Nanthakumar, P. Gregor, M. Kuhar, and G. Uhl, *Science* **254,** 576 (1991).
[12] R. T. Fremeau, M. G. Caron, and R. D. Blakely, *Neuron* **8,** 915 (1992).
[13] G. Pines, N. C. Danbolt, M. Bjoras, Y. Zhang, A. Bendahan, L. Eide, H. Koepsell, J. Storm-Mathisen, E. Seeberg, and B. I. Kanner, *Nature* **360,** 464 (1992).
[14] Y. Kanai and M. A. Hediger, *Nature* **360,** 467 (1992).
[15] T. Storck, S. Schulte, K. Hofmann, and W. Stoffel, *Proc. Natl. Acad. Sci. USA* **89,** 10955 (1992).
[16] J. A. Clark and S. G. Amara, *BioEssays* **15,** 323 (1993).
[17] S. G. Amara and J. L. Arriza, *Curr. Biol.* **3,** 337 (1993).

Interestingly, the gene family encoding excitatory amino acid transporters bears little if any sequence relationship to the gene family encoding GABA, glycine, and biogenic amine (DA, NE, Epi, 5HT) transporters. The GABA/amine transporters are further discriminated by their dependence on both extracellular Na^+ and Cl^-.[4,17,22–24] Both gene families, however, comprise topologically complex integral membrane proteins where canonical N-glycosylation and phosphorylation sites suggested multiple opportunities for posttranslational processing.

Transporter substrate specificity and antagonist selectivity are likely to be attributes of amino acid side chains that comprise the transmitter permeation pathway (See E. L. Barker and R. D. Blakely, this volume[25]). However, the mechanisms by which transporter proteins fold to achieve the requisite structure for substrate recognition, how efficiently they reach the cell surface, how long they remain resident in the plasma membrane, and how efficiently they catalyze neurotransmitter transport appear to also require specific features of transporter structure, some of which are affected by posttranslational mechanism.[16,19,26–28] Thus, point mutations introduced in transporter proteins *in vitro* sometimes reduce cell surface expression or compromise transport efficiency with little or no alteration in substrate recognition.[29–33] These studies reveal sensitive sites in transporters whose mutation *in vivo* might contribute hereditary determinants to brain disease.

---

[18] Q. Liu, B. Lopez-Corcuera, S. Mandiyan, H. Nelson, and N. Nelson, *J. Biol. Chem.* **268,** 2106 (1993).

[19] B. Borowsky, E. Mezey, and B. J. Hoffman, *Neuron* **10,** 851 (1993).

[20] J. L. Arriza, M. P. Kavanaugh, W. A. Fairman, Y. Wu, G. H. Murdoch, R. A. North, and S. G. Amara, *J. Biol. Chem.* **268,** 15329 (1993).

[21] B. I. Kanner, *FEBS Lett.* **325,** 95 (1993).

[22] B. I. Kanner, *Biochem. Soc. Trans.* **19,** 92 (1991).

[23] S. G. Amara and M. J. Kuhan, *Annu. Rev. Neurosci.* **16,** 73 (1993).

[24] D. M. Worrall and D. C. Williams, *Biochem. J.* **297,** 425 (1994).

[25] E. L. Barker and R. D. Blakely, *Methods Enzymol.* **296,** [33], 1998 (this volume).

[26] K. E. Smith, L. A. Borden, P. R. Hartig, T. Branchek, and R. L. Weinshank, *Neuron* **8,** 927 (1992).

[27] Q. Liu, B. Lopez-Corcuera, S. Mandiyan, H. Nelson, and N. Nelson, *J. Biol. Chem.* **268,** 22802 (1993).

[28] K. Kim, S. F. Kingsmore, H. Han, T. L. Yang-Feng, N. Godinot, M. F. Seldin, M. G. Caron, and B. Giros, *Mol. Pharmacol.* **45,** 608 (1994).

[29] C. G. Tate and R. D. Blakely, *J. Biol. Chem.* **269,** 26303 (1994).

[30] H. E. Melikian, J. K. McDonald, H. Gu, G. Rudnick, K. R. Moore, and R. D. Blakely, *J. Biol. Chem.* **269,** 12290 (1994).

[31] H. E. Melikian, S. Ramamoorthy, C. G. Tate, and R. D. Blakely, *Mol. Pharmacol.* **50,** 266 (1996).

[32] T. T. Nguyen and S. G. Amara, *J. Neurochem.* **67,** 645 (1996).

[33] L. Olivares, C. Aragon, C. Gimenez, and F. Zafra, *J. Biol. Chem.* **270,** 9437 (1995).

Mutations of this nature are well described in other transport systems. For example, the most prominent mutation of cystic fibrosis transmembrane conductance regulator (CFTR) results in retention of CFTR protein in intracellular organelles.[34–36] As work proceeds to define the consequences of site-directed mutagenesis,[37–40] evaluate heritable structural variants,[41–43] and determine the pathways for regulated neurotransmitter transporter surface expression,[31–33,44,45] knowledge of basic methods to evaluate biosynthetic pathways and transporter localization should be useful. These methods become important again as one seeks to elucidate regulatory pathways by which transporter half-life and surface expression may be regulated in the absence of structural mutations. For example, we have returned to cell-surface biotinylation techniques that we first employed to study the role of N-glycosylation in NET function[31] for an evaluation of how activators of protein kinase C (PKC) regulate the transport capacity of norepinephrine and serotonin transporters (NET and SERT, respectively).[46,47]

Below we illustratc methods we have employed to characterize the synthesis, N-glycosylation, and surface expression of antidepressant-sensitive NETs and SERTs in mammalian cells and discuss the resulting data in the context of recent studies of transporter posttranslational processing and trafficking. Our specific goals in these studies was to determine (1)

[34] S. H. Cheng, R. J. Gregory, J. Marshall, S. Paul, D. W. Souza, G. A. White, C. R. O'Riordan, and A. E. Smith, *Cell* **63,** 827 (1990).

[35] P. J. Thomas, P. Shenbagamurthi, J. Sondek, J. M. Hullihen, and P. L. Pedersen, *J. Biol. Chem.* **267,** 5727 (1992).

[36] G. M. Denning, M. P. Anderson, J. F. Amara, J. Marshall, A. E. Smith, and M. J. Welsh, *Nature* **358,** 761 (1992).

[37] B. Giros, Y. M. Wang, S. Suter, S. B. McLeskey, C. Pifl, and M. G. Caron, *J. Biol. Chem.* **269,** 15985 (1994).

[38] K. J. Buck and S. G. Amara, *Proc. Natl. Acad. Sci. USA* **91,** 12584 (1994).

[39] E. L. Barker, H. L. Kimmel, and R. D. Blakely, *Mol. Pharmacol.* **46,** 799 (1994).

[40] E. L. Baker and R. D. Blakely, *Mol. Pharmacol.* **50,** 957 (1996).

[41] A. Heils, A. Teufel, S. Petri, G. Stober, P. Riederer, D. Bengel, and K. P. Lesh, *J. Neurochem.* **66,** 2621 (1996).

[42] G. Stöber, A. Heils, and K. P. Lesch, *Lancet* **347,** 1340 (1996).

[43] K. P. Lesch, D. Bengel, A. Heils, S. Z. Sabol, B. D. Greenberg, S. Petri, J. Benjamin, C. R. Müller, D. H. Hamer, and D. L. Murphy, *Science* **274,** 1527 (1996).

[44] H. H. Gu, J. Ahn, M. J. Caplan, R. D. Blakely, A. I. Levey, and G. Rudnick, *J. Biol. Chem.* **271,** 18100 (1996).

[45] C. Perego, A. Bulbarelli, R. Longhi, M. Caimi, A. Villa, M. J. Caplan, and G. Pietrini, *J. Biol. Chem.* **272,** 6584 (1997).

[46] Y. Qian, A. Galli, S. Ramamoorthy, S. Risso, L. J. DeFelice, and R. D. Blakely, *J. Neurosci.* **17,** 45 (1997).

[47] S. Apparsundaram, A. Galli, L. J. DeFelice, H. C. Hartzell, and R. D. Blakely, *J. Pharm. Exp. Ther.* in press, 1998.

whether mutations introduced into NET and SERT cDNAs disrupt protein synthesis or stability, (2) whether NET and SERT utilize canonical N-glycosylation sites found in the TMD 3-4 loop, (3) whether N-glycosylation plays an important role in transporter protein biosynthesis and surface trafficking, and (4) whether kinase-mediated signal transduction pathways regulate transporter surface expression.

Cell Culture and Transfections

Cell lines are maintained in 6- or 24-well tissue culture plates in a humidified atmosphere of 5% (v/v) CO_2 at 37°. HeLa cells, HEK 293 (human embryonal kidney) cells, and COS-1 (monkey kidney) cells are grown in Dulbecco's modified Eagle's medium (DMEM), and LLC-NET cells are grown in minimum essential medium (MEM) (Life Technologies, Gaithersburg, MD). HEK-293 cells are grown on poly(D-lysine) (0.1 mg/ml) coated plates to improve adherence. All media are supplemented with 5–10% fetal bovine serum, 2 m*M* glutamine, 100 U/ml penicillin, and 100 μg/ml streptomycin. Transporter cDNAs cloned in pBluescript SKII$^-$ (Stratagene, La Jolla, CA), pRC/CMV (InVitrogen), or pcDNA3 (InVitrogen, Carlsbad, CA) are introduced into mammalian cells (HeLa, COS, LLC-PK1, HEK-293) using liposome (Lipofectin, Life Technologies)-mediated transfection at a lipid/DNA ratio of 3:1. For transient expression of transporters, we make use of the vaccinia virus–T7 expression system[48] or the replication-enhanced COS cell transient system.[49] The vaccinia virus system provides for rapid, high-level expression (6–18 hr) of transport activities due to the cytosolic expression of T7 RNA polymerase,[50–52] along with viral factors to assist in focused translation of viral-transcribed mRNAs, whereas the COS system affords high-level transient expression because of the replication (up to 10,000-fold) of episomal DNA containing the SV40 replication origin.[49] Stable cell lines (LLC-PK1 and HEK-293) are derived from single cells after 2-week initial culture in G418 (250–1000 μg/ml) to select for neomycin resistance. Stable LLC-hNET cells were a gift from Gary Rudnick (Yale University, New Haven, CT). Stable HEK-293hSERT cells are produced in our laboratory.[46]

[48] R. D. Blakely, J. A. Clark, G. Rudnick, and S. G. Amara, *Anal. Biochem.* **194,** 302 (1991).
[49] Y. Gluzman, *Cell* **23,** 175 (1997).
[50] T. R. Fuerst, E. Niles, F. W. Studier, and B. Moss, *Proc. Natl. Acad. Sci. USA* **83,** 8122 (1986).
[51] T. R. Fuerst, P. L. Earl, and B. Moss, *Mol. Cell Biol.* **7,** 2538 (1987).
[52] O. Elroy-Stein, T. R. Fuerst, and B. Moss, *Proc. Natl. Acad. Sci. USA* **86,** 6126 (1989).

Transport and Radioligand Binding Assays

For NE and 5HT transport assays, medium is removed by aspiration, and the cells are washed with 2 ml of Krebs–Ringer–HEPES (KRH) buffer containing 130 m*M* NaCl, 1.3 m*M* KCl, 2.2 m*M* $CaCl_2$, 1.2 m*M* $MgSO_4$, 1.2 m*M* KH_2PO_4, 1.8 g/liter glucose, and 10 m*M* HEPES, pH 7.4. Cells are preincubated in KRH buffer containing 100 μM pargyline and 100 μM ascorbic acid (to block catechol *O*-methyltransferase) for 10 min at 37° and the uptake assays are initiated by the addition of 1-[7,8-^{3}H]norepinephrine ([^{3}H]NE), [7,8-^{3}H]dopamine ([^{3}H]DA), or 5-hydroxy[^{3}H]tryptamine ([^{3}H]5-HT) for 5–10 min at 37° and terminated by three rapid washes (2 ml each) with cold KRH buffer. Studies also suggest that inclusion of catechol-*O*-methyltransferase (COMT) inhibitors (e.g., tropolone, U-0251) can limit the degradation of catecholamines, particularly DA, and increase accumulation of radiolabel in transport assays.[53] Cells are then solubilized in a small volume (0.5–1.0 ml) of 1% sodium dodecyl sulfate (SDS) and a detergent-tolerant scintillation fluor is added to lysed samples to allow ^{3}H accumulation to be quantitated by liquid scintillation spectrometry. Specific uptake is determined by subtracting the amount of accumulated radiolabel observed in the presence of low concentrations of tricyclic antidepressants (NET and SERT uptake inhibitors) or utilizing transport observed in vector transfected cells or parental cell lines. Antagonists are typically preincubated with cells for 10 min prior to addition of substrates. Competitive substrates are added at the same time as labeled substrates. Figure 1A illustrates the consequences on uptake of NE if hNET-transfected LLC-PK1 cells are treated with the N-glycosylation inhibitor tunicamycin (TM) for 24 hr at 10 μg/ml.[30] Transport capacity is significantly reduced, though the K_m for NE is unchanged. The residual transport activity retained after TM treatment has normal antagonist sensitivity (data not shown). These data suggest a role for proper N-glycosylation in either NET stability, establishing transporter surface expression, or NE transport efficiency, though the effects could be mediated, at this point, but an associated, N-glycosylated protein other than hNET.

Transporters can also be evaluated by assessing the presence and characteristics of high-affinity binding sites for radiolabeled antagonists. This is particularly useful to assess transporter density irrespective of transport activity and to verify an absence of gross structural changes in transporter structure introduced by mutations. These are conducted either as whole-cell assays or using membranes harvested from lysed cells. Caution is required in

[53] A. J. Eshleman, E. Stewart, A. K. Evenson, J. N. Mason, R. D. Blakely, A. Janowsky, and K. A. Neve, *J. Neurochem.* **69,** 1459 (1997).

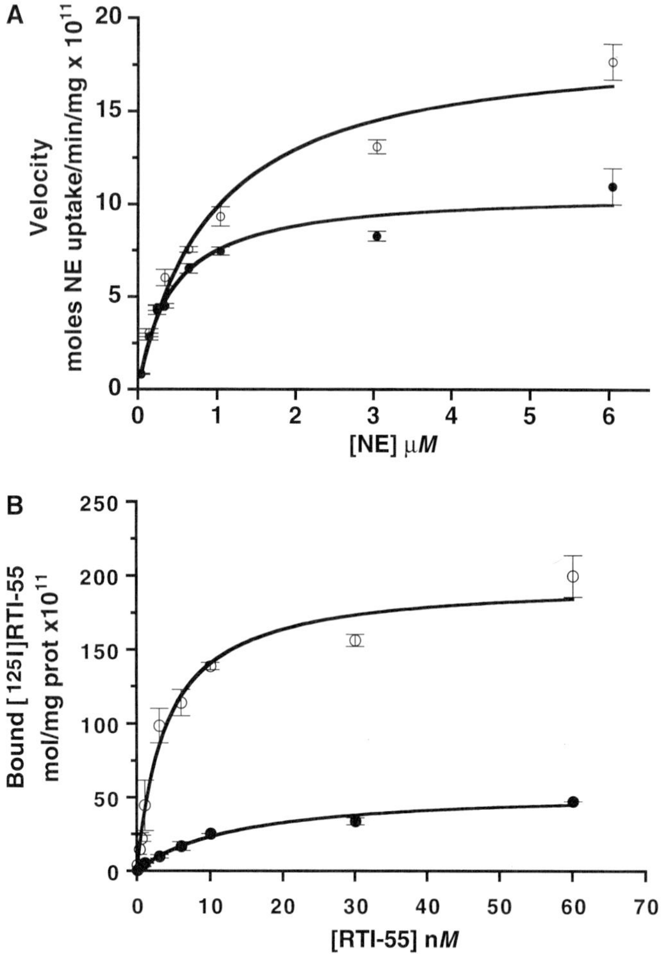

FIG. 1. Effect of tunicamycin (TM) treatment on NE transport (A) and ligand binding (B) of hNET in LLC-NET cells. Substrate velocity profiles were obtained from NE transport studies conducted under initial rate conditions with vehicle (○) or TM-treated (10 μg/ml, 24 hr) (●) LLC-NET cells. Results are the means ± S.E. for assays conducted in triplicate, with values adjusted for nonspecific transport, defined as the accumulation of NE in the presence of 1 μM desipramine. Saturation binding of [^{125}I]RTI-55 was performed on membranes prepared from vehicle (○) or TM-treated (10 μg/ml, 24 hr) (●) LLC-NET cells. Results are expressed as pmol/mg of protein, with derived values adjusted for nonspecific binding, defined as [^{125}I]RTI-55 binding in the presence of 1 μM desipramine. [Reprinted with permission from Melikian *et al.*, *J. Biol. Chem.* **269,** 12290 (1994).]

interpreting binding to intact cells, particularly as a measure of surface expression, since many if not all of the available antagonist radioligands are lipophilic and can presumably bind to sequestered, nonfunctional transporters. In either case it is important to first determine times at which binding reaches equilibrium and to utilize structurally distinct antagonists, vector-transfected cells, or parental cells to define nonspecific binding. We have utilized [3β-(4-iodophenyl)tropane-2β-carboxylic acid methyl ester tartrate ([^{125}I]RTI-55) (also known as β-CIT) as a ligand for NET and SERT, since the agent binds to both transporters, though with higher affinity to SERT.[54,55] For whole-cell assays, washed cells are incubated with a range of concentrations of radioligands or antagonists in KRH buffer on ice (equilibrium ~1 hr) to limit internalization of radioligand. Unbound radiolabel is then removed by aspiration and washing and the cells solubilized as above; iodinated samples are transferred to a gamma counter for quantiation.[56] Membrane binding assays are perfomred using membranes prepared from transfected cells by homogenization in ice-cold phosphate-buffered saline (PBS) [Polytron tissue homogenizer (Brinkmann Instruments, Inc., NY), 18,000 rpm, 30 sec] followed by centrifugation at 20,000*g* for 30 min at 4°. Membranes (1–100 μg protein) are incubated with radioligands in borosilicate glass tubes to reach equilibrium (~1 hr) and then rapidly filtered over 0.3% (v/v) polyethyleneimine (PEI)-coated glass fiber filters (GFB, Whatman, Clifton, NJ) using a Brandel cell harvester (Gaithersburg, MD). Membranes are washed three times within 10 sec and the dried membranes transferred directly to counting tubes for quantitation of bound isotope. Figure 1B demonstrates the effects of TM on [^{125}I]RTI-55 binding to membranes from LLC-NET cells. As with transport measurements, biosynthetic blockage of N-glycosylation results in a loss of hNET activity, revealed as a loss in maximal capacity for antagonist binding. Radioligand affinity is unchanged and can be inhibited with appropriate concentrations of unlabeled antagonists, suggesting a loss of transporter proteins. The greater loss of binding sites than transport capacity (see Fig. 1A) is consistent with a slow turnover of plasma membrane pools of uptake sites and a significant intracellular concentration of hNET protein that can bind radioligand, both suspicions borne out in subsequent studies.[29,30,32]

[54] J. W. Boja, W. M. Mitchell, A. Patel, T. A. Kopajtic, F. I. Carroll, A. H. Lewin, P. Abraham, and M. J. Kuhar, *Synapse* **12,** 27 (1992).

[55] H. Gu, S. C. Wall, and G. Rudnick, *J. Biol. Chem.* **269,** 7124 (1993).

[56] K. R. Moore and R. D. Blakely, *Biotechniques* **17,** 130 (1994).

Immunoblotting and Immunoprecipitation of Transporter Protein

Measurements of transport kinetics and antagonist binding fail to present the entire picture of *how* and *where* transporter proteins become modified during the course of normal biosynthesis or as a result of mutations or regulatory events. Detection of transporter protein in native tissues and transfected cells is typically achieved by immublotting of detergent-solubilized extracts using antisera raised against discrete epitopes (conjugated peptides, fusion proteins).[30,32,57,58] We prepare extracts from cells and tissues using radioimmunoprecipitation assay (RIPA) buffer (10 m*M* Tris-HCl, pH 7.4, 150 m*M* NaCl, 1 m*M* EDTA, 0.1% SDS, 1% Triton X-100, 1% sodium deoxycholate) supplemented with protease inhibitors [1 mg/ml soybean trypsin inhibitor, 1 m*M* *o*-phenanthroline, 1 μg/ml leupeptin, 1 m*M* iodoacetamide, 1 μM pepstatin A, 250 μM phenylmethylsulfonyl fluoride (PMSF), and 1 μg/ml of aprotinin] for 25 min at 4° with constant shaking (200 rpm). Samples are centrifuged at 20,000*g* for 30 min at 4° to pellet insoluble material and supernatants are electrophoresed on a denaturing 8–10% SDS–polyacrylamide gel electrophoresis (PAGE) gel and transferred to ECL (enhanced chemiluminescence) nitrocellulose membranes (Hybond, Amersham, Arlington Heights, IL) by electroblotting. Gels can be stained with Coomassie Brilliant Blue or the membranes stained with 0.1% Ponceau S following transfer to verify complete transfer. Membranes are blocked in blocking solution (5% Carnation nonfat dried milk, 0.5% Tween 20, PBS) for 45–60 min at 22° and then incubated with affinity purified antibodies, typically at 0.5–1 μg/ml in blocking solution for 45–60 min at 22°. Volumes of incubations are kept minimized so as not to waste antibody; incubation solutions can often be reused if stored at 4° in azide (0.1%). Blots are then washed (2 × quick rinse, 1 × 15 min, 2 × 5 min) in PBS with 0.5% Tween 20 at 22° and incubated with horseradish peroxidase-conjugated goat antirabbit secondary Ab at 1 : 3000 to 1 : 10,000 dilution in blocking solution for 45 min at 22°. Blots are then washed as above, and the immunoreactive bands detected by enhanced chemiluminescence of HRP substrates (Amersham). For transfected cells, it is useful to run nontransfected cell extracts in adjacent lanes to document specificity of antibodies. Bands should be absent when blotting is performed with (1) preimmune sera, (2) primary antibody preincubated with immunizing peptide or fusion protein, or (3) an irrelevant antibody. An example of an immunoblot of hNET protein expressed in stably transfected LLC-PK1

[57] Y. Qian, H. E. Melikian, D. B. Rye, A. I. Levey, and R. D. Blakely, *J. Neurosci.* **15,** 1261 (1995).

[58] M. Brüss, R. Hammermann, S. Brimijoin, and H. Bönisch, *J. Biol. Chem.* **270,** 9197 (1995).

cells is shown in Fig. 2. Two bands are seen in the LLC-NET cells that are absent in parental LLC-PK1 cells, one at ~56 kDa and the other a broader, more abundant band at ~80 kDa. To examine whether differences in the mobility of hNET polypeptides are accounted for by proteolysis or differential N-glycosylation, LLC-NET cells are treated with TM prior to immunoblots or LLC-NET RIPA extracts are subjected to enzymatic deglycosylation using PNGase F, an enzyme that removes all N-linked glycosylation regardless of sugar content, followed by visualization of core hNET protein on immunoblots. An immunoblot of TM-treated cells (Fig. 3) reveals the appearance of a new 46-kDa band, a loss of the 56-kDa band, and a gradual reduction of the 80-kDa band. These data suggest that the 80-kDa form is a more highly N-glycosylated form of hNET protein, that the 56-kDa form also carries N-glycosylation, and that the 56-kDa form is an intermediate in the synthesis of the 80-kDa form. The retention

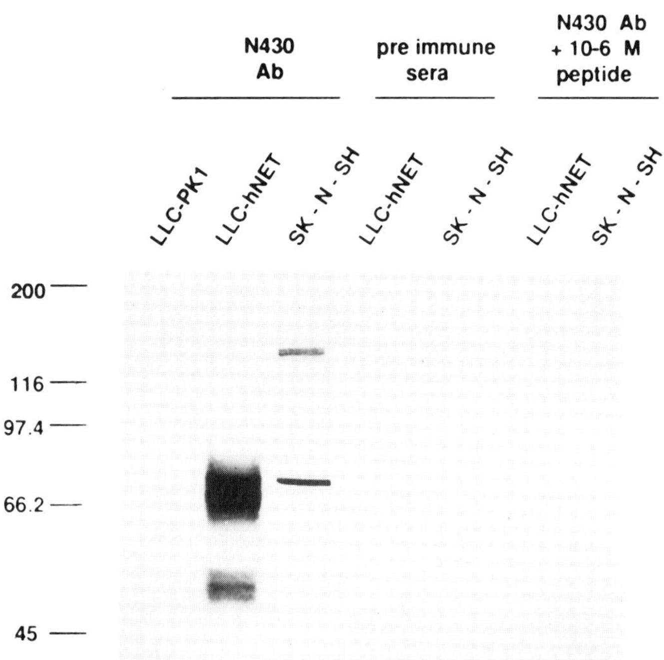

FIG. 2. Immunoblot of hNET expression in stably transfected LLC-PK1 cells. RIPA extracts of total protein from LLC-PK1 (25 μg) and LLC-NET cells (25 μg) were subjected to SDS–PAGE, electroblotted, and probed with NET specific antibody (N430). The numbers on the left of the blot indicate the positions of molecular mass marker in kDa. [Reprinted with permission from Melikian *et al.*, *J. Biol. Chem.* **269,** 12290 (1994).]

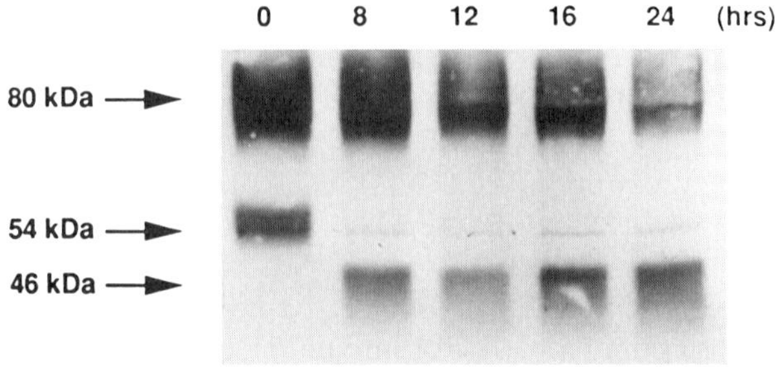

FIG. 3. Effects of TM treatment on hNET protein expression in LLC-NET cells. LLC-NET cells were incubated with TM (10 μg/ml, 37°) for the times indicated and RIPA extracts immunoblotted after SDS–PAGE with NET specific antibody N430. [Reprinted with permission from Melikian *et al.*, *J. Biol. Chem.* **269,** 12290 (1994).]

of 80-kDa material is likely to reflect the presence of material that was synthesized and N-glycosylated prior to TM treatment. To explore these conclusions, extracts are treated directly to remove N-glycosylation. LLC-NET cell extracts prepared with RIPA as described above are incubated with 0.1 volumes of PNGase F denaturation buffer (New England Biolabs) for 5 min, 22°, followed by addition of 0.1 volume of each of 10% Nonidet P-40, 10× PNGase F buffer, and PNGase F (40 mU/ml), and the samples are incubated for 16 hr at 37°. PNGase F treatment results in the loss of both 80- and 54-kDa immunoreactive bands, and the appearance of the lower molecular mass doublet 46- and 44-kDa bands (Fig. 4). We should note that these digestion conditions are rather harsh and may limit opportunities to assay functions (binding) of treated proteins, though some sucesses have been described.[59] On the basis of the migration of hNET synthesized in the presence of tunicamycin as well as from glycosylation site-mutated hNET (see later text), the 46-kDa band is concluded to be the non-N-glycosylated core polypeptide and the 44-kDa fragment presumably a proteolytic fragment generated in the rather longer incubation required for complete PNGase treatment.

These studies with TM and PGNAse F reveal that hNET protein is N-glycosylated as predicted by the appearance of canonical N-glycosylation sites in the TMD 3–4 loop. The loss of transport activity and antagonist binding sites suggest that an inability to N-glycosylate NET would alter the stability of the transporter, cell surface expression, or turnover rate. To this point, however, we have not documented that the N-glycosylation states observed reflect sequential processing variants, that processing coin-

[59] E. Nunez and C. Aragon, *J. Biol. Chem.* **269,** 16920 (1994).

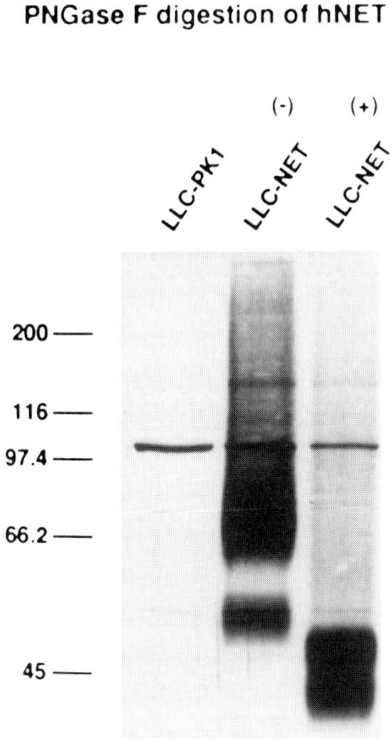

FIG. 4. Deglycosylation of hNET protein by PNGase F. Total extracts of LLC-NET cells (50 μg) were incubated with (+) or without (−) PNGase F for 16 hr, at 37°, subjected to SDS–PAGE, and immunoblotted with NET specific antibody N430. As a control, LLC-PK1 cell extracts were blotted in parallel. Arrows indicate the M_r of hNET species as determined by parallel electrophoresis of molecular mass standards. [Reprinted with permission from Melikian *et al.*, *J. Biol. Chem.* **269,** 12290 (1994).]

cides with cell surface expression, or how lack of N-glycosylation affects transport function. The maturation of transporter protein and its movement through different cellular compartments can be readily explored by immunoprecipitation of transporters metabolically labeled with [^{35}S]methionine and [^{35}S]cysteine. Pulse-chase metabolic labeling involves a brief (10-min pulse) labeling period followed by removal of medium and replenishment by normal medium containing unlabeled methionine and cysteine (chase). Cells are harvested at timed intervals after initiation of the chase with unlabeled amino acids and transporter protein immunoprecipitated as described for steady-state labeling. The pulse-chase procedure results in visualization of proteins synthesized only during the labeling period, thus permitting a kinetic evaluation of changes in transporter SDS–PAGE mobility and subcellular localization. Stable cell lines are seeded on 6-well cell culture plates at 400,000 cells/well for 48 hr. Cells are washed three times with prewarmed methionine/cysteine-free media and incubated further with the 2 ml of methionine/cysteine-free DMEM media for 30 min at 37° to deplete intracellular methionine and cysteine pool. Deficient media are removed and replaced with 500 μl of methionine/cysteine-free DMEM media supplemented Tran^{35}S-label (^{35}S *Escherichia coli* hydrolyzate labeling regent, ICN, Costa Mesa, CA) to a final concentration of 500 μCi/ml and labeled for 10 min at 37°. Cells are washed three times with cold sodium phosphate-buffered saline, pH 7.4, and incubated with complete media until solubilization in 300 μl of RIPA. A variant of this procedure, where transporters are incubated for 2–24 hr (steady-state labeling) in labeled medium without chase, is often convenient for identifying all biosynthetic products and establishing their relative abundance in cells, though this has no particular attractiveness over direct immunoblotting (however, some antibodies immunoprecipitate but will not immunoblot). Lysates are collected in a 1.5 ml microcentrifuge tubes and centrifuged at 20,000*g* for 30 min at 4° to remove particulate material and nucleic acids. Labeled total cell extracts may be precleared by incubation with protein A-Sepharose alone or preimmune sera plus protein A-Sepharose for 2–4 at 4° by end-over-end rotation. In addition, protein A-Sepharose beads can be preblocked with unlabeled cell extracts or with BSA prior to use for immunocomplex isolation. These treatments help to minimize nonspecific background interaction of labeled protein with immunoprecipitation reagents and are optional or required, depending on the expression system and specificity of reagents. Immunoprecipitations are then performed by incubating transporter antibody (typically 5–25 μl serum/500 μl extract) for 1 hr at 22° or overnight at 4° with continuous mixing. We use protein A-Sepharose beads that are preblocked by incubation with unlabeled, parental cell extract for 1 hr at 22° followed by three washes in RIPA buffer and stored at 4° as gel slurry suspension at 30 mg/ml in RIPA buffer. Blocked beads (100 μl) are added to samples and incubated for either 1 hr at 22° or overnight at 4° with continuous mixing on a rocking platform. After incubation, the immunocomplexes are collected by low-speed centrifugation (10,000*g*, 4 min, 4°) and washed three times with cold RIPA buffer. Immunoprecipitated proteins are eluted from the beads with Laemmli sample buffer (0.0625 *M* Tris-HCl, pH. 6.8, 10% (v/v) glycerol, 2% (w/v) SDS, 5% (v/v) 2-mercaptoethanol, 0.00125% (w/v) bromphenol blue) for 20 min at 22°. In our experiences, boiling beads at this stage to elute samples induces transporter protein aggregation and is to be avoided. Beads are centrifuged (10,000*g*, 4 min, 22°) and supernatants are subjected to electrophoresis on 8–10% denaturing SDS–polyacrylamide gels.[60] Gels are then soaked in Entensify enhancing solution (Du Pont

[60] U. K. Laemmli, *Nature* **227,** 680 (1970).

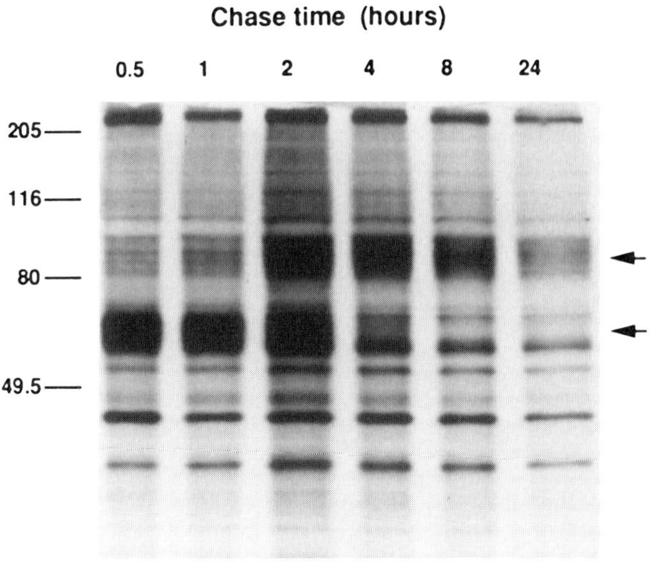

FIG. 5. Pulse-chase analysis of hNET biosynthesis in LLC-NET cells. LLC-NET cells were incubated with 500 μCi/ml [35S] Met/Cys label in Met/Cys-free medium for 10 min (pulse) rapidly washed, and incubated in unlabeled serum containing media (chase) for the times indicated. RIPA extracts were immunoprecipitated with NET specific antibody N430. Immunoisolated hNET was subjected to SDS–PAGE, treated with Entensify enhancing solution, dried and exposed for autoradiography. Arrows on right indicate the migration of the 80- and 54-kDa products synthesized in LLC-NET cells. [Reprinted with permission from Melikian *et al.*, *J. Biol. Chem.* **269,** 12290 (1994).]

NEN) or 1 *M* sodium salicylate,[61] dried, and exposed to X-ray film at −80° or a Phosphoimager (Molecular Dynamics) plate for 2–24 hr. Quantitation is performed by scanning densitometry of films using serial dilutions fo Tran^{35}S-label as a standard to establish the linear range of the film or directly (Phosphoimager) using ImageQuant software (Molecular Dynamics). As with immunoblots, control immunoprecipitations are performed with (1) preimmune sera, (2) an irrelevant primary antibody at the same dilution, (3) antisera preblocked with the corresponding antigen used to raise the antibody or with an irrelevant antigen, (4) protein A-Separose lacking antisera, or (5) nontransfected parental cell line extracts. These control experiments are essential to interpretation of immunoprecipitation experiments since often spurious bands arise from a lack of absolute specificity of antibodies or nonspecific adherence of cellular proteins to beads. Figure 5 shows the result of a pulse-chase experiment performed in LLC-NET cells.

[61] J. Chamberlain, *Anal. Biochem.* **98,** 132 (1979).

Note the initial appearance of a 54-kDa band and its gradual disappearance which coincides with the appearance of the 80-kDa form. These data confirm the precursor–product relationship of the two hNET N-glycosylated bands in LLC-NET cells. It should be noted that vaccinia T7 expressed hNET in Hela cells and in COS cells exists predominately in the 54-kDa form.[30,31] Differential N-glycosylation of transporters is a common occurrence across different cell lines and has been observed also to vary with development and brain region.[30,31,58,62–64] Studies performed using site-directed mutagenesis and cell-surface biotinylation, described later, suggest that differential N-glycosylation may affect protein stability and cell surface expression and also have on impact on the kinetic efficiency of substrate translocation.

Site-Directed Mutagenesis of Canonical N-Glycosylation Sites in hNET

Although the previous data presented show hNET proteins to be N-glycosylated, they fail to identify where N-glycosylation is occuring, nor do they demonstrate how N-glycosylation affects transport activity expressed in intact cells. In addition, the effects of TM may be indirect and depress catecholamine uptake through nonspecific mechanisms. To advance this analysis, we mutated the three canonical N-glycosylation sites (N184, N192, N198Q) in the loop connecting TMDs 3–4[6] (Fig. 6A) and expressed the protein in transiently transfected cells.[31] We changed the nucleic acid sequence to substitute Gln for Asn in the three Asn-X-Ser/Thr motifs to provide as conservative a change as possible and block N-glycosylation. In HeLa cells, the protein synthesized from the N184N192N198Q mutant hNET migrates as a polypeptide of 46-kDa just as seen when cells expressing the wild-type hNET cDNA are treated with TM or PNGAse F (Fig. 6B). A similar shift is observed when the N-glycosylation sites in the serotonin transporter are mutated.[29] Since only the TMD 3–4 loop bears N-glycosylation sites, the 46-kDa protein observed reflects the core hNET polypeptide lacking N-glycosylation. Consistently, these peptides migrate almost 20-kDa faster in SDS–PAGE gels than is predicted by their M_r (68 kDa), most likely because of the highly hydrophobic nature of the protein and anomalous absorption of SDS.

To determine why cells expressess transporter protein that lacks N-glycosylation exhibit reduced uptake in activity, we repeated the pulse–chase paradigm using either the TM treated cells or the N-glycosylated

[62] R. Lew, R. Vaughan, R. Simantov, A. Wilson, and M. J. Kuhar, *Synapse* **8,** 152 (1991).
[63] A. Patel, G. Uhl, and M. J. Kuhar, *J. Neurochem.* **61,** 496 (1993).
[64] A. P. Patel, C. Cerruti, R. A. Vaughan, and M. J. Kuhar, *Dev. Brain Res.* **83,** 53 (1995).

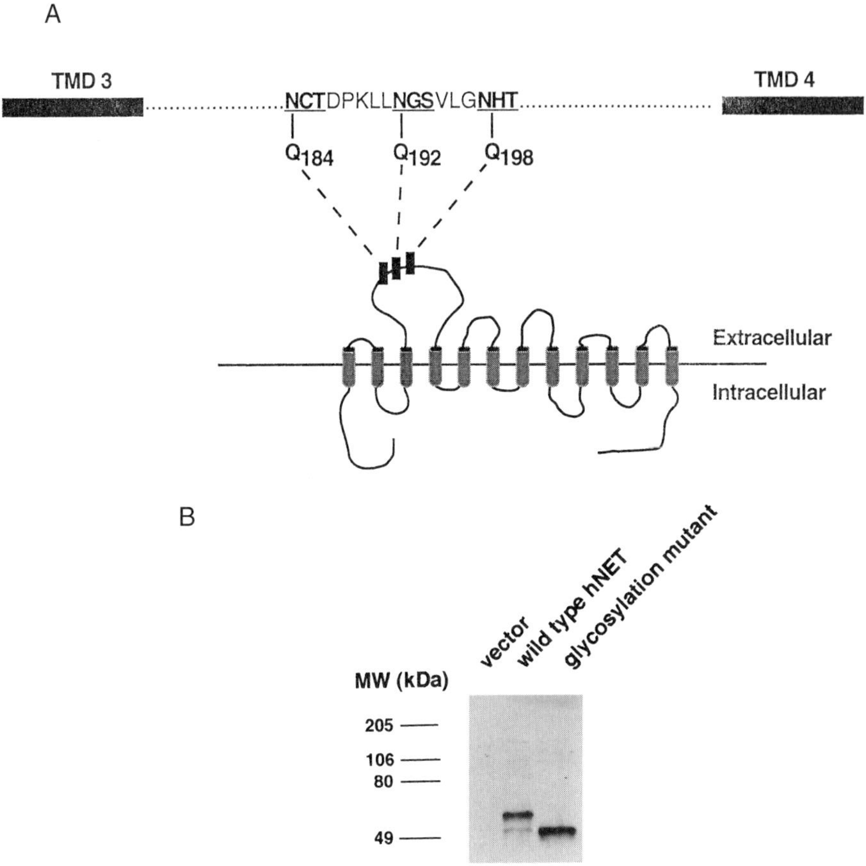

FIG. 6. (A) Diagrammatic presentation of the location of canonical N-glycosylation sites in topological model for hNET protein (Pacholczyk *et al.*, *Nature* **350,** 350, 1991). Barrels represent putative TMDs; black boxes denote the relative location of N-glycosylation sites between TMDs 3 and 4; underline reflects canonical N-glycosylation motifs modified in hNETN184,192,198Q. (B) Electrophoretic mobilities of wild-type hNET and mutated hNETN184,192,198Q in HeLa cells. Cell extracts were prepared from HeLa cells transiently transfected with wild-type hNET and mutated hNETN184,192,198Q cDNA, subjected to SDS–PAGE, and immunoblotted with NET specific antibody N430. [Reprinted with permission from Melikian *et al.*, *Mol. Pharmacol.* **50,** 266 (1996).]

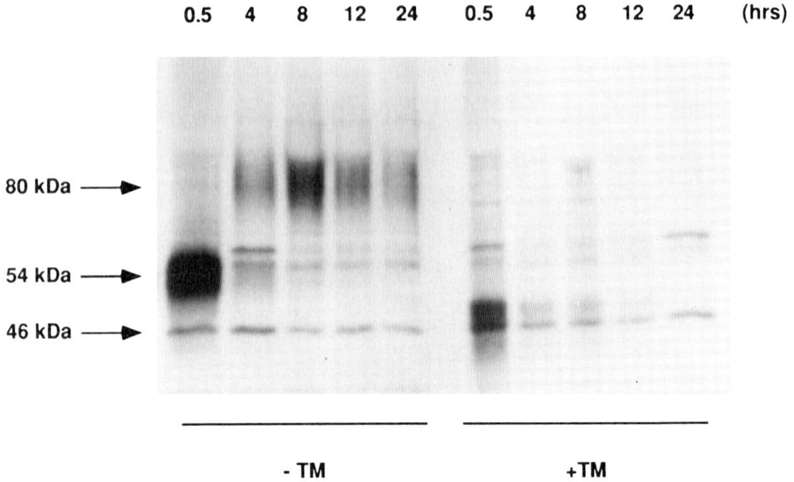

FIG. 7. Pulse-chase experiments on the effects of TM treatment on hNET protein expression in LLC-NET cells. Immunoprecipitation of LLC-NET cells pulse-chase labeled in the absence (−TM) or presence (+TM) of TM. LLC-NET cells were incubated with vehicle or TM (10 μg/ml) in complete media for 1 hr, then changed to Met/Cys-free medium containing vehicle or TM for 30 min, followed by pulse chase labeling with [^{35}S]Met/Cys (300 μCi/ml) for 10 min. Label was removed and cells incubated with complete media containing vehicle or TM. Cells were harvested at the times indicated followed by RIPA extraction, immunoisolation with NET specific antibody N430, and SDS–PAGE and fluorography. Sizes of hNET species indicated with arrows. [Reprinted with permission from Melikian *et al.*, *J. Biol. Chem.* **269,** 12290 (1994).]

mutant, expressed in transiently transfected cells. As shown in Fig. 7, 46-kDa nonglycosylated hNET evident after TM treatment is completely lost after 4 hr, whereas in nontreated cells the protein matures to a stable 80-kDa glycosylated form. These data suggest that N-glycosylation affects transporter protein stability. In support of these findings, the half-life of hNET lacking N-glycosylation sites expressed in HeLa cells is reduced by 50% as determined by pulse–chase/immunoisolation studies (Fig. 8). This greater instability of nonglycosylated hNET could thus result in a reduced level of expression of protein at the cell surface reducing apparent transport capacity. Notably, loss of canonical N-glycosylation sites does not appear to affect recognition by substrates or antagonists.[29,31,32]

Cell-Surface Biotinylation of Transporter Proteins

A greater instability of transporter protein could be solely responsible for the loss in transport capacity observed after either TM treatment or mutation of N-glycosylation sites. However, trafficking and substrate trans-

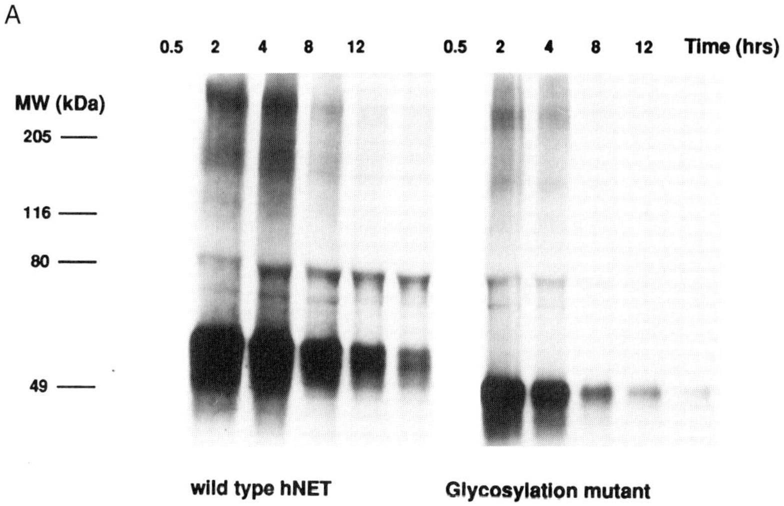

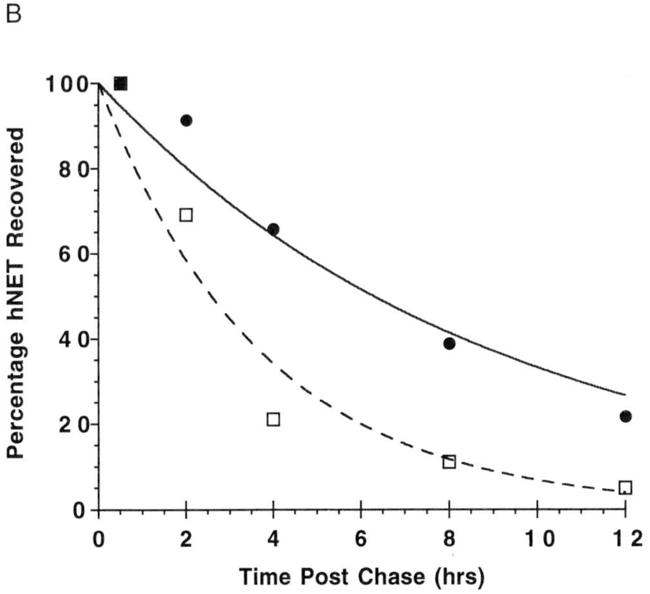

FIG. 8. Half-life estimation of hNET and hNETN184,192,198Q in transfected HeLa cells evaluated by pulse-chase methods. (A) Pulse-chase experiments were conducted as described in Fig. 7, except in HeLa cells, which were transfected with either wild-type hNET or hNETN184,192,198Q using vaccinia T7 expression methods. (B) Plot of band intensities, expressed as a percentage of initial time point, plotted as a function of time and fit by exponential decay curve. Densities across the experiment were expressed as a fraction remaining after chase with unlabeled medium. Average half-life for hNET (closed circles) and hNETN184N192N198Q (open squares) proteins was determined to be 5.0 ± 0.6 and 2.8 ± 0.9 hr, respectively. [Reprinted with permission from Melikian *et al.*, *Mol. Pharmacol.* **50,** 266 (1996).]

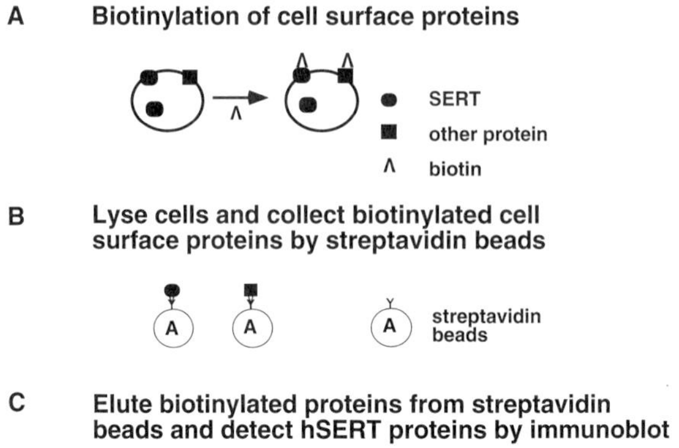
A
Biotinylation of cell surface proteins
SERT
other protein
biotin
B
Lyse cells and collect biotinylated cell surface proteins by streptavidin beads
A
A
A
streptavidin beads
C
Elute biotinylated proteins from streptavidin beads and detect hSERT proteins by immunoblot

FIG. 9. Schematic representation of biotinylation procedure to label cell surface transporters in intact cells. Flow charts shows the general steps in biotinylating surface SERT proteins stably expressed in HEK-293 cells.

location rates may also require proper glycosylation of the TMD 3–4 loop. Transport activity is a property of the protein at the cell surface and to this point, we have not actually distinguished cytoplasmic from surface pools. Our concern that available radiolabeled antagonists are membrane permeant brought us to adopt a different method, cell-surface biotinylation,[24,31,32] for quantiation of transporter resident in the plasma membrane. As we will describe, these methods allow us not only to evaluate the delivery of transporter protein in the study of N-glycosylation,[31–33] but also to reveal that endogenous protein kinases regulate transport capacity through modulated cell-surface expression.[46] These methods are also appropriate for establishing polarity of expression in polarized epithelial cells.[44,65]

The general scheme for surface biotinylation of NET is outlined in Fig. 9. The principle of this method is the use of biotin coupled to highly reactive *N*-hydroxysuccinimide ester group which is membrane impermeant and thus can be used to covalently tag any surface-resident protein with free amino groups. Biotinylated (cell-surface) protein can be isolated using avidin-agarose (Pierce Chemical, Rockford, IL) and then eluted Laemmli sample buffer (0.0625 *M* Tris-HCl, pH. 6.8, 10% glycerol, 2% SDS, 5% 2-mercaptoethanol, 0.00125% bromphenol blue) for 20 min at 22° for SDS–PAGE analysis, followed by immunoblot detection with transporter-specific antibodies. Thus, we washed LLC-NET cells twice in cold PBS supplemented with 0.1 m*M* $CaCl_2$ and 1 m*M* $MgCl_2$ (PBS/Ca/Mg) and then treated cells with sulfosuccinimido-NHS-biotin (Pierce) added in PBS/Ca/Mg

[65] H. H. Gu, J. Ahn, M. J. Caplan, R. D. Blakely, A. I. Levey, and G. Rudnick, *J. Biol. Chem.* **271,** 18100 (1996).

buffer at a final concentrations of 1 mg/ml (1 ml/well) and incubated for 25 min at 4°. All solutions and the complete steps should be carried out on ice to limit endocytosis of the labeling reagent and surface proteins. Biotinylating reagents are also available (Pierce) that possess an extended spacer arm to reduce steric hindrance and increase opportunity for capture on avidin agarose. Free biotinylating reagent is removed by washing with cold PBS/Ca/Mg containing 100 m*M* glycine and incubated for a further 20 min in this buffer to quench all of the unreacted biotin, at which time the cell monolayer is washed again with PBS/Ca/Mg and solubilized with cold RIPA buffer. Solubilized biotinylated cell surface proteins, after centrifugation (20,000*g*, 30 min, 4°) to remove insoluble debris, are separated from nonbiotinylated (intracellular plus surface nonbiotinylated) proteins by batch affinity chromatography using monomeric avidin beads (250 μl beads/300 μg protein). Prior to use, irreversible biotin binding sites on these beads are blocked with 2 m*M* biotin in PBS, followed by washing with 0.1 *M* glycine, pH. 3.0. Beads are then washed and reequilibrated with PBS. Equal amounts of total cell protein are incubated with pretreated beads for 1 hr at room temperature by end-over-end rotation. Beads are centrifuged (10,000*g*, 4 min; 4°) and samples of the supernatant are kept for analysis (nonbiotinylated protein pool). Beads are washed four times in 1 ml of cold RIPA buffer by centrifugation and resuspension and samples of the final wash are kept for subsequent analysis (to determine washing efficiency).[66,67] Biotinylated proteins are eluted from beads by incubation with 50 μl of Laemmli buffer for 20 min at room temperature with shaking (see previous note regarding boiling of transporter protein). Aliquots of the total cell lysate, the nonabsorbed lysate after incubation with avidin beads, the last wash, and the total bead eluate are resolved by SDS–PAGE in parallel with prestained molecular weight markers (Bio-Rad). Electrophoresed gels are soaked in transfer buffer (50 m*M* Tris, 250 m*M* glycine, 0.01% SDS, 20% methanol) for 20 min and electroblotted at 30 V, 16 hr, 4° to Hybond ECL nitrocellulose membranes. Biotinylated proteins are detected using horseradish peroxidase (HRP)-conjugated streptavidin (1:5000), and transporter protein in fractions is visualized with specific antibodies [for hNET affinity purified N430 antibody (0.5 μg/ml)]. Blots are incubated for 1 hr at room temperature or overnight at 4° in blocking buffer (5% Carnation nonfat dried milk, 0.5% Tween 20, PBS) followed by incubation with primary antibodies in the same buffer for 45 min at 22°. The blots are rinsed (two times in PBS, 0.5% Tween 20) prior to incubation

[66] M. Sargiacomo, M. Lisante, A. Le Bivic, and E. Rodriguez-Boulan, *J. Membr. Biol.* **107,** 277 (1989).

[67] C. J. Gottardi, L. A. Dunbar, and M. J. Caplan, *Am. J. Physiol.* **268,** F285 (1995).

with HRP-conjugated goat anti-rabbit secondary antibody (1 : 10,000) for 1 hr at room temperature, washes are repeated as above, and immunoreactive bands are visualized by ECL on Hyperfilm ECL (Amersham). To assess whether intracellular proteins could be inappropriately biotinylated and recovered in avidin beads, the same blot is stripped in 100 m*M* 2-mercaptoethanol, 62.5 m*M* Tris pH 6.7, 2% SDS for 30 min at 50°, washed, preblocked as above, and probed with a monoclonal antibody (MAb) against the intracellular marker actin (0.5 μg/ml; Boehringer-Mannheim Biochemicals, Indianapolis, IN) and detected using HRP-conjugated sheep antimouse secondary antibody (1 : 10,000) as described above. We have also found that anticalnexin (Stressgen Biotechnologies Corp., Victoria, Canada) can also be used as a reliable intracellular marker and lacks the adherance to membranes observed for actin in some preparations. Quantitiative densitometry of hNET immunoreactivity is performed across all fractions, normalizing to the actin band in the total extract to allow separate experiments to be averaged.

Using this procedure, we tested the effect of TM on cell surface abundance of hNET in LLC-NET cells (Fig. 10A). Biotinylated proteins could be detected in the whole-cell extract and the bead eluent, but not in wash or avidin bead supernatant (intracellular pool) fractions, indicating and efficient and quantitative separation of biotinylated protein from nonbiotinylated proteins by the avidin beads. Reprobing the blot with anti-actin showed that actin segregates primarily with the nonbiotinylated fraction. An immunoblot of the biotinylated proteins from control and TM treated LLC-NET cells are shown in Fig. 9B. N430 antibody detected both the 80- and 54-kDa hNET proteins in the biotinylated total cell extract, indicating that biotinylation per se has no effect on hNET detection by N430. However, analysis of the avidin bead eluate lane shows the 80-kDa hNET form to be the predominant cell surface species, whereas the 54-kDa form is predominates in the nonbiotinylated fraction consistent with its less mature status in the biosynthetic pathway. Moreover, TM treatment of LLC-NET cells results in a marked reduction in the 80-kDa band in the surface fraction, whereas the nonglycosylated 46-kDa hNET form is not detected in the surface fraction, suggesting that the nonglycosylated hNET synthesized in the presence of TM cannot contribute to transport activity. Quantitative analyses of surface trafficking in COS cells, transiently transfected with wild-type and N-glycosylation mutant hNETs, reveals that the reduced delivery to the cell surface is largely a consequence of reduced protein in the intracellular pool available for plasma membrane insertion.[31] In COS cells, nonglycosylated hNET can populate the cell surface at an efficiency comparable to that of N-glycosylated protein. However, the mutant protein does not catalyze as much catecholamine transport as it should if all the

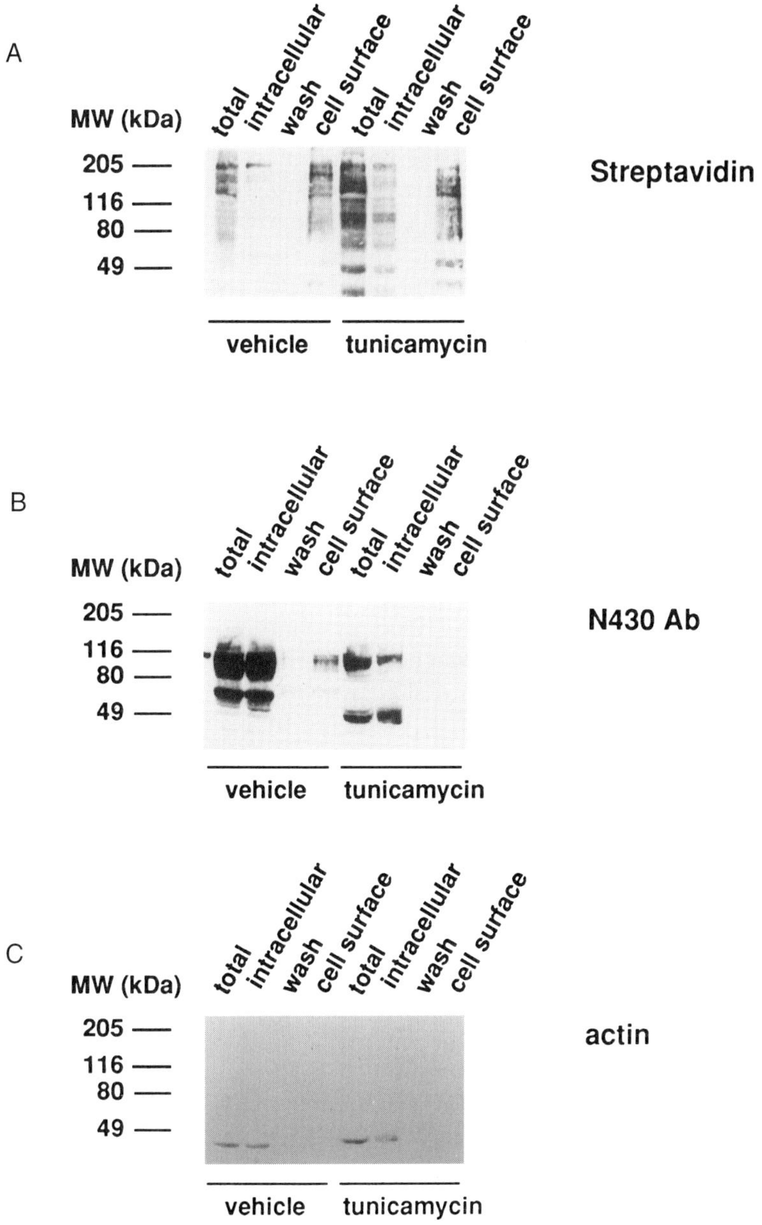

FIG. 10. Effect of TM on cell surface abundance of hNET in LLC-NET cells. After treatment with TM (10 μg/ml, 24 hr) or vehicle, LLC-NET cells were biotinylated, subjected to 10% SDS–PAGE, blotted, and probed with (A) HRP-conjugated streptavidin (1:5000), (B) N430 Ab (0.5 μg/ml), or (C) actin Ab (0.5 μg/ml). Loading in total and intracellular (supernatant) lanes for vehicle and TM was balanced for protein content (33 μg), whereas half of the entire surface fraction recovered after avidin chromatography was split between the N430/actin blot and a parallel streptavidin blot. Actin detection was performed on the N430 blot after stripping. [Reprinted with permission from Melikian *et al.*, *Mol. Pharmacol.* **50,** 266 (1996).]

surface transporters were as efficient at catecholamine translocation as wild-type protein. Interestingly, chimera studies in the TM 3 and 4 loop also reveal kinetic changes imposed by modification of this domain.[68] Reduced transport efficiency occurs despite no change in substrate recognition (as revealed by substrate K_m). From these studies we conclude that N-glycosylation is required for normal transporter protein stability and may also contribute to substrate translocation efficiency. Similar, although not identical, conclusions have been reached in studies of the role of N-glycosylation in glycine transporters.[33]

Regulation of Transporter Biosynthesis and Surface Expression by Protein Kinases

The tools described above for assessing the biosynthesis, N-glycosylation, and surface expression of transporter proteins may be useful in regulatory contexts. For example, we have employed biotinylation techniques to evaluate the basis for protein kinase C-mediated regulation of NET[69] and SERT[46] activity. Activation of PKC by phorbol esters leads to a rapid reduction in transport activity for both transporters in native and transfected cells. In hSERT stably transfected HEK-293 cells, we have shown that the reduction in 5HT uptake capacity induced by β-PMA is paralleled by a loss of hSERT protein from the cell surface with no change in the total cellular content of hSERT protein (Fig. 11A). This loss is blocked by the PKC inhibitor staurosporine (Fig. 11C) and coincides with a loss of hSERT-associated currents. Similar findings have been generated for NET and have been extended to include distinct serine and tyrosine kinase-linked pathways triggered by native receptors (Apparsundaram and Blakely, manuscript in preparation).[47,70] We have also demonstrated that SERT proteins are phosphorylated in parallel with reductions in surface abundance (Ramamoorthy and Blakely, manuscript in preparation).[71,72] Together, these data suggest that one major mechanism for the regulation of transporter activity *in vivo* is kinase-mediated modulation of surface expression. The consequence of this modulation would be to regulate the amount of transmitter clearance achieved and modify the temporal and

[68] M. M. Stephan, M. A. Chen, K. M. Y. Penado, and G. Rudnick, *Biochem.* **36,** 1322 (1997).
[69] S. Apparsundaram and R. D. Blakely, *Soc. Neurosci.* (abstract) **22,** 450 (1997).
[70] S. Apparsundaram and R. D. Blakely, "Insulin modulation of sympathetic function: Regulation of human L-norepinephrine transporter" (1997, unpublished).
[71] S. Ramamoorthy, E. Giovanetti, and R. D. Blakely, *Soc. Neurosci.* (abstract) **22,** 450 (1997).
[72] S. Ramamoorthy, E. Giovanetti, Y. Qian, and R. D. Blakely, *J. Biol. Chem.* **273,** 2458 (1998).

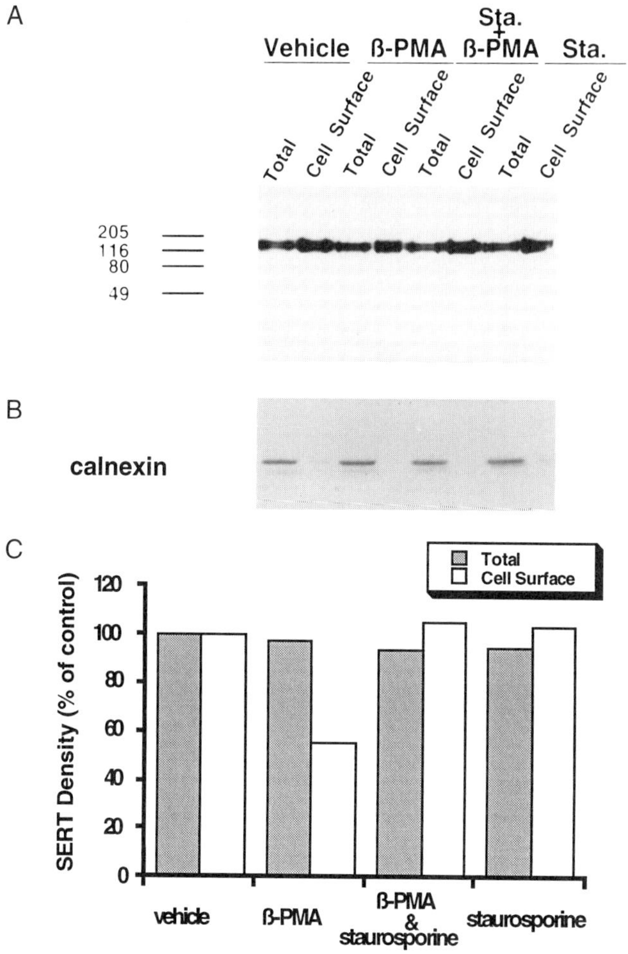

FIG. 11. Determination of protein kinase C mediated regulation of hSERT surface density using cell surface biotinylation techniques. (A) 293-hSERT cells were exposed to 1 μM β-PMA, β-PMA + staurosporine (μM), staurosporine or vehicle for 40 min, cells were then washed and surface proteins were biotinylated using sulfo-NHS-biotin. Biotinylated proteins were isolated by streptavidin agarose followed by SDS–PAGE and electroblotting. hSERT proteins were visualized by with SERT specific antibody CT-2. (B) The same blot in A was stripped and probed with anti-calnexin antibody as a marker for intracellular protein to validate the efficiency of surface biotinylation. Calnexin immunoreactivity was observed only in the total fraction, but not in streptavidin agarose eluate (cell surface). (C) Quantitation of the total and cell surface hSERT. CT-2 immunoreactive bands were scanned and density values were normalized for calnexin immunoreactivity in the total cell extracts. Data are expressed as a percentage of values obtained with vehicle-treated cells (Qian *et al.*, *J. Neurosci.* **17,** 45, 1997).

spatial impact on target receptors. Studies suggest that chronic antagonist treatment may reduce cellular expression of transporter protein, and thus methods to evaluate transporter half-life are also likely to be important.[73] Together these methods described should allow for a dissection of mechanisms involved in posttranslational processing of transporters and how such modifications contribute to regulated changes in transport capacity.

Acknowledgments

NIH Awards DA07390 and NS33373 to R.D.B. and a NARSAD Young Investigator Award to S.R. provided support during completion of this review.

[73] M. Yang Shu and G. A. Ordway, *J. Neurochem.* **68,** 134 (1997).

[25] Expression of Neurotransmitter Transport Systems in Polarized Cells

By JINHI AHN, GRAZIA PIETRINI, THEODORE R. MUTH, and MICHAEL J. CAPLAN

Introduction

In order to subserve their many and varied physiologic functions, each of the neurotransmitter transport systems must be localized to a specific subdomain of the surface membrane of a cell. In the case of the neuronal transporters which function to terminate synaptic transmission, this requirement is readily understandable. Concentration of transport systems in the presynaptic plasmalemma, near the site of transmitter release, facilitates their ability both to inactivate signaling rapidly and to permit efficient reutilization of transmitter.[1] Similarly, postsynaptic and glial transport systems must also be positioned close to the synapse if they are to be effective in controlling the duration of synaptic communication. Those members of the neurotransmitter transport system family which are expressed in epithelial cells are likewise targeted to specific regions of the plasmalemma. The betaine transporter, for example, is expressed exclusively at the basolateral surface of renal medullary epithelial cells.[2] This distribution allows the

[1] S. G. Amara and M. J. Kuhar, *Annu. Rev. Neurosci.* **16,** 73 (1993).
[2] A. Yamauchi, H. M. Kwon, S. Uchida, A. S. Preston, and J. S. Handler, *Am. J. Physiol.* **261,** F197 (1991).

0076-6879/98 $25.00

cell to import betaine from the plasma into its cytosol, where it can fulfill its osmoprotective function.

Although the teleologic necessity for targeting transporters to specific subcellular destinations may be clear, the mechanisms through which this targeting is achieved remain largely unelucidated. It is widely accepted that polarized cells must possess sorting machinery which is capable of recognizing specific classes of membrane proteins and directing them to the appropriate domains of their plasmalemmas. Furthermore, embedded within the structures of membrane proteins must be the targeting information or sorting signals which specify the proteins' sites of ultimate functional residence.[3] Evidence suggests that neurons and epithelial cells may employ similar signals and machinery to allocate polypeptides selectively to their particular plasmalemmal domains.[4–6] Furthermore, closely related members of the neurotransmitter transporter family appear to be spatially segregated from one another in both polarized cell types.

To study the cellular and molecular correlates of these sorting processes, it is necessary to develop methods through which neurotransmitter transporters can be expressed in polarized cells in culture. Techniques must also be established through which the subcellular distributions of the exogenous transporters can be assessed and compared. This review will discuss procedures for transporter expression in polarized epithelial cells and hippocampal neurons in culture. It will also describe several different approaches to the problem of determining the localization of expressed proteins in both cell types.

Exogenous Transporter Expression in Polarized Epithelial Cells

Polarized Epithelial Cell Lines

Studies of membrane protein sorting often take advantage of the existence of numerous well-characterized epithelial tissue culture lines.[7] Although derived from diverse tissues, these cell types share several properties which underlie their utility. The most important of these is their capacity

[3] M. J. Caplan and K. S. Matlin, *in* "Functional Epithelial Cells in Culture" (K. S. Matlin and J. Valentich, Eds.), pp. 71–130. A. R. Liss, New York, 1989.

[4] C. G. Dotti and K. Simons, *Cell* **62,** 63 (1990).

[5] G. Pietrini, Y. J. Suh, L. Edelmann, G. Rudnick, and M. J. Caplan, *J. Biol. Chem.* **269,** 4668 (1994).

[6] J. Ahn, O. Mundigl, T. R. Muth, G. Rudnick, and M. J. Caplan, *J. Biol. Chem.* **271,** 6917 (1996).

[7] M. J. Caplan and E. Rodriguez-Boulan, *in* "Handbook of Physiology" (J. F. Hoffman and J. D. Jamieson, Eds.), pp. 665–688. Oxford University Press, New York, 1997.

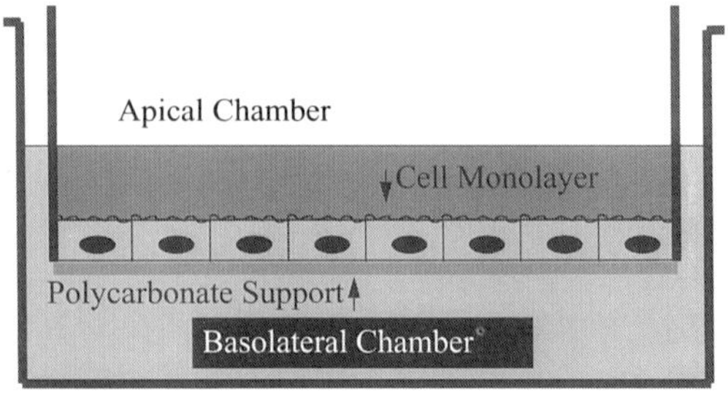

FIG. 1. Epithelial cells grown on permeable filter supports. Tissue culture filter inserts allow epithelial cells to grow on permeable polycarbonate membranes. When it reaches confluence, an epithelial cell monolayer forms a barrier between two medium-filled compartments. The medium within cup of the filter insert bathes the cells' apical surfaces (Apical Chamber). Intercellular tight junctions separate this solution from the medium filling the well in which the insert is suspended (Basolateral Chamber). This fluid has access to the basolateral plasma membranes of the epithelial cells through the pores of the filter.

to form morphologically, biochemically, and functionally polarized monolayers upon reaching confluence. Like their *in vivo* counterparts, these polarized epithelial cell lines organize their plasma membranes into two distinct domains. The basolateral domain includes the portion of the plasmalemma which rests upon the substrate, as well as those regions which participate in cell–cell contact. The apical domain is often referred to as the "free" surface and is bathed by the overlying medium (Fig. 1). Basolateral membranes are frequently extensively infolded, whereas apical membranes are often decorated with microvilli or microplicae.[3]

The barrier between the apical and basolateral surfaces is defined by the zonula occludens, or tight junction. This structure prevents diffusional mixing of the protein components of the two domains, thus maintaining their biochemical identities. It also impedes the paracellular movement of fluid and solutes and accounts for the electrical resistance which confluent epithelial monolayers develop. The magnitude of this resistance varies widely among different cell lines, reflecting intrinsic differences in tight junctional properties and architecture. In many epithelial cell lines, functional polarity and the presence of tight occluding junctions is dramatically illustrated by the capacity to mediate transmonolayer fluid and solute fluxes. When grown on solid supports, epithelia which actively transport in the apical to basolateral direction form domes or blisters, in which small regions of the monolayer are separated from the substratum by the basolateral accumulation of transported fluid.[8]

[8] M. Cereijido, E. S. Robbins, W. J. Dolan, C. A. Rotunno, and D. D. Sabatini, *J. Cell Biol.* **77,** 853 (1978).

Most of the commonly used epithelial cell lines are derived from mammalian kidney or gut.[7–12] The best known of these include the MDCK (Madin–Darby Canine Kidney) cell line, which is derived from canine distal nephron and bears some morphologic and functional resemblance to that tissue. Similarly, the LLC-PK_1 cell line was propagated from pig kidney and is similar in many respects to the renal proximal tubule. Although Caco2; HT-29, and T84 cells all originated from tumors of the human intestinal epithelium, their distinct properties strongly suggest that they arose from different cell types. The list of useful polarized cell lines is by no means limited to the five presented here. On the contrary, an ever-growing catalog of epithelial lines, with interesting characteristics and diverse origins, is available for adapation to specific experimental situations. For the purposes of exogenous membrane protein expression, however, these five lines are among the most heavily used.

Of these, the most popular by far is MDCK. Its popularity can be attributed, in part, to the MDCK cell lines' stable phenotype, desirable growth characteristics, and relative ease of transfectability. It is also somewhat self-perpetuating, since the wide use of MDCK cells ensures that they are readily available and that the relevant transfection protocols and assay techniques are well established and thoroughly documented. Since our own work on transporter expression in polarized cells makes exclusive use of MDCK cells, the remainder of this section of the review will focus primarily on this cell type. It should be noted, however, that for sorting studies it may be extremely important to employ more than one epithelial cell type. Studies indicate that the targeting of neurotransmitter transport systems may change dramatically when they are expressed in different epithelial cell lines.[13]

Cell Growth and Transfection

Two different subclones of MDCK cells are frequently used in membrane protein sorting studies. The strain II, or low resistance cell line, typically generates transepithelial electrical resistances on the order of 300 Ω cm^2. In contrast, the strain I, or high resistance cell line, is capable of mounting resistances as high as 3000 Ω cm^2.[14] Both lines polarize apical

[9] E. Rodriguez-Boulan and D. D. Sabatini, *Proc. Natl. Acad. Sci. USA* **75,** 5071 (1978).
[10] C. J. Gottardi and M. J. Caplan, *J. Cell Biol.* **121,** 283 (1993).
[11] A. LeBivic, A. Quaroni, B. Nichols, and E. Rodriguez-Boulan, *J. Cell Biol.* **111,** 1351 (1990).
[12] G. M. Denning, L. S. Ostedgaard, S. H. Cheng, A. E. Smith, and M. J. Welsh, *J. Clin Invest.* **89,** 339 (1992).
[13] H. H. Gu, J. Ahn, M. J. Caplan, R. D. Blakely, A. I. Levey, and G. Rudnick, *J. Biol. Chem.* **271,** 18100 (1996).
[14] S. D. Fuller and K. Simons, *J. Cell Biol.* **103,** 1767 (1986).

and basolateral membrane proteins with equal fidelity. MDCK cells import the majority of their nutrients through basolaterally disposed transport systems.[15–17] When MDCK cells are grown to confluence on nonporous substrates, only the apical surface is bathed in nutrient-supplying medium. Consequently, growth on solid supports such as plastic or glass does not favor the aquisition of the fully polarized phenotype. Furthermore, because the high transepithelial resistance they develop further reduces nutrient access to the basolateral surface, the strain I cells can be difficult to propagate on standard plastic. For optimal growth, strain I cells should be cultured on porous filter supports (see later discussion). It is easier, therefore, to create stably transfected lines in strain II cells.

Culture on permeable filter supports allows the epithelial cells to import nutrients from the medium directly bathing their basolateral plasmalemmas. Hence, they are able to polarize more completely in this configuration.[14] For these reasons, immunocytochemical analysis of transporter distribution is generally best performed on filter grown cells (see later discussion). As can be seen in Fig. 1, formation of a confluent monolayer on the surface of a filter can physically separate the apical from the basolateral media compartments. This permits simultaneous and independent experimental manipulations to be performed at the two surface domains. Techniques such as cell surface biotinylation (see below) rely on this feature of filter culture. Similarly, growth on fiters allows the apical and basolateral neurotransmitter transport activities to be measured independently. As discussed below, transmitter uptake is a very useful technique for quantifying the polarity of transporter distribution. Filter-grown cells can also be used for measurement of electrical parameters, such as transepithelial resistance, potential, and short circuit current.

For routine culture, MDCK cells are grown in 75 cm^2 plastic tissue culture flasks (Corning). The cells are maintained in Eagle's minimal essential medium supplemented with Earle's salts (EMEM) (GIBCO-BRL, Gaithersburg, MD), 10 m*M* HEPES, 5% fetal bovine serum (FBS), and antibiotic solution containing penicillin and gentamicin in a 5% (v/v) CO_2, 37° incubator. The cells are passaged with a solution of 0.25% trypsin, 2 m*M* EDTA (GIBCO). Fresh vials of cells should be thawed after 15 passages. Backup stocks of parent and transfected lines should be frozen for storage at −120° in medium containing 20% fetal bovine serum (FBS) and 10% dimethyl sulfoxide (DMSO) at a cell density of $0.5–1 \times 10^6$/ml.

[15] J. E. Lever, B. G. Kennedy, and R. Vasan, *Arch. Biochem. Biophys.* **234,** 330 (1984).
[16] W. S. Pascoe, K. Inukai, Y. Oka, J. W. Slot, and D. E. James, *Am. J. Physiol.* **271,** C547 (1996).
[17] P. Zlatkine, C. Leroy, G. Moll, and C. Le Grimellec, *Biochem. J.* **315,** 983 (1996).

Our experiments requiring MDCK cells grown on permeable supports employ monolayers cultured on 0.45-μm pore size polycarbonate filters mounted in plastic chambers (Transwell, Corning-Costar). Cells are plated on the filters at approximately 10^5 cells/cm^2 of filter surface area, which corresponds to approximately one-tenth of confluent density. Cells are allowed to reach confluence over the course of 7–10 days, with twice-weekly media changes, prior to being used. Plating at low density allows an even, uniform monolayer to form over the entire surface of the monolayer. Because polycarbonate filters with 0.45-μm pores are not transparent, it is difficult to assess visually the state of filter plated cells. We often plate cells at the same density in a plastic dish of the same size in order to monitor the density of our cultures. It is also possible to monitor the integrity of filter grown monolayers by (1) measuring electrical resistance,[8] (2) measuring monolayer permeability to small soluble molecules,[18] and (3) assessing the polarity of [^{35}S]methionine incorporation.[14]

We find that studies of the synthesis and sorting of exogenous membrane transport proteins expressed in MDCK cells are best performed on stably transfected cell lines. Polarity tends not to be fully developed in transiently transfected epithelial cells and their sorting behavior is not always reproducible. Furthermore, the high levels of overexpression which can be attained in transient transfection can overwhelm normal protein processing and sorting pathways. Finally, transfection of epithelial cells is notoriously inefficient, and hence the pool of transfected cells is generally extremely small in transient expression experiments. Vectors employing the cytomegalovirus (CMV) promoter to drive exogenous protein expression work well in MDCK cells.[19] Several such vectors are commercially available. Most of these also incorporate sequences encoding the bacterial phosphotransferase gene as a dominant selectable marker, since production of this enzyme renders MDCK cells resistant to the antibiotic G418. Addition of G418 to the growth media kills all untransfected cells and allows the identification and expansion of stably transfected cell lines.

The methods of Puddington *et al.*[20] provide a relatively easy and effective protocol for MDCK cell transfection. These investigators achieve a transformation frequency of $\sim 10^{-4}$ using calcium phosphate DNA precipitation enhanced through the presence of 100 μM chloroquine and followed by a 5-min 15% glycerol shock. Following 2 days of growth in normal medium (EMEM, 5% FBS), cells treated in this manner can be exposed to medium containing 0.4 mg/ml G418 (effective concentration). It should be noted

[18] M. J. Caplan, H. C. Anderson, G. E. Palade and J. D. Jamieson, *Cell* **46,** 623 (1986).
[19] C. B. Brewer, *Methods Cell Biol.* **43,** 233 (1994).
[20] L. Puddington, C. Woodgett, and J. K. Rose, *Proc. Natl. Acad. Sci. USA* **84,** 2756 (1987).

that different epithelial cell lines (and even different MDCK cell subclones) manifest markedly distinct sensitivities to G418. It is useful, therefore, to identify the minimal lethal dose of G418 by carrying out a titration, or "kill curve," prior to transfecting any new epithelial cell line. After ~9 days, colonies of cells which survive this selection are identified and isolated through the placement of cloning rings. Transfected lines are expanded, characterized, and frozen. Transfected cell lines of interest should be recloned by limiting dilution so that cells expressing the exogenous protein are not lost through overgrowth of more rapidly dividing nonexpressing cells. Expression of the gene of interest by transfected MDCK cell clones can be assessed by the techniques discussed in the next section. If necessary, levels of exogenous protein under the control of the CMV promoter can be increased by incubating the transfected cells for 12 hr with 10 m*M* sodium butyrate.

Evaluation of Protein Distribution

Immunocytochemistry, cell surface biotinylation, and tracer flux assays have all proven useful in evaluating the distributions of neurotransmitter transport proteins expressed by transfection in polarized cells.[5,6,13,21] Application of the first two of these techniques depends on the availability of specific antibodies which recognize the exogenous transporter protein and do not cross-react with any endogenous epithelial cell proteins. When specific antibodies are not available, adding short sequences encoding epitope tags to the cDNA specifying the transporter allows well-characterized commercially available antibodies directed against narrowly defined antigenic sites to be employed. Commonly used epitopes include short sequences derived from the influenza hemagglutinin (HA) protein and c-*myc*. Although such tags can save investigators the often prodigious effort involved in generating or scrounging high-quality antibodies, their use carries important risks which must be recognized. Addition of even a short and seemingly inoccuous stretch of sequence to a protein of interest can affect the folding, function, or sorting of that protein in unanticipated ways. We have found, for example, that appending the 10 amino acid *c-myc* tag to the C terminus of the GAT-3 isoform of the α-aminobutyric acid (GABA) transporter completely randomizes its targeting in MDCK cells.[22] We have subsequently determined that residues in the GAT-3 C terminus appear to carry important sorting information whose interpretation is sterically impeded through the addition of the tag. In light of this observation, it cannot be assumed that the subcellular distribution achieved by a tagged transporter protein

[21] C. Perego, A. Bulbarelli, R. Longhi, M. Caimi, A. Villa, M. J. Caplan and G. Pietrini, *J. Biol. Chem.* **272,** 6584 (1997).

[22] T. R. Muth, J. Ahn and M. J. Caplan, *J. Biol. Chem.*, in press.

will be identical to that associated with the native protein. It is desirable, therefore, to compare the distributions of native and tagged transporters through an antibody-independent assay, such as measurement of transmitter uptake.

Immunofluorescence Microscopy. Subcellular localization of the transporter molecules can be readily accomplished by immunofluorescence microscopy. For screening purposes—that is, for experiments designed to examine a large number of transfected clones and identify those which are expressing detectable levels of the exogenous transporter protein—immunofluorescence can be performed on MDCK cells grown on Lab-Tek (Nunc, Naperville, IL) plastic microscope slides. Examination of clonal stable lines to establish transporter distribution is best performed with cells grown on Transwell filters (Corning-Costar, Cambridge, MA). Since the effects of various fixation protocols on individual antigen–antibody interactions cannot be predicted *a priori*, it is useful to try more than one approach. We have had reasonable success in staining confluent monolayers which have been fixed either with 100% methanol prechilled to $-20°$ or with 4% formaldehyde dissolved in PBS containing 100 μM $CaCl_2$ and 1 mM $MgCl_2$. In either case, fixation is followed by permeabilization (to allow antibodies access to intracellular antigens), which entails a 5-minute wash with PBS containing 0.2% (v/v) Triton X-100. Next, nonspecific antibody binding is prevented by incubating the cells for 30 min in a "blocking" solution containing goat serum [16% (v/v) goat serum (Sigma), 0.3% (v/v) Triton X-100, 0.45 M NaCl, 20 mM NaP_i, pH 7.4].[23] Incubations with primary and secondary antibodies are also carried out in this goat serum dilution buffer (GSDB) for 2 hr at room temperature or overnight at 4° in a humidified chamber. After each incubation, unbound antibodies are removed through three washes in PBS–BSA [PBS-Ca-Mg containing 0.1% (w/v) bovine serum albumin (BSA) and 0.3% (v/v) Triton X-100] for 5 min each at room temperature. We generally use goat anti-rabbit or mouse immunoglobin G (IgG) (whole molecule) (Sigma, St. Louis, MO) conjugated to rhodamine or fluorescein as our secondary probes. After the final washes, a drop of mounting solution (0.1% *p*-phenylenediamine (Sigma), 70% glycerol, 150 mM NaCl, 10 mM NaP_i, pH 7.4) is placed on the surface of a stained Lab-Tek slide or filter (which in turn is placed face up on a glass microscope slide). A coverslip is placed over the sample, with care being taken to avoid trapping air bubbles. Excess mounting solution is gently aspirated from around the edges of the coverslip, which are then sealed to the slide with nail polish. Immunofluorescence patterns can be observed with a conventional epifluorescent photomicroscope or with a laser scanning confocal microscope. Whereas the conventional microscope permits only an *en face* view

[23] P. L. Cameron, T. C. Sudhof, R. Jahn and P. De Camilli, *J. Cell Biol.* **115,** 151 (1991).

of the monolayer epithelium, confocal microscopy allows both *en face* and cross-sectional images to be generated from the same specimen preparations (Fig. 2).

Cell Surface Biotinylation. Cell surface biotinylation has been applied with great success to determine the distributions of a large number of membrane proteins in polarized epithelial cells.[24–26] *N*-Hydroxysuccinamidylbiotin (NHS-biotin) is a membrane-impermeant reagent which covalently incorporates into the free amino groups of proteins exposed at the cell surface. Addition of NHS-biotin to the medium bathing the apical or the basolateral plasma membranes of epithelial cells grown on Transwell filters results in the covalent modification of many of the polypeptides exposed at the treated surface. Since biotin binds to the protein avidin with extremely high afinity, this interaction can be exploited to recover biotin-tagged proteins. Cells are solubilized in a detergent-containing buffer and the resulting lysates are incubated with avidin immobilized on agarose beads. Material bound to the beads can be analyzed by immunoprecipitation or by Western blotting employing antibodies directed against specific membrane proteins. In this manner, the size of the pools of a particular polypeptide which are susceptible to apical versus basolateral biotinylation can be determined (see Fig. 3). It must be noted that there are several technical caveats which may affect the ability of the cell surface biotinylation technique to provide quantitatively meaningful results. These considerations, as well as protocols designed to address them, are presented in detail elsewhere.[27]

Transport Assays. Measuring neurotransmitter uptake provides a relatively easy and reproducibly quantitative technique for assessing the polarized distributions of neurotransmitter transport proteins. Transport assays can be carried out as described in Ahn *et al.*, with minor modifications. MDCK cells stably expressing the transporter of interest are plated at an initial density of 4.0×10^4 cells per 6.5 mm Costar Transwell filter support. Cells are allowed to grow to confluence under standard conditions for 1 week. On the day of the assay, cells are rinsed three times with PBS-Ca-Mg (150 m*M* NaCl, 1 m*M* $MgCl_2$, 0.1 m*M* $CaCl_2$, 10 m*M* NaP_i, pH 7.4) to thoroughly remove the culture medium. For assays of GABA transport, PBS-Mg-Ca containing [^{3}H]GABA at a final concentration of 50 n*M* (0.33

[24] M. Sargiacomo, M. Lisanti, L. Graeve, A. Le Bivic and E. Rodriguez-Boulan, *J. Membr. Biol.* **107,** 277 (1989).

[25] M. P. Lisanti, A. Le Bivic, A. R. Saltiel and E. Rodriguez-Boulan, *J. Membr. Biol.* **113,** 155 (1990).

[26] M. P. Lisanti, A. Le Bivic, M. Sargiacomo and E. Rodriguez-Boulan, *J. Cell Biol.* **109,** 2117 (1989).

[27] C. J. Gottardi, L. A. Dunbar, and M. J. Caplan, *Am. J. Physiol.* **268,** F285 (1995).

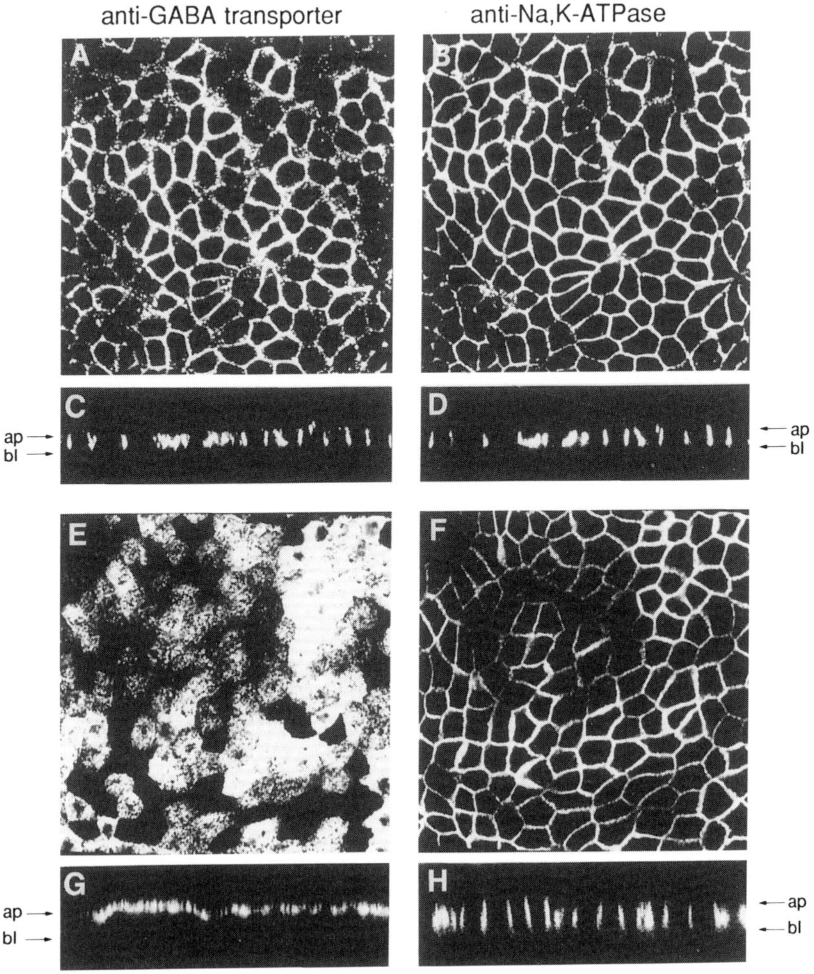

FIG. 2. Immunolocalization of exogenous GABA transporters expressed by transfection in MDCK cells. MDCK cells transfected with the GAT-2 and GAT-3 GABA transporters were processed for immunofluorescence labeling as described in the text. A confocal microscope was used to generate both *en face* (A, B, E, and F) and xz cross-section images (C, D, G, and H). Like the endogenous Na,K-ATPase (B, D, F, and H), the GAT-2 transporter (A and C) is restricted to basolateral domains of the plasmalemma. This basolateral localization is revealed in the "chicken wire" pattern seen in the *en face* views as well as in the linear staining at sites of cell–cell contact detected in cross sections. In contrast, GAT-3 (E and G) is present at the apical surface, as demonstrated by the "ground glass" appearance of the *en face* staining pattern and the linear stripe along the apical border seen in the cross sections. The apical and basal surfaces of the cells in the cross sections are indicated by the arrows labeled ap and bl, respectively.

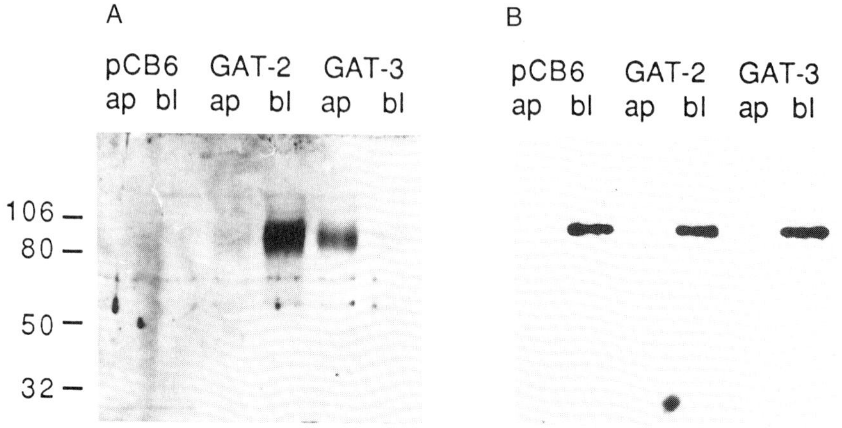

FIG. 3. Assessment of GABA transporter polarity by cell surface biotinylation. MDCK cells transfected with the GAT-2 and GAT-3 GABA transporters were processed for cell surface biotinylation as described in the text. Following NHS-biotin labeling of the apical or the basolateral plasma membrane, epithelial monolayers were solubilized and biotin-tagged proteins were recovered on streptavidin-conjugated agarose beads. After separation by SDS–PAGE, proteins were transferred to nitrocellulose and probed with antibodies directed against the GABA transporter (A) or the α-subunit of the Na^+, K^+-ATPase (B). The Na^+, K^+-ATPase is available to biotinylation only from the basolateral surface in mock transfected cells (pCB6) and in GAT-2 and GAT-3 expressing cell lines. A band corresponding to GABA transporter is not detected in mock transfected cells. While the GAT-2 protein is only isolated when the basolateral surface is labeled, GAT-3 is preferentially accessible to biotin added at the apical surface. These results are in good agreement with those obtained by immunocytochemistry (Fig. 2).

μCi/ml) is placed in the medium compartment (apical or basolateral) from which transport activity is to be determined. The same buffer containing 5 μM unlabeled GABA is placed in the contralateral medium compartment, to block the uptake of any labeled GABA which may leak across the monolayer. Filters are incubated with the labeled transmitter for 12 min at room temperature, after which they are rinsed three times in ice-cold PBS-Mg-Ca to remove any noninternalized counts. Prior to solubilizing the monolayers for scintillation counting, it is necessary to excise the filters from the plastic chamber support. This step is important, since cells can grow up the inner wall of the chamber support. Because the basolateral surfaces of the cells on the walls are not in contact with the basolateral medium compartment, this subpopulation of cells will only import labeled transmitter from the apical medium compartment. Consequently, including these cells in the analysis can substantially bias the measurement of apical versus basolateral transporter activity. Following excision with a razor blade, filters are placed in scintillation vials with 800 μl of 1% SDS and incubated for 10 minutes at room temperature with gentle agitation on an orbital shaker. A 200 μl aliquot of the lysate is removed from each vial for use in determining the total protein concentration. The remaining 600 μl is mixed by vortexing with 5 ml of EcoLume scintillation fluid (ICN, Costa Mesa, CA) and radioactivity is determined in a scintillation counter. Each assay is performed in quadruplicate. Background transport is measured by including a 100-fold excess of unlabeled GABA with the [^{3}H]GABA during the uptake assay. Background is measured for both apical and basolateral transport and these numbers are subtracted from the respective uptake values. Transport activity is then calculated as picomoles of GABA transported normalized either to microns squared of cell monolayer or to milligrams of cell protein.

Transient Expression of Transporters in Neurons

Although expression of neurotransmitter transport proteins in polarized epithelial cells can provide valuable insights into their sorting and regulation, it is clearly desirable to study the cell biologic properties of the transporters in their native cell types. Relatively few techniques have been developed, however, which permit exogenous proteins to be expressed in neurons, in culture or *in situ,* under conditions in which the complex architecture and properties of these cells are preserved. Transient transfection methodologies exploiting replication-deficient recombinant neurotrophic viruses have become available. This approach, which has been applied quite successfully with herpes simplex virus (HSV) and Semliki Forest virus (SFV)-based expression systems, is described in detail elsewhere.[28,29] Although potentially extremely powerful, the applicability of viral-based expression protocols may be limited by a number of factors. Viral infection can produce varying degrees of cytotoxicity, which can prevent extended study of transfected cells or alter their critical features (this is especially true of SFV-based systems). Construction of recombinant viruses can be time consuming and necessitates culture of live viral stocks, which may require institutional approval. For these reasons, we explored a number of methods for expressing exogenous proteins in neurons which could take advantage of standard plasmid expression vectors.

Unlike the MDCK cell line, which is rapidly dividing and immortal, neurons do not undergo mitosis and can be maintained in culture for short

[28] A. M. Craig, R. J. Wyborski, and G. Banker, *Nature* **375,** 592 (1995).

[29] V. M. Olkkonen, P. Liljestrom, H. Garoff, K. Simons, and C. G. Dotti, *J. Neurosci. Res.* **35,** 445 (1993).

periods of time (up to 5 weeks). Neurons are more sensitive to culture conditions and survive poorly when exposed to standard transfection reagents (i.e., calcium phosphate and DEAE-dextran). Even the milder cationic lipid reagents did not prove to be successful in our hands. When embryonic hippocampal neurons were transfected with commercial cationic lipids, no protein expression could be detected; either the transfection efficiency with these methods was too low (less than 1 in 100) or the neurons were no longer viable and actively synthesizing proteins after treatment.

Many laboratories use microinjection as a method for introducing macromolecules into the nucleus or cytoplasm of different cell types, including neurons.[30–33] Hippocampal neurons are smaller than average, but they, too, can be injected and survive the procedure. We have developed microinjection-based techniques for expressing exogenous proteins in hippocampal neurons in culture. We have found that this method generates reasonable levels of expression without perturbing neuronal morphology, thus permitting studies of sorting of exogenous neurotransmitter transport proteins. In this section we discuss in detail both the setup procedure for microinjecting hippocampal neurons with plasmid DNA and the types of analysis which can be applied to assess the results of the transfection.

Production of Micropipettes

A glass microcapillary tube (Narishige, Tokyo, Japan; model GD-1) is heated and pulled using a Narishige micropipette puller to produce two pipettes, each having a tip with an approximate diameter of 0.5 μm (Fig. 4). With a felt pen, two marks are made, 1.5 and 2.5 cm, respectively, from the pulled end (tip). The glass tube is trimmed at the outer mark with a diamond pen, leaving the pipette that is approximately 2.5 cm long. (The inner mark is used as a guide when inserting the open end of the pipette into the holder of the micromanipulator.) It is best to make the pipettes just before use; however, the pipettes may be stored, covered, for a week or longer. If stored for later use, the pipettes should be checked under the microscopic before they are filled with DNA solution. Place the pipette in the micromanipulator and focus on the tip with the microscope. At high magnification (40×), the tip should appear as a dark pinpoint. If the tip appears flat or circular, the opening is too large and the pipette should be discarded.

[30] M. R. Capecchi, *Cell* **22,** 479 (1980).

[31] D. Bar-Sagi and J. R. Feramisco, *Cell* **42,** 841 (1985).

[32] I. A. Muslimov, E. Santi, P. Homel, S. Perini, D. Higgins, and H. Tiedge, *J. Neurosci.* **17,** 4722 (1997).

[33] J. Wang, W. Yu, P. W. Baas, and M. M. Black, *J. Neurosci.* **16,** 6065 (1996).

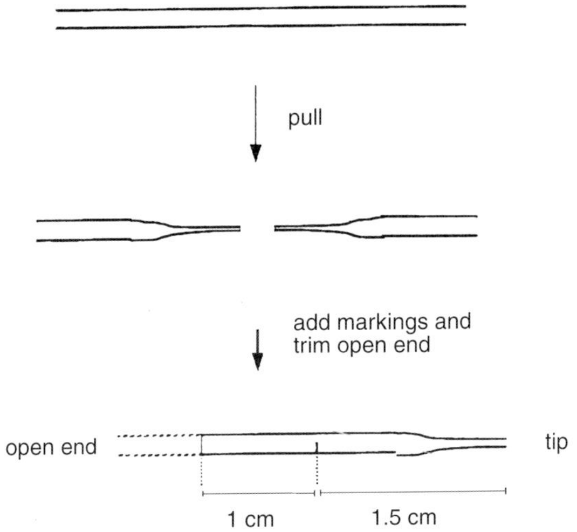

FIG. 4. Production of micropipettes. Micropipettes are pulled from capillary tubes, scored, and cut to size as described in the text.

Preparation of DNA

In all of our studies, the gene of interest is inserted into an expression vector carrying the cytomegalovirus (CMV) promoter. The construct is amplified in bacteria and plasmid DNA is isolated. We use Qiagen columns to purify the plasmid because this method is quick and simple for producing transfection-quality DNA. The DNA is ethanol precipitated and suspended in water at about 1 μg/ml. The DNA solution must be passed through a 0.2 μm Spin-X membrane (Costar) by centrifugation to remove particles which may clog the injection pipette. The concentration of the filtered DNA should be at least 0.1 μg/ml. The filtered DNA may be stored at $-20°$.

Cell Culture

Neurons are isolated from the hippocampus of 17-day rat embryos by the method described by Goslin and Banker.[34] The trypsin-dissociated neurons are plated at 3000 cells/cm^2 on treated 18-mm round poly(L-lysine)-treated coverslips and cocultured with rat glial cells in neuronal medium [0.1% bovine albumin, 2 m*M* L-glutamine, 1% (v/v) HL-1 supplement (Ventrex Laboratories, Portland, ME) in MEM]. One-half of the conditioned medium is replaced with fresh neuronal medium every 4–5 days. It is

[34] K. Goslin and G. A. Banker, *in* "Culturing Nerve Cells" (G. A. Banker and K. Goslin, Eds.), pp. 251–282 MIT Press, Cambridge, Massachusetts, 1990.

desirable to wait until the neural outgrowths are morphologically and biochemically distinguishable as axons or dendrites before transfecting the neurons (6–11 days). This ensures that the neurons are truly polarized and that the newly synthesized proteins as well as the endogenous markers (MAP2 and synapsin in our case) are properly distributed when the neurons are examined 2 days after transfection.

Up to 1 hr before microinjection, each coverslip is transferred, cell side up, to a 35-mm dish containing 1.5 ml of fresh neuronal medium buffered with 20 m*M* NaHEPES, pH 7.4. The neurons to be injected are kept in the incubator [37°, 5% (v/v) CO_2] while the pipettes are loaded with DNA solution and the injection apparatus is set up. When ready, the dishes are taken out of the incubator and placed on the stage of the injection microscope, one at a time, for 15–20 min (which is the time it takes to inject 100 cells) before being returned to the incubator to reequilibrate. The coverslips with injected neurons are transferred, cell side down, to their original dishes containing glial cultures and incubated for 48 hr before protein expression is monitored.

Loading Pipettes and Microinjection

If cells are to be coinjected with a fluorescent dye to facilitate the subsequent search for injected cells, the filtered DNA is mixed with one-third volume of 40 mg/ml lysine-fixable fluorescein/dextran (molecular weight 10,000, Molecular Probes, Eugene, OR, dissolved in 10 m*M* Tris-HCl, pH 7.5, and filtered through 0.2-μm Spin-X membrane). The resulting yellow DNA-dextran solution (2.5 μl) is transferred to the open end of the glass micropipette using a Seque/pro pipette tip (Bio-Rad Laboratories, Richmond, CA). The open end of the filled pipette is placed into the micromanipulator (to the 1.5-cm mark on the pipette) and the position of the pipette is adjusted laterally so that the pipette tip is centered in the microscopic field.

An inverted Olympus microscope equipped with a Narishige micromanipulator is used to place the neurons and pipette in position and deliver the DNA solution into the nucleus of the neuron (Fig. 5). The open end of the pipette is attached to a syringe which can be finely tuned, if necessary, to adjust the water pressure in the pipette. We try to keep the following variables constant: the water pressure in the pipette (equal to or only slightly greater than the pressure in the nucleus), the depth of penetration of the pipette into the nucleus (just enough to create a hole in the nuclear membrane), and time of contact between the pipette and nuclear membrane (0.5 sec). These values are determined empirically after numerous trials to find conditions that produce moderate nuclear enlargement without causing

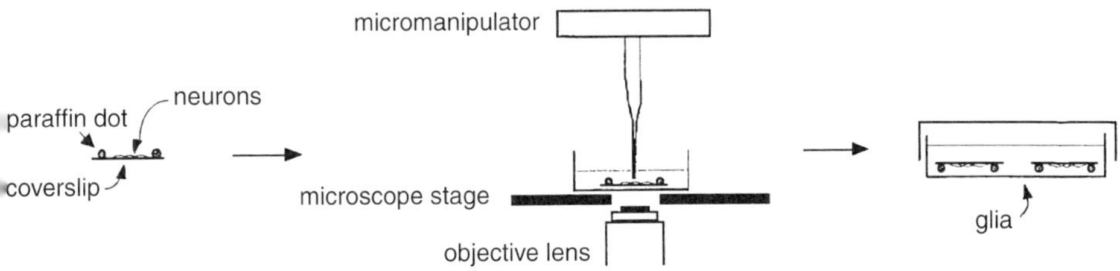

FIG. 5. Microinjection procedure. Coverslips with neurons are transferred to a petri dish and placed on the stage of an inverted microscope. A micromanipulator controls the movement of the pipette which introduces DNA into the neurons. Following injection, the neurons are returned to their original dishes containig glia and conditioned medium.

catastrophic nuclear explosions. When the injection apparatus is set up, a 35-mm dish containing a single coverslip is placed on the stage of the injection microscope and the microscope is focused on the neurons. The pipette is carefully lowered and adjusted so that it comes to a stop when its tip just pierces the nucleus. The pipette is raised, the stage is moved, and another neuron is brought to the center. The pipette is lowered until it injects the nucleus, then quickly raised. The pipette is repeatedly lowered and raised over each neuron until 100 cells have been injected. It is usually possible to detect a slight swelling of the nucleus and a change in light reflection when a neuron is injected. However, to be sure that the pipette is not blocked, it is advisable that a neuron be sacrificed occasionally by extending the injection time to 2–3 sec, during which time, the nucleus will rapidly fill with injection fluid, swell, and burst open. It is also advised that before attempting to inject DNA solutions, one first practices injecting water or fluorescein–dextran alone. The coverslip can be mounted onto a microscope slide immediately following microinjection and scanned under a fluorescence microscope to see if any cells are labeled with fluorescein. An injected cell has a bright green nucleus with some cytoplasmic staining since the fluorescein–dextran is able to diffuse through the nuclear pores.

Analysis of Exogenous Protein Expression by Fluorescence Microscopy

The coverslips with the transfected neurons are transferred, cell side up, to a 6-well plate containing warm PBS-Ca-Mg, rinsed, and fixed in freshly prepared 4% paraformaldehyde in PBS for 25 min at room temperature. The cells are washed three times in 10 m*M* glycine in PBS-Ca-Mg and once in PBS-Ca-Mg to remove excess fixative. Before labeling with antibodies, the cells are permeabilized in Buffer A (PBS-Ca-Mg containing 0.3% Triton X-100 and 0.1% bovine albumin), blocked for 30 min in Buffer B (16% goat serum, 0.3% Triton X-100, 0.45 *M* NaCl, 20 m*M* sodium phosphate, pH 7.4), 15 min in avidin/Buffer B (avidin solution from Vector Laboratories diluted 1:1 in 2× Buffer B) followed by a 5-min rinse in PBS-Ca-Mg, then 15 min in biotin/Buffer B (biotin solution from Vector Laboratories diluted 1:1 in 2× Buffer B). Primary and secondary antibodies and avidin conjugates are diluted in Buffer B. The volume of antibody solution required to cover the specimen could be reduced to 30 μl by placing the coverslip, cell side down, over a drop of the diluted antibody resting on a sheet of Parafilm. Excess reagents are removed by three 5-min washes in Buffer A. All incubations are carried out at room temperature in a humidified light-protected box. The labeled cells are rinsed in 5 m*M* sodium phosphate, pH 7.4, and mounted, cell side down, over a tiny drop of Vectashield mounting solution (Vector Laboratories) onto a microscope slide (Fig. 6).

Each mounted slide is examined under a Zeiss Axiophot microscope and the number of fluorescent neurons are counted. Cells which look interesting are further analyzed and recorded on a Zeiss laser scanning confocal microscope. Images are generated using the following excitation/emission wavelengths: 488 nm/515–560 nm (fluorescein), 568 nm/590–640 nm (Texas Red), 647 nm/650–710 nm (Cy5). On average, 5 out of 100 cells receiving injections were labeled with fluorescein dextran. Of these cells which received DNA, one of two were found to express protein at a level that could be detected by indirect immunofluorescence (Fig. 7). This is the limitation of transfecting neurons by microinjection. It is practical if protein expression is to be measured in single cells using techniques such as immunofluorescence, patch clamping, monitoring ion concentrations, or electron microscopy. However, this method of transfection is not feasible for producing large quantities of recombinant protein for biochemical analysis. Overall, because of its simplicity and safety, microinjection may be preferred over

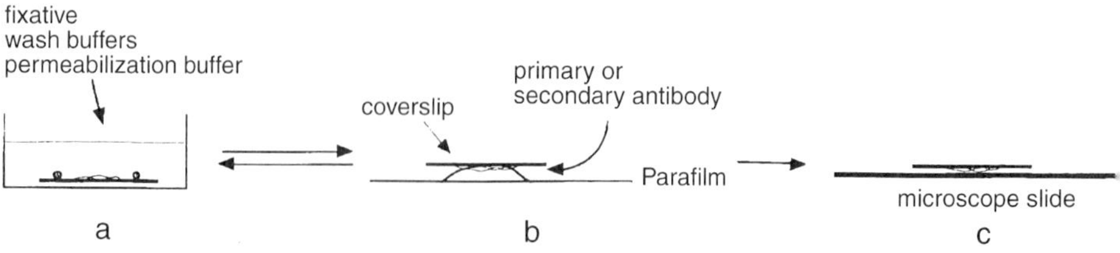

FIG. 6. Preparing cells for immunofluorescence microscopy. (a) Before labeling with antibodies, cells are fixed and washed in 1–2 ml of buffer. The paraffin dots on coverslips should be removed. (b) To incubate cells in 30–100 μl of antibody or avidin–biotin blocking solutions, the coverslips are placed over a drop of the solution on parafilm. (c) Coverslips are mounted onto microscope slides.

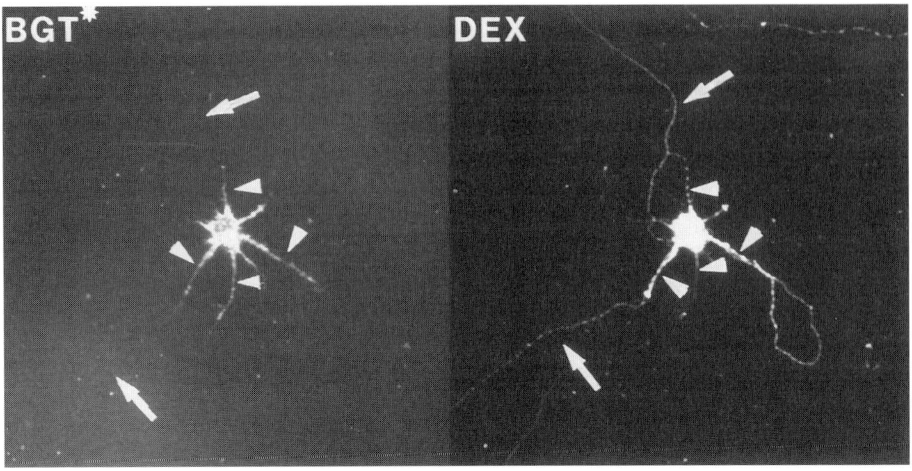

FIG. 7. Expression of BGT-1 in cultured hippocampal neurons. Hippocampal neurons in culture were microinjected with 1 μm/μl pCB6-BGT-1 (tagged with c-*myc*) and fluorescein isothiocyanate (FITC)–dextran, as described in the text. Cells were labeled for immunofluorescence microscopy with anti-c-*myc* (Oncogene Science, Manhasset, NY; diluted 1:50, v/v) monoclonal antibody followed by biotinylated antimouse IgG (Vector Laboratories, 10 μg/ml) and avidin–Texas Red (Molecular Probes, Eugene, OR; 1:200, v/v). While FITC-dextran (DEX) labels both dendritic (arrowheads) and axonal (arrows) processes, c-*myc*-tagged BGT-1 (BGT*) is found only in the dendrites.

more efficient transfection methods (i.e., use of inactivated viruses, which have side effects and can be detrimental to the neuron).

Conclusions

Members of the neurotransmitter transport protein family are normally expressed in polarized cells. In order to understand fully their cell biologic properties, therefore, it is necessary to study these polypeptides in polarized cell types and to examine the mechanisms through which they attain their specific subcellular distributions. Cultured epithelial cell lines, such as MDCK, can be readily transfected with cDNAs encoding transporter proteins. When expressed in this setting, the neurotransmitter transport proteins are targeted to specific subdomains of the plasma membrane. Their distributions can be evaluated by immunocytochemical and biochemical assays as well as through transport assays. Transporters can also be transiently expressed in neurons in culture. Once again, they appear to be segregated to the specific subcellular structures which correlate with their predicted functions. In the future, studies of chimeric neurotransmitter transport proteins composed of complementary portions of closely related family members normally subject to differential targeting should allow the molecular signals responsible for this sorting to be dissected. Identifying and characterizing these signals may lead to an enhanced understanding of the mechanisms which regulate transporter function.

Acknowledgments

We are grateful to Drs. M. Roth, C. Brewer, P. De Camilli, R. Jahn, B. Kanner, G. Rudnick, and L. Borden for generously sharing reagents. Work in the author's laboratory is supported by NIH GM-42136 and a National Young Investigator Award from the NSF.

[26] Localization of Transporters using Transporter-Specific Antibodies

By N. C. Danbolt, K. P. Lehre, Y. Dehnes, F. A. Chaudhry, and L. M. Levy

Introduction

Neurotransmitter transporters in the plasma membranes of neurons and glial cells play essential roles in the nervous system by removing the transmitters from the extracellular fluid surrounding the receptors. This is required for securing a high signal-to-noise ratio and for avoiding overstimulation of receptors. Two different families of plasma membrane neurotransmitter transporters have been identified by molecular cloning. One family includes the transporters for γ-aminobutyric acid (GABA), noradrenaline, dopamine, serotonin, proline, choline, and glycine. The other family includes the four different glutamate transporters and a transporter for neutral amino acids.[1] Some of the neurotransmitter transporters are regulated[2–5] and highly differentially localized.[6–11] In order to determine their

[1] N. C. Danbolt, *Prog. Neurobiol.* **44,** 377 (1994).

[2] M. Casado, A. Bendahan, F. Zafra, N. C. Danbolt, C. Aragón, C. Giménez, and B. I. Kanner, *J. Biol. Chem.* **268,** 27313 (1993).

[3] J. L. Corey, N. Davidson, H. A. Lester, N. Brecha, and M. W. Quick, *J. Biol. Chem.* **269,** 14759 (1994).

[4] L. M. Levy, K. P. Lehre, S. I. Walaas, J. Storm-Mathisen, and N. C. Danbolt, *Eur. J. Neurosci.* **7,** 2036 (1995).

[5] K. Sato, R. Adams, H. Betz, and P. Schloss, *J. Neurochem.* **65,** 1967 (1995).

[6] F. A. Chaudhry, K. P. Lehre, M. van Lookeren Campagne, O. P. Ottersen, N. C. Danbolt, and J. Storm-Mathisen, *Neuron* **15,** 711 (1995).

0076-6879/98 $25.00

physiological and pathophysiological roles, it is essential to know their exact localizations, not only qualitatively, but also quantitatively. This article describes methods used by our laboratory for the immunocytochemical investigation of glutamate and glycine transporters. These techniques depend on specific antibodies, optimal tissue and antigen preservation, and, above all, on proper controls.

Preparation of Antibodies

The traditional way of producing antibodies is by immunizing animals with the antigen of interest.[12] The antigen may, for instance, be a purified protein, synthetic peptides corresponding to parts of the deduced amino acid sequence encoded in a cloned cDNA, polysaccharides, or even free amino acids conjugated with glutaraldehyde to carrier proteins.[13] Monoclonal antibodies may be obtained by fusing the antibody-producing plasma cells with myeloma cells, thereby obtaining stable antibody-producing cell lines. Nontraditional methods include genetic immunization, which is based on the injection of a suitable mammalian expression vector containing the gene of the desired antigen rather than the antigen itself.[14]

Selection and Preparation of Antigens

The main difficulty in the first production of anti-GABA transporter[15,16] and anti-glutamate transporter[17] antibodies was the availability of the antigens. In these cases, methods had first to be developed to purify the pro-

---

[7] C. Freed, R. Revay, R. A. Vaughan, E. Kriek, S. Grant, G. R. Uhl, and M. J. Kuhar, *J. Comp. Neurol.* **359,** 340 (1995).

[8] K. P. Lehre, L. M. Levy, O. P. Ottersen, J. Storm-Mathisen, and N. C. Danbolt, *J. Neurosci.* **15,** 1835 (1995).

[9] A. Minelli, N. C. Brecha, C. Karschin, S. Debiasi, and F. Conti, *J. Neurosci.* **15,** 7734 (1995).

[10] F. Zafra, C. Aragon, L. Olivares, N. C. Danbolt, C. Gimenéz, and J. Storm-Mathisen, *J. Neurosci.* **15,** 3952 (1995).

[11] C. E. Ribak, W. M. Y. Tong, and N. C. Brecha, *J. Comp. Neurol.* **367,** 595 (1996).

[12] D. Drenckhahn, T. Jöns, and F. Schmitz, *Meth. Cell Biol.* **37,** 7 (1993).

[13] J. Storm-Mathisen, A. K. Leknes, A. T. Bore, J. L. Vaaland, P. Edminson, F.-M. S. Haug, and O. P. Ottersen, *Nature* **301,** 517 (1983).

[14] J. J. Donnelly, J. B. Ulmer, and M. A. Liu, *J. Immunol. Meth.* **176,** 145 (1994).

[15] R. Radian, A. Bendahan, and B. I. Kanner, *J. Biol. Chem.* **261,** 15437 (1986).

[16] R. Radian, O. P. Ottersen, J. Storm-Mathisen, M. Castel, and B. I. Kanner, *J. Neurosci.* **10,** 1319 (1990).

[17] N. C. Danbolt, J. Storm-Mathisen, and B. I. Kanner, *Neuroscience* **51,** 295 (1992).

teins.[15,18,19] Now, after the cloning of a number of different transporters, the situation is completely different. It is no longer necessary to isolate the proteins, and the investigator can choose which part(s) of the sequence to use for immunization. Initially it must be decided if the antigens are to be produced as fusion proteins or as synthetic peptides. The fusion proteins have the advantage that they can be made long enough to cover several possible antigenic sites, thereby increasing the probability that at least one of them will give rise to good antibodies. The disadvantage is that there is some work involved in cloning of the cDNA encoding the protein or part of it into a suitable expression vector, expressing it in cells, and purifying it from a lysate of these cells. Peptides, on the other hand, have the advantage of being easily synthesized without any work on the part of the investigator. A disadvantage is that the peptides are relatively short, usually in the range 8–30 residues, because of the costs and difficulties involved in the synthesis of larger peptides. This raises the question of how to select an antigenic sequence.

Unfortunately, there is still no perfect way to predict the most antigenic part(s) of a sequence and there appear to be no absolute rules. Try to avoid hydrophobic parts of the sequence (e.g., putative transmembrane regions) and sites that may be glycosylated or lipidated *in vivo*. It is usually (but not always[20]) recommended to convert the C-terminal carboxyl group of the peptide to an amide unless it is free in the native protein. Cysteines inside the sequence may create problems during synthesis. If a cysteine is wanted to enable directional coupling (via its SH group), it may be best to let it be the last residue added. To select the peptides, a computer program (such as the program Protean in the Lasergene program package from DNASTAR) that can run different algorithms for calculating hydophobicity, immunogenicity, etc., is very helpful. It is, however, important to realize that these programs only give an indication. Sequence stretches that are suggested to be highly immunogenic may not be immunogenic at all or may give rise to high titers of antibodies that recognize the peptide, but not the native protein.

The synthetic peptides are usually too small to be antigenic in themselves and must be coupled to a carrier. This can be done in different ways:

1. The peptide may be synthesized directly onto a carrier, making coupling reagents redundant. The most common is the so-called multiple antigenic peptide (MAP) system which has been presented as a novel and efficient approach for eliciting antibodies to peptides. The system is based

[18] R. Radian and B. I. Kanner, *J. Biol. Chem.* **260,** 11859 (1985).

[19] N. C. Danbolt, P. Pines, and B. I. Kanner, *Biochemistry* **29,** 6734 (1990).

[20] B. Maillère and M. Hervé, *Mol. Immunol.* **34,** 1003 (1997).

on a small immunogenically inert core matrix of branched polylysine[21] and may be designed to contain both defined B- and T-cell epitopes.[21,22] The peptides have their N termini free and are attached by their C termini. Because the C terminus is blocked, the MAP system is not recommended for production of antibodies to the extreme C terminus.[23]

2. Still the most common procedure is to synthesize the peptide in free form and then couple some of it to a carrier. The frequently used carriers include keyhole limpet hemocyanin (KLH), bovine serum albumin (BSA), and thyroglobulin, but other proteins may also be used. Common coupling procedures are based on *m*-maleimidobenzoyl-*N*-hydroxysuccinimide ester (MBS) and glutaraldehyde, but a variety of other methods and reagents are described in the literature.

MBS is a heterobifunctional reagent and is used for directional coupling of peptides containing a free SH group to carrier protein in a two-step procedure.[12] It contains one maleimide group (which is SH group reactive) and one *N*-hydroxysuccinimide group (which is amino reactive). The first step is the reaction of MBS with the amino groups of the carrier (activation of the carrier) and the removal of free unreacted MBS. Then, the activated carrier (which now has a number of maleimide groups) is reacted with free SH groups on the peptide. (*Note:* The peptides usually have to be freshly reduced with dithiothreitol (DTT), 2-mercaptoethanol, or sodium borohydride prior to coupling.) The MBS coupling procedure results in only three possible combinations: peptide–MBS–carrier, unreacted peptide, and carrier–MBS.

Glutaraldehyde reacts primarily with amino groups, but also with sulfhydryl groups and to some extent with the phenolic and the imidazole rings of tyrosine and histidine.[24] The glutaraldehyde coupling procedures usually involve the mixing or carrier protein and peptide together with glutaraldehyde (G). This results in a random mixture of carrier–G–peptide, carrier–G–carrier, peptide–G–peptide, carrier–G, and peptide–G conjugates, as well as unreacted glutaraldehyde. The mixture will be even more complicated if the peptide contains more than one glutaraldehyde reactive group.

It is not possible to know in advance which method will be the best, as this may vary with the peptide. Most of our antibodies have been made using the glutaraldehyde method, but we have also had success with the MBS method and the MAP system. The selection of an antigenic peptide

[21] J. P. Tam, *J. Immunol. Meth.* **196,** 17 (1996).

[22] M. Christodoulides and J. E. Heckels, *Microbiology* **140,** 2951 (1994).

[23] J.-P. Briand, C. Barin, M. H. V. van Regenmortel, and S. Muller, *J. Immunol. Meth.* **156,** 255 (1992).

[24] A. F. S. A. Habeeb and R. Hiramoto, *Arch. Biochem. Biophys.* **126,** 16 (1968).

is more important than the coupling procedure, type of carrier, etc. The undefined nature of the glutaraldehyde procedure does not make it intellectually appealing and it may initiate the production of antibodies that only recognize the modified protein or the coupling reagent. However, since the antibodies are intended for use on aldehyde-fixed tissue and chemical modification of the sequence may help overcome the tolerance of the immune defenses of the animals, the glutaraldehyde method should not be forgotten.

The MBS coupling method[12] is somewhat tricky to set up. If a small number of peptides are to be coupled, it may be worthwhile to have them prepared commercially. The simpler glutaraldehyde methods, however, are best done in the laboratory. The procedures, which should be performed in the cold room, are as follows:

Procedure A

1. Dissolve the carrier protein in water and dialyze overnight against 1000 volumes 10 m*M* Na-4-(2-hydroxymethyl)-1-piperazineethanesulfonic acid (HEPES), pH 7.5, and freeze in aliquots. (Do not be in a hurry when dissolving the KLH and do not try to make it more concentrated than 40 mg/ml. A practically insoluble mass may result.)

2. Mix 1–3 peptides (1 mg of each) and 1 mg dialyzed carrier protein in 1 ml of a solution containing 50 m*M* NaP_i (sodium phosphate buffer pH 7.4) and 300 m*M* NaCl (omit the salt and add dimethyl sulfoxide (DMSO) or *N*, *N*-dimethylformamide (DMF) if the peptides are too hydrophobic to stay in solution). Use a 50-ml tube in order to contain the foam that forms at stage 3. Put a stirring bar in the tube and place the tube on a magnetic stirrer in the cold room. Add 1 ml ice-cold 2% (w/v) glutaraldehyde and continue stirring overnight.

3. Stop reaction and reduce double bonds (thereby making the coupling irreversible) by adding 220 μl 10% (w/v) sodium borohydride freshly dissolved in water. Incubate for 1 hr. (Most of the color produced by addition of glutaraldehyde should disappear.) Make sure the solution is still clear. Gel filter to remove aldehyde, free peptide, and salts on columns equilibrated with 50 m*M* NaP_i. Azide or other toxic preservatives should of course not be added. Store the 3 ml of filtrate (the conjugate) at −20°. Fifty to one hundred microliters of this solution will be enough for one intradermal immunization dose.

Procedure B

Take 1–3 peptides (1 mg of each) in an Eppendorf tube with 1 mg carrier protein (see Procedure A, step 1) and bring the volume to 400 μl with 0.2 *M* NaPi pH 7.4. Make sure the components are soluble. Add 15 μl 25% (w/v) glutaraldehyde and mix quickly. Incubate end-over-end (2 hours or overnight, 4°). Store at −20°. For each immunization, use 50 μl of the mixture (which may contain precipitates).

Immunization

The selection of the species to immunize will depend on the required amounts of antibodies and to some extent on the source of the antigen. (The demands will often be far greater than originally thought.) Theoretically, it is not a good idea to try to make antibodies to rabbit proteins in rabbits, although this sometimes works with intracellular proteins or intracellular parts of membrane proteins (e.g., the C-terminal of the EAAC glutamate transporter, which is identical in rabbit, humans, and rat).

For production of polyclonal antibodies, rabbits will usually be the first choice. They are easy to handle and to bleed (large ears with superficial veins) and are good responders. Chinchilla rabbits can be bled 3 × 10 ml/kg each month (without reducing the hematocrit below 30 at any time). A 5-kg rabbit may then give 70–80 ml serum/month. If larger amounts of serum are required or immunoglobulin (Ig) from a different species, sheep are often a good alternative. It is our experience that rabbits respond after the second or third immunization, whereas sheep need four to five immunizations to respond. Large amounts of immunoglobulins can be obtained by collecting eggs of immunized chickens. There may be more than 100 mg of immunoglobulin (IgY) in one egg yolk. Unfortunately, IgY is not as stable as IgG[25] and is more difficult to purify.[26] Mice and rats are interesting only if production of monoclonal antibodies is intended.

When deciding doses it should be kept in mind that larger animals do not need more antigen than smaller animals when the antigen is given by intradermal injections. In our experience, which is based mainly on intradermal injections, Freund's adjuvant, and glutaraldehyde-coupled peptides, doses around 10–100 μg peptide give better antibodies than higher doses.

When producing antipeptide antibodies, it may be desirable to try several different peptides, carrier proteins, and immunization protocols. One animal can be immunized with a number of peptides simultaneously; some of the best antibodies from our laboratory have been obtained after immunizing the same animal with up to nine different peptides. In these cases,

[25] M. Shimizu, H. Nagashima, K. Sano, K. Hashimoto, M. Ozeki, K. Tsuda, and H. Hatta, *Biosci. Biotech. Biochem.* **56,** 270 (1992).

[26] J. C. Jensenius, I. Andersen, J. Hau, M. Crone, and C. Koch, *J. Immunol. Meth.* **46,** 63 (1981).

all the peptides given to one animal have been different parts of the same protein. The antibodies obtained to each individual peptide are easily separated by affinity purification (see later discussion).

Several excellent descriptions on alternative adjuvants and the handling of the animals are available[12,27] and will therefore not be described here. One practical tip concerning the bleeding of rabbits is worth mentioning. We routinely give the rabbits 0.1–0.2 ml Hypnorm (Janssen, Buckinghamshire, UK) intramuscularly (i.m.) per kg body weight before ear bleeding. Hypnorm, which contains fentanyl citrate (0.315 mg/ml) and fluanisone (10 mg/ml), has a strong sedative and analgesic effect, but more importantly, it also has a strong vasodilating effect. This combination makes the bleeding of rabbits painless, easy, and quick.

Testing of Antisera to Transporters

The importance of proper specificity testing for antibodies for immunocytochemistry cannot be overemphasized. It is not enough to demonstrate reactivity towards the desired antigen. The absence of reactivity toward other antigens in the preparation under study must also be shown. When a tissue section is immunostained, both antibodies with the desired specificity and antibodies with unwanted reactivity will give rise to staining. On Western blots,[28] different reactivity might be distinguished by the differences in molecular mass of the labeled polypeptides. No such information is obtained from tissue sections. Labeling of both glial and neuronal cell membranes could mean that the transporter is expressed in both types of cells, or it could mean that the transporter is expressed in one of them and that unwanted reactivity causes the labeling of the other. From looking at the tissue sections, there is no way one can tell the difference. If, on the other hand, cell nuclei are stained, one might be suspicious about the specificity of the antibodies because the neurotransmitter transporters are not expected to be expressed there. Labeling of cytoplasm could represent transporter protein in the granular endoplasmic reticulum or in vesicles, but could also represent spurious reactivity. Used in sufficiently high concentrations, virtually any antibody preparation will label aldehyde-fixed brain tissue. Further, autoantibodies to intracellular proteins are frequently observed in rabbits.[12] It must therefore be realized that the proof of antibody specificity is primarily biochemical. This may, however, not be enough and it may be necessary to compare the histochemical labeling obtained with different supposedly specific antibodies to the same protein.

[27] J. W. Goding, "Monoclonal Antibodies: Principles and Practice," 2nd ed., Chapter 8. Academic Press, London, 1986.

[28] H. Towbin, T. Staehelin, and J. Gordon, *Proc. Natl. Acad. Sci. USA* **76,** 4350 (1979).

Antibodies recognizing unwanted proteins are commonly called "unspecific." This is terribly misleading because some of these antibodies can be highly specific in that they recognize a well-defined (although unknown and unwanted) epitope. They could actually also be specific for an epitope on the right protein, but the epitope might not be specific for the protein. A protein sequence (e.g., phosphorylation consensus sites) may be shared by many different proteins. During the screening for a monoclonal antibody to GLT,[29] some clones producing polyreactive antibodies were isolated (L. M. Levy and N. C. Danbolt, unpublished). One of them (7G4) produced a mouse IgG_{2b} antibody that recognizes a huge number of different proteins (Fig. 1, lane 1) as compared to the monospecific IgM anti-GLT (9C4) antibody (Fig. 1, lane 2).

The antibodies should be tested on immunoblots made from crude sodium dodecyl sulfate (SDS) extracts of the tissue under study. It is important that labeling of a single polypeptide can be demonstrated on a blot containing as many of the tissue antigens as possible. However, the specificity testing done on immunoblots of fresh unfixed SDS-denaturated tissue proteins is not necessarily valid on tissue sections of aldehyde-fixed tissue. It must also be tested whether the antibody preparation contains antibodies recognizing aldehyde-treated proteins. Several of our rabbits have such antibodies even in the preimmune serum. This should be tested not only with crude serum, but also with affinity-purified antibodies (see below). Another control[8] is to preabsorb the antibodies with the antigen (e.g., the free peptide) and see if this competitively inhibits the labeling both of blots and tissue sections.

Very often there will be weak extra bands and the temptation may be strong to ignore them. If these antigens are evenly distributed in the tissue, they will probably not affect the overall picture. If, however, they should be concentrated in certain small areas, the local concentration may be high enough to give a significant signal. To control for this possibility, it is necessary to have another antibody, produced in a different animal against a different epitope of the same protein. If the staining patterns of the two different antibodies are the same, this will strongly support the initial interpretation. If the labeling pattern is different, one has to decide if one of the antibodies has some unwanted reactivity or if the protein exists in two different forms.

Further, these tests have to be performed for each tissue studied. Antibodies specific for transporters in the brain may not be specific for the same transporters in another tissue. The anti-B493 antibodies to GLT (rabbit 84946) may serve as an example (N. C. Danbolt, unpublished).

[29] L. M. Levy, K. P. Lehre, B. Rolstad, and N. C. Danbolt, *FEBS Lett.* **317,** 79 (1993).

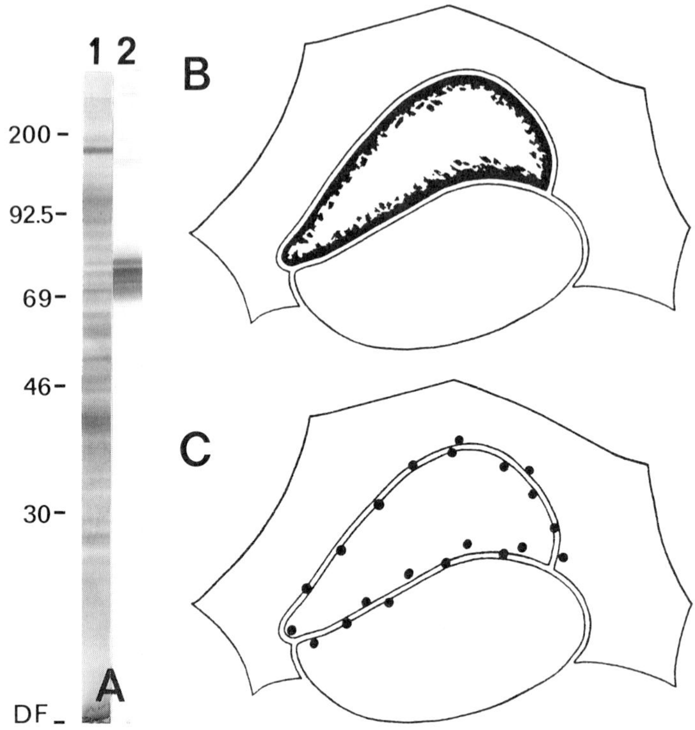

FIG. 1. Demonstration of antibody polyreactivity (A) and the different appearance of electron micrographs after labeling with preembedding peroxidase (B) and postembedding immunogold (C) immunocytochemistry. (A) Proteins from rat forebrain brain were separated by SDS–PAGE and immunoblotted with cell culture supernatants containing mouse monoclonal antibodies: (lane 1) a polyreactive IgG_{2b} antibody and (lane 2) an IgM antibody to a glutamate transporter.[29] Note that multiple bands are labeled by a monoclonal antibody. (B) Diagram of an electron microscopic picture shows the cell membranes of three different cellular compartments. In this example the antigen is located on the inside of the cell membrane of the central compartment. With the preembedding technique, the immunoreagents have reached the antigen by diffusion, limited by the cell membranes in the tissue, from the surface of the thick (30–40 μm) Vibratome section. The peroxidase reaction is formed on the inside and remains on the inside since the cell membrane represents a diffusion barrier. Some of the reaction product may diffuse a short distance into the cytoplasm. After termination of the peroxidase reaction, the tissue has been embedded and the ultrathin section cut. (C) The same structure labeled with the postembedding immunogold technique. Here, the antibodies have been applied onto the flat surface of the ultrathin (0.05–0.1 μm) sections. Movement of the immunoreagents is therefore not restricted by the membranes. The length of the gold particle–secondary antibody–primary antibody–antigen complex is such (10–20 nm) that some of the gold particles may be on the outside even though the antigen is on the inside.

When used on immunoblots or sections of adult rat brain, they appear specific for GLT although they also react with another protein of about the same molecular mass as GLT. The reaction with this protein is not easily recognized when the antibodies are tested on adult brain because the concentration of the GLT protein is much higher. However, when the antibodies are tested on tissue with little or no GLT (e.g., newborn rat brain, placenta, or heart), this other protein is clearly seen and the labeling may be wrongly attributed to GLT. Thus, a polyclonal antibody is not universally specific. It must be tested for each new application.

When trying out a new serum, the first question is whether it recognizes the protein of interest. If no signal is obtained with a crude serum or an immunoblot of a whole tissue protein extract, there are three possibilities: (1) the serum does not contain antibodies to the wanted protein, (2) the antigen is expressed in very low concentration in the tissue or for any other reason present in low amounts on the blots, or (3) the serum contains a lot of unwanted reactivity toward antigens present in high concentrations on the blot, making it difficult to recognize any labeling of the desired antigen. (Theoretically, it is possible that the antibodies only recognize the nondenaturated form of the protein, but this is rather unlikely when a synthetic peptide corresponding to a small piece of the sequence has been used as the immunogen.) Therefore, one should not judge a serum useless before it has been affinity purified and tested on a preparation containing a high concentration of the desired protein. A partially purified preparation of most neurotransmitter transporters may be obtained by isolating membrane glycoproteins by affinity chromatography on wheat germ agglutinin (lectin)-agarose.[17]

When immunizing with synthesis peptides corresponding to parts of a protein, antibodies to the peptides are virtually always obtained (we have antibodies to 55 out of the 56 peptides we have tried so far), but only a fraction of these antibodies recognize the native protein specifically. Most peptide antisera contain a mixture of antibodies recognizing both the peptide and the native protein and antibodies recognizing only the peptide. There may be several reasons for this. The peptides may have a different conformation than the native protein. The selected part of the sequence may not be exposed on the surface of the native protein. The antibodies may be directed to one of the termini of the peptide which may not exist free in the native protein.[30] The peptides may be different from the native protein because of peptide bond racemization,[31] side reactions during syn-

[30] T. C. Liang, W. P. Luo, J. T. Hsieh, and S. H. Lin, *Arch. Biochem. Biophys.* **329,** 208 (1996).

[31] G. I. Szendrei, H. Fabian, H. H. Mantsch, S. Lovas, O. Nyeki, I. Schon, and L. Otvos, *Eur. J. Biochem.* **226,** 917 (1994).

thesis,[32,33] deletions (peptides with missing residues could be more immunogenic), incompletely removed protecting groups or modification of antigenicity due to coupling, C-terminal amidation,[20] chemical instability of the peptide, etc. It is usually easier to immunize with a new peptide or the same peptide in a different way than to try to figure out the exact cause(s).

Purification of Sera

Antibodies recognizing the antigen may be isolated by so-called "affinity purification." The serum is passed through a column containing covalently immobilized antigen. After washing, the bound antibodies are eluted (detailed below). It should be realized that the procedure isolates molecules that bind to the antigen, the coupling reagent used to immobilize the antigen, or the column material itself. It is not a good idea to couple the peptide conjugate used for immunization to the column. If, for example, a peptide–glutaraldehyde–KLH conjugate has been used for immunization, it is better to avoid both glutaraldehyde and KLH and couple the free unconjugated peptide to, for instance, *N*-hydroxysuccinimide-activated agarose. In principle, only the peptide antibodies and not the carrier–aldehyde-reactive antibodies should bind to the column. Quite often, this is the only purification needed. However, sometimes unwanted antibodies, such as aldehyde and polyreactive antibodies, do bind. These antibodies will give background and often cell nuclei staining. Their removal may require absorption of the serum.

Affinity Purification of Antibodies

1. Peptide is coupled to agarose (at about 4°) in the following way: Take 1 mg free peptide in 1 ml 1 *M* Na-HEPES, pH 8, in a 50-ml tube (e.g., Falcon 2070). If the peptide is poorly soluble in water, add 1 ml dimethyl sulfoxide or dimethylformamide. Wash Affi-Gel 15 (*N*-hydroxysuccinimide-activated agarose, Bio-Rad, Richmond, CA) for about 10 min with some 25 volumes of cold distilled water adjusted to about pH 4 with HCl. Do not allow air to come into the gel. After washing, suspend the gel in about one volume of water. Immediately take 5 ml gel (10 ml of suspension) to the above tube and start end-over-end mixing immediately for 2 hr or overnight. Add NaN_3 to 0.1% (w/v) and store the gel at 4° until needed. The gels prepared in this way have a binding capacity around

[32] A. A. Mazurov, S. A. Andronati, T. I. Korotenko, V. Y. Gorbatyuk, and Y. E. Shapiro, *Int. J. Peptide Prot. Res.* **42,** 14 (1993).

[33] P. Rovero, S. Pegoraro, F. Bonelli, and A. Triolo, *Tetrahedron Lett.* **34,** 2199 (1993).

1–5 mg antibodies per ml gel. Higher concentrations of immobilized peptide will not necessarily increase the binding capacity, but may in fact decrease it.

2. All the subsequent steps are performed at room temperature. A column with 5 ml peptide–Affi-Gel is washed (to remove free peptide) with 10 column volumes washing buffer consisting of 50 m*M* Tris-HCl pH 7.4, 0.3 *M* NaCl, 1 m*M* ethylenediaminetetraacetic acid (EDTA), 0.05% (v/v) Tween 20 and 0.1% (w/v) NaN_3.

3. Pass 50–150 ml serum with 1 m*M* phenylmethylsulfonyl fluoride (PMSF) and 5 m*M* EDTA through the column for about 1 hr. Collect the serum as there may be more antibodies left. Wash the column with 10 column volumes washing buffer (10 min).

4. Elute bound antibodies with 20 ml 0.2 *M* glycine hydrochloride buffer (pH 2.5) with 0.15 *M* NaCl and 0.1% NaN_3. Apply the buffer on the surface of the gel. Elute at a flow rate of 0.5–1 column volumes per min. (Various other buffers are reported in the literature for the elution of bound antibodies, but the glycine hydrochloride buffer described above has been the most reliable buffer in the author's hands.) Neutralize the eluate immediately by 2 *M* Na-HEPES, pH 8.5, or 2 *M* Tris-HCl, pH 8.5. (Add the neutralization buffer to the tubes prior to elution and mix during elution.) Use indicator paper to check for a pH of 7–8.

5. Determine the protein concentration spectrophotometrically by absorption at 280 nm. Add PMSF (1 m*M*).

Absorption of Crude Serum or Affinity-Purified Antibodies

1. Aldehyde–carrier protein–agarose is prepared as follows: dialyzed carrier protein is coupled to Affi-Gel as described for peptides above using 10 mg of protein per ml of Affi-Gel. After coupling, the gel is rinsed briefly with water, suspended in several volumes of 2% (w/v) glutaraldehyde freshly diluted in 0.1 *M* NaP_i, incubated (2 hr, room temperature) end-over-end, washed with 10 m*M* NaP_i to remove free aldehyde, and transferred to a large Erlenmeyer flask. Solid sodium borohydride [to 1% (w/v)] is added and the gel is incubated (1 hr or overnight). Wash the gel in NaP_i, block unspecific binding sites with a "neutral" serum, wash the gel with 3 volumes of the above glycine hydrochloride buffer, and neutralize with a few milliliters of 2 *M* Tris-HCl, pH 8.5. Finally, wash the gel with the washing buffer (see step 2 above) and store it in this buffer (4°) until needed.

2. Let 1 ml crude serum or the affinity purified antibodies (isolated from 50–150 ml serum) slowly pass through a 10 ml glutaraldehyde-carrier protein-AffiGel column. BSA should be added to the purified antibodies to minimize loss of antibodies due to unspecific binding to the column. (*Note:* In this step collect the proteins that do not bind to the column.)

3. Wash out remaining antibodies with 2–3 column volumes of washing buffer. Do not worry about the large volume, as the antibodies can be concentrated on protein A (or G)-Sepharose. (See later discussion.)

4. Regenerate the column with the glycine hydrochloride buffer, neutralize, and equilibrate with washing buffer. Discard the antibodies that are bound to the column.

Isolation of IgG from Affinity-Purified Antibodies

Use protein A-Sepharose Fast Flow (Pharmacia, Piscataway, NJ) when purifying rabbit antibodies and protein G-Sepharose Fast Flow when purifying sheep antibodies. Equilibrate a 0.5 ml column with the above washing buffer. Apply the antibody solution over about 1 hr. (Do not discard the flowthrough, as there may be more antibodies.) Wash with 6 column volumes (2–3 min) 0.1 *M* NaP_i with 0.02% (w/v) NaN_3. Elute rabbit IgG from protein A with 2.5 ml 0.2 *M* sodium-citrate, pH 3.7, with 0.1% (w/v) NaN_3 at a flow rate of 0.5–1 column volume per min. Elute sheep IgG from protein G with the above glycine hydrochloride buffer. Neutralize the eluate immediately with 2 *M* Na-HEPES, pH 8.5. Check that the pH is pH 7–8. Determine the protein concentration and add PMSF (1 m*M*) and EDTA (5 m*M*).

Immunocytochemistry

The immunocytochemical techniques require thin sections of tissue with preserved antigenicity and structure. The purpose of the investigations will determine the necessary degree of preservation of structural detail. Even poor preservation is satisfactory when the aim is to reveal regional distribution, whereas only optimal preservation will suffice for the most demanding studies at high resolution.

Several different tissue-cutting devices are available. The freezing microtome consists of a microtome with an attached cooling unit making it possible to keep the specimen holder at subzero temperature while the microtome itself, including the knife and tissue, is exposed to room-temperature air. The sections (20–50 μm) will thaw immediately after cutting and are wiped off the knife and into buffer with a fingertip or a fine brush. The tissue will disintegrate if it is not fixed prior to cutting. The cryostat is a microtome located inside a freezer. Here the sections will remain frozen after cutting, and it is possible to cut unfixed tissue. Sections can be cut as thin as 5 μm. For glutaraldehyde-fixed tissue, drying of the sections may cause cross-linking of the tissue, making staining difficult. Neither the freezing microtome nor the cryostat are recommended for electron microscopy, but are excellent for quickly producing a number of sections for initial testing of antibodies and regional labeling differences. Better tissue preservation is obtained by cutting wet fixed unfrozen tissue. This can be done on a Vibratome, which consists of a small tray and a razor blade-like knife. The knife is moved slowly toward the tissue horizontally in a sideways vibrating fashion. The tissue (approximately 5 mm thick slice) is glued by cyanoacrylate glue onto a wooden chuck for mounting in a holder at the bottom of the tray. The tray is filled with cold (2–8°) isotonic buffer to keep the tissue firm and to prevent it from drying. When properly adjusted, it is possible to cut 30 to 40-μm-thick sections of a whole fixed rat brain. After cutting, the sections are stored (4°, 12 hr–weeks) in 0.1 *M* NaP_i with NaN_3 (0.2–1 mg/ml). For electron microscopy ultrathin (0.05–0.1 μm) sections are needed. These are cut on an ultramicrotome equipped with a glass or diamond knife. The cutting requires embedding (see below for details) of the tissue in a hard plastic-like material (e.g. Lowicryl, Epon, or Durcupan).

Thc immunocytochemical labeling may be done either before (preembedding) or after embedding and sectioning (postembedding). For preembedding electron microscopy, the tissue is usually cut on a Vibratome and processed as for light microscopy (see later discussion). Both enzyme (peroxidase) and gold-labeled reporter molecules may be used. After the labeled sections have been evaluated light microscopically, small pieces of tissue are cut out from the sections and embedded (see below). For postembedding electron microscopy, the tissue is first embedded and cut, then incubated with primary antibodies followed by labeled (usually gold-labeled) reporter molecules. It is therefore essential that the antigenicity be preserved after embedding. This may be very difficult to achieve. At present, the best way is to perform freeze substitution and embedding at very low temperature. However, even this procedure only works for some of the antibodies that have proved useful for preembedding.

When the relatively (30–40 μm) thick Vibratome sections are stained without Triton and without ever letting the tissue dry prior to labeling, the preembedding method gives information on the topology of the membrane proteins.[8,34] The immunoreagents will diffuse into the tissue from the surface, but will be restricted by the cell membranes (Fig. 1B,C). This implies (1) that the peroxidase reaction product or gold particles will be on the same side of the membrane as the enzyme and thereby the antigen itself and (2) that there will be a variable penetration of the reagents. A structure may be unlabeled either because it does not contain the antigen or because

[34] E. Molnar, R. A. J. McIlhinney, A. Baude, Z. Nusser, and P. Somogyi, *J. Neurochem.* **63,** 683 (1994).

the antigen has been inaccessible to the immunoreagents. The method is therefore not quantitative. The postembedding immunogold method[6] is less sensitive, but it gives quantitative information and offers higher resolution than the preembedding peroxidase method.

Fixation

The purpose of fixation is to preserve the tissue components of interest after death and during the cutting and labeling procedures. Preservation of tissue for electron microscopy (i.e., preservation of ultrastructure) requires very rapid fixation as the autolytic degradation starts almost immediately postmortem. Several different fixatives are available. The most common is formaldehyde (HCHO), which binds to and thereby modifies amino groups and forms covalent bonds between adjacent amino groups. Formaldehyde may be used alone or in combination with other compounds, notably glutaraldehyde (pentane 1,5-dial). Other fixatives, such as carbodiimides which modify carboxyl groups, may be used when free amino groups are required for antibody reactivity.

Both formaldehyde and glutaraldehyde have a tendency to polymerize. Since the polymers have a much slower tissue penetration than the monomers, the monomers are required for the quickest and most efficient fixation. Formaldehyde monomers can be obtained by depolymerizing commercially available paraformaldehyde shortly before use.

Depolymerization of Paraformaldehyde

Perform the whole procedure in a fume hood. To make 1000 ml of a 4% solution in 0.1 *M* NaP_i, dissolve 11.5 g (water-free; molecular weight 142) Na_2HPO_4 in about 800 ml water in an Erlenmeyer flask. Add 40 g paraformaldehyde and heat to 60° while stirring until dissolved. This may take 30 min. The solution will not be completely clear. Then add 2.28 g (water-free, molecular weight 120) NaH_2PO_4 and water to make 1000 ml. Let the solution cool down to room temperature and filter it through filter paper. The formaldehyde is now ready. The most demanding applications may require that the solution be used within a couple of hours, but for routine perfusion fixation of rat brain satisfactory results are obtained with formaldehyde solution which is a few days old, provided that the fixative is filtered before use. If glutaraldehyde or picric acid are to be added to the fixative, this is done immediately before use. The standard fixative used in our laboratory for light and preembedding electron microscopical visualization of neurotransmitter transporters is a mixture of (4%/0.2%/

0.05%) formaldehyde/picric acid/glutaraldehyde in 0.1 *M* NaP_i.[35] For post-embedding, we often use other fixatives.[6,36]

Perfusion Fixation of Rats

Thin pieces of tissue, such as 2- to 300-μm-thick hippocampal slices or isolated pieces of retina, may be adequately fixed simply by dropping them into the fixative solution and leaving them there for 20–60 min at room temperature. Thicker pieces of tissue are best fixed *in situ* by perfusion[8,37] through the heart (left ventricle, aorta). The perfusion liquids are delivered by a peristaltic pump (50 ml/min; adult rats). The tubing is filled with 25 ml ice-cold washing solution (0.1 *M* NaP_i with dextran 70, 40 mg/ml) and connected to the bottle of fixative without introducing air bubbles. (This arrangement allows the fixation to start immediately after most of the blood has been washed out from the animal.) Then the rats are given a lethal dose of pentobarbital (100 mg/kg). When deeply anesthetized, but before cessation of spontaneous respiration, do the following quickly in this order: Open the thorax, start the pump, push the cannula (OD 2–3 mm blunted) through the apex of the heart into the left ventricle, and cut open the right atrium. Continue to perfuse for 15 min (room temperature) with fixative. The procedure is performed in a fume hood. The brains are postfixed (2–3 hr) in the same fixative and stored (4°) in a storage solution (1 part fixative and 9 parts NaP_i) until they are processed for immunocytochemistry.

Immunocytochemical Labeling Procedure

The whole procedure is performed at room temperature when not stated otherwise. To increase penetration of antibodies and thereby labeling intensity, 0.5% (v/v) Triton X-100 may be added to the solutions in steps 1–8 (below). This is recommended for light microscopical evaluation of regional differences. Inclusion of Triton does, however, make light microscopical evaluation of fine details more difficult. Since it destroys the cell membranes, it should not be used with sections intended for electron microscopy.

1. Rinse the (20–40 μm thick) sections in 0.1 *M* NaP_i and quench aldehyde groups present in the tissue after aldehyde fixation by incubation (30 min) in 1 *M* ethanolamine (adjusted to pH 7.4 with HCl and 0.1 *M* sodium dihydrogen phosphate).

[35] P. Somogyi and H. Takagi, *Neuroscience* **7,** 1779 (1982).

[36] Y. Dehnes, F. A. Chaudhry, K. Ullensvang, K. P. Lehre, J. Storm-Mathisen, and N. C. Danbolt, *J. Neurosci.* **18,** 3606 (1998).

[37] O. P. Ottersen, *Anat. Embryol.* **178,** 407 (1988).

2. Wash (3 × 1 min) in PBS (0.135 *M* NaCl, 10 m*M* NaP_i) and incubate (1 hr, room temperature) in a "blocking" solution consisting of TBS (0.3 *M* NaCl, 0.1 *M* Tris-HCl, pH 7.4) with NaN_3 (1 mg/ml) and either 10% v/v) newborn calf serum or 4% (w/v) nonfat dry milk.

3. Incubate (12–48 hr, room temperature or 4°) with primary antibodies diluted in blocking solution.

4. Wash (first 3 × 1 min and then 2 × 10–20 min) in washing solution consisting of TBS with either 1% (v/v) newborn calf serum or 1% (w/v) nonfat dry milk.

5. Incubate (1 hr) with secondary antibody in washing solution. We routinely use biotinylated antirabbit. If peroxidase is to be used as marker, it is important to remember to avoid NaN_3 in all steps from this onwards. Peroxidase is reversibly, but potently, inhibited by NaN_3, NaN_3 must therefore be removed by extensive washing before the enzyme activity will be recovered.

6. Wash (3 × 1 min and 2 × 10–20 min) in washing solution.

7. Incubate (1 hr) with streptavidin–biotinylated horseradish peroxidase complex in washing solution.

8. Wash (first 3 × 1 min and then 2 × 10–20 min) in washing solution.

9. Wash (3 × 1 min) in PBS.

10. Preincubate (6 min) the sections in diaminobenzidine (DAB) solution and then incubate (5 min) in DAB solution containing 0.01% (w/v) hydrogen peroxide. Stop the reaction with 0.1 *M* NaP_i (2 × 5 min). To prepare the DAB-solution, dissolve one 10 mg DAB tablet (Sigma) in 20 ml 0.1 *M* NaP_i in a 50-ml tube with cap (e.g., Falcon 2070) and centrifuge (3000 rpm, 5 min). Divide the supernatant in two aliquots. Use the first for the preincubation and add 3.3 μl 30% hydrogen peroxide to the other. *Note that DAB is harmful and should be handled with care.* For this reason we prefer to use DAB tablets rather than powder. (For safe disposal, see ref. 38.)

11. Store sections (4°) in NaP_i with NaN_3 (0.2–1 mg/ml) until they can be either mounted (e.g., in glycerol–gelatin) or processed for electron microscopy.

Preembedding Electron Microscopy

Interesting parts of the above labeled Vibratome sections are cut out and treated (30–45 min, room temperature) with osmium tetroxide (OsO_4, 10 mg/ml) in 0.1 *M* NaP_i. *Note that osmium tetroxide is toxic and evaporates*

[38] G. Lunn and E. B. Sansone, *TIG* **8,** 7 (1992).

easily. It is supplied in glass ampules which are supposed to be broken while immersed in the buffer.

After osmication, the tissue sections are washed (3 × 1 min) in 0.1 *M* NaP_i and dehydrated in graded ethanols (1 × 5 min in each of 50, 70, 80, and 96% ethanol; 3 × 10 min in 100% ethanol) followed by propylene oxide (2 × 5 min). *Note that propylene oxide is harmful.* Do the work in a fume hood. Then the sections are immersed in Durcupan ACM (Fluka, Ronkonkoma, NY), incubated (56°, 30 min), and transferred for overnight infiltration in a new aliquot of Durcupan. Finally, the sections are transferred to a mold with fresh Durcupan, oriented to optimize cutting, and hardened at exactly 56° for exactly 48 hr.

Cut ultrathin sections at right angles to the thick (40-μm Vibratome) ones in order to be able to study the parts of the tissue that have been in immediate contact with the reagents. The sections are contrasted (10 mg/ml uranyl acetate, 10–15 min; 3 mg/ml lead citrate, 1–2 min) and examined in a transmission electron microscope. Some ultrathin sections are left uncontrasted for better evaluation of low levels of immunoreactivity.

Freeze-Substitution Postembedding Immunocytochemistry

Freeze Substitution

The procedures are based largely on the work of van Lookeren Campagne *et al.*[6,39] To allow quick substitution and infiltration, small blocks of perfusion fixed brain tissue (less than 1 mm^3) are cut with a razor blade and cryoprotected with increasing concentrations of glycerol: 1 hr in 10% and 20% and overnight in 30%. To achieve quick freezing and avoid crystallization of the water, the specimens are plunged into liquid propane cooled by liquid nitrogen (−190°) using a rapid-freeze apparatus (KF80; Reichert Jung, Wien, Austria). *Note that liquid propane is highly flammable and should be handled with care.* The specimens are then moved to a precooled (−90°, for at least 1 day) chamber of an automatic freeze substitution apparatus (CS auto; Reichert-Jung, Wien, Austria). The tissue is immersed in anhydrous methanol containing 0.5% uranyl acetate for protection of phospholipid membranes, overnight at −90°; then the temperature is slowly (4°/hr) increased to −45° (methanol gives low lipid extraction at such low temperatures). Excess uranyl acetate is removed by washing the specimens several times with anhydrous methanol. The specimens are infiltrated by

[39] M. van Lookeren Campagne, A. B. Oestreicher, T. P. Van der Krift, W. H. Gispen, and A. J. Verkleij, *J. Histochem. Cytochem.* **39,** 1267 (1991).

filling the chamber with a solution of methanol and Lowicryl HM20 resin (Chemische Werke Lowi, Waldkraiburg, Germany) at $-45°$ in three stages with increasing ratios of resin to methanol: 1/3, 2/3, and 1.[40] The HM20 was chosen because of its hydrophobic properties, its weak denaturating effect on proteins, and its low viscosity at low temperatures. The samples are then transported to an embedding mall filled with pure resin at $-45°$. The resin is polymerized (48 hr at $-45°$C followed by 24 hr at room temperature) by UV radiation (360 nm). Ultrathin sections (about 80 nm) are cut by a diamond knife on an ultramicrotome and mounted on nickel grids.

Postembedding Immunocytochemistry

For immunocytochemistry, TBST-buffer (50 m*M* Tris-HCl, pH 7.4, 0.15 *M* NaCl, and 1% Triton X-100) is used in all steps. The detergent apparently increases the specific labeling.[41] The grids with the specimen are incubated (10 min) in droplets of sodium borohydride (1 mg/ml; freshly prepared) and 50 m*M* glycine in TBST, pH 7.4, to inactivate aldehyde groups. After washing (3×10 min) in blocking solution (TBST with either 2% human serum albumin or 0.1–4% nonfat dry milk), the grids are incubated (overnight, room temperature) with primary antibodies diluted in blocking solution. They are then washed in TBST, pH 7.4 (3×10 min), and incubated with goat antirabbit immunoglobulin coupled to colloidal gold particles (1:20, 15 nm, Amersham) diluted in blocking solution. Finally, the specimens are contrasted with 1% uranyl acetate (10 min) and 1% lead citrate (1 min) and washed with ultrapure water (2×5 min).

Concluding Remarks

Antibodies remain powerful tools in protein research. By means of immunocytochemical techniques it is possible to obtain quantitative data on the localizations of the neurotransmitter transporters. The quality of the data obtained depends on the specificity of the antibodies used. The specificity of each antibody preparation should therefore be documented for each application.

[40] M. Müller, T. Marti, and S. Kriz, *Proc. 7th Eur. Congr. Electron Microsc.* **2,** 720 (1980).
[41] K. D. Phend, R. J. Weinberg, and A. Rustioni, *J. Histochem. Cytochem.* **40,** 1011 (1992).

Acknowledgments

This work was supported by the Norwegian Research Council (fellowships awarded to FAC, KPL, and NCD), Anders Jahres fond, Nansenfondet, Langfeldts fond, Schreiners fond, Bruuns legat, and EU BIOMED II (contract no. BMH4-CT95-0571). A private donation made it possible to buy custom-synthesized peptides. We thank Jon Storm-Mathisen for discussions and for critical reading of the manuscript.

[27] Generation of Transporter-Specific Antibodies

By GARY W. MILLER, MICHELLE L. GILMOR, and ALLAN I. LEVEY

1. Introduction

Neurotransmitter transporters play an integral role in the regulation of synaptic and intracellular neurotransmitter levels. Various methods have been employed to study the biology of these transporters. Radioligand uptake and binding assays provide functional data on the kinetics of the transporter; however, many compounds are substrates or ligands at different transporters, making it difficult to distinguish among closely related proteins. For example, dopamine can be transported by several neurotransmitter transporters, including the plasma membrane dopamine transporter, the norepinephrine transporter, the serotonin transporter, and the vesicular monoamine transporter. Autoradiography can provide useful data regarding the distribution of transporters, but relies on the specificity of the radioligand. The molecular cloning of many of the neurotransmitter transporters has permitted the use of oligonucleotide probes in conjunction with *in situ* hybridization to identify the cells that express the transporter mRNA. However, the mRNA tends to localize in the soma, thus providing little data regarding the localization of the transporter in the nerve terminals and dendrites, the regions of the cell most likely to express the functional transporter. Furthermore, the level of mRNA expression does not necessarily correlate with protein expression. Antibodies generated against specific regions of the individual transporters offer complementary approaches to study the function, localization, and biochemistry of these closely related molecules.

There are two distinct methods for producing transporter-specific antibodies. The first, anti-peptide antibody generation, uses a relatively short peptide as the target sequence. The production of anti-peptide antibodies has become very common with numerous companies and universities providing services to produce both the peptides and antibodies. Briefly, a short

0076-6879/98 $25.00

peptide sequence (≈8–20 amino acids) is selected from the known sequence of the protein. The peptide is then synthesized and linked to separate carrier molecules, one for immunization and one for affinity purification. Common carrier molecules include keyhole limpet hemocyanin (KLH), bovine serum albumin (BSA), or thyroglobulin. As discussed below, it is of utmost importance to select a region that is unique to that protein, antigenic, and exposed on the native transporter, which within the same or related gene families can be especially difficult using only <20 amino acids. Because of these limitations we have turned to an alternative method of antibody production that usually provides excellent reactivity and discrimination among closely related proteins.

Fusion protein antibody generation involves cloning a targeted sequence of the protein, much longer than typically used for peptide antibody synthesis. The larger protein sequence (30–150 amino acids) presents more potential epitopes and may represent the native conformation of the protein. In addition, the same fusion protein (carrier molecule attached to the target sequence) can be used for both immunization and affinity purification. The first, and potentially most important, step in generating fusion protein antibodies to transporter proteins is selecting the region of the protein to be used for antibody production. Generally, hydrophilic regions, often the N terminus, nonmembrane-bound intra- or extracellular loops, and the C terminus, are preferred to hydrophobic domains. These regions tend to yield fusion proteins which are soluble and more accessible to the generated antibodies since they are not buried within the membrane or hidden within the protein itself. However, it is of utmost importance to select regions that are unique with respect to other neurotransmitter transporter proteins. A computer search of a sequence database program, such as GenBank, is often helpful in this respect. Careful selection of the target region allows one to distinguish not only among different transporter types, but also among the same protein in different species.

The use of large amino acid sequences increases the number of potential epitopes. This is important since some of the epitopes contained within the fusion protein may be inaccessible in the native molecule *in vivo* because of posttranslational modifications, such as glycosylation or phosphorylation. These modifications may not occur during the production of the fusion protein even though it contains the identical sequence. Furthermore, some epitopes may not be recognized under denaturing conditions such as occur in immunoblot analysis. The potential for numerous epitopes is one of the primary advantages between fusion protein antibodies and peptide antibodies, as the relatively short sequence used for peptide antibody production can limit the number of potential epitopes and ultimate antigenicity of the peptide. This chapter will focus on the steps involved in generating fusion protein antibodies, including construction of the expression plasmid, production and purification of the fusion protein containing a portion of the transporter, immunization of animals for antisera production, and purification and characterization of antibodies, although many of these procedures can be employed in the production and characterization of antipeptide antibodies.

2. Recombinant pGEX-2T Plasmid Construction

pGEX Expression System

Several protein expression systems are commercially available for fusion protein production. We have successfully used the pGEX vectors available from Pharmacia (Piscataway, NJ) to produce numerous fusion proteins.[1] In the pGEX vector system, the target cDNA sequence is ligated to the *Schistosoma japonicum* gene encoding the 26-kDa protein glutathione *S*-transferase (GST). Produced fusion proteins are typically soluble on nondenaturing lysis of the bacteria, and the high affinity of GST for glutathione permits simple affinity purification of the fusion proteins from bacterial lysates with glutathione-agarose beads. The pGEX vectors also have recognition sites for simple proteolytic cleavage of the transporter fragment from GST (e.g., pGEX-2T has a recognition site for thrombin).

Design Oligonucleotide Primers

Design sense and antisense oligonucleotide primers to the cDNA sequence(s) to be amplified by the polymerase chain reaction (PCR). Restriction sites for *Bam*HI and *Eco*RI or other endonucleases should be incorporated into the primers to facilitate directional cloning into the cloning site of the pGEX-2T vector. A further six base pairs should be added outside the restriction site for optimal cutting. The primers should be approximately 20 base pairs in length. If the cDNA of interest contains either a *Bam*HI or *Eco*RI restriction site, a *Sma*I site can be used in place of one of these. Alternatively, nondirectional cloning or another pGEX vector can be used.

Amplification of cDNA Template by PCR

The annealing temperature should be selected based on the composition of the primers. A general rule is to add 4° for every GC base pair, 2° for every AT base pair, and subtract from that value 5° for every mismatched base pair (often the result of insertion of a restriction site). Many other

[1] D. B. Smith and K. S. Johnson, *Gene* **67,** 31 (1988).

factors can influence the annealing of the oligonucleotides to the template, and the reader is encouraged to refer to a manual on molecular biology for tips on optimizing PCR conditions.[2] The length of the PCR fragment can be confirmed by agarose gel electrophoresis by running a small quantity of the PCR reaction.

Ligation of Fragment and pGEX 2T Vector

Digest the PCR product and the pGEX-2T vector with *Bam*HI and *Eco*RI. This can be done sequentially or concomitantly. Confirm successful digestion by agarose gel electrophoresis. Purify the PCR product and vector by running the digested samples on a low-melt agarose gel and extract using glass binding or column chromatography according to manufacturer's instructions. Samples of the purified fragment and vector should be run on an agarose gel to estimate the quantity. Ligate the vector and a five-fold molar excess of fragment with T4 DNA ligase according to manufacturer's instructions. A good starting point is 50 ng of vector.

Expression of Recombinant Plasmid

Transform competent RR1 *Escherichia coli* (or other appropriate strain) with the recombinant plasmid, plate on agar containing 50 μg/ml ampicillin, and grow overnight at 37°. Perform minipreparation DNA purification on the transformants and screen minipreparations by restriction digest analysis. This is done by cutting the DNA with a restriction enzyme that will result in an obvious shift in molecular weight in the plasmid banding pattern if the fragment has been incorporated. Selected digests can then be used to produce microgram quantities of DNA using a Qiagen kit according to manufacturer's instructions. To ensure the proper insertion of the fragment and the absence of mutations in the fragment which may have occurred during PCR, the sequence must be verified by double-stranded dideoxy sequencing.

3. Expression and Purification of Fusion Protein

After transformation of the pGEX expression plasmid, the protein synthesis machinery of *E. coli* is used to drive expression of the fusion protein. The BL21(DE3) strain of *E. coli* is well suited for protein expression from introduced plasmids. This strain lacks many proteases which could degrade the produced fusion protein, and also carries the lambda DE3 lysogen which suppresses basal synthesis of endogenous proteins, but allows for

[2] J. Sambrook, E. F. Fritsch, and T. Maniatis, "Molecular Cloning: A Laboratory Manual," 2nd ed. Cold Spring Harbor, 1989.

protein synthesis from the pGEX plasmid when induced by isopropylthio-β-D-galactoside (IPTG). The isolated fusion protein is used for immunization, antibody purification, and characterization; thus, large quantities of fusion protein should be produced. In addition to cultures of fusion protein, cultures of BL21 transformed with the parent pGEX plasmid should also be generated for subsequent use in purification.

Transformation of Competent BL21 Escherichia coli

Transform BL21 cells with either the recombinant plasmid or the parent pGEX-2T vector and plate on 2YT agar plates containing 50 μg/ml ampicillin and 150 μg/ml chloramphenicol. The lysate of the parent pGEX-2T vector expression will be used during affinity purification to remove antibodies generated to the GST or nonspecific bacterial components of the expression system. Incubate at 37° overnight. Select a small isolated colony and grow overnight in 3 ml of 2YT broth and antibiotics on shaker at 37°. Inoculate 400 ml fresh broth with antibiotics with 100 μl of overnight culture and grow on shaker (37°) until the OD_{600} is 0.4 to 0.5 (approximately 6–8 hr).

Induction of Bacteria

The transporter gene in pGEX-2T is under the control of the tac promoter and can be induced with IPTG. Remove a 100-μl sample for subsequent SDS–PAGE analysis of preinduction. Add 400 μl of 1 *M* IPTG and grow for 4 hr. Remove 10 μl of culture at various time points (e.g., every hour) for postinduction analysis.

Confirmation of Fusion Protein Expression

Run the pre- and postinduction samples on a 12% SDS–PAGE gel followed by Coomassie staining. The fusion protein should appear as an obviously new protein and is often the most abundant protein in the postinduction sample. Isolate the bacteria by centrifugation at 8000*g* for 15 min at 4°. Remove the supernatant and resuspend the pellet in 10 ml buffer A (50 m*M* Tris, pH 8.0, 25% sucrose, 10 m*M* EDTA). Transfer to 50-ml tube and centrifuge at 5000*g* for 10 min at 4°. Decant supernatant. These washed pellets can be frozen at −80° for future purification. Resuspend the pellet in buffer A and add 40 mg lysozyme and protease inhibitors (2 μg/ml leupeptin, pepstatin, and aprotinin) to solution and incubate on ice for 1 hr to digest cell walls. Freeze the sample in dry ice/ethanol bath and thaw two times. Add one-tenth volume of 10% Triton X-100 (v/v) and freeze thaw two more times. Add 2 μg/ml DNase I and 3 ml 1 *M* $MgSO_4$. Incubate at room temperature with gentle shaking until viscosity decreases (~30–60 min). Centrifuge at 3400*g* for 10 min at 4°. Transfer the supernatant into a 50-ml tube and save. Resuspend the pellet in 50 m*M* Tris, 10 m*M* EDTA, pH 8.0 (TE; same volume as removed supernatant). Analyze the supernatant and pellet fraction by SDS–PAGE and Coomassie staining. The soluble fusion protein should be in the supernatant fraction. If the fusion protein is in the pellet and therefore insoluble, it will need to be processed as described in Section 7, Insoluble Fusion Proteins.

Purification of Fusion Protein

Add 15 ml of buffer C [20 m*M* HEPES, pH 7.6, 100 m*M* KCl, 0.2 m*M* EDTA, 20% (v/v) glycerol, 1 m*M* dithiothreitol (DTT); immediately prior to use add 1 μg/ml of aprotinin, leupeptin, and pepstatin) to the supernatant fraction. Add 4 ml of a 1 : 1 slurry of glutathione agarose beads, which have been washed several times in buffer C and refrigerated overnight. Shake at 4° for 1–2 hr (can be left overnight). Centrifuge at 800*g* for 5 min at 4°. Carefully remove and save the supernatant from the soft pellet of beads. Check for unabsorbed fusion protein in the supernatant by SDS–PAGE and Coomassie staining. There should only be a small amount of unabsorbed protein as the majority of the protein should be bound to the beads. Wash the beads 3× in buffer C. Elute the fusion protein with 6 × 2 ml of 50 m*M* reduced glutathione (GSH) in buffer C (GSH should be made fresh at pH 7.5). The fusion protein can be tracked by checking the OD_{280} (OD_{280} of 1 = ~1 mg/ml protein). Combine elutions 1–3 (typically containing the most fusion protein) and 4–6 and centrifuge at 5000*g* at 4° to remove any remaining beads. The glutathione agarose beads can be recycled by overnight washing in 3 *M* NaCl followed by several washes in buffer C. Dialyze elutions into large volumes of phosphate-buffered saline (PBS, 1–2 liters), changing the dialysis buffer at least three times (residual GSH can interfere with many assays; this complete removal is critical). Estimate the concentration and yield of the fusion protein by protein assay and store at −80°.

4. Immunization and Serum Collection

Immunization of New Zealand White Rabbits

The purified fusion protein is used to immunize New Zealand (NZ) White rabbits. Prior to starting injections take a serum sample from an ear vein to be used as a control for testing of antisera. The initial immunization of a soluble fusion protein should contain 200–500 μg of purified fusion protein emulsified in Freund's complete adjuvant immediately prior to immunization. Monthly boost injections should consist of half the amount of purified fusion protein emulsified in Freund's incomplete adjuvant. For insoluble fusion proteins the same amount of purified fusion protein (200–500 μg) is used, but the alum is used as the adjuvant for all immunizations.

Collection of Antisera

Ten and 24 days after each immunization, a serum sample (15–25 ml) should be taken from the ear vein. The sample should be allowed to clot overnight at 4° and clarified by centrifugation. Aliquots should be stored at −80°. Screen the various test bleeds to determine that they contain antibodies to the protein of interest by immunoblotting or immunocytochemistry. The unpurified antiserum should recognize the transporter, although significant background may result. These results should be compared to those found with the preimmune test bleed. It is not necessary to immediately screen all of the test bleeds as they can be stored at −80°. The immunization procedure can be performed commercially at a reasonable cost, avoiding the housing and maintenance of animals. In addition, rats or mice can also be immunized for monoclonal antibody production. This method involves isolating the splenocytes, which are then fused to a hybridoma cell line according to standard procedures. The protocols necessary for monoclonal antibody production can be found in lab manuals devoted to the subject.[3]

5. Affinity Column Production

Two separate affinity columns are used to purify the antibodies. The first is a BL21/GST column, which contains all of the proteins in the parent vector transformed *E. coli,* including GST, but not the target region protein. Antibodies that were formed to residual bacterial proteins or GST will be removed in this step. To specifically isolate antibodies to the transporter portion of the fusion protein, a column expressing the fusion protein is constructed to extract the remaining antibodies of interest from the sera. Any unrelated proteins will run through the column, and then after rinsing, the antibodies are eluted off the column, yielding purified antibodies against the transporter protein only.

Protein Binding to Affinity Resin

The columns are constructed by attaching the BL21/GST lysate or fusion protein to an affinity resin, such as Affi-Gel 10 or 15 (Bio-Rad,

[3] E. Harlow and D. Lane, "Antibodies: A Laboratory Manual." Cold Spring Harbor, 1988.

Richmond, CA). To determine which affinity resin provides the better binding to the fusion protein, incubate a small quantity of fusion protein with a 100 μl bed volume of the potential resins and compare it with an equivalent amount of nonabsorbed fusion protein by SDS–PAGE and Coomassie staining. Use the resin that removes the greater amount of fusion protein from the sample.

Place affinity resin slurry into 15-ml tubes (1–2 ml slurry for fusion protein column; 12–16 ml slurry for BL21/GST column). Wash bead three times with ice-cold deionized water, pelleting the resin at 800*g* at 4°, carefully aspirating water. Add up to 25 mg of fusion protein or the BL21/GST lysate from a 400-ml culture to the respective tubes, add water for a final volume of 15 ml, and incubate overnight at 4°. Be sure to reserve pre- and postsamples of both the fusion protein and BL21/GST lysate to determine efficiency of protein binding to resin by SDS–PAGE and Coomassie staining. Wash resin in 15 ml of 0.1 *M* glycine, pH 3.0, followed by 0.1 *M* Tris, pH 8.0; 0.1 *M* triethylamine (TEA), pH 11.0; and 0.1 *M* Tris, pH 8.0.

Packing of Chromatography Columns

Fill 2-ml and 10-ml double-frit columns (Pierce, Rockford, IL) with water. After determining that water passes through the column, cap the end of tube while the water level is above the neck of the column. Float the first frit on the water and gently push the frit to the bottom of the column with the wide end of a Pasteur pipette, making sure no air bubbles are present in the column. Uncap the column and transfer the resin/beads to the column with a pipette tip that has been enlarged by cutting the end off with a razor blade. Be sure to rinse the tip thoroughly in 0.1 *M* Tris to prevent the loss of any beads. Allow the beads to settle in the column. Fill the column with 0.1 *M* Tris and cap the bottom while the fluid volume is above the neck. Slowly place the second frit into the column, leaving a small space between the frit and beads. Air bubbles will prevent proper function of the column and should be removed by centrifugation. To do this fill the column with 0.1 *M* Tris, cap both ends, place in a 50 ml tube, and spin at 800*g* for 10 min. Wash the column with 30 ml of 0.1 *M* Tris at 4°. The OD_{280} of the eluant should be zero. The column should be stored with 0.1 *M* Tris above the second frit, both ends capped, at 4°.

5. Purification of Antibodies by Affinity Chromatography

The unpurified serum should produce positive immunoreactivity and may be suitable for some purposes; however, affinity purification selectively concentrates the antibodies of interest, reducing background staining. To ensure that the serum does indeed contain antibodies to the transporter protein, test a small quantity by immunocytochemistry at a dilution of 1:5000 to 1:20,000 prior to purification. All of the following procedures should be performed at 4° to prevent degradation of the antibodies.

BL21/GST Affinity Column

Remove a small aliquot (5 ml) of serum to be purified. Add 50 U of heparin per ml of serum to inhibit endogenous thrombin activity, thereby preventing cleavage of the GST fusion protein. It is often desirable to dilute the serum 2- to 4-fold with Tris pH 8.0 to improve flow over the column. Pass the serum over the BL21/GST column and collect effluent in a 50-ml tube and place on ice. Wash column with 10 bed volumes of 0.1 *M* glycine, pH 3.0, to remove bound antibodies, and discard wash. The initial fractions can be collected to monitor the elution of the GST and BL21 antibodies. The column is then neutralized with 10 bed volumes of 0.1 *M* Tris, pH 8.0, which is discarded. Confirm that eluant pH is 8.0 using pH paper. Reintroduce the collected serum and washes over the BL21/GST column and collect in a new 50-ml tube(s). The column is stripped as before and the serum/washes run over the BL21/GST column for a total of 3–4 passes to ensure complete removal of antibodies to GST or BL21 proteins, with the column stripped between each run. Check that the eluant OD_{280} is near 0 to confirm adequate removal of GST and bacterial antibodies.

Fusion Protein Column

Pass the collected serum and washes from the BL21/GST column over the fusion protein column to isolate antibodies specific for the transporter protein. Make sure to collect this serum as all of the antibodies may not be removed upon the first pass. Wash the column with 20 ml of 10 m*M* Tris to remove unbound protein and discard. Place 50 μl of 1 *M* Tris, pH 8.0, into five microfuge tubes, labeled 1–5. The 1 *M* Tris will neutralize the elution buffer. Elute antibodies with 5 × 500 μl of 0.1 *M* glycine into the respective tubes. Vortex briefly and measure antibody concentration at OD_{280}. Additional 500-μl eluants can be collected if the OD is still high after the first five. If the OD has returned close to baseline, a tail elution of 7.5 ml 0.1 *M* glycine should be collected in a 15-ml tube containing 750 μl 1 *M* Tris. Wash the column with 20 ml 0.1 *M* Tris. The serum should be passed over the column one or two more times following the above procedure. Pool the fractions containing the peak of the antibody elution curve. This may be two to three of the first pass elutions, and only one fraction from subsequent passes. Store the antibody at 0.1 mg/ml or greater in 50% glycerol at −20°. Working aliquots should be treated with sodium azide and stored at 4°.

7. Insoluble Fusion Proteins

One of the purposes of using a GST fusion protein for antibody production is to have a large source of a soluble fusion protein. The soluble protein is easy to isolate and produces a relatively pure immunogen. However, occasionally the GST fusion protein will be insoluble. This can occur because the protein is incorporated into inclusion bodies or the target region, conferring hydrophobicity to the fusion protein. Fortunately, this does not mean that the fusion protein is not suitable for antibody production; rather, some special steps need to be taken during the procedures.

Solubilization of Insoluble Fusion Protein

Determine the amount of fusion protein present following the expression, lysing, and washing of the fusion protein by subjecting the resuspended pellet fraction to SDS–PAGE and Coomassie staining. Place a volume of the resuspended pellet that will be sufficient for all immunizations into a centrifuge tube. Spin at 3400*g* for 10 min and discard supernatant. Resuspend the pellet in an equal volume of buffer C and centrifuge at 3400*g* for 10 min. Remove supernatant and resuspend pellet fraction in 10% SDS at 2 μl SDS solution/100 μg protein. Boil for 2 min. Examine the solution to ensure that the pellet material has solubilized. If it has not, more SDS can be added, but the absolute minimum amount of SDS should be used since SDS can interfere with the ability of the protein to bind to the affinity resin and can be toxic to the animal. Dilute the solubilized pellet solution 10-fold in PBS. Aliquot solution for immunizations and freeze until use. Immunize as described in Immunization of Animals. Alternatively, insoluble fusion proteins can be purified by SDS–PAGE, and the gel slice containing the protein used for immunizations.

Production of Affinity Columns with Insoluble Fusion Protein

For the BL21/GST column, grow a 400-ml culture of BL21 *E. coli* transfected with the pGEX parent vector and purify to the point of DNase treatment, stopping prior to the glutathione-agarose bead purification step. Be sure to save both the supernatant and pellet fractions. Solubilize the pellet fraction as described above. Bind the supernatant and solubilized pellet fraction to separate aliquots of affinity resin 4-ml bed volume, as described in the section on affinity column production. Either two separate columns can be made, or the beads can be stacked on top of each other in one column. For the fusion protein column, solubilize the pellet fraction from one 400-ml culture and use as described in the section on affinity column production.

8. Characterization of Antibodies

Once antibodies have been generated, it is essential that their specificity be confirmed. This is of key concern with neurotransmitter transporters, as there is a great deal of homology within transporter families. We employ three procedures in the characterization of antibodies: immunoprecipitation, immunoblotting, and immunocytochemistry. Each of these techniques provides key information regarding the utility and specificity of the generated antibodies.

Immunoprecipitation

Antiserum can be tested for its ability to recognize the native conformation of the protein by assaying how efficiently it removes the transporter from a solubilized tissue sample after. This is achieved by precipitating the antibody–antigen complex using protein A bound to beads. Briefly, we equilibrate 300 μl of protein A-Sepharose CL-4B with equilibration buffer (50 m*M* Tris, 300 m*M* NaCl, 10 m*M* EDTA, pH 7.5) and incubate with 2–6 μl of antiserum for 2 hr at 4° in a final volume of 600 μl. After centrifugation at 50*g* for 1 min, the supernatant is removed and the pellet washed three times in equilibration buffer. The pellets can then be incubated with 300 μl of digitonin (1%) solubilized membranes prepared from an area known to contain binding activity of the transporter of interest in the range of 100–500 fmol of ligand overnight at 4° with gentle shaking. Then residual binding can be assayed by standard binding methods. An example of immunoprecipitation with DAT (dopamine transporter) antibodies is shown in Fig. 1.[4] Controls consist of sample incubated with protein A-Sepharose beads coupled with preimmune sera.

Immunoblotting

In combination with SDS–PAGE, immunoblotting provides an opportunity to assess the ability of an antibody to specifically recognize denatured proteins as well as provide information on the molecular weight of the recognized species. The ability of the purified antibody to recognize the fusion protein (GST + target sequence), but not the protein produced from

[4] B. J. Ciliax, C. Heilman, L. L. Demchyshyn, Z. B. Pristupa, E. Ince, S. M. Hersch, H. B. Niznik, and A. I. Levey, *J. Neurosci.* **15,** 1714 (1995).

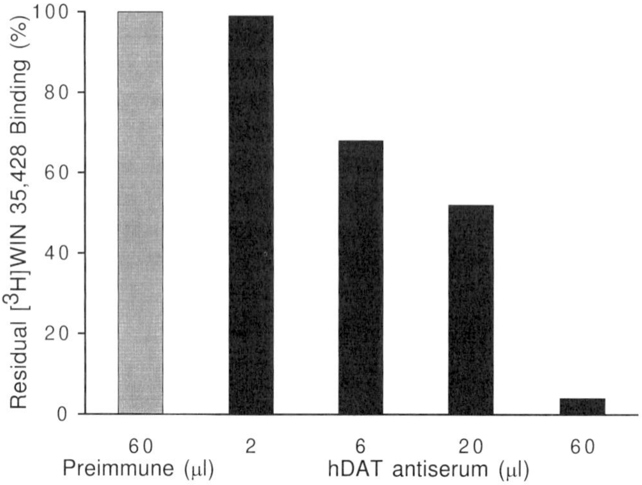

FIG. 1. Immunoprecipitation of [^{3}H]WIN 35,428 binding activity with anti-human DAT antiserum (hDAT antiserum). Human caudate membranes were solubilized with 1% digitonin and incubated with various volumes of either preimmune sera or hDAT antiserum coupled with protein A-Sepharose for 12 hr at 4°. Following centrifugation, supernatants were assayed for [^{3}H]WIN 35,428 binding activity as previously described.[4] Values represent the means of duplicate experiments from two separate experiments. 100% binding activity was approximately 200 fmol/ml of [^{3}H]WIN 35,428 binding activity to solubilized membranes.

the parent vector (GST only) or other unrelated fusion proteins helps confirm the specificity of the antibody; however, this does not prove that the antibody will recognize the full-length protein. To address the ability of the antibody to recognize the full-length protein and to discriminate among the related members within neurotransmitter transporter gene families, we suggest that immunoblot analysis be performed on cell lines expressing the full-length transporters from within the family of interest. For example, to confirm the specificity of our DAT-Nt antibodies we screened mammalian cell lines expressing DAT, SERT, or NET (Fig. 2).[5] We also recommend that Western blots be performed on brain tissue selected from different brain regions that are known or suspected to express the transporter, as well as regions known to be devoid of the transporter (Fig. 2). Control blots using preimmune sera and affinity purified antibody preabsorbed with fusion protein should also be performed, and the bands specific for the transporter of interest should be absent in both control conditions. Monospecificity on Western blots using native tissue is the single

[5] G. W. Miller, J. K. Staley, C. J. Heilman, J. T. Perez, D. C. Mash, D. B. Rye, and A. I. Levey, *Ann. Neurol.* **41,** 530 (1997).

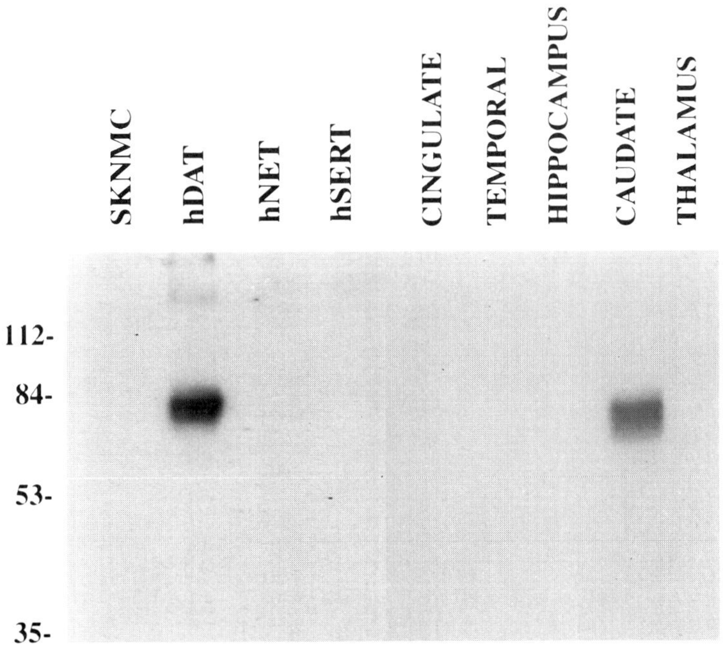

FIG. 2. Molecular specificity of DAT antibodies by immunoblot analysis. Homogenates from untransfected SK-N-MC cell line (SKNMC), stable SK-N-MC cell line expressing DAT (hDAT), stable HeLa cell line expressing human norepinephrine transporter (hNET), stable HeLa cell line expressing human serotonin transporter (hSERT), cingulate gyrus (CINGULATE), temporal lobe (TEMPORAL), HIPPOCAMPUS, CAUDATE, and THALAMUS were subjected to SDS–PAGE (10%), transferred to Immobilon-P membrane, subjected to immunoblot analysis with hDAT antibody, and developed with enhanced chemiluminescence. [Reproduced with permission from G. W. Miller, J. K. Staley, C. J. Heilman, J. T. Perez, D. C. Mash, D. B. Rye, and A. I. Levey, *Ann. Neurol.* **41,** 530 (1997).]

most valuable characteristic to demonstrate antibody specificity, particularly to validate immunocytochemical applications.

It is possible under the denaturing conditions of SDS–PAGE that potential epitopes may be destroyed. Thus, although the antibodies may not work in immunoblotting, they still may be suitable for immunoprecipitation and immunocytochemistry. We therefore suggest that immunocytochemistry be performed to determine if there are antibodies present. Caution must be exercised when using antibodies that do not work in immunoblotting, since it is not known if the antibody recognizes a single protein species.

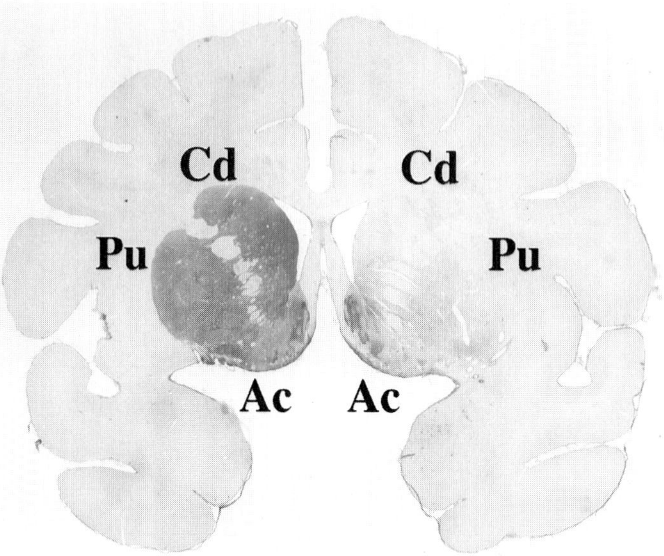

FIG. 3. DAT immunocytochemistry in a hemiparkinsonian monkey. The dopaminergic neurotoxin 1-methyl-4-phenyl-1,2,3,6-tetrahydropyridine (MPTP) was administered via a unilateral intracarotid injection. After clinical verification of lesion, DAT immunocytochemistry was performed as previously described.[5] DAT immunoreactivity is robust in the untreated caudate (Cd) and putamen (Pu) (left-hand side), but nearly absent in the lesioned side (right-hand side). DAT immunoreactivity appears to be preserved in the nucleus accumbens (Ac) of the lesioned side.

A necessary control for such antibodies is performing immunocytochemistry on mock-transfected cells and cells stably expressing the protein of interest.

Immunocytochemistry

Immunocytochemistry is very useful in determining the localization of transporter proteins, and the reader is referred to a chapter in this volume for a detailed description on this subject.[6] We also view immunocytochemistry as an excellent method of screening antibodies and aiding in their characterization. Here, we briefly describe the methods we employ in the characterization of our antibodies and examples are shown in Figs. 3 and 4. Animals are transcardially perfused with 4% paraformaldehyde in phos-

[6] N. C. Danbolt, K. P. Lehre, Y. Debres, F. A. Chandhry, and L. M. Levey, *Methods Enzymol.* **296,** [26], (1998).

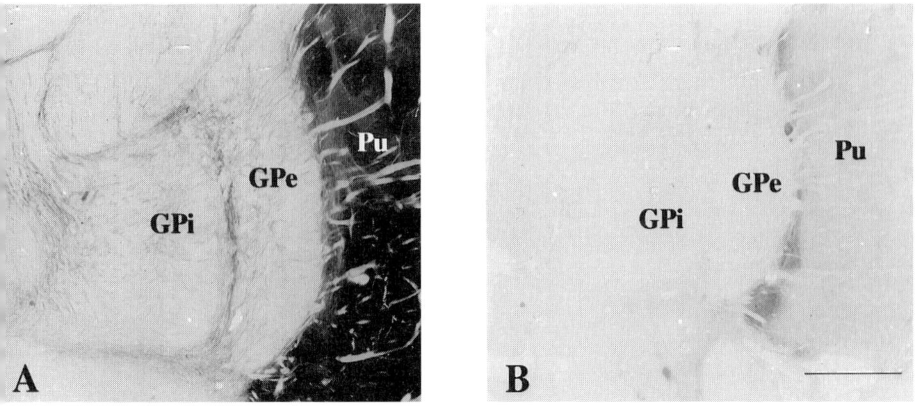

FIG. 4. DAT immunoreactivity in control and Parkinson's diseased striatum. Postmortem tissue was processed as previously described.[5] (A) Control (83-year-old male). (B) Parkinson's disease (75-year-old male). Note loss of DAT immunoreactivity in putamen (Pu) and DAT-immunoreactive fibers in the internal (Gpi) and external (GPe) pallidum and the internal medullary lamina (iml). Bar: 0.5 cm. [Reproduced with permission from G. W. Miller, J. K. Staley, C. J. Heilman, J. T. Perez, D. C. Mash, D. B. Rye, and A. I. Levey, *Ann. Neurol.* **41,** 530 (1997).]

phate-buffered saline (PBS) followed by 10% sucrose (for rats 200 ml over 10 min for each solution). Alternatively, brains can be postfixed by submersion in 4% paraformaldehyde followed by 10% sucrose. Brains are then removed and cryopreserved in 30% sucrose at 4°. Sections (40–50 μm) are cut on a freezing sliding microtome and collected in cold Tris-buffered saline pH 7.2 (TBS). Sections should be treated with 4% normal goat serum (NGS) (v/v) in TBS containing 0.5–1.0% Triton X-100 and 3% hydrogen peroxide (v/v) for 20 min to permeabilize the membranes and block endogenous peroxidase, respectively. After thorough rinsing in cold TBS (6 $\times$ 10 min), sections are then incubated in primary antibody solution containing 0.1% Triton X-100 and 4% serum (v/v), and the affinity purified antibody (the optimal concentration of each antibody should be determined by titration, but final concentration of affinity purified antibody of approximately 0.5 μg/ml usually produces excellent results) for 24–48 hr at 4°. The sections are then rinsed 3 $\times$ 10 min followed by incubation in secondary antibody solution containing 4% NGS, 0.1% Triton X-100, and 1% (v/v) goat anti-rabbit antibody for 1 hr. After 3 $\times$ 10 min rinsing in cold TBS, the sections are incubated in 4% NGS in TBS containing 0.5% rabbit peroxidase antiperoxidase complexes for 1 hr at 4°. After 3 $\times$ 10 min rinsing in cold TBS, the sections are developed in 0.1 *M* Tris, pH 8.0, and 0.05% diaminobenzidine and 0.01% hydrogen peroxide for 5–10 min. After 3 $\times$ 10 min rinsing in TBS, the sections are then mounted on to subbed slides, air dried, dehydrated, and covered with coverslips. Other detection methods such as avidin-biotin based protocols (ABC) produce good results, but the improved sensitivity is often accompanied by increase background staining.

Section V

Expression Systems and Molecular Genetic Approaches

Articles 28 through 36

[28] Homologies and Family Relationships among Na^+/Cl^- Neurotransmitter Transporters

By HOLGER LILL and NATHAN NELSON

Introduction

Synaptic transmission involves the release of a neurotransmitter into the synaptic cleft, interaction with the postsynaptic receptor, and subsequent removal of the transmitter from the cleft. The majority of transmitters are removed from the cleft by rapid sodium-dependent uptake systems in the plasma membrane of the presynaptic cells and their surroundings. These reuptake systems are catalyzed by transporters specific for the various neurotransmitters in the brain. A defining moment in the advancement of our knowledge in neurotransmitter transport came in 1990 when the first cDNAs encoding neurotransmitter transporters were cloned and sequenced.[1–8] A family of transporters with a common structure of presumably 12 transmembrane helices has been defined. Sequence analysis revealed that this family of transporters contains four subfamilies of monoamine, γ-aminobutyric acid (GABA), amino acid, and "orphan" (NTT4) transporters.[9]

GABA is a major inhibitory neurotransmitter in the mammalian brain and is widely distributed throughout the nervous system.[10–12] Molecular cloning studies have resulted in the isolation and characterization of

[1] J. Guastella, N. Nelson, H. Nelson, L. Czyzyk, S. Keynan, M. C. Miedel, N. Davidson, H. A. Lester, and B. I. Kanner, *Science* **249,** 1303 (1990).
[2] H. Nelson, S. Mandiyan, and N. Nelson, *FEBS Lett.* **269,** 181 (1990).
[3] T. Pacholczyk, R. D. Blakely, and S. G. Amara, *Nature* **350,** 350 (1991).
[4] S. Shimada, S. Kitayama, C.-L. Lin, A. Patel, E. Nanthakumar, P. Gregor, M. Kuhar, and G. Uhl, *Science* **254,** 576 (1991).
[5] J. E. Kilty, D. Lorang, and S. G. Amara, *Science* **254,** 578 (1991).
[6] B. J. Hoffman, E. Mezey, and M. J. Brownstein, *Science* **254,** 579 (1991).
[7] T. B. Usdin, E. Mezey, C. Chen, M. J. Brownstein, and B. J. Hoffman, *Proc. Natl. Acad. Sci. USA* **88,** 11168 (1991).
[8] R. D. Blakely, H. E. Berson, R. T. Fremeau, Jr., M. G. Caron, M. M. Peek, H. K. Prince, and C. C. Bradley, *Nature* **354,** 66 (1991).
[9] N. Nelson and H. Lill, *J. Exp. Biol.* **196,** 213 (1994).
[10] L. L. Iversen and J. S. Kelly, *Biochem. Pharmacol.* **24,** 933 (1975).
[11] B. I. Kanner and S. Schuldiner, *CRC Crit. Rev. Biochem.* **22,** 1 (1987).
[12] B. I. Kanner, *Curr. Opin. Cell Biol.* **1,** 735 (1989).

0076-6879/98 $25.00

cDNAs encoding four different GABA transporters, GAT1, GAT2, GAT3, and GAT4,[13] as well as creatine,[14] betaine,[15] and taurine[16,17] transporters.

The subfamily of monoamine transporters contains the transporters of noradrenaline, serotonin, and dopamine.[9,18] Cloning of a cDNA encoding a noradrenaline transporter revealed that all the fundamental properties of noradrenaline uptake in the brain are encoded by this single cDNA species.[3] Dopamine transporter appears to be the most important site for the behavioral effects of amphetamines and cocaine.[19,20] The identity between the GABA and noradrenaline transporters was used for designing degenerative oligonucleotide probes corresponding to the regions of greatest identity. The attempt resulted in the isolation of a cDNA clone encoding the serotonin transporter in rat brain.[8] Simultaneously, using expression cloning, the serotonin transporter from rat basophilic leukemia cells was also cloned.[6] Expression of both cDNAs in nonneuronal cells generated sodium-dependent serotonin uptake ability which was sensitive to antidepressants.

The subfamily of amino acid transporters contains the transporters of glycine and proline. The amino acid glycine is a classical inhibitory neurotransmitter localized in the spinal cord, brain stem, and retina.[21] cDNAs encoding the amino acid transporters of glycine (GLYT1 and GLYT2) and proline were cloned and sequenced.[22–26] GLYT2 differs from GLYT1 in

[13] Q.-R. Liu, B. López-Corcuera, S. Mandiyan, H. Nelson, and N. Nelson, *J. Biol. Chem.* **268,** 2106 (1993).

[14] C. Guimbal and M. W. Kilimann, *J. Biol. Chem.* **268,** 8418 (1993).

[15] A. Yamauchi, S. Uchida, H. M. Kwon, A. S. Preston, R. B. Robey, A. Garcia-Perez, M. B. Burg, and J. S. Handler. *J. Biol. Chem.* **267,** 649 (1992).

[16] S. Uchida, H. M. Kwon, A. Yamauchi, A. S. Preston, F. Marumo, and J. S. Handler, *Proc. Natl. Acad. Sci. USA* **89,** 8230 (1992).

[17] Q.-R. Liu, B. López-Corcuera, H. Nelson, S. Mandiyan, and N. Nelson, *Proc. Natl. Acad. Sci. USA* **89,** 12145 (1992).

[18] G. Rudnick and J. Clark, *Biochim. Biophys. Acta* **1144,** 249 (1993).

[19] M. C. Ritz, R. J. Lamb, S. R. Goldberg, and M. J. Kuhar, *Science* **237,** 1219 (1987).

[20] J. Bergman, B. K. Madras, S. E. Johnson, and R. D. Spealman, *J. Pharmacol. Exp. Ther.* **251,** 150 (1989).

[21] E. C. Daly, *in* "Glycine Transmission" (O. P. Ottersen and J. Storm-Mathisen, eds.), p. 25. John Wiley, New York, 1990.

[22] Q.-R. Liu, H. Nelson, S. Mandiyan, B. López-Corcuera, and N. Nelson, *FEBS Lett.* **305,** 110 (1992).

[23] J. Guastella, N. Brecha, C. Weigmann, H. A. Lester, and N. Davidson, *Proc. Natl. Acad. Sci. USA* **89,** 7189 (1992).

[24] K. E. Smith, L. A. Borden, P. R. Hartig, T. Branchek, and R. L. Weinshank, *Neuron* **8,** 927 (1992).

molecular structure, tissue specificity, and pharmacological properties. The cDNA of GLYT2 encodes for 799 amino acid residues with an extended amino-terminal peptide of 200 amino acids containing potential phosphorylation sites for protein kinase C, cAMP-dependent kinase, and calmodulin-dependent kinase.[26] Although a specific proline receptor has not yet been identified, it was argued that this amino acid also functions as a neurotransmitter.[25] It was shown that radiolabeled L-proline is released from brain slices and synaptosomes in a Ca^{2+}-dependent manner following K^+-induced depolarization.[27] These and other experimental findings supported a synaptic role for L-proline in specific excitatory pathways in the brain.

The subfamily of "orphan" (NTT4) transporters consists of four gene products that appear to encode transporters which differ structurally from the three other subfamilies.[28,29] They display large second and fourth extracellular domains with sites for N-linked glycosylation in both large domains. Their function is not known and attempts to identify their substrates thus far have been unsuccessful.[30]

Genomic cloning of genes encoding neurotransmitter transporters and search for sequences without known function in the GenBank, revealed that insects and *Caenorhabditis elegans* contain genes encoding potential neurotransmitter transporters.[31] Therefore, it was assumed that this family of transporters evolved from a common ancestor about 1 billion years ago.[9] We have identified and cloned a bacterial gene with relatively high homology to neurotransmitter transporters.[32] This gene is present in a potential tryptophan operon of *Symbiobacterium thermophilum* that contains the gene encoding the enzyme tryptophanase. The entire genomic sequences of *Haemophilus influenzae* and *Methanococcus jannaschii* have

[25] R. T. Fremeau, Jr., M. G. Caron, and R. D. Blakely, *Neuron* **8,** 915 (1992).

[26] Q.-R. Liu, B. López-Corcuera, S. Mandiyan, H. Nelson, and N. Nelson, *J. Biol. Chem.* **268,** 22802 (1993).

[27] V. J. Nickolson, *J. Neurochem.* **38,** 289 (1982).

[28] G. R. Uhl, S. Kitayama, P. Gregor, E. Nanthakumar, A. Persico, and S. Shimada, *Mol. Brain Res.* **16,** 353 (1992).

[29] Q.-R. Liu, S. Mandiyan, B. López-Corcuera, H. Nelson, and N. Nelson, *FEBS Lett.* **315,** 114 (1993).

[30] S. E. Mestikawy, B. Giros, M. Pohl, M. Hamon, S. F. Kingsmore, M. F. Seldin, and M. G. Caron, *J. Neurochem.* **62,** 445 (1994).

[31] Q.-R. Liu, S. Mandiyan, H. Nelson, and N. Nelson, *Proc. Natl. Acad. Sci. USA* **89,** 6639 (1992).

[32] H. Nelson, S. Mandiyan, S. Horinouchi, and N. Nelson, in preparation (1998).

been determined.[33,34] Each genome contains a single gene homologous to the mammalian family of neurotransmitter transporters. These findings indicate that the family of neurotransmitter transporters is rooted to the onset of life and is present not only in eukaryotes, but also in Eubacteria and Archaea.

Evolutionary trees were constructed for the genes that were available at the time of the particular studies. Moreover, an attempt was made to calculate evolutionary trees for individual exons in the various transporters of mammalian origin.[9] Here we describe the construction of evolutionary trees that include the various mammalian transporters together with the recently discovered bacterial transporters.

Construction of Evolutionary Trees of Neurotransmitter Transporters

All sequences used in the evolutionary calculations are available on the Internet (GenBank at http://www2.ncbi.nlm.nih.gov/genbank, EMBL

[33] R. D. Fleischmann, M. D. Adams, O. White, R. A. Clayton, E. F. Kirkness, A. R. Kerlavage, C. J. Bult, J.-F. Tomb, B. A. Dougherty, J. M. Merrick, K. McKenney, G. Sutton, W. FitzHugh, C. A. Fields, J. D. Gocayne, J. D. Scott, R. Shirley, L.-I. Liu, A. Glodek, J. M. Kelley, J. F. Weidman, C. A. Phillips, T. Spriggs, E. Hedblom, M. D. Cotton, T. R. Utterback, M. C. Hanna, D. T. Nguyen, D. M. Saudek, R. C. Brandon, L. D. Fine, J. L. Fritchman, J. L. Fuhrmann, N. S. M. Geoghagen, C. L. Gnehm, L. A. McDonald, K. V. Small, C. M. Fraser, H. O. Smith, and J. C. Venter, *Science* **269,** 496 (1995).

[34] C. J. Bult, O. White, G. J. Olsen, L. Zhou, R. D. Fleischmann, G. G. Sutton, J. A. Blake, L. M. FitzGerald, R. A. Clayton, J. D. Gocayne, A. R. Kerlavage, B. A. Dougherty, J.-F. Tomb, M. D. Adams, C. I. Reich, R. Overbeek, E. F. Kirkness, J. F. Weidman, J. L. Fuhrmann, D. Nguyen, T. R. Utterback, J. M. Kelley, J. D. Peterson, P. W. Sadow, M. C. Hanna, M. D. Cotton, K. M. Roberts, M. A. Hurst, B. P. Kaine, M. Borodovsky, H.-P. Klenk, C. M. Fraser, H. O. Smith, C. R. Woese, and J. C. Venter, *Science* **273,** 1058 (1996).

[35] W. Mayser, P. Schloss, and H. Betz, *FEBS Lett.* **305,** 31 (1992).

[36] D. J. Vandenberg, A. M. Persico, and G. R. Uhl, *Mol. Brain Res.* **15,** 161 (1992).

[37] L. A. Borden, K. E. Smith, P. R. Hartig, T. A. Branchek, and R. L. Weinshank, *J. Biol. Chem.* **267,** 21098 (1992).

[38] K. E. Smith, L. A. Borden, C.-H. Wang, P. R. Hartig, T. Branchek, and R. L. Weinshank, *Molec. Pharmac.* **45,** 563 (1992).

[39] K. E. Smith, S. G. Fried, M. M. Durkin, E. L. Gustafson, L. A. Borden, T. A. Branchek, and R. L. Weinshank. *FEBS Lett.* **357,** 86 (1995).

[40] J. C. Wasserman, E. Delpire, W. Tonidandel, R. Kojima, and S. R. Gullans, *Am. J. Physiol.* **267,** F688 (1994).

[41] F. Jurksy, S. Tamura, A. Tamura, S. Mandiyan, H. Nelson, and N. Nelson, *J. Exp. Biol.* **196,** 283 (1994).

[42] B. López-Corcuera, Q.-R. Liu, S. Mandiyan, H. Nelson, and N. Nelson, *J. Biol. Chem.* **267,** 17491 (1992).

TABLE I
SEQUENCES USED FOR EVOLUTIONARY CALCULATIONS

Abbreviation	Name	Substrate	Accession	Refs.
BETAT-D	Dogncbta	Betaine	M80403	15
CRETR-R	Cretr	Creatine	X67252	14, 35
DOPAT-B	Bovdopatr	Dopamine	M80234	7
DOPAT-H	Humdoptra	Dopamine	M95167	36
DOPAT-R	Ratdoper	Dopamine	M80570	4
GABAT1-H	Hsgat1mr	γ-Aminobutyrate	X54673	2
GABATI-M	Gat1	γ-Aminobutyrate	M92378	31
GABAT2-M	Gat2	γ-Aminobutyrate	M97632	42
GABAT3-M	Gat3	γ-Aminobutyrate	L04663	13
GABAT4-M	Gat4	γ-Aminobutyrate	L04662	13
GABAT4-R	Rat3gat	γ-Aminobutyrate	M95763	37
GLYTI-M	Glyt1	Glycine	X67056	22
GLYT2-R	Glyt2	Glycine	L21672	26
HAEM1		?	U32703	33
METHAN1		?		34
NORAT-B	Ntt1	Noradrenaline	U09198	41
NORAT-H	Humnortr	Noradrenaline	M65105	3
NTT4	Ntt4r	?	S52051	29
NTT73	Ntt73	?	L22022	28
NTTB21	rB21a	?	S76742	39
PROT-R	Protr	Proline	M88111	25
ROSI-T	ROSIT	?	U12973	40
SEROT-R	Rsertran	Serotonin	X63253	8
STRYP1	Sat1	?	—	32
TAUT-D	Dognacltau	Taurine	M95495	16
TAUT-M	Taurt	Taurine	L03292	17
TAUT-R	Ratttransp	Taurine	M96601	38

at http://www.ebi.ac.uk, SwissProt at http://expasy.hcuge.ch). In Table I, we compiled the names, synonyms, substrates, and accession numbers of all the sequences employed, if available. Prior to the evolutionary calculations, the protein sequences have been aligned by means of the programs PILEUP and LINEUP of the GCG package.[43] Three regions, showing a relatively high degree of similarity between all sequences and only few gaps in the alignment have been chosen for further analysis (Fig. 1). The respective DNA sequences were aligned and then run with evolutionary tree building software of the PHYLIP package developed by Felsenstein[44] on a DEC 3000 workstation under OpenVMS ALPHA 1.5. The DNAML program

[43] J. Devereux, P. Haeberli, and O. Smithies, *Nucleic Acid Res.* **12,** 387 (1984).
[44] J. Felsenstein, "PHYLIP 3.2 Manual." University of California Herbarium, Berkeley, 1989.

```
BETAT-D    51 GEIIGLGNVWRFPYLCYKNGGGAFFIPYFIFFFTCGIPVFFLE
CRETR-R    64 GFAVGLGNVWRFPYLCYKNGGGVFLIPYVLIALVGGIPIFFLE
DOPAT-B    75 GFAVDLANVWRFPYLCYKNGGGAFLVPYLFFMVVAGVPLFYME
DOPAT-H    75 GFAVDLANVWRFPYLCYKNGGGAFLVPYLLFMVIAGMPLFYME
DOPAT-R    75 GFAVDLANVWRFPYLCYKNGGGAFLVPYLLFMVIAGMPLFYME
GABAT1-H   59 GYAIGLGNVWRFPYLCGKNGGGAFLIPYFLTLIFAGVPLFLLE
GABAT1-M   59 GYAIGLGNVWRFPYLCGKNGGGAFLIPYFLTLIFAGVPLFLLE
GABAT2-M   51 GEIIGLGNVWRFPYLCYKNGGGAFFIPYFIFFFSCGIPVFFLE
GABAT3-M   47 GEIIGLGNVWRFPYLCYKNGGGAFFIPYLIFLFTCGIPVFFLE
GABAT4-M   60 GEIIGLGNVWRFPYLCYKNGGGAFLIPYVVFFICCGIPVFFLE
GABAT4-R   60 GEIIGLGNVWRFPYLCYKNGGGAFLIPYVVFFICCGIPVFFLE
GLYT1-M    42 GYAVGLGNVWRFPYLCYRNGGGAFMFPYFIMLIFCGIPLFFME
GLYT2-R   208 GYAVGLGNVWRFPYLAFQNGGGAFLIPYLMMLALAGLPIFFLE
HAEM1      26 GSAVGLGNIWRFPYTTYENGGGAFIIPYLIALLTAGIPLLFLD
METHAN1    21 GSAIGLGNIWRFGYMVYTNGGGAFLIPYIVALLCVGIPLMILE
NORAT-B    69 GFALDLANVWRFPYLCYKNGGGAFLIPYTLFLIIAGMPLFYME
NORAT-H    71 GFAVDLANVWRFPYLCYKNGGGAFLIPYTLFLIIAGMPLFYME
NTT4       75 GFSVGLGNIWRFPYLCQKNGGGAYLVPYLVLLIIIGIPLFFLE
NTT73      76 GFSVGLGNVWRFPYLCQKNGGGAYLLPYLILLLVIGIPLFFLE
NTTB21     44 SYAVGLGNVWRFPYLCQMYGGGSFLVPYLIMLIVEGMPLLYLE
PROT-R     52 GYCVGLGNVWRFPYRAYTNGGGAFLVPYFLMLAICGIPLFFLE
ROSI-T     32 GFAVGLGNIWRFPYLCHTHGGGAFLIPYFIALVFEGIPLFYIE
SEROT-R    94 GYAVDLGNIWRFPYICYQNGGGAFLLPYTIMAIFGGIPLFYME
STRYP1     54 GAAVGLGNFWRFPFMAYQNGGGAFLLPYFVALLTAGVPLMILE
TAUT-D     56 GGFVGLGNVWRFPYLCYKNGGGAFLIPYFIFLFGGGLPVFFLE
TAUT-M     56 GGFVGLGNVWRFPYLCYKNGGGAFLIPYFIFLFGSGLPVFFLE
TAUT-R     56 GGFVGLGNVWRFPYLCYKNGGGAFLIPYFIFLFGSGLPVFFLE

BETAT-D       VALGQYTSQGSVTAWRKICPLLQGIGLASVVIESYLNIYY
CRETR-R       ISLGQFMKAGSINVWN-ICPLFKGLGYASMVIVFYCNTYY
DOPAT-B       LALGQFNREGAAGVW-KICPILRGVGYTAILISLYIGFFY
DOPAT-H       LALGQFNREGAAGVW-KICPILKGVGFTVILISLYVGFFY
DOPAT-R       LALGQFNREGAAGVW-KICPVLKGVGFTVILISFYVGFFY
GABAT1-H      CSLGQYTSIGGLGVW-KLAPMFKGVGLAAAVLSFWLNIYY
GABAT1-M      CSLGQYTSIGGLGVW-NWAPMFKGVAVAAAVLSFWLNIYY
GABAT2-M      VALGQYSSQGSVTAWRKICPLLQGIGMASVVIESYLNIYY
GABAT3-M      TALGQYTNQGGITAWRRICPIFEGIGYASQMIVSLLNVYY
GABAT4-M      TALGQFTSEGGITCWRRVCPLFEGIGYATQVIEAHLNVYY
GABAT4-R      TALGQFTSEGGITCWRRVCPLFEGIGYATQVIEAHLNVYY
GLYT1-M       LSFGQFASQGCLGVWR-ISPMFKGVGYGMMVVSTYIGIYY
GLYT2-R       VSLGQFASQGPVSVW-KAIPALQGCGIAMLIISVLIAIYY
HAEM1         YAIGHRHRGGAPLSYRRFSPHFEVFGWWQMMVNVIIGLYY
METHAN1       FAIGHYTKKSAPLALEKLHKGSEWTGWFAVISGFIITSYY
NORAT-B       LALGQYNREGAATVW-KICPFFKGVGYAVILIALYVGFYY
NORAT-H       LALGQYNREGAATVW-KICPFFKGVGYAVILIALYVGFYY
NTT4          LAVGQRIRRGSIGVWHYVCPRLGGIGFSSCIVCLFVGLYY
NTT73         LSVGQRIRRGSIGVWNYISPKLGGIGFASCVVCYFVALYY
NTTB21        LAVGQRMRQGSIGAWRTISPYLSGVGVASVVVSFFLSMYY
PROT-R        LSLGQFSSLGPLAVWK-ISPLFKGAGAAMLLIVGLVAIYY
ROSI-T        LAIGQRLRRGSIGVWKTISPYLGGVGLGCFSVSFLVSLYY
SEROT-R       LALGQYHRNGCISIWRKICPIFKGIGYAICIIAFYIASYY
STRYP1        FGFGHKMRTATITAFKKLNRRFEWIGWWQITVPVVVVTFY
TAUT-D        VIIGQYTSEGGITCWEKICPLFSGIGYASIVIVSLLNIYY
TAUT-M        VIIGQYTSEGGITCWEKICPLFSGIAYASIVIVSLLNVYY
TAUT-R        VIIGQYTSEGGITCWEKICPLFSGIGYASIVIVSLLNVYY
```

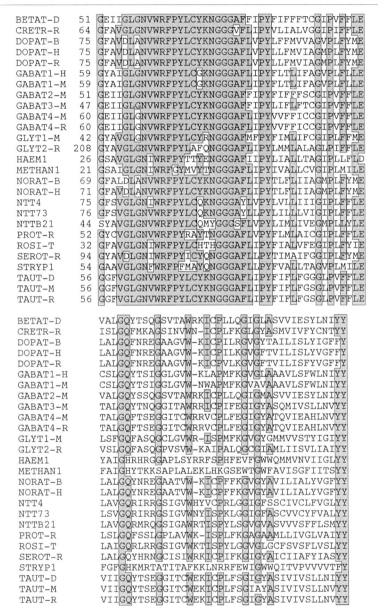

FIG. 1. Amino acid sequence alignment of the three regions in the various neurotransmitter transporters used for calculating the evolutionary trees. The abbreviations of the various transporters are listed in Table I. The source of some transporters is indicated by the last letter: D, dog; H, human; B, bovine, R, rat; M, mouse. HAEM1 is a gene of the *Haemophilus influenzae* genome,[33] MRTHAN1 is a gene of the *Methanococcus jannaschii* genome,[34] and STRYP1 is a gene of the *Symbiobacterium thermophilum* genome.[32]

```
BETAT-D    220  KVVYFTATFPYLMLVILLIRGITLPGAYQGVIYYLKPDLLRLKDPQVWMDA
CRETR-R    242  KIVYFTATFPYVVLVVLLVRGVLLPGALDGIIYYLKPDWSKLGSPQVWIDA
DOPAT-B    244  KVVWITATMPYVVLFALLLRGITLPGAVDAIRAYLSVDFHRLCEASVWIDA
DOPAT-H    247  KVVWITATMPYVVLTALLLRGVTLPGAIDGIRAYLSVDFYRLCEASVWIDA
DOPAT-R    246  KVVWITATMPYVVLTALLLRGVTLPGAMDGIRAYLSVDFYRLCEASVWIDA
GABAT1-H   221  KVVYFSATYPYIMLIILFFRGVTLPGAKEGILFYITPNFRKLSDSEVWLDA
GABAT1-M   220  KVVYFSATYPYIMLIILFFRGVTLPGAKEGILFYITPNFRKLSDSEVIFDA
GABAT2-M   220  KVVYFTATFPYLMLIILLIRGVTLPGAYQGIVFYLKPDLLRLKDPQVWMDA
GABAT3-M   215  KVVYFTATFPYLMLVVLLIRGVTLPGAAQGIQFYLYPNITRLWDPQVWMDA
GABAT4-M   230  KVVYVTATFPYIMLLILLIRGVTLPGASEGIKFYLYPDLSRLSDPQVWVDA
GABAT4-R   230  KVVYVTATFPYIMLLILLIRGVTLPGASEGIKFYLYPDLSRLSDPQVWVDA
GLYT1-M    224  KVVYFTATFPYVVLTILFVRGVTLEGAFTGIMYYLTPQWDKYLEAKVWGDA
GLYT2-R    405  KVVYFTATFPYVVLVILLIRGVTLPGAGAGIWYFITPKWEKLTDATVWKDA
HAEM1      164  KVSSVLMPVLVVMFMVLVIYSLFLPGAAKGLDALFTPDWSKLSNPSVWIAA
METHAN1    158  KANKIMIPFLLFLIILLVLNALTLPGALTGIEWYLTPDFSALFNYNVWLSA
NORAT-B    242  KVVWITATLPYLVLFVLLVHGITLPGASNGINAYLHIDFYRLKEATVWIDA
NORAT-H    244  KVVWITATLPYFVLFVLLVHGVTLPGASNGINAYLHIDFYRLKEATVWIDA
NTT4       234  KVMYFSSLFPYVVLACFLVRGLLLRGAVDGILHMFTPKLDKMLDPQVWREA
NTT73      235  KIMYFSSLFPYVVLICFLIRSLLLNGSIDGIRHMFTPKLEMMLEPKVWREA
NTTB21     201  KVVYFTALMPYCVLIIYLVRGLTLHGATNGLMYMFTPKIEQLANPKAWINA
PROT-R     224  KVVYFTATFPYLILLMLLVRGVTLPGAWKGIQFYLTPQFHHLLSSKVWIEA
ROSI-T     189  KVIYFTALFPYLVLTIFLIRGLTLPGATEGLTYLFTPNMKILQNSRVWLDA
SEROT-R    262  KVVWVTATFPYIVLSVLLVRGATLPGAWRGVVFYLKPNWQKLLETGVWVDA
STRYP1     174  KACKIMTPFLIVAMLIFDIRGITLPGATYGLNYFLNPDFSKIMDPGVWVAA
TAUT-D     227  KVVYFTATFPFAMLLVLLVRGLTLPGAGAGIKFYLYPDISRLEDPQVWIDA
TAUT-M     227  KVVYFTATFPFAMLLVLLVRGLTLPGAGEGIKFYLYPDISRLGDPQVWIDA
TAUT-R     227  KVVYFTATFPFAMLLVLLVRGLTLPGAGEGIKFYLYPNISRLEDPQVWIDA

BETAT-D         GTQIFFSFAICQGCLTALGSYNKYHNNCYRDSIALCFLNSATSFAAGFVVF
CRETR-R         GTQIFFSYAIGLGALTALGSYNRFNNNCYKDAIILALINSGTSFFAGFVVF
DOPAT-B         AIQICFSLGVGLGVLIAFSSYNKFTNNCYRDAIITTSVNSLTSFSSGFVVF
DOPAT-H         ATQVCFSLGVGFGVLIAFSSYNKFTNNCYRDAIVTTSINSLTSFSSGFVVF
DOPAT-R         ATQVCFSLGVGFGVLIAFSSYNKFTNNCYRDAIITTSINSLTSFSSGFVVF
GABAT1-H        ATQIFFSYGLGLGSLIALGSYNSFHNNVYRDSIIVCCINSCTSMFAGFVIF
GABAT1-M        ATQIFFSYGLGLGSLIALGSYNSFHNNVYRDSIIVCCINSCTSMFAGFVIF
GABAT2-M        GTQIFFSFAICQGCLTALGSYNKYHNNCYRDSIALCFLNSATSFVAGFVVF
GABAT3-M        GTQIFFSFAICLGCLTALGSYNKYHNNCYRDCIALCILNSSTSFMAGFAIF
GABAT4-M        GTQIFFSYAICLGCLTALGSYNNYNNNCYRDCIMLCCLNSGTSFVAGFAIF
GABAT4-R        GTQIFFSYAICLGCLTALGSYNNYNNNCYRDCIMLCCLNSGTSFVAGFAIF
GLYT1-M         ASQIFYSLGCAWGGLITMASYNKFHNNCYRDSVIISITNCATRLYAGFVIF
GLYT2-R         ATQIFFSLSAAWGGLITLSSYNKFHNNCYRDTLIVTCTNSATSIFAGFVIF
HAEM1           YGQIFFSLSIGFGIMVTYASYLKKESDLTGSGLVVGFANSSFEVLAGIGVF
METHAN1         FSQIFFSLSLGFGILIAYASYLPKKSDLTINAVTVSLLNCGFSFLAGFAVF
NORAT-B         ATQIFFSLGAGFGVLIAFASYNKFDNNCYRDALLTSTINCVTSFISGFAIF
NORAT-H         ATQIFFSLGAGFGVLIAFASYNKFDNNCYRDALLTSSINCITSFVSGFAIF
NTT4            ATQVFFALGLGFGGVIAFSSYNKQDNNCHFDAALVSFINFFTSVLATLVVF
NTT73           ATQVFFALGLGFGGVIAFSSYNKRDNNCHFDAVLVSFINFFTSVLATLVVF
NTTB21          ATQIFFSLGLGFGSLIAFASYNEPSNDCQKHAVIVSVINSSTSIFASIVTF
PROT-R          ALQIFYSLGVGFGGLLTFASYNTFHQNIYRDTFIVTLGNAITSILAGFAIF
ROSI-T          ATQIFFSLSLAFGGHIAFASYNQPRNNCEKDAVTIALVNSMTSLYASITIF
SEROT-R         AAQIFFSLGPGFGVLLAFASYNKFNNNCYQDALVTSVVNCMTSFVSGFVIF
STRYP1          YSQVFFSTTLAVGVMIAYASYVPEDSDLANNAFITVFANSSFDFMAGLAVF
TAUT-D          GTQIFFSYAICLGAMTSLGSYNKYKYNSYRDCMLLGCLNSGTSFVSGFAIF
TAUT-M          GTQIFFSYAICLGAMTSLGSYNKYKYNSYRDCMLLGCLNSGTSFVSGFAIF
TAUT-R          GTQIFFSYAICLGAMTSLGSYNKYKYNSYRDCMLLGCLNSGTSFVSGFAIF
```

FIG. 1. (*continued*)

```
BETAT-D    384  FFIMLIFLGLDSQFVCVECLVTASMDMFPSQLRKSGRRELLILAIAVFCYLAGLFLVT
CRETR-R    406  FFFMLLLLGLDSQFVGVEGFITGLLDLLPASYYFRFQREISVALCCALCFVIDLSMVT
DOPAT-B    408  FFVMLLTLGIDSAMGGMESVITGLADEF--QLLHR-HRELFTLLVVLATFLLSLFCVT
DOPAT-H    411  FFIMLLTLGIDSAMGGMESVITGLIDEF--QLLHR-HRELFTLFIVLATFLLSLFCVT
DOPAT-R    410  FFLMLLTLGIDSAMGGMESVITGLVDEF--QLLHR-HRELFTLGIVLATFLLSLFCVT
GABAT1-H   385  FFSMLLMLGIDSQFCTVEGFITALVDEYPRLLRN--RRELFIAAVCIISYLIGLSNIT
GABAT1-M   384  FFSMLLMLGIDSQFCTVEGFITALVDEYPRLLRN--RRELFIAAVCIVSYLIGLSNIT
GABAT2-M   384  FFIMLLFLGLDSQFVCMECLVTASMDMFPQQLRKSGRRDVLILAISVLCYLMGLLLVT
GABAT3-M   379  FFFMVVLLGLDSQFVCVESLVTALVDMYPRVFRKKNRREVLILIVSVISFFIGLIMLT
GABAT4-M   394  FFMMLIFLGLDSQFVCVESLVTAVVDMYPKVFRRGYRRELLILALSIISYFLGLVMLT
GABAT4-R   394  FFMMLIFLGLDSQFVCVESLVTAVVDMYPKVFRRGYRRELLILALSIVSYFLGLVMLT
GLYT1-M    388  FFFMLILLGLGTQFCLLETLVTAIVDEVGNEWILQ-KKTYVTLGVAVAGFLLGIPLTS
GLYT2-R    569  FFLMLLTLGLDTMFATIETIVTSISDEFPK-YLRT-HKPVFTLGCCICFFIMGFPMIT
HAEM1      329  FFGSLTFAALTSFISVIEVIISAIQDKIRIS-----RGKVTFIVGVPMMLVSVILFGT
METHAN1    323  FFLALVFAGISSAVSIVEASVSAIIDKFSLS-----RKKALLAVLALF-IIISPIFTT
NORAT-B    406  FFIMLLALGIDSSMGGMEAVITGLADDF--QVLKR-HRKLFTFAVSFGTFLLALFCIT
NORAT-H    408  FFVMLLALGLDSSMGGMEAVITGLADDF--QVLKR-HRKLFTFGVTFSTFLLALFCIT
NTT4       462  FFLMLINLGLGSMIGTMAGITTPIIDTF------KVPKEMFTVGCCVFAFFVGLLFVQ
NTT73      463  FFLMLINLGLGSMFGTIEGIITPVVDTF------KVRKEILTVICCLLAFCIGLMFVQ
NTTB21     419  YFFMLLMLGMGSMLGNTAAILTPLTDS--KVISSYLPKEAISGLVCLINCAVGMVFTM
PROT-R     388  FFFMLLTLGLDSQFAFLETIVTAVTDEFPY-YLRP-KKAVFSGLICVAMYLMGLILTT
ROSI-T     405  FFGMLFTLGLSSMFGNMEGVITPLFDM--GILPKGVPKETMTGVVCFICFLSAICFTL
SEROT-R    427  FFLMLITLGLDSTFAGLEGVITAVLDEFPHIWAK--RREWFVLIVVITCVLGSLLTLT
STRYP1     339  FFSALLLAGISSSISQMESFASAVIDRFGVD-----RKK-LLGWFSLIGFAFSALFAT
TAUT-D     391  FFIMLLLLGLDSQFVEVEGQVTSLVDLYPSFLRKGFRREIFIAFMCSISYLLGLSMVT
TAUT-M     409  FFIMLLLLGLDSQFVEVEGQITSLVDLYPSFLRKGYRREIFIAILCSISYLLGLTMVT
TAUT-R     409  FFIMLLLLGLDSQFVEVEGQITSLVDLYPSFLRKGYRREIFIAIVCSISYLLGLTMVT

BETAT-D         GGMYIFQLFDYYASSGICLLFLAMFEVICISWVYGADRFYDNIEDMIGYRPWPLVKIS
CRETR-R         GGMYVFQLFDYYSASGTTLLWQAFWECVVVAWVYGADRFMDDIACMIGYRPCPWMKWC
DOPAT-B         GGIYVFTLLDHFAA-GTSILFGVLMEVIGVAWFYGVWQFSDDIKQMTGRRPSLYWRLC
DOPAT-H         GGIYVFTLLDHFAA-GTSILFGVLIEAIGVAWFYGVGQFSDDIQQMTGQRPSLYWRLC
DOPAT-R         GGIYVFTLLDHFAA-GTSILFGVLIEAIGVAWFYGVQQFSDDIKQMTGQRPNLYWRLC
GABAT1-H        GGIYVFKLFDYYSASGMSLLFLVFFECVSISWFYGVNRFYDNIQEMVGSRPCIWWKLC
GABAT1-M        GGIYVFKLFDYYSASGMSLLFLVFFECVSISWFYGVNRFYDNIQEMVGSRPCIWWKLC
GABAT2-M        GGMYIFQLFDYYASSGICLLFLSLFEVICIGWVYGADRFYDNVEDMIGYRPWPLVKIS
GABAT3-M        GGMYVFQLFDYYAASGMCLLFVAIFESLCVAWVYGAGRFYDNIEDMIGYKPWPLIKYC
GABAT4-M        GGMYIFQLFGSYAASGMCLLFVAIFECVCIGWVYGSNRFYDNIEDMIGYRPLSLIKWC
GABAT4-R        GGMYIFQLFDSYAASGMCLLFVAIFECVCIGWVYGSNRFYDNIEDMIGYRPLSLIKWC
GLYT1-M         AGIYWLLLMDNYAA-SFSLVVISCIMCVSIMYIYGHRNYFQDIQMMLGFPPPLFFQIC
GLYT2-R         GGIYMFQLVDTYAA-SYALVIIAIFELVGISYVYGLQRFCEDIEMMIGFQPNIFWKVC
HAEM1           TGLPMLDVFDKFVN-YFGIVAVAFASLIAIVANEKLGLLGNHLNETSSFKVGFF----
METHAN1         AGLYYLDIIDHFAS-GYLLPIAAILEIIIAIWLFGGDKLREHVNKLSEIKLGVW----
NORAT-B         GGIYVLTLLDTFAA-GTSILFAVLMEAIGVSWFYGVDRFSNDIQQMMGFKPGLYWRLC
NORAT-H         GGIYVLTLLDTFAA-GTSILFAVLMEAIGVSWFYGVDRFSNDIQQMMGFRPGLYWRLC
NTT4            SGNYFVTMFDDYSAT-LPLTVIVILENIAVAWIYGTKKFMQELTEMLGFRPYRFYFYM
NTT73           SGNYFVTMFDDYSAT-LPLLIVVILENIAVSFVYGIDKFLEDLTDMLGFAPSKYYYYM
NTTB21          AGNYWFDIFNDYAAT-LSLLLIVLVETIAVCYVYGLRRFESDLRAMTGRPLNWYWKAM
PROT-R          GGMYWLVLLDDYSA-SFGLMVVVITTCLAVTRVYGIQRFCRDIHMMLGFKPGLYFRAC
ROSI-T          SGSYWLEIFDSFAAS-LNLIIFAFMEVVGVIHVYGIKRFCDDIEWMTGRRPSLYWQVT
SEROT-R         GGAYVVTLLEEYAT-GPAVLTVALIEAVAVSWFYGITQFCSDVKEMLGFSPGWFWRIC
STRYP1          AGVHILDIVDHFVG-SYAIAILGLVEAIVLGYIMGTARIREHVNLTSDIRVGMW----
TAUT-D          GGMYVFQLFDYYAASGVCLLWVAFFECFVIAWIYGSDNLYDGIEDMIGYRPGPWMKYS
TAUT-M          GGMYVFQLFDYYAASGVCLLWVAFFECFVIAWIYGGDNLYDGIEDMIGYRPGPWMKYS
TAUT-R          GGMYVFQLFDYYAASGVCLLWVAFFECFVIAWIYGGDNLYDGIEDMIGYRPGPWMKYS
```

FIG. 1. (*continued*)

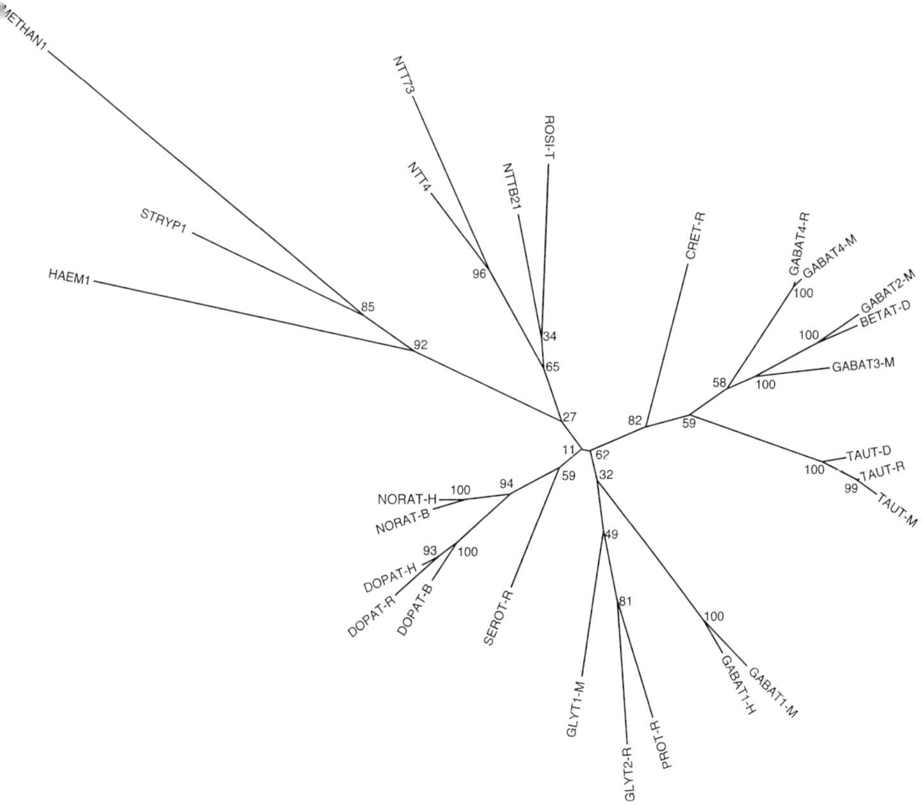

FIG. 2. Three evolutionary trees constructed from the amino acid sequences shown in Fig. 1.

included in the package builds trees according to the maximum likelihood algorithm[45] with branch lengths proportional to the relative number of base changes. Eight runs of the program were carried out on each dataset, always jumbling the input order of the sequences according to random numbers and thereby avoiding the possibility of introducing artifacts by an unfavorable order of sequence input. After finishing a run, the program was directed to perform global rearrangements of the trees found (i.e., to remove subtrees from the tree and put them back on in all possible ways so as to have a better chance of finding an optimal tree). Out of the eight results, the tree with the best likelihood score was finally selected. We also performed a statistical analysis on the same data sets. The sequences were first bootstrapped by the SEQBOOT program, creating 100 new data sets by sam-

[45] J. Felsenstein, *J. Molecular Evolution* **17,** 368 (1981).

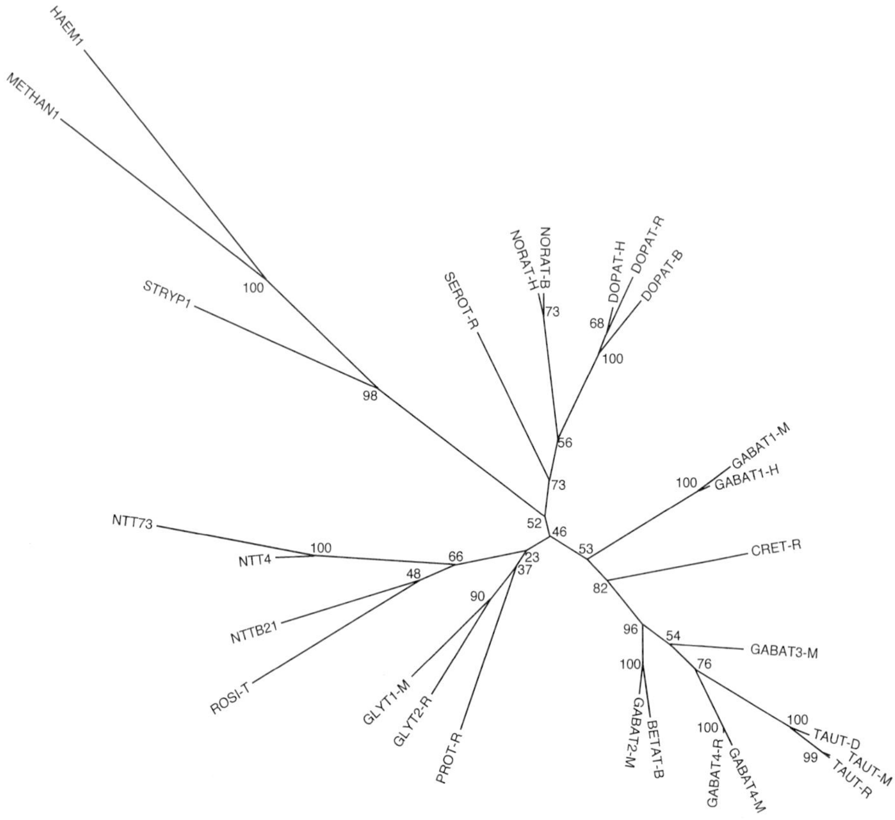

FIG. 2. (*continued*)

pling the characters randomly with replacement. The resulting data sets have the same size as the original, but some characters have been left out and others have been duplicated. The random variation of the results from analyzing these bootstrapped data sets can be shown statistically to be typical of the variation that one would get from collecting new data sets.[46] The 100 sets were used as input to DNAPARS, a program calculating evolutionary trees according to the parsimony algorithm.[47] The 100 trees resulting from the DNAPARS runs were used as input to the CONSENSE program to calculate one consensus tree and to examine the statistical relevance of the branchings. We were pleased to learn that the trees showed

[46] J. Felsenstein, *Evolution* **39,** 783 (1985).

[47] A. G. Kluge and J. S. Farris, *Systematic Zoology* **18,** 1 (1969).

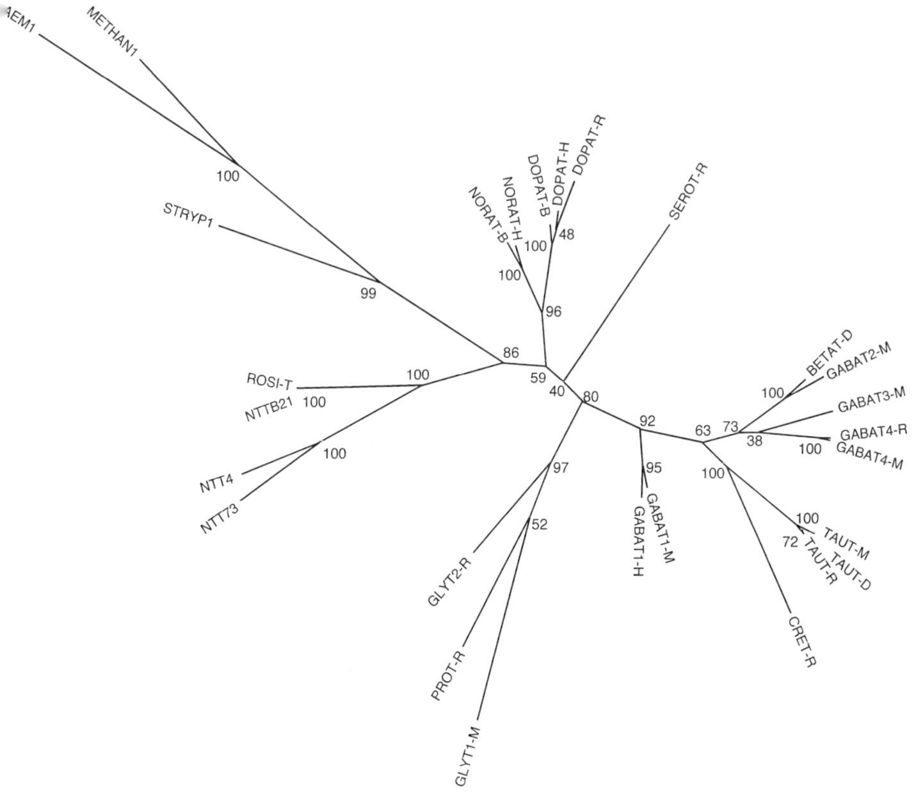

FIG. 2. (*continued*)

very high similarity. Only minor differences occurred in the length of certain branches of trees calculated by the maximum likelihood method, and the six trees obtained with the maximum likelihood as well as with the bootstrapping method differed only slightly in the innermost branching order. The trees obtained with the three fragments are shown in Fig. 2. The branch lengths were taken from the maximum likelihood analysis and show the relative genetic diversity between species. The numbers depict the results of the consensus analysis, indicating how many times the group which consists of the species to the right of the fork occurred.

By dividing the various transporter sequences into three fragments that contain the most homologous sequences, not only did we put the method to a rigorous test, but we also obtained additional insight into the relations among the various transporters. In general, the four subfamilies of neurotransmitter transporters are maintained. This is quite remarkable because three bacterial transporters were added to the calculation of the evolutionary trees. The bacterial transporters are loosely grouped with the "orphan" (NTT4) transporters. The only striking deviation was recorded with the GABA transporter GAT1, which in fragment 1 is grouped with the amino acid transporters of glycine and proline. It should be noted that the amino acid sequence of GAT1 is the most distant of all the other members of the subfamily of GABA transporters. This property of GAT1 is in line with its substrate specificity, which is unique among all the other members of this subfamily.

[29] Vaccinia Virus–T7 RNA Polymerase Expression System for Neurotransmitter Transporters

By SUE L. POVLOCK and SUSAN G. AMARA

Introduction

Biochemical analyses of transport activities in heterogeneous preparations such as brain slices, synaptosomes, and plasma membrane vesicles have been complicated by the presence of multiple carrier subtypes and the frequent overlap of their substrate specificities. Furthermore, the low abundance of many of the carrier subtypes and the difficulties encountered in protein purification have made efforts at isolating these activities a challenging undertaking. Ligand affinity and binding techniques have contributed to the successful purification of at least one transporter, the human platelet plasma membrane serotonin transporter (SERT).[1] Transporters for which high-affinity binding ligands are unavailable have required more difficult reconstitution assays for purification. Carriers purified by this approach include a rat brain γ-aminobutyric acid (GABA) transporter,[2] a porcine brain stem glycine transporter,[3,4] and a rat glial glutamate transporter.[5] As a way of circumventing the difficulties associated with the purification of carrier proteins, the cloning of genes encoding neurotransmitter transporters not only has offered insight into the diversity of carrier

[1] J.-M. Launay, C. Geoffroy, V. Mutel, M. Buckle, A. Cesura, J. E. Alouf, and M. Da Prada, *J. Biol. Chem.* **267,** 11344 (1992).
[2] R. Radian, A. Bendahan, and B. I. Kanner, *J. Biol. Chem.* **261,** 15437 (1986).
[3] B. Lopez-Corcuera, J. Vazquez, and C. Aragon, *J. Biol. Chem.* **266,** 24809 (1991).
[4] C. Aragon and B. Lopez-Corcuera, *Methods Enzymol.* **296,** [1], (1998) (this volume).
[5] N. C. Danbolt, G. Pines, and B. I. Kanner, *Biochemistry* **29,** 6734 (1990).

0076-6879/98 $25.00

subtypes, but also has allowed new approaches for examining their functional properties and pharmacological specificities.

The availability of cDNAs encoding neurotransmitter carriers offers the advantages of high level expression of single activities in heterologous cells, in the absence of vesicular storage compartments and free from the confounding influences of other transport pathways. A variety of heterologous expression systems have proven useful in characterizing neurotransmitter transporters. *Xenopus laevis* oocytes have been used both to study activities encoded by endogenous brain mRNAs and to identify and characterize transporter cDNAs by functional expression. To facilitate the investigations into the basic structural and functional properties of transporter proteins, cDNAs have been stably transfected into cell lines such as HeLa, LLC-PK_1 (porcine kidney), COS (monkey kidney), and MDCK (Madin–Darby canine kidney) using a variety of standard transfection and selection techniques. For transient expression of cloned transporter cDNAs, mammalian cell systems such as COS cells[6] offer significant advantages over *Xenopus laevis* oocytes in terms of ease and reproducibility. As will be considered in this chapter, a vaccinia virus/T7 polymerase expression system has also been used as a convenient system for characterizing cloned transporter cDNAs.[7–12]

The vaccinia virus/T7 polymerase expression system has been used to direct the expression of cDNAs encoding a variety of cellular genes in mammalian cells. In contrast to the more standard vaccinia virus expression systems, it does not require the generation of recombinant vaccinia virus containing the gene of interest. The only prerequisite is that the gene must be inserted into a plasmid vector under the control of a T7 promoter. The strain of vaccinia used in this approach has been engineered to express T7 RNA polymerase,[13] and when a cDNA driven by the T7 promoter is transfected into cells infected with this virus, the encoded protein is expressed at high levels. Vaccinia virus replicates in the cytoplasm and has a relatively broad host range, allowing cDNAs to be expressed in a variety of cellular environments. Furthermore, the gene of interest is transcribed

[6] Y. Gluzman, *Cell* **23,** 175 (1981).

[7] R. D. Blakely, J. A. Clark, G. Rudnick, and S. G. Amara, *Anal. Biochem.* **194,** 302 (1991).

[8] T. Pacholczyk, R. D. Blakely, and S. G. Amara, *Nature* **350,** 320 (1991).

[9] R. D. Blakely, H. E. Berson, R. T. Fremeau, Jr., M. G. Caron, M. M. Peek, H. K. Prince, and C. C. Y. Bradley, *Nature* **354,** 66 (1991).

[10] J. Kilty, D. Lorang, and S. G. Amara, *Science* **254,** 578 (1991).

[11] K. J. Buck and S. G. Amara, *Proc. Natl. Acad. Sci. USA* **91,** 12584 (1994).

[12] T. T. Nguyen and S. G. Amara, *J. Neurochem.* **67,** 645 (1996).

[13] T. R. Fuerst, E. G. Niles, F. W. Studier, and B. Moss, *Proc. Natl. Acad. Sci. USA* **83,** 8122 (1986).

by the virally encoded T7 polymerase within the cytoplasm, circumventing the loss in efficiency that can occur when the DNA must reach the nucleus to be expressed. Multiple host cell lines can be screened to identify those lines with low background uptake, high expression levels, and optimal transport kinetics. This system is well suited for the expression of plasma membrane carriers and has been useful for studying the cloned norepinephrine (NET),[8] serotonin (SERT),[9] and dopamine (DAT)[10] transporters for structure–function analyses, for addressing cellular targeting, and for studying the regulation of transport activity. It has also been adapted for the cloning of transporter cDNAs by functional expression.[7] The advantages of this system include its high sensitivity and ease of expression (assays can be done on a single well of transfected cells in a microtiter dish). The combined viral infection/liposome-mediated transfection procedure has been found to be more efficient and less time consuming than the DEAE-dextran or calcium phosphate precipitation methods.[7] The applicability of the system to studies of the kinetics and pharmacology of the cloned Na^+-dependent glucose and GABA transporters has been considered in greater detail elsewhere.[7]

This chapter will describe a procedure for the liposome-mediated transfection of the human DAT gene under the control of the T7 promoter into HeLa cells infected with a recombinant vaccinia virus expressing the T7 polymerase gene. A brief description of the background of this method of transfection is given here in lieu of in-depth discussions provided elsewhere.[13–16] When a strain of vaccinia virus engineered to express the gene for bacteriophage T7 RNA polymerase is added to cells, it will infect the cells and synthesize the polymerase in the cytoplasm. The enzyme then acts on T7 promoter-driven target sequences to produce a rapid, high level of expression of a gene inserted downstream of a T7 promoter. Plasmids containing the target DNA are introduced into the cells using Lipofectin, a cationic lipid that spontaneously forms liposomes which complex with the DNA, fuse with the plasma membrane, and ultimately allow the expression of the desired gene. A schematic of the approach is shown in Fig. 1.

Materials and Methods

Maintenance of HeLa Cells

HeLa cells are maintained in Dulbecco's modified Eagle's medium (DMEM), supplemented with 10% fetal calf serum (FCS, GIBCO, Grand

[14] P. L. Felgner, T. R. Gadek, M. Holm, R. Roman, H. W. Chan, M. Wenz, J. P. Northrop, G. M. Ringold, and M. Danielson, *Proc. Natl. Acad. Sci. USA* **84,** 7413 (1987).
[15] M. Mackett, G. L. Smith, and B. Moss, *J. Virol.* **49,** 857 (1984).
[16] T. R. Fuerst, P. L. Earl, and B. Moss, *Mol. Cell. Biol.* **7,** 2538 (1987).

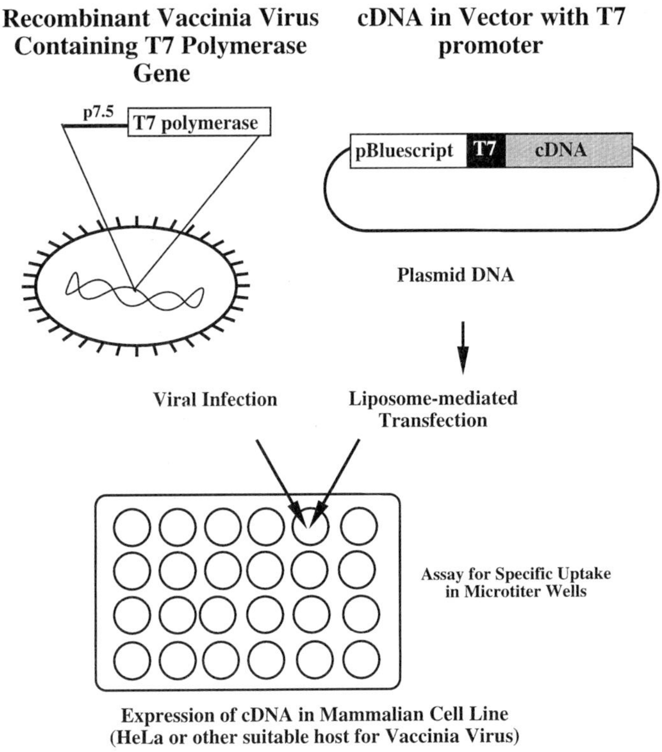

FIG. 1. Vaccinia virus/T7 polymerase transient expression system.

Island, NY), 10 U/ml penicillin, and 10 μg/ml streptomycin (GIBCO) (DMEM$^+$) at 37° in 5% CO_2. The cells grow slowly, with a doubling time of roughly 36 hr. HeLa cells are most robust during subconfluent stages ($\leq$70%). Allowing the cells to reach confluency may result in alterations of their basic properties.

Virus Preparation

The following protocol to prepare the VTF-7, a recombinant vaccinia virus that expresses the T7 RNA polymerase, was adapted from methods outlined by Mackett *et al.*[17] Grow HeLa cells to near confluency (10^7 cells/plate) in two 15-cm tissue culture plates in DMEM$^+$. Remove the media from the plates and wash once with 25 ml of sterile phosphate-buffered saline (PBS) or serum-free DMEM (DMEM$^-$) to remove any fetal calf

[17] M. Mackett, G. L. Smith, and B. Moss, *in* "DNA Cloning" (D. M. Glover, Ed.), Vol. 2, p. 191, IRL, Oxford, 1985.

serum that may adsorb the virus. Infect with one plaque-forming unit (pfu)/10 cells in the smallest volume (~5 ml) of DMEM$^-$ possible. Allow the virus to adsorb for 30 min at 37° in 5% (v/v) CO_2 with frequent rotation of the plates to ensure complete adsorption. Remove the virus, wash once with DMEM$^-$, and add DMEM$^+$ to the plates. Incubate the cells for 48 hr under normal HeLa growth conditions.

To harvest the cells, remove the medium and wash once with 25 ml of PBS. Using a sterile rubber policeman, scrape the cells from the plate into 10 ml PBS and pellet in a tabletop centrifuge at 1000*g* at room temperature for 5 min. Resuspend cell pellet in 2.5 ml ice-cold 10 m*M* Tris (pH 9.0) and homogenize with 10 strokes in a Dounce homogenizer. Centrifuge at 1000*g* for 5 min at 4° and place the supernatant on ice. Aspirate the PBS, resuspend the cell pellet in 2.5 ml ice-cold 10 m*M* Tris (pH 9.0), and homogenize and spin as before. Combine the supernatants and subject them to at least three rounds of freeze–thawing by alternately placing them in liquid nitrogen and a 37° water bath. Layer the supernatant over 36% sucrose in 10 m*M* Tris (pH 9.0) and centrifuge at 90,000*g* for 20 min at 4°. Discard the supernatant. Resuspend the pellet in 5 ml 10 m*M* Tris (pH 9.0), aliquot into 25- to 100-μl volumes, and store at −80°.

To titer the virus, prepare dilutions of the stock (10^2 to 10^6) in 10 m*M* Tris (pH 9.0). Grow HeLa cells in two 6-well plates to 70% confluency. Wash with DMEM$^-$ and add 1 μl of each dilution to each well in 500 μl of DMEM$^-$. Allow virus to adsorb for 30 minutes in the incubator at 37°, 5% CO_2, with frequent rotation to ensure complete absorption. Remove virus and replace with DMEM$^+$. Incubate for 36–48 hr until plaques are clearly visible. Count the number of plaques to determine the plaque-forming units of the original stock solution.

Infection/Transfection Procedure

HeLa cells, cultured to 70% confluency, are plated at a density of 2–4 × 10^5 cells/well. Twelve to twenty-four hours after plating, the cells will be ready to begin the infection/transfection process (approximately 70% confluency). This is an optimal confluency, although the infection/transfection procedure will be successful if the confluency is ≤90%. At least 15 min prior to beginning the infection process, add 3 μl Lipofectin (GIBCO) to 135 μl DMEM$^-$ for each well. Wash the plated HeLa cells once with 500 μl DMEM$^-$. Add the VTF-7 (10 pfu/100 μl DMEM$^-$/well) to the cells. Incubate the cells with the virus at 37° in 5% (v/v) CO_2 for 30 min, rotating often to ensure complete absorption of the virus. During this incubation period, add 1.0 μg recombinant DAT plasmid to 135 μl DMEM$^-$ for each well. Fifteen minutes after starting the virus infection period, mix the Lipofectin/DMEM$^-$ and plasmid/DMEM$^-$ solutions together, and incubate at room temperature to allow formation of micelles. After the virus incubation, add 270 μl of the Lipofectin/plasmid/DMEM$^-$ solution to the cells and return the plates to the incubator. Incubate cells for 12–16 hr before assaying for expression and functional characterization.

Optimizing Expression

There are several modifications in this protocol that may be considered to maximize expression of the gene of interest, and these variations should be tested whenever the system is being adapted for use with new host cell types. First, the amount of DNA added to each well will affect the efficiency of transfection and the expression level of the protein. Quantities between 0.5 and 3 μg of plasmid DNA per well have yielded acceptable levels of expression for a variety of plasma membrane carrier proteins. Generally, DNA to lipid ratios of 1 μg DNA per 3 μl of lipofectin give the most optimal transfection efficiencies. Other cationic lipid reagents such as dimethyldioctadecylammonium bromide,[18] LipofectAMINE (GIBCO), and LipofectAMINE Plus (GIBCO) have been tested with HeLa cells and give transfection efficiencies comparable to those obtained using Lipofectin (GIBCO). However, for some cell types these alternative reagents should be considered as they may be more efficient and less toxic. If toxicity of either the cationic lipid reagents or the virus becomes a concern, cell viability can be visually assessed using the trypan blue exclusion method. Cell lines that we have tested with the vaccinia/T7 polymerase system include HeLa cells (ATCC, Rockville, MD), Intestinal 407 (Int 407; ATCC) and baby hamster kidney cells (BHK; ATCC).

Another variable crucial to obtaining maximum expression is the length of time between completing the transfection procedure and performing the functional assays. The time course of expression can vary with different activities, and the function of some carriers may be more or less sensitive to cellular changes that occur during the later stages of viral infection.[7] Infection with vaccinia virus leads to a rapid repression of the translation of cellular mRNAs, which can lead to alterations in endogenous transport activities. This can have advantages in that background transport activities may be reduced, but there is also the risk that with time there will be a decline in the abundance and activities of membrane proteins involved in maintaining the ion gradients needed to sustain transport. For these reasons a time course of expression should be determined for each carrier subtype

[18] J. K. Rose, L. Buonocore, and M. A. Whitt, *Biotechniques* **10,** 520 (1991).

examined. The possible consequences of viral infection underscore the importance of determining background transport values in vector-transfected, virally infected cells at the same time postinfection.

Other approaches for improving expression of specific cDNAs have included the introduction of additional sequences to the T7 promoter-containing plasmid.[19,20] In our hands a modest improvement in expression has been obtained for some inserts[7] through the use of a vector, pET-3,[21] which introduces an f10 stem–loop structure and a Tf transcriptional termination signal at the 5′ and 3′-noncoding ends of the insert, respectively. We have not found this to be important for most of the carrier cDNAs we have examined, but it is worth considering for cDNAs that are difficult to express.

Transport Assay

At the appropriate time posttransfection, aspirate the virus/Lipofectin/DNA solution from the cells and wash cells once with 1 ml 37° KRH buffer (120 m*M* NaCl, 4.7 m*M* KCl, 2.2 m*M* $CaCl_2$, 1.2 m*M* $MgSO_4$, 1.2 m*M* KH_2PO_4, 10 m*M* HEPES, pH 7.4). Add 0.5 ml 37° KRH and incubate from 15–45 min at 37°. In some experiments, the subsequent uptake of substrate is more robust when longer preincubation times are used. During this incubation period, prepare the stock [^{3}H]dopamine (50–100 n*M* final concentration) in KRH. Unlabeled dopamine may be added to this solution to vary the concentration of dopamine. All dopamine solutions should be made fresh daily and kept on ice to prevent the oxidation of dopamine over the duration of the experiment. However, ascorbate (100 μM final concentration) may also be added to the dopamine solutions to prevent degradation from occurring. After the preincubation period, add [^{3}H]dopamine, with or without inhibitors, in a total volume of 500 μl and incubate cells for 20 min at 37°. The time of the transport assay should fall within the linear portion of the uptake versus time curve. This profile may be different for each transporter protein and should be predetermined. The transport assay is terminated by placing cells on ice and washing them three times with 500 μl ice-cold KRH. Lyse cells in 500 μl 0.1% SDS for at least 60 min with shaking. Determine lysate radioactivity by liquid scintillation counting and protein concentration using the BCA Protein Assay (Pierce, Rockford, IL). Transport velocity is expressed as fmol dopamine/min/mg protein.

[19] O. Elroy-Stein, T. R. Fuerst, and B. Moss, *Proc. Natl. Acad. Sci. USA* **86,** 6126 (1989).
[20] T. R. Fuerst and B. Moss, *J. Mol. Biol.* **206,** 333 (1989).
[21] F. W. Studier, A. H. Rosenberg, J. J. Dunn, and J. W. Dubendorff, *Methods Enzymol.* **185,** 60 (1990).

Acknowledgments

The authors acknowledge support from NIDA, DA07595 (S. G. Amara), T32 DA07262 (S. L. Povlock), and the Howard Hughes Medical Institute. Appreciation is also extended to Gwynn Daniels and Gursharn Khatra for their contributions to this chapter.

[30] Baculovirus-Mediated Expression of Neurotransmitter Transporters

By CHRISTOPHER G. TATE

Introduction

Na^+/Cl^--dependent neurotransmitter transporters constitute a small family of integral membrane proteins that couple the uptake of small molecules (monoamine neurotransmitters and amino acids) to the movement of ions (Na^+, Cl^-, and sometimes K^+) down their concentration gradients (reviewed in refs. 1–3). The transporters contain between 581 and 799 amino acid residues, of which about 99 amino acid residues are absolutely conserved. All members of this family are predicted to have 12 transmembrane regions, with both the N and C termini in the cytoplasm, and they contain a large N-glycosylated loop between putative transmembrane regions 3 and 4. N-Linked glycosylation is not required for transport or inhibitor binding per se, but glycosylation does seem to be important for cell-surface expression of transporters and for their stability in the cell.[4–7]

The determination of the three-dimensional structure of at least one member of the neurotransmitter transporter family is essential to understanding how they transport molecules into the cell and how inhibitors work. Unfortunately, the majority of these transporters are normally expressed at low levels and only the γ-aminobutyric acid (GABA) transporter (GAT),[8]

[1] G. Rudnick and J. Clark, *Biochim. Biophys. Acta* **1144,** 249 (1993).
[2] D. M. Worrall and D. C. Williams, *Biochem. J.* **297,** 425 (1994).
[3] B. Borowsky and B. J. Hoffman, *Int. Rev. Neurobiol.* **38,** 139 (1995).
[4] C. G. Tate and R. D. Blakely, *J. Biol. Chem.* **269,** 26303 (1994).
[5] L. Olivares, C. Aragon, C. Gimenez, and F. Zafra, *J. Biol. Chem.* **270,** 9437 (1995).
[6] H. E. Melikian, S. Ramamoorthy, C. G. Tate, and R. D. Blakely, *Mol. Pharmacol.* **50,** 266 (1996).
[7] T. T. Nguyen and S. G. Amara, *J. Neurochem.* **67,** 645 (1996).
[8] R. Radian, A. Bendahan, and B. I. Kanner, *J. Biol. Chem.* **261,** 15437 (1986).

0076-6879/98 $25.00

serotonin transporter (SERT),[9] and glycine transporter[10] have been purified from natural sources. No reports have been published describing the successful purification of the milligram quantities of fully functional transporter necessary for crystallization. It is, therefore, essential to devise a heterologous overexpression system for these transporters. The structural characteristics of the neurotransmitter transporters (12 transmembrane regions, glycosylation essential for good expression) make them among the most poorly overexpressed membrane proteins, regardless of the expression system used (reviewed by Grisshammer and Tate[11]). All of the commonly used expression systems have been tried for the expression of SERT. *Escherichia coli* and yeast gave poor functional expression of SERT[12] compared with expression in stable cell lines derived from mammalian cells[13] and in HeLa cells using the vaccinia virus expression system.[14] The baculovirus system has given good levels of expression that could be scaled up to produce milligrams of protein.[4] This paper describes how to construct recombinant baculoviruses expressing neurotransmitter transporters and their subsequent characterization.

Principle

The *Autographa californica* multicapsid nuclear polyhedrosis virus (AcMNPV) is an enveloped baculovirus that infects the larvae of only a few invertebrate species (reviewed in refs. 15 and 16). It has a two-phase life cycle. During the first phase, the immediate early, delayed early and late baculovirus-encoded genes are expressed, resulting in the budding of new viruses from the infected cell about 12 hours postinfection (p.i.). In the second phase, the very late genes are expressed (polyhedrin and p10), resulting in the production of occlusion bodies in the nucleus. Occlusion bodies contain viruses embedded in a matrix of polyhedrin; this form of the baculovirus can survive in the environment for years until ingested by another insect larva. The important feature of this life cycle is that the

[9] J.-M. Launay, C. Geoffroy, V. Mutel, M. Buckle, A. Cesura, J. E. Alouf, and M. Da Prada, *J. Biol. Chem.* **267,** 11344 (1992).

[10] B. Lopez-Corcuera, J. Vazquez, and C. Aragon, *J. Biol. Chem.* **266,** 24809 (1991).

[11] R. Grisshammer and C. G. Tate, *Q. Rev. Biophys.* **28,** 315 (1995).

[12] C. R. Baker, D. C. Williams, F. Titgemeyer, and G. T. Robillard, *J. Neurochem.* **66,** S43 (1996).

[13] H. Gu, S. C. Wall, and G. Rudnick, *J. Biol. Chem.* **269,** 7124 (1994).

[14] S. Ramamoorty, H. E. Melikian, Y. Qian, and R. D. Blakely, *Methods Enzymol.* **296,** [24], 1998 (this volume).

[15] G. W. Blissard, *Cytotechnology* **20,** 73 (1996).

[16] M. J. Fraser, *Curr. Top. Microbiol. Immunol.* **158,** 131 (1992).

second phase is not required for the propagation of baculovirus in the laboratory, and hence the viral proteins polyhedrin and p10 are redundant. Replacement of the polyhedrin gene with a cDNA results in a recombinant virus that expresses the cDNA-encoded protein in infected cells. The polyhedrin promoter is extremely strong and is turned on about 20 hr p.i., i.e., after virus budding has commenced, thus allowing the overexpression of potentially cytotoxic, integral membrane proteins.

The baculovirus expression system has become one of the most popular eukaryotic expression systems currently in use. This is partly due to the high levels of expression obtained, and also to the commercialization of the system, which has ensured the ease and reproducibility of recombinant virus construction. The insect cells used for expression (Sf9, Sf21, *Spodoptera frugiperda,* fall armyworm ovary cells) are also relatively easy to grow, even in large-scale bioreactors if scale-up is required.

Recombinant baculoviruses are constructed in two steps.[17] First, a bacterial plasmid containing the gene of interest downstream from the viral polyhedrin promoter is made. Numerous plasmids containing up to five polyhedrin and/or p10 promoters are commercially available. The plasmid is cotransfected with linearized baculovirus DNA into insect cells where *in vivo* recombination between homologous regions in the plasmid and the viral DNA produces the recombinant baculovirus. The success of this step has been dramatically improved by deleting an essential gene during the generation of the linearized baculovirus DNA; this gene is present on the bacterial plasmid and, therefore, only recombined baculoviruses can reproduce, resulting in greater than 90% of progeny viruses being recombinant.[18]

Methods

The steps required to produce a high-titer recombinant baculovirus suitable for characterization of neurotransmitter transporter expression in insect cells are listed below. (1) Construction of a transfer vector with the neurotransmitter transporter cDNA downstream from the polyhedrin promoter. (2) Cotransfection of the transfer vector with linearized baculovirus DNA into insect cells to produce the recombinant baculovirus. (3) Plaque purification of the recombinant baculovirus to ensure a homogeneous virus population. (4) Amplification of the baculovirus in the plaque to make 5 ml of first passage (1°) virus. (5) Titration of the 1° virus to

[17] D. R. O'Reilly, L. K. Miller, and V. A. Luckow, "Baculovirus Expression Vectors: A Laboratory Manual." W. H. Freeman, New York, 1992.

[18] P. A. Kitts and R. D. Possee, *Biotechniques* **14,** 810 (1993).

determine the number of plaque-forming units (pfu) per ml. (6) Amplification of the 1° virus to product 100 ml of second passage (2°) virus. (7) Titration of the 2° virus stock. The methods described below for the propagation of cells and baculovirus are derived from Summers and Smith.[19]

Maintenance of Cell Lines

All cell lines are grown at 27° in TNM-FH medium (Sigma, St. Louis, MO) supplemented with 10% fetal bovine serum (HyClone), 2.5 μg/ml Fungizone (GIBCO-BRL, Gaithersburg, MD) and 200 μg/ml penicillin/streptomycin (GIBCO-BRL). The quality of the fetal bovine serum is critical for good cell growth. The surfactant Pluronic F-68 (Sigma) is included in media for suspension cultures at a final concentration of 0.1% (w/v). Sf9 and Sf21 cells (Invitrogen, Leek, the Netherlands) are routinely grown in spinner flasks (Techne, Cambridge, UK); 100 ml cultures are grown in 250 ml spinners until they reach 2×10^6 cells/ml, when they are split to 0.5×10^6 cells/ml. The doubling time for the cells is roughly 24 hr. Cells to be left over the weekend are split to 0.3×10^6 cells/ml. Once a week (Mondays) the cells are spun down (4 min, 800*g*, 21°) and resuspended in fresh medium and replaced in a clean spinner bottle. Insect cells have a high oxygen demand, especially when infected with baculovirus, and it is therefore essential not to overfill spinner bottles and to have caps open a quarter-turn. A healthy cell culture suitable for baculovirus infection should contain at least 97% viable cells as determined by trypan blue exclusion staining (add 100 μl 0.4% trypan blue in PBS to 900 μl cells; nonviable cells are stained blue). Cells are harvested, for use in the protocols below, when they are at a density of between 1 and 2 million cells per milliliter.

Two other commercially available insect cell lines are used for expression studies (High Five and MG1 cells, Invitrogen). These are adherent cell lines and were used only to test whether expression of neurotransmitter transporters was better than in Sf9 cells. High Five and MG1 cells are grown in 75 cm^2 flasks containing 15 ml of medium and are split when about 90% confluent. Fresh flasks are reseeded with a 1 in 5 dilution of cells. High Five cells are removed from the flask by washing with 5 ml of 250 m*M* EDTA, followed by washing the cells once in fresh medium (4 min, 800*g*, 21°). MG1 cells require a mild trypsin treatment to remove them from the flask (5 ml of 0.25 μg/ml trypsin in 250 m*M* EDTA, 5 min room temperature).

[19] M. D. Summers and G. E. Smith, "A Manual of Methods for Baculovirus Vectors and Insect Cell Culture Procedures." Texas Agricultural Experiment Station, Bulletin No. 1555, Texas A & M University, Texas, 1987.

Construction of Recombinant Baculovirus

The cDNAs encoding the human norepinephrine transporter (NET),[20] rat SERT,[21] and human GAT[22] are all altered to remove the 5′ untranslated regions and to place a consensus Kozak sequence at the initiator Met codon (CCACCATGG).[23] A unique restriction site is included 5′ of the Kozak sequence for cloning. This avoids any potential effects of the 5′ untranslated region on the rate of translation and produces an *Nco*I site at the initiator Met codon that can be used to insert DNA encoding purification tags. The Kozak sequence does not correspond to the most favored sequence found around initiator Met codons in baculovirus genes (CAAATGAA),[24] but it has been found to be satisfactory for expression of neurotransmitter transporters, because factors other than translation are found to limit the level of expression. Removal of the 3′ untranslated region does not affect the expression levels of SERT or NET. Mutations are performed by standard PCR (polymerase chain reaction) protocols, and all the DNA generated by PCR is subsequently checked by DNA sequencing.[25]

The cDNAs encoding the neurotransmitter transporters are cloned downstream from the polyhedrin promoter in the plasmid pVL1393 (Pharmingen, San Diego, CA). Any of the other commercially available plasmids containing either a polyhedrin or p10 promoter would be equally suitable. Plasmid is prepared from a 200 ml bacterial culture (Wizard maxiprep, Promega, Madison, WI) for cotransfection with linearized baculovirus DNA (Baculogold, Pharmingen). The purity of plasmid DNA is crucial to the success of transfections. Other companies (Clontech, Novagen) also supply linearized baculovirus DNA which should give identical results to Baculogold. The manufacturer's protocol is followed exactly for making the recombinant virus. All manipulations using automatic pipetters are performed using sterile, aerosol-resistant tips, and automatic pipetters that are dedicated to tissue culture work and regularly cleaned with 70% (v/v) ethanol.

Plaque Purification

The cotransfection of linearized baculovirus DNA and plasmid DNA results in 5 ml of culture supernatant containing a mixture of recombinant

[20] T. Pacholczyk, R. D. Blakely, and S. G. Amara, *Nature* **350,** 350 (1991).

[21] R. D. Blakely, H. E. Berson, R. T. Fremeau, Jr., M. C. Caron, M. M. Peck, H. K. Prince, and C. C. Bradley, *Nature* **354,** 66 (1991).

[22] H. Nelson, S. Mandiyan, and N. Nelson, *FEBS Lett.* **269,** 181 (1990).

[23] M. Kozak, *J. Biol. Chem.* **266,** 19867 (1991).

[24] M. D. Ayres, S. C. Howard, J. Kuzio, M. Lopez-Ferber, and R. D. Possee, *Virology* **202,** 586 (1994).

[25] J. Sambrook, E. F. Fritsch, and T. Maniatis, "Molecular Cloning: A Laboratory Manual," 2nd ed. Cold Spring Harbor Laboratory, Cold Spring Harbor, New York, 1989.

virus clones. To ensure that a homogeneous virus population is used in subsequent experiments, a single round of plaque purification is performed. First, 1.5×10^6 cells in 3 ml medium are allowed to attach to 60×15 mm tissue culture dishes (Easy Grip, Falcon) for 10 min. The medium is removed and replaced with 1 ml of baculovirus supernatant diluted in medium to give 10^{-2} to 10^{-6} dilutions in steps of 10. The dishes are left for 1 hr. A sterile solution of 2% (w/v) agarose (SeaPlaque, FMC BioProducts, Rockland, ME) in water is melted in a microwave oven and placed with 2× TNM-FH medium containing 20% (v/v) fetal bovine serum in a 37° water bath for 15 min. Immediately before use, equal volumes of the two solutions are mixed at room temperature. The virus dilutions are aspirated from the cells and replaced with 4 ml agarose–medium by gently pipetting the solution at the side of the dish and allowing it to flow slowly over the cells. The agarose is allowed to set without any disturbance for 30 min at room temperature, and then the dishes are placed upright in a sealed sandwich box with a damp paper towel in the bottom to provide a humid atmosphere. The dishes are incubated for 5 days at 27° until plaques appear. Plaques are identified by using an inverted microscope at low magnification and marked with a pen. Plaques are only picked from dishes containing fewer than 10 plaques which are widely separated (only one plaque per field of view). Plaques are picked using a sterile pipetter tip and placed in 1 ml fresh medium, which is subsequently added to 1×10^6 cells in 4 ml medium in a 25 cm^2 flask and incubated for 3 days at 27°. The supernatant is decanted from the flasks and centrifuged at 1000*g* for 5 min. The supernatant [first passage (1°) virus stock] is transferred to a fresh tube. Aliquots are frozen at −70° for long-term storage and the remainder is stored at 4°.

Generally, five independent plaques are picked for each transfection and 1° virus stocks produced. To determine whether a recombinant virus is present, 100 μl of the first passage virus is used to infect 5×10^6 cells in 5 ml medium in a 25 cm^2 flask. The cells are harvested after 2 days, and membranes are prepared and analyzed by Western blotting using antibodies raised against the neurotransmitter transporter (see below). Of the five plaques picked, usually all five express the desired protein. Recombinant baculoviruses that expressed SERT, NET, and GAT were called bvSERT, bvNET, and bvGAT, respectively.

Virus Titration

The number of virus particles in a culture supernatant is determined by the end-point dilution method. Sf9 cells are diluted to $0.25–0.3 \times 10^6$ cells per milliliter. A serial dilution of baculovirus from 10^{-2} to 10^{-9} in steps of 10 is set up (200 μl per tube). An equal volume of diluted cells is added to each tube and mixed by shaking. Ten microliters of each mixture is placed in a well of a 60 × 10 μl well microtiter plate (Nunclon, GIBCO-BRL), ensuring that the tubes are well mixed before removing aliquots. For each dilution, 10 wells (one row) are set up. Plates are incubated in a humidified sandwich box at 27° for 6 days. The number of infected wells at each dilution is determined using an inverted microscope and the titer is calculated.[17,26] An "infected well" had at least some cells that showed signs of baculovirus infection (increased diameter, enlarged nucleus, grayish in color, cell lysis). Virus titers vary from 5×10^8 pfu/ml for amplified 2° stocks to about 5×10^7 pfu/ml for 1° stock.

Virus Amplification

To prepare high titer second passage virus, 1×10^8 Sf9 cells in 10 ml medium in a sterile 50 ml tube are infected with sufficient 1° virus to give a multiplicity of infection (MOI) of 0.2. The cells arc left at room temperature for 1 hr and then diluted to 100 ml with fresh medium (1×10^6 cells per ml) and incubated in a 250 ml spinner bottle at 27° for 2 days. Cells are decanted into two sterile 50 ml tubes and centrifuged for 5 min at 1000*g*, room temperature. The supernatant is filtered through a 0.2 μm filter and stored at 4°. Before use, the virus titer is determined as above. Second passage virus is used for characterization of all recombinant viruses. Higher passages are used to generate liters of baculovirus for large-scale fermentation. No virus is passaged more than four times, because decreased expression levels are commonly observed after serial passaging.[27]

Infection of Cells for Expression Studies

Expression of neurotransmitter transporters is generally performed in 250 ml spinner bottles containing 100 ml of medium. Exponentially growing Sf9 cells are pelleted (800*g*, 4 min, 21°) and resuspended in 10 ml fresh medium containing the recombinant baculovirus at MOI 5 in a sterile 50 ml tube. After 1 hr at room temperature, the cells are added to 90 ml medium in a 250 ml spinner bottle and stirred at 60 rpm, 27°. Ten milliliters of cells are removed at specific time points and a crude membrane preparation is made and analzyed by Western blotting and inhibitor binding (see below). The optimal MOI for infection is determined empirically by infecting 1×10^7 cells in a 75 cm^2 flask with MOI 2, 5, or 10 of second passage baculovirus. The cells are added to the flask in a total of 10 ml of medium and left to adhere for 15 min. The medium is removed and replaced with 10 ml medium

[26] L. J. Reed and H. Muench, *Amer. J. Hygiene* **27,** 493 (1938).
[27] P. J. Krell, *Cytotechnology* **20,** 125 (1996).

containing the desired amount of baculovirus, and left for 1 hr at room temperature. The virus-containing medium is then removed and replaced with 15 ml of fresh medium. The flasks are incubated at 27° for 2 days. The cells are removed from the surface of the flask by sharply banging the flask on a flat surface twice and are then pelleted by centrifugation at 1000*g* for 5 min at 4°. A crude membrane preparation is made from the cell pellet and analyzed by an inhibitor binding assay (see below).

Expression of neurotransmitter transporters in MGI and Hi5 cells is performed in 75 cm^2 flasks as described above for optimum MOI determination.

Preparation of Crude Cell Membranes

Ten milliliters of baculovirus-infected cells are pelleted (1000*g*, 5 min, 4°) and resuspended in 1 ml of ice-cold 0.1 × PBB + protease inhibitors [5 m*M* HEPES, pH 7.4, 12 m*M* NaCl, 0.5 m*M* KCl, 200 μ*M* phenymethylsulfonyl fluoride (PMSF), 1 μ*M* leupeptin, 1 m*M* EDTA].[28] Cells are sheared by seven passages through a 26-gauge hypodermic needle and pelleted (full speed in a microcentrifuge, 5 min, 4°). The crude membranes are washed once with 0.1 × PBB, once with 1 × PBB, and finally resuspended in 1 ml 1 × PBB. Protease inhibitors are present throughout the preparation at the concentration given above. Protein contents are determined using the method of Schaffner and Weissmann (1973),[29] which is unaffected by the presence of up to 100 μg of lipid.[30]

Western Blot Analysis

Cell membranes are solubilized in SDS-sample buffer (20 m*M* Tris, pH 6.8, 1% SDS, 1 *M* glycerol, 1 mg/ml bromphenol blue) by incubation at 37° for 15 min, and the proteins are resolved on a 9% sodium dodecyl sulfate–polyacrylamide gel (SDS–PAGE). Proteins are transferred to a nitrocellulose membrane (Hybond ECL, Amersham) using a semidry blotter (120 mA, 45 min, for a 100 cm^2 gel) with a Tris–glycine transfer buffer (25 m*M* Tris, 190 m*M* glycine, 20% (v/v) methanol, pH 8.3). Blotted proteins are probed with antibodies to the rat SERT,[31] human NET,[32] and rat GAT.[33]

[28] W. A. Wolf and D. M. Kuhn, *J. Biol. Chem.* **267,** 20820 (1992).

[29] W. Schaffner and C. Weissman, *Anal. Biochem.* **56,** 502 (1973).

[30] C. G. Tate, unpublished results, 1995.

[31] Y. Qian, H. E. Melikian, D. B. Rye, A. I. Levey, and R. D. Blakely, *J. Neurosci.* **15,** 1261 (1995).

[32] H. E. Melikian, J. K. McDonald, H. Gu G, K. R. Moore, and R. D. Blakely, *J. Biol. Chem.* **269,** 12290 (1994).

[33] N. J. Mabjeesh and B. I. Kanner, *J. Biol. Chem.* **267,** 2563 (1992).

Blots are developed using the enhanced chemiluminescence system (ECL, Amersham).

Three immunoreactive proteins of different molecular weight are presented in cells expressing each neurotransmitter transporter. These represent the unglycosylated transporter, glycosylated transporter and an SDS-resistant aggregate, respectively (Fig. 1). Glycosylated forms are identified by the absence of bands after treatment of membranes with endoglycosidase H (removes high-mannose N-linked oligosaccharides from proteins) and when cell membranes are prepared from cells that are grown in the presence of 20 μg/ml tunicamycin (inhibits N-linked glycosylation). The aggregate is unlikely to represent a biologically important form of the transporter, because its formation can be induced by boiling membranes in SDS–sample buffer prior to gel electrophoresis. The unglycosylated transporters consistently run at a lower molecular weight on SDS polyacrylamide gels than would be predicted from their cDNA sequences; this is commonly observed for integral membrane proteins. One piece of experimental evidence supporting the assumption that the bands represent full-length transporter is that anti-Myc and anti-FLAG antibodies recognize the same protein when

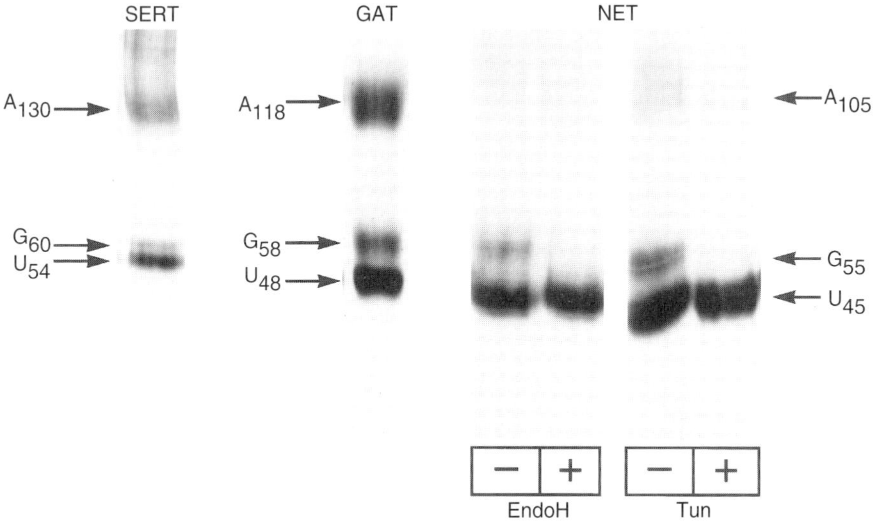

FIG. 1. Western blots of membranes containing SERT, NET, and GAT. Transporters were identified on blots by probing with transporter-specific antibodies. No endogenous Sf9 cell membrane proteins cross-reacted with the antisera (results not shown). Different forms of the transporters are indicated by arrows: U, unglycosylated; G, glycosylated; A, aggregate. Molecular masses are given as subscripts. The effect of treatment of membranes with endoglycosidase H (endoH) and the effect of growing cells in the presence of tunicamycin (Tun) are shown for NET.

the c-Myc and FLAG epitope tags are present at the N and C termini, respectively, of SERT.[4]

Uptake of Neurotransmitters

First, 5×10^7 exponentially growing Sf9 cells are pelleted (800*g*, 4 min, 21°) and resuspended in 5 ml fresh medium, recombinant baculovirus is then added to give an MOI of 5. The cells are left standing at room temperature for 1 hr and then made up to a final volume of 50 ml and placed in a 100 ml spinner bottle. The cells are grown at 27° for 24 hr and then harvested as above. The cells are resuspended in 0.25 *M* sucrose in KRH buffer (120 m*M* NaCl, 4.7 m*M* KCl, 2.2 m*M* $CaCl_2$, 1.2 m*M* $MgSO_4$, 1.2 m*M* KH_2PO_4, 10 m*M* HEPES pH 7.4, 10 m*M* glucose)[34] at room temperature and used immediately for uptakes. Trypan blue staining of the resuspended cells shows that greater than 90% of the cells are unstained, i.e., intact. Ten microliters of [^{3}H]serotonin binoxalate (DuPont NEN, Stevenage, UK) (10–100 n*M*, carrier free) is added to 700 μl of cells in a 1.5 ml microcentrifuge tube and the tube is immediately mixed by inversion. Aliquots (100 μl) are removed at specific time points (0.5–10 min) and rapidly filtered through nitrocellulose filters (0.65 μm DAWP, Millipore, Bedford, MA), which are immediately washed with 5 ml of ice-cold KRH/sucrose. Background counts are determined by using insect cells infected with the recombinant baculovirus bv-pVL, constructed from plasmid pVL1392. Filters are placed into scintillation vials with 5 ml scintillant and left overnight before counting.

Uptake of serotonin by SERT in Sf9 cells is clearly detected,[4] but norepinephrine transport by NET is consistently extremely poor. Neurotransmitter uptake is dependent on expression of the transporter in the plasma membrane of the insect cell, so these results suggest that NET is expressed primarily in intracellular membranes (a proportion of NET is correctly folded as indicated by mazindol binding; see below). The measurement of neurotransmitter uptake into baculovirus-infected cells has to be performed between 20 and 36 hr p.i. when some of the transporter is expressed and the cell is still intact. Uptakes performed on day 2 p.i. are unsuccessful, despite higher expression levels of SERT, because the cells are too leaky. Expression of SERT in Sf9 cells accelerates cell death (Fig. 2). SERT-induced cell death seems to be independent of SERT activity since the presence of the inhibitor RT155 (Research Triangle Institute, Research Triangle Park, NC) in the cell medium does not slow down the rate of cell death. Coexpression of the apoptosis inhibitor BCL2 also has

[34] R. D. Blakely, J. A. Clark, G. Rudnick, and S. G. Amara, *Anal. Biochem.* **194,** 302 (1991).

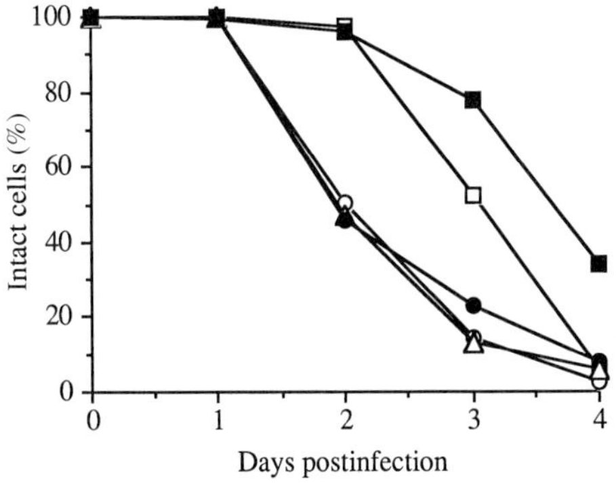

FIG. 2. Survival of Sf9 cells during the expression of SERT, Sf9 cells were infected with either bvSERT (○) or bv-pVL (control, □) and grown in a 100 ml suspension culture. Cells were removed at time points and the percentage of cells that were intact was determined by trypan blue exclusion. (●) Coinfection with bvSERT and bvBCL2. (△) Addition of RT155 to the cell culture containing bvSERT-infected cells. (■) Infection with bvBCL2 alone.

no effect, even though BCL2 slows the rate of death of Sf9 cells infected with wild-type baculovirus.[35]

Inhibitor Binding Assays

Insect cells are infected with recombinant baculovirus at an MOI of 5 as described for uptake studies. Cells are grown at 27° for 2 days and then harvested. A crude cell membrane preparation is prepared as above; typically 1 ml of membranes with a protein concentration of 250–500 μg/ml is prepared from 5 ml cells (1×10^6 cells/ml). Membranes are diluted 1 in 5 in PBB before use.

For the determination of the K_d and B_{max} of [^{125}I]RT155 binding by SERT, a binding assay is performed in triplicate at 13 different [^{125}I]RT155 concentrations (0.0–2.0 nM, carrier free). Each assay contains 5 μl membranes and [^{125}I]RT155 (DuPont NEN) in a final volume of 50 μl of PBB in a 1.5 ml microcentrifuge tube. To determine nonspecific counts, an identical set of reactions is set up containing 100 μM serotonin (Sigma, freshly made from solid). The tubes are incubated at room temperature for 1 hr and then centrifuged for 5 min in a microcentrifuge at full speed. The supernatant is carefully removed with an automatic pipetter and the pellet is resuspended in 0.1 M formic acid. The formic acid is removed and

[35] E. S. Alnemri, N. M. Robertson, T. F. Fernandes, C. M. Croce, and G. Litwack, *Proc. Natl. Acad. Sci. USA* **89,** 7295 (1992).

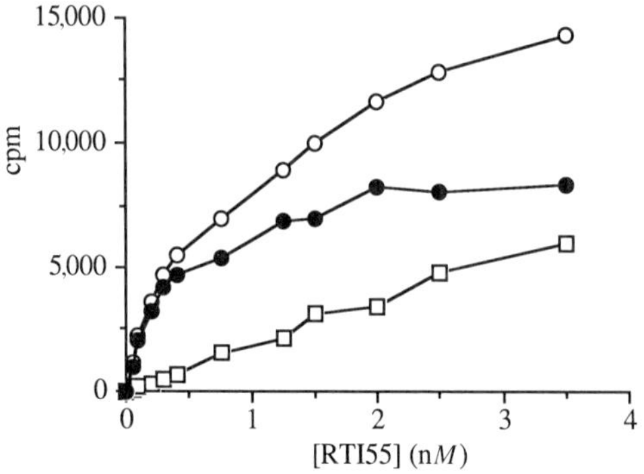

FIG. 3. RT155 binding assay for bvGST-SERT. The serotonin transporter was fused to the C terminus of glutathione *S*-transferase (GST) to allow the purification of the fusion protein on immobilized glutathione (C. G. Tate, unpublished work, 1995). The ability of the SERT portion of the fusion protein to fold correctly was tested with a [^{125}I]RT155 binding assay: (○) total bound counts (no serotonin); (□) nonspecifically bound counts (serotonin present); (●) specific counts (total counts minus nonspecific counts). The K_d for binding was 0.2 n*M*, which is identical to that measured for native SERT, suggesting that GST–SERT is folded in a native conformation.

counted in a γ-counter (Fig. 3). The binding data is analyzed by the computer program EBDA and LIGAND.[36,37]

If an approximate determination of expression levels is required in a routine preparation of the serotonin transporter, a single concentration (2 n*M*, i.e., 10 times the K_d) is used along with the serotonin-containing control; if specific counts are below 10% of the total counts added, it is assumed that the number of molecules of [^{125}I]RT155 per cell equals the number of molecules of the serotonin transporter. Ligand binding assays for the norepinephrine transporter are performed essentially as described for SERT, using [^{3}H]mazindol (DuPont NEN) in a final volume of 1 ml, and using 1 m*M* nortryptiline (Sigma) to determine the specificity of mazindol binding.

Inhibitor binding assays are performed to ascertain the levels of expression of active SERT and NET in baculovirus-infected insect cells. The assumption is made that if the transporter can bind the inhibitor, and this binding is prevented by the inclusion of substrate or another inhibitor, then it is likely that the transporter is correctly folded and would be capable of performing transport if it is present at the cell surface. This is particularly important for NET, because the uptake of norepinephrine by infected

[36] P. J. Munson and D. Rodbard, *Anal. Biochem.* **107,** 220 (1980).
[37] G. A. McPherson, *J. Pharmacol. Methods* **14,** 213 (1985).

cells is exceedingly low. The assays for NET are complicated by the presence of an endogenous Sf9 cell membrane protein that binds [^{3}H]mazindol in a nortryptiline-protectable manner. This protein is present at about 2 pmol/mg membrane protein and decreases to about 1.3 pmol/mg during virus infection. No endogenous membrane proteins in Sf9 cells bind [^{125}I]RT155 (up to 3.5 nM) in a serotonin-protectable manner. [^{125}I]RT155 is used in preference to [^{3}H]paroxetine, because the latter shows very high nonspecific binding to membranes. Expression levels in Sf9 cells of SERT and NET are about 500,000 copies per cell (9 pmol/mg membrane protein) and 300,000 copies per cell (4 pmol/mg membrane protein), respectively. Expression levels (as measured by inhibitor binding) of both SERT and NET are lower in High Five and MG1 cells, although more immunoreactive protein is detected on Western blots. Sf21 cells are larger than Sf9 cells and consequently express more copies per cell of SERT and NET, although this represents slightly lower levels when calculated in terms of picomoles of functional transporter per milligram membrane protein.

Concluding Remarks

The baculovirus expression system has successfully been used to express SERT and NET in an active form and in sufficient quantities for purification. It is likely that poor glycosylation of the neurotransmitter transporters in Sf9 cells during baculovirus expression contributes significantly to the relatively low levels of active transporter present in relation to the total amount of transporter protein expressed. Future modifications of the baculovirus genome and the selection of improved cell lines may enable complete glycosylation of expressed neurotransmitter transporters and hence give higher expression levels.

Acknowledgments

I am indebted to R. Blakely for providing cDNAs encoding hNET and rSERT and antibodies to both of the transporters, to N. Nelson for providing the cDNA to hGAT, to B. Kanner for providing the anti-GAT antibody, and to F. Carroll for providing RT155. I am grateful to J. Ellis, J. Tucker, and L. Tierney for helpful comments during the preparation of this manuscript. C.G.T. is the recipient of an MRC Career Development Award.

[31] Dopamine Transporter Mutants, Small Molecules, and Approaches to Cocaine Antagonist/Dopamine Transporter Disinhibitor Development

By GEORGE UHL, ZHICHENG LIN, THOMAS METZGER, and DALIT E. DAR

Introduction

Dopamine transporters (DAT) terminate dopaminergic neurotransmission by Na^+- and Cl^--dependent reaccumulation of dopamine into presynaptic neurons.[1,2] They thus play important roles in the spatial and temporal buffering of released dopamine, and key roles in its recycling.

A number of substantial lines of evidence have suggested that the actions of cocaine in inducing its rewarding and reinforcing properties are largely due to its inhibition of the DAT, especially the transporter expressed on terminals of mesolimbic–mesocortical dopaminergic neurons whose cell bodies lie in the ventral tegmental area (VTA) of the basal midbrain.[3,4] Striking reduction in psychostimulant reward after lesions of these VTA neurons, the ability of psychostimulants to robustly enhance dopamine spillover from their terminal areas in nucleus accumbens, the good correlations between the relative potencies of cocaine analogs in tests of behavioral reward and their potencies at the dopamine transporter, and results in transgenic animals in which dopamine transporter is overexpressed in catecholaminergic neurons or deleted have each been interpreted as fitting with the idea that the role of cocaine in reward could largely depend on its action at the dopamine transporter. These transporters are thus currently identified as the principal brain cocaine receptors related to drug abuse.[3–6]

Opportunities for elucidation of the molecular details of binding of cocaine and dopamine to the DAT were spurred considerably by cloning of dopamine transporter cDNAs from several species.[7–11] Successful cloning

[1] L. L. Iversen, "Handbook of Psychopharmacology" (L. L. Iversen, S. J. Iversen, and S. H. Snyder, Eds.), Vol. 381. Plenum, New York (1976).

[2] A. S. Horn, *Prog. Neurobiol.* **34,** 397 (1990).

[3] M. Ritz, R. Lamb, S. Goldberg, and M. Kuhar, *Science* **237,** 1219 (1987).

[4] J. Bergman, B. Madras, S. Johnson, and R. Spealman, *J. Pharmacol. Exp. Ther.* **251,** 150 (1989).

[5] B. Giros, M. Jaber, S. Jones, R. Wightman, and M. Caron, *Nature* **379,** 606 (1996).

[6] D. Roberts and G. Koob, *Pharmacol. Biochem. Behav.* **17,** 901 (1982).

[7] S. Shimada, S. Kitayama, C.-L. Lin, A. Patel, E. Nanthakumar, P. Gregor, M. Kuhar, and G. R. Uhl, *Science* **254,** 576 (1991).

[8] J. Kilty, D. Lorang, and S. G. Amara, *Science* **254,** 578 (1991).

METHODS IN ENZYMOLOGY, VOL. 296
0076-6879/98 $25.00

of DAT cDNA by our research group used expression systems developed for this purpose, including assays monitoring whether radiolabeled dopamine could be taken up in *Xenopus* oocyte and other cell types expressing DAT mRNA and/or cDNAs. These efforts also relied on cDNA homologies with other members of the sodium-dependent neurotransmitter transporter, a family recently cloned at the time.[12–16] The rat and human DAT cDNAs encode proteins of 619–620 amino acids with 12 putative transmembrane hydrophobic segments, sites for N-linked glycosylation and for possible intramolecular disulfide bond formation in the putative extracellular domains, and sites for phosphorylation in putative intracellular domains.[15–21] Molecular models for this transporter have been developed, using analogies with other protein families and data from mutagenesis studies.[22,23] However, no exact assignments for the topology of any of the amino acids of the DAT can be made unequivocally in the absence of any crystallographic or high resolution nuclear magnetic resonance (NMR) structural data from it, or from any member of the gene family in which it resides. Topological arguments made in this chapter, and in all current discussions of these proteins, must thus be couched in tentative terms.[23]

Despite these limitations, cloning and expression of the DAT cDNAs from several species in several cellular expression systems did clearly indicate that the single gene product could confer virtually all of the properties of the wild-type transporter onto several expressing cell systems, including avid sodium- and chloride-dependent uptake of dopamine, and high-affinity

[9] T. Usdin, E. Mezey, C. Chen, M. Brownstein, and B. Hoffman, *Proc. Natl. Acad. Sci. USA* **88,** 11168 (1991).

[10] B. Giros, S. el Mestikawy, L. Bertrand, and M. Caron, *FEBS Lett.* **295,** 149 (1992).

[11] D. Vandenbergh, A. Persico, and G. R. Uhl, *Molecular Brain Res.* **15,** 161 (1992).

[12] G. R. Uhl, B. O'Hara, S. Shimada, R. Zaczek, J. DiGiorgianni, and T. Nishimor, *Molecular Brain Res.* **9,** 23 (1991).

[13] T. Pacholczyk, R. D. Blakely, and S. G. Amara, *Nature* **350,** 350 (1991).

[14] J. Guastella, N. Nelson, N. Nelson, L. Czyzyk, S. Kenyan, M. C. Midel, N. Davidson, and H. Lester, *Science* **249,** 1303 (1990).

[15] G. R. Uhl, *TINS* **15,** 265 (1992).

[16] G. R. Uhl and P. R. Hartig, *TIPS* **13,** 421 (1992).

[17] R. Lew, D. Grigoriadis, A. Wilson, J. Boja, R. Simantov, and M. Kuhar, *Brain Res.* **539,** 239 (1991).

[18] J.-B. Wang, A. Moriwaki, and G. R. Uhl, *J. Neurochem.* **64,** 1416 (1995).

[19] J.-B. Wang, A. Patel, M. Kuhar, and G. R. Uhl, *Molecular Brain Res.,* submitted (1997).

[20] R. Huff, R. Vaughan, M. Kuhar, and G. R. Uhl, *J. Neurochem.* **68,** 225 (1997).

[21] R. Vaughan, R. Huff, G. R. Uhl, and M. J. Kuhar, *J. Biol. Chem.,* submitted (1997).

[22] O. Edvardsen and S. G. Dahl, *Molecular Brain Res.* **27,** 265 (1994).

[23] C. Suratt, J.-B. Wang, S. Yuhasz, M. Amzel, H. Kwon, J. Handler, and G. R. Uhl, *Curr. Opin. Nephrol. Hypertens.* **2,** 744 (1993).

cocaine analog binding with appropriate pharmacological profiles, including high- and lower-affinity cocaine analog binding sites.[7–11,24] These data, along with data confirming DAT mRNA and protein expression at high levels in midbrain and other dopaminergic neurons, suggested that the dopamine transport and cocaine recognition features ascribed to the pharmacologically defined dopamine transporter were in fact likely to reside in a single expressed gene product.[25,26]

This information had implications for anticocaine medications development, a major goal for drug abuse therapeutics because of the current virtual absence of clearly effective anticocaine pharmacotherapeutics. One major possible strategy for development of such a pharmacotherapeutic approach was to assess the feasibility of identifying a cocaine blocker that would still allow the transporter to recognize and transport dopamine. No clear-cut terminology to unambiguously describe such a pharmacologic property currently exists. Cocaine itself, in a sense, functions as an "antagonist" at the dopamine transporter, blocking uptake with generally competitive kinetic properties. Despite this, several investigators have used the term "cocaine antagonist" to describe the properties of such a blocker of dopamine transporter sites important for cocaine recognition but less essential for dopamine transport. Alternatively, compounds that can act by such a mechanism not previously well described in widely recognized classic pharmacology could be termed "disinhibitors" to describe antagonists of transport blockers that are not themselves transport blockers. A complete "antagonist" or "disinhibitor" would allow unimpeded movement of dopamine in the joint presence of dopamine, cocaine, and the anticocaine medication, and thus represent a strong candidate treatment for cocaine intoxication. A partial "antagonist" would possibly exert mild dopaminergic blockade at concentrations that could block cocaine. Analogies from work in antiopiate addiction pharmacotherapeutics suggested that some mild rewarding properties might be beneficial for recruitment and retention of drug-dependent individuals to treatment settings. However, available literature provided scant precedent for compounds that could block any inhibitor and spare neurotransmitter transport function by DAT or any member of the sodium- and chloride-dependent neurotransmitter transporter family to which it belongs.

How could one find such compounds? We reasoned that this process would be facilitated by identification of transporter regions necessary for

[24] J. W. Boja, L. Markham, A. Patel, G. R. Uhl, and M. J. Kuhar, *NeuroReport* **3,** 247 (1992).

[25] S. Shimada, S. Kitayama, D. Walther, and G. R. Uhl, *Molecular Brain Res.* **13,** 359 (1992).

[26] C. Freed, R. Revay, R. A. Vaughan, E. Kriek, S. Grant, G. R. Uhl, and M. J. Kuhar, *J. Compar. Neurol.* **359,** 340 (1995).

cocaine recognition but not for dopamine transport (and vice versa). Studies detailed elsewhere in this volume indicate that construction of chimeric transporter molecules employing parts of dopamine, norepinephrine, and serotonin transporters provide substantial insights into sequences important for recognition of many antidepressant molecules. This strategy, however, has provided fewer definitive data concerning sequences selectively important for cocaine recognition. Many such constructions, indeed, fail to express properly. An alternative approach to this problem that we have employed utilized much smaller molecular alterations, made through site-directed mutagenesis techniques, to establish transporter structure–function relationships. These studies have especially sought structural alterations that could differentially affect dopamine and cocaine analog recognition.

In a second and parallel approach, we also used data from transporter mutagenesis results and transporter models to seek selective compounds that could provide cocaine antagonism with less potency in dopamine transport blockade. We hoped that data from these two lines of approach would reinforce each other to provide increasingly rich ideas about the possibility of identifying a dopamine-sparing cocaine antagonist, the nature of compounds that might be able to exert such selective actions, the sites on the transporter that might recognize such selective compounds, and even the nature of lead small-molecule structures that could serve such functions. Below, we detail the current status of this combined approach.

Structure–Function Studies: Dopamine Transporter

Studies of the structure–function relationships of cocaine analogs available in the early 1990s revealed models in which several polar and at least one hydrophobic interaction were involved in cocaine's recognition by a putative DAT recognition site.[27] Similarly, studies of catecholamine recognition by their G-linked, seven-transmembrane (TM7) domain receptors revealed that polar interactions with amine and catechol hydroxyl residues were reasonable candidates to play major roles in recognition.[28] Our initial studies modifying the putative sites for N-linked glycosylation and sites possibly implicated in disulfide bond formation displayed substantial effects on levels of functional transporter expression, but little evidence for selectivity of effect on cocaine recognition when compared to dopamine transport.[1–19] Similarly, evidence for modification of transport function by phosphorylation, especially when stimulated by enhancers of protein kinase C

[27] F. I. Carroll, Y. Gao, A. Rahman, P. Abraham, K. Parham, A. Lewin, J. Boja, and M. Kuhar, *J. Med. Chem.* **34,** 883 (1991).

[28] C. D. Strader, I. Siegel, and R. Dixon, *FASEB J.* **3,** 1825 (1989).

activity, showed selective impact on transport, rather than on cocaine analog recognition.[20,21] Accordingly, our group began a series of studies evaluating the results of alanine substitutions that removed amino acid side chains from interesting charged, polar, and aromatic amino acids in the DAT.[29–32] Reasoning that at least some of the selective effects on uptake vs. transport were likely to take place through actions in transmembrane domains, and finding that amino acids in these domains were among the most conserved in the monoamine transporter subfamily, each of whose members recognized both a monoamine and cocaine, we have focused most attention on residues modeled as lying in transmembrane domains. The results of these mutagenesis studies fall into several groups. A number of amino acids side chains could be eliminated without substantial effect on either cocaine analog recognition or dopamine transport.[29–32] Changes in many other amino acids reduced both dopamine transport and cocaine analog binding, in some cases to such low levels that they effectively eliminated the functional expression of the transporter.[29–32] As previously reported, mutations in TM7 serines resulted in reduced dopamine transport with preservation of cocaine analog binding.[29]

Mutations in the DAT transmembrane domain 4 displayed reduced cocaine analog recognition with preserved dopamine transport.[23] These features were noted in initial mutants that changed multiple polar amino acids to alanine, and in single mutants changing the tyrosine located toward the middle of this putative transmembrane sequence, Y251, to alanine or to phenylalanine.[31] These data supported a stronger and more essential role for this mixed polar/aromatic residue in DAT recognition of cocaine analogs than in any DAT feature necessary for dopmaine transport. Several other amino acids whose replacement by alanine led to a more that threefold greater loss of affinity for the cocaine analog carboxyfluorotropane (CFT) than for dopamine are illustrated in Fig. 1.

Aromatic DAT transmembrane domain (TM) residues could also play roles in DAT recognition of catecholamine catechol groups, cocaine phenyl groups, and/or the halogenated phenyl group of CFT. Alternatively, they could also serve to orient DAT transmembrane regions, helping to direct the correct amino acids toward ligand- and substrate-recognizing pockets. To address possible roles played by TM aromatic residues in such interactions, we have substituted alanines for most of the phenylalanines that lie

[29] S. Kitayama, S. Shimada, H. Xu, L. Markham, D. Donovan, and G. R. Uhl, *Proc. Natl. Acad. Sci. USA* **89,** 7782 (1992).

[30] S, Kitayama, J.-B. Wang, and G. R. Uhl, *Synapse* **15,** 58 (1993).

[31] J.-B. Wang, S. Davis, and G. R. Uhl, in preparation.

[32] Z.-C. Lin and G. R. Uhl, in preparation.

in postulated transmembrane domains.[32] Transmembrane domains 1 and 2 phenylalanine substitutions reduced dopamine uptake and [^{3}H]CFT binding.[32] In transiently expressing COS cells, these receptor mutants as well as mutants in other TM domains yielded the perinuclear patterns of DAT immunostaining that appear to indicate abnormal cellular trafficking of these mutant DATs. Alanine substitutions for phenylalanines found in transmembrane domains 4, 7, 8, and 11, however, produced nearly normal patterns of cellular immunostaining. Conceivably, these studies could be elucidating essential aromatic amino acid contributions to the correct DAT folding necessary for plasma membrane targeting. The data may also identify residues essential for cocaine recognition and/or dopamine uptake (Fig. 1). Indeed, several of these phenyl amino acid side chains may contribute more to cocaine analog recognition than to dopamine affinity. These are distributed through several of the domains of the primary structure of the transporter in a fashion that suggests the possibility that several recognition subsites that might selectively interact with cocaine might even be recognized by several distinct classes of small-molecule compounds. Alternatively, this scattering of sites could also be pointing out the limitations of our current understanding of the true topology of the assembled dopamine transporter in its membrane environment.

Structure–Function Studies: Small Molecules

These studies of the dopamine transporter mutants have been complemented by pharmacological studies that examined drugs in several classes. These drugs appeared to manifest charge and hydrophobicity distributions that might parallel those of cocaine, and/or those for which some dopamine transporter activities had been suggested.[33] We have screened compounds selected from several chemical classes for selective activities in cocaine analog recognition blockade. Tested drugs from several different chemical classes displayed effects on DAT that fell into three categories.[33] Compounds including cocaine, dopamine, and tyramine showed virtually identical potencies in inhibiting dopamine uptake and CFT binding. Compounds in a second group, such as cloperastine and diphenhydramine, were substantially more potent in inhibiting dopamine uptake then in blocking CFT binding. Finally, compounds including diphenhydramine, tetracaine, procaine, amelioride, and tolperisone showed greater potencies in inhibiting cocaine analog binding than in reducing dopamine uptake. Interestingly, the anticholinergic compound trihexyphenidyl (see Fig. 2) was able to

[33] D. Dar, S. Kitayama, L. Miner, F. I. Carroll, and G. R. Uhl, in preparation.

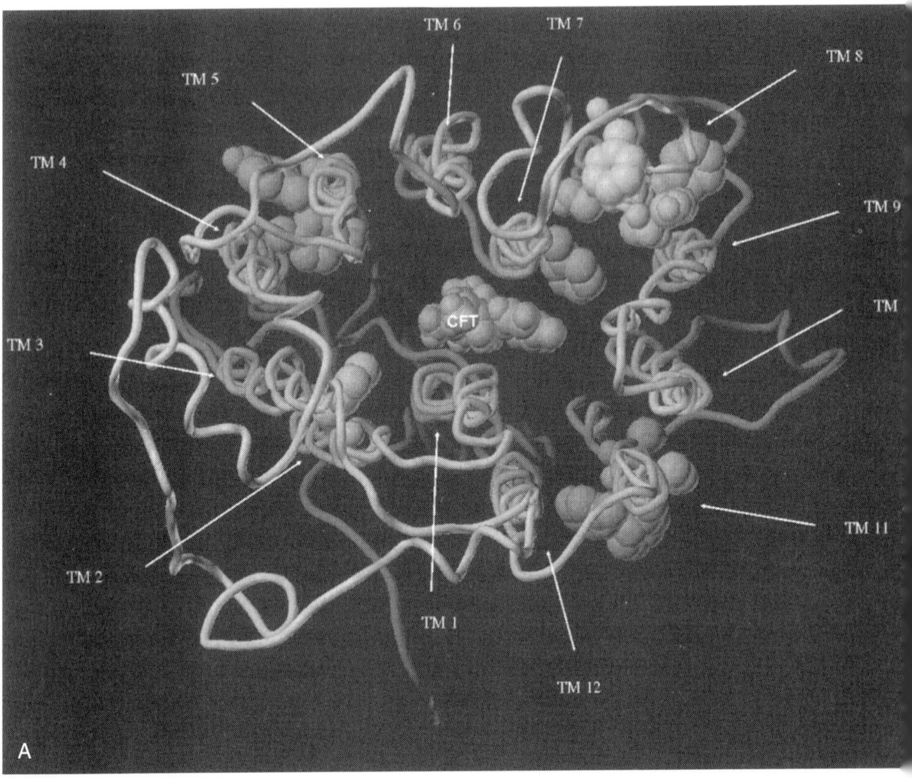

FIG. 1. Molecular models of the dopamine transporter showing helical transmembrane domains [according to O. Edvardsen and S. G. Dahl, *Molecular Brain Res.* **27,** 265 (1994).] Space-filling side chains indicate amino acids which, when substituted with alanine, result in selective loss of affinity for [^{3}H]carboxyfluorotropane (CFT) (affinity reduction for CFT > three-fold greater than affinity reduction for dopamine; see text for references). (A) View from "extracellular" side of membrane, with transmembrane domains indicated, and CFT included for size comparison only. (B) Side view, with extracellular domains on top, approximate borders of transmembrane domains indicated by white lines, and intracellular domains located at the bottom.

substantially inhibit cocaine analog recognition at concentrations virtually ineffective in blocking dopamine uptake.[33]

None of the agents that manifested selective effects on cocaine analog recognition is likely to act only on this site; the potency of trihexyphenidyl in blocking muscarinic cholinergic receptors is substantially greater than its activity in selectively inhibiting DAT recognition of cocaine analogs.[33] However, use of the compounds identified here as lead structures provided

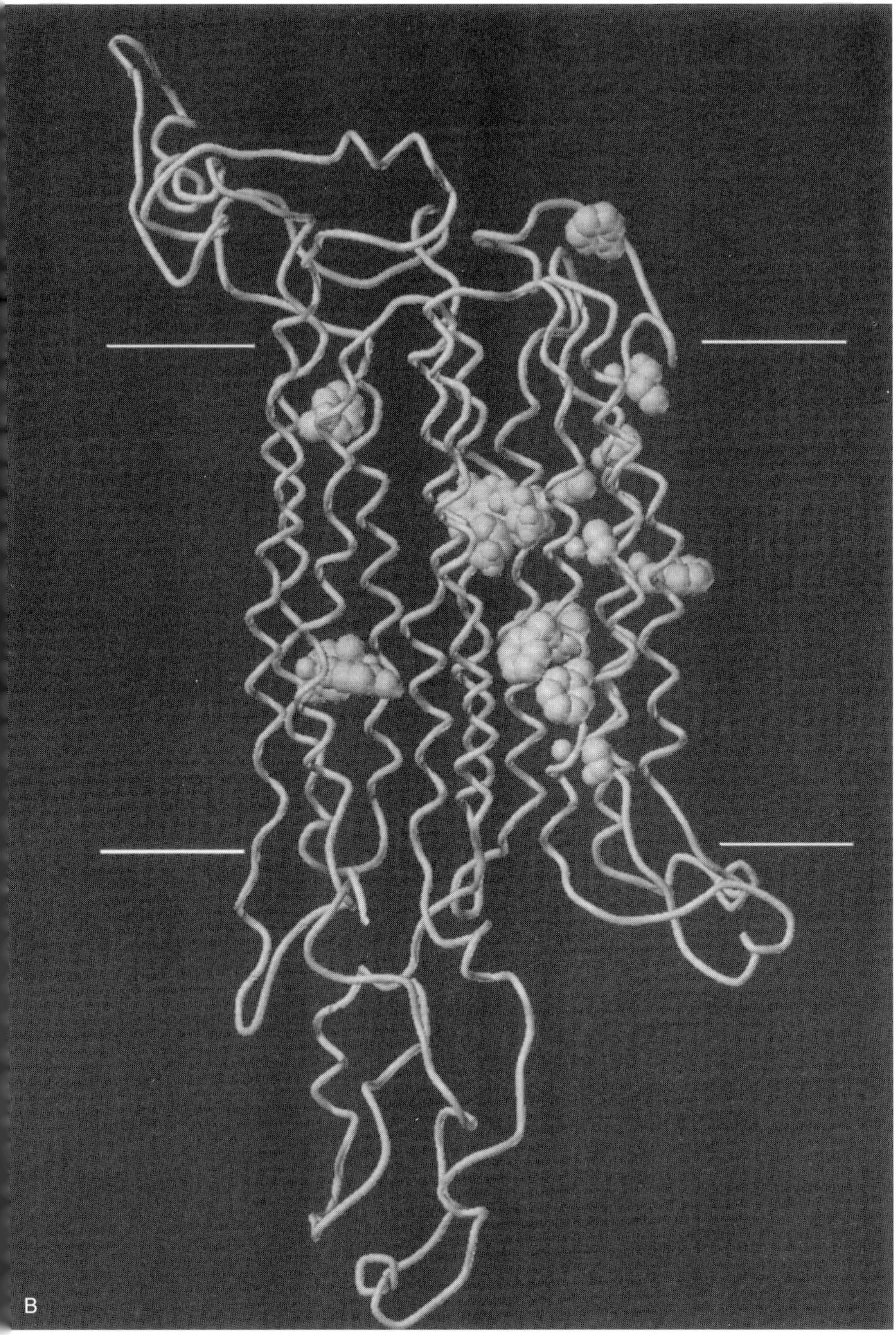

FIG. 1. (*continued*)

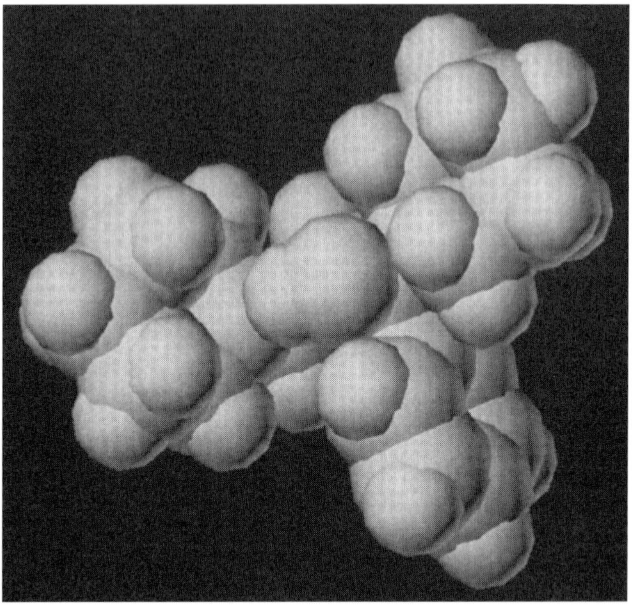

FIG. 2. Space-filling model for trihexyphenidyl.

the first step toward development of potent and selective agents that would allow testing in intact animals of the extent to which a dopamine transport-sparing cocaine antagonist disinhibitor could exert effects that might even prove clinically useful.

In order to test these ideas, we have worked to develop small-molecule variants of the trihexyphenidyl lead that might display sufficiently selective actions at dopamine transporters and could be used for *in vivo* testing. Identification of a compound that is many thousandfold less potent than trihexyphenidyl at muscarinic cholinergic receptors, but that retains the selective ability to antagonize cocaine analog recognition by the DAT *in vitro,* encouraged us to seek *in vivo* cocaine antagonism by the drug, coded as RTI 026. Interestingly, the compound has been able to block cocaine-induced locomotor activity increases in mice at doses that do not, in themselves, augment location.[33] These data support the feasibility of the idea that a cocaine disinhibitor or antagonist that was efficacious *in vitro* might display desirable properties *in vivo.* Conceivably, further derivatives of these or other leads might prove clinically useful in combined behavioral and pharmacologic strategies for combating cocaine abuse.

Conclusions

Two convergent lines of evidence now demonstrate the feasibility of a major promising path for anticocaine pharmacotherapeutic development. A "disinhibitor" or partial antagonist at the DAT can block transporter recognition of the competitive transport inhibitor cocaine and even possibly allow unimpeded movement of the substrate dopamine in the joint presence of dopamine, cocaine, and the disinhibitor. Mutagenesis studies in each DAT transmembrane domain identify mutants in polar and aromatic amino acids which spare dopamine transport but dramatically reduce cocaine analog affinity. Testing compounds on the expressed DAT reveals that triphenylhexidyl and several analogs can block most cocaine analog binding at concentrations ineffective in inhibiting dopamine uptake. These experiments define the feasibility of a strategy that can allow DAT function in the presence of cocaine, identify transporter sites where novel drugs might act to display such activities, and even provide a lead compound with these properties.

These pharmacologic and mutagenesis results, taken together, provide strong evidence that, even though the same molecule is the target of both dopamine and cocaine, strategies for anticocaine drug development that exploit selectivity of the regional recognition of compounds by these large, multidomain proteins appear possible. These sorts of approaches could even be applied to other transporters, to selectivity for neurotoxin uptake, to ion selectivity, and to other molecular functional features that could be of specific therapeutic importance in the multiple neuronal systems, if they depend on these transporters for spatial and temporal control of extracellular concentrations of a number of neurotransmitters of several different chemical families.

Acknowledgments

We gratefully acknowledge support for this work from the Intramural Research Program, National Institutes of Drug Abuse, NIH.

[32] *In Vivo* Generation of Chimeras

By KARI JOHNSON BUCK and SUSAN G. AMARA

Introduction

Chimeras between related proteins, in which similar sequence domains and distinct functional properties are exchanged, can be used to assign important pharmacological and mechanistic properties of the parental proteins to specific domains within their primary structures. Unlike other conventional methods of mapping functional domains, such as analyses of site-directed or deletion mutants in which the function of interest is frequently impaired or destroyed, chimeras between homologous proteins can provide an assayable phenotype that allows positive inferences to be drawn from particular functional properties associated with distinct protein domains. Here, an efficient *in vivo* method for generating chimeras between homologous parental cDNAs is described. We have used this method to generate a series of chimeras between two carrier proteins with extensive functional and sequence similarities, the human norepinephrine transporter (NET) and the rat dopamine transporter (DAT).[1,2] These transporters terminate catecholamine-mediated neurotransmission at synapses by high-affinity reuptake into presynaptic terminals and are the initial sites of action for a variety of therapeutic antidepressants and drugs of abuse. NET and DAT cDNAs predict highly similar protein sequences of 617 and 618 amino acids (64% identity),[3,4] and hydropathy analyses indicate 12 hydrophobic regions proposed to represent transmembrane (TM) domains.[5] Chimeras were generated *in vivo* in bacteria transformed using linearized plasmids that contained a single copy of the two parental cDNAs (DAT and NET) cloned in tandem in a tail-to-head configuration. Generation of chimeras within the host bacterium is believed to involve partial exonuclease digestion of the linear plasmid and base pairing between exposed ends of complementary DNA within regions of high sequence homology, followed by repair and ligation to recircularize the chimeric plasmids. Chimeras were identified based on cDNA size, diagnostic restriction digests, and nucleotide sequencing to determine the precise location of each chimera junction and confirm

[1] K. J. Buck and S. G. Amara, *Proc. Natl. Acad. Sci. USA* **91,** 12584 (1994).
[2] K. J. Buck and S. G. Amara, *Mol. Pharmacol.* **48,** 1030 (1996).
[3] T. Pacholczyk, R. D. Blakely, and S. G. Amara, *Nature* (*London*) **350,** 350 (1991).
[4] J. E. Kilty, D. Lorang, and S. G. Amara, *Science* (*Washington, DC*) **254,** 578 (1991).
[5] S. G. Amara and M. J. Kuhar, *Annu. Rev. Neurosci.* **16,** 73 (1993).

0076-6879/98 $25.00

that the junction was in-frame. The expression plasmids contained a T7 RNA polymerase promoter and the vaccinia virus T7-encoded RNA polymerase, so that chimeric proteins could be directly examined for expression in mammalian cells. Using this approach, we generated 59 chimeric cDNAs with junctions positioned throughout much of the transporter molecule in regions of sequence homology between DAT and NET. Functional screening demonstrated that most junctions within conserved regions of DAT and NET were not disruptive of transporter expression or function.

Materials and Methods

Clones and Plasmids

The cDNAs of the human norepinephrine transporter (NET) and rat dopamine transporter (DAT) are previously cloned into pBluescript SKII$^-$ (pBSK; Stratagene, La Jolla, CA).[3,4] Chimeric transporters are generated using an *in vivo* method developed by R. Reed (personal communication, 1993) which generates chimeras that junction in regions of sequence homology (Fig. 1). The T7DAT-NET tandem plasmid used to generate the DN chimera cDNAs (i.e., DAT at the 5′ end, and NET at the 3′ end) is engineered by cloning the 1.9 kb NET cDNA into the *Not*I site of DAT/pBSK, in order to leave at least two unique sites within the pBluescript SKII$^-$ polylinker between the DAT and NET cDNA inserts; the tandem plasmid (2 μg) is then linearized using two unique restriction sites between

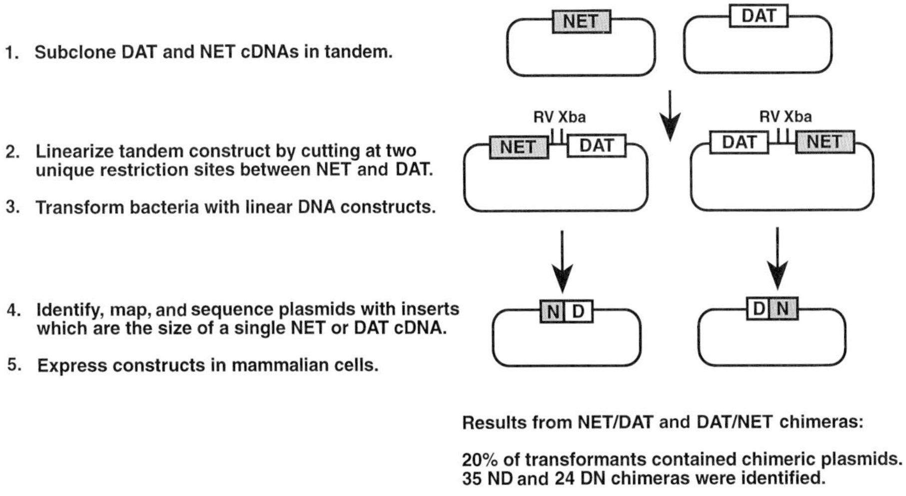

FIG. 1. Method for generating chimeric transporters.

the DAT and NET cDNA sequences, *Eco*RI and *Xba*I (1 U each enzyme per μg plasmid, 37°C, 4 hr). The complementary T7NET-DAT tandem plasmid used to generate the ND chimera cDNAs (i.e., NET at the 5′ end, and DAT at the 3′ end) is constructed by cloning the 1.9 kb DAT cDNA into the *Xba*I site of NET/pBSK. The T7NET-DAT tandem plasmids are then linearized using two unique restriction sites between the NET and DAT cDNA inserts, *Eco*RV and *Xba*I. Linearized tandem constructs are separated from any remaining supercoiled plasmid on a 1% agarose gel and purified using the Qiagen Plasmid Maxi Kit (Qiagen Inc., Chatsworth, CA).

Bacterial Transformation using Linearized Tandem Plasmids

Approximately 50–100 ng of purified linearized T7DAT-NET plasmid is transformed by heat shock (42°, 60 sec) into 10 μl of high efficiency competent DH5α bacteria (Life Technologies, Gaithersburg, MD), which ultimately generates the DAT-NET (or DN) chimera series. Similarly, competent DH5α bacteria are transformed using 50–100 ng of purified linearized T7NET-DAT plasmid, which ultimately generates the NET-DAT (or ND) chimera series. Repeated transformations using the linearized T7NET-DAT or T7DAT-NET constructs resulted in 298 carbenicillin (50 μg/ml)-resistant colonies. Each carbenicillin-resistant colony is streaked on another LB plate containing 50 μg/ml carbenicillin, and its DNA immediately subjected to alkaline lysis and electrophoresis on 0.8% agarose gels.[6] Colonies are identified as representing plasmids with tandem sized inserts (i.e., recircularized parental constructs), plasmids with monomer sized inserts (i.e., possible chimeras), or plasmids with inserts of intermediate size (e.g., due to inappropriate rearrangement).

Analysis of Monomer-Sized cDNAs

Bacteria carrying inserts representing possible chimeras are grown overnight in 4 ml of LB broth containing 50 μg/ml carbenicillin and their plasmid DNA purified using the Magic Minipreps DNA Purification System (Promega, Madison, WI). Each plasmid minipreparation is then subjected to diagnostic digests to ascertain insert size and chimeric status. Bacteria carrying recombinant monomer-sized inserts (i.e., chimeras) are grown overnight in 250 ml of LB broth containing 50 μg/ml carbenicillin, and their plasmid DNA is purified using the Plasmid Maxi Kit (Qiagen, Chatsworth, CA). Each purified plasmid is then subjected to additional diagnostic digests to ascertain the approximate location of the chimera junction. For

[6] D. E. Titus, "Promega Protocols and Applications Guide," 2nd ed. Promega Corp., Madison, Wisconsin, 1991.

example, DAT contains one *Cla*I site 1.0 kb from the 5′ end of the insert and one *Pst*I site at 1.8 kb, whereas NET contains one *Hinc*II site at 1.2 kb. A chimeric DAT-NET cDNA with a *Cla*I site at 1.0 kb and a *Hinc*II site at base 1.2 kb, but no *Pst*I site at 1.8 kb, would have its junction somewhere between bases 1.0 and 1.2 kb. Dideoxynucleotide sequencing (Sequenase Version 2.0, United States Biochemical Corporation, Cleveland, OH) of this region determined the precise location of the chimera junction and confirmed that the junction was in-frame.

Transport and Binding Assays: Analysis of Chimera Protein Expression

Expression in the plasma membrane is directly assessed through transport assays using mammalian cells transiently transfected with purified chimera plasmids. Wild-type and chimeric transporter cDNAs are expressed in HeLa cells using the vaccinia/T7 transient expression system as previously described.[1,7] The method uses a recombinant vaccinia virus strain (VTF_{7-3}) which expresses a T7 RNA polymerase.[8] In host cells, VTF_{7-3} replicates in the cytoplasm, resulting in rapid high-level expression of proteins encoded by plasmids bearing T7 promoters. Cells are plated at 2×10^5 per well in 24-well tissue culture plates and transfected the following day. Thirty minutes prior to transfection, cells are infected at 10 pfu (plaque-forming units) per cell in 100 μl of Dulbecco's modified Eagle's medium. Plasmids are added in liposome suspension (1 μg of DNA and 3 μg of Lipofectin; Life Technologies, Gaithersburg, MD) in a total volume of 350 μl per well. Sixteen hours after transfection, the virus/liposome suspension is removed by aspiration, and the cells are washed once with 1 ml of KRTH (120 m*M* NaCl, 4.7 m*M* KCl, 2.2 m*M* $CaCl_2$, 1.2 m*M* $MgSO_4$, 1.2 m*M* KH_2PO_4, 5 m*M* Tris, 10 m*M* HEPES, pH 7.4), and preincubated for 15 min at 37° in 500 μl of KRTH. Uptake is initiated by the addition of 10 n*M* [^{3}H]NE or [^{3}H]DA (DuPont NEN Research Products, Boston, MA) with or without various concentrations of unlabeled substrate (Research Biochemicals, Natick, MA) diluted in KRTH containing L-ascorbate (100 μM). Each substrate concentration is tested in quadruplicate, and well-to-well variability is typically <10% of the mean. Uptake is terminated after 20 min at 37° by washing twice with 1 ml of cold KRTH, cells are solubilized with 0.5 *M* NaOH, and the accumulated radioactivity is determined by scintillation spectrometry. Nonspecific transport is determined by assays of cells transfected with the pBluescript $SKII^-$/plasmid vector on the same plate and subtracted from the data. Kinetic parameters are determined by nonlinear least-squares fit of uptake–velocity profiles using

[7] R. D. Blakely, J. A. Clark, G. Rudnick, and S. G. Amara, *Anal Biochem.* **194,** 302 (1991).
[8] T. R. Fuerst, P. L. Earl, and B. Moss, *Mol. Cell. Biol.* **7,** 2538 (1987).

the INPLOT data analysis program. Functional screening has demonstrated that 46 encoded chimeric proteins of DAT and NET were expressed and inserted into the plasma membrane in a conformation that recognized and translocated a variety of substrates.

Thirteen chimeras showed attenuated transport activity relative to DAT or NET. All 13 chimeras junction in a region spanning TM5–TM8. Expression of these chimeric proteins is examined using whole cell binding of [^{125}I]RTI-55, an antagonist that recognizes DAT and NET, in HeLa cells transfected with wild-type or chimeric transporters using the vaccinia/T7 expression system as described above. Sixteen hours after transfection, the cells are gently washed twice with 1 ml of KRTH and preincubated in 450 μl KRTH for 20 min at 22°. Then, [^{125}I]RTI-55 (DuPont NEN Research Products, Boston, MA; final 0.5 n*M*) is added in a volume of 50 μl, and the plates are gently rocked for 60 min at 22° for equilibrium binding. The cells are washed twice with 1 ml ice-cold KRTH and solubilized in 0.5 *N* NaOH; bound radioligand is then measured using a gamma counter. Nonspecific binding is defined using cells on the same plate that were transfected with the pBluescript SKII$^-$ vector and is subtracted from total binding to obtain specific binding. Although transport in these chimeras is perturbed, specific [^{125}I]RTI-55 binding demonstrates that the encoded chimeric proteins are expressed, and that antagonist recognition is retained.

Results and Discussion

Generation and Analysis of Chimeric cDNAs

In order to establish a potentially general approach to study the superfamily of sodium and chloride ion dependent carriers, we generated and expressed a series of functional recombinant chimeric transporters, in which similar sequence domains and distinct functional properties of catecholamine transporters are exchanged.[1,2] Chimeric transporters between DAT and NET were generated and expressed in order to delineate the structural domains responsible for differential selectivity of catecholamine transporters for a variety of substrates and uptake inhibitors. Using a method for *in vivo* generation of chimeras between related proteins, we rapidly and efficiently generated a series of 59 chimeras between DAT and NET. Chimeric cDNAs were generated in bacteria transformed using linearized plasmids that contained a single copy of the two parental cDNAs cloned in tandem in a tail-to-head configuration. Transformation of DH5α bacteria with the linearized T7DAT-NET or T7NET-DAT construct resulted in 298 carbenicillin-resistant colonies. Most of the colonies (238/298) corresponded in size to the recircularized parental tandem plasmid, despite the use of constructs that were linearized using two restriction enzymes. Only one colony (1/298) represented a plasmid of intermediate size, indicative of a very low incidence of inappropriate rearrangement between NET and DAT. A higher incidence of plasmids too large (or too small) to represent chimeras formed at appropriate locations was observed following *in vivo* generation of chimeras between NET and the rat serotonin transporter,[9] or between two excitatory amino acid transporters,[10] perhaps because these transporters show much less sequence similarity relative to NET and DAT. Fifty-nine colonies represented plasmids with monomer-sized inserts (i.e., possible chimeras). Nucleotide sequence analyses demonstrated that all 59 plasmids represented in-frame chimeric cDNAs between DAT and NET, representing a chimera efficiency of about 20%. All of the chimeras junction at single sites, and no deletions or insertions of nucleotide sequence were observed. Chimera junctions were positioned throughout much of the transporter molecule in regions of sequence homology between DAT and NET. Alignments of sequence surrounding the chimeric junctions indicated DNA sequence identity of 50–90% flanking the junctions between DAT and NET sequences. The structures of 14 representative members of this series of chimeras, their apparent substrate affinity constants for uptake of norepinephrine and dopamine, and their selectivity for inhibition by desipramine or GBR 12935 are compared relative to wild-type DAT and NET in Fig. 2.

Functional Screening for Expression of Chimeric Transporter Protein Expression

Transport assays performed in parallel with the parental DAT and NET transporters identified 46 chimeras that showed robust transport activity. To provide an indication of the robust activity of these chimeras, the V_{max} of representative chimeras is compared to DAT activity observed in a parallel experiment (Table I, normalized V_{max}). All of these chimeric transporters junction in or before TM4 (e.g., DN1, DN2, DN3, ND1, ND3, ND4) or in or after TM9 (e.g., DN9, DN10, ND10, ND11). Additional transport assays showed that these chimeras demonstrate robust translocation of a variety of substrates (e.g., dopamine, norepinephrine; see Table I and Fig. 2) and show high affinity for uptake inhibitors (e.g., desipramine and GBR 12935; see Fig. 2). These analyses demonstrate that these chimeras are expressed in the plasma membrane in a conformation in which structural integrity of the transporter molecule is maintained.

[9] K. R. Moore and R. D. Blakely, *Biotechniques* **17,** 130 (1994).
[10] R. J. Vandenberg, J. L. Arriza, S. G. Amara, and M. P. Kavanaugh, *J. Biol. Chem.* **270,** 17668 (1995).

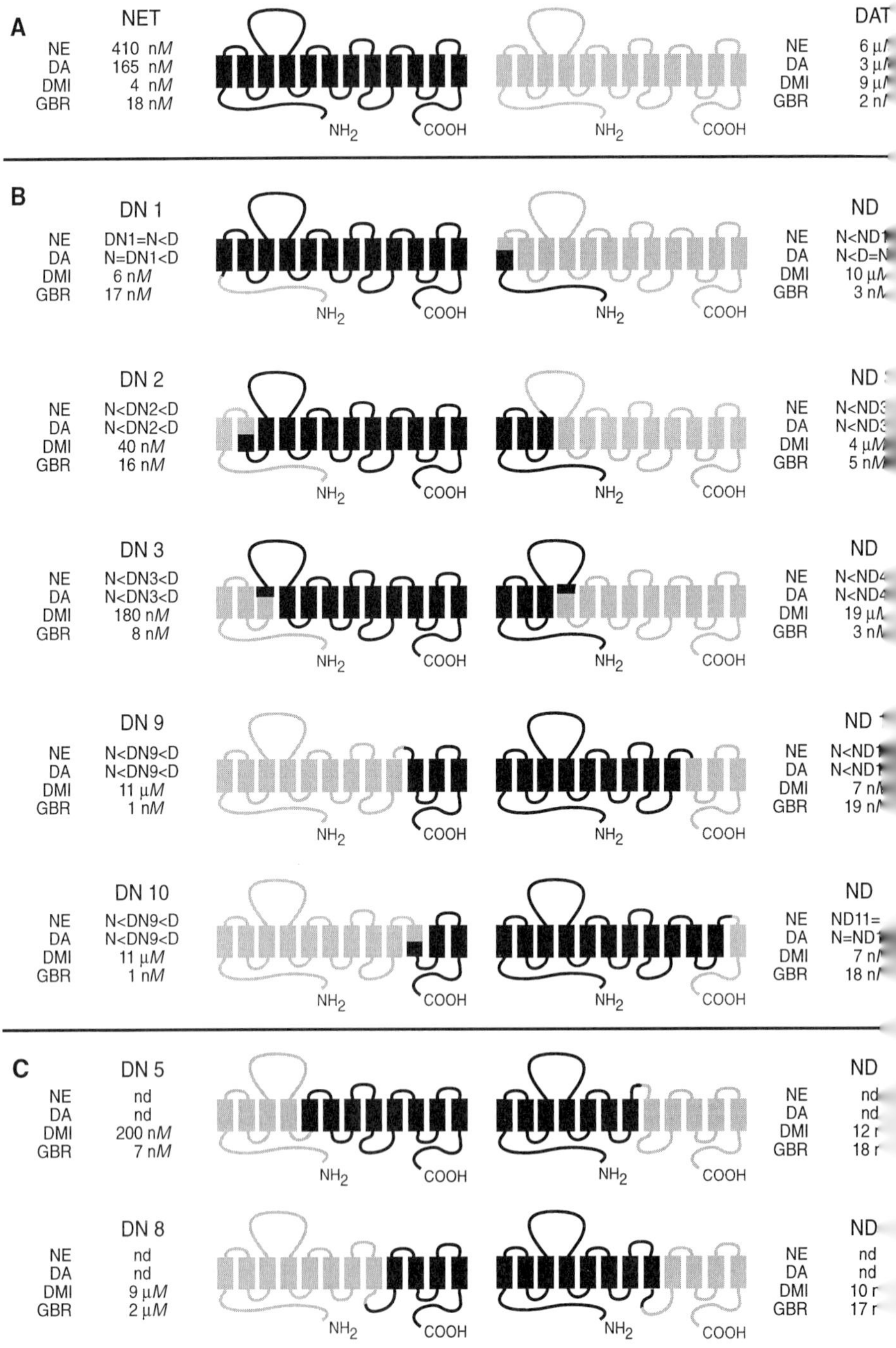
A
NET
NE 410 nM
DA 165 nM
DMI 4 nM
GBR 18 nM
NH2
COOH
DAT
B
DN 1
NE DN1=N<D
DA N=DN1<D
DMI 6 nM
GBR 17 nM
DN 2
NE N<DN2<D
DA N<DN2<D
DMI 40 nM
GBR 16 nM
DN 3
NE N<DN3<D
DA N<DN3<D
DMI 180 nM
GBR 8 nM
DN 9
NE N<DN9<D
DA N<DN9<D
DMI 11 μM
GBR 1 nM
DN 10
NE N<DN9<D
DA N<DN9<D
DMI 11 μM
GBR 1 nM
C
DN 5
NE nd
DA nd
DMI 200 nM
GBR 7 nM
DN 8
NE nd
DA nd
DMI 9 μM
GBR 2 μM

[^{125}I]RTI-55 Binding of Chimeras with Attenuated Transport Function

Thirteen chimeras, all functioning in a region spanning TM5-TM8 (including DN5, DN8, ND7, and ND8), and show attenuated transport activity as compared to DAT or NET. For example, chimera DN5 demonstrated a dramatic reduction in transport (i.e., V_{max} <10% for dopamine uptake by DAT) with no decrease in apparent dopamine affinity relative to DN3 and DN9. Expression of these chimeras was examined using whole-cell [^{125}I]RTI-55 binding.[11] RTI-55 is a structural analog of cocaine with high affinity for DAT,[12] which also recognizes NET. Cells transfected with these chimeras demonstrated specific [^{125}I]RTI-55 binding that was approximately 40–50% of binding to whole cells transfected with NET. These assays indicate that these chimeras are expressed, and that antagonist recognition is maintained. These data suggest that a region spanning TM5 through TM8 may be important for substrate translocation, or for appropriate processing and insertion of transporters into the plasma membrane. Immunoprecipitation represents an alternative method to assess the expression of chimeras with null phenotypes. For example, immunoprecipitation assays

[11] K. J. Buck, D. Lorang, and S. G. Amara, in "Molecular Approaches to Drug Abuse Research" (T. H. H. Lee, Ed.), Vol. III. USDHHS, Rockville, Maryland, 1996.

[12] J. W. Boja, A. Patel, F. I. Carroll, M. A. Rahman, A. Philip, A. H. Lewin, T. A. Kopajtic, and M. J. Kuhar, *Eur. J. Pharmacol.* **194,** 133 (1991).

FIG. 2. Wild-type and chimeric catecholamine transporters: structures and summary of pharmacological selectivity for substrates and uptake inhibitors. (A) Diagram of wild-type NET (in black) and DAT (in gray). Twelve hydrophobic domains are modeled as membrane-spanning domains. The NH_2 and COOH termini are presumed to lie in the cytoplasm (intracellular). Beside each transporter is a summary of its pharmacological characteristics. The following properties are summarized: the apparent affinity constant (K_T) for transport of the substrates norepinephrine (NE) and dopamine (DA); and the potency (K_i) with which desipramine (DMI) and GBR 12935 (GBR) inhibit [^{3}H]dopamine transport. (B) Properties of chimeric transporters. The functional characteristics of each chimera are summarized relative to wild-type NET (N) and DAT (D). Chimeras are referred to as ND (i.e., NET/DAT) or DA (i.e., DAT/NET) and numbered to reflect the transmembrane domain nearest their junction. Uptake inhibition by desipramine and GBR 12935 was examined using [^{3}H]dopamine (10 nM). Uptake inhibition constants (K_i) were determined using INPLOT. Data represent the mean $K_i \pm$ SEM determined from 4–6 independent experiments performed in triplicate. (C) Apparent substrate affinities were not determine (ND) for several chimeras in which transport was significantly attenuated, e.g., DN5, DN8, ND7, and ND8, all of which junction within a region spanning TM5 through TM8. For these chimeras, apparent dopamine transport affinities were not readily determined. IC_{50} values determined using 200 nM substrate in the presence of increasing concentrations of unlabeled inhibitor are reported for these chimeras. [Reprinted with permission from K. J. Buck and S. G. Amara, *Proc. Natl. Acad. Sci. USA* **91,** 12584 (1991). Copyright 1991 National Academy of Sciences, U.S.A.]

TABLE I
FUNCTIONAL PROPERTIES OF DA/NE TRANSPORTER CHIMERAS[a]

Transporter	Junction	K_T (μM) (apparent affinity constant) NE	DA	V_{max} (uptake efficacy) Relative (NE:DA)	DA uptake % wild type DAT
NET	none	0.4 ± 0.1	0.2 ± 0.1	2:1	160
ND 11	NET F550	0.4 ± 0.1	0.3 ± 0.1	1:1	130
ND 10	NET F474	0.9 ± 0.4	0.6 ± 0.3	1:1	110
ND 4	NET L237	1.6 ± 0.1	1.5 ± 0.5	1:1	130
ND 3	NET L163	1.7 ± 0.5	0.8 ± 0.2	1:1	60
ND 1	NET N78	2.6 ± 0.8	2.6 ± 0.7	1:4	100
DAT	none	5.7 ± 2.7	3.2 ± 0.4	1:9	100
DN 10	DAT F484	2.4 ± 0.8	2.2 ± 0.8	1:20	180
DN 9	DAT V469	2.3 ± 0.6	2.1 ± 0.5	1:17	170
DN 3	DAT F154	2.0 ± 0.6	1.1 ± 0.4	1:5	80
DN 2	DAT L113	2.1 ± 0.3	1.8 ± 0.2	4:5	90
DN 1	DAT W63	0.4 ± 0.1	0.2 ± 0.1	1:2	70

[a] Kinetic analysis of substrate transport in HeLa cells expressing chimeric transporters or wild-type NET or DAT. Chimeras are identified as ND or DN, and numbered to reflect the transmembrane domain near which they junction. The position of the junction is identified by the amino acid residue immediately preceding the junction. The kinetic constants, K_T (apparent affinity constant for transport) and V_{max} (maximal uptake rate at steady-state) for each of the substrates were obtained by nonlinear least-squares fits of substrate–velocity profiles using the data analysis program INPLOT. The apparent affinity constant (K_T) for transport of the substrates norepinephrine (NE) and dopamine (DA) is reported as the mean K_T ± SEM as determined from three to six independent experiments performed in quadruplicate. The relative V_{max} is reported as a rank order for DA and NE transport in cells transfected and assayed in parallel. The relative maximal transport rate is expressed as a percent of the V_{max} for DA uptake by the wild-type DAT.

using whole cell extracts from transfected cells showed that chimeras generated between NET and the rat serotonin transporter were expressed, although four of the seven chimeras had null phenotypes for substrate transport and antagonist binding.[9] Biotinylation of carriers on the cell surface is yet another approach that can provide insight into basis for a loss of function observed with a mutated or chimeric transporter. To determine the relative abundance of a chimeric carrier at the plasma membrane compared to the unaltered carrier, transporters can be biotinylated with the hydrophilic membrane-impermeant NHS-SS-biotin, then affinity purified on an avidin resin, and the purified proteins can be immunoblotted with anti-NET or anti-DAT sera.[13]

[13] G. M. Daniels and S. G. Amara, *Methods Enzymol.* **296,** [21], (1998) (this volume).

Chimeras constructed using *in vivo*[1,2,9,10] or traditional methods[14,15] have been generated from several neurotransmitter transporters. Here, we describe the successful *in vivo* formation of a large series of recombinant chimeras between two related transporters, DAT and NET. This approach generated chimeras that junction at single sites, in regions of conserved nucleic acid sequence, and typically resulted in chimeras that are expressed and inserted into the plasma membrane in a conformation in which functional integrity is retained. These chimeras provided an assayable phenotype which allowed us to assign a number of pharmacological and mechanistic properties of cocaine- and antidepressant-sensitive catecholamine transporters to specific domains within their primary structures.

Acknowledgments

We thank Dr. Randall Reed (Johns Hopkins University, Baltimore, MD) for providing the details of the unpulished method for *in vivo* formation of chimeras, and Dr. Robert J. Vandenberg for contributions to this chapter. This chapter was prepared with support from National Institutes of Health grants DA05228 and AA10760 (KJB), and DA07595 (SGA).

[14] B. Giros, Y.-M. Wand, S. Suter, S. B. McLeskey, C. Pifl, and M. G. Caron, *J. Biol. Chem.* **269,** 15985 (1994).

[15] E. L. Barker, H. L. Kimmel, and R. D. Blakely, *Mol. Pharmacol.* **46,** 799 (1994).

[33] Structural Determinants of Neurotransmitter Transport using Cross-Species Chimeras: Studies on Serotonin Transporter

By ERIC L. BARKER and RANDY D. BLAKELY

Introduction

With the cloning of the γ-aminobutyric acid (GABA) and norepinephrine transporters (GAT and NET, respectively),[1,2] new molecular tools and approaches based on homology became available to identify structural determinants of function and specificity for these Na^+- and Cl^--dependent transporter proteins. With this structural information, a functional dissec-

[1] J. Guastella, N. Nelson, H. Nelson, L. Czyzyk, S. Keynan, M. C. Miedel, N. Davidson, H. A. Lester, and B. I. Kanner, *Science* **249,** 1303 (1990).

[2] T. Pacholcyzk, R. D. Blakely, and S. G. Amara, *Nature* **350,** 350 (1991).

0076-6879/98 $25.00

tion of the transporters has proceeded using mutagenesis and chimeric protein strategies attempting to link specific amino acids with one or more of the transporter's properties.[3–14] For members of the GAT/NET gene family of transporters, the functional properties of interest include shared traits such as Na^+ and Cl^- dependence of transport and more defining characteristics such as recognition of specific substrates or binding of selective antagonists. The availability of cDNAs for individual transporters allows for functional comparisons in heterologous expression systems (i.e., transiently or stably transfected mammalian cells, *Xenopus* oocytes, etc.) where experimental conditions can be tightly controlled and influences and functional contributions from other receptors, ion channels, and transporters can be eliminated. Furthermore, the functional properties of closely related proteins can be compared in the same cellular context, providing a foundation for further molecular analysis of transporter characteristics. Focusing on the work from our laboratory with species variants of the serotonin (5-hydroxytryptamine; 5-HT) transporter (SERT), this chapter describes the formation and analysis of cross-species chimeras illustrating how species variants of SERT have served as an important guide for molecular structure–function studies. The ultimate goal of structure–function studies of SERT and related neurotransmitter transporters is to gain a better understanding of transport mechanisms to provide insight into how the proteins may be altered in disease states and regulated by endogenous or therapeutic agents.

SERTs are most closely related to the transporters for the monoamine transmitters norepinephrine and dopamine (DATs), all of which are modeled to possess 12 putative transmembrane domains with cytoplasmic amino and carboxyl tails.[15–17] SERTs, NETs, and DATs are of particular interest

[3] E. L. Barker, H. L. Kimmel, and R. D. Blakely, *Mol. Pharmacol.* **46,** 799 (1994).
[4] E. L. Barker and R. D. Blakely, *Mol. Pharmacol.* **50,** 957 (1996).
[5] K. J. Buck and S. G. Amara, *Mol. Pharmacol.* **48,** 1030 (1995).
[6] K. J. Buck and S. G. Amara, *Proc. Natl. Acad. Sci. USA* **91,** 12584 (1994).
[7] B. Giros, Y. M. Wang, S. Suter, S. B. McLeskey, C. Pifl, and M. G. Caron, *J. Biol. Chem.* **269,** 15985 (1994).
[8] S. Kitayama, S. Shimada, H. Xu, L. Markham, D. M. Donovan, and G. R. Uhl, *Proc. Natl. Acad. Sci. USA* **89,** 7782 (1992).
[9] S. Kitayama, J. B. Wang, and G. R. Uhl, *Synapse* **15,** 58 (1993).
[10] B. I. Kanner, A. Bendahan, S. Pantanowitz, and H. Su, *FEBS Lett.* **356,** 191 (1994).
[11] N. Kleinberger-Doron and B. I. Kanner, *J. Biol. Chem.* **269,** 3063 (1994).
[12] A. Bendahan and B. I. Kanner, *FEBS Letts.* **318,** 41 (1993).
[13] S. Pantanowitz, A. Bendahan, and B. I. Kanner, *J. Biol. Chem.* **268,** 3222 (1993).
[14] N. J. Mabjeesh and B. I. Kanner, *J. Biol. Chem.* **267,** 2563 (1992).
[15] S. Ramamoorthy, A. L. Bauman, K. R. Moore, H. Han, T. Yang Feng, A. S. Chang, V. Ganapathy, and R. D. Blakely, *Proc. Natl. Acad. Sci USA* **90,** 2542 (1993).

as they have been implicated in multiple psychiatric diseases as well as being the molecular targets for many antidepressants and drugs of abuse.[17] Attempts to map specificity for substrates and antagonists have generally followed two strategies. In one case, residues are identified as candidates for specificity based on homology across the gene family or within subsets of transporters which exhibit overlapping drug sensitivites. In the other strategy, chimeras are formed between related genes and transported behavior, and antagonist sensitivity of the resulting chimeras is used to track molecular recognition to discrete domains and residues. Lessons from transporter chimeras will be the primary focus of this chapter; however, mutagenesis guided by amino acid residues conserved within a subfamily of transporters provides an important context to profile less constrained chimera studies. For example, sequence comparisons across GAT/NET gene family members identifies several residues that are absolutely conserved only in the monoamine transporter family of proteins. One such residue is an aspartate (D) located in TMD1 (Fig. 1A). Based on the proposed role of a D residue for interactions with the amine side chain of catecholamines at adrenergic receptors[18,19] and the absolute conservation of this residue in NET, DAT, and SERT, this residue was hypothesized to coordinate the interactions of the amine group with the transporter. Indeed, mutagenesis of this residue in DAT caused a marked disruption of transport activity and ligand binding.[8] We have confirmed these studies with NET and SERT where mutation of this D residue again caused a disruption of transporter function[20] (Fig. 1B). Interestingly, whereas the NET D75E substitution is nonfunctional, the SERT D98E mutant retains significant transport activity, suggesting that requirements for NET recognition of substrates are more sensitive to perturbation than those for SERT's recognition of indoleamine substrates. Using cell-surface biotinylation techniques,[21-23] we have verified

[16] R. D. Blakely, H. E. Berson, R. TJ. Fremeau, M. G. Caron, M. M. Peek, H. K. Prince, and C. C. Bradley, *Nature* **354,** 66 (1991).

[17] E. L. Barker and R. D. Blakely, *in* "Psychopharmacology: The Fourth Generation of Progress" (F. Bloom and D. Kupfer, Eds.), p. 321. Raven Press, New York, 1995.

[18] C. D. Strader, I. S. Sigal, and R. A. Dixon, *FASEB J.* **3,** 1825 (1989).

[19] C. D. Strader, T. Gaffney, E. E. Sugg, M. R. Candelore, R. Keys, A. A. Patchett, and R. A. Dixon, *J. Biol. Chem.* **266,** 5 (1991).

[20] H. E. Melikian, K. R. Moore, Y. Qian, H. L. Kimmel, S. B. Taylor, R. W. Gereau, A. Levey, and R. D. Blakely, *Neurosci. Abs.* **19,** 407.8 (1993).

[21] Y. Qian, A. Galli, S. Ramamoorthy, S. Risso, L. J. DeFelice, and R. D. Blakely, *J. Neurosci.* **17,** 45 (1997).

[22] H. E. Melikian, S. Ramamoorthy, C. G. Tate, and R. D. Blakely, *Mol. Pharmacol.* **50,** 266 (1996).

[23] S. Ramamoorthy, H. E. Melikian, Y. Qian, and R. D. Blakely, *Methods Enzymol.* **296,** [24], (1998) (this volume).

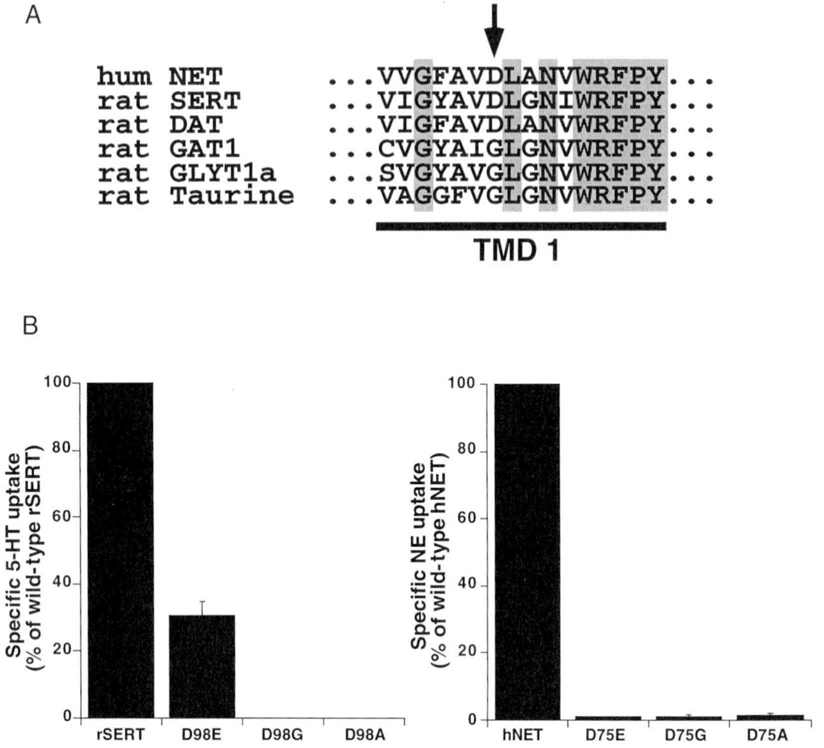

FIG. 1. (A) Amino acid sequence alignments through the TMD1 region of the human NET, and rat SERT, DAT, GAT1 (GABA), GLYT1 (glycine), and taurine transporter. Shaded residues indicate areas of conserved sequence identity. The arrow indicates the NET D75 and SERT D98 position. Whereas this position is always a D residue in SERTs, NETs, and DATs, other members of the gene family possess a G residue at this location. (B) Uptake assays comparing SERT D98 and NET D75 mutants. Hela cells transiently transfected with either the wild-type or mutant transporters were incubated with either 40 n*M* [^{3}H]5-HT (for SERT) or [^{3}H]norepinephrine (for NET) for 10 min. Uptake for mutant transporters is expressed as percentage of wild-type transport with data representing means ± S.E.M. from three separate experiments performed in triplicate.

that all nonfunctional rat SERT D98 and human NET D75 mutants have comparable surface expression to the wild-type transporters, ruling out mutation-induced disruptions of trafficking mechanisms as an explanation for the lack of transport activity.[24] Regardless, the lack of transport activity by the NET mutants hinders efforts to further understand the molecular role of this residue. The recent characterization of the activity retained by

[24] E. L. Barker and R. D. Blakely, unpublished work (1996).

the SERT D98E mutation using the tryptamine derivative gramine [3-(dimethylaminomethyl)indole] has generated data consistent with an interaction between D98 and the amine side chain.[24] Nonetheless, the generation of nonfunctional, and hence nontestable, mutants is a major impediment to using only sequence-based comparisons to guide functional studies. Alternatively, species variants, where sequence variations occur in the context of normal transport activity, could be conceived of as nature's mutagenesis experiments leading to potentially interesting and informative mutagenesis studies that can test variations in sequence for functional implications.

Chimera Strategies

Chimeric proteins formed between functionally distinct members of the same gene family (e.g., NET and SERT, NET and DAT) as well as species homologs of the same protein (e.g., human, rat, and *Drosophila* SERTs) have been instrumental in identifying discrete domains involved in specific pharmacological and biophysical properties.[3,5–7,25] The advantage for generating chimeras between two homologous proteins is based on detecting fundamental differences in functional characteristics of the two proteins and then using chimeras to localize the domains involved in those transporter-specific properties. Whereas the functional differences between species homologs can be relatively subtle, chimera formation between transporters with more divergent structure such as SERT and NET or NET and DAT might yield greater information regarding structural determinants of transporter properties. For example, in addition to recognition of distinct substrates, SERT and NET also demonstrate markedly different sensitivities to antagonists such as fluoxetine and imipramine. NET/SERT chimeras with exchanged intracellular amino- and carboxyl-tail domains appear to carry properties of the membrane-embedded polypeptide, suggesting that the tail domains contribute little to substrate and antagonist recognition.[25] Unfortunately, attempts at analyzing additional SERT/NET chimeras have revealed hybrid proteins that are largely nonfunctional.[25,26] These studies revealed that a nonfunctional protein was formed when the chimera junction point existed anywhere internal to either the amino- or carboxyl-tail domains, demonstrating the strict structural and conformational requirements that must be maintained for transport to occur. Functional NET/DAT chimeras have been successfully generated, providing much information about domains involved in substrate and antagonist recognition.[5–7] Thus, NET/DAT chimeras suggest that TMDs 1–3 contribute primary

[25] R. D. Blakely, K. R. Moore, and Y. Qian, *Soc. Gen. Physiol. Ser.* **48,** 283 (1993).

[26] K. R. Moore and R. D. Blakely, *Biotechniques* **17,** 130 (1994).

determinants for substrate recognition and secondary determinants of antagonist recognition, TMDs 5–8 are involved in substrate translocation and primary influences for antagonist affinity, and TMDs 9–12 contain modulatory sites for substrate affinity. Presumably, these NET/DAT chimera studies will guide future mutagenesis strategies to identify specific residues involved in transporter properties.

Cross-species chimeras using evolutionary variants of the *same* protein have been extremely useful in identifying functional domains of cell-surface receptors and ion channels and, more recently, SERTS.[3,4,27–31] For species variants of transporters, successful strategies will be dependent on a thorough characterization of the two parental species homologs. As with any chimera or mutagenesis approach, assaying multiple functional attributes is critical to identifying domains involved not only in transport activity and antagonist sensitivity, but also in proper protein folding and trafficking. Protein analysis is facilitated by either specific antibodies or the introduction of epitope tags[32] to assist in assessing perturbations of transporter protein structure. In our case, the availability of the cDNAs for SERT species variants allowed functional analyses to be performed using heterologous expression systems, where comparisons of wild-type transporters could be achieved in the same cellular context, removing concerns about nonspecific influences of different native tissue preparations. We have determined transport kinetics and ion dependence of transport, as well as the pharmacological profile, of human,[15] rat,[16,33] and *Drosophila*[34,35] SERTs using HeLa cells transiently transfected with the vaccinia virus–T7 expression system (Table I).[36] Previous studies in native preparations revealed that the tricyclic antidepressants imipramine and clomipramine were more potent at human platelet SERTs relative to rat

[27] J. M. Hall, M. P. Caulfield, S. P. Watson, and S. Guard, *Trends. Pharmacol. Sci.* **14,** 376 (1993).

[28] D. Oksenberg, S. A. Marsters, B. F. ODowd, H. Jin, S. Havlik, S. J. Peroutka, and A. Ashkenazi, *Nature* **360,** 161 (1992).

[29] H. T. Kao, N. Adham, M. A. Olsen, R. L. Weinshank, T. A. Branchek, and P. R. Hartig, *FEBS Lett.* **307,** 324 (1992).

[30] M. A. Gray, E. L. Barker, L. Demchyshyn, H. B. Niznik, and R. D. Blakely, *Exp. Biology '95 Abs.* (1995).

[31] E. L. Barker, M. A. Gray, E. A. Camp, and R. D. Blakely, *Neurosci. Abs.* **207,** 6 (1996).

[32] P. A. Kolodziej and R. A. Young, *Methods Enzymol.* **194,** 508 (1991).

[33] B. J. Hoffman, E. Mezey, and M. J. Brownstein, *Science* **254,** 579 (1991).

[34] L. L. Demchyshyn, Z. B. Pristupa, K. S. Sugamori, E. L. Barker, R. D. Blakely, W. J. Wolfgang, M. A. Forte, and H. B. Niznik, *Proc. Natl. Acad. Sci. USA* **91,** 5158 (1994).

[35] J. L. Corey, M. W. Quick, N. Davidson, H. A. Lester, and J. Guastella, *Proc. Natl. Acad. Sci. USA* **91,** 1188 (1994).

[36] R. D. Blakely, J. A. Clark, G. Rudnick, and S. G. Amara, *Anal. Biochem.* **194,** 302 (1991).

TABLE I

SERT SPECIES VARIANTS EXHIBITING DISTINCT PHARMACOLOGICAL PROFILES[a]

DRUG	K_i or K_m (nM) Rat SERT	Human SERT	*Drosophila* SERT
5-HT	629 ± 116	499 ± 89	490 ± 35
Imipramine	46 ± 9.3	8.2 ± 3.4	1450 ± 280
Despiramine	209 ± 45[b]	54 ± 4.2	580 ± 147[b]
Paroxetine	0.09 ± 0.03	0.05 ± 0.01	3.4 ± 1.6[b]
Citalopram	4.9 ± 1.1	4.7 ± 0.7	88 ± 17[b]
Fluoxetine	7.3 ± 3.3	3.1 ± 0.6	73 ± 5.6[b]
Sertraline	0.9 ± 0.2	1.3 ± 0.4	ND
Cocaine	540 ± 47	611 ± 66	464 ± 31
RTI-55	0.26 ± 0.12	0.11 ± 0.04	444 ± 64[b]
d-Amphetamine	5500 ± 400[b]	11500 ± 1500	33800 ± 4200[b]
Fenfluramine	1000 ± 100	1990 ± 320	4550 ± 490[c]
Mazindol	ND	98 ± 1.7	3.9 ± 0.02[b]

[a] K_i and K_m values for the inhibition of [^{3}H]5HT uptake for various compounds to HeLa cells transiently transfected with the cloned rat, human, or *Drosophila* SERTs. Drugs are grouped according to therapeutic classification. Data represent means ±SEM of triplicate determinations from 2–3 separate experiments. ND, not determined. Means were compared to human SERT values using a two-sided Student's *t*-test.

[b] $P < 0.01$ (GraphPad InStat v.2.0, Intuitive Software for Science).

[c] $P < 0.05$.

preparations.[37] Analysis of the cloned human and rat SERTs revealed that the tricyclic antidepressants were indeed more potent at human as compared to rat SERT, whereas *d*-amphetamine exhibited a higher potency for rat SERT versus human SERT (Table I). These observations are unlikely to reflect structural anomalies of SERT resulting from heterologous expression since other antagonists, such as fluoxetine, paroxetine, and cocaine, were equipotent at both species and both human and rat SERT transported 5-HT with similar kinetics.

Having identified a functional difference between these two SERT species homologs, we constructed a series of rat/human SERT chimeras to track the higher potency interactions of the tricyclic antidepressants with human SERT and *d*-amphetamine with rat SERT to specific regions of the transporter. These chimeras identified the TMD12 domain of human SERT as conferring the higher potency interactions of the tricyclics (Fig. 2) and also the lower potency interactions with *d*-amphetamine.[3] As discussed

[37] M. Wielosz, M. Salmona, G. de Gaetano, and S. Garattini, *Naunyn Schmiedeberg's Arch. Pharmacol.* **269,** 59 (1976).

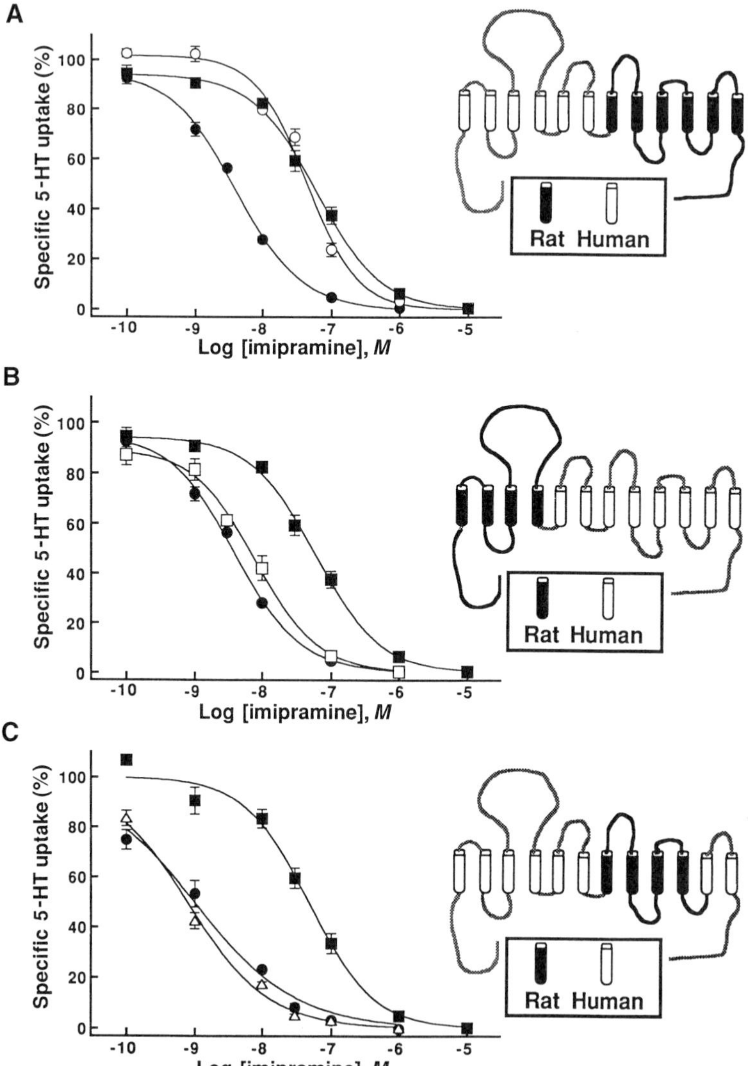

FIG. 2. Imipramine inhibition of 5-HT uptake at parental human SERT (●), rat SERT (■), and chimeras (A, ○; B, □; C, △). [^{3}H]5-HT uptake assays were performed on transiently transfected HeLa cells with increasing concentrations of imipramine added 10 min before the addition of 20 n*M* [^{3}H]5-HT. Data were plotted as percentage of specific 5-HT uptake and represent means ± standard errors of triplicate determinations, representative of three (A and B) or two (C) separate experiments. Chimeras indicated that the region distal to TMD11 was involved in the more potent interactions of imipramine with human SERT (C). An additional chimera [see E. L. Barker and R. D. Blakely, *Mol. Pharmacol.* **50,** 957 (1996)]

previously, mutagenesis based upon sequence conservation across a gene family may provide multiple logical mutagenesis targets, whereas functional data provided by cross-species chimeras may target only a handful of candidate residues. For example, rat and human SERT share 92% overall sequence identity. Once chimeras identified the TMD12 domain as involved in the species-specific properties, sequence comparison showed only four amino acids differing between rat and human SERT in the region, limiting the targets for site-directed mutagenesis. Subsequently, site-directed mutagenesis strategies identified a single residue, phenylalanine-586, as being responsible for the higher potency of the tricyclics with human SERT.[4] No single residue could be identified as influencing *d*-amphetamine potency, suggesting more complex interactions with amphetamine that may include interactions with multiple transporter domains.

We have extended this comparative approach to analysis of more distant relatives within the SERT gene family, specifically the *Drosophila melanogaster* SERT and human SERT. Despite human and *Drosophila* SERTs only showing 49% sequence identity, the two molecules exhibit essentially equivalent transport kinetics for 5-HT; however, vastly different pharmacological profiles are observed (Table I). Chimeras between *Drosophila* and human SERT have been generated and have yielded insights into functional domains of SERT. These *Drosophila*/human SERT chimeras have identified the TMDs 1–2 region as being potentially involved in recognition of the antagonists mazindol and citalopram as well as interactions with the cotransported Na^+ ions.[31] Mutagenesis studies are currently underway seeking to identify specific residues that may mediate some or all of these species-selective properties; however, the advantages of cross-species comparisons are already evident since our chimeras produced from fairly divergent SERT proteins are functional. Since both parental transporters recognize 5-HT as their primary substrate and transport 5-HT with similar kinetics, there is a strong likelihood that rat/human or *Drosophila*/human SERT chimeras should demonstrate 5-HT uptake properties similar to those of the parental SERTs. This reduces the fear of producing nonfunctional and hence, hybrid transporters that cannot be studied. The strength of cross-species chimeras is their ability to track alterations, not disruptions, in transporter function to specific domains. For example, our studies with SERT species variants have identified domains influencing antagonist po-

indicated that the carboxyl-tail domain was not involved in this effect, narrowing the targeted region to the TMDs. In the region of TMDs 11 and 12, only four amino acids differ between human and rat SERT, all located in or near TMD12. [This figure used by permission from E. L. Barker, H. L. Kimmel, and R. D. Blakely, *Mol. Pharmacol.* **46,** 799 (1994).]

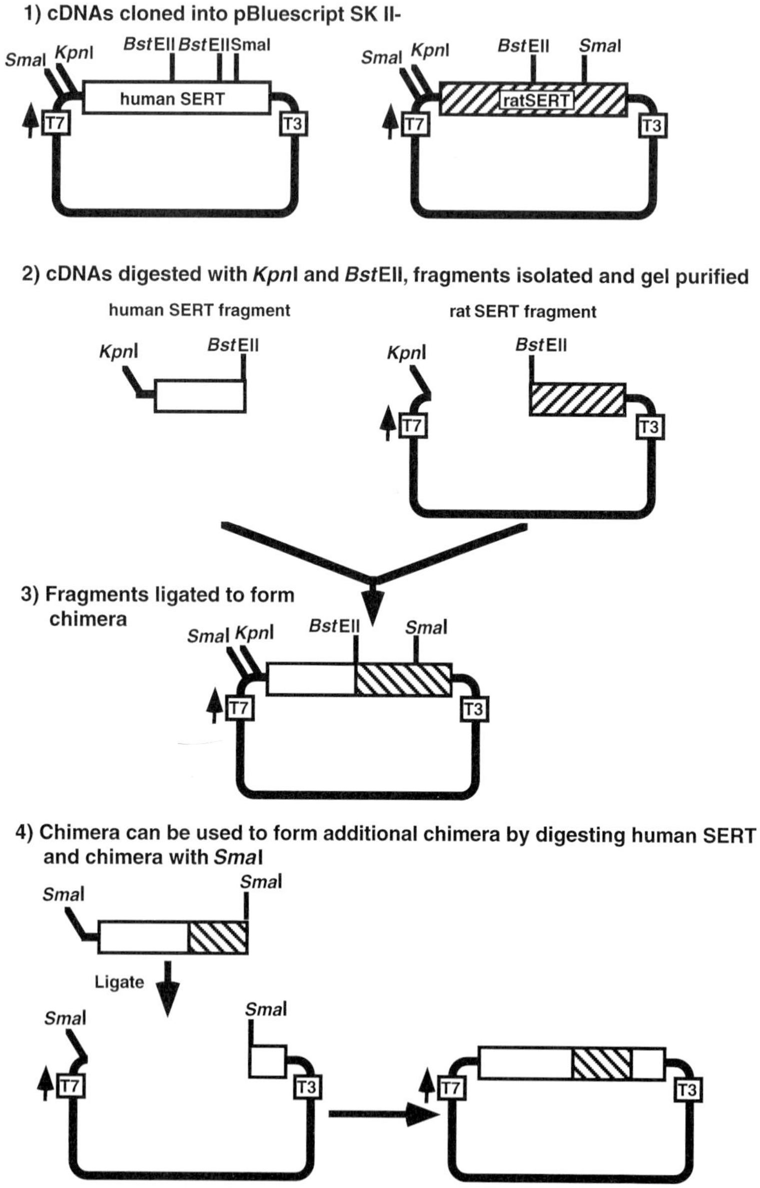

FIG. 3. Schematic representation of chimeras formed by exploiting shared compatible restriction enzyme sites. This method was used to generate chimeras between rat and human SERT. More details regarding these constructions can be found in E. L. Barker, H. L. Kimmel, and R. D. Blakely, *Mol. Pharmacol.* **46,** 799 (1994). In brief, the *Kpn*I and *Bst*EII fragment

tencies. As such, chimeras and mutants can test for reciprocal effects on antagonist potency and can lead to further mechanistic and structure–activity studies revealing the molecular role of a specific residue in transporter function. Below, we describe methods involved in the construction of transporter chimeras.

Chimera Formation

The three most common methods of chimera formation are described along with a brief discussion of the main advantages and disadvantages to each method.

Restriction Enzyme-Based Chimeras

This method is perhaps the easiest and most straightforward of the approaches discussed. Relying on shared unique restriction enzyme sites, the two cDNAs are digested and fragments are interchanged to create chimeras (Fig. 3). This is the method that we used to construct our NET/SERT tail chimeras[25] and the rat and human SERT chimeras.[3] Basic restriction site and transformation methodologies can be obtained from commercial vendors of relevant enzymes or from standard molecular biology methods manuals (Sambrook *et al.*, "Molecular Cloning: A Laboratory Manual," "Current Protocols in Molecular Biology," etc.).

Method

1. Analyze restriction enzyme maps seeking to identify enzymes sites that are shared by the two cDNAs. This means that one focuses primarily on enzymes with unique digestion sites; however, enzymes targeting two sites can be also useful when attempting to construct chimeras (Fig. 3). *Note:* Be sure to check the expression vector map, as well to verify lack of sites in the vector itself which would complicate construction strategies, although often sites within the vector sequence, either in the multiple cloning site or in other regions of the DNA, can be incorporated into chimera constructions (Fig. 3, see *Kpn*I site).

2. Once appropriate sites have been identified, completely digest plasmids to liberate desired fragments. Fragments must then be gel purified

---

of human SERT was substituted for the corresponding fragment of removed rat SERT, creating the chimera represented in (3). (4) represents the formation of a second chimera using the first chimera as a base molecule. The *Sma*I fragment of chimera 1 was isolated and substituted for the corresponding fragment removed human SERT.

using method of choice (i.e., Qiaex resin, Qiagen Inc., Chatsworth, CA; GENECLEAN, Bio101, Vista, CA; tombstoning, etc.[38]).

3. Ligate fragments to generate chimeric molecule.

4. Transform into competent *Escherichia coli* and subsequently screen colonies for inserts.

5. Purify positive colonies using plasmid purification methods.

6. Chimeric constructs can now be expressed and analyzed in comparison with the two parental transporters.

Advantages and Disadvantages. The main advantage to this strategy is the relatively simple molecular manipulations that are required to form the chimeras. The approach is fast and requires little or no DNA sequencing since there are no error-prone polymerase reactions. The strategy also requires no oligonucleotide primers for polymerase chain reaction (PCR), site-directed mutagenesis, or sequencing, making this probably the least costly method. Also, it should be noted that the chimeras formed can sometimes be used to generate further hybrid proteins, as the chimera created has a unique restriction enzyme map that may prove advantageous in later constructions (Fig. 3). If multiple chimeras are needed, this method can become time consuming as each chimera requires separate construction; however, the method can be useful for generating a few initial chimeras to guide further chimera constructions. Because the method relies on shared restriction enzyme sites, chimera junction points are limited to these rather arbitrary points that may not be ideal or important. Many times, convenient restriction digestion sites do not even exist, thus hindering the ability to form chimeric transporters. This method works best with proteins that share a high degree of sequence identity both at the amino acid and nucleotide level, such as the rat and human SERTs. For more divergent proteins such as the human and *Drosophila* SERTs, which share only 49% amino acid identity, this method has not been particularly useful because of the lack of shared restriction enzyme sites.[34,35] These caveats can be overcome by taking advantage of degeneracy in the genetic code using site-directed mutagenesis techniques to introduce restriction enzyme sites into one or both cDNAs without altering protein sequence. These new sites can then be used to form chimeras; the application of this method is discussed below.

Restriction Site Introduction and cDNA Cassette Swapping

The introduction of compatible restriction enzyme digestion sites into two parental transporter cDNAs using site-directed mutagenesis can be an

[38] F. M. Ausubel, R. Brent, R. E. Kingston, D. D. Moore, J. G. Seidman, J. A. Smith, and K. Struhl, "Current Protocols in Molecular Biology." Wiley, New York, 1994.

extremely valuable tool for construction of interspecies chimeras, particularly for variants whose nucleotide structures are highly divergent, thus lacking many common restriction sites. This method essentially allows the investigator to determine where chimera junction points will exist and then exploit the degeneracy in the genetic code to use molecular approaches for introducing compatible restriction sites into the parental cDNAs for subsequent digestion and ligation. This method has been used to construct chimeras between the NET and DAT[7] and between human and *Drosophila* SERT.[24]

Method

1. Restriction enzyme digestion maps for the two parental transporter cDNAs are compared to compile a list of restriction enzymes that either cut the cDNA in a single location or do not digest the sequence at all. This list of "single cutters" can be correlated with the amino acid translation to determine the "cut" sites on the protein sequence corresponding to potential chimeric junction points. Typically, it is easier to choose a restriction enzyme that cuts one of the parental cDNAs so that a site has to be introduced in only the order cDNA; if no convenient single cutters exist, introduction of sites from the "no cutter" list will have to be performed on both cDNAs. At this point, a decision is made as to where chimera junction points will be generated. Areas of the protein sequence with conserved identity will most likely yield the areas where the simplest molecular manipulations can be performed to add restriction sites. The most advantageous strategy is perhaps to plan on 2–3 internal chimera switch points, thus allowing for multiple "drop and swap" steps to form various chimeric transporters (Fig. 4). Starting with a site somewhere in the middle of the sequence allows one to split the transporter essentially in half and begin to target which domains are involved in certain properties, and hence guide future chimera constructions.

2. Once a decision has been made as to the general area where restriction sites are to be introduced, one should use commercial sequence analysis software (i.e., MacVector, GeneWorks, LaserGene, etc.) to identify nucleotide sequences where the desired restriction site can be added without altering the amino acid translation (Fig. 4) Following these analyses, oligonucleotide primers can be designed and synthesized for use in the mutagenesis method of choice (we use either the Stratagene Chameleon Kit or the QuikChange kit; however, PCR-based mutagenesis techniques can be employed as an alternative to the commercially available methods).

3. Using the oligonucleotide primers and the mutagenesis method, the restriction site is introduced. This new species of cDNA can then be used as the template to introduce further restriction sites until the two wild-type

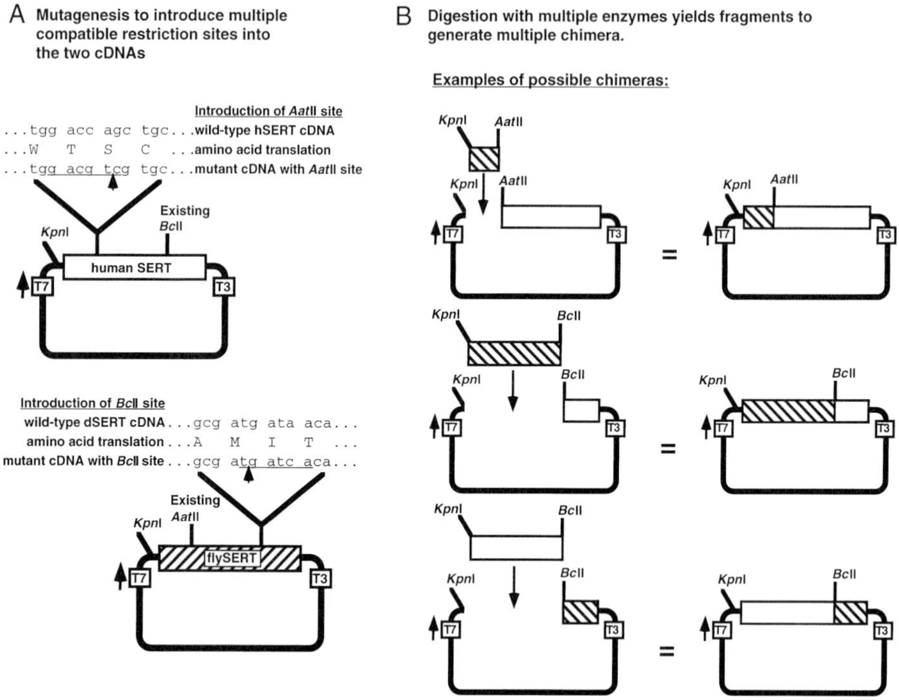

FIG. 4. Chimera formation by introducing compatible restriction sites to create cDNA cassettes. (A) Mutagenesis is used to introduce compatible *Aat*II and *Bcl*I sites in the human and *Drosophila* SERTs. Restriction enzyme recognition site is underlined on the cDNA with the cut site indicated by the arrow. (B) Combinations of digestions using *Kpn*I, *Aat*II, or *Bcl*I can be used to liberate cDNA cassettes that can be interchanged to create multiple chimeras. Three examples of the many possible chimeras are represented.

transporter cDNAs contain a series of compatible sites. DNA sequencing is performed to confirm that the only mutation in the cDNA is the addition of the restriction enzyme recognition sequence.

4. Using the newly compatible sites as well as sites in the multiple cloning site, the cDNAs are digested to create multiple cDNA cassettes that can be ligated together as described above (Fig. 4).

5. After transformation into competent bacteria, the positive clones are purified and chimeric formation verified by restriction digestion. Subsequently, cDNAs are ready for expression and analysis.

Advantages and Disadvantages. By introducing compatible restriction enzyme digestion sites, this method allows one to dictate where chimera junction points will be formed and is not hindered by the limits of existing restriction sites. Thus, cDNA cassettes bordered by the introduced restriction sites can be swapped between the two parental transporters to form multiple chimeric species. This is particularly useful in creating complementary chimeras (see Fig. 4) for testing the reciprocity of a specific domain's role in a functional test. In the NET/DAT chimera examples,[7] when the DAT TMDs 6–8 domain are introduced into the NET backbone, the chimeras exhibit low potency for the NET antagonists nortriptyline and desipramine. Conversely, the functional chimeras containing NET TMDs 6–8 show the more characteristic higher affinity interactions with these antagonists. The reciprocal effects of the TMDs 6–8 domain strongly suggest that major determinants of tricyclic antidepressant recognition exist in these domains. Again, for cross-species chimera construction, this method has the greatest utility when the desired switch point exists in a region where no common restriction sites exist.

The main disadvantages to this method of chimera construction are the expense and time required to introduce multiple unique sites. Each chimera construction requires a separate digestion and ligation step, which can be labor and time intensive. The other major disadvantage of this method lies in the bias of the investigator when choosing chimera junction points, a process that tests the limits of one's understanding of viable and informative chimeric junctions. Since each chimera requires separate construction before expression, extensive time and effort may be expended before chimera analysis can occur, and unfortunately, there is no guarantee that the hybrid transporters will possess measurable transport activity. For example, the one human and *Drosophila* SERT chimera that we have generated using this method does not demonstrate 5-HT uptake activity despite chimeric protein translation and expression on the cell surface.[24] The NET/DAT chimeras constructed using this method further demonstrate sensitivity to chimera constructions as half of the chimeric transporters failed to transport either norepinephrine or dopamine.[7] This illustrates the strict conformational requirements of the transporters for uptake activity to be maintained and how chimeric constructions even between closely related molecules can lead to disruption of this complex activity. Another potential problem with this method is that introduction of restriction sites might not be allowable at chosen points, since degeneracy in the amino acid code and restriction enzyme digestions may not maintain in-frame translation at the chimera junction. Finally, placement of restriction sites limit and govern future chimera constructions, and if the chimeras fail to function, the utility of the introduced sites will be limited as well.

Restriction Site-Independent Chimera Formation

The methods described above all depend on the existence of common restriction enzyme sites for the ultimate formation of chimeric transporter proteins. The construction of multiple chimeras are limited to these restriction sites, and formation is time consuming and costly as procedures are repeated for each new chimera. A PCR overlap extension method of cDNAs has been used to create chimeric proteins without the need for compatible restriction sites[39,40]; although this method can be quite useful, creation of oligonucleotides, amplification and construction steps, and extensive DNA sequencing for each chimera make this method rather laborious and costly. It is also biased toward a specific chimera junction predetermined by the investigator. Consequently, we have described another restriction site-independent method of chimera formation that can be used to rapidly generate a library of chimeric transporters.[26] This method uses transformation of competent bacteria with a linearized vector construct containing the two homologous transporter cDNAs in a "tail-to-head" orientation downstream of the T7 RNA polymerase promoter. Following transformation, multiple in-frame chimeric neurotransmitter transporter cDNAs can be isolated with determination of chimera junction point by restriction enzyme digestion patterns and DNA sequencing. This method has been successfully used to generate chimera between human NET and rat DAT,[5,6] human NET and rat SERT,[26] and human SERT and *Drosophila* SERT.[30,31]

Method

1. For the sake of clarity, we describe the method used to generate NET/SERT chimeras.[26] The SERT cDNA is liberated from its plasmid using restriction enzyme digestion and is ligated into the NET plasmid downstream of the NET cDNA (Fig. 5). For the purpose of linearing the dimer construct, it is advantageous to leave 30–50 bp of the vector polylinker, which contains multiple restriction sites, between the two cDNAs.

2. The dimer construct is linearized using two unique restriction sites found in between the two parental cDNAs. The linearized DNA should be gel purified.

3. Fifty to 100 ng of the purified DNA is transformed by heat shock into a competent *E. coli* strain. For some plasmids, it may be necessary to transform with up to 500 ng of linearized DNA. Also, any single cut or nondigested DNA contaminating the purified double digested DNA sample leads to transformation of the dimer construct and yield multiple false positives. Therefore, it is imperative that all transformed DNA be double digested. *Note:* Digestion can be performed essentially anywhere in the

[39] R. M. Horton, Z. L. Cai, S. N. Ho, and L. R. Pease, *Biotechniques* **8,** 528 (1990).
[40] R. M. Horton, H. D. Hunt, S. N. Ho, J. K. Pullen, and L. R. Pease, *Gene* **77,** 61 (1989).

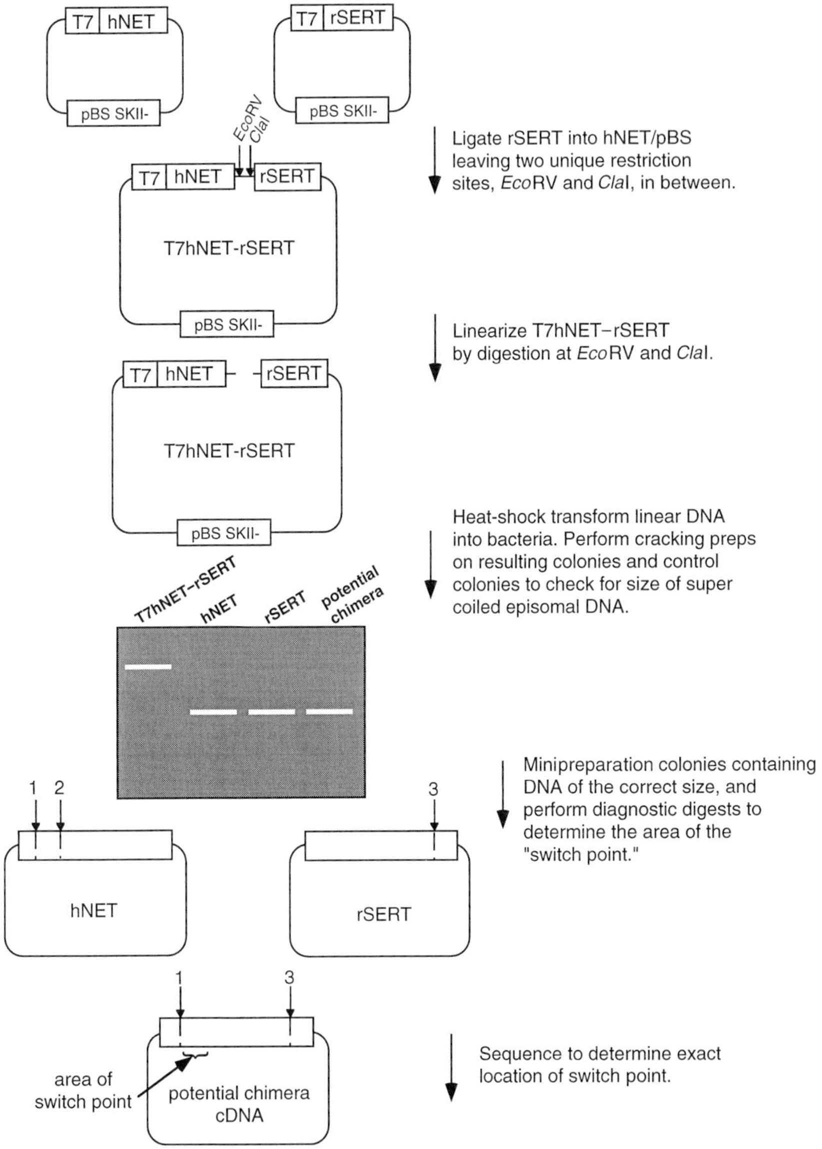

FIG. 5. Schematic representation of NET/SERT chimeras generated using the restriction site-independent method. [Reprinted with permission from K. R. Moore and R. D. Blakely, *Biotechniques* **17,** 130 (1994).]

parental cDNAs to restrict the region for potential chimeric junction formation.

4. Alkaline lysis or "cracking" minipreps can be performed on resistant colonies to determine relative sizes of inserts screening for only monomer-sized cDNA inserts indicative of chimera formation.

5. Colonies containing monomer-sized inserts are isolated and a minipreparation of the DNA performed. The plasmid DNA is digested with restriction enzymes that have unique patterns of digestion on the parental transporters. By screening using a set of enzymes will sites proceeding from amino to carboxyl termini, one can ascertain the general region of the chimera switch point. For example, human NET contains two *Hind*III sites at 200 and 500 bp (sites 1 and 2 on Fig. 5), and rat SERT contains one *Hind*III site at nucleotide 1800 (site 3 on Fig. 5). If digestion of potential chimeric cDNAs with *Hind*III were to liberate a fragment of 1600 bp, the chimera junction point would be implicated in the region between the two native NET *Hind*III sites.

6. DNA sequencing is performed to identify the exact location of the switch point and verify that the chimeric molecule is in-frame with respect to the amino acid coding sequence of the parental transporters.

7. The plasmids are purified and expressed in heterologous expression systems for analysis.

Advantages and Disadvantages. This method of chimera formation has shown great utility in generating multiple chimeras in a random and rapid fashion. In generating NET/SERT chimeras, although only a small number of colonies resulted from the transformation of the linearized dimer construct, all colonies containing monomer-sized inserts represented in-frame transporters. Because the junction points are randomly determined, any experimental bias on the part of the investigator as to where switch points should exist is removed. Furthermore, because the cDNA dimer construction is essentially the only molecular manipulation of the cDNAs, the method requires little time and is relatively inexpensive, particularly if multiple, unique chimeras are generated. Thus, the greatest strength of this method lies in the analysis of proteins where little structural information is available to guide selection of chimera switch points by restriction enzyme sites and when a library of chimeras is needed to provide a thorough structural dissection of the transporter.

Although the power of this approach to generate chimeras is potentially great, there are a number of weaknesses to the approach that potential users should recognize. The mechanism for actual bacterial-mediated chimera formation is not well understood. The most likely possibility links bacterial exonuclease digestion of the linearized dimer DNA with alignment and annealing of conserved cDNA sequences present throughout the two transporter clones, since it occurs in *recA* mutant strains of *E. coli* and appears to bias chimera junction formation in regions of maximum nucleotide homology. Thus, chimeric junctions arise most often in regions where the two cDNAs share sequence identity limiting the use of the method in more divergent members of a gene family. However, as interspecies variants of transporters typically demonstrate a high degree of distributed sequence identity, this method can be particularly powerful in generating interspecies chimeras. Again, this method does not allow the investigator to choose the location of the chimera junction and hence cannot be used to address questions about specific domains; these chimeras must be made using a restriction enzyme-dependent method. Finally, we have had difficulty in generating chimeras with switch points in the carboxyl-terminal half of the transporter. Our experience has shown that the cDNA closest to the T7 RNA polymerase promoter (i.e., in the more 5′ location) in the dimer construct typically contributes much less to the chimeras than the second parental cDNA (in the more 3′ location). As such, we have been unable to generate a chimeric transporter with a switch point distal to the TMD 7 domain, making exploration of these more carboxyl terminal domains difficult. This effect is possibly due to reduced sequence homology between the parental transporters in the region distal to TMD7. Buck and Amara,[5,6] however, have used this method to generate a library of NET/DAT chimeras with junctions evenly dispersed throughout the transporter molecule, suggesting that if one has the perseverance to screen additional colonies, chimeric junctions can be obtained throughout the molecule.

Analysis of Chimeras

Once a series of chimeric transporters has been generated with the species variants, a thorough functional characterization must occur to identify domains involved in species-specific properties. We have found that characterization of the parental and chimeric transporters is facilitated by performing constructions in plasmids containing the T7 RNA polymerase promoter (like pBluescript, Stratagene, La Jolla, CA) where the vaccinia virus–T7 expression system[36] can subsequently be used for rapid expression of cloned transporters without the need for subcloning cDNAs into mammalian expression vectors. Chimera analysis will, in general, be guided by the results of the characterization of the parental transporters and will seek to track the species-specific properties identified; however, properties that are shared by the two parental species variants also should be tested as positive controls, showing that the chimeras share appropriate characteristics with the wild-type transporters.

Specific Tests

Transport Activity

Substrate transport kinetics should be determined to compare substrate K_m and V_{max} values. Whereas K_m values can reveal information regarding relative affinity of a substrate for transporters, differences in V_{max} values for different clones can be more difficult to interpret as a decrease in V_{max} could result from a decreased cell-surface expression level of a particular clone and not be indicative of an alteration in an intrinsic property of the transporter. We have found that determining transport kinetics of multiple substrates can be useful providing an opportunity for comparisons using V_{max} ratios. These ratios normalize the data for each transporter and can reveal subtle differences in the rates of transport for substrates.

Ion Dependence of Transport

As all members of the GAT/NET gene family of transporters depends on ion gradients (Na^+ and Cl^-) to energize the transport process, the ion dependence of transport for each parental and chimera should be determined. For example, both human and *Drosophila* SERT exhibit Na^+ and Cl^- dependence for 5-HT uptake. Whereas the stoichiometry of 1 Cl^- : 1 5-HT for transport appear to be shared by the two species homologs, the stoichiometry for Na^+ appears distinct. Human SERT shows a Na^+ stoichiometry of 1 Na^+: 1 5-HT for transport, but *Drosophila* SERT displays a 2 Na^+: 1 5-HT coupling.[31] We are presently using the *Drosophila*/human SERT chimeras to identify the domains of *Drosophila* SERT that mediate the additional interactions with Na^+. It is interesting that our NET/SERT chimeras were nonfunctional as these two transporters demonstrate different external and internal ion dependencies that may have been compromised by chimera formation.[17,41] However, our ability to form fly/human SERT chimeras, where external Na^+ requirements appear to be different, argues that this limitation may be relaxed *within* species variants.

Pharmacological Profiles

Thorough pharmacological profiles of the cloned transporters performed in side-by-side comparisons in the same cellular context are critical for identifying species differences in antagonist sensitivities. We have identified significant differences in antagonist potencies for the three SERT variants (Table I) and have exploited these differences to track domains in-

[41] G. Rudnick and J. Clark, *Biochim. Biophys. Acta* **1144,** 249 (1993).

volved in antagonist recognition.[3,4,34] Some of the most powerful and convincing data from chimera studies will come from "gain of function" effects; thus, particular attention should be paid to situations where an antagonist gains potency for a chimera as a result of the introduction of only a small domain from the parental transporter with greater potency for that antagonist. The ideal situation would allow for the reciprocal effect by introducing the same domain from lower potency transporter into the transporter with higher potency for the antagonist and observing a loss of potency. Regardless, because multiple nonspecific effects could account for loss of potency, a leftward shift in potency for an antagonist induced by chimera formation will always provide strong evidence for the involvement of a domain in contributing to the molecular interactions of the antagonist with the transporter.

Transporter-Mediated Currents

An ion conducting state associated with translocation of substrates has been reported for various members of the GAT/NET gene family of transporters.[35,42–46] This property allows one to study the behavior of substrates without the requirement for radiolabeled compounds. Thus, substrates induce transporter-associated currents, whereas antagonist block substrate-induced currents, but fail to induce currents themselves. In our hands, even the closely related human and *Drosophila* SERT demonstrate differences in this conducting state, providing another opportunity to investigate structural features of the SERTs using the interspecies chimeras. For example, transport of 5-HT by *Drosophila* SERT appears to be voltage dependent, but human SERT exhibits voltage-independent transport activity.[35] Also, the magnitude of current associated with *Drosophila* SERT is 5–10 times larger than that observed for human SERT. We suspect that the human/*Drosophila* SERT chimeras will reveal domains involved in these properties, providing important information about the ions responsible for the currents as well as direct evidence of residues lining the permeation pathway.

[42] A. Galli, L. J. DeFelice, B. J. Duke, K. R. Moore, and R. D. Blakely, *J. Exp. Biol.* **198,** 2197 (1995).

[43] S. Mager, C. Min, D. J. Henry, C. Chavkin, B. J. Hoffman, N. Davidson, and H. A. Lester, *Neuron* **12,** 845 (1994).

[44] A. Galli, R. D. Blakely, and L. J. DeFelice, *Proc. Natl. Acad. Sci. USA* **93,** 8671 (1996).

[45] S. Risso, L. J. DeFelice, and R. D. Blakely, *J. Physiol.* **490,** 691 (1996).

[46] M. S. Sonders, S. J. Zhu, N. R. Zahniser, M. P. Kavanaugh, and S. G. Amara, *J. Neurosci.* **17,** 960 (1997).

Radioligand Binding

Classically, radioligand binding has been a major tool in the characterization of membrane receptors, ion channels, and transporters. Selective high-affinity radioligands for many of the transporters are commercially available and can be used in the study of transporter chimeas.[47] Coupled with functional studies using uptake of radiolabled substrates, radioligand binding can provide useful information regarding the affinity of antagonists for transporters. Binding studies may also have use in the study of chimeras that lack transport activity. Loss of function for chimeric transporters is typically defined by a lack of substrate translocation ability; however, the inability to transport a substrate does not necessarily prevent a radiolabeled antagonist from binding to the transporter. Thus, if a particular radioligand retains affinity for a chimera, characterization using competition binding studies could proceed and provide useful information not attainable using uptake studies. For example, mutation of serine residues in TMD7 of DAT cause a marked reduction in DA transport activity; however, these TMD7 mutants retain affinity for the radiolabeled cocaine analog, CFT, allowing for characterization using binding paradigms.[8] The lack of transport activity certainly suggests that the active conformation of the transporter protein has been disrupted; hence, any data derived from binding studies on nonfunctional chimeras or mutants carries a level of doubt that should be considered before any conclusions are drawn. Nevertheless, the utility of radioligand binding studies in the characterization of wild-type and chimeric transporters can provide unique information about the interactions of ligands with the transporter.

Protein Analysis

The availability of antibodies directed to transporter-specific epitopes allows for specific analysis of transporter protein synthesis and trafficking. Western blot and immunoprecipitations can be particularly useful in verifying the proper synthesis of chimeric molecules in terms of quantity, appropriate molecular mass, and cellular processing. For nonfunctional chimeras or chimeras with markedly reduced transport activity, determination of cell-surface expression levels of the proteins can reveal chimera-induced disruptions of trafficking mechanisms. We have successfully used a cell-surface biotinylation technique[21–23] to demonstrate that a nonfunctional human/*Drosophila* SERT chimera has cell-surface expression levels equiva-

[47] E. L. Barker and R. D. Blakely, *in* "The RBI Handbook of Receptor Classification and Signal Transduction" (K. J. Watling, J. W. Kebabian, and J. L. Neumeyer, Eds.), p. 20. Research Biochemicals International, Natick, Massachusetts, 1995.

lent to those of the parental transporters.[24] This confirms that the lack of transport activity is due to a disruption in the transporter itself and not in cellular trafficking mechanisms responsible for targeting the transporter to the cell surface. Such analysis of proteins are invaluable in a thorough characterization of both functional and nonfunctional chimeric transporters.

Conclusions

Much structural information can be obtained by comparing species variants of transporters and using interspecies chimeras to track functional properties to specific transporter domains as demonstrated by our studies with human, rat, and *Drosophila* SERT. The power of interspecies comparisons is completely dependent on the comprehensive characterization of the parental transporters to identify species-specific properties that the chimeras can localize. One must then analyze the parental transporter data with an extremely critical eye, as species differences can be quite subtle. Our studies with human and rat SERTs revealed a 5 to 10-fold difference in tricyclic antidepressant potency between the two SERTs, but the distinction for *d*-amphetamine was less obvious at approximately three-fold. Once species-specific properties have been identified and cross-species chimeras have implicated a domain involved, site-directed mutagenesis can be initiated seeking to identify the residue(s) that mediate the species differences. Amino acid sequence analysis between the species variants provide a list of residues that differ between the two parental transporters as potential candidates. In our studies of rat and human SERT, chimeras allowed us to narrow our targeted region to the TMD12 domain where only four residues differed between the two SERTs (Fig. 2)[3,4] We then substituted the identities from human SERT into rat SERT, attempting to create a rat SERT mutant that possessed the higher potency interactions with the tricyclic compounds that were observed with the parental human SERT. This strategy was chosen because it tested for the gain of potency, which is a more powerful indicator of effects than mutation-induced loss of potency. These studies identified a single residue, phenylalamine-586, responsible for the higher potency interactions of the tricyclics with human SERT. Had our initial approach been to mutate distinct residues between the species variants, we would have had to mutate them *all* in a single construct to verify that *one* residue accounted for observed differences. Chimeras allowed us not only to narrow the mutagenesis effort, but also to establish the importance of a single residue for functional distinctions carried out by the entire protein.

After a residue has been implicated in mediating a specific functional property, the final step is to determine the molecular role for that residue in the function of the transporter. We suggest two approaches in this final set of studies. Additional mutations can be made at the site of interest to explore the nature of the residue's interactions with a ligand. In our studies, we substituted position 586 with the aromatic tyrosine, acidic aspartate, and basic arginine. Only the negatively charged aspartate mimicked the effect of phenylalanine, suggesting that the native role of phenylalanine may involve cation–π electron interactions either with ligand or through intramolecular interactions stabilizing the ligand binding pocket.[4,48] If possible, mutation studies addressing substrate or antagonist recognition sites should be coupled with structure–activity studies using ligand derivatives to identify chemical modifications that either mimic or are insensitive to the effects of transporter mutations. Strader and co-workers have used this approach in the study of the G-protein-coupled receptors to confirm the interactions of ligand functional groups with receptor residues.[19,49] Similar studies can be performed with derivatives of transporter ligands to define molecular information linking a specific transporter residue with either a substrate or antagonist functional group. We have described the use of tryptamine derivatives in the analysis of the SERT D98 mutation (see above). Similar studies using various tryptamine analogs could be performed on SERT species variants as well as SERT chimeras to identify domains involved in the recognition of transporter substrates. The use of interspecies differences and chimeras to associate domains with specific transporter properties can be a very useful method to initiate structure–function studies, ultimately leading to the identification of specific residues involved in antagonist and substrate binding, ion interactions, and protein trafficking.

[48] D. A. Dougherty, *Science* **271,** 163 (1996).
[49] T. M. Fong, M. A. Cascieri, H. Yu, A. Bansal, C. Swain, and C. D. Strader, *Nature* **362,** 350 (1993).

[34] Molecular Cloning of Neurotransmitter Transporter Genes: Beyond Coding Region of cDNA

By DAVID J. VANDENBERGH

Introduction

Molecular cloning of genes encoding the neurotransmitter transporters opened up many new avenues of study into evolution, transcriptional regulation, and genetics of these molecules that are important for neural func-

0076-6879/98 $25.00

tion. Cloning of cDNAs of the transporters allowed questions of pharmacology and physiology to be addressed via analysis of protein function, but could not contribute directly to these other areas. These transporters exhibit specific patterns of expression on selected sets of neuronal and, in some cases, nonneuronal cells that are essential to their proper functioning. There are also characteristics of the transporters that make them candidates for a genetic basis for neuropsychiatric disorders when they do not function properly. Study of expression patterns and genetic disorders requires isolating clones of the neurotransmitter transporter genes, including exons, introns, and flanking regions. For example, a deletion within a tandem repeat in the 5′ flank of the serotonin transporter gene has been associated with anxiety-related personality traits.[1] This tandem repeat plays a role in serotonin transporter expression in JAR cells, an immortalized mast cell line, and lies approximately one kilobase pair (kb) upstream of the start site of transcription.[2] A comparison of human and mouse dopamine transporter (DAT) genes shows a high density of potential transcription factor binding sites within the 200 base pairs 5′ of the start site of transcription.[3] This short region appears to allow brain specific expression of a marker gene (β-galactosidase), but with variable patterns of brain region expression between transgenic mouse lines.[4] With the information generated by efforts to map and sequence the entire genome of several organisms, such as the Human Genome Project, it will become increasingly helpful to have genomic DNA of the transporter genes in hand.

Since 1992, our group has focused attention on the structure and regulation of the DAT gene, as well as genetic differences between individuals in this gene. The following methods describe the techniques used to isolate and characterize DAT genomic clones, but these methods should apply to any gene for which a cDNA clone is available. Variants of commonly used techniques such as Southern blot hybridizations with radioactively labeled primers are described. In addition, application of newer techniques, including Long and Accurate PCR (polymerase chain reaction),[5,6] that have

[1] K.-P. Lesch, D. Bengel, A. Heils, S. Z. Sabol, B. Greenberg, S. Petri, J. Benjamin, C. R. Muller, D. H. Hamer, and D. L. Murphy, *Science* **274,** 1527 (1996).

[2] A. Heils, A. Teufel, S. Petri, G. Stober, P. Riederer, D. Bengel, and P. K. Lesch, *J. Neurochem.* **66,** 2621 (1995).

[3] D. M. Donovan, D. J. Vandenbergh, M. P. Perry, G. S. Bird, R. Ingersoll, E. Nanthakumar, and G. R. Uhl, *Mol. Brain Res.* **30,** 327 (1995).

[4] D. J. Vandenbergh, G. S. Bird, R. Ingersoll, E. Nanthakumar, and G. R. Uhl, *Soc. Neurosci. Abs.* **20,** 381.7.

[5] W. Barnes, *Proc. Nat. Acad. Sci. USA* **91,** 2216 (1994).

[6] R. Saiki, D. H. Gelfand, S. Stoffel, S. J. Scharf, R. Higuchi, G. T. Horn, K. B. Mullis, and H. A. Erlich, *Science* **239,** 487 (1988).

proved useful in isolating and cloning regions of the DAT gene not found in two genomic DNA libraries are described. Approaches to optimize the use of information from other transporter genes both to speed the characterization of the new gene and to prevent spending effort on clones of a gene already identified are discussed.

Identifying Genomic Clones

Library Plating and Screening

Escherichia coli lambda (λ) phage are a frequently used vector for genomic DNA libraries because they can accommodate insert DNA of 15 to 20 kb with minimal rearrangement or deletions of the insert, and because the phage themselves are stable. An outgrowth of the Human Genome Initiative is the generation of several vectors that can accommodate much larger fragments of DNA than are found in lambda [up to 1000 kb for a yeast artificial chromosome (YAC)]; however, the large insert DNA is more difficult to handle and suffers a higher frequency of rearrangements than lambda inserts. Little detail is provided in this section in order to focus on characterization of the isolated clones, which is more specific to the transporter genes, and because detailed instructions accompany the libraries or can be found in several molecular cloning manuals[7,8] and this series.[9] It is worth noting that fresh *E. coli* host bacteria and an accurate titer of plaques in the library are the two critical elements to screening a lambda phage library that allow plating of the phage at a density high enough to maximize the chance of finding positives, but not so dense that individual clones do not grow large enough to give a detectable signal.

Isolating Phage DNA

Once a lambda phage has been purified to homogeneity, DNA can be isolated by growing a large-scale preparation of the phage by either liquid culture or plate lysate methods. The latter is generally preferred because it is rapid and less demanding of accurate phage concentrations and timing.

[7] J. Sambrook, E. F. Fritsch, and T. Maniatis, "Molecular Cloning, A Laboratory Manual," 2nd ed. Cold Spring Harbor Laboratory Press, 1989.

[8] F. M. Ausubel, R. Brent, R. E. Kingston, D. D. Moore, J. G. Seidman, J. A. Smith, and K. Struhl, "Current Protocols in Molecular Biology." Wiley, New York, 1994.

[9] J. L. Slightom, *Methods Enzymol.* **224,** 251 (1993).

Four to five 15 cm agar plates are warmed to 37° for each phage isolate to be purified. For each plate, 10^5 phage and 0.6 ml of *E. coli* that have been grown overnight and resuspended to a density of OD_{600} = 0.5 in 10 m*M* $MgSO_4$ are mixed in a test tube and allowed to bind at 37° for 30 min. Six milliliters of molten top agarose (0.7%) that has been tempered to 47° in a separate test tube is poured into a tube containing phage and *E. coli,* and the whole solution poured back into the first tube for mixing before pouring onto the agar plate. Agar should not be used for the top solution because it contains contaminants that copurify with DNA and inhibit most enzymes used in subsequent steps. The plates are allowed to solidify and then grown overnight until the plaques are large enough to form a relatively clear plate with only a weblike pattern of lysed bacteria. Ten milliliters of Lambda Diluent (10 m*M* Tris, pH 7.5, 10 m*M* $MgSO_4$) is pipetted on each plate and incubated at 4° with gentle shaking for several hours. The solution is then removed and gently mixed in a polypropylene centrifuge tube with 0.1 ml of chloroform to lyse remaining bacteria. After centrifugation at 4000*g* for 10 min at 4° to pellet bacterial debris and agarose particles, the supernatant is removed to a fresh tube and incubated with RNase A and DNase I (60 μg/ml and 20 μg/ml final concentrations, respectively) at 37° for 30 min to degrade bacterial nucleic acids. The phage particles are precipitated by the addition of 2 ml of ice-cold Lambda Precipitation buffer [30% (w/v) polyethylene glycol 6000, 3 *M* NaCl, 10 m*M* Tris pH 7.5, 10 m*M* $MgSO_4$], incubated on ice for 1 hr, and centrifuged at 10,000*g* for 10 minutes. The phage pellet is resuspended in buffer M2 (500 m*M* guanidinium isothiocyanate, 10 m*M* MOPS buffer, pH 6.5, 1% Triton X-100) and heated to 65° for 20 min to break open the phage and free the DNA. This solution is ready for loading onto an ion exchange column such as the Tip-20 from Qiagen (Chatsworth, CA; Buffer M2 was suggested by technical services at Qiagen and has solved DNA degradation problems encountered in other protocols). The DNA recovered from the column is ready for analysis as described below.

Characterizing Genomic Clones

There are six questions involved in characterization of genomic clones: (1) Is the clone a true positive? (2) Where are convenient restriction sites located that will serve as landmarks within the large insert DNA? (3) Which fragments generated by restriction enzyme digestion (restriction fragments) contain an exon or exons? (4) What are the intron sizes? (5) What is the exact position of an intron with respect to the mRNA (cDNA) sequence? (6) Does the map reflect the gene structure found in native genomic DNA? These questions are addressed below.

1. Confirmation That the Phage DNA Contains the Gene of Interest. Once the phage has been isolated, it should be characterized to ensure that it contains a genomic DNA insert that hybridizes to the cDNA used for the initial screening. Restriction enzyme digestions and Southern blotting of the phage DNA are used for this last stage of screening. Many lambda vectors are engineered so that the insert is flanked by restriction sites that cut genomic DNA infrequently, for example, *Xho*I and *Sac*I in the lambda vector EMBL-3 SP6/T7. Digestion with one of these enzymes will "pop out" the insert and allow its size to be determined, as well as allowing isolation of the insert in one or a few pieces. These flanking-site enzymes, as well as several others that have six base recognition sites, for example, *Eco*RI, *Bam*HI, *Hind*III, *Pst*I, and *Xba*I, should be used to digest approximately 1 μg of phage DNA for each enzyme. The digested DNA is then size fractionated by agarose gel electrophoresis. Minigel units work well and the finished gel can be used to prepare duplicate Southern blots in the following manner: The gel is soaked in at least 5× the gel volume of Transfer Solution (1 *M* NaOH/1.5 *M* NaCl) for 30 min. Two nylon-based filters (e.g., Nytran, Schleicher & Schuell, Keene, NH) are cut to gel size and wetted in H_2O and then soaked in Transfer Solution. The gel is placed on a glass plate, the nylon filter is overlayed, and a pipette is rolled across the filter to force out any trapped air between the gel and filter. The procedure is repeated for each of three sheets of gel-sized pieces of heavy filter paper (Whatman 3mm, Whatman, Clifton, NJ), and then a small stack of dry paper towels is placed on top. The whole setup is turned upside down on the laboratory bench and the glass is slid carefully off of the gel. The second filter, filter papers, and paper towels are set on top of the gel, and transfer allowed to occur overnight. Hybridization of the resulting Southern blots with primers specific to the desired transporter gene provides a rapid way of confirming the identity of the genomic DNA from the purified phage. The primer must contain enough nucleotide differences between it and exon sequences of related but undesired transporter genes to prevent a false positive signal. In general, five nucleotide differences in a primer of 15 to 20 nucleotides will greatly reduce cross hybridization. The use of primers for hybridization will also begin to indicate which portion of the cDNA is encoded by the genomic clone of the Southern blot and which fragments of the phage insert contain exons.

The use of primers as probes on a Southern blot is very similar to using a full cDNA as a probe, but can be completed in 1 day. The first step is to start the prehybridization of one Southern blot per primer. The easiest method is to place a filter into a sealable plastic bag (e.g., Seal-a-Meal, available at discount stores) with the bag and filter close fitting on three sides, but with 10 cm or more of extra space on the fourth side ot allow several sealings of the bag. Prehybridization solution [5 × Denhardt's, 6 × SSC, 0.5% SDS, 100 μg/ml single-stranded DNA (from herring or salmon sperm)] is added to the bag at approximately 1 ml/15 cm^2 of filter. The prehybridization solution is allowed to wet the filter, and then the remaining air is forced out of the bag and the bag is sealed. Bubbles that are trapped on the filter are dislodged by pulling the bag over an edge of a laboratory bench, trapping the air in the extra space in the bag, and allowing the solution to fall down to the filter. The bag with filter is placed in a plastic box with the trapped bubbles propped against one side of the box, and placed in a shaking water bath with a weight on top of the box to keep it shaking rather than floating. The temperature of the bath should be 2–5° below the T_m of the primer (see Primer Selection for information on calculating the T_m of a primer) during both the prehybridization and hybridization steps. After 15 min, the bag is removed from the bath, any bubbles that have formed on the filter are dislodged as described above, and prehybridization is continued for a minimum of 2 hr.

During the prehybridization step the primer chosen for hybridization is radioactively labeled in 10 μl total volume consisting of 50 mM Tris, pH 7.6, 10 mM $MgCl_2$, 5 mM dithiothreitol (DTT), 0.1 mM spermidine, 0.1 mM EDTA, 2 μM primer, 3 μCi [γ-^{32}P]ATP (not dATP) (approximately 3 μl of 3000 Ci/mmole specific activity) and 10 units of T4 polynucleotide kinase. The reaction is mixed gently and incubated at 37° for 30 min and is terminated by bringing the volume to 100 μl with Stop Solution (0.1% SDS/5 mM EDTA). One microliter of this solution is removed for scintillation counting to measure the total radioactivity in the reaction. The labeled primer is then separated from free nucleotide through a spin column. Many columns are commercially available, but one can be constructed by packing a small plug of silanized glass wool in a 1000 μl disposable pipetter tip, and filling the tip with Sephadex G-25 until the Sephadex is within 1 cm of the top when liquid stops flowing from the tip. A 100 μl volume of Stop Solution (see above) with 1 μg of tRNA as a nonspecific blocker is added to the column for a prespin. A collar to hold the tip can be made by cutting off the bottom of a 0.5 ml microfuge tube and sliding it inside a 1.5 ml tube. The tip is placed into the collar so that the 1.5 ml tube catches the eluate, and the whole assembly is placed inside a polypropylene tube (Falcon tube #2059 or equivalent). The prespin only needs to be fast enough to force the liquid from the spin column; 1 min at 1000 rpm in a swinging bucket of a table-top centrifuge is sufficient. The eluate is removed from the 1.5 ml tube, the column is reassembled, and the labeling reaction is added and centrifuged under the same conditions as the prespin. Scintillation counting of 1 μl of this eluate represents incorporated radioactivity. Any reaction that yields less than 20% incorporation should not be used and the problem causing poor incorporation should be determined (usually inactive enzyme or old buffer) before the labeling reaction is repeated.

The hybridization and washing steps with a primer probe are much shorter than those of longer probes because of the rapid rates of annealing of short DNA fragments. The recovered probe from the spin column is added to the bag containing the filter and prehybridization solution to a final concentration of 1–5 $\times$ 10^6 cpm/ml of solution. In order to minimize the spreading of radioactivity to everything in harm's way, the bag is sealed at the top edge, trapping some air and all of the radioactivity inside. The bag is inverted to mix the contents and then the air is forced to the top of the bag by pulling the bag over an edge of the box (behind a radioactivity shield). Once the solution has settled down to the filter, the bag can be resealed close to the edge of the filter, preventing air bubbles from interfering with hybridization. The bag is returned to the water bath and hybridization allowed to proceed for 30 min. Longer times will cause significant background problems. After hybridization, the bag is opened and the filter transferred to a plastic box containing at least 100 ml of 2 $\times$ SSC. The hybridization solution can be poured into a 50 ml conical tube with a cap and stored at room temperature or below for at least 2 weeks for reuse. The SSC solution is removed as soon as the probe is dispensed with, and the blot is washed once for 15 min by shaking at room temperature with at least 200 ml of blot washing solution (0.1 $\times$ SSC, 0.5% SDS). It is quite easy to believe that the blot is too hot and inadvertently wash off all of the true hybrids during a second wash. An exposure of 5–10 min to film or a phosphor storage system is sufficient to detect hybridization. If the filter is wrapped moist in plastic wrap or a plastic bag, it can be washed again if the first wash did not remove enough of the nonspecific hybridization. Figure 1A shows an agarose gel of lambda DNA clones. Note that the inserts have internal *Xho*I sites; thus, there are several bands in addition to the two lambda vector fragments at 23 and 8 kb. An autoradiogram of the Southern blot of the gel from Fig. 1A that has been probed with a primer from exon 1 of the human dopamine transporter is shown in Fig. 1B.

2. Restriction Site Mapping and Ordering of Subclones Generated from Some of the Restriction Sites. If the purified DNA hybridizes to a probe from the cDNA used in the initial screening, a restriction map of one or a few restriction enzyme sites should be generated to help serve as landmarks for subsequent analysis. There are two methods of generating a restriction map from the genomic clone. The first, by partial digestion of

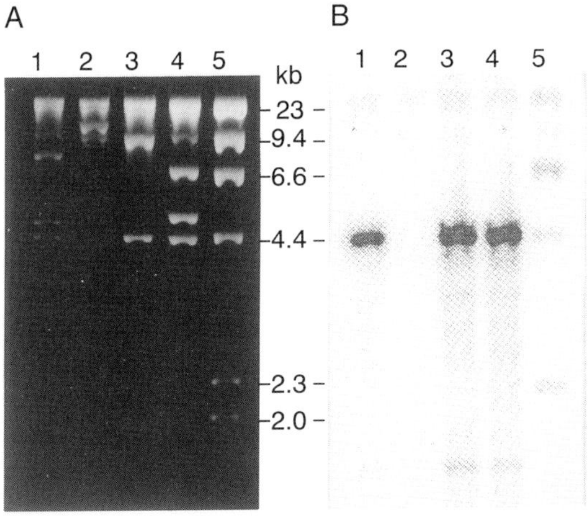

FIG. 1. Lambda phage inserts containing sequences from the dopamine transporter gene. (A) Ethidium bromide stained agarose gel showing lambda DNA digested with *Xho*I (lanes 1–4) and size marker DNA (lane 5) of lambda cut with *Hind*III. Sizes in kilobase pairs, (kb) are shown on the side. (B) Autoradiogram of a Southern blot of the same gel in (A), probed with a primer specific to exon 1 of the dopamine transporter gene. Note that the clone in lane 2 overlaps with the other three clones, but does not hybridize to exon 1.

the insert DNA, can be carried out while subcloning is underway, but it requires large amounts of DNA and it can be difficult to optimize the partial digestion (see Ref. 7 for details).

The second method, based on partial sequencing, is less demanding technically and is shown in Fig. 2. This method generates sequence data that is useful in later steps of characterization and also detects very small fragments that might not be seen by partial digestion, or even missed in subcloning. The first step is to purify the genomic insert by cutting the phage DNA with a restriction enzyme (F) that only cuts the lambda vector DNA at the flanks of the insert (site F in Fig. 2A) and then gel isolating the insert DNA (a few bands if the enzyme F cuts within the insert). This purified insert DNA can then be further digested with one of the enzymes determined in Part 1 to give a few bands, especially in the 1–6 kb size range (X in Fig. 2B). These subfragments are then subcloned into a plasmid vector to allow easier isolation of large amounts of DNA (Fig. 2C). Sequencing with primers that bind to the flanks of the plasmid cloning site (arrows in Fig. 2C) generates partial data from each end of the subclones. These primers are commercially available and are specific to the RNA polymerase binding sites that have been engineered into many plasmids; the most

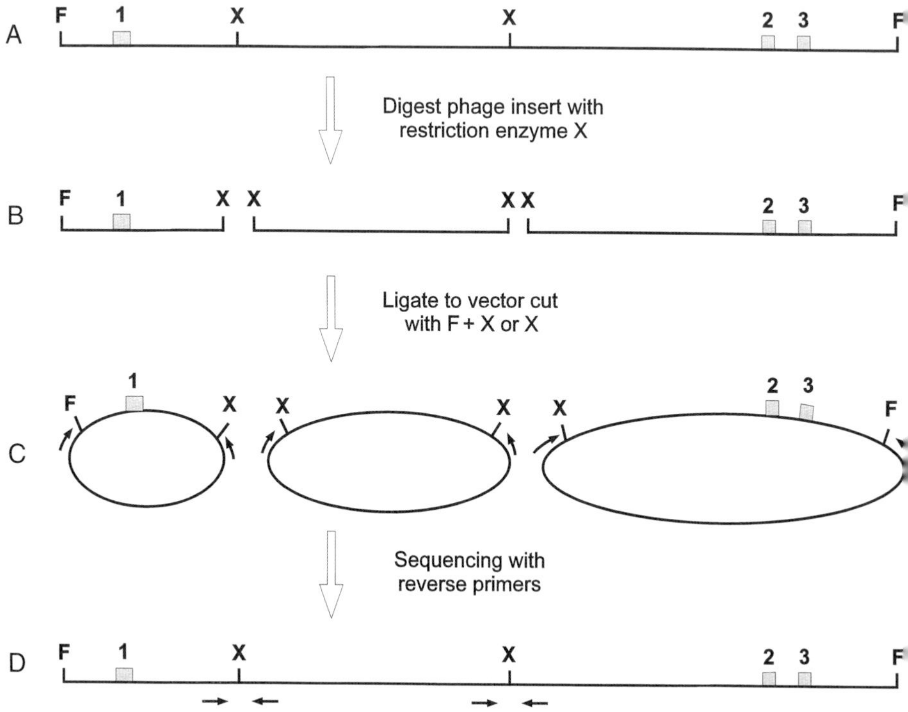

FIG. 2. Schematic of restriction site mapping by subcloning and sequencing. (A) Purified phage insert with restriction sites F (flanking) and X (internal) is shown along with exons 1, 2, and 3 (shaded boxes). (B) Restriction fragments that are generated by enzyme recognizing site X. (C) Subclones generated from the restriction fragments. The arrows show the plasmid-derived primer binding sites for generating sequence of the ends of each subclone in order to make reverse primers. (D) Position of the reverse primers (arrows) on the original phage insert for sequencing or PCR.

common of the sites are those known as SP6, T3, and T7. From the sequence of the ends of the clones, "reverse" primers are picked from the opposite strand (heading 5′ to 3′ toward the flanking restriction site). The choice of sequence for a primer is described in more detail below under Primer Selection. As shown in Fig. 2D, these reverse primers generate sequence from the original phage insert across the restriction site X from one subclone to its neighboring subclone. The lack of a match indicates that a fragment of DNA was missed in the subcloning. This problem occurs when a fragment is too small to be seen when subcloning. In such a case, the sequences align only up to the restriction enzyme site (X in Fig. 2) that defines the end of the subclone. The alignment may begin again if the region of nonmatching sequence is short enough to be spanned in the sequencing experiment, and should start with another restriction site X. It is also possible that two restriction fragments of the phage insert are the same size and comigrate on the agarose gel. If the two fragments are copurified but only one is subcloned and sequenced, then there will be a gap in the sequence alignment. The undetected fragment can be recovered by more screening of subclone ligation reactions. The size of the phage insert should match up to the sum of the sizes of its subclones, but it is easy to underestimate the size of the phage insert, and thus miss a fragment.

If the phage insert is in limited supply, PCR reactions using 1–10 ng of insert or intact phage DNA as template can serve to identify adjacent subclones. If all possible combinations of the reverse primers are used in separate reactions, only those primer pairs from subclones that abut one another will generate DNA fragments. The resulting PCR fragments can be sequenced to verify the identity of the subclones.

3. Identification of Exon-Containing Subclones. The positions of exon/intron boundaries for other transporter genes should be used as a guide for those of the newly identified gene. With a few exceptions, these boundaries are well conserved among the transporter gene family, and each transmembrane domain of the protein is encoded on a separate exon. The large number of exons, 14 or 15 per transporter gene, makes for small exons of 100 to 300 base pairs (bp) that can be separated by introns of 10 kb or more. Identifying all of the subclones that contain exons has already begun under Part 1 by determining the fragment size that hybridizes to a specific primer, and thus the corresponding subclone generated in Part 2. Primer probing is fast, but requires multiple blots or a series of single experiments.

4. Determining the Position of an Exon within a Subclone and the Size of Introns. With the map order of the clones from Part 2 and with the knowledge of which end of each subclone is 5′ or 3′, PCR can be used to position an exon within a clone. In fact, these experiments can be started with the entire phage insert, but may require more effort if the exons are more than 3 or 4 kb from a phage insert's flank (see Long and Accurate PCR, below). Two PCR reactions are set up as in Section 2, using one primer that anneals to a transporter exon for both reactions, and in each reaction, one of the two primers that anneal to vector sequence (e.g., SP6, T3, or T7 primers). The amount of time for the extension step depends on the length of insert that the *Taq* polymerase must traverse. As a starting point, assume an enzyme synthesis rate of 1 kb per minute, and that the polymerase must traverse three-quarters of the entire insert each cycle. Usually a PCR product can be detected by gel electrophoresis of one-quarter to one-half of the reaction. Even with a subclone, which is highly purified DNA of low complexity compared to genomic DNA, the amplification of a large fragment can be difficult. If a reaction fails to yield any product, the concentrations of primers, template, or magnesium can be altered as first attempts to increase the yield of reaction product. The PCR fragments generated in this analysis are often useful as probes themselves and can be cloned for future experiments.

5. *Defining the Exact Exon/Intron Junctions.* Once it is known that a primer hybridizes to a subclone, the primer can be used to generate sequence from the exon into the flanking intron. The Sequenase Kit (USB, Cleveland, OH), or similar ones from other vendors, works well and provides consistency that is worth its expense. Preparation of the template DNA and primer is simplified by the following: The subclone (4–5 μg) is mixed with 1 μl of 1 μM primer in a total volume of 9 μl, boiled for 3 min, incubated at room temperature for 10 min, and then briefly centrifuged to force any condensation back to the bottom of the tube. While the mix is annealing at room temperature, a reaction mix is prepared and the remainder of the procedure follows the manual provided in the kit. There are several chapters of other volumes of this series that describe sequencing and troubleshooting in more detail, and there are several commercial sequencing services or university molecular biology core facilities that can sequence by contract. Determination of the junction between exon and intron can sometimes be ambiguous when only one end of an intron is sequenced; for example, when the first two nucleotides of an exon are GT (which is the same as the first two nucleotides of an intron). By obtaining sequence from both sides of an intron, the exact location can be pinpointed. The conserved elements that compose an intron/exon junction are described by Senapthy *et al.*[10] and greatly help in the analysis.

6. *Comparison of Data Derived from the Clone to Native Genomic DNA.* It is possible for pieces of genomic DNA to be deleted or rearranged by recombination during propagation in bacteria. This problem is not common, but does arise and can be detected easily with a sample of genomic DNA blotted to a filter (Genomic Southern). A Southern blot prepared with restriction digested genomic DNA (5–20 μg of DNA per lane) and probed with a ^{32}P-labeled subclone insert should produce the same size fragment on the blot as in the cloned DNA. If the genomic DNA has been cut with different restriction enzymes, it is possible to test the relative positions of restriction sites that have been mapped on the genomic clones. The distance measured between two restriction sites is limited by the resolution of fragments by gel electrophoresis—15–20 kb, and longer if pulsed field gel electrophoresis is used. Thus, a short fragment from a subclone can verify

[10] P. Senapathy, M. B. Shapiro, and N. L. Harris, *Methods Enzymol.* **183,** 252 (1990).

a large region of surrounding DNA. It is also possible to detect deletions by PCR, but only those between the two primers used.

One important caution is that intron DNA may contain repetitive elements, such as Alu repeats, that are scattered randomly at more than 100,000 sites in the genome. If such a repeat is present in a probe, the result is a blot with radioactivity along the entire length of the lanes of DNA. These repetitive elements hybridize at a significantly faster rate than the remainder of genomic DNA, which can be used to block repeats in probe DNA. DNA that consists of repeated elements is sold as COT-1 DNA by GIBCO-BRL (Gaithersburg, MD). When nonradioactive COT-1 DNA is mixed with single-stranded probe the repeat sequences hybridize in solution, thus preventing their hybridization to the Southern blot. If COT-1 DNA is not available, it can be established whether a subclone contains a repeat by a "reverse blot." A Southern blot with subclones on it is probed using genomic DNA that has been digested with any restriction enzyme and then radioactively labeled. Single-copy DNA is present at such lower abundance than repetitive elements that single-copy subclones hybridize very weakly if at all, whereas those subclones that have repetitive DNA give strong signals.

The procedures described above generate a series of genomic clones with a restriction map, and the positions of the exons of the gene. These reagents and information can then be used in experiments to examine genetic and molecular biological questions of the transporter gene.

Isolating Genomic Regions Poorly Represented in Libraries

Long and Accurate PCR

There are sequences in mammalian genomic DNA that are particularly unstable in the bacteria or yeast that serve as hosts for the maintenance and propagation of genomic libraries. These sequences are underrepresented or entirely absent from the libraries, and unlikely to be isolated with the techniques described above. A combination of Long and Accurate PCR and specialized bacterial strains that have low rates of recombination may allow successful cloning of these sequences if they are present in a transporter gene. Barnes and colleagues[5] developed a technique that allows PCR amplification of long fragments of DNA, greater than 4 kb in length, as well as generating high yields of the amplified fragment. An example of a gap between genomic clones that contain exons 3 and 4 of a transporter gene, but do not overlap, is shown in Fig. 3. The first step is to design primers, a forward primer from exon 3 and a reverse primer from exon 4 (see below). Long primers with high melting temperatures that allow an-

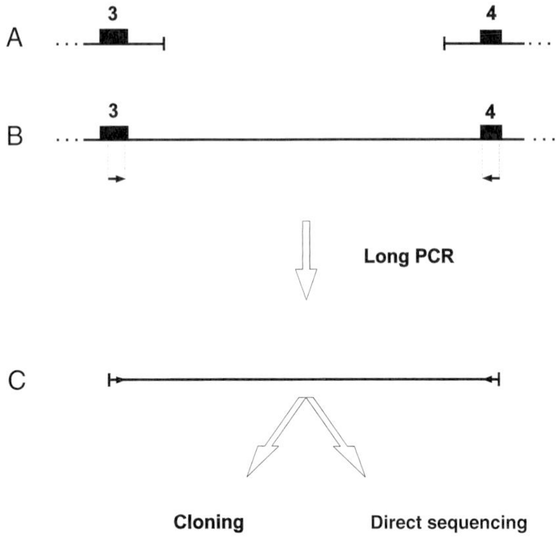

FIG. 3. Long PCR scheme to clone introns and gaps. Lambda clones that do not overlap are shown in (A) with exons 3 and 4 as shaded boxes. A forward primer that anneals to exon 3 and a reverse primer that anneals to exon 4 (arrows) are shown in (B) with the intron to be amplified. The PCR product shown in (C) contains the intron sequences missing from the lambda phage clones.

nealing to the target DNA in the range of 68–72° have a higher success rate than primers with lower melting temperatures in long PCR reactions. Several of the components, as well as the conditions, of a PCR reaction can be optimized if DNA amplification does not occur. The following setup is a good starting point and has been designed to use a technique known as "hot start" to prevent mispriming at sites other than those intended, which can generate background.

Reagent	Bottom mix	Top mix
Water	15.5 μl	4.0 μl
10 × Reaction Buffer	4.5 μl	0.5 μl
All 4 dNTPs (2.5 m*M*)	5.0 μl	—
Forward Primer (20 μ*M*)	5.0 μl	—
Reverse Primer (20 μ*M*)	5.0 μl	—
Genomic DNA (10 ng/μl)	10.0 μl	—
Long *Taq* Polymerase (5 U/μl)	0 μl	0.5 μl
	45.0 μl	5.0 μl Total volumes

The Bottom Mix can be prepared as a master mix without genomic DNA and then aliquotted into individual tubes so that different genomic DNA samples can be tested individually. After the Bottom Mix and genomic DNA are combined in the PCR tube, a drop of mineral oil is added as an overlay to prevent evaporation, and the samples are started in the first round of thermal cycling. When the temperature has reached 80° or slightly higher, the machine is placed on hold to maintain a constant temperature. The Top Mix is added directly to the PCR tube without removing the tube from the machine by pipetting the mix under the mineral oil, and mixing very briefly with the tip. Once all tubes have had Top Mix added, the cycling is resumed so that the reactions now heat to the first denaturing step. The denaturing step requires temperatures of 94–98° for 10 sec to 1 min, followed by a combined annealing and extension step at 66–70° if the primers have melting temperatures that allow binding in this range. A separate annealing step should be used if the primers require annealing below 66°. A total of 35 cycles of denaturation and annealing/extension should give enough PCR product to be detected on an agarose gel. Figure 4 shows the result of amplification of intron 14 of the human DAT gene. Cloning and sequencing, or direct sequencing, of the product verifies that the correct region of genomic DNA has been amplified.

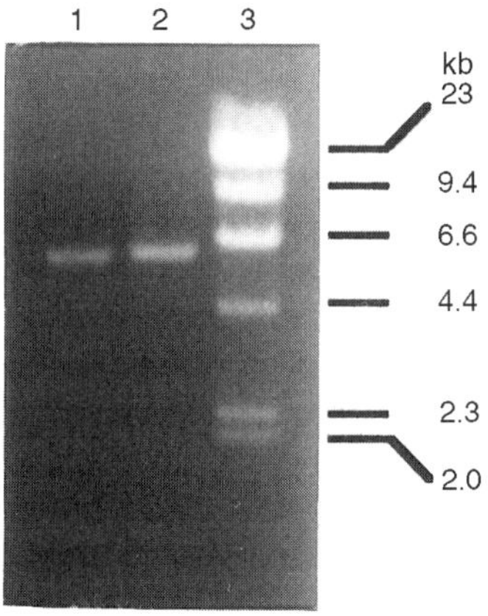

FIG. 4. Long PCR amplification of intron 14 from the dopamine transporter gene. Primers specific to exon 14 (forward primer) and exon 15 (reverse primer) were used to amplify genomic DNA from two individuals (lanes 1 and 2) for subsequent isolation and subcloning. The intron 14 DNA migrates at approximately at 6 kb. Lane 3 shows size markers of lambda DNA digested with *Hind*III.

Primer Selection

The choice of two primers for the amplification of any desired region of DNA is probably the most important decision that has to be made for successful amplification, and the complexity of mammalian genomic DNA magnifies the importance of this decision. The success of a pair of primers in amplification of DNA is still something of a mystery despite the heavy use of PCR since its invention.[11] It is strongly recommended that a computer program be used to aid in this part of an experiment. A number of commercial DNA analysis software packages contain sophisticated programs to aid in making a choice of a primer pair. "Primer," free software written by Steve Lincoln, Mark Daly, and Eric Lander at MIT, has been used very successfully to choose primer pairs. Primer can be downloaded via anonymous ftp from "genome.wi.mit.edu" (IP address 18.157.0.135).

Several characteristics of a primer pair are described below that should be checked even when software is used in primer design.

(a) The melting temperature (T_m) is probably the most frequently thought-of parameter and should be around 68–70° as described above for two-step cycles. Primers with T_m values as low as 50–55° can be used successfully but with greater risk of mispriming. If software is unavailable or does not calculate T_m values, then the 2 + 4 rule can be used as a quick estimation: $T_m = 2(A + T) + 4(G + C)$. This method tends to underestimate the T_m and actually serves as a starting point for the annealing temperature of a PCR reaction.

(b) Primer length can be variable, but for work with genomic DNA the length should be a minimum of 18 nucleotides, and the maximum being that necessary to achieve the correct T_m (usually 30 or less), with an average primer being 20–22 nucleotides long. Shorter primers, thus with lower T_m values, were used initially because of the cost and low purity of long primers when PCR was first developed. It is now possible to purchase primers inexpensively with high enough purity that primers of 30 bases in length rarely need further purification before use. The complexity, or presence of many different sequences, in a mammalian genome also requires that long primers be chosen to prevent unexpected binding of the primers at multiple sites along the genome.

(c) There should be no more than four of any single base in succession, particularly at the 3′-end of the primer where polymerization is initiated.

(d) The presence of simple sequence repeats, such as CACACA or

[11] S. F. Altschul, W. Gish, W. Miller, E. W. Meyers, and D. J. Lipman, *J. Mol. Biol.* **215,** 403 (1990).

GTGTGT, should be avoided to prevent binding of the primer to simple repeats present at many sites in the genome.

(e) The primer should end in a G or C "clamp" to allow tighter binding, and thus more efficient polymerization, than with a terminal A or T.

(f) A primer should not be able to form a hairpin structure, or bind to itself or the second primer of the pair. Due to their small size, primers hybridize at a higher rate, and this fact coupled with the high molar excess of primers over template during a reaction may allow the primers to serve as a template for polymerization. Such a side reaction product is known as a "primer dimer" and can become the predominant PCR product in a reaction to the exclusion of the desired product. The dimer products tend to be small, running below 200 bp in size on an agarose gel.

(g) A primer that has been chosen from a gene's introns or flanking DNA, or even untranslated regions of a cDNA, must be screened for the ability of the primer to bind to repetitive elements such as Alu repeats and L1 elements. The potential for binding to a repetitive element is best screened by using a computer search algorithm such as BLAST[12] or FASTA[13] that compares the sequence of the primer for matches in the GenBank sequence database. This analysis may also reveal a significant match to a nonrepetitive sequence that might still compete for primer annealing. This type of analysis is limited to known sequences that are in the database, but is particularly important to do when isolating members of a gene family, such as the transporters, which have regions of high sequence match.

Conclusion

With gene sequences in hand, it is possible to ask questions concerning transporter function and expression that are difficult or impossible to address with only the transporter protein. Screening the cloned DNA for repeated elements, such as simple sequence repeats or tandem repeats, may uncover polymorphic markers to trace inheritance of traits that may be caused by variants of the transporter.[14] Gene knockout, which depends on gene cloning, has begun to clarify the role of the dopamine transporter in amphetamine and cocaine response in the intact mouse. Finally, the relationship of a transporter gene to other genes nearby on the chromosome may provide insight into gene regulation and evolution, as is the case with the vesicular choline transporter, which is contained entirely within the

[12] W. Pearson, *Methods Enzymol.* **183,** 63 (1990).

[13] GenBank, Release No. 97.0, Oct 15, 1996.

[14] D. J. Vandenbergh, A. M. Persico, A. L. Hawkins, C. A. Griffin, X. Li, E. W. Jabs, and G. R. Uhl, *Genomics* **14,** 1104 (1992).

first intron of the choline acetyltransferase gene.[15,16] These and perhaps other aspects of transporter function not yet imagined can be addressed, once the initial isolation of the transporter genes is completed.

Acknowledgments

For support, I would like to thank George Uhl, in whose laboratory the work on gene cloning was carried out, and Donna Walther, Geoffrey Bird, and Elisabeth Bendahhou, with whom I have enjoyed the efforts of isolating the dopamine transporter gene. The support of the National Institute on Drug Abuse Intramural Research Program and the Frederick Biomedical Supercomputer Center of the Frederick Cancer Research Facility, National Cancer Institute, NIH, and the DNA Analysis Facility, Johns Hopkins University Medical School, particularly Roxann Ingersoll, is gratefully acknowledged.

[15] J. D. Erickson, H. Varoqui, M. K. H. Schafer, W. Modi, M. F. Diebler, E. Weihe, J. Rand, L. E. Eiden, T. I. Bonner, and T. B. Usden, *J. Biol. Chem.* **269,** 21929 (1994).

[16] S. Bejanin, R. Cervini, J. Mallet, and S. Berrard, *J. Biol. Chem.* **269,** 21944 (1994).

[35] Use of Antisense Oligodeoxynucleotides to Inhibit Expression of Glutamate Transporter Subtypes

By LYNN A. BRISTOL and JEFFREY D. ROTHSTEIN

Introduction

A family of glutamate transporters has been identified and all share the characteristic of high-affinity transport, although their ion requirements vary. Immunohistochemical and immunoelectron microscopic studies have also detailed the cellular localization of these proteins.[1] In spite of the ability of these techniques to elegantly detail the precise localization of these proteins, they provide no information on the quantitative and functional contribution of the proteins to normal and abnormal synaptic transmission. Some investigators have attempted to utilize immunoprecipitation and protein isolation methods to estimate the relative amounts of each transporter subtypes, but these studies provide no true information on the physiological contribution of each transporter subtype to synaptic and extracellular clearance of glutamate. So how can one study the role of each transporter

[1] J. D. Rothstein, L. Martin, A. I. Levey, M. Dykes-Hoberg, L. Jin, D. Wu, N. Nash, and R. W. Kuncl, *Neuron* **13,** 713–725 (1994).

0076-6879/98 $25.00

subtype in the extracellular clearance of glutamate? In this chapter we provide methodological information on the use of a highly specific and powerful method to parcel out the individual role of each glutamate transporter subtype to total tissue glutamate transport. As will be shown, this method can provide a reliable and relevant method to understand the contribution of each glutamate transporter subtype in the clearance of glutamate.

This chapter outlines a highly specific method to study the biology of glutamate transporter subtypes using antisense oligodeoxynucleotide (ODN) inhibition of individual transporter proteins. As will be discussed, this method can be employed for both *in vitro* and *in vivo* paradigms. An essential requirement for the use of this method is knowledge of the nucleotide sequence for the transporter of interest. To date, five glutamate transporters have been cloned in animals, along with their human counterparts: EAAT1 (GLAST), EAAT2 (GLT-1), EAAT3 (EAAC1), EAAT4, and EAAT5.

General Considerations in the Use of Antisense Oligodeoxynucleotides

Antisense oligodeoxynucleotides (ODN) offer several important advantages when studying the function of selected proteins. (1) They are highly specific, and given the fact that no highly specific glutamate transporter subtype-specific inhibitor exists, they offer the optimum method to study transporter subtypes in normal brain tissue. (2) Their effect can be reversible. If they are only briefly administered, they can also be used to produce transient and reversible transporter synthesis inhibition. (3) They can be used for both *in vitro* and *in vivo* paradigms. (4) Depending on the experimental paradigm, they offer a low cost method to "knock out" a gene product (as opposed to the high cost and long delay in producing null mutation mice). They require no specialized equipment to use, other than the availability of facilities to synthesize them. Most major universities now have core facilities capable of oligodeoxynucleotide synthesis; furthermore, many commercial sources exist for low cost, rapid synthesis of oligodeoxynucleotide reagents.

In spite of these distinct advantages, antisense oligodeoxynucleotides also have some clear drawbacks which one must consider in designing experiments: they can have sequence-independent effects (e.g., phosphorothioate oligodeoxynucleotides can be nonspecifically cytotoxic); repeated treatments may be necessary because of degradation; and they may not completely inhibit synthesis of the target protein.

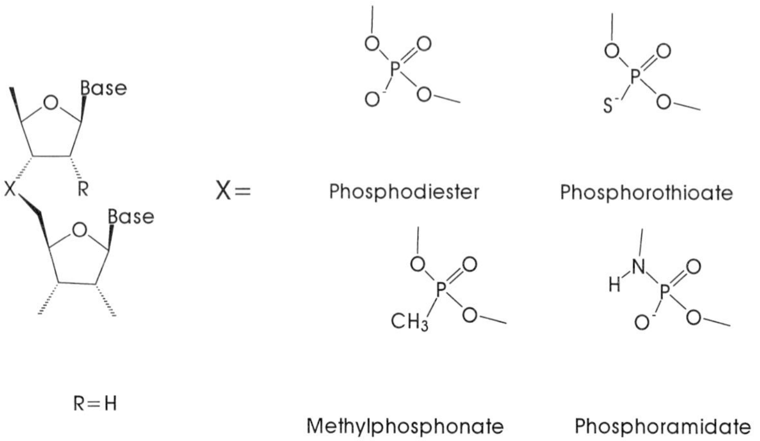

FIG. 1. Various modifications of oligonucleotides used for antisense inhibition of protein synthesis.

Mechanism of Action

There are a number of mechanisms proposed to account for the inhibition of protein synthesis following antisense ODN use: (1) hybrid arrest, the hybridization of the antisense to mRNA to prevent translation and protein synthesis; (2) degradation by RNase H which involves cleavage of heteroduplexes of the mRNA and antisense ODN; and (3) secondary effects on gene expression. A number of excellent reviews concerning the mechanism of antisense action are available.[2,3]

Types of Oligodeoxynucleotides

Standard oligodeoxynucleotides have a naturally occurring phosphodiester linkage between ribose sugars. Because of the rapid degradation of these molecules, many modifications have been studied in order to stabilize the molecule and increase its antisense actions. The most widely used modification is the phosphorothioate substitution in the linkage (see Fig. 1). This appears to decrease degradation *in vivo* and *in vitro*, thus increasing their half-life to 12–24 hr. Many other modifications are being developed, including methyl phosphonates and phosphoroamidation, as well as base modifications.

[2] S. T. Crooke and C. F. Bennett, *Ann. Rev. Pharmacol. Toxicol.* **36,** 107–129 (1996).
[3] C. Helene and J. J. Toulme, *Biochim. Biophys. Acta* **1049,** 99–125 (1990).

Design of Antisense Oligodeoxynucleotides

Sequence Selection

In general, the choice of the antisense sequence should be near the ATG start site of the nucleotide sequence for the transporter subtype. Nevertheless, in theory, any site within the coding sequence should be effective at inhibiting the ultimate formation of the protein. However, practical experience has shown that the best inhibition of synthesis often occurs when the chosen sequence spans the ATG site or resides in the early 5′ sequence. The actual length of the sequence should be 17- to 21-mer. This is not absolute; it has been calculated that at the genomic level any sequence of 17-mer or more is unique. If the sequence is too long, entry into cells may be impaired. It is very important to analyze the chosen sequence with a sequence program, e.g., Oligo (NBI, Plymouth, MN), in order to avoid potential hairpin loops which could affect binding to the target RNA sequence. In addition, the antisense sequences can form duplexes with themselves (self-complementarity) (see example, Fig. 2) and these programs can predict the likelihood of such interactions. As a general

```
Current Oligo   24-mer [1]:

Current + Oligo: the most stable 3'-dimer: 2 bp, -1.5 kcal/mol
5'  ATATTGTTGGCACCCTCGGTTGAT  3'
    ::        :     :      ||
3'  TAGTTGGCTCCCACGGTTGTTATA  5'

Current - Oligo: the most stable 3'-dimer: 4 bp, -4.0 kcal/mol
5'  ATCAACCGAGGGTGCCAACAATAT  3'
                        ||||
                    3'  TATAACAACCGTGGGAGCCAACTA  5'

Current + Oligo: the most stable dimer overall: 3 bp, -4.4 kcal/mol
5'  ATATTGTTGGCACCCTCGGTTGAT  3'
               |||    :::
        3'  TAGTTGGCTCCCACGGTTGTTATA  5'

Hairpin: ΔG = 0.80 kcal/mol, Loop = 3 nt, Tm = 6 °C
5'  ATATTGTTGGCACCC┐
               ||| T
        3'  TAGTTGGC┘
```

FIG. 2. Evaluation of the suitability of an antisense sequence. Oligo can be used to determine if a chosen oligodeoxynucleotide sequence will be suitable for use by determining the likelihood of either duplex dimers or hairpin formation. Both the sense and antisense orientations (i.e., + or − Oligo) are automatically evaluated.

rule, to avoid this problem, the ΔG should be less than −5 kcal/mol (e.g., −1 to −4 kcal/mol). Finally, the sequence should be analyzed against known sequences in a genome data bank to be certain that the chosen sequence is unique. This is commonly performed by BLAST analysis, available via the internet and the National Institutes of Health (www.ncbi.nih.gov/BLAST). Examples of sequences that can be used to inhibit glutamate transporter subtypes are shown in Table I. Importantly, sequences are generally species specific. Although the peptide sequences for the transporters subtypes are very similar between species (e.g., human and rat), the nucleotide sequences are often sufficiently different that antisense ODN can only be synthesized based a species-specific sequence.

Control Sequences

Because antisense ODN are known to cause nonspecific effects in a number of experimental systems, the use of controls is critical in the experimental design. Controls need to take into account several factors: the size of the antisense ODN, the actual composition of the nucleotides in each antisense, and the effects of the ODN on a particular cell type. A standard control must include ODN that corresponds to the appropriate *sense* sequence. In addition, since the sense stand does not actually contain the same proportion of A, T, G, and C nucleotides as the antisense ODN, a control which consists of random ODN with the exact same proportion of the four nucleotides in the antisense ODN should be designed. It is important to check the random sequence in a genome database to ensure that it is truly random and does accidentally code for another protein. An additional control that can be considered, which is the equivalent of peptide blocking experiments used in immunohistochemistry, is to preincubate sense and antisense strands together, before administration. Finally, as an additional control for the increased formation of intracellular RNA–oligonucleotide duplexes that could occur in astrocytes following the administration of GLT-1 or GLAST antisense oligodeoxynucleotides, animals can be treated with antisense to other astroglial specific proteins, such as glial fibrillary acidic protein (GFAP). A suitable antisense sequence for GFAP is 5′-CAGAGGCGAGGTAGAACG-3′.

Modifications

Standard antisense oligonucleotides have a phosphodiester linkage which exactly mimics the natural biochemical structure of nucleotide sequences. However, this linkage is highly susceptible to endogenous nucleases present in tissue or in tissue culture conditions. For this reason, many investigators have used modified linkage structures in the design of

TABLE I

Antisense and Sense Sequences for Inhibition of Glutamate Transporter Subtype Synthesis in Rat

Transporter subtype	Sense ODN	Antisense ODN	Random ODN
rEAAT1 (GLAST)	5′GAAAGATAAAATATGACAAAAAGCAAC-3′ (nucleotides −12 to 15)	5′-GTTGCTTTTTGTCATATTTTATCTTTC-3′	5′-TGTCGTTTTGTTATCTATATTCTTTCT-3′
rEAAT2 (GLT-1)	5′-ATCAACCGAGGGTGCCAACAATAT-3′ (nucleotides 6 to 29)	5′-ATATTGTTGGCACCCTCGGTTGAT-3′	5′-AATTGTGTTAGCCCCCTCTGTTGA-3′
rEAAT3 (EAAC1)	5′-GCTCGGGATGCGACTGGC-3′ (nucleotides 17 to 34)	5′-GCCAGTCGCATCCCGAGC-3′	5′-GCGGATCCGTACGCCCAG-3′

the antisense ODN in order to prevent degradation of the nucleotides. Typically, thioate derivatization of the oligonucleotides increases the half-life (e.g., 12–24 hr). Phosphorothioate derivatized oligonucleotides can be prepared with replacement of the entire phosphodiester linkage using thioate, or can be "end-capped" with thioate only at the end linkages. End-capping can effectively inhibit endonuclease degradation and serves to limit the number of thioate linkages. This alternative can be utilized if the thioate derivatized oligonucleotides appear to cause excessive nonspecific toxicity. There are now newer modifications of the basic oligonucleotides, often referred to as "second" and "third" generation oligonucleotides. These modifications, as shown in Fig. 1, include methyl phosphonates and phosphoramidate derivatizations.

Because oligonucleotides have a relatively limited ability to enter cells, alternative methods have also been developed to enhance delivery of the molecules. Researchers have been able to increase the efficacy of antisense protein inhibition by more than 100-fold by attaching oligonucleotides to the Antennapedia homodomain protein. This protein can translocate across biological membranes to serve as an excellent vector for delivery of ODN to cells. Researchers have found that it can enhance the potency of ODN effects by 100-fold.[4]

Other Methods

Two other methods can also be employed to inhibit the synthesis of glutamate transporters, hammerhead ribozyme mRNA cleavage and triple helix. Hammerhead ribozymes can be engineered to contain a partial sequence of a glutamate transporter subtype. These can bind to the mRNA resulting in immediate cleavage of the RNA, thereby inhibiting the synthesis of the target mRNA. Triple helix approaches require knowledge of the promoter site sequence, but can also be employed to inhibit synthesis of glutamate transporter subtypes.

Preparation of Antisense Oligodeoxynucleotides

Oligonucleotides can be synthesized from university core molecular biology/genetics facilities or can be purchased from many commercial sources. For the large amounts used for *in vitro* and *in vivo* experiments, orders of 1–3 mg are more cost effective than the small orders typically required for PCR (polymerase chain reaction) primers or other molecular

[4] C. M. Troy, D. Derossi, A. Prochiantz, L. A. Greene, and M. L. Shelanski, *J. Neurosci.* **16,** 253–261 (1996).

biology experiments. The antisense should be purified by HPLC (usually performed by the supplier). However, before delivery into animals or into culture media, there are two important procedures necessary to minimize specific toxic effects of all ODN used. It is necessary to dialyze them to remove trace chemical impurities, since they will be delivered to central nervous system (CNS) cultures or brain, which are quite sensitive to pyrogens, and should be filtered to remove contaminating bacteria.

Following are the methods necessary for the preparation of antisense ODN for use in chronic infusion paradigms.

Preparation of Oligonucleotides for Chronic Intraventricular Infusion

1. Preparation of Vehicle for Oligodeoxynucleotides: Artificial Cerebrospinal Fluid

	For 1 liter
NaCl	8.4 gm
$MgCl_2 \cdot 6H_2O$	0.244
$CaCl_2 \cdot 2H_2O$	0.185
KCl	0.25
$NaC_2HPO_4 \cdot 7H_2O$	0.129
$NaHCO_3$	1.764
Glucose	0.61
Urea	0.13

Adjust pH to 7.4 with 1 *N* HCl. Keep refrigerated (4°C).[5]

The above has been corrected for hydrated chemicals. The original formula for each ionic species is as follows:

	Mol/liter
Na^+	0.141
Ca^{2+}	0.00125
K^+	0.0033
Mg^{2+}	0.0012
Cl^-	0.152
HPO_4^-	0.00048
HCO_3^-	0.021
Glucose	0.0034
Urea	0.0022

[5] Modified from J. K. Merlis, *Am. J. Physiol.* **131,** 67–72 (1940).

2. Preparation of Antisense Oligonucleotides for Use

(a) Add artificial CSF to purified oligonucleotides, at about 2.5 mg/ml.

(b) Dialyze newly dissolved oligonucleotides to remove impurities, pyrogens, and endotoxins. This is a critical step that appears to markedly reduce nonspecific toxicity associated with both *in vitro* or *in vivo* ODN use. Dialysis tubing should be rinsed and soaked for at least 30 min in double distilled H_2O prior to use. It is important to use dialysis tubing with a low molecular-weight cutoff, typically around 2000 [e.g., Spectra/Por CE, (Spectrum, Los Angeles, CA), cellulose ester, molecular weight cutoff 2000, flat width 10 mm]. Dialysis should be performed overnight in 2 l of artificial CSF at 4°. Very carefully remove dialyzed solution and sterilize it by filtration through a 0.22 μm culture filter (e.g., Millipore, Bedford, MA Millex-GV4 for solutions of 1 ml or less).

(c) Fill Alzet (Palo Alto, CA) miniosmotic pump. The size of the osmotic pump varies with the amount of delivery time required. For a 7- to 9-day delivery period, the Alzet model is suitable for subcutaneous placement in a rat, and it can hold 200 μl of solution. This pump has a delivery rate of approximately 1 μl per hr, a volume quite suitable for intrathecal or intraventricular delivery.

Dosing and Delivery

In Vitro. When administering antisense ODN *in vitro,* generally a full range of concentrations needs to be studied to determine the most effective dose to inhibit synthesis of the transporter. There are variables that can be factored into the dosing regimen: what is the effective dose, how to administer the antisense, for how long, and should membrane permeabilization aids be used?

For *in vitro* studies, we generally recommend a large range: 0.5–20 μm, added to culture medium for up to 1 week. The length of addition to culture medium depends, in part, on the half-life of the transporter proteins. Our studies suggest, for permanent cell lines (L-cells), transporter subtypes EAAT2 and EAAT3 have a half-life of about 16 hr. For organotypic cultures, new antisense ODN should be added twice weekly, along with the standard culture media changes. Various cationic lipid preparations are also commercially available to enhance delivery of phosphorothioate ODN into cells (e.g., Lipofectin, GIBCO-BRL, Gaithersburg, MD). However, the drawbacks to their use are (1) direct cytotoxicity with chronic use, and (2) cell-to-cell variability in the efficacy of intracellular delivery.

In Vivo. For *in vivo* inhibition of transporter synthesis, multiple variables need to be considered: rate and mode of delivery, concentration of the antisense, length of delivery period, distribution and or location of delivery, vehicle for delivery. The location for delivery depends on the region of interest. The antisense molecules do not pass through the blood–brain barrier and thus cannot be administered via intraperitoneal or intravenous delivery; they must be directly injected into the brain. Furthermore, ODN do not diffuse large distances when delivered directly to the intraventricular space. For example, as shown in Fig. 3, after intraventricular delivery of GLAST antisense ODN, the effect on protein levels decreased dramatically within 5 mm of the site of delivery.

There are two basic methods for *in vivo* delivery—direct, acute intracerebral and intraventricular injections or chronic infusions. Since treatment with antisense ODN may take several days before there are detectable changes in the steady-state levels of transporter proteins, it requires multiple daily or chronic infusions of the ODN to achieve an effect. Because multiple daily direct injections often leads to CNS infections, we do not recommend this method for chronic delivery. Instead, we recommend chronic infusions using indwelling intracerebral or intraventricular cannulas along with mini-osmotic pumps commercially available from Alzet. These pumps have calibrated rates of delivery and deliver the ODN over 3, 7, 14, or 30 days. Alternatively, single pumps can be used, e.g., 7 days, and replaced repeatedly in the same animal.

Chronic delivery of antisense ODN can be performed via indwelling stainless steel cannulas into the lateral ventricles, directly into cerebral parenchyma, or via thin polyethylene tubing intrathecally for delivery to the spinal cord. For intraventricular delivery cannulas are generally placed at the following stereocoordinates: −0.8 mm anterior–posterior; −1.5 mm lateral; approximately −4.8 mm dorsoventral in the right lateral ventricle. Correct placement of cannulas should be confirmed in the animals at the time of sacrifice. The concentration of antisense ODN to use for chronic delivery is entirely empirical. A range of 1–50 nmol delivered over 24 hr (e.g., 2 nmol/μl/hr) has been effective at decreasing transporter synthesis.[6]

Assessing Efficacy of Synthesis Inhibition

The effectiveness of the antisense ODN to alter synthesis of glutamate transporter proteins is best determined by immunoblot analysis of the

[6] J. D. Rothstein, M. Dykes-Hoberg, C. A. Pardo, L. A. Bristol, L. Jin, R. W. Kuncl, Y. Kanai, M. Hediger, Y. Wang, J. Schielke, and D. F. Welty, *Neuron* **16,** 675–686 (1996).

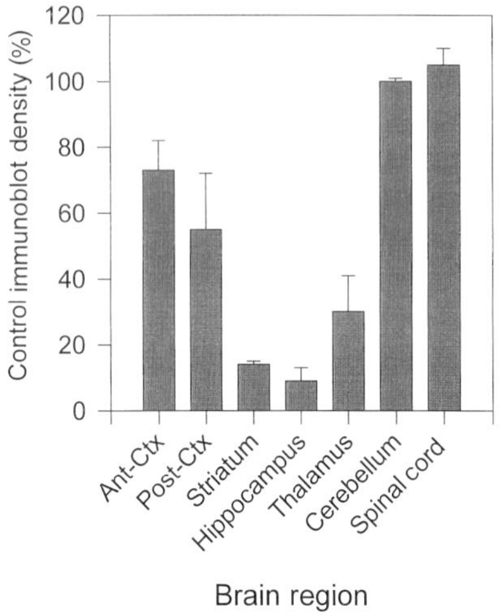

FIG. 3. Regional decrease in GLAST protein following intraventricular infusions of GLAST antisense ODN. Rats received chronic infusions of GLAST antisense (2 nmol/hr) for 7 days. After sacrifice, brains were removed on ice and 1 mm coronal sections were collected anterior and posterior to the intraventricular cannula site. Brain regions were dissected from the slices; the tissue was homogenized, and then immunoblotted with affinity-purified anti-GLAST antibody (see footnote 6). Synthesis of GLAST protein was less affected at distances far from the infusion site (e.g., cerebellum, spinal cord, and anterior cortex) compared to sense ODN treated animals. Ctx, Cortex.

tissue. Samples of culture homogenates or of tissue surrounding the infusion site can be subjected to standard sodium dodecyl sulfate–polyacrylamide gel (SDS–PAGE) and immunoblot analysis. Immunoblots should always compare sense, antisense, random oligonucleotide treated tissue as well as untreated tissue. The degree of inhibition of transporter synthesis may not reach 100%; depending on the half-life of existing protein, the length of treatment, and the efficacy of synthesis inhibition. Furthermore, for *in vivo* experiments, the actual tissue sampled may affect the interpretation; if intraventricular delivery is used and large amounts of tissue are analyzed that lie far from the infusion site, the immunoblots will reflect a combination of affected and unaffected tissue and thus artificially diminish the degree of synthesis inhibition. For this reason, tissue should be sampled at defined intervals (e.g., 1–5 mm) from the infusion site (see Fig. 3).

An additional test for the effect of synthesis inhibition should be the actual measurement of sodium-dependent glutamate transport. This is particularly important for transporters which are active when inserted into the plasma membrane—immunoblots alone only estimate total transporter

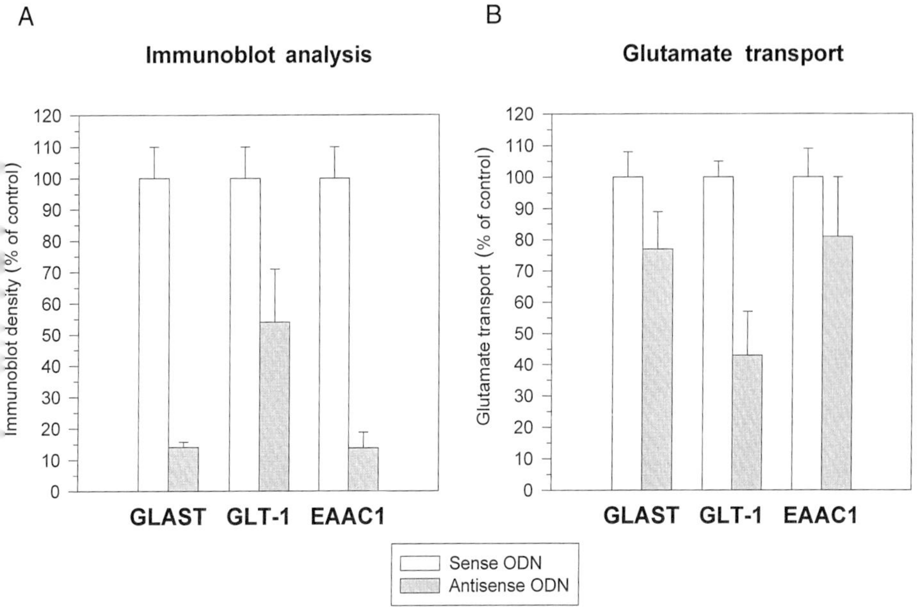

FIG. 4. Analysis of the effects of chronic intraventricular infusion of antisense or sense ODN for each glutamate transporter subtype (GLAST, GLT-1, or EAAC1) on tissue levels of each transporter protein. Striatal tissue was collected after 7 days of chronic infusions individual ODN. Tissue was then slit for immunoblot analysis and measurement of dependent glutamate transport. **(A)** Each antisense treatment produced a substantial decrease in transporter protein, although not always complete (e.g., GLT-1). **(B)** The effect of the loss of each transporter subtype on total tissue glutamate transport can be analyzed. In spite of the incomplete loss of GLT-12 protein in the tissue homogenates, it appears to contribute the most to total glutamate transport.

proteins—both membrane and cytoplasmic pools. As shown in Fig. 4, the loss of the transporter protein may not correlate with changes in tissue glutamate transporter since tissue measurements reflect the combined total of all sodium-dependent transporters. Widely used methods for sodium-dependent glutamate transport are available[7] and are outlined elsewhere in this volume and at the end of this chapter. Finally, it is important to be certain that the effect of the antisense is specific for the transporter subtype; if the inhibition of one subtype affects the expression of other subtypes, interpretation of results can be complicated. Therefore, immunoblots should routinely be performed for multiple transporter subtypes even if

[7] J. D. Rothstein and B. Tabakoff, *Biochem. Pharm.* **34,** 73–79 (1985).

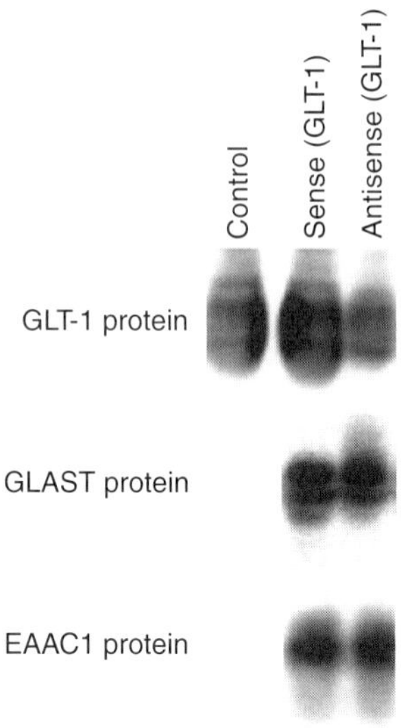

FIG. 5. Controls for potential nonspecific effects of antisense should be included in all experimental designs. The immunoblots from spinal cord cultures treated for 7 days with GLT-1 antisense revealed a large loss of GLT-1 protein; sense ODN had no effect. As shown in this immunoblot study, the GLT-1 antisense treatment had no effect on glutamate transporter subtypes GLAST or EAAC1.

only one has been targeted for antisense inhibition. As shown in Fig. 5, antisense inhibition of GLAST can be specific, without effects on either EAAC1 or GLT-1 proteins. The specificity of an antisense ODN can also be modified by the chronicity of the synthesis inhibition. For example, acute loss of GLT-1 by antisense inhibition could be compensated days later by an increased synthesis of other transporter subtypes.

Summary

Antisense ODN are highly effective tools to selectively inhibit synthesis of glutamate transporter subtypes. They have been used to evaluate the biology of individual transporter subtypes *in vivo* and *in vitro*. Appropriate use of antisense ODN, however, requires a number of important controls to validate the specificity of their effects. Ultimately, the efficacy at inhibiting the synthesis of transporters proteins reflects several variables: the actual efficacy of the antisense to directly inhibit synthesis, the rate of degradation of the ODN, the rate of intracellular penetration of the ODN, the rate of new protein synthesis for the transporter subtype, and the actual penetration of the ODN into CNS tissue. The methods described in this chapter will help one to optimize each of these parameters.

Appendix: Glutamate Uptake Assay

The procedures described below are for a standard assay for sodium-dependent, high-affinity glutamate transport in crude tissue preparations.

A. Tissue Preparation

1. Homogenize in tissue in 20 volumes 0.32 *M* buffered sucrose (Tris-HCl 0.05 *M*, pH 7.4) with a Teflon/glass homogenizer.
2. Centrifuge at 1000*g* for 10 min, collect supernatant.
3. Centrifuge supernatant at 20,000*g* for 30 min (P2 pellet).
4. Homogenize P2 pellet by vortexing in original 20 volume of sodium-free Krebs buffer.
5. Repeat step 4 up to 2 times to remove extracellular glutamate.
5. Check protein concentration of homogenate.

B. Assay

1. Prepare assay buffer (Krebs and sodium-free Krebs), tissue buffer, glutamate assay solution.
2. Prepare tissue as detailed above; keep resuspended tissue on ice.
3. Assay is performed in a total of 1 ml final volume in miniscintillation vials (4–7 ml).
4. Add in the following sequence: assay buffer (calculated to make final volume of all ingredients 1 ml), glutamate (always 100 μl), then add tissue aliquot (typically 50–100 μl).
5. *Immediately* shake tubes; place in shaking incubator.
6. Incubate at 37° for 4 min.
7. Place in an ice–water bath at end of incubation.
8. Immediately and rapidly add 1 ml ice-cold Krebs buffer to each tube (e.g., using Eppendorf repeating pipetter).
9. Place tubes in refrigerated centrifuge (4°); spin at least 5000*g* for 2–4 min.
10. Decant supernatant, clean inside of tubes with cotton swabs; avoid touching pellet.
11. Add 0.5 ml tissue solubilizer; let solution rest for 30 min (at 37–50°).
12. Add scintillation cocktail, shake, count radioactivity in standard β-counter.

Assay Reagents

A. Krebs Buffer

Prepare 1 *M* stock solutions. Use fresh solutions for all assays.

Chemical	Molecular weight	g/100 ml (1 *M*)
NaCl	58.4	5.84
$NaHCO_3$	84.01	8.4
KCl	74.6	7.46
$CaCl_2$	147	14.7
KH_2PO_4	136.09	13.61
$MgSO_4$	120.4	12.04
Choline chloride	139.6	13.96
Tris	121.1	12.1 (adjust to pH 7.4 with HCl)

B. Standard Krebs Solution

Chemical	Final concentration (m*M*)	Volume of 1 *M* stock (ml)
NaCl	120	12
$NaHCO_3$	25	2.5
KCl	5	0.5
$CaCl_2$	2	0.2
KH_2PO_4	1	0.1
$MgSO_4$	1	0.1

Add distilled water to 10 g D-glucose with final volume 100 ml. Adjust pH to 7.4.

C. Sodium-Free Krebs Solution

Chemical	Final concentration (m*M*)	Volume of 1 *M* stock (ml)
Choline chloride	120	12
Tris	25	2.5
KCl	5	0.5
$CaCl_2$	2	0.2
KH_2PO_4	1	0.1
$MgSO_4$	1	0.1

Add distilled water to 10 g D-glucose with final volume 100 ml. Adjust to pH 7.4.

Note: Add choline, Tris, KCl, KH_2PO_4, $MgSO_4$; bring almost to 100 ml volume; then add glucose, mix. Add $CaCl_2$ last to avoid precipitation.

D. Tissue Buffer

0.05 *M* Tris, 0.32 *M* sucrose:
Tris (molecular weight 121.1): 3.03 g/500 ml
Sucrose (molecular weight 342.3): 54.8 g/500 ml
Add Tris, sucrose in distilled water, adjust to pH 7.4

E. Glutamate

Final concentration/assay tube is 0.5 μM with 0.5 μCi ^{3}[H]glutamate.

Glutamate (molecular weight 169.1): (always prepare new solutions prior to assay):

Glutamate stock: prepare 1 m*M* solution (33.8 mg/200 ml).
Glutamate assay solution: take 5 μl of 1 m*M* glutamate stock solution, add to 10 ml Krebs (or sodium-free Krebs) along with 5 μl [^{3}H]glutamate (~54.7 Ci/mmole; 1 μCi/ml).
Plan to use 100 μl of Glutamate assay solution for uptake assay.

[36] Using *Caenorhabditis elegans* to Study Vesicular Transport

By JAMES B. RAND, JANET S. DUERR, and DENNIS L. FRISBY

Introduction

In this chapter, we explore ways in which studies using the nematode *Caenorhabditis elegans* can add to our knowledge of vesicular transporters and vesicular transport. We discuss four general areas where the use of *C. elegans* offers advantages over other approaches: the use of *C. elegans* mutants to determine cellular and behavioral requirements for transporter function, the use of *C. elegans* for structure–function studies of vesicular transporters, the use of antibodies and mutants to explore protein targeting, and the use of *C. elegans* genetics to identify interacting genes and proteins. These studies rely on the strengths of *C. elegans* as a research organism: a simple nervous system in which all cells are identified and synaptic connectivity is known and reproducible; a large collection of mutants and powerful

0076-6879/98 $25.00

methods of genetic analysis; simple methods for the generation and analysis of transgenic animals; and a number of relatively straightforward quantifiable behaviors. Thus, the major uses of *C. elegans* in this field are for analysis of the *functional* (i.e., behavioral) role(s) of transporters, and the analysis of cellular and subcellular localization of these proteins.

We conclude this chapter with a brief discussion of plasma membrane transporters of *C. elegans*. Although little research has been performed to date on these proteins in *C. elegans,* apparent homologs of the mammalian genes have been identified by the *C. elegans* Genome Sequencing Project and by other investigators. We explain how the strategies and methods described in this chapter can be applied to this set of transporters.

Caenorhabditis elegans Biology and Methodology

As the result of the combined efforts of many laboratories, *C. elegans* has been developed from relative obscurity to a remarkably well-characterized system. This effort has been motivated by two central features of the organism, originally recognized and capitalized on by Sydney Brenner[1]: (1) genetic manipulability and (2) cellular simplicity. A 2.5-day generation time, numerous progeny, small size, and ease of growth make genetic experiments relatively fast and straightforward. Among metazoans possessing nervous systems, only in *Drosophila melanogaster* are there comparable tools for classical and molecular genetic analyses. Cellularly, *C. elegans* is remarkably simple; the adult contains 959 somatic cells, of which 302 are neurons. All of the cell divisions, including those in the embryo, have been carefully observed using Nomarski optics[2,3]; the cell lineages for the entire adult are known and are extremely reproducible. In addition, reproducibility of neuron morphology and connectivity for most portions of the nervous system has been demonstrated by serial section electron microscopy.[4–7]

Caenorhabditis elegans is particularly suited for studying neural function. Its nervous system has many similarities to those of mammals, both morphologically and biochemically. Most proteins identified as important for neuro-

[1] S. Brenner, *Genetics* **77,** 71 (1974).

[2] J. E. Sulston, E. Schierenberg, J. G. White, and J. N. Thomson, *Dev. Biol.* **100,** 64 (1983).

[3] J. E. Sulston and H. R. Horvitz, *Dev. Biol.* **56,** 110 (1977).

[4] D. G. Albertson and J. N. Thomson, *Phil. Trans. R. Soc. Lond. B* **275,** 299 (1976).

[5] J. G. White, E. Southgate, J. N. Thomson, and S. Brenner, *Phil. Trans. R. Soc. Lond. B* **275,** 327 (1976).

[6] J. G. White, E. Southgate, J. N. Thomson, and S. Brenner, *Phil. Trans. R. Soc. Lond. B* **314,** 1 (1986).

[7] D. H. Hall and R. L. Russell, *J. Neurosci.* **11,** 1 (1991).

transmitter release have homologs in both *C. elegans* and mammals.[8,9] In addition, because hermaphrodites can self-fertilize and produce progeny without mating, it is possible to maintain strains of worms carrying severe neural defects which disrupt movement or other behaviors.

The genome is relatively small, consisting of 10^8 base pairs on six chromosomes, and has been extensively mapped and sequenced.[10–12] The *C. elegans* Genome Sequencing Project expects to complete (virtually) sequencing of the genome by late 1998. *Caenorhabditis elegans* can maintain injected DNA as extrachromosomal arrays, and it is therefore possible to express cloned genes by transformation.[13,14]

A collection of *C. elegans* methods has been published as a volume in *Methods in Cell Biology*,[15] which contains all of the basic techniques of *C. elegans* growth and maintenance. In addition, a compendium of anatomical, genetic, behavioral, developmental, and molecular information about *C. elegans* has been published,[16] and any investigator contemplating using this organism is strongly advised to consult this volume.

Vesicular Transporters in *Caenorhabditis elegans*

So far, two vesicular transporters have been identified and cloned in *C. elegans:* the vesicular acetylcholine transporter (VAChT) is encoded by the *unc-17* gene[17] and a vesicular monamine transporter (VMAT) is en-

[8] T. C. Südhof, *Nature* **375,** 645 (1995).

[9] K. G. Miller, A. Alfonso, M. Nguyen, J. A. Crowell, C. D. Johnson, and J. B. Rand, *Proc. Natl. Acad. Sci. USA* **93,** 12593 (1996).

[10] A. Coulson, J. Sulston, S. Brenner, and J. Karn, *Proc. Natl. Acad. Sci. USA* **83,** 7821 (1986).

[11] R. Wilson, R. Ainscough, K. Anderson, C. Baynes, M. Berks, J. Bonfield, J. Burton, M. Connell, T. Copsey, J. Cooper, A. Coulson, M. Craxton, S. Dear, Z. Du, R. Durbin, A. Favello, A. Fraser, L. Fulton, A. Gardner, P. Green, T. Hawkins, L. Hillier, M. Jier, L. Johnston, M. Jones, J. Kershaw, J. Kirsten, N. Lassiter, P. Latreille, J. Lightning, C. Lloyd, B. Mortimore, M. O'Callaghan, J. Parson, C. Percy, L. Rifken, A. Roopra, D. Saunders, R. Shownkeen, M. Sims, N. Smaldon, A. Smith, M. Smith, E. Sonnhammer, R. Staden, J. Sulston, J. Thierry-Mieg, K. Thomas, M. Vaudin, K. Vaughan, R. Waterston, A. Watson, L. Weinstock, J. Wilkinson-Sproat, and P. Wohldman, *Nature* **368,** 32 (1994).

[12] J. Hodgkin, R. H. A. Plasterk, and R. H. Waterston, *Science* **270,** 410 (1995).

[13] C. C. Mello, J. M. Kramer, D. Stinchcomb, and V. Ambros, *EMBO J.* **10,** 3959 (1991).

[14] C. Mello and A. Fire, *in* "*Caenorhabditis elegans:* Modern Biological Analysis of an Organism" (H. E. Epstein and D. C. Shakes, Eds.), pp. 451–482. Academic Press, San Diego, 1995.

[15] H. E. Epstein and D. C. Shakes, Eds. "*Caenorhabditis elegans:* Modern Biological Analysis of an Organism." Academic Press, San Diego, 1995.

[16] D. L. Riddle, T. Blumenthal, B. J. Meyer, and J. R. Priess, Eds. "*C. elegans* II." Cold Spring Harbor Laboratory Press, Cold Spring Harbor, New York, 1997.

[17] A. Alfonso, K. Grundahl, J. S. Duerr, H.-P. Han, and J. B. Rand, *Science* **261,** 617 (1993).

coded by the *cat-1* gene.[18] In addition, there is some evidence that the *unc-47* gene may encode a vesicular γ-aminobutyric acid (GABA) transporter (VGAT; see ref. 19).

VAChT. Mutations in the *unc-17* gene (so named because mutations conferred *unc*oordinated locomotion) were first described by Brenner.[1] In addition to the locomotion deficit, *unc-17* mutants with a slight amount of residual gene function are slow-growing, small, and resistant to inhibitors of cholinesterase.[1,20] Mutations which eliminate all gene function (null mutations) are lethal; mutant homozygotes are able to complete embryogenesis and hatch, but the young larvae are unable to move or feed normally, and they shrink and die within a few days.[17] This is the same phenotype displayed by *cha-1* mutants, which lack the enzyme choline acetyltransferase (ChAT) and are totally deficient for acetylcholine synthesis.[21]

The *unc-17* gene was cloned and sequenced by Alfonso *et al.*,[17] who showed that the predicted UNC-17 protein was associated with synaptic vesicles and was expressed only in cholinergic neurons (Fig. 1; see color insert). This was the first VAChT to be cloned; examination of the sequence demonstrated that VAChTs and VMATs are members of a gene family. An unusual aspect of *unc-17* was its genomic organization: it is part of a complex transcription unit that also includes the *cha-1* gene. The *cha-1* and *unc-17* transcripts appear to be derived by alternative splicing, with a common 5′-exon and the remainder of the *unc-17* gene nested within the long first intron of *cha-1*.[22] A similar genomic organization was subsequently identified in mammals: the mammalian VAChT gene is also nested within the first intron of the ChAT gene.[23,24] Thus, in both mammals and nematodes, the synthesis and the vesicular transport of acetylcholine are coupled at the genomic level. This suggests that the organization of this "cholinergic locus" is somehow important for its function. (Note that a similar organization is not found for VMAT.)

VMAT. The *cat-1* gene (abnormal *cat*echolamines) was first characterized by Sulston *et al.*[25] *cat-1* mutants were isolated on the basis of an altered

[18] J. S. Duerr, D. L. Frisby, J. D. Erickson, A. Duke, L. E. Eiden, and J. B. Rand, unpublished (1997).

[19] S. L. McIntire, E. Jorgensen, and H. R. Horvitz, *Nature* **364,** 334 (1993a).

[20] M. Nguyen, A. Alfonso, C. D. Johnson, and J. B. Rand, *Genetics* **140,** 527 (1995).

[21] J. B. Rand, *Genetics* **122,** 73 (1989).

[22] A. Alfonso, K. Grundahl, J. R. McManus, J. M. Asbury, and J. B. Rand, *J. Mol. Biol.* **241,** 627 (1994).

[23] S. Bejanin, R. Cervini, J. Mallet, and S. Berrard, *J. Biol. Chem.* **269,** 21944 (1994).

[24] J. D. Erickson, H. Varoqui, M. K.-H. Schäfer, W. Modi, M.-F. Diebler, E. Weihe, J. Rand, L. E. Eiden, T. I. Bonner, and T. B. Usdin, *J. Biol. Chem.* **269,** 21929 (1994).

[25] J. Sulston, M. Dew, and S. Brenner, *J. Comp. Neurol.* **163,** 215 (1975).

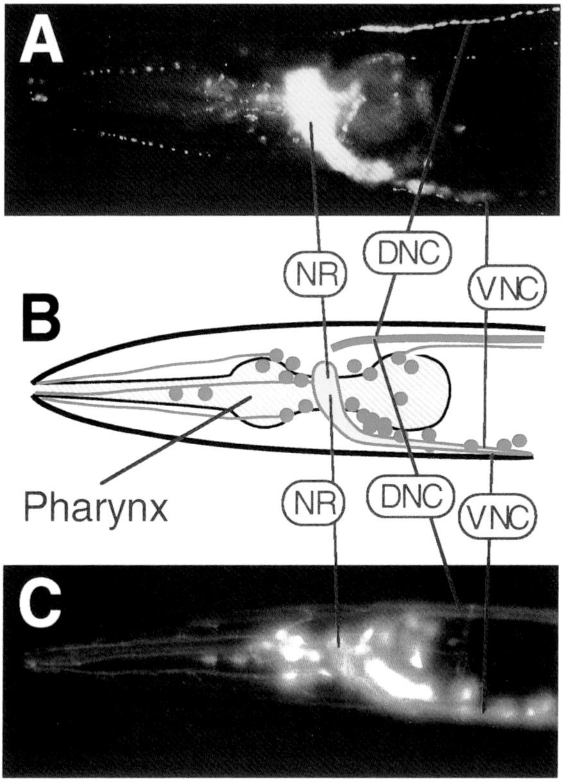

FIG. 1. Expression of VAChT in the nervous system of *C. elegans.* (A) Indirect immunofluorescence with antibody against UNC-17/VAChT. The antibody labels synaptic regions of putative cholinergic neurons; very little protein is detected in cell bodies. Distinct puncta believed to correspond to synapses can be seen in individual processes in the anterior region and in the dorsal nerve cord (DNC). More numerous synapses are present in the major neuropil, the nerve ring (NR), and the ventral nerve cord (VNC). (B) A diagram of the same region with the area of most intense anti-UNC-17/VAChT immunoreactivity shown in yellow. Also shown are cells (green circles) and processes (green lines) which express green fluorescent protein (GFP) with the construct shown in the lower panel. The black indicates the outer cuticle of approximately the anterior third of a larval *C. elegans* whereas the pink demarcates the pharynx. (C) The head of a transgenic *C. elegans* in which 3.2 kb of genomic DNA upstream from the *unc-17* (VAChT) coding sequence was used in vector pPD95.67 [C. Mello and A. Fire (1995). *In* "*Caenorhabditis elegans:* Modern Biological Analysis of an Organism" (H. E. Epstein and D. C. Shakes, Eds.), pp. 451–482. Academic Press, San Diego] to drive expression of GFP. This genomic region is sufficient to cause expression of the GFP marker protein in most of the neurons which normally express UNC-17/VAChT, including putative cholinergic neuron cell bodies (seen as small whitish circles) and processes (thin green processes). The pattern of fluorescence appears quite different from that seen with anti-UNC-17/VAChT antibodies because of the distinct subcellular locations of the endogenous protein (synaptic VAChT) versus the marker protein (cytoplasmic GFP).

tion of the phenotype in the adult. In the simplest case, the ablation of the cell(s) leads to the same phenotype as nonexpression of the transporter in that cell. However, sometimes a cell ablation leads to a different phenotype from nonexpression of the transporter. Such results might be obtained, for example, if the ablated neuron contained another transporter with an overlapping substrate specificity, or if the ablated neuron formed gap junctions or released additional neurotransmitters or neuromodulators. All of these cases can provide new insights into synaptic mechanisms; however, the structure–function experiments presented in the following section are best interpreted using "clean" phenotypes.

Use of *Caenorhabditis elegans* for Structure–Function Studies of Vesicular Transporters

C. elegans offers several advantages for *in vivo* structure–function studies of vesicular transporters. Once a set of behavioral functions has been ascribed to a transporter, it is then possible to design a series of experiments in which regions of the transporter gene are systematically altered or deleted, the altered gene is then used to generate a transgenic strain, and the behavioral consequences are determined. These experiments are best performed in conjunction with biochemical and kinetic studies of the altered transporter (generally using a mammalian cell culture system). Together, these studies can provide valuable insights into the role(s) of specific parts of the transporter protein in its biochemical and behavioral functions.

The fundamental paradigm for these studies must meet three requirements: (1) existence of an experimentally altered transporter gene (mutated *in vitro*, a chimeric construct, etc.); (2) a *C. elegans* mutant strain deficient in the transporter to be analyzed; and (3) some method(s) for assessing the extent of phenotypic "rescue" by the transgene. In the following sections, we discuss constructing, injecting, and assaying behavioral function of modified transporter genes. We first present the tools and methods of *C. elegans* transgenic technology, followed by the methods to assess the behavioral function conferred by a transgenic construct. As discussed in a subsequent section, an extension of this approach can be used to study subcellular targeting of vesicular transporter proteins.

Transgenic Methods for Caenorhabditis elegans

Fire Expression Vectors. An extensive set of "modular" plasmid vectors has been designed for expressing RNA or protein products in *C. elegans*. These have been constructed in the laboratory of Andrew Fire at the

Carnegie Institution.[27] The canonical vector, pPD49.26, has a modular design with three multiple cloning sites: the first is for inserting promoter and/or enhancer elements, the second is for inserting the coding sequence to be expressed, and the third is for inserting additional downstream regulatory elements (if needed). The plasmids also contain a small synthetic *C. elegans* intron and the 3′ region from a body muscle myosin gene—both of these have been shown to improve the efficiency of mRNA processing.[14] A large series of *C. elegans* promoters is available for these plasmids, including so-called "minimal promoters" into which enhancers or control regions can be placed, many different tissue-specific promoters, and inducible promoters (i.e., heat-shock promoters). Also available are a set of reporter modules containing β-galactosidase and/or green fluorescent protein (GFP; see Fig. 1); these may be useful for evaluating the expression pattern of promoter constructs, or for making chimeras (see below) with a gene of interest. Using β-galactosidase as a reporter necessitates killing the animal for staining, whereas GFP can be monitored in living animals. For more information on design and use of these vectors the reader is referred to a review by Mello and Fire.[14]

Translational Fusions. It may also be useful to produce functional translational fusions to tag a transporter protein. Modular vectors (of the same type described above) are available for constructing chimeras with reporter genes. GFP is currently the reporter of choice, although β-galactosidase is also quite useful. For tagging purposes, the *gfp* gene can be inserted at either end of a cloned gene of interest. Functional translational fusions have been constructed for a number of *C. elegans* transmembrane proteins, including VAMP/synaptobrevin[28] and a putative cyclic nucleotide-gated channel.[29]

Transformation Methods. DNA transformation in *C. elegans* is generally conducted by microinjection directly into the syncytial gonad of the adult hermaphrodite. Injected DNA can be maintained in the progeny for several generations.[30] By coinjecting a selectable or scorable marker along with the DNA of interest, stable transgenic lines can be maintained indefinitely. The most commonly used methods are modified from studies by Mello *et al.*[13]

Transformed DNA usually assembles into large extrachromosomal arrays containing several to hundreds of copies of the injected DNA. Coin-

[27] A. Fire, S. W. Harrison, and D. Dixon, *Gene* **93,** 189 (1990).

[28] E. M. Jorgensen, E. Hartwieg, K. Schuske, M. L. Nonet, Y. Jin, and H. R. Horvitz, *Nature* **378,** 196 (1995).

[29] C. M. Coburn and C. I. Bargmann, *Neuron* **17,** 695 (1996).

[30] D. T. Stinchcomb, J. E. Shaw, S. H. Carr, and D. Hirsh, *Mol. Cell. Biol.* **5,** 3484 (1985).

jected DNAs are almost always incorporated into the same array. Chromosomal integration of the arrays can be achieved most easily by gamma-irradiating a transgenic population, then selecting cloned progeny that give rise to 100% expressing populations. Another method is to include a single-stranded oligonucleotide in the injection mixture; this reduces the frequency of extrachromosomal transformation at least tenfold, permitting recovery of integrated lines in a single step.[13]

Perhaps the most popular, and in many cases the most convenient, marker for DNA transformation in *C. elegans* is *rol-6*(*su1006*).[31] This mutant cuticular collagen gene causes an easily scorable dominant phenotype: the animals twist and move around in circles. However, because the "roller" phenotype distorts the shape and movement of the animal, it is less attractive for studies involving the nervous system and behavior.

In our laboratory, the selectable marker *pha-1*[32] has proved useful for transformation experiments. *pha-1*(*e2123*) is a recessive, temperature-sensitive lethal allele; animals homozygous for this mutation are arrested in embryonic development at 25°, but develop and grow normally at 16°. Transformants are obtained by coinjecting a wild-copy of the *pha-1* gene along with the DNA of interest into *pha-1*(*e2123*) recipients and shifting the injected animals to the nonpermissive temperature. A disadvantage of the *pha-1* method is that if one wishes to transform animals containing a particular mutation, one must first construct a double mutant with *pha-1*. However, a major advantage of this system is that it permits easy selection for animals maintaining the transformed DNA.

Functional Assays for Vesicular Transport

General Considerations. Here we are concerned not with biochemical and kinetic functions of transporters (which are best measured using an *in vitro* assay system), but with their contribution to neuronal function and behavior. Although the measurement of animal behavior is clearly an indirect measure of protein function, it is often the most biologically meaningful. Useful information about biological function can be obtained from "yes or no" analysis; however, in practice much more information is obtained from a quantitative analysis of behavior. Any easily quantifiable behavior that is significantly altered in the mutant may be used for these assays. This permits assay of strains transformed with plasmids containing transporters (as described above) to allow quantitative measurement of the relative degree of rescue of behavior provided by different transporter constructs. It is desirable (but not absolutely necessary) to use a behavior which has

[31] J. M. Kramer, R. P. French, E.-C. Park, and J. J. Johnson, *Mol. Cell. Biol.* **10,** 2081 (1990).
[32] M. Granato, H. Schnabel, and R. Schnabel, *Development* **120,** 3005 (1994).

been shown to be dependent on at least one cell (or cell type) expressing the transporter.

All of the behavioral assays described below may be performed by direct observation through a high-quality dissecting stereomicroscope, although for some a video camera attached to the microscope allows analysis of the behavior at a slower speed. It is often informative to use several different assays, since it is quite possible that different assays may have differential requirements for transporter function(s).

Assays for VAChT Function. There are approximately 100 cholinergic neurons in the *C. elegans* nervous system, and most of them appear to be excitatory motor neurons.[33] We present three assays that may be used to measure different types of cholinergic function: measurements of pharyngeal pumping and body "thrashing" are relatively quick and quantifiable, and the rescue of lethality is extremely sensitive. All of these assays are suitable for animals carrying transgenes.

The two easiest behaviors to measure are the rates of pharyngeal pumping and body thrashing (Fig. 2). These behaviors involve different neural circuits and muscle types working at different frequencies. For these experiments, the *unc-17(e245)* allele is quite useful as a "recipient" of transgenes—animals homozygous for this allele are extremely uncoordinated, yet are relatively healthy. The assays work best on young adult animals—approximately 3 days old for wild-type animals and 4 days for *unc-17(e245)* mutants. Both of these assays are temperature dependent and should therefore be performed in a (relatively) controlled temperature environment.

The pumping of the pharynx is a stereotypic motor program,[34,35] normally occurring 200 times per minute in wild-type adults but only 41 times per minute in *unc-17(e245)* homozygotes (Fig. 2). The pumping assay involves counting the number of closings and openings of the terminal bulb of the pharynx (not necessarily the jerking of the pharynx back into the gut) for 1 min in each of 50 animals. The animals should be undisturbed and feeding actively.

When animals are immersed in liquid, they initiate a rapid swimming behavior ("thrashing"), which normally involves approximately 115 body flexions per minute for wild-type adults, but barely 1 per minute in *unc-17(e245)* homozygotes (Fig. 2). The thrashing assay involves flooding a growth plate with a small amount of buffer (M9 or PBS), waiting 1 min,

[33] J. S. Duerr, H.-P. Han, and J. B. Rand, unpublished (1997).
[34] L. Avery and H. R. Horvitz, *Neuron* **3,** 473 (1989).
[35] D. M. Raizen and L. Avery, *Neuron* **12,** 483 (1994).

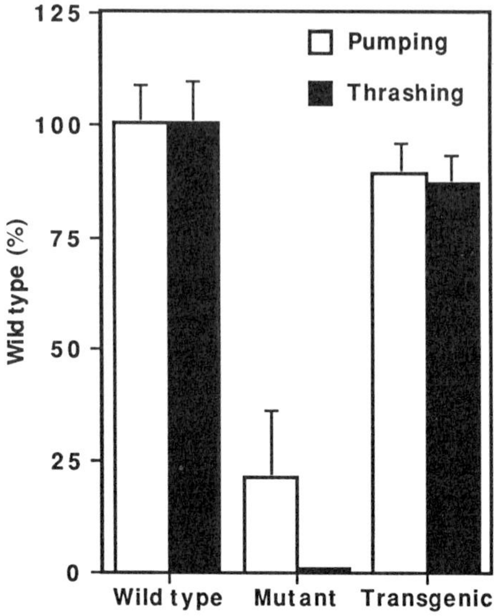

FIG. 2. Contribution of VAChT to pharyngeal pumping and body thrashing behaviors. Young adult *C. elegans* hermaphrodites were assayed for pharyngeal pumping rate (white bars) or body thrashing (black bars). For pumping measurements, 10 nematodes of each strain were placed on individual lawns of bacteria and the number of pharyngeal pumps in 1 min was determined for each animal. Thrashing was measured by placing 10 worms individually into microtiter wells containing M9 buffer on top of a bed of agarose and counting the number of body thrashes over a 3-min interval. For all assays, the means are presented as the percent of wild-type function and the error bars represent the standard deviation. The first set of bars represent wild-type animals, and the second set of bars represent measurements of *unc-17* (*e245*) mutants. The third set of bars represent measurements obtained from a transgenic line of *C. elegans,* in which *unc-17*(*e245*) animals were transformed with a 9-kb genomic fragment containing the complete wild-type *unc-17* coding sequence and regulatory region. It is clear that pumping and thrashing behaviors are almost completely restored in this particular transgenic line.

then counting the changes in the direction of flexion at the midbody for a 3-min period for 20–40 animals.

The most sensitive measure of VAChT function is the rescue of lethality. Total elimination of VAChT (*unc-17* null mutations) leads to larval lethality, but there are *unc-17* missense mutants with less than 5% of wild-type VAChT (judged by VAChT immunoreactivity) that can hatch and grow, albeit poorly.[17] Thus, a transgene construct with even minimal transporter function can probably provide enough VAChT to permit survival of the recipient strain. Null *unc-17* alleles (e.g., *md64*)[17] are usually maintained on a chromosome containing a linked *dpy-13* mutation (when homozygous, this confers a "dumpy" or short, fat appearance) and balanced with the *nT1* balancer chromosome[36]; thus, the genotype of such a strain would be *unc-17(null) dpy-13/nT1*.

Normally, one-quarter of the self-progeny derived from such animals are homozygous for the *unc-17(null)* allele and die soon after hatching. However, if the mutant homozygotes carry a transgene conferring some VAChT function, they may be able to grow slowly (although they may be uncoordinated). These animals can be readily identified because they will also be homozygous for the *dpy-13* mutation, and therefore dumpy. The growth rate and behavior of the transgenic animals are expected to be dependent on the degree of VAChT function conferred by the transgene. This type of assay does not easily lend itself to quantitation, but, as indicated above, it is probably the most sensitive assay for VAChT function.

Assays for VMAT Function. Approximately 25 of the 302 neurons in *C. elegans* hermaphrodites are VMAT-immunoreactive.[18] Several behaviors appear to be modulated by biogenic amines, including foraging (J. Kaplan, personal communication, 1996), response to bacteria (E. Sawin, personal communication, 1996), egg-laying,[37] and mating.[25,38] We present two methods for monitoring the function of aminergic neurons and their transporters: sensation of food and male mating efficiency.

Nematodes normally slow down when they encounter a streak of bacteria on an agar dish. Eight dopamine-containing sensory cells are involved in detection of bacteria (E. Sawin, personal communication, 1996). When nematodes are starved prior to exposure to the bacterial food source, they slow down even more on encountering bacteria. This starvation response involves serotonergic cells, including a pair of neurons in the pharynx (E. Sawin, personal communication, 1996).

The response to food has been monitored in different ways, including the following. Groups of *C. elegans* are grown on agar plates completely covered with bacteria. Young adult hermaphrodites are transferred to the side of an agar plate with only a central streak of bacteria. After 1 min has elapsed, animals encountering the bacterial food source are observed to determine their response. Generally, when a wild-type nematode's head enters the bacterial streak, it responds by significantly slowing its forward progress. The time to enter food is defined as the interval from when the tip of the nose first enters the bacterial streak until the tip of the tail enters the streak. Increasing starvation times up to at least 6 hr cause a steady increase in response to the bacteria. As shown in Fig. 3, *cat-1* mutants

[36] E. L. Ferguson and H. R. Horvitz, *Genetics* **110,** 17 (1985).

[37] C. Desai, G. Garriga, S. L. McIntire, and H. R. Horvitz, *Nature* **336,** 638 (1988).

[38] C. M. Loer and C. J. Kenyon, *J. Neurosci.* **13,** 5407 (1993).

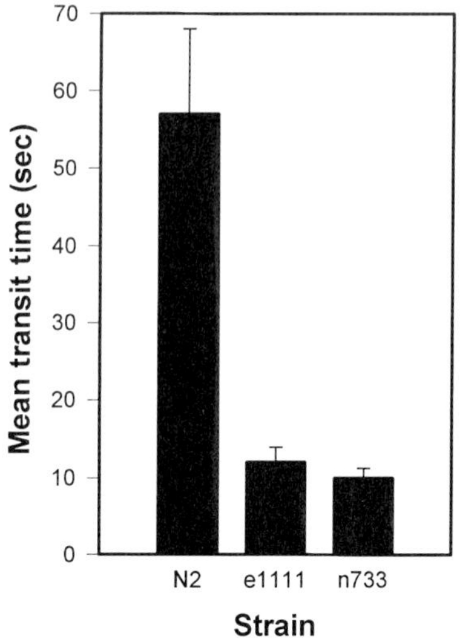

FIG. 3. VMAT-mediated feeding behavior. Groups of nematodes were transferred from plates covered with bacteria and tested immediately as described in the text. Individuals that took more than 2 min to completely enter the bacteria were assigned a value of 120 sec. To control for the nonnormal distribution of measured times, groups of 10 worms were tested and the mean time to enter food was calculated. The graphs show the average of three to four means. Error bars indicate standard errors of the mean. Data are presented for wild-type (N2) animals, as well as for two *cat-1* mutants, *e1111* and *n733*. [J. S. Duerr, D. L. Frisby, J. D. Erickson, A. Duke, L. E. Eiden, and J. B. Rand (1997). Unpublished.]

show a reduced response to the presence of bacteria compared with wild-type animals.

Males (XO) arise spontaneously in populations of hermaphrodites (XX) at a low frequency (less than 0.2%).[39] This frequency is increased by heat shock or by a number of mutations which affect meiotic nondisjunction (called *him* for high incidence of males).[39] Half of the cross progeny of a hermaphrodite and a male are male. Males have a more elaborate nervous system than hermaphrodites; many of the extra neurons in males are involved in copulation. Approximately 40 of 381 neurons in males, including approximately 15 male-specific neurons, are VMAT-immunoreactive. Aminergic neurons are important for normal mating behavior,[38,40] and several mating assays can be performed with varying degrees of ease.

A simple assay is to monitor the percent successful matings and the

[39] J. A. Hodgkin, H. R. Horvitz, and S. Brenner, *Genetics* **91,** 67 (1979).
[40] K. S. Liu and P. W. Sternberg, *Neuron* **14,** 79 (1995).

number of progeny per mating. To perform these assays, it is convenient to use *tra-2*(*e2046*) females. This mutation transforms XX hermaphrodites into females, which produce eggs but no sperm.[41] These females are therefore not self-fertile, so that all progeny from these females are cross progeny. Single immature males to be tested are placed with single immature *tra-2* females on an agar plate with a spot of bacteria. Pairs are kept on the same plate at 20° for at least 48 hr after maturity. The number of pairs with progeny and the number of progeny per pair are counted. Typically, with wild-type males, more than half such pairings result in cross progeny, with more than 100 progeny per pair. Mutant *cat-1* males mate successfully approximately 10% of the time, and the progeny per mating pair is generally lower than for wild-type males.[18]

For more detailed analysis of mating behavior, the movement of the males as they encounter and attempt to copulate with females or hermaphrodites can be analyzed.[38,40] For example, the dopamine-containing ray neurons R5, R7, and R9[25] and the serotonin-containing CP motor neurons[38] are necessary for the typical turning behavior of males as they back up along the body of the hermaphrodite while attempting to locate the vulva.[40] Analysis of the role of VMAT in particular neurons involved in these behaviors is now possible.

Assays for VGAT Function. Although the vesicular GABA transporter(s) has not yet been characterized molecularly, GABA localization and function have been studied,[19,42] and it is therefore possible to speculate on the phenotype of a VGAT-deficient mutant.

The most specific and quantifiable GABA-mediated behavior so far reported is the expulsion step of the defecation cycle.[42] The defecation motor program is initiated approximately every 45 sec in feeding wild-type adults by a contraction of the posterior muscles of the body wall; the final step of the program is the contraction of the posterior enteric muscles.[43] Typically, 10 consecutive defecation cycles are monitored in each of 10 undisturbed, feeding animals, and the frequency with which the defecation cycle, once initiated, fails to go to completion is measured. The failure rate of the enteric muscle contraction is 0–5% in wild-type adults and 88% in GABA-deficient *unc-25* mutants.[19] Published values for *unc-47* are 82–87% failure rate,[19] so if *unc-47* does turn out to encode VGAT, this assay will clearly be suitable for behavioral analysis of VGAT function.

[41] J. A. Hodgkin and S. Brenner, *Genetics* **86,** 275 (1977).
[42] S. L. McIntire, E. Jorgensen, J. Kaplan, and H. R. Horvitz, *Nature* **364,** 337 (1993b).
[43] J. H. Thomas, *Genetics* **124,** 855 (1990).

Assays for VGluT Function. There are no mutations in *C. elegans* which are known to eliminate all glutamate-dependent neurotransmission. Therefore, we cannot be certain which behaviors will be altered in mutants that lack vesicular glutamate transport. The *C. elegans* Genome Sequencing Project has identified at least one putative metabotropic and several putative ionotropic glutamate receptors. We might expect nematodes with a deficit in the glutamate vesicular transporter to show a phenotype similar to the sum of effects observed when single receptors are eliminated. Unfortunately, if a large number of cells and behaviors are involved, the behaviors seen with individual receptor knockouts may not be totally additive. However, it is possible that selective expression of the transporter may allow correlation of function in particular neurons with particular behaviors.

Mutations which eliminate the function of an AMPA-type ionotropic glutamate receptor (*glr-1*) have been described.[44,45] Mutations in this receptor disrupt the sensitivity of the nematode to light touch of the "nose" (the anterior tip of their head) as well as harsh touch of the body. The response to nose touching is easy to observe; the response to harsh touch can only be observed in conjunction with a mutation in a second gene. To observe the nose-touch response, an eyelash is placed in the path of an advancing nematode. Normal animals withdraw their heads upon contacting the eyelash with their noses,[46] whereas *glr-1* mutants continue to advance.[45]

Another behavior which might be altered in glutamate vesicular transport mutants is pharyngeal pumping. Glutamate probably acts through a gated chloride channel[47] to change the timing of the relaxation phase of pharyngeal pumping.[48] Assays of this behavior would be best done by videotaping of pumping behavior or with the use of an electropharyngeogram[35]; both methods are far more complicated than the simple touch response described above.

Use of Antibodies and Mutants to Explore Targeting

As an extension of the structure–function experiments described above, one can use *C. elegans* to analyze the targeting of vesicular transporters. The basic experimental structure is the same—transformation of *C. elegans*

[44] A. V. Maricq, E. Peckol, M. Driscoll, and C. I. Bargmann, *Nature* **378,** 78 (1995).

[45] A. C. Hart, S. Sims, and J. M. Kaplan, *Nature* **378,** 82 (1995).

[46] J. M. Kaplan and H. R. Horvitz, *Proc. Natl. Acad. Sci. USA* **90,** 2227 (1993).

[47] D. F. Cully, D. K. Vassilatis, K. K. Liu, P. S. Paress, L. H. T. Van der Ploeg, J. M. Schaeffer, and J. P. Arena, *Nature* **371,** 707 (1994).

[48] L. Avery and J. H. Thomas, *in* "*C. elegans* II" (D. L. Riddle, T. Blumenthal, B. J. Meyer, and J. R. Priess, Eds.). Cold Spring Harbor Laboratory Press, Cold Spring Harbor, New York, 1997.

with a set of mutated or chimeric transporter genes, followed by evaluation of the phenotype. In this case, however, the phenotype to be evaluated is the subcellular localization of the transporter. If one is interested only in site(s) on the transporter molecule required for proper subcellular localization, then a heterologous construct may be transformed into a recipient strain mutant for the *pha-1* gene, but without any transporter mutation. If one is interested in the functional deficits associated with deficient or altered targeting, then one may use a recipient strain containing a transporter mutation as described above.

The targeting of the transgenic transporter to particular cells and subcellular regions may be assessed either by indirect immunofluorescence or by use of a functional translational fusion of the transporter with a reporter protein. The first approach requires the generation of antibodies against that transporter and the use of those antibodies *in situ*. The second approach requires the construction of a fully functional translational fusion transgene.

Antibodies can be generated against peptides (for more specificity) or against fusion proteins (generally more robust staining); we have used both types of antigens. A problem particular to *C. elegans* is the difficulty of penetrating the impermeable cuticle for *in situ* immunofluorescence. Accessibility of the tissues to antibody has been increased as follows.[33,49] Mixed populations of nematodes are compressed between two polylysine-coated slides and placed on a piece of dry ice. After 20 min, the frozen slides are separated. The top layers of cuticle adhere to the top slide, while the bottom slide holds the rest of the nematodes. The bottom slide is immediately placed in ice-cold fix (2 min in methanol, then 4 min in acetone). Primary and secondary antibody incubations follow standard procedures. (This "whole mount" procedure is followed because the hard cuticle of *C. elegans* makes cryostat sectioning very difficult, and the animals' small size and turgor pressure makes most dissections of the nervous system impractical.)

Once antibodies have been generated, they may be used for a number of purposes. For example, they may be used to examine whether particular mutations in the transporters disrupt transporter stability *in vivo*. In this case, mutants without endogenous transporter would be transformed with a mutant transporter whose location and abundance could be monitored with antibodies. Antibodies can also be useful in analyzing the effects of suppressor genes (discussed below) or transporter abundance and localization.

Studies with antibodies can be complemented by studies using GFP translational fusions (described above; see Fig. 1). Antibodies against multiple epitopes give the most unambiguous representation of the localization

[49] D. G. Albertson, *Dev. Biol.* **101,** 61 (1984).

of the protein of interest. However, production of robust antibodies may be time-consuming and expensive, whereas transformation with translational fusion proteins can be done rapidly and the GFP product is bright and readily visualized. To be sure of the accuracy of the observed GFP localization, it is necessary to have a functional fusion that can provide rescue of a mutant phenotype. Without such functional rescue, one cannot be sure that the fusion protein is properly expressed or localized. If such a functional fusion protein is available, it would also be very useful in suppressor screens as discussed below.

Use of *Caenorhabditis elegans* Methods to Identify Interacting Genes and Proteins

In order to understand better the biological functions of a particular transporter, it is often desirable to identify interacting proteins. Such proteins may be needed for proper folding and/or localization of the transporter, maintenance of vesicle structure, or regulation of transporter function. We present two basic strategies that have been successfully employed with *C. elegans*.

One strategy involves screening for new mutations that suppress a phenotype of a known transporter mutant. The basic rationale is that if two proteins interact physically, a mutational disruption of one of the proteins can sometimes be "corrected" by a compensatory alteration in the interacting protein.[50] This approach takes advantage of some of the major strengths of *C. elegans* as a research organism: the ability to screen large numbers of mutagenized animals, isolate mutants quickly, and analyze them (relatively) easily.

The other method takes advantage of the two-hybrid system in yeast.[51,52] This is a fast, powerful, and now well-established method for identifying interacting proteins. The strength of the genetic (suppressor) approach is that it identifies *functional* interactions between proteins. The advantages of the two-hybrid system are that it is quick and that it provides a cloned cDNA, so that molecular analysis of the new gene is relatively fast and simple. Although the two-hybrid system can in principle be used equally well to identify non-nematode genes and proteins, the overall analysis of such genes (cellular and subcellular localization of the protein, behavioral consequences of knockout, etc.) is usually much quicker in *C. elegans*. Together, these approaches can identify additional proteins important in the transport of neurotransmitters.

[50] P. E. Hartman and J. R. Roth, *Advan. Genet.* **17,** 1 (1973).

[51] S. Fields and O.-K. Song, *Nature* **340,** 245 (1989).

[52] C.-T. Chien, P. L. Bartel, R. Sternglanz, and S. Fields, *Proc. Natl. Acad. Sci. USA* **88,** 9578 (1991).

Suppressor Screens

The choice of the starting mutation(s) is an important consideration before beginning a suppressor screen. In general, null mutations (complete absence of function) usually fail to identify an interacting protein, but allow recovery of bypass-of-function mutations. Starting with a reduction-of-function mutation, however, often identifies mutations in genes whose products interact with the protein of interest. Thus, in a search for proteins that interact with a transporter, the starting mutation should confer some residual activity (i.e., a "leaky" mutation). In addition, it is important to choose a starting mutant with a strong and/or easily recognizable phenotype, so that a "revertant" is obvious.

In order to recover stable suppressor mutant lines, it is necessary to obtain germ-line (heritable) mutations. A variety of mutagenic treatments, including chemical and irradiation techniques, may be used.[53] The basic approach is to mutagenize late larval stage animals (which contain the maximum number of germ-line nuclei) and screen for phenotypic revertants in subsequent generations. Mutations that are recovered in the F_1 generation are dominant, whereas mutations recovered in the F_2 or subsequent generations may be either dominant or recessive. Typically, it is possible to screen 10^6 animals (200–300 large petri plates, each containing 3000–5000 animals) in two person-days. Once phenotypic revertants have been recovered, it is necessary to outcross them to segregate the suppressor mutation from the mutagenized background. The outcrossed mutant can then be mated back to the parent strain. In doing so, it is possible to determine whether the suppressor mutation is linked or unlinked to the starting mutation, whether it is dominant or recessive, and if it has a characteristic phenotype of its own.

From Mutant to Clone

Standard *C. elegans* genetic analysis (i.e., mapping and complementation testing; see ref. 54) can be used to determine if the suppressor mutations are in a previously characterized gene—if so, there is a significant chance that the gene has already been cloned. However, even if the suppressor locus turns out to be a new gene or a gene previously identified but not yet cloned, there are a number of approaches to cloning the mutated gene.

[53] P. Anderson, *in* "*Caenorhabditis elegans:* Modern Biological Analysis of an Organism" (H. E. Epstein and D. C. Shakes, Eds.), pp. 31–58. Academic Press, San Diego, 1995.

[54] R. K. Herman and H. R. Horvitz, *in* "Nematodes as Biological Models," Vol. 1 (B. Zuckerman, Ed.), pp. 228–261. Academic Press, New York, 1980.

Positional cloning is now the commonly accepted method for the cloning of a gene for which there is a mutant phenotype. This takes advantage of several powerful *C. elegans* tools: the availability of the soon-to-be-completed sequence of the *C. elegans* genome, the virtually complete coverage of the genome by an overlapping set of mapped cosmids, and the large number of genes which have been placed both on the genetic map (by mutant analysis) and on the cosmid map (by cloning). From the position of the locus on the genetic map, one can usually identify a small set of overlapping cosmids which must contain the wild-type copy of the mutated gene. The transformation methods described above are then used to determine the cosmid or cosmid fragment capable of "rescuing" the mutant phenotype. It is sometimes possible, by examining the genome sequence of the region, to identify a putative "candidate gene" and then perform the transformation rescue experiments for confirmation.

Two-Hybrid Screen

The two-hybrid system was initially developed to examine protein–protein interactions in yeast.[51,52] It has since been employed to identify interacting proteins in a number of other experimental systems, including *C. elegans* (e.g., refs. 55 and 56). As indicated above, the advantage of using *C. elegans* genes and libraries to identify interacting proteins is the ease with which the molecular analysis of a gene can move to the mutational (i.e., functional) analysis of the gene product. Often, the cDNAs isolated by two-hybrid analysis are derived from a gene for which mutants have already been described. However, even if such mutants have not yet been described (or even if the gene lies in a portion of the genome not yet sequenced), a number of strategies for "reverse genetics" are available to isolate mutants with insertions or deletions in the gene of interest.[57]

Plasma Membrane Transporters

Although the plasma membrane transporters of mammals were well characterized before the cloning and analysis of synaptic vesicle transporters, in *C. elegans* the pattern has been reversed. Thus, molecular, mutational, and cellular analysis of VAChT and VMAT (and soon, VGAT) have preceded characterization of any of the plasma membrane neurotransmitter transporters. However, when the complete *C. elegans* genome sequence

[55] E. J. A. Hubbard, Q. Dong, and I. Greenwald, *Science* **273,** 112 (1996).

[56] D. Wu, H. D. Wallen, and G. Nuñez, *Science* **275,** 1126 (1997).

[57] R. H. A. Plasterk, *in* "*Caenorhabditis elegans:* Modern Biological Analysis of an Organism" (H. E. Epstein and D. C. Shakes, Eds.), pp. 59–80. Academic Press, San Diego, 1995.

is available, all plasma transporters of known classes should be readily identifiable. Inspection of the genomic sequences so far submitted to GenBank suggests that *C. elegans* contains a significant number of genes with homology to plasma membrane neurotransmitter transporters. However, to date there have been only two published reports of the cloning of transporter homologs, and these are both putative glutamate transporters.[58,59]

It is reasonable to expect that the next 2 or 3 years will see a large number of *C. elegans* transporter genes cloned. What will then be needed for these genes will be mutants and cell-specific and/or transmitter-specific phenotypes. Once these mutants and phenotypes have been described, the corresponding mammalian genes can be studied in a manner that is conceptually identical to what we have described above for the vesicular transporters.

Note Added in Proof

The identification of the *unc-47* gene as the vesicular GABA transporter and the isolation of homologous mammalian clones has recently been published.[60]

Acknowledgments

Many members of our laboratory (past and present) have been involved in the analysis of *C. elegans* vesicular transporter genes, proteins, and mutants. These include Aixa Alfonso, Tony Crowell, Angie Duke, Kiely Grundahl, He-Ping Han, and John McManus. Ken Miller has helped greatly in the development of behavioral assay strategies. We also wish to acknowledge our collaborators Lee Eiden and Jeff Erickson at the National Institute of Mental Health, who have helped with the biochemical and kinetic studies of *C. elegans* VAChT and VMAT. Our research in this area has been supported by research grants from the National Institute of General Medical Sciences and the Oklahoma Center for the Advancement of Science and Technology.

[58] T. Kawano, K. Takuwa, and T. Nakajima, *Biochem. Biophys. Res. Commun.* **228,** 415 (1996).
[59] A. D. Radice and S. Lustigman, *Mol. Biochem. Parasitol.* **80,** 41 (1996).
[60] S. L. McIntire, R. J. Reimer, K. Schuske, R. H. Edwards, and E. M. Jorgensen, *Nature* **389,** 870 (1997).

Section VI

Application of Electrophysiological Techniques to Neurotransmitter Carriers

Articles 37 through 43

[37] Measurement of Transient Currents from Neurotransmitter Transporters Expressed in *Xenopus* Oocytes

By SELA MAGER, YONGWEI CAO, and HENRY A. LESTER

Introduction

Removal of neurotransmitter from the synaptic cleft by ion-coupled transporters helps to terminate synaptic transmission. Energy for this process is provided by the electrochemical potentials of Na^+ and other ions. To understand the transport process at the molecular level, one must study individual steps in the transport cycle, such as binding and dissociation of ions and the neurotransmitter, and the conformational changes of the proteins that allow the translation of substrate from one side of the membrane to the other. The turnover rate of transport is on the order of 1–10 sec^{-1}; thus, the study of individual steps requires measurements with time resolutions of less than a second. Traditionally, transporter kinetics have been studied using flux measurements. These measurements, however, cannot resolve changes of less than 1 sec; indeed, most of these flux measurements are done in equilibrium or steady-state conditions.

Electrophysiological measurements have been employed to study transporters with high time resolution. Transporters are amenable to electrophysiological studies primarily because, in addition to the fact that the energy for transport is provided by the ion electrochemical gradient, the transport process is electrogenic.[1,2] For example, at the γ-aminobutyric acid (GABA) transporter GAT1, for each transport cycle, one zwitterionic GABA molecule is cotransported with two Na^+ ions and one Cl^- ion, giving one net charge per transport cycle. Surprisingly, in many cases transporters also display conducting states similar to those of ionic channels. A conducting state, in the absence of an organic substrate, gives rise to a "leakage current"; a different conducting state, in the presence of an organic substrate, gives rise to a "transport-associated current."[3,4] The nature of these currents, how they relate to transport activity, and what we can learn from them about transporter functions are subjects of intensive research.

[1] H. A. Lester, S. Mager, M. W. Quick, and J. L. Corey, *Annu. Rev. Pharmacol. Toxicol.* **34,** 219 (1994).
[2] B. I. Kanner and S. Schuldiner, *CRC Crit. Rev. Biochem.* **22,** 1 (1987).
[3] H. A. Lester, Y. Cao, and S. Mager, *Neuron* **17,** 807 (1996).
[4] M. K. Sonders and S. G. Amara, *Curr. Opin. Neurobiol.* **6,** 294 (1996).

0076-6879/98 $25.00

Electrophysiological measurements can also illuminate individual steps in the transport cycle. Although one cannot yet resolve the charge movements due to one or two ions binding to a single transporter protein, synchronized binding to many transporters generates measurable currents. Transporters can be synchronized by a rapid change in one of the experimental conditions that affect binding, such as ion concentration and membrane potential. Correlation of the charge movements with steady-state measurements, uptake of the organic substrate, and binding of high-affinity ligands provides new insights into transporter kinetics. In addition, electrophysiological measurements reveal new and somewhat peculiar behaviors of neurotransmitter transporters that have yet to be resolved.

Transporter Expression in Oocytes

The specific mechanism by which neurotransmitter transporters can generate currents are discussed later in this chapter. At this time, we note that each transporter generates a current 1000 to 1,000,000 times smaller than most ionic channels; thus, a high expression level (on the order of 1000 transporters per μm^2 of membrane) is required. In general, there are two methods for heterologous expression in oocytes. The first is based on nuclear injection of plasmid DNA: after the transcription occurs in the nucleus, the mRNA is transported to the cytoplasm and translated to protein.[5] The second is based on cytoplasmic injection of *in vitro* transcribed mRNA. We use the latter because it seems to give a higher expression. Several plasmids have been optimized for high-level expression in *Xenopus* oocytes. Most of these plasmids contain 3′ poly(A) and, in addition, upstream and downstream sequences that facilitate ribosome binding and/or increase the stability of the RNA (see Table I). We use pAMV-PA, a plasmid which is derived from pBluescript KS(+) and contains 42 nucelotides from the alfalfa mosaic virus[6] upstream, and a poly(A_{50}) tail downstream. After linearization of the plasmid with *Not*I, RNA is transcribed with bacteriophage T7 RNA polymerase. We find that the mMessage mMachine kit from Ambion (Austin, TX) is simple to use and gives reliable results. Oocytes are prepared and maintained as described previously.[7] RNA (10–25 ng) is injected into each oocyte. Oocytes are incubated at 19°

[5] D. Bertrand, E. Cooper, S. Valera, D. Rungger, and M. Ballivet, in "Methods in Neuroscience" (P. M. Conn, Ed.), p. 174. Academic Press, San Diego, 1991.

[6] S. A. Jobling and L. Gehrke, *Nature* **325,** 622 (1987).

[7] M. W. Quick and H. A. Lester, in "Ion Channels of Excitable Cells" (T. Narahashi, Ed.), p. 261. Academic Press, San Diego, 1994.

TABLE I
EXPRESSION VECTORS

Vector	Provider (location)	Base vector	5′ UTR	3′ UTR	Sites	Use	Ref[a]
pBluescript	Strategene (La Jolla, CA)		None	None	Nice polylinker		
pAMV-PA	Labarca & Co.	pBluescript KS(+)	(AMV) Avian myeloblastosis virus	poly(A)	Rudimentary	rSERT GAT1	1 2
pSPORT1	GIBCO-BRL (Gaithersburg, MD)		None	None		ROMK1	3
pGEMHE	E. Liman	pGEM3Z	UTR from *Xenopus* globin	UTR from *Xenopus* globin	Nice polylinker	Kv1.1, 1.5	4
pMXT	L. Salkoff, (Wash University)	pBluescript KS II$^+$	UTR from *Xenopus* globin	UTR from *Xenopus* globin		GIRK1, GIRK2, GIRK4	5,6
pFROGy	A. Connolly, S. Coughlin, (UCSF)	pCDM8	UTR from *Xenopus* globin	UTR from *Xenopus* globin		Thrombin receptor, $G_{\beta 1}$, $G_{\gamma 2}$	7 8
pGEM3	Promega (Madison, WI)		None	None		M2 muscarinic receptor	9
pOTV	S. Amara, (Vollum Institute)	pBluescript SKII$^-$	UTR from *Xenopus* globin	UTR from *Xenopus* globin		Glutamate transporters	10
pSP64/pSP65	Promega		None	None		β_2-Adrenergic receptor	11
pAlter1	Promega		None	None		CFTR	12

[a] *Key to references:* (1) F. Lin, H. A. Lester, and S. Mager, *Biophys. J.* **71,** 3126 (1996); (2) S. Mager, N. Kleinberger-Doron, G. I. Keshet, N. Davidson, B. I. Kanner, and H. A. Lester, *J. Neurosci.* **16,** 5405 (1996); (3) P. Kofuji, C. A. Doupnik, N. Davidson, and H. A. Lester, *J. Physiol.* **490,** 633 (1996); (4) E. R. Liman, J. Tytgat, and P. Hess, *Neuron* **9,** 861 (1992); (5) A. Wei, C. Solaro, C. Lingle, and L. Salkoff, *Neuron* **13,** 671 (1994); (6) P. Kofuji, M. Hofer, K. J. Millen, J. M. James H. Millonig, N. Davidson, H. A. Lester, and M. E. Hatten, *Neuron* **16,** 941 (1996); (7) T.-K. H. Vu, D. T. Hung, V. I. Wheaton, and S. R. Coughlin, *Cell* **64,** 1057 (1991); (8) C. A. Doupnik, N. F. Lim, P. Kofuji, N. Davidson, and H. A. Lester, *J. Gen. Physiol.* **106,** 1 (1995); (9) P. Kofuji, N. Davidson, and H. A. Lester, *Proc. Natl. Acad. Sci. USA* **92,** 6542 (1995); (10) J. L. Arriza, W. A. Fairman, J. I. Wadiche, G. H. Murdoch, M. P. Kavanaugh, and S. G. Amara, *J. Neurosci.* **14,** 5559 (1994); (11) N. F. Lim, N. Dascal, C. Labarca, N. Davidson, and H. A. Lester, *J. Gen. Physiol.* **105,** 421 (1995); and (12) S. McDonough, N. Davidson, H. A. Lester, and N. A. McCarty, *Neuron* **13,** 623 (1994).

for 4–10 days for translation. In many cases, 1–5% horse serum is added to the incubation medium to increase the lifetime of the oocytes.[8]

Functional Measurements

Uptake Measurements

Removing released neurotransmitter is the main function of transporters; therefore, any functional measurements must eventually be compared to neurotransmitter uptake. Radiolabeled substrate is most commonly used to measure uptake, but in some specific cases fluorescent analogs of neurotransmitters are available.[9] In a typical uptake experiment, oocytes are placed in 100–300 μl of incubation solution containing ^{3}H-labeled neurotransmitter (in a 24- or 48-well plate). The uptake is terminated by multiple washes and oocytes are then solubilized in 1% sodium dodecyl sulfate (SDS). Uptake is determined by liquid scintillation counting. To study the effect of membrane potential on uptake rate or to compare the uptake to the current, uptake measurements can be performed under two-electrode voltage clamp as follows: the oocyte membrane is clamped and the flow is stopped; the neurotransmitter is then added to the chamber and mixed with a micropipette; and uptake is terminated by restarting the perfusion.

Ligand Binding

To prepare membranes, oocytes are homogenized in a buffer containing 0.32 *M* sucrose, 10 m*M* sodium phosphate, and 1 m*M* EGTA (pH 7.4). The homogenate is centrifuged at 1000*g* for 30 min and the supernatant is saved. The pellet is resuspended in the same buffer and recentrifuged. The supernatants are combined and centrifuged at 100,000*g* for 30 min. The pellet is resuspended in the binding buffer and aliquots are incubated with various concentrations of the ligand. The binding is terminated by rapid filtration through Whatman (Clifton, NJ) GF/B filters previously soaked with polyethyleneimine (using a filter manifold). The filters are washed three times and radioactivity is counted.[10]

Electrophysiology

Two-Electrode Voltage Clamp. A standard two-electrode voltage clamp setup is used to record currents associated with the expression of transport-

[8] M. W. Quick, J. Naeve, N. Davidson, and H. A. Lester, *BioTechniques* **13,** 358 (1992).

[9] D. Bruns, F. Engert, and H. D. Lux, *Neuron* **10,** 559 (1993).

[10] S. Mager, C. Min, D. J. Henry, C. Chavkin, B. J. Hoffman, N. Davidson, and H. A. Lester, *Neuron* **12,** 845 (1994).

ers. For voltage-jump experiments, intracellular electrodes typically have resistances of less than 1 mΩ to enable fast charging of the oocyte membrane capacitance. For recordings at constant membrane potentials, sharper electrodes with higher resistances (1–2 mΩ) are frequently used to increase the stability of the recording. When the experiment requires changing the perfusion composition, the reference electrodes should be connected via agar/KCl bridges to eliminate junction potentials.

Macropatches. Macropatches (6–12 pF membrane capacitance) of oocyte membranes have two major advantages over the two-electrode voltage clamp: First, they enable a rapid exchange of the solution from one side of the membrane (usually the inner side in the inside-out configuration). Second, the low resistance of the patch electrode compared to the small capacitance of the patch enables faster control of the membrane potential. This method is described in detail elsewhere[11] and has been employed to study various transporters and channels. In one study, capacitance measurements from macropatches excised from oocytes expressing the GABA transporter GAT1 were used to study Cl^- binding at the intracellular side of the transporter.[12]

Small Patches. The discovery that, in some cases, neurotransmitter transporter ionic currents are not coupled to the uptake of organic substrate[10] suggests that transporters might have channel-like activity. Patch clamp recordings from the 5-hydroxytryptamine (5-HT) transporter expressed in oocytes[13] and the norepinephrine and GABA transporters expressed in mammalian cells[14,15] did reveal single-channel currents. The channel is believed to be formed directly by the expressed transporter protein because, in both cases, the open probability of the channel is affected by the organic substrate, and the channel activity is eliminated by specific uptake inhibitors. Also, a point mutation at the 5-HT transporter altered the single channel conductance.[13] A detailed description of patch-clamp recordings in *Xenopus* oocytes can be found elsewhere.[16]

Concentration Jumps

The study of the interaction between ions and transporters requires rapid changes in the ion composition at the surface of the oocyte. In this

[11] A. Collins, A. V. Somlyo, and D. W. Hilgemann, *J. Physiol.* **454,** 27 (1992).

[12] C.-C. Lu, A Kabakov, V. S. Markin, S. Mager, A. Frazier, and D. W. Hilgemann, *Proc. Natl. Acad. Sci. USA* **92,** 11220 (1996).

[13] F. Lin, H. A. Lester, and S. Mager, *Biophys. J.* **71,** 3126 (1996).

[14] A. Galli, R. D. Blakely, and L. J. DeFelice, *Proc. Natl. Acad. Sci. USA* **93,** 8671 (1996).

[15] J. N. Cammack and E. A. Schwartz, *Proc. Natl. Acad. Sci. USA* **93,** 723 (1996).

[16] W. Stühmer, *Methods Enzymol.* **207,** 319 (1992).

section we discuss the advantages and disadvantages of two different methods for changing the perfusion solution.

Multiple-Position Valve. This system is based on the use of a valve with 6–12 input tubes and a single output tube. Changing the position of the valve selects one of the input tubes. The advantage of this system is that it allows application of a large number of solutions with a convenient manual switch. There are also disadvantages to this system: (1) There is a relatively large volume between the input tubes and the chamber containing the oocytes. This volume includes the mixing chamber of the valve and the output tube. As a result, there is a delay between switching the valve and changing the solution—the concentration change takes more than a second. (2) The sequential access to the input tubes puts constraints on the experimental design. In general, this system is convenient for manual experiments that require a large number of solutions but do not require time-resolved measurements.

Separate Valves and Manifold. In this system, each of the input tubes is controlled by an electrical Teflon valve (Auto-Mate Scientific, San Francisco, CA), and all the tubes are connected to the manifold just before the oocyte chamber. The small volume between the manifold and the oocyte allows for rapid switching among several input tubes. This switching can be controlled by digital outputs from a computer. Thus, the system allows for synchronized digital storage of membrane currents for data analysis and manipulation. Two major factors that affect the solution exchange rate are the volume between the manifold and the oocyte and the flow rate. The oocyte chamber has a U-shaped cross section and is 1.5 mm wide and 1.5 mm deep. Two pins are placed in front of the chamber to brace the oocyte during fast perfusion. The system allows for solution change in 400–700 ms. The major disadvantage of the system is that there is a small solution exchange between the mixing chamber of the manifold and the incoming tubes. This is especially problematic if one of the tubes contains an inhibitor with a high affinity or if there are large differences in concentration among different tubes.

Internal Ion Substitution

The transport cycle of ion-coupled transporters involves ion binding on one side of the membrane and dissociation on the other side. Thus, transporter activity strongly depends on ion composition on both sides of the membrane. Moreover, some of the currents carried by the transporter are mediated by the ion channel mode; thus, the reversal potential of the current depends on ion concentrations on both sides of the membrane. The internal ion concentration can be changed by microinjection of the desired ion into the oocyte[17] or by incubating the oocytes in the desired ion composition. To study the Cl^- conductance of the glutamate transporter, internal Cl^- was depleted by overnight incubation of the oocyte in a Cl^--free solution.[18] In another experiment, depletion of internal Cl^- by 2 hr incubation in a Cl^--free solution (acetate substitution) increased the apparent affinity of the GABA transporter GAT1 for Na^+ (S. Mager, unpublished observation). To study H^+ permeation at the 5-HT transporter, internal H^+ concentration has been elevated by incubating the oocytes in a low pH Na^+-free Ringer's solution.[19]

Several Ionic Currents Mediated by Neurotransmitter Transporters

Transport Current vs Transport-Associated Current

For all neurotransmitter transporters tested, a current is induced by organic substrate. This current might reflect an electrogenic transport cycle. For example, according to the present model of the GABA transporter GAT1, for each transport cycle, two positively charged Na^+ ions are cotransported with one negatively charged Cl^- ion and one zwitterionic GABA molecule. This stoichiometry yields a net transfer of one elementary charge per transport cycle. Nevertheless, the discovery of a channel-like mode at different transporters requires a reexamination of the electrogenicity of the transporters. Transport current is a result of a stoichiometrically coupled transport cycle with a net charge transfer. Uncoupled current induced by organic substrate is called "transport-associated current" and seems to result from the channel mode of operation.[13,14]

Is it a transport current? Several factors help to determine whether ion current induced by the substrate is stoichiometrically coupled to the transport of organic substrate: (1) If the uptake of neurotransmitter does not depend on the membrane potential but current is induced by the organic substrate, the current is probably not directly coupled to the uptake. This occurs with the mammalian serotonin transporter.[10,13] (2) If the current is carried by ions that have no effect on uptake, it is not a transport current. An example of this is the Cl^- current at several glutamate transporters.[18] (3) If there is a large discrepancy between the ionic current and the uptake of radiolabeled organic substrate, a large portion of the current is probably

[17] K.-W. Wang, K.-K. Tai, and S. A. N. Goldstein, *Neuron* **16,** 571 (1996).
[18] J. I. Wadiche, S. G. Amara, and M. P. Kavanaugh, *Neuron* **15,** 721 (1995).
[19] Y. Cao, Mager, and H. Lester, *J. Neurosci.* **17,** 2257 (1997).

not coupled. This occurs with both serotonin and norepinephrine transporters.[10,14,20]

Leakage Current

Leakage current is defined as the ionic current mediated by a transporter in the absence of organic substrate. Although uptake depends on the cotransport of Na^+ in all known neurotransmitter transporters, the leakage current can be carried by one of several alkali metal ions.[10,20,21] In many cases the current is larger in the presence of external Li^+ than in the presence of Na^+. For the serotonin transporter expressed in oocytes,[13] and for the GABA transporter GAT1 expressed in mammalian cells,[15] it has been shown that the leakage current is a result of the channel mode. Nevertheless, in other cases the current could be a result of carrier mode, whereby permeation of each ion requires conformational changes of the proteins.

Is it a leakage current? Leakage current is revealed by a decrease in membrane current due to the application of specific uptake inhibitors to oocytes expressing the transporter.

Charge Movements

In cells expressing the GABA,[21,22] norepinephrine,[20] or glutamate transporters,[23] transient current develops in response to voltage jumps to negative or positive potentials. Current with a similar charge transfer but in the opposite direction develops when the membrane potential is stepped back to the original potential. For the GABA and glutamate transporters, this current is affected by Na^+ concentration and is absent when external Na^+ is absent. The current is therefore believed to reflect either the binding and dissociation of Na^+ within the membrane electric field, or the conformational changes of the proteins induced by ion binding. For the GABA transporter, transient current also develops in response to a rapid change in ion concentration. When Na^+ is removed from the perfusion medium, an outward transient current develops, just as if Na^+ ions were leaving the membrane.[21] However, transient current in the norepinephrine transporter is not Na^+dependent.[20]

[20] A. Galli, L. J. DeFelice, B.-J. Duke, K. R. Moore, and R. D. Blakely, *J. Exper. Biol.* **198,** 2197 (1995).

[21] S. Mager, N. Kleinberger-Doron, G. I. Keshet, N. Davidson, B. I. Kanner, and H. A. Lester, *J. Neurosci.* **16,** 5405 (1996).

[22] S. Mager, J. Naeve, M. Quick, J. Guastella, N. Davidson, and H. A. Lester, *Neuron* **10,** 177 (1993).

[23] J. I. Wadiche, J. L. Arriza, S. G. Amara, and M. P. Kavanaugh, *Neuron* **14,** 1019 (1995).

Is it charge movement? The major characteristic of charge movement current is the symmetry between the "on" response (the change from state A to state B) and the "off" response (the change back to state A). States A and B can be either two different membrane potentials or two different ion concentrations. The symmetry reflects the reversibility of the process that generates the current. Transient current can also occur with ion permeation, but the "on" and "off" currents are usually not symmetric. For example, in oocytes expressing the rat or human serotonin transporter, a large transient inward current develops in response to voltage jumps to negative potentials. However, "off" current is observed when the potential is stepped back to the initial voltage. Thus, this current reflects ion permeation and not binding.[10]

The charge movement of transporters can also be distinguished from membrane capacitance. As for the gating current in the ion channel, charge movements in the transporters are not linear with voltage, tend to saturate at extreme membrane potentials, and are slower than the membrane capacitance transients, which are limited only by the speed of the clamping circuit.[20,23]

Data Analysis

Isolation of Charge Movements Induced by Voltage and Concentration Jumps

As discussed previously, many ion-coupled transporters display capacitive-like transient currents in response to a step change in the membrane potential. Several approaches can be utilized to isolate this transient current from the membrane capacitance as well as from the currents endogenous to the oocyte. First, the charge movement of ion-coupled transporters (like the gating current of voltage-dependent ion channels) behaves as a nonlinear capacitor. Capacitance reaches a maximum at a specific voltage and reaches zero at more negative or positive potentials. Thus, membrane capacitance, which is the linear component of the transient current, can be subtracted from the total current. This method is described in detail by Heinemann *et al.*[24] Second, in some cases, the charge movement depends on the presence of external Na^+. Thus, recordings made in the absence of Na^+ can be subtracted from recordings made in its presence. The disadvantage of this method is that Na^+ permeates several endogenous channels and transporters in the oocyte membrane; therefore, the elimination of Na^+ also affects the endogenous currents. Third, charge movements are

[24] S. H. Heinemann, F. Conti, and W. Stühmer, *Methods Enzymol.* **207,** 353 (1992).

blocked by nonsubtrate uptake inhibitors and records in the presence of the inhibitor can be subtracted. It should be noted that, in many cases, uptake inhibitors are a competitive substrate for the transporter; therefore, they not only block the transient current, but also activate the transport current. The major obstacle in using inhibitors is that, at low Na^+ concentrations, the inhibitor does not efficiently block the transient current. In a typical experiment, a specific voltage protocol is applied, and the current is digitized and saved. The same voltage protocol is then repeated in the presence of the inhibitor or in a Na^+-free solution. The second recording is then subtracted from the first recording to yield the transient component (Fig. 1B).

A similar procedure is used for concentration-jump experiments. First, a specific concentration jump is applied and the current is digitized. The inhibitor is then applied in the presence of Na^+. The same protocol is repeated in the presence of the inhibitor (Fig. 1A).

Time Constant of Charge–Movement Current

The transient charge–movement current is fitted to an exponential equation of the form

$$I(t) = I_0 \exp(-t/\tau) \tag{1}$$

where $I(t)$ is the current as a function of time t, I_0 is the initial current, and τ is the time constant of the current decay. If the waveform contains more than a single time constant, a second exponent can be added to the equation. The effect of membrane potential, ion concentration, or other experimental conditions can provide important information about the molecular mechanism that underlies the charge movement. For example, in a single binding–dissociation reaction of the form

$$\mathrm{T} + n\mathrm{X} \underset{k_{-1}}{\overset{k_1}{\rightleftharpoons}} \mathrm{T}n\mathrm{X}$$

T is an empty transporter and TnX is a transporter occupied by n ions X. The first-order unidirectional rate constants that govern the transitions are $k_1(\mathrm{X}, V) = \mathrm{X}^n k_1(0) \exp[V/v_1]$ and $k_{-1}(V) = k_{-1}(0) \exp[V/v_{-1}]$. The time constant of this reaction is

$$\tau(\mathrm{X}, V) = 1/(k_1 + k_{-1}) \tag{2}$$

The predicted time constant increases when [X] is decreased (Fig. 2C). However, for the GABA transporter GAT1, the data show a decreased

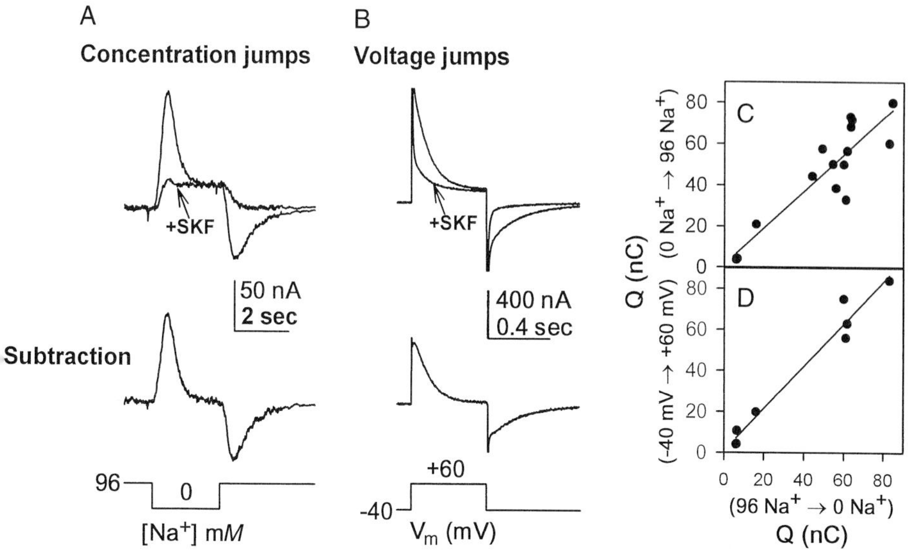

FIG. 1. Na^+ concentration jumps and voltage jumps in a single oocyte expressing GAT1. (A) The membrane potential was held at −40 mV during perfusion with a NaCl Ringer's solution. At the indicated time, the perfusion was changed to NMDG-Cl Ringer's solution and returned to NaCl Ringer's solution. (B) After an additional 5 sec, the membrane potential was stepped to +60 mV and back to −40 mV. The oocyte was then perfused with a Na^+ Ringer's solution containing 30 μM SKF-100330A and the procedures in (A) and (B) were repeated. The records in the presence SKF-100330A have been subtracted from those in the absence of the drug in order to isolate GAT1 currents. (C) Comparison of charge movements for the Na^+ concentration jumps (96 → 0 vs 0 → 96 m*M* Na^+) at a holding potential of −40 mV. (D) Charge movements during concentration jumps (96 → 0 m*M* Na^+) vs. voltage jumps (−40 → +60 mV).

time constant at lower $[Na^+]$ (Fig. 2A). A model that predicts such behavior incorporates sequential binding of 2 Na^+,

$$T + 2\,Na^+ \underset{k_{-1}}{\overset{k_1}{\rightleftharpoons}} T\,Na^+ + \overset{K_2}{\longleftrightarrow} T2\,Na^+$$

For simplicity, the binding dissociation of the second Na^+ ion is assumed instantaneous and voltage-independent; thus, $K_2 = [Na^+][T\,Na^+]/[T2\,Na^+]$ is an equilibrium constant. The time constant of this reaction is

$$\tau = 1 \Big/ \left(k_1[Na^+] + k_{-1}\left(\frac{K_2}{K_2 + [Na^+]}\right)\right) \tag{3}$$

The best fit of the data to the model is given in Fig. 2B. As usual, even good curve fitting only supports the model; it does not prove it.

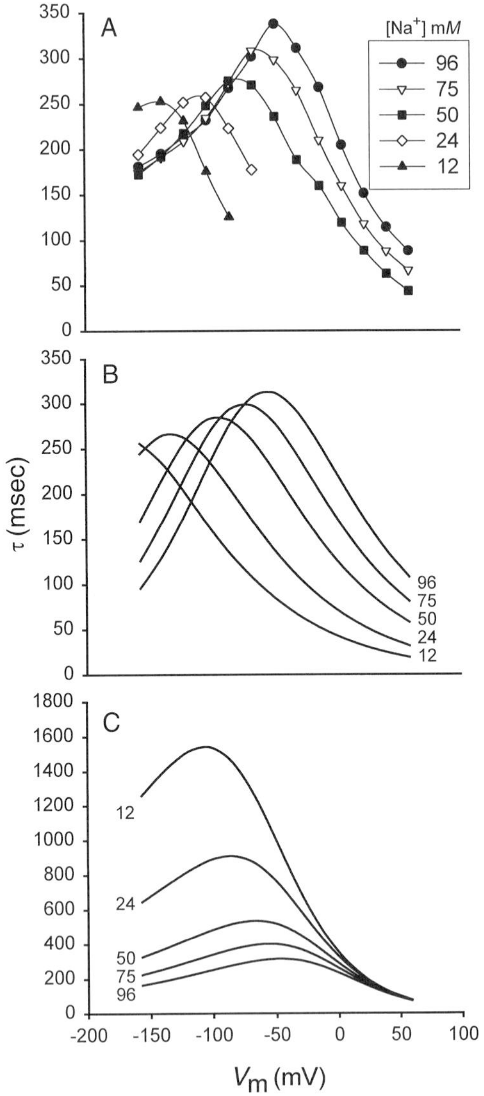

FIG. 2. Time constant of charge movements measured from voltage jumps at varying $[Na^+]$. (A) Data from oocytes expressing the GABA transporter GAT1. Currents were isolated by subtraction of traces in the presence of the GABA uptake inhibitor NO-05-711 and were fitted to Eq. (1). (B) Curve fitting of the data in (A) to Eq. (3), ($k_1 = 4.4 \times 10^{-6}$ ms^{-1} mM^{-1}, $k_{-1} = 6.8 \times 10^{-2}$ ms^{-1}, $V_1 = -50$ mV, $V_{-1} = 73$ mV, $K_2 = 6.2$). (C) Much less satisfactory predictions of time constants from the simpler two-state kinetic model. Initially, the curve in (A) that corresponds to 96 m*M* Na^+ was fitted to Eq. (2) ($k_1 = 1.7 \times 10^{-5}$ ms^{-1} mM^{-1}, $k_{-1} = 2.65 \times 10^{-3}$ ms^{-1}, $V_1 = -120$ mV, $V_{-1} = 37$ mV). Curves for lower $[Na^+]$ were then calculated.

Current Integration

To obtain the charge movement, the transient current is integrated over time. It has been shown that the charge movements are equal for the "on" and "off" currents, corresponding to the jump from the holding potential to a specific voltage and back to the holding potential, respectively (see Fig. 1). Nevertheless, most of the oocyte's endogenous current is carried by Cl^- and, because in many oocytes the Cl^- conductance changes with time, it is therefore more reliable to hold the membrane potential near the Cl^- reversal potential and to integrate the "off" current.

Curve Fitting

The charge movement, as a function of membrane potential, is fitted to a Boltzmann distribution of the form

$$\Delta Q(V_h \rightarrow V) = Q(V) - Q(V_h) \quad (4)$$

where $Q(V) = Q_{\max}/[1 + \exp(-(V - V_{1/2})q\ \delta n_{\rm H}/kT)]$ and $\Delta Q(V_h \rightarrow V)$ is the charge movement for a jump from the holding potential V_h to a final voltage V. The quantity $Q(V)$ equals the charge that moves between an extreme hyperpolarized potential and the test potential V. $Q_{\max}$ corresponds to the complete movement of charges between the membrane and the medium (for instance, from an extreme hyperpolarized potential to an extreme depolarized potential), $V_{1/2}$ is the voltage at which charge movements are half-complete, $n_{\rm H}$ is the charge on the particle moving, δ is the fraction of the membrane field through which the charge moves, and q is the elementary charge. k and T have their usual values.

At lower $[Na^+]$, the $Q(V)$ curve is shifted to the left, as if a more negative potential is required to bind the Na^+ to the transporter. For the GABA and glutamate transporters, the only parameter in Eq. (4) affected by the $[Na^+]$ is $V_{1/2}$. The "field-dependent ion binding equation," a modified Hill/Boltzmann distribution, can be used to describe the effects of both Na^+ and membrane potential on the charge movement[20]:

$$Q(V, {\rm Na}^+) = \frac{Q_{\max}}{1 + \{[{\rm Na}^+]/K_{\rm Na} \exp(-Vq\ \delta/kT)\}^{n_{\rm H}}} \quad (5)$$

In Eq. (5), $K_{\rm Na}$ is a zero-voltage, intrinsic, Na^+-binding constant of the transporter in the Hill equation, and the midpoint of the Boltzmann distribution is

$$V_{1/2} = \frac{kT}{q\delta} \ln\left(\frac{[{\rm Na}^+]}{K_{\rm Na}}\right) \quad (6)$$

For a typical experiment, the measured $\Delta Q(V_h \rightarrow V)$ at different $[Na^+]$ is fitted to Eq. (4) (Fig. 3A). $Q(V_h)$ is then added to each curve, and the curves are superimposed (Fig. 3B). Equation (5) can be applied if (1) $V_{1/2}$ is the only parameter significantly affected by the $[Na^+]$, and (2) there is a linear relation between $\log[Na^+]$ and $V_{1/2}$ (Fig. 3C). The detailed analysis presented above can be applied for systematic comparisons between different transporters or to analyze the effect of mutations for structure–function studies. In one paper, a single mutation in the GABA transporter GAT1

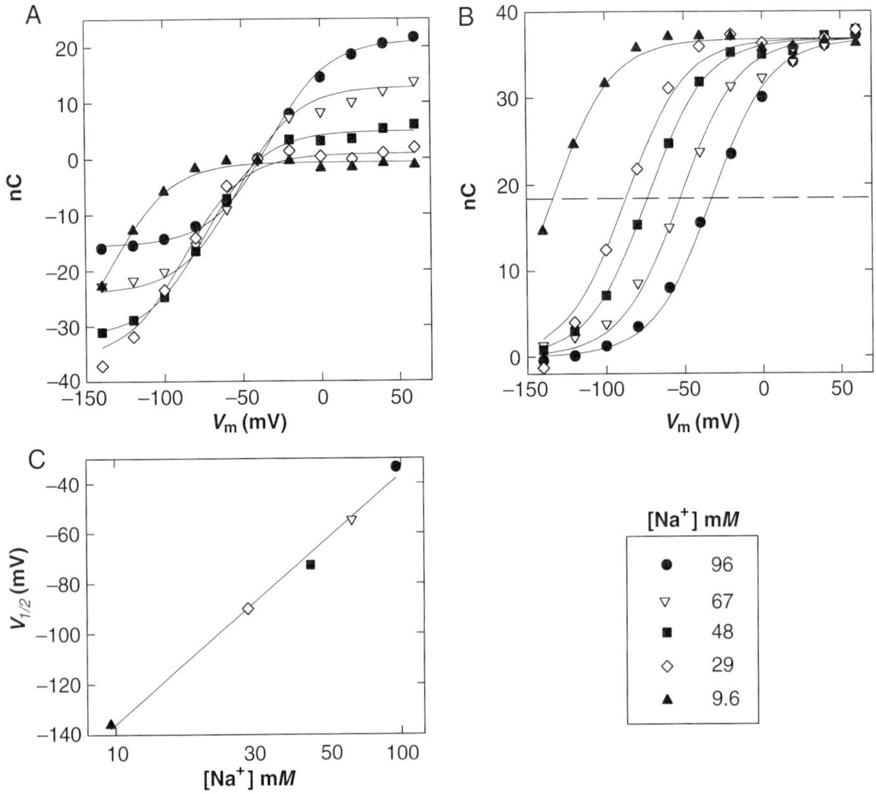

FIG. 3. Voltage-jump charge movements at varying external $[Na^+]$. (A) Charge movements $-\Delta Q(V \rightarrow -40)$ in a single cell for various values of V. (B) Data in A have been fit to Eq. (4) and translated vertically so that the charge movements superimpose at positive potentials. This procedure reveals that the curves also shift along the voltage axis toward more hyperpolarized potentials at reduced $[Na^+]$. The dashed horizontal line intersects each curve at half-completion of charge movements [corresponding to $V_{1/2}$ in Eq. (4)]. (C) Fitted values for $V_{1/2}$ are plotted vs. $[Na^+]$ on a semilogarithmic scale (solid line). Least-squares fit corresponds to an *e*-fold change in $V_{1/2}$ for 43 mV.

decreased K_{Na}, but did not significantly affect any other parameter in Eq. (5).[21]

Turnover Rate and Stoichiometry

The transporter turnover rate is simply the ratio between flux rate and the number of transporters per oocyte. In the simplest form, the uptake rate is measured by the flux of radiolabeled organic substrate; the number of transporters is measured by the binding of specific, high-affinity ligands to transporters at the cell membrane. Electrophysiology can provide an alternative estimate of the turnover rate. The number of transporters in the membrane is estimated by converting the charge movement to the number of elementary charges, and the flux is estimated from the transport current. The ratio of the transport current at specific conditions to Q_{max} is the turnover rate of the transporter (Fig. 4).

Summary

Electrophysiological measurements add new dimensions to the study of neurotransmitter transporters. (1) One can perform measurements with

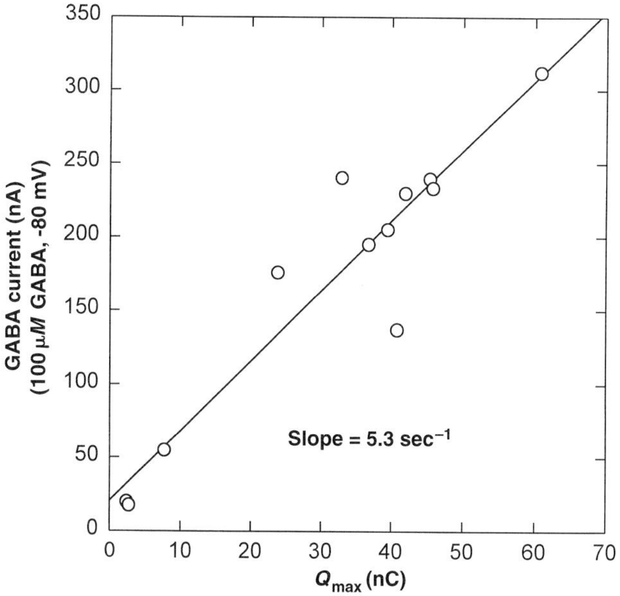

FIG. 4. Turnover analysis of the GABA transporter GAT1. Maximal GABA currents at −80 mV are plotted versus Q_{max}. Least-squares fit line represents the turnover rate of the transporter and has a slope of 5.3 sec^{-1}.

high temporal resolution (however, uptake of radioactive substrate is limited in that it cannot resolve events that occur within 1 sec, which is greater than the time of a single transport cycle). (2) Electrophysiology provides information about partial steps in transport cycles, including the fact that ion binding and dissociation at transporters can generate currents, which provides new insights about ion–transporter interaction. (3) Electrophysiology provides information about single transporter molecules, from patch-clamp recordings of single-channel activity of neurotransmitter transporters.

At present, little is known about the molecular mechanisms that underlie transport. Electrophysiological measurements of ion binding and permeation contribute to the analysis of mutations that affect transport.[13,21] Electrophysiology may help to identify amino acids and domains in neurotransmitter transporters that participate in specific ways in the transport process, such as ion neurotransmitter binding, permeation pathways, voltage sensors, and gates. In combination with spectroscopic measurements,[25] it may also be possible to identify the actual conformational changes of the proteins that enable substrate translocation.

[25] L. M. Mannuzzu, M. M. Moronne, and E. Y. Isacoff, *Science* **271,** 213 (1996).

[38] Fluorescence Techniques for Studying Cloned Channels and Transporters Expressed in *Xenopus* Oocytes

By Albert Cha, Noa Zerangue, Michael Kavanaugh, and Francisco Bezanilla

In this article we describe some applications of fluorescence to the study of transporters and channels expressed in *Xenopus laevis* oocytes. Two general approaches are covered. The first section deals with measurement of changes in surface fluorescence associated with conformational transitions, or gating, of exogenously expressed channels and transporters. This approach is based on changes in the fluorescence of covalently incorporated probes resulting from perturbation of the probe microenvironment during a process such as channel opening. The second part of the article deals with the use of ion-sensitive fluorescent dyes to analyze solute movements during transport. Methods for combining each technique with measurement of membrane currents under voltage clamp are also described. A useful

0076-6879/98 $25.00

source of general information on fluorescence techniques can be found in Volume 278 of this series.[1]

Surface Fluorescence Measurements from *Xenopus* Oocytes *in Situ*

Valuable information about conformational changes in transmembrane proteins can be obtained from *Xenopus* oocytes *in situ* with site-directed fluorescent labeling. After using site-directed mutagenesis to substitute cysteine for a particular residue in extracellularly exposed regions of the protein, an extrinsic sulfhydryl-reactive fluorescent probe can be covalently attached to this residue. By using a fluorescent probe whose quantum yield is highly dependent on its environment, changes in the environment of the labeled residue resulting from conformational changes in the protein can be monitored via changes in the fluorescence intensity and spectrum of the probe from the surface of *Xenopus* oocytes. This technique has been used to examine the voltage dependence and kinetics of conformational changes in different regions of the voltage-gated *Shaker* potassium channel.[2,3] Although structural models of gating in transporters are generally less refined than for voltage-gated channels, we anticipate that future studies will lead to the development of more detailed conformational models in this class of proteins.

Optical Measurements

The epifluorescence optics of the experimental setup utilize same light path for excitation of the fluorophore and for fluorescence emission (Fig. 1a, b). These light components are separated with a dichroic mirror (F), which reflects the excitation light while transmitting the emitted fluorescence. In an inverted microscope, a fiber optic (A) couples the light from an objective (B) to the surface of the oocyte (C). The excitation light is filtered (D) from a tungsten–halogen lamp (E), is reflected off of a dichroic mirror (F), and passes through the objective and fiber optic to the oocyte surface. The emitted fluorescence is transmitted from the oocyte surface to the fiber optic, through the objective and dichroic mirror, is filtered (G), and travels to one of two different detection systems. For measurements of fluorescence intensity, the fluorescence is focused onto the active area of a photodiode (H), which is attached to the headstage of a patch clamp amplifier (K). For spectral measurements, the fluorescence

[1] L. Brand and M. L. Johnson, Eds., *Methods Enzymol.* **278** (1997).
[2] L. M. Mannuzzu, M. M. Moronne, and E. Y. Isacoff, *Science* **271,** 213 (1996).
[3] A. Cha and F. Bezanilla, *Neuron* **19,** 1127 (1997).

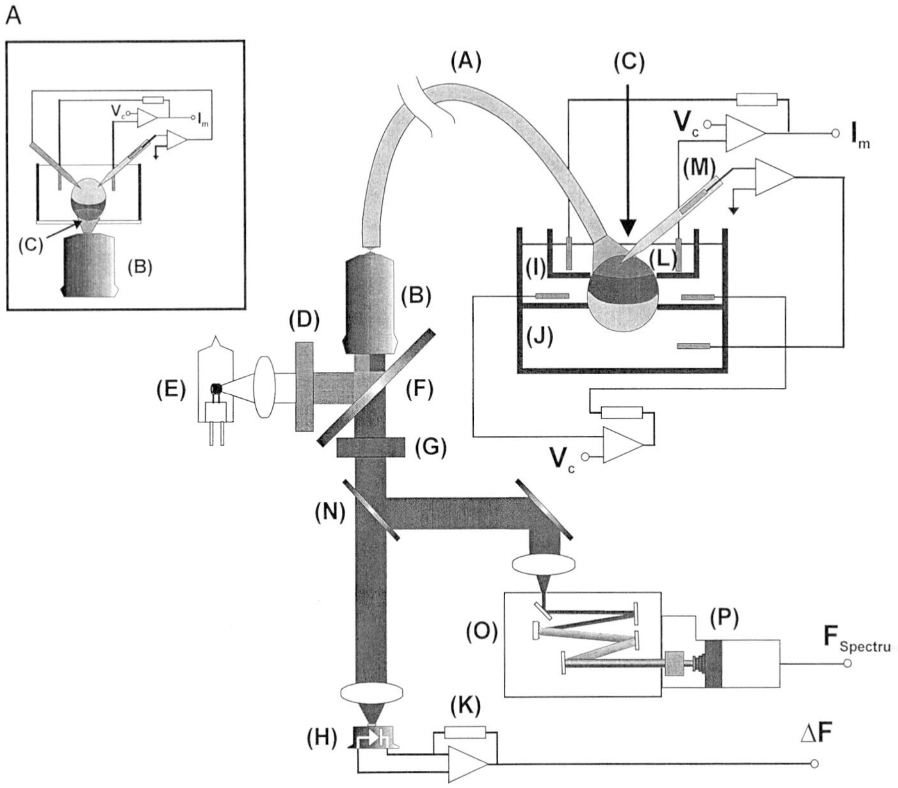

FIG. 1. (a) Schematic, (b) photograph of cut-open oocyte fluorescence technique. The epifluorescence setup utilizes a fiber optic (A) to couple the light from an objective (B) to the surface of the oocyte (C). The excitation light is filtered (D) from a tungsten–halogen lamp (E), is reflected off of a dichroic mirror (F), and passes through the objective and fiber optic. The emitted fluorescence is transmitted through the fiber optic, objective, and dichroic mirror and is filtered (G). This light can be switched (N) between being focused onto the surface of a photodiode (H), which is attached to the headstage of a patch clamp amplifier (K) and being focused onto the slit of a spectrograph (O) attached to an array of charge-coupled devices, or CCD (P). The voltage clamp setup is composed of an upper, middle, and bottom chamber. The bottom chamber (J) contains the portion of the oocyte which is permeabilized with saponin so that current can be injected directly into the oocyte. The middle chamber (I) serves as an electronic guard, and the top chamber (L), which is painted black, contains the portion of the oocyte membrane from which the fluorescence changes and gating or ionic currents are measured. The voltage electrode (M) measures the membrane potential inside the oocyte and is part of the feedback loop that holds the interior of the oocyte at virtual ground. *Inset:* The alternative experimental setup utilizes the two-microelectrode voltage clamp. In this setup, the objective (B) is focused directly on the bottom surface of the oocyte (C). The components below the objective are the same as those in the main figure. [Adapted with permission from *Neuron* **19,** 1127 (1997), copyright 1997 Cell Press.]

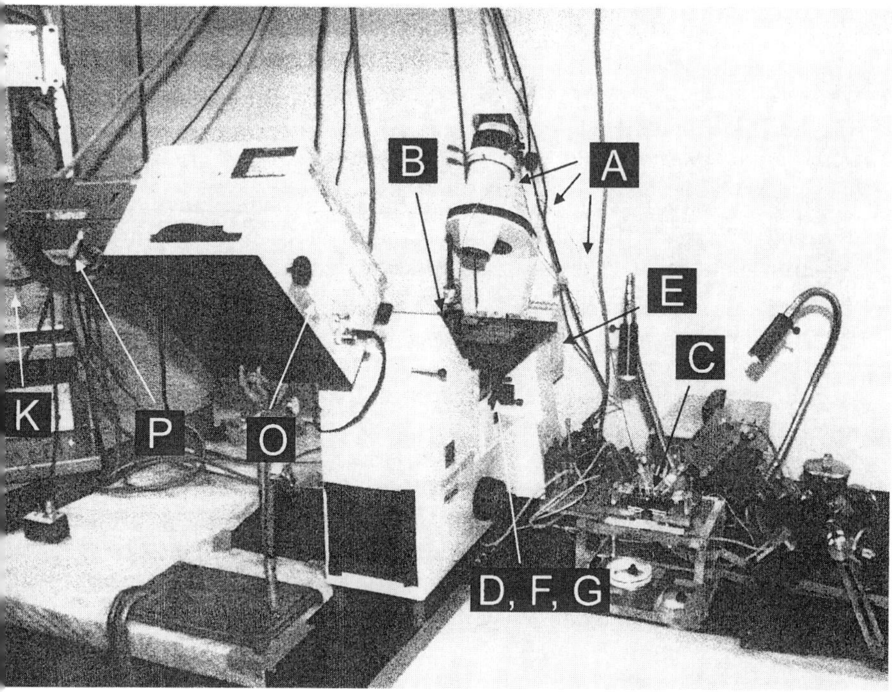

FIG. 1. (*continued*)

is focused onto a slit at the entrance of a spectrograph (O), which separates the fluorescence into its component wavelengths and projects it onto an array of charge-coupled devices, or CCD (P). A mechanical switch in the microscope allows the light to be switched between the epifluorescence port, where the photodiode is attached, and the eyepiece port, where the CCD is attached, so that fluorescence intensities and spectra can be measured during the same experiment (N).

The objective's field of view is effectively extended to the oocyte surface with the fiber optic, which is positioned with a hydraulic manipulator to focus the light onto the oocyte. The fluorescence and electrical currents are measured from a large, 600 μm-diameter area of membrane located in the top chamber (L), which generally contains at least 10^9 channels. Using an opaque top chamber ensures specific fluorescence excitation of the same membrane from which the electrical currents are measured.

The oocyte should be positioned in the chamber so that the animal pole, which is the dark side of the oocyte, is visible in top chamber for fluorescence measurements. This side contains a pigment layer beneath the plasma membrane which absorbs light, preventing cytoplasmic autofluorescence from interfering with measurements of fluorescence from labeled proteins.

Photodiode Measurements of Fluorescence Intensity

When a photon of an energy greater than the band gap of silicon hits the active area of the photodiode, an electron–hole pair forms. These electron–hole pairs generate a potential across the semiconductor P–N junction, as well as a photocurrent proportional to the amount of light absorbed. For lowest noise operation, the silicon photodiode should be operated in photoconductive mode, where a small reverse voltage bias is applied to the photodiode terminals. In order to detect small changes in the fluorescence intensity on the order of picowatts (pW), which generates a photocurrent in picoamperes (pA), it is critical that a low-noise amplifier be used to amplify the photocurrent. The patch clamp amplifier is ideal for this task because it is designed to amplify very small electrical currents. In order to achieve the best signal-to-noise ratio, the photodiode and headstage must be properly shielded from electromagnetic interference. This can be accomplished by ensuring that there are no ground loops between any of the measurement equipment, including the computers used for acquisition, and by placing a grounded aluminum foil shield around the photodiode and its attachment to the headstage. By attaching the photodiode to the headstage of a patch clamp amplifier and carefully shielding the photodiode and headstage, changes in fluorescence of ~0.1 pW can be measured with signal averaging.

The choice of photodiode is also important. With silicon-based photodiodes, the smaller the active area, the lower the noise and capacitance associated with the signal. Thus, the best photodiode is one with a very small active area of less than 0.50 mm^2. However, the light must be focused extremely accurately onto this active area in order to maximize the measured fluorescence. With an infinity-corrected optical system, the light can be focused onto the extremely small active area by using a lens with a short focal length, such as a microscope condenser lens, and mounting the photodiode in a manipulator for positioning. Without infinity-corrected optics, this can be done without the lens as long as the photodiode is located in the focal plane. In this manner, almost all of the fluorescence from the oocyte can be focused onto the surface of the photodiode.

Avalanche photodiodes utilize a cascade of electrons generated from a single electron–hole pair at high voltages for internal amplification of the photocurrent. Because of this internal gain, avalanche photodiodes may be useful for measuring extremely low levels of fluorescence. However, the problem with avalanche photodiodes is that the amplification process also amplifies the noise of the avalanche process. In this system, the fluorescence signals measured with avalanche photodiodes are no better than those seen with regular photodiodes using a patch clamp amplifier.

It is also important to note that large amounts of light generate photocurrents which are capable of saturating the capacitor in an integrating headstage during a single data sweep. This can be handled in one of two different ways. One method requires the addition of an adjustable voltage source in series with a high-impedance resistor of at least 10 gigohms (GΩ), which is fed into the summing junction of the current-to-voltage converter in the headstage. By adjusting the voltage, the current fed into the summing junction can offset the amount of photocurrent from a large background light signal, preventing the capacitor from saturating during a single sweep. The other method utilizes a resistive headstage to avoid the problem of capacitor saturation, and low-noise amplification can still be attained with a very high (>10 GΩ) resistance as long as a booster circuit is added to restore the high-frequency response of the system.

Measurements of Fluorescence Spectra

Fluorescence spectra from the surface of the oocyte are measured with a spectrograph attached to the face of a CCD array. This spectrograph has a narrow slit at its entrance, a diffraction grating which separates the incoming light into its component wavelengths, and a combination of mirrors which image onto the surface of the CCD. The CCD integrates the fluorescence intensity along one of its two dimensions to obtain the total intensity of light as a function of wavelength. In order to optimize the light entering the spectrograph, a cylindrical lens was used to focus the light onto the entrance slit. This increased the spectral signals severalfold without changing the spectral characteristics. Using a gated image intensifier also increases the sensitivity of the CCD without substantially increasing the noise level.

Voltage-dependent steady-state spectra of the probe are obtained with the CCD while the oocyte is held at different potentials in voltage clamp. For spectral kinetics, the CCD is triggered externally from a control line from the acquisition system at various points during the sweep. An electronically triggered shutter which opens only during signal acquisition avoids unnecessary photobleaching of the fluorescent probe between acquisitions.

In order to obtain an accurate fluorescence spectrum from CCD measurements, the transfer function of the system must be measured. This can be done by transilluminating the objective with the tungsten–halogen light source with and without the dichroic mirror and emission filter. After subtracting the dark noise, dividing these two spectra determines the transfer function, which can be used to correct the spectral shape of CCD measurements.

Electrophysiological Measurements

By measuring changes in fluorescence from the surface of the oocyte while clamping the transmembrane voltage, gating and ionic currents have been measured simultaneously with changes in fluorescence intensity and fluorescence spectra.[3] The voltage clamp for recording electrical currents from the surface of the oocyte is based on the cut-open oocyte technique.[4–6] The voltage clamp setup is composed of an upper, middle, and bottom chamber (see Fig. 1). The bottom chamber (J) contains the portion of the oocyte which is permeabilized with saponin so that current can be injected directly into the oocyte. The middle chamber (I) serves as an electronic guard, and the top chamber (L), which is painted black, contains the portion of the oocyte membrane from which the fluorescence changes and electrical currents are measured. The voltage electrode (M) measures the membrane potential inside the oocyte and is part of the feedback loop that holds the interior of the oocyte at virtual ground.

The cut-open oocyte technique is ideal for these measurements for several reasons. During the experimental protocol, the bottom surface of the oocyte is permeabilized with saponin, which generates a low access resistance path into the interior of the oocyte and also enables solution changes of the internal medium. External solution changes are performed by exchanging solutions in the top and middle chamber. By injecting current into the oocyte through the bottom chamber with an agar bridge while clamping the interior to virtual ground, series resistance errors are substantially reduced. In addition, the guard voltage clamp minimizes problems of spatial voltage heterogeneity. As a result, this clamp is capable of resolving capacitive transients in 50–100 μsec, which permits the examination of very fast kinetic changes.[5] In addition, for transporter studies, access to the intracellular compartment allows introduction of substrates such as glucose, permitting measurement of bidirectional transport.[7]

The two-microelectrode oocyte voltage clamp can also be used to measure fluorescence changes without the use of a fiber optic.[2] This optical setup is illustrated in Fig. 1a (inset). The advantage with this approach is

[4] E. Stefani and F. Bezanilla, *Methods Enzymol.* **293,** [17], (1998).

[5] E. Stefani, L. Toro, E. Perozo, and F. Bezanilla, *Biophys. J.* **66,** 996 (1994).

[6] M. Taglialatela, L. Toro, and E. Stefani, *Biophys. J.* **61,** 78 (1992).

[7] X.-Z. Chen, M. J. Coady, F. Jackson, A. Berteloot, and J.-Y. Lapointe, *Biophys. J.* **69,** 2405.

that no light is lost through the fiber optic coupling to the oocyte surface, increasing the total light signal. However, there are a few disadvantages with using the two-microelectrode voltage clamp. First of all, clamping with two microelectrodes does not uniformly clamp the entire oocyte to the same voltage, as a result of the distributed series resistance within the oocyte. Thus, the voltage seen by channels from which fluorescence is measured may be different from the clamped voltage. In addition, the cut-open oocyte voltage clamp has substantially faster time resolution than the two-microelectrode voltage clamp, although this may not be an issue when measuring currents with relatively slow kinetics. Finally, using the two-microelectrode voltage clamp prevents direct access to the internal solution of the oocyte.

In order to simultaneously measure the electrophysiological data with the fluorescence, an acquisition system capable of two-channel recording must be used. It is also advantageous to have a trigger which fires before the acquisition of fluorescence, so that the integrating headstage can be reset, preventing premature saturation of the headstage capacitor during acquisition.

An example of simultaneously recorded gating currents and fluorescence changes is shown in Fig. 2. The gating currents (Fig. 2A) and fluorescence changes (Fig. 2B) were measured from a cut-open *Xenopus* oocyte expressing a *Shaker* potassium channel with a cysteine substitution at residue 356 near the fourth transmembrane segment. This oocyte was labeled with tetramethylrhodamine-5-maleimide for 40 min with the protocol described below.

Oocyte Labeling

Xenopus oocytes are injected with cRNA transcripts of the protein of interest. Because the *Shaker* channel does not contain any naturally occurring, externally accessible cysteines, external cysteines introduced with site-directed mutagenesis[8] can be specifically labeled with extrinsic, membrane-impermeant fluorophores. In this manner, sites in the extracellular portions of the second and fourth transmembrane segments, as well as the putative pore region, have been successfully labeled with the maleimide-based fluorescent probes tetramethylrhodamine, fluorescein, and Oregon Green (Molecular Probes; Eugene, OR).[3] By incubating these oocytes in a solution containing 5 μM of the probe from 40 min to an hour at room temperature, substantial voltage-dependent fluorescence changes can be

[8] D. D. Moore (1994). *In* "Current Protocols in Molecular Biology," F. M. Ausubel, R. Brent, R. E. Kingston, D. D. Moore, J. G. Seidman, J. A. Smith, and K. Struhl, Eds., pp. 8.2.8–8.2.11. John Wiley & Sons, New York.

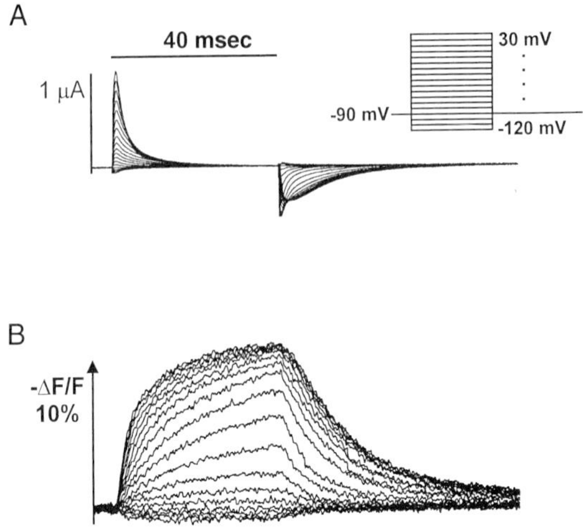

FIG. 2. Data obtained with the cut-open oocyte fluorescence technique. (A) Gating currents measured from a *Xenopus* oocyte mounted in the cut-open oocyte voltage clamp. These currents were measured from a *Shaker* H4IR channel construct with N-type inactivation removed (Δ6–46), ionic conduction blocked (W434F), a cysteine substitution at site 356 (M356C), and labeled with tetramethylrhodamine-5-maleimide. The voltage protocol (*inset*) consisted of pulses of 40 msec from a holding potential of −90 millivolts (mV) to potentials ranging from −120 mV to +50 mV in increments of 10 mV. (B) Fluorescence traces measured simultaneously during the gating current in (A), displayed on the same time scale. The scale arrow indicates the direction and magnitude of the decrease in fluorescence intensity. Each record is an average of three sweeps.

obtained from these fluorescently labeled oocytes. The optimal labeling conditions should be empirically determined to ensure saturation of channel labeling for each dye and cysteine construct combination. In addition, comparing the extent of labeling with a hyperpolarizing solution (i.e., 0 m*M* potassium) to labeling with a depolarizing solution (i.e., 120 m*M* potassium) can determine at which potential the residue is most accessible to labeling. Covalent modification of cysteine residues in transporters can also be achieved using similar techniques, either with naturally occurring cysteines or by using site-directed mutagenesis protocols.

Specifications

The inverted epifluorescence microscope is an Axiovert 100TV (Carl Zeiss; Thornwood, NY) with an excitation filter (D), dichroic mirror (F), and emission filter (G) appropriate to the fluorescent dye being used. The tungsten lamp provided with the microscope (E) is used as the light source and contains a 150-watt filament (Osram BRJ EVB), powered by a 6286A power supply (Hewlett-Packard; Palo Alto, CA). The lamp output is interrupted with a TTL-triggered VS25 shutter (Vincent Associates; Rochester, NY) to minimize photobleaching of the probe. The fiber optic cable (A) is an FSUS760 (PolyMicro Technologies; Phoenix, AZ) with a numerical aperture of 0.65 and core diameter of 760 μm. Fluorescein-5-maleimide and Oregon Green-5-maleimide (Molecular Probes) measurements are made with a 470DF35 excitation filter (Omega Optical; Brattleboro, VT), a 505DRLP dichroic mirror (Omega Optical), and a 535DF50 emission filter (Chroma Technologies; Brattleboro, VT). Tetramethylrhodamine-5-maleimide (Molecular Probes) measurements are made with a 535DF50 excitation filter (Chroma Technologies), a 570DRLP dichroic mirror (Omega Optical), and a 565EFLP emission filter (Omega Optical).

Fluorescence intensity measurements are made with a PIN-020A photodiode (H) (UDT Technologies; Torrance, CA) mounted on an FP-1 fiber optic manipulator (Newport Corporation; Irvine, CA), which is attached to the microscope's epifluorescence port. The avalanche photodiode used is the C-30902S (EG&G Optoelectronics; Santa Clara, CA). The patch clamp amplifier is an Axopatch-1B (Axon Instruments; Foster City, CA), used either with a custom-built resistive headstage or an IHS-1 integrating headstage (Axon Instruments). The microscope objective (B) is an infinity-corrected 20× quartz objective (Partec; Münster, Germany) with a numerical aperture of 0.65. The fluorescence is focused onto the photodiode active area using a microscope condenser lens with a focal distance of 1 cm.

Fluorescence spectra are obtained with an Multispec 257i spectrograph (Oriel Instruments; Stratford, CT) with a 1200 l/mm grating, which is attached to an Instaspec V CCD with image intensifier (Oriel Instruments) cooled to −20°. The spectrograph and CCD are attached to the eyepiece port of the epifluorescence microscope to enable measurements with the photodiode and the CCD during the same experiment.

Voltage clamp of the oocyte is performed with a CA-1 cut-open oocyte clamp (Dagan Corporation; Minneapolis, MN). Gating, ionic, and fluorescence currents are acquired with a PC44 board (Innovative Technologies; Moorpark, CA), which interfaces with a Pentium-based computer via an IBM-compatible AT slot. The fluorescence and electrophysiology are simultaneously acquired on two analog-to-digital converters and transferred to two separate channels of the PC44.[8a]

Intracellular Ion Measurements with Fluorescent Dyes

Steady-state current measurements under voltage clamp is a powerful technique for analyzing transporter function, but it does not provide direct

[8a] A new version of this setup, which uses an upright microscope and 40× water immersion objective, has been developed which increases the fluorescence signal by around an order of magnitude (Cha and Bezanilla, submitted).

information about individual ionic species. Determining the possible role of a particular ion in co- or countertransport can be complicated by factors including the involvement of multiple ions and the presence of thermodynamically uncoupled currents which are frequently observed in cloned transporters. This section provides an overview of the use of fluorescent dyes.[9] in *Xenopus* oocytes to directly monitor changes in intracellular ion concentrations. Combining ion-sensitive fluorescent dye measurements in oocytes with voltage clamp allows the experimenter to control the electrochemical gradients of a particular ion which is coupled to transport of other solutes. The existence and direction of ion fluxes can be monitored under various conditions, providing insight into the thermodynamics of the transporter being studied.[10] This technique can be applied to demonstrate coupled flux of a particular inorganic ion (which may be either cotransported or countertransported with substrate) by demonstrating that the transporter can mediate movement of the ion of interest against its electrochemical gradient under conditions in which the net driving force for all coupled solutes is favorable.

Ion-Sensitive Dye Loading

A large number of ion-sensitive fluorescent dyes and filter sets are commercially available for monitoring commonly transported ionorganic ions, including Na^+, K^+, Cl^-, Ca^{2+}, and H^+. A convenient sourcebook can be obtained from Molecular Probes (www.probes.com/handbook/toc.html). Oocytes are loaded with a membrane-permeant form of the appropriate dye, commonly an acetoxymethyl ester form which can pass into the cytoplasm where it is trapped following hydrolysis by endogenous esterases. The conditions of dye loading should be empirically determined for each case. For BCECF-AM (Molecular Probes), a 3 mM stock solution is prepared in anhydrous DMSO and stored at $-20°$. Oocytes are incubated in a Ringer's solution at room temperature containing 3 μM dye for approximately 1 hr in the dark. Longer loading times can be employed to increase the fluorescence signal if necessary.

Fluorescence and Voltage Clamp Setup

A two-microelectrode voltage clamp setup with a standard inverted epifluorescence microscope is utilized for simultaneous voltage clamp and intracellular dye fluorescence recording (see Fig. 1a, inset). In order to facilitate positioning of the microelectrodes, a stereoscopic zoom microscope may be mounted above the chamber by attachment to the condenser

[9] R. Y. Tsien, *Methods Cell Biol.* **30,** 127 (1989).
[10] N. Zerangue and M. P. Kavanaugh, *Nature* **383,** 634 (1996).

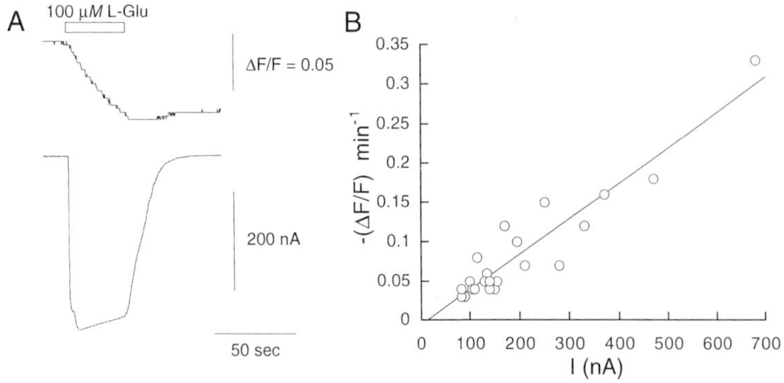

FIG. 3. BCECF fluorescence changes associated with glutamate transport. (A) Simultaneous recording of fluorescence change and electrical current in an oocyte expressing the human glutamate transporter clone EAAT3. The 490/440 nm ratio of emission intensity (monitored at 540 nm) decreases during application of 100 μM glutamate ($V_{mem} = -80$ mV), reflecting a decrease in intracellular pH. (B) The slope of the fluorescence decrease is correlated ($r = 0.94$) with the magnitude of the ionic flux measured as current at −80 mV in 22 cells expressing different levels of EAAT3.

support or by means of a free-swinging arm. Alternatively, the electrodes may be simply positioned using the inverted optics as described by Stühmer.[11] Illumination is provided by a 75 W xenon lamp. A set of variable neutral density filters is useful for optimizing the fluorescence signal intensity. For ratiometric dyes such as BCECF which have variable excitation spectra, signals at appropriate wavelengths are recorded by switching the excitation filter and ratios are computed off-line. Depending on the kinetic requirements of the experiment, a simple manual filter switch may be used for many applications, including steady-state current measurements. Fluorescence is detected by a photomultiplier tube attached to the side port of the microscope. The setup in use consists of a Diaphot 200 with a 200× objective and a P100S photomultiplier tube controlled by a P101 photometer (Nikon). The output from the controller is directed to an A/D converter such as a Maclab (ADInstruments Milford, MA) or a Digidata 1200 (Axon Instruments) via a BNC connection so that the signals can be stored on computer.

After dye loading, the oocyte is positioned in a recording chamber directly on top of a glass coverslip with the vegetal (light) pole oriented down, facing the light beam. The cell is impaled with microelectrodes and the chamber is perfused at a flow rate of ~5 mL/min. The photomultiplier output is monitored while the objective is focused onto a plane between

[11] W. Stühmer, *Methods Enzymol.* **207,** 319 (1992).

the bottom of the oocyte and the equator which results in the maximum fluorescence signal. Fluorescence signals are typically >100 times larger than background measured in unloaded oocytes. The cell is voltage clamped and current and fluorescence signals are simultaneously recorded. After stable signals are obtained, the chamber solution is changed to one containing the transporter substrate and current and fluorescence changes are recorded. Figure 3 illustrates an example of this technique in oocytes expressing the human glutamate transporter EAAT3[10] in which transport causes an increase in intracellular proton concentration monitored as a decrease in fluorescence of the dye BCECF. The change in dye signal is proportional to the flux of monitored ion, as indicated by the correlation of the transport current magnitude with slope of the fluorescence signal (Fig. 2B).

[39] Fluctuation Analysis of Norepinephrine and Serotonin Transporter Currents

By LOUIS J. DEFELICE and AURELIO GALLI

Introduction

The uptake of norepinephrine (NE) by cocaine- and antidepressant-sensitive NE transporters (NETs) in presynaptic terminals requires the cotransport of Na^+ and Cl^- ions.[1,2] Therefore, NE uptake is associated with an ionic current.[3–5] In order to study the relationship between uptake and current, it is convenient to establish stable lines. We have constructed a human norepinephrine transporter (hNET) cell line that is suitable for NE uptake studies, radioligand binding studies, and voltage-clamp recording.[6] Stably transfected hNET into HEK-293 (human embryonal kidney) cells (hNET-293) have an average of 10^6 transporter copies per cell. hNET-293 cells display micromolar affinity for NE, Na^+-dependent uptake of NE, and inhibition by NET-specific antagonists similar to that observed in native

[1] L. L. Iversen, *in* "Handbook of Psychopharmacology," L. L. Iversen, Ed. Plenum Press, New York, 1975.
[2] G. Rudnick and J. Clark, *Biochim. Biophys. Acta* **1144,** 249 (1993).
[3] R. D. Blakely, L. J. DeFelice, and H. C. Hartzell, *J. Exp. Biol.* **196,** 263 (1994).
[4] L. J. DeFelice and R. D. Blakely, *Biophys. J.* **70,** 579 (1996).
[5] M. Sonders and S. G. Amara, *Curr. Opin. Neurobiol.* **6,** 294 (1996).
[6] A. Galli, L. J. DeFelice, B. J. Duke, K. R. Moore, and R. Blakely, *J. Exp. Biol.* **198,** 2197 (1995).

METHODS IN ENZYMOLOGY, VOL. 296
0076-6879/98 $25.00

membranes.[6] In addition, hNET-293 cells also demonstrate Na^+-dependent and antidepressant/cocaine-sensitive NE-induced currents that are absent in parental HEK-293 cells. These stable cell lines therefore provide a useful model in which to investigate the relationship between NE uptake and NE-induced current. hNET-293 cells also possess a leak current that is mediated by hNET but is independent of NE. The leak current is revealed by hNET antagonists applied to transfected cells. hNET agonists and antagonists have no similar effects in nontransfected cells.[6,7] This review focuses on the NE-induced currents, not the NE independent currents. We include a similar analysis of serotonin (5-HT) transport and 5-HT-induced currents in serotonin transporters (SERTs).

The initial problem that concerns us is to compare the current expected from alternating access models to the current measured under voltage clamp. We then concentrate on the contribution of fluctuation analysis to understanding these currents. From uptake studies on synaptosomes, primary culture cells, and resealed membranes, it is known that the rate of NE transport saturates at micromolar concentrations of NE and millimolar concentrations of Na^+ and Cl^-.[8–14] These studies predict fixed stoichiometric cotransport of NE with Na^+ and Cl^-. For example, the transport model derived from uptake studies in PC-12 (rat pheochromocytoma) cells imply a transport coupling ratio of 1 NE/1 Na^+/1 Cl^-.[13] Because NE is a monovalent cation at physiological pH, this coupling ratio implies a net charge transfer of one electronic charge unit (e) per transport cycle. Thus, NE transport is said to be electrogenic.

Transporter Currents

Studies on membrane vesicles indicate a NET transport rate of 2.5 cycles/sec.[12] Assuming this rate, we can estimate the current associated with NE transport in hNET-293 cells. If there are $N = 10^6$ transporters cycling at a rate $\nu = 2.5$/sec with a net charge transfer of $q = 1e$, where $e = 1.6 \times 10^{-19}$ coulomb (C), then the transport current is

$$I = N\nu q$$

[7] A. Galli, R. Blakely, and L. J. DeFelice, *Proc. Natl. Acad. Sci. USA* **93,** 8671 (1996).
[8] L. L. Iversen and E. A. Kravitz, *Mol. Pharm.* **2,** 360 (1966).
[9] D. F. Bogdanski and B. B. Brodie, *Life Sci.* **5,** 1563 (1966).
[10] S. Sanchez-Armáss and F. Orrego, *Life Sci.* **20,** 1829 (1977).
[11] R. Harder and H. Bönisch, *J. Neurochem.* **45,** 1154 (1985).
[12] H. Bönisch and R. Harder, *Arch. Pharm.* **334,** 403 (1986).
[13] U. Friedrich and H. Bönisch, *Arch. Pharm.* **333,** 246 (1986).
[14] T. Pacholczyk, R. D. Blakely, and S. G. Amara, *Nature* **350,** 350 (1991).

Therefore,

$$I = 10^6 \times 2.5/\text{sec} \times 1.6 \times 10^{-19}\,\text{C} = 0.4\,\text{pA}$$

By this estimate, 1 million transporters are expected to generate only a fraction of a picoampere.

There are several features to notice about this calculation. First, the formula considers only the net rate of uptake. Thus, ν is calculated from $V_{\max}/B_{\max}$, where $V_{\max}$ is the initial rate of uptake at saturating NE concentrations and $B_{\max}$ is the number of transporters (N) on the surface. $B_{\max}$ is measured from binding assays. Therefore, the calculation above is actually

$$I_{\max} = V_{\max} q$$

which represents the saturation current $I_{\max}$, for N transporters under specific conditions. We have written this formula as $I = N\nu q$ for comparison with formulas below. Note that neither representation explicitly includes voltage dependence, which we might expect for a current. Measured transporter currents are voltage dependent, and we seek a physical explanation for the steady-state $I(V)$ curves. Many models exist for transporter current as a function of voltage, and we shall not review them here. Rather, we consider the general structure of the problem as a preamble to our discussion of fluctuation analysis. For the alternating access model, an empty transporter T_o facing the outside binds to a substrate S_o in solution. S_o is the concentration of NE in combination with the concentration of the required cotransported ions Na^+ and Cl^-. Thus, in the case we are considering, three ions must bind to the transporter, and $S_o = [NE]_o\,[Na^+]_o\,[Cl^-]_o$. On binding these ions, the substrate–transporter complex ST_o orients to the inside at the rate α to achieve the inward-facing conformation ST_i. The substrate and cotransported ions may then unbind and be released to the internal solution as S_i. The dissociation constants for the two binding reactions on either side of the membrane are K_o and K_i. We may also consider K_o and K_i as ratios of rate constants for these reactions, but here we define them as the disassociation constants. After releasing the substrates and cotransported ions, the empty transporter returns to the outside face at the rate **a**. All of the chemical reactions and transporter conformational changes in this scheme are reversible. Figure 1 summarizes this model, which we refer to as the four-state model.

Hansen *et al.*[15] have provided a useful approach to analyzing the four-state model and to comparing the model with measured steady-state $I(V)$ curves. They show that a four-state reaction scheme, such as that shown

[15] U. P. Hansen, D. Gradmann, D. Sanders, and C. L. Slayman, *J. Memb. Biol.* **63,** 165 (1981).

$$\begin{array}{ccc} & \alpha & \\ ST_o & \Leftrightarrow & ST_i \\ & \beta & \\ K_o \Uparrow & & \Downarrow K_i \\ & a & \\ T_o & \Leftrightarrow & T_i \\ & b & \end{array}$$

FIG. 1. Four-state alternating access model for neurotransmitter transporters. This model describes a mechanism for the transport of a transmitter and the cotransported ions (S) via a neurotransmitter transporter (T). ST stands for the bound state of substrate and ions to the transporter, the Ks are the disassociation constants, and T refers to the empty transporter on the inside (i) and the outside (o) of the membrane.

in Fig. 1, can be represented by a two-state reaction scheme in the special case that the transport cycle contains one voltage-dependent transition. In the case we are considering, the conformational change $ST_o \rightleftharpoons ST_i$ is the voltage-dependent step. The voltage dependence of this transition is introduced through the rate constants: $\alpha = \alpha_o \exp(qV/2kT)$ and $\beta = \beta_o \exp(-qV/2kT)$, where q has its previous meaning and α_o and β_o are voltage-independent constants that set the amplitude of the reactions. In the Hansen model, $V/2$ is the voltage drop across a symmetric barrier in the membrane, k is the Boltzmann constant, and T is the temperature. The voltage-independent transitions in the four-state scheme, namely, $ST_o \rightleftharpoons T_o$, $ST_i \rightleftharpoons T_i$, and $T_o \rightleftharpoons T_i$ are grouped together in the two-state scheme. In this way the four-state model reduces to an equivalent two-state model. This simplification results because by assumption the net current depends only on the voltage-dependent transitions $ST_o \rightleftharpoons ST_i$. Figure 2 summarizes the two-state model.

In the two-state model, two fictitious states, ST_o and ST_i, are connected by composite rate constants α_f and β_f. In this case, the current is given by

$$I = N(\beta_f - \alpha_f)q$$

$$\begin{array}{ccc} & \alpha_f & \\ ST_o & \Leftrightarrow & ST_i \\ & \beta_f & \end{array}$$

FIG. 2. The equivalent two-state scheme. This model is an alternative representation of the four-state model in Fig. 1. In the two-state model, the rate constants α_f and β_f are fictitious representations of the rate constants and other parameters in the four-state model (see text).

where as before N is the number of transporters, and q is the net charge transported per cycle. Whereas in the four-state model, α and β depended only on voltage, in the equivalent two-state model, α_f and β_f are functions of substrate and cotransported ion concentrations (S_o and S_i), the dissociation constants (K_o and K_i), and the transmembrane voltage (V). Thus, we may think of the four-state scheme as having been compressed into a two-state scheme with more complex rate constants. It is informative to compare the above equation with the simpler equation $I = N\nu q$. In the alternating access model, the net current is the difference between an outward current, given by $N\beta_f q$, and an inward current, given by $N\alpha_f q$. The turnover rate (ν) that we first discussed therefore represents the difference between an outward rate and an inward rate:

$$\nu = (\beta_f - \alpha_f)$$

which we see depends on concentration and voltage. Although $\nu = V_{max}/N$ is measured at saturating concentrations, it is not always evident that V_{max} is determined at saturating voltages. Therefore, the previous estimate of the current may be an underestimate.

Gadsby and Nakao[16] have applied the two-state model to steady-state $I(V)$ curves generated from Na^+,K^+-ATPase in cardiac myocytes at 36°. $I(V)$ curves that saturate at both positive and negative voltages are required for a complete analysis. However, any voltage range that includes a saturation of the current is useful. Their experiments on voltage-clamped heart cells revealed (1) a monotonic increase of pump current with increased voltage, and (2) a saturation of the outward current at sufficiently positive voltages, which is sufficient to calculate specific parameters of the model. For the Na pump, the maximum moveable charge in response to a voltage step is $Q_{max} = 20$ nC/μF. If a single charge q moves per transporter, and if this is the only charge that contributes to Q_{max}, then the number of transporters is given by

$$N = Q_{max}/q$$

and therefore,

$$N = (20\ \text{nC}/\mu\text{F})(1\ \mu\text{F/cm}^2)/(1.6 \times 10^{-19}\ \text{C}) = 1250/\mu\text{m}^2$$

The saturating pump current occurred near V of 0 mV, where I_{max} is 1 μA/μF. The turnover rate is given by

$$\nu = I_{max}/Nq$$

[16] D. C. Gadsby and M. Nakao, *J. Gen. Physiol.* **94,** 511 (1989).

and therefore,

$$\nu = (1\ \mu A/\mu F)/(1250/\mu m^2 \times 1.6 \times 10^{-19}) = 50/\text{sec}$$

This result is obtained more directly as $\nu = I_{max}/Q_{max}$. In these calculations we used the symbol I_{max} to represent the current saturation with voltage. However, this current was not measured under saturating substrate concentrations (in this case the substrates are Na^+, K^+, and ATP). The maximum turnover rate under conditions of saturating substrate concentrations and saturating voltage is ν = 80/sec. This example illustrates the importance of considering both the chemical and the electrical driving forces in defining transport parameters. Mager *et al.*[17] have applied a similar analysis to GAT1, a γ-aminobutyric acid (GABA) cotransporter in the same gene family as hNET. These data yielded a density of GAT1 expressed in *Xenopus* oocytes of N = 7200/μm^2 and a turnover rate of ν = 5/sec under saturating GABA/Na concentrations (22°).

Fluctuation Models

Läuger[18] has developed a theory of the current fluctuations associated with electrogenic ion pumps. This theory is based on the idea that, at the microscopic level, individual reaction steps in the transport process are probabilistic events. The current that is generated by pumps at a particular voltage therefore fluctuates around a mean value, which we have designated as $I(V)$. Using the four-state model in Fig. 1, Läuger's theory takes the following form. We assume that binding and unbinding of the substrate and cotransported ions to the transporter (described by the dissociation constants K_i and K_o) are much faster reactions than the conformational transitions associated with transport ($ST_o \rightleftharpoons ST_i$ and $T_o \rightleftharpoons T_i$). In this case, the current spectral density $S_I(f)$ from the four-state reaction cycle in Fig. 1 is given by

$$S_I(f) = \Sigma A_i/(1 + \omega^2\tau_i^2) + S_\infty$$

in units of amperes squared/hertz. Σ stands for the sum over three dispersion factors characterized by time constants τ_1, τ_2, τ_3, and $\omega = 2\pi f$ (the frequency f is in units of hertz). The τ and A values in this equation are frequency-independent, as is S_∞, the high frequency limit of $S_I(f)$. However, the τ, A, and S_∞ values vary with the constants of the reaction scheme and the

[17] S. Mager, J. Naeve, M. Quick, J. Guastella, N. Davidson, and H. A. Lester, *Neuron* **10,** 177 (1993).

[18] P. Läuger, *Eur. Biophys. J.* **11,** 117 (1984).

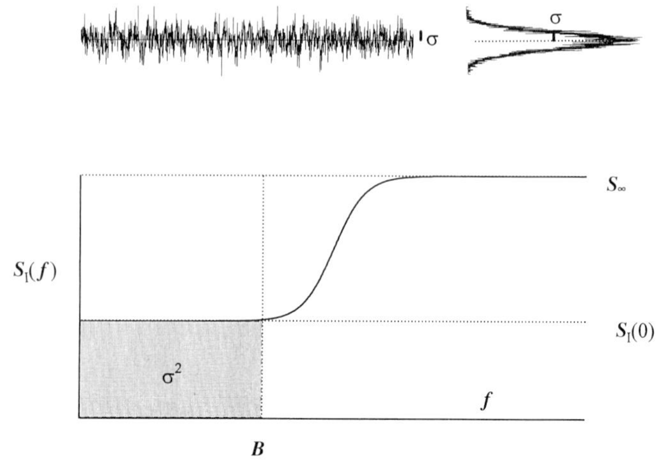

FIG. 3. Spectral density for the alternating access model. The theoretical spectral density for the four-state (or the equivalent two-state) model, $S_I(f)$, is shown as a function of the frequency (f) on an arbitrary scale. The plot shows the relationship between the variance (σ^2), the bandwidth (B), and the spectral density. The noise trace above the plot, and the accompanying amplitude histogram, illustrates the relationship between the standard deviation of the fluctuations (σ) and the variance. In this hypothetical example, the variance represents the pure fluctuations associated with the transporter. In a practical situation the transporter fluctuations must be separated from the background fluctuations (see Fig. 4).

concentrations of the substrates and the cotransported ions. The factors A are negative, because of negative correlations in the reaction cycle. Thus, in contrast to the spectral density of ion channel currents,[19] the spectral density of the transporter currents is a sum of inverted Lorentzians that are flat or low frequencies and plateau at the maximum value S_∞. The shape of the current spectral density $S_I(f)$ for the four-state reaction scheme (or the two-state equivalent) is shown in Fig. 3.

Consider the situation where $\alpha_f \gg \beta_f$, i.e., the electrochemical driving force for uptake is much larger than for efflux. Läuger has shown that in this case the low frequency limit of the spectral density $S_I(f)$ reduces to the constant value

$$S_I(0) = 2qI$$

where, as before, q is the net charge of the transported substrates and the cotransported ions, and I is the mean current generated by transport. $S_I(0)$ is indicated in Fig. 3. The variance of the current fluctuations, σ^2, is the

[19] L. J. DeFelice, "Introduction to Membrane Noise." Plenum Press, New York, 1981.

area under the spectral density in any bandwidth. In the low frequency limit where the spectral density is flat, $\sigma^2 = S_I(f)B$ and

$$\sigma^2 = 2qIB$$

Figure 3 shows the relationship between the variance, σ^2, and the standard deviation, σ, which is the characteristic width of the Gaussian distribution of the fluctuations. In electrical engineering, $2qI$ is called the "shot noise" due to the movement of discrete charges q (shots) through a conductor.[19] As a numerical example, Läuger[18,20] considered the outward current that is generated by 5×10^8 Na^+/K^+ pumps in a cardiac myocyte voltage clamped at $V = +100$ mV. The mean current in this example was $I =$ 18 nA. Using the measured value for pump parameters, the low-frequency limit of the spectral density has the predicted value $S_I(0) = 0.004$ pA^2/Hz. Thus, in a band width of 1000 Hz we would expect $\sigma^2 = 4$ pA^2 and $\sigma =$ 2 pA. It is informative to compare the predicted Na^+,K^+-ATPase pump noise with measured ACh-induced noise from the nicotinic receptor.[21] From an excised patch that contains only 150 ACh-R channels, the variance in 2000 Hz bandwidth at a mean current of 100 pA was 200 pA^2, or $\sigma =$ 14 pA. This comparison illustrates that current fluctuations expected from coupled transport in an alternating access model are considerably less than the fluctuations from a ligand-gated ion channel. Below we consider an application of the shot noise theory to NE transporters transfected into HEK-293 cells.

Brief Methods for Stable Mammalian Cell Lines and Electrophysiology

An *Xho*I/XbaI fragment containing complete hNET cDNA is released from pBluescript SKII^{-14} and subcloned into *Xho*I/*Xba*I-digested pcDNA3 (Invitrogen), placing hNET expression under control of the cytomegalovirus (CMV) and T7 RNA promoters. To generate stable transfected cells, hNET/pcDNA3 is transfected by Lipofectin (Life Technologies, Gaithersburg, MD) into HEK-293 cells plated at 50–60% confluence. The culture medium is Dulbecco's modified Eagle's medium with 10% heat-inactivated fetal bovine serum, 100 μg/ml penicillin, and 100 units/ml streptomycin. After 3 days parental and transfected cells are switched to a medium containing 250 μg/ml geneticin (G418). Resistant colonies are isolated from transfected plates. Individual cells are used to generate clonal lines. Stably transfected HEK-293 cells have advantages for electrophysiological studies,

[20] P. Läuger, "Electrogenic Ion Pumps." Sinauer, Sunderland, MA, 1991.
[21] J. P. Dilger and R. S. Brett, *Biophys. J.* **57,** 723 (1990).

compared with transiently transfected HeLa cells.[22] They express uniformly, have resting potentials in the range of −30 to −50 mV, and they form tighter seals (greater than 20 GΩ) for whole-cell voltage clamp. Prior to recording, cells are plated at a density of 10^5 per 35 mm culture dish and washed three times with bath solution at room temperature. All experiments are done at 37°. The bath solutions used for electrophysiology contain (in m*M*): 130 NaCl, 1.3 KCl, 1.3 KH_2PO_4, 0.5 $MgSO_2$, 1.5 $CaCl_2$, 10 HEPES, and 34 dextrose. The solution is adjusted to pH 7.35 and 300 mOsm with 1 *M* NaOH and dextrose. For Na^+ replacement, LiCl or Tris-Cl replaces NaCl in the bath. For the whole-cell recording, pipette solutions contain (in m*M*): 130 KCl, 0.1 $CaCl_2$, 2 $MgCl_2$, 1.1 EGTA, 10 HEPES, and 30 dextrose adjusted to pH 7.35 and 270 mOsm. Free Ca^{2+} in the pipette is 0.1 μ*M*. To reduce oxidation of NE, solutions contain 100 μ*M* pargyline and 100 μ*M* ascorbic acid. Patch electrodes (5 MΩ) are pulled from borosilicate glass (Corning 7052) with a programmable puller (Sachs-Flaming, PC-84). An Axopatch 200A amplifier band-limited at 5000 Hz is used to measure current fluctuations. Additional filtering to specific bandwidths *B* is done off-line. The series conductance is 0.1 μS or greater, and cell capacitance ranges between 25 to 80 pF, implying 2500 to 8000 μm^2 surface area.

NE-Induced Inward Currents

If the level of amine substrate in the bath is increased incrementally from 0 to 30 μ*M*, and if the voltage is clamped at negative potentials, the NE-induced current saturates.[6] The induced current is sensitive to desipramine. In these experiments the voltage is constant at −120 mV and the Na^+ concentration is 130 m*M*. These data are fit by nonlinear regression to the equation

$$I = I_{max}[NE]^n/(K_m^n + [NE]^n)$$

with values K_m of 600 ± 40 n*M*, *n* of 1.02 ± 0.06 (3 cells for each data point). If we assume the stoichiometry 1 net charge translocated per cycle and 10^6 transporters, the predicted cycle time is 340/sec, which is more than 300 times greater than predicted from V_{max} measured in these same cells. The magnitude of the NE-induced current thus suggests a much higher turnover rate than previously measured by others. There are several possible explanations for this discrepancy. One is that the estimate for V_{max} is too low because the voltage is not controlled in the cells in which it is measured. Alternatively, the alternating access model may be an incomplete

[22] S. Risso, L. J. DeFelice, and R. D. Blakely, *J. Physiol.* **490.3,** 691 (1996).

description of the NE transporter. Based on these data, we have proposed a model in which hNET binds to its substrate (NE) and to obligatory ions (Na^+ and Cl^-) with fixed stoichiometry. Such binding initiates NE uptake; however, this transport mechanism exists in parallel with a channel pathway that allows other ions to move through the transporter. Thus, in addition to the traditional transporter mode associated with the coupled translocation of neurotransmitter, we propose that NETs contain a mode of conduction similar to ligand-gated ion channels.[7] Other neurotransmitter transporters demonstrate comparable properties.[4,5,23] If NETs have the property of ligand-gated ion channels, it should be possible to investigate the size of the underlying elementary events using fluctuation analysis. We can then compare these channel-like events with expectations of the alternating access model.

NE-Induced Current Fluctuations

The increase in the NE-induced current fluctuations is defined as $\sigma^2 = \sigma^2_{\mathrm{NE}} - \sigma^2_{\mathrm{DS}}$. The variance σ^2_{DS} represents the fluctuations in the presence of the hNET antagonists desipramine (DS). The difference variance from whole-cell recordings is a function of the concentration of NE according to the following equation:

$$\sigma^2 = \sigma^2_{\max}[\mathrm{NE}]^{\mathrm{n}}/(K^n_m + [\mathrm{NE}]^n)$$

In the case that V is -120 mV, and in 30 μM NE, $\sigma^2_{\max}$ is 31 ± 3 pA^2 (from whole-cell recordings), K_m is 0.6 ± 0.13 μM, and n is 0.8 ± 0.15. We see that the variance of the NE-induced fluctuations and the magnitude of the NE-induced current have similar dependence on NE concentration. Therefore the ratio of the variance to the mean is approximately independent of the NE concentration.[7] If we assume that hNET acts exclusively as a transporter with the fixed stoichiometry 1 NE/1 Na^+/1 Cl^-, then each event should carry net charge $q = 1e$. By the shot noise theory online above,

$$q = \sigma^2/2BI$$

Using this equation, we can test the prediction of the coupled transport theory. Experiments reveal that for $B = 2000$ Hz, $q \sim 0.5 \times 10^{-16}$ coulomb at -20 mV.[7] This means that approximately 300 electronic charges make up each unitary event, which contradicts the assumed stoichiometry. Before discussing the implications of this result, there are two points to consider. First, the transporter-associated events underlying the NE-induced current may have a distribution of open times, like an ion channel, each translocat-

[23] H. A. Lester, Y. Cao, and S. Mager, *Neuron* **17,** 807 (1996).

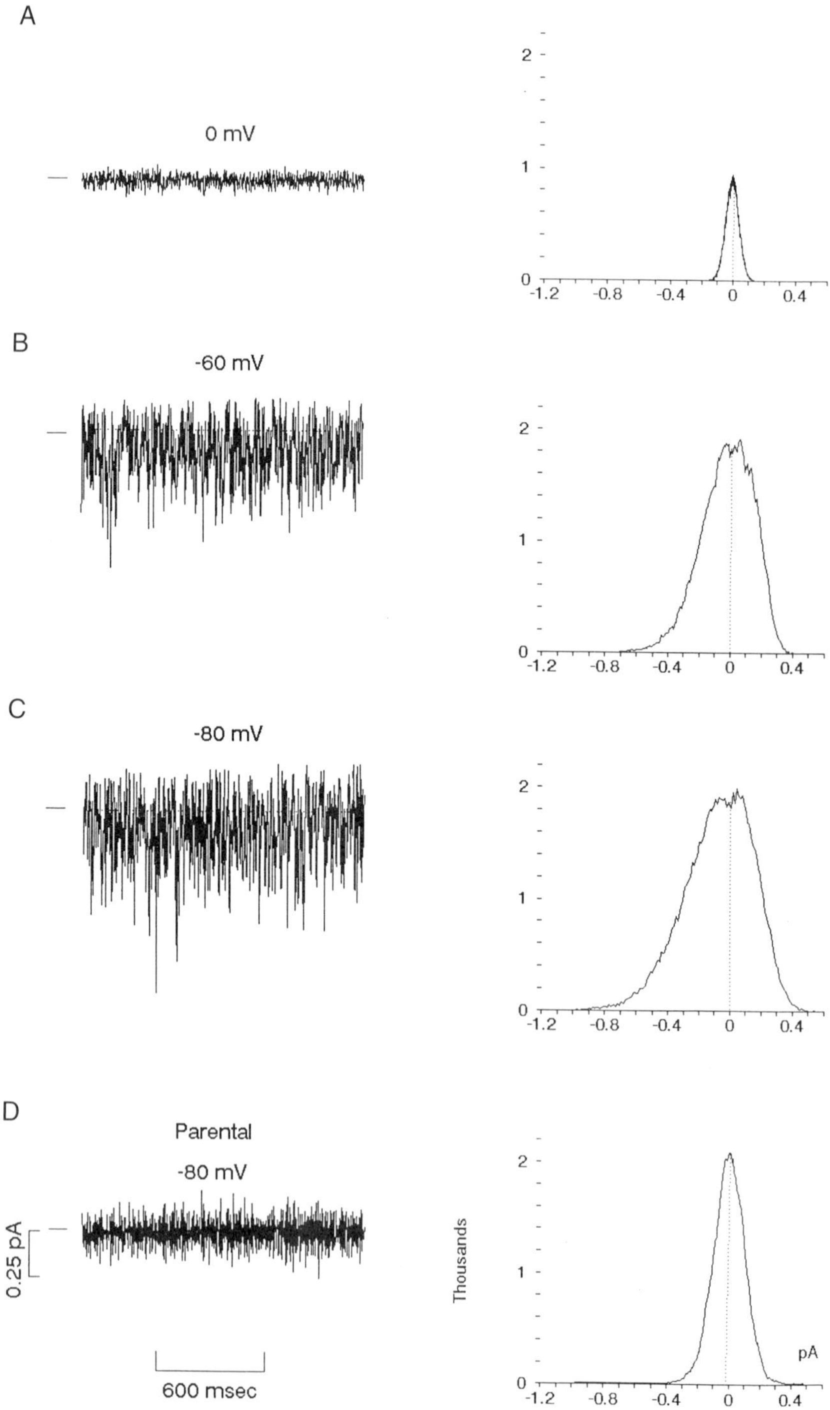
A
0 mV
B
-60 mV
C
-80 mV
D
Parental
-80 mV
0.25 pA
600 msec
Thousands
pA
2
1
0
-1.2
-0.8
-0.4
0
0.4

ing a different total charge. If the elementary events are not exactly the same, as assumed in the shot noise theory, then the estimate $q = 300e$ represents an average charge per event. Second, this estimate of q is approximately independent of the bandwidth if $S_I(f)$ is flat. Figure 3 illustrates this point for the alternating access model in the low frequency range: decreasing the bandwidth decreases the variance, but the ratio σ^2/B theoretically remains the same. We have verified this expectation for hNETs in the range 200 to 2000 Hz. The bandwidth does, however, limit the ability to resolve brief openings. For $B = 2000$ Hz, the briefest events that one may observe are approximately $\Delta t = 1/2\pi B \sim 1$ msec. Events with open times below 1 msec contribute to the variance, but they are not resolved as complete openings in this bandwidth. Larsson *et al.*[24] have measured current fluctuations in glutamate transporters in photoreceptors, for which the spectral density is flat only up to about 20 Hz. The substrate-induced events in the glutamate transporter are apparently much slower than the corresponding events in hNET, and they give rise to channel-like Lorentzian spectral densities[19] in the frequency range below 2000 Hz. In contrast, Cammack and Schwartz[25] have reported for a γ-aminobutyrate transporter an inverted Lorentzian spectral density similar to the rising phase in Fig. 3.

To look for NE-induced unitary events, we performed experiments on cell-attached and inside-out patches on transfected cells.[7] The inward currents from such patches show an inwardly directed asymmetry in the amplitude histograms which becomes larger at negative membrane potentials. These inwardly directed fluctuations, which are shown in Fig. 4, are eliminated by the addition of 20 μM desipramine to the bath.

[24] H. P. Larson, S. A. Picaud, F. S. Werblin, and H. Lecar, *Biophys. J.* **70,** 733 (1996).
[25] J. N. Cammack and E. A. Schwartz, *Proc. Natl. Acad. Sci. USA* **93,** 723 (1996).

FIG. 4. Current fluctuations from cell-attached patches in hNET-transfected cells. These data are from cell-attached patches on hNET-293 cells. The pipette contained 30 μM guanethidine (GU), a nonoxidizable substrate for hNET. (A) Trace at 0 mV. B and C are at −60 and −80 mV, respectively. Amplitude histograms, shown to the right of each trace, are from 9 sec of similar data. In these histograms, the mean current was set arbitrarily to zero at each voltage. Similar data were obtained with 30 μM NE. The histograms in (A)–(C) show that in transfected cells with GU in the pipette (or NE), patch noise has an asymmetric distribution of amplitudes that indicates the presence of inward channel-like openings. These openings are buried in the noise, but they are evident in the asymmetry of the traces and in the left asymmetry of the amplitude histograms. (D) Sample trace from similar experiment on a parental cell stepped to $V = -80$ mV. The variance associated with the DS-blocked fluctuations (see text: $\sigma^2 = \sigma^2_{NE} - \sigma^2_{DS}$) increases with voltage from 0.05 ± .011 pA^2 at 0 mV to 0.45 ± .04 pA^2 at −140 mV (four-cell–attached patches). In parental cells, with or without NE or GU in the pipette, patch noise has a symmetric distribution of amplitudes, as indicated in the histogram in (D).

Although an analysis of amplitude histograms suggest the presence of channels of the size predicted from fluctuation analysis, we rarely observe channels as discrete openings for a B of 2000 Hz. Nevertheless, from such measurements we have estimated that the NE-induced single-channel conductance is about 3 pS between −120 and −30 mV.

The probability of opening the transporter in channel mode is probably small, perhaps the order of 10^{-3}. It is difficult to arrive at a solid estimate of this probability because the number of transporters on the particular cell under voltage clamp is unknown. Furthermore, the NE uptake in the exact cell used for electrophysiology is unknown. Thus, in the mammalian expression system we have been considering, it is not possible to accurately compare the NE-induced current to the NE flux. In particular, it would be informative to know the number of transport events that occur per channel opening. By measuring transmitter uptake and transmitter-induced current fluctuations from the same cell, it is possible to answer this question. However, this requires using *Xenopus* oocytes as an expression system.[26]

Brief Methods for *Xenopus* Oocyte Expression and Electrophysiology

The 2.3 kb dSERT *Bgl*II fragment is cloned into the *Bam*HI site of pBluescript KS^+ plasmid (Stratagene, La Jolla, CA) downstream of the T7 promoter.[27] Stage V and VI defolliculated *Xenopus* oocytes were injected with 42 nl (1 mg/ml) dSERT cRNA and incubated at 18° in Ringer's solution (pH 7.6) supplemented with 1% penicillin–streptomycin stock, 10,000 U/ml (GIBCO-BRL, Gaithersburg, MD) and 5% horse serum (GIBCO-BRL; see ref. 28). Peak expression of dSERT varied from 4 to 9 days. Current records have been obtained using the two-microelectrode voltage clamp technique (GeneClamp 500, Axon Instruments, Foster City, CA) band-limited at 2000 Hz. All the data are obtained at room temperature. The recording solutions are as follows (in mM): Na Ringer's solution contains 96 NaCl, 2 KCl, 5 $MgCl_2$, 5 HEPES, and 0.6 $CaCl_2$. Current measurements and flux assays are simultaneously measured while perfusing voltage-clamped oocytes with 1.8 μM 5-HT containing 200 nM [^{3}H]5-HT. Experiments are terminated by washing the oocytes in Ringer's solution three times. The oocytes are then solubilized in 1% SDS, and the 5-HT uptake is determined by liquid scintillation counting.

[26] A. Galli, C. I. Petersen, M. deBlaquiere, R. D. Blakely, and L. J. DeFelice, *J. Neurosci.* **17,** 3401 (1997).

[27] L. L. Demchyshyn, Z. B. Pristupa, K. S. Sugamori, E. L. Barker, R. B. Blakely, W. J. Wolfgang, M. A. Forte, and H. B. Niznik, *Proc. Natl. Acad. Sci. USA* **91,** 5158 (1994).

[28] M. W. Quick, J. Naeve, N. Davidson, and H. A. Lester, *Biotechniques* **13**(3), 357 (1992).

5-HT Induced Inward Currents

Oocytes injected with cRNA for dSERT exhibit an inward current when oocytes were held at negative potentials and 5-HT is added to the bath. To verify the oocyte expression system for studying the electrophysiological characteristics of dSERT, we have compared the ionic selectivity of 5-HT uptake with previous studies.[29] Replacing Na^+ with either Li^+ or NMDG, or replacing Cl^- with acetate, lowers 5-HT uptake and 5-HT-induced current proportionally. As with NETs, the amplitude of the 5-HT-induced current in dSERT-expressing oocytes saturates with increasing 5-HT concentration. These data are fit by nonlinear regression to the equation

$$I = I_{\max}[5\text{HT}]^n/(K_m^n + [5\text{HT}]^n)$$

which gives K_m of 1.26 ± 0.20 and n of 1.19 ± 0.19 for voltages between −100 and −60 mV.

Current to Flux Ratios

To correlate current and uptake at different voltages, each oocyte was voltage clamped and bathed for 50 sec in ^{3}H-labeled 1.8 μM 5-HT Ringer's solution. Integration of the 5-HT-induced current measures the total charge movement (Q) at particular voltage, and scintillation counting of lysed oocytes measures the 5-HT charge movement ($Q_{5\text{HT}}$) from the same oocyte at the same voltage. We define

$$\rho(V) = Q/Q_{5\text{HT}}$$

Because $\rho(V)$ is a ratio, it is independent of expression level. Thus, $\rho(V)$ can then be obtained from different oocytes held at different voltages. For frog oocytes expressing dSERT we found that $\rho(V)$ varies between ρ of 17 at 20 mV and ρ of 67 at −80 mV. Note that the measurement of Q also includes the charge of the transmitter; therefore, at 20 mV, 5-HT induces the translocation of 16e for every 5-HT molecule that passes through dSERT. If we assume that one 5-HT translocation represents one transport cycle, then $\rho(V) = 17$ implies that $16e + 1$ (5-HT) are translocated per cycle. This result is at variance with standard models in which 5-HT is translocated at a fixed stoichiometry with cotransported ions. First, the stoichiometry is much higher than previously indicated. Second, the number of ions moving with 5-HT varies with voltage. Another way to cast this data is to calculate

[29] Y. L. Corey, M. W. Quick, Q. N. Davidson, H. A. Lester, and J. Guastella, *Proc. Natl. Acad. Sci. USA* **91,** 1188 (1994).

the current that is carried by 5-HT alone. The absolute current carried by 5-HT, which is a monovalent cation at physiological pH, is given by

$$I_{5\mathrm{HT}} = I(V)/\rho(V)$$

$I(V)$ is measured as the difference between the 5-HT-induced current and the control current without 5-HT. The result of this calculation is that the pure 5-HT current is 5% of the total 5-HT-induced current. This result is for the case $V = -20$ mV, which is selected because the leak current is zero at this potential and cannot contribute to the fluctuations. At $V = -20$ mV, $\rho(V)$ has approximately the same value as at +20 mV, namely, $\rho(V) = 17$. In the same oocytes, held at −20 mV, we also measured the charge q that is carried by a typical elementary event underlying the 5-HT-induced current. We obtained q from a measurement of $\sigma^2/2BI$ as described previously for NE transporters. We can now combine the results from uptake and fluctuation analysis to calculate

$$q_{5\mathrm{HT}} = q(V)/\rho(V)$$

As we argue below, the experimentally determined value of $q_{5\mathrm{HT}}$ has two interpretations. Before discussing these, note that it is also possible to determine q by fitting a linear regression to the equation $\sigma^2/I = i_o + 2qB$, where B is a variable. Using sharp filters to vary the upper cutoff frequency between 100 and 1000 Hz, such a plot yielded $i_o = 0.102$ pA and $q = 6755e$. Using the entire bandwidth to estimate q yielded the value $q = 8250e$. In the following discussion we use the larger value.

Models

Based on such measurements, it is possible to determine parameters for specific models for neurotransmitter transporters. If 5-HT translocation occurs exclusively through coupled transport, $q_{5\mathrm{HT}}$ estimates of the number of transport cycles that occur (n_ν) for every 5-HT-induced channel opening. Dividing $q = 8250e$ by $\rho = 17$ gives $458e$. Assuming that one 5-HT translocates per cycle implies that $n_\nu = 458$. Thus the combination of flux data and noise analysis provides a novel measure of transporter cycles per channel opening, leading to the conclusion that channel openings are rare compared to transport cycles. However, this interpretation is not unique. These data are also consistent with a model in which transport and channel openings are not separate states of the transporter. In this case, $q_{5\mathrm{HT}}$ has a different interpretation, namely, the number of 5-HT molecules that translocate per channel opening.

Summary

Findings from an electrophysiological analysis of neurotransmitter transporters show that transmitter-induced currents are associated with these transporters: For charged transmitters, such as NE and 5-HT, a fraction of the total current is carried by the transmitter itself; however, the transmitter also induces an extra current in analogy to an ligand-gated ion channel. An additional conductance not discussed in this article is the so-called leak, in which neurotransmitter transporters generate an ionic current in the absence of transmitter. Using a combination of flux measurements, voltage clamp, and fluctuation analysis has shown that, for norepinephrine and serotonin transporters, the transmitter-induced current greatly exceeds the transmitter current. Such data can provide an exact measure of the ratio of these charge movements to transmitter translocation at the molecular level, suggesting new strategies to understand neurotransmitter transporters.

Acknowledgment

This work was supported by NIH Grant NS-34075 and NARSAD Investigator Awards LJD and AG.

[40] Serotonin Transport in Cultured Leech Neurons

By DIETER BRUNS

Introduction

Fast removal of transmitter from the synaptic cleft terminates chemical neurotransmission and enables transient and rapid signaling between neurons. At cholinergic synapses, transmitter clearance is thought to be regulated by integral membrane transport proteins that mediate the rapid uptake of transmitter into presynaptic terminals or surrounding glia. Over the past few years, several members of a family of Na^+-dependent neurotransmitter cotransporters have been cloned and examined under voltage-clamp conditions in various expression systems.[1–3] When translocating substrates, all

[1] H. A. Lester, S. Mager, M. W. Quick and J. L. Corey, *Annu. Rev. Pharmacol. Toxicol.* **34,** 219 (1994).
[2] M. S. Sonders and S. G. Amara, *Curr. Opin. Neurobiol.* **6,** 294 (1996).
[3] H. A. Lester, Y. Cao, and S. Mager, *Neuron* **17,** 807 (1996).

0076-6879/98 $25.00

transporters elicit currents and are thus considered to be electrogenic. Characterization of transporter activity in the presynaptic neuron is essential for our understanding of synaptic function. However, recordings of transport-associated currents in the intact neuron have so far only been reported from a limited number of preparations.[4–8] Here we describe the use of the cultured Retzius P cell synapse of the leech (*Hirudo medicinalis*) as a model system to study serotonin uptake. When Retzius cells are paired with P cells, a stable serotonergic connection with well-defined physiological properties is established.[9–12] Furthermore, the large Retzius cell forms the synaptic contact with the postsynaptic P cell over a short distance,[13] which makes this preparation well-suited for examination of electrical events in the presynaptic terminal region.

This chapter describes electrophysiological and fluorescence imaging techniques that have been used to characterize electrogenic serotonin uptake in the cultured Retzius P cell synapse. Furthermore, techniques that enable stimulation of phasic transmitter release using flash photolysis of Ca^{2+} caging 1-(4,5-dimethoxy-2-nitrophenyl) EDTA (DM-nitrophen™, Calbiochem Corp.) are described. These techniques can be used to study the activation of electrogenic uptake at the synaptic site.

Synapse Formation between Cultured Leech Neurons

Since the early 1980s, the pioneering work of John Nicholls and coworkers has formed the foundation for studying serotonin release from the cultured Retzius P cell synapse of the leech. Details regarding the techniques (aspiration method) for removing cells from leech ganglia have been described elsewhere.[10] Here, a brief review of the preparation and the handling of the neurons is given. Retzius and P cells are isolated from desheathed ganglia using suction pipettes after mild enzymatic treatment with collagenase–dispase (2 mg/ml for 60 min; Boehringer Mannheim, Germany) in culture medium. Suction pipettes are fabricated from soft glass under a microforge and are heat-polished in a tip opening of 80 to

[4] R. P. Malchow and H. Ripps, *Proc. Natl. Acad. Sci. USA* **87,** 8945 (1990).

[5] K. Kaila, B. Rydquist, M. Pasternack, and J. Viopo, *J. Physiol. (London)* **453,** 627 (1992).

[6] D. Bruns, F. Engert, and H. D. Lux, *Neuron* **10,** 559 (1993).

[7] S. A. Picaud, H. P. Larsson, G. B. Grant, H. Lecar, and F. S. Werblin, *J. Neurophysiol.* **74,** 1760 (1995).

[8] G. B. Grant and J. E. Dowling, *J. Neurosci.* **15,** 3852 (1995).

[9] P. A. Fuchs, L. P. Henderson, and J. G. Nicholls, *J. Physiol. (London)* **323,** 195 (1982).

[10] I. D. Dietzel, P. Drapeau, and J. G. Nicholls, *J. Physiol. (London)* **372,** 191 (1986).

[11] D. Bruns and R. Jahn, *Nature* **377,** 62 (1995).

[12] D. Bruns, S. Engers, C. Yang, R. Ossig, A. Jeromin, and R. Jahn, *J. Neurosci.* **17,** 1898 (1997).

[13] Y. Liu and J. G. Nicholls, *Proc. R. Soc. Lond. (B)* **236,** 253 (1989).

100 μm for handling of Retzius cells. The smaller-sized P cells are best handled with pipettes having a tip operating of 50 to 60 μm. The culture medium is Leibowitz L-15 medium (GIBCO, Gaithersburg, MD) supplemented with 6% (v/v) fetal bovine serum (GIBCO), glucose, 6 mg/ml and gentamicin, 0.1 mg/ml (Sigma, St. Louis, MO). The cells are maintained in culture medium for 24 hr to allow removal of adhering cell debris from the membrane. Maneuvering single neurons with suction pipettes is technically difficult and needs some practice. Individual Retzius cells and P cells are paired, so that their axon stumps are brought in close apposition. This step is performed on nonadhesive culture dishes (Falcon 3001, 35 mm, Becton Dickinson, Lincoln Park, NJ) and should be visualized on the stage of an inverted microscope (Nikon-TMS). Alternatively, pairing of the cells can be performed in dishes coated with polylysine or other substrates, as described by Liu and Nicholls.[13] However, we found that manipulation of the cells into "stump-to-stump" contact is more readily achieved on nonadhesive dishes. After 24 hr the connection between the cells is sufficiently stable to allow the transfer of the cell pairs to dishes coated with polyornithine (Sigma, molecular weight 15,000–30,000) for long-term culturing. For this, dishes (Falcon 3001, 35 mm) are filled with serum-free culture medium to facilitate cell adhesion to the substrate. Plating of the cell pairs is visualized under a stereo microscope (Nikon-SM-1B). The cell pair is carefully released from the suction pipette and maneuvered onto the bottom of the dish in order to ensure that both cells attach with soma and axon stump to the substrate. One hour after plating the culture medium is supplemented with fetal bovine serum as described above. This procedure results in synapse formation with high reliability (success rate >80%). Under these culture conditions neurite extension is largely prevented, minimizing space clamp problems. For the same reason it is advisable to reject Retzius cells with thin and elongated axon stumps. The cells are distinguished easily by their different sizes (P cell diameter 40–50 μm, Retzius cell diameter (70–90 μm). Culture dishes are incubated at room temperature in a humidity chamber. Cell pairs can be used for recordings on days 2 to 8 of culture (Fig. 1).

5-Hydroxytryptamine Transporter Currents Induced by External Application of Transmitter

Neurotransmitter transporters have a specific requirement for sodium. Indeed, even similar ions, such as lithium, cannot substitute for sodium.[14–16]

[14] G. Rudnick and J. Clark, *Biochim. Biophys. Acta* **1144,** 249 (1993).
[15] R. D. Blakely, L. J. de Felice, and H. C. Hartzell, *J. Exp. Biol.* **196,** 263 (1994).
[16] B. I. Kanner, *J. Exp. Biol.* **196,** 237 (1994).

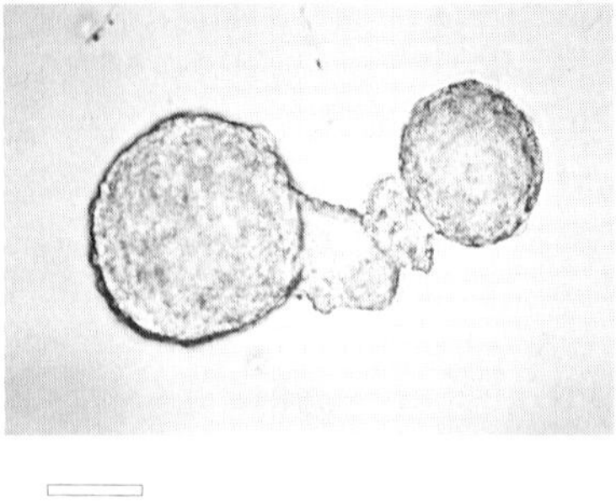

FIG. 1. Retzius P cell pair plated on polyornithine. The axon stumps of the cells were manipulated into direct contact. Bar: 40 μm. [Reproduced with permission from D. Bruns, S. Engers, C. Yang, R. Ossig, A. Jeromin, and R. Jahn, *J. Neurosci.* **17,** 1898 (1997).]

The identification of a neurotransmitter uptake process in the presynaptic neuron requires the demonstration that the electrogenic process is activated by application of neurotransmitter as well as being dependent on extracellular sodium.

Retzius cells respond *in vitro* to 5-hydroxytryptamine (5-HT) application mainly with the activation of a Na^+ and a Cl^- conductance.[6,17–19] The Na^+-dependent current is best isolated in external and internal solutions devoid of Cl^- ions (see Appendix). In contrast, to the mammalian 5HT transporter,[20,21] uptake of 5-HT in the leech[6] is not dependent on extracellular Cl^-. Interestingly, serotonin transport in *Drosophila melanogaster*[22,23] is facilitated by, but is also not absolutely dependent on, external Cl^-.

Transporter-associated currents in presynaptic Retzius cells have amplitudes ranging from 100 pA to 3 nA. They are best studied under voltage clamp, using patch pipettes (see Appendix) in the whole-cell mode[24] and a List EPC-7 amplifier. The large Retzius cell allows for penetration with an additional microelectrode, which can be connected to a conventional bridge amplifier. This makes it possible to verify effective clamp control by recording the actual membrane potential. The Retzius cell should be

[17] P. Drapeau, E. Melinyshyn, and S. Sanchez-Armass, *J. Neurosci.* **9,** 2502 (1989).

[18] V. Lessmann and I. D. Dietzel, *J. Neurosci.* **11,** 800 (1991).

[19] V. Lessmann and I. D. Dietzel, *J. Neurosci.* **15,** 1496 (1995).

[20] P. J. Nelson and G. Rudnick, *J. Biol. Chem.* **257,** 6151 (1982).

[21] S. Mager, C. Min, D. J. Henry, C. Chavkin, B. J. Hoffman, N. Davidson, and H. A. Lester, *Neuron* **12,** 845 (1994).

[22] L. L. Demchyshyn, Z. B. Prestupa, K. S. Sugamori, E. L. Baker, R. D. Blakely, W. J. Wolfgang, M. A. Forte, and H. B. Niznik, *Proc. Natl. Acad. Sci. USA* **91,** 5185 (1994).

[23] E. L. Barker, M. A. Gray, E. A. Camp, and R. D. Blakely, *Soc. Neurosci. Abstr.* **207,** 509 (1996).

[24] O. P. Hamill, A. Marty, E. Neher, B. Sakmann, and F. J. Sigworth, *Pfluegers Arch.* **391,** 85 (1981).

penetrated with the microelectrode before establishment of the whole-cell recording configuration in order to prevent disruption of the seal formed between patch pipette and membrane.

In our experiment,[6] superfusion of presynaptic Retzius cells with 100 μM serotonin has resulted in a steady inward current at a membrane potential of -80 mV. The current amplitude decreases with membrane depolarization and assumes values near zero at potentials of $+60$ mV. Substitution of external Na^+ with *N*-methyl-D-glucamine ($NMDG^+$) or with other monovalent cations, such as lithium or cesium, reduced the current over the whole voltage range.

The unfavorable ratio of the cell volume (400–800 pl) of Retzius cells and the maximum tip opening (3–5 μm) of patch pipettes that can be used for whole-cell recording with reasonable success rate allows only a slow equilibration between the patch pipette and cell content. However, on replacement of external Na^+ with various substituents (lithium, cesium, or NMDG) only a small residual inward current (~5%) was observed that is probably due to efflux of remaining intracellular Cl^- through 5-HT–activated Cl^- channels. A fraction of Retzius cells (10–30%) also responded to 5-HT application with activation of a K^+ conductance (See ref. 18). K^+ currents can be identified in Na^+-free solution by their reversal near the calculated K^+ reversal potential. In order to facilitate the isolation of transporter-associated currents and to be able to examine their voltage dependence, these cells should be excluded from the experiments.

Changing External Solutions

For most purposes, the best method for exchanging external solutions is to expose the cell to the outflow of solutions from a multibarreled superfusion pipette.[25] The pipette is assembled from several hypodermic needles mounted on a plastic holder (Fig. 2). The tip of the aligned needles is glued into a glass capsule 6–9 mm in length to create a common outlet for the superfusion solutions. The capsule is prepared under a microforge from soft glass and made with a tip opening of 100–150 μm. Driven by gravity, each capillary is fed by a reservoir above the bath. Switching of superfusion solutions can be accomplished by triggered electromagnetic valves (The Lee Co., Westbrook, CT). The outlet of the pipette is positioned 200–300 μm from the cell. For fluid level control a separate suction pipette can be used. Changes in fluid level will cause changes in the electrode shunt capacitance and require readjustment of capacitance compensation to avoid oscillation of the amplifier, particularly when close to a critical setting. The

[25] E. Carbone and H. D. Lux, *J. Physiol.* (*London*) **386,** 547 (1987).

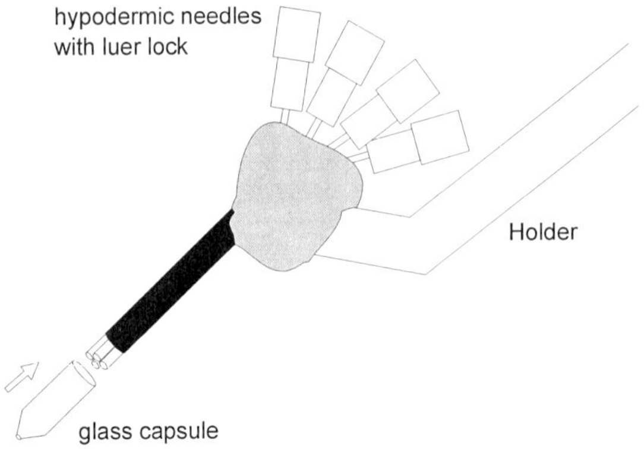

FIG. 2. Schematic illustration of the multibarreled superfusion pipette. Hypodermic needles are aligned with a heat-shrink tubing (black) and are mounted with glue on a plastic holder. The glass capsule fits snugly over the capillary bundle. The dead space of the pipette is reduced when the needle tips are glued together.

dead space created by the common outlet of the pipette should be as small as possible to reduce the time needed for changing solutions. Filtered solutions should be used to prevent blocking of the pipette by particles. The proper flow of saline-containing solution from each capillary can be readily visualized using phase-contrast optics in a dish containing distilled water.

Monitoring Uptake Activity with Autofluorescent 5-Hydroxytryptamine Analog 5,7-Dihydroxytryptamine

A critical test for the activation of an electrogenic serotonin transporter is the demonstration that transmitter molecules are taken up during charge influx. 5-HT uptake in leech neurons has been demonstrated by accumulation of either radiolabeled 5-HT[26] or autofluorescent transmitter analogs of 5-HT, such as 5,7-dihydroxytryptamine[6,18,27] (5,7-DHT). Using single cultured neurons, a selective accumulation of [^{3}H]5-HT in Retzius cells, but not in P cells, has been described by Henderson.[26] 5,7-DHT is transported, as the native transmitter, by active uptake mechanisms into serotonin-utilizing neurons.[28,29] Similarly, in the ganglia of the leech, serotonergic neurons

[26] L. P. Henderson, *J. Physiol.* (*London*) **339,** 309 (1983).
[27] C. M. Lent and H. M. Dickinson, *Brain Res.* **300,** 167 (1984).
[28] G. H. Baumgarten, H. P. Klemm, and L. Lachenmayer, *Ann. NY Acad. Sci.* **305,** 3 (1978).
[29] J. Rohrbacher, K. Krieglstein, S. Honerkamp, A. Lewen, and U. Misgeld, *Neurosci. Lett.* **199,** 207 (1995).

were selectively stained after 5,7-DHT treatment.[18,27] 5,7-DHT fluorescence imaging in conjunction with current recordings enables repetitive and almost simultaneous measurements of uptake and charge transfer.[6] Furthermore, fluorescence imaging can be used to study whether uptake occurs selectively in the presynaptic element of the synapse.

Current Measurements in Conjunction with 5,7-DHT Fluorescence Imaging

Microelectrodes instead of patch pipettes should be used in these experiments in order to minimize intracellular perfusion which would otherwise interfere with the fluorescence measurement. Currents are isolated in external Cl^--free saline and are measured with a single-electrode voltage clamp (SEC01, NPI-Instruments, Tamm, Germany). The Retzius cell should be voltage clamped at a membrane potential of −70 mV near the K^+ reversal potential, assuming an intracellular K^+ concentration of approximately 100 m*M*.[30] 5,7-DHT–containing solutions can be applied with a superfusion system, as described above. For these experiments, Retzius P cell pairs should be cultured on polyornithine-coated Petriperm dishes (Bachhofer, Reutlingen, Germany), which show less UV light adsorption than plastic petri dishes. Figure 3 illustrates schematically the optical setup used for fluorescence imaging. Excitation wavelengths of 360 ± 6 nm for 5,7-DHT fluorescence are selected from the emission spectrum of a xenon lamp (Osram XBO 75 W) by using a combination of two dichroic mirrors, DM-400 (Nikon) and FT 390 (Zeiss, Germany), and a bandpass (BP 360, Melles Girot, Darmstadt, Germany). The excitation light is directed onto the cells via the optical path of an inverted microscope (Zeiss Axiovert 35; objective 25× Plan-Neofluar 0.8 Imm. Korr., Zeiss). The 5,7-DHT fluorescence (maximum emission wavelength 420 nm; Silva *et al.*[31]) is filtered with a bandpass BP 420 (Zeiss). Images should be obtained before and after application of 5,7-DHT while cells are superfused with control solution. This allows correction for background fluorescence that results mainly from endogenous fluorophores excited by light in the UV range[32] (for example, pyridine nucleotides such as NADH and NADPH) and enables repetitive measurements of the same cell pair. On washing with control solution no residual fluorescence should be detected in noncell areas. In calibration tests, 5,7-DHT fluorescence intensity was found to increase linearly within the concentration range of 10 to 500 μM.[6] We use an ordinary hemocytome-

[30] J. W. Deitmer and W. R. Schlue, *J. Exp. Biol.* **91,** 87 (1981).
[31] N. L. Silva, A. P. Mariani, N. L. Harrison, and J. L. Barker, *Proc. Natl. Acad. Sci. USA* **85,** 7346 (1988).
[32] J. Aubin, *J. Histochem. Cytochem.* **27,** 36 (1979).

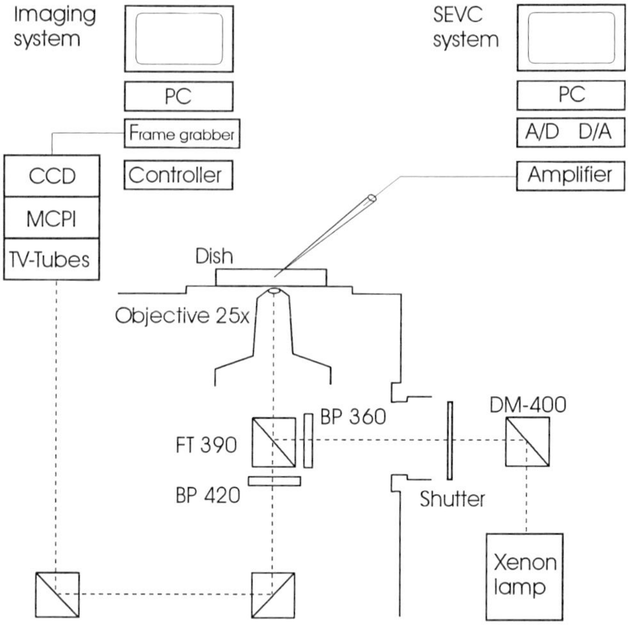

FIG. 3. A schematic representation of the experimental setup used for 5,7-DHT fluorescence detection in conjunction with SEVC (single-electrode voltage clamp)-recordings. The UV illumination unit is equipped with an electronically controlled shutter. For fluorescence imaging an intensified CCD camera system is mounted on the TV port of the microscope. The imaging system consists of a computer-based device for acquisition, display and storage of data and a controller unit for UV exitation. 5,7-DHT induced currents are recorded with a SEVC, digitized, and stored on a separate computer.

ter filled with various concentrations of 5,7-DHT to test linearity of the fluorescence. Images can be obtained with a Peltier-cooled S/W CCD camera (HTMC-87, B&M Spectronic, Germany) in combination with a multichannel-plate light intensifier (MCPI, B&M, Germany, Hahnel *et al.*[33]). The dynamic range of intensified charge-coupled device (CCD) cameras is not as large as that of photomultipliers. Consequently, saturation effects are more easily observed. The camera gain should be adjusted so that the majority of the fluorescence signal is recorded within the linear part of the camera response. This can be done at the beginning of an experiment by focusing on the tip of the superfusion pipette while solutions containing various 5,7-DHT concentrations are applied. The video image analysis can be performed with an image processor (ITEX 151, Imaging Technology,

[33] C. Hahnel, G. Brändle, K. Gottmann, and H. D. Lux, *in* "Proc. 20th Neurobiol. Conference, Göttingen" (Elsner and Richter, Ed.), p. 114 (1992).

Woburn, MA) and objected-oriented 32-bit C^{++} software written by G. Brändle (Max Planck Institute for Psychiatry, Munich, Germany). Changes in cellular fluorescence can be displayed by digital subtraction of the images. Because of the large size of the cells, spatially resolved information for fluorescence changes in axonal and somatic regions can be readily obtained. To minimize fading of 5,7-DHT fluorescence, shutter-controlled minimum exposure times should be used, which are a function of excitation intensity, fluorescence efficeincy, and concentration. Furthermore, it is good practice to perform these experiments in darkened rooms and to add ascorbic acid (100–200 μM) to all external solutions in order to reduce autoxidation of 5,7-DHT.

Measurements of Reuptake Currents during Serotonergic Transmission

The measurement of synaptic signaling is a challenging task, particularly when invasive stimulation and detection methods are used. This section of the chapter describes techniques that were used to detect presynaptic reuptake currents and related postsynaptic responses on Ca^{2+}-evoked transmitter release. Calcium, entering the presynaptic terminal thorugh voltage-gated Ca^{2+} channels, is essential for triggering the release of neurotransmitter into the synaptic cleft.[34] The two-electrode voltage clamp configuration (2-EVC) would be the method of choice for measuring presynaptic Ca^{2+} currents which easily reach amplitudes of 50 nanoamperes in these neurons.[35] Activation of transporter currents, however, with mean amplitudes in the picoampere range, would be obscured by the magnitude of the Ca^{2+} current and the noise generated by the 2-EVC configuration. In order to evoke transmitter release and to separate transporter currents from other voltage gated conductances, transient rises in presynaptic intracellular Ca^{2+} concentration can be elicited by flash photolysis of the caged Ca^{2+} compound DM-nitrophen.[36] In the past few years, caged Ca^{2+} compounds have proven to be valuable tools for studying transmitter release and accompanying processes in various secretory cells (see ref. 37 for review). We have successfully used this approach to stimulate transient transmitter release from Retzius cells.[6] Photolytic "uncaging" of Ca^{2+} in the cytosol of Retzius cells evokes postsynaptic currents that resemble those elicited by

[34] B. Katz, *in* "The Release of Neural Transmitter Substances." (C. C. Thomas, Ed.). Liverpool University Press, Liverpool, 1969.

[35] R. R. Stewart, J. G. Nicholls, and W. B. Adams, *J. Exp. Biol.* **141,** 1 (1989).

[36] J. H. Kaplan and G. C. R. Ellis-Davies, *Proc. Natl. Acad. Sci. USA* **85,** 6571 (1988).

[37] R. Zucker, *Methods Cell Biol.* **40,** 31 (1994).

presynaptic action potentials. More importantly, postsynaptic signals are paralleled by a fast activating presynaptic reuptake current (Fig. 4B) that is highly dependent on extracellular sodium. The serotonin transporter acts as a rapid transmitter removal system and affects the time course of the postsynaptic response.

Isolation of Pre- and Postsynaptic Currents Induced by Flash Photolysis

Presynaptic and postsynaptic currents are recorded using patch pipettes and two List EPC-7 amplifiers (Fig. 4A). Released 5-HT activates a Cl^-

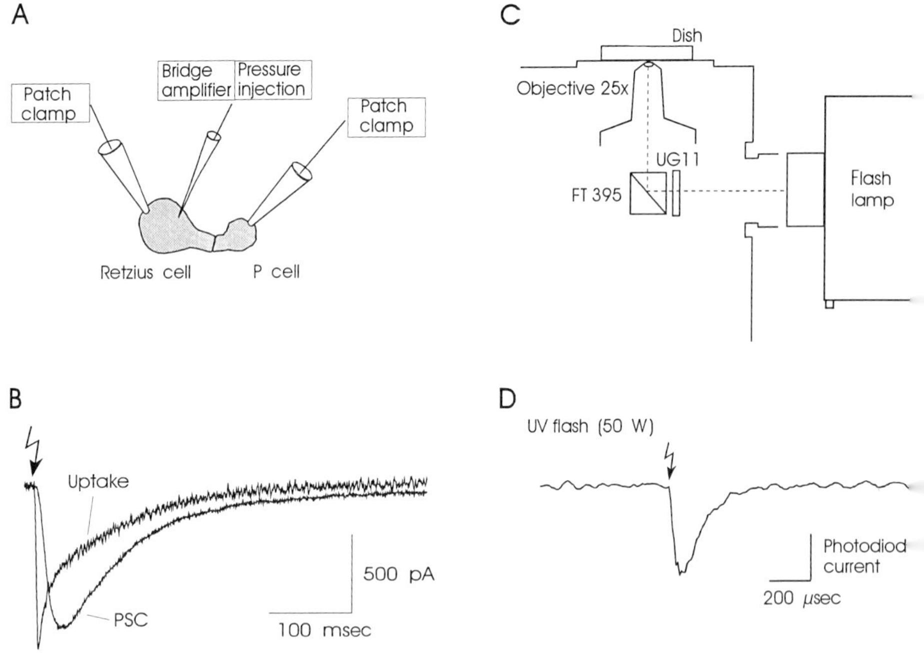

FIG. 4. (A) Schematic illustration of the electrode arrangement used for caged Ca^{2+} photolysis experiments. Pre- and postsynaptic currents are recorded with patch pipettes in the whole-cell configuration. The Retzius cell is penetrated with an additional microelectrode, which serves for DM-nitrophen pressure injection and measures clamp control. (B) Simultaneous recording of the presynaptic reuptake current and the related postsynaptic response (PSC) on flash-evoked transmitter release (currents superimposed). The short UV-light flash (50 W · s electrical energy) is indicated by the arrow. (C) Schematic representation of an inverted microscope system for caged Ca^{2+} photolysis. The UV flash is guided through the epifluorescent port and the optical path of the microscope onto the cells. (D) Photodiode current evoked by a flash of 50 W · s.

conductance in the membrane of the postsynaptic P-cell.[38] We found that the Cl^- mediated current was more stable when Cl^- ions were present in both external and internal solutions. A robust 5-HT induced inward current can be recorded when the postsynaptic patch pipette contains a high Cl^- concentration solution. The presynaptic pipette solution should be composed to give a common reversal potential (e.g., −70 mV) for Cl^- and K^+ conductances, in order to facilitate the isolation of Na^+ dependent reuptake currents. This procedure restricts the measurement of transporter currents to a certain holding potential (−70 mV), but minimizes their possible contamination by 5-HT–activated Cl^- and K^+ conductances or by the Ca^{2+}-activated K^+ conductance.[35] Because the Ca^{2+}-activated K^+ conductance generates large currents in these neurons the K^+-channel blocking agent, tetraethylammonium (TEA) should be added to the external saline.[35] When 5-HT release is diminished, it is possible to test whether channels or transporters that are directly activated by intracellular Ca^{2+} contribute to the observed charge influx in the Retzius cell. For example, vesicular serotonin can be efficiently depleted by culturing Retzius P cell pairs in the presence of reserpine.[6,17] Another approach is offered by clostridial neurotoxins, such as tetanus toxin. This neurotoxin impairs transmitter release by specifically cleaving the synaptic vesicle protein synaptobrevin[39,40] (also referred to as VAMP) and efficiently inhibits synaptic vesicle exocytosis in Retzius cells.[11,12]

Photochemistry of DM-nitrophen

It is important to understand the chemistry of caged Ca^{2+} compounds in order to interpret the synaptic response to illumination in cells injected with DM-nitrophen. A number of excellent reviews have appeared in the past few years that cover the chemistry and application of DM-nitrophen.[37,41,42] DM-nitrophen[36] is a photolabile chelator which binds calcium and magnesium with affinities similar to those of its parent compound ethylenediaminetetraacetic acid (EDTA). Photolytic cleavage of DM-nitrophen, by exposure to UV light, results in the conversion of the high-affinity Ca^{2+} buffer [dissociation constant (K_d) 5 n*M*, 22°, pH 7.1, 150 m*M* ionic strength] to photo products that bind calcium with a low

[38] P. Drapeau and S. Sanchez-Armass, *J. Neurosci.* **8,** 4718 (1988).

[39] G. Schiavo, F. Benfenati, B. Poulain, O. Rosetto, P. Polverino de Laureto, B. DasGupta, and C. Montecucco, *Nature* **359,** 832 (1992).

[40] E. Link, L. Edelmann, J. H. Chou, T. Binz, S. Yamasaki, U. Eisel, M. Baumert, T. C. Südhof, H. Niemann, and R. Jahn, *Biochem. Biophys. Res. Com.* **189,** 1017 (1992).

[41] J. H. Kaplan, *Annu. Rev. Physiol.* **52,** 897 (1990).

[42] A. M. Gurney, *in* "Microelectrodes Techniques" (D. Odgen, Ed.), The Company of Biologists Limited, Cambridge, 1994.

affinity (K_d ~3 mM). Following Ca^{2+} liberation, which occurs with submillisecond time constants, free Ca^{2+} reequilibrates with unphotolyzed free chelator and native cytoplasmic calcium buffers.[43] Experiments by Kaplan and co-workers[44] have demonstrated that partial photolysis of DM-nitrophen, preloaded and nonsaturating Ca^{2+} concentrations, produces a rise of Ca^{2+} concentration that peaks at a concentration of greater than 100 μM in less than 100 μsec, but then declines to steady state within 1 to 2 msec. Thus, DM-nitrophen can be used to generate spikes of intracellular Ca^{2+}[43–45] which are probably similar in magnitude and duration to submembrane Ca^{2+} transients generated by Ca^{2+} influx through voltage-gated Ca^{2+} channels.[46–48]

Microinjection of DM-nitrophen

Because of the large size of the Retzius cell, Ca^{2+}-loaded DM-nitrophen (75 mM DM-nitrophen, 30% loaded with calcium) is best delivered by pressure injection via an additional microelectrode (Fig. 4A). This procedure offers distinct advantages: First, the presynaptic neuron can be filled with the chelator–Ca^{2+} complex more rapidly by pressure injection than by equilibration with the patch pipette solution. This is important, because DM-nitrophen binds Mg^{2+} with considerable affinity (K_d of 2.5 μM) and, therefore, cytosolic magnesium (0.46 mM Mg^{2+} in Retzius cells[49]) tends to displace calcium until the concentration of free DM-nitrophen is higher than the concentration of free magnesium. This scenario has been shown to cause a transient rise in $[Ca^{2+}]_i$ and uncontrolled secretion in the squid giant synapse and chromaffin cells.[45,50] Second, the microelectrode serves as independent monitor for clamp control during measurements of presynaptic reuptake currents and can be used to stimulate action potentials in Retzius cells when spike-evoked transmitter release is studied. We find it essential to compare spike- and flash-evoked transmitter release in the same cell pair because the strength of synaptic signaling varies between cell pairs. Furthermore, it is helpful to monitor spike-evoked transmitter throughout the experiment in order to unmask possible deteriorating effects caused by pressure injection or external solution exchange.

[43] R. S. Zucker, *Cell Calcium* **14,** 87 (1993).
[44] G. R. C. Ellis-Davies, J. H. Kaplan, and R. J. Barsotti, *Biophys. J.* **70,** 1006 (1996).
[45] E. Neher and R. S. Zucker, *Neuron* **10,** 21 (1993).
[46] S. M. Simon and R. R. Llinas, *Biophys. J.* **48,** 485 (1985).
[47] R. S. Zucker and R. A. Folgeson, *Proc. Natl. Acad. Sci. USA* **83,** 3032 (1986).
[48] R. Llinas, M. Sugimori, and R. B. Silver, *Science* **256,** 677 (1992).
[49] D. Günzel and W. R. Schlue, *J. Physiol. (London)* **491,** 3 (1996).
[50] K. R. Delaney and R. S. Zucker, *J. Physiol. (London)* **426,** 473 (1989).

DM-nitrophen containing solution is injected by applying controlled pressure pulses of nitrogen gas (20 psi for 2–3 sec, PLI-100, Medical Systems, Greenvale, NY). A rough estimate of the injected volume and, based on this, of the resulting intracellular concentration (~5–10 mM DM-nitrophen in our experiments) can be obtained by test injections into an oil drop. Following DM-nitrophen injection into Retzius cells, we occasionally observed a significant increase (70–120%) in the action potential evoked postsynaptic response. This may have been caused either by a Ca^{2+} leak into the neuron as a result of injection damage or by Mg^{2+}-mediated displacement of DM-nitrophen–bound Ca^{2+}, leading to an increase in resting calcium before photolysis. The latter scenario may be bypassed in future studies by using, instead of DM-nitrophen, the caged Ca^{2+} compound NP-EGTA,[51] which has a significantly reduced affinity for Mg^{2+}. The presynaptic current recording patch pipette can be filled with solution containing Ca^{2+} free DM-nitrophen to prevent resealing, as we frequently encountered after flash-induced rise in $[Ca^{2+}]_i$. Flashes (interval ~30 sec) should be delivered shortly after injection (within 4 to 5 min) to minimize loss of Ca^{2+}-loaded DM-nitrophen into the patch pipette. Flash evoked signals in the presynaptic and postsynaptic cell tend to decline to repetitive flashes owing mainly to progressive destruction of the intracellular caged Ca^{2+} compound. When flashes are adjusted to small intensities (e.g., 50 W·s) the reproducibility of elicited signals is improved. Usually, five to seven responses could be evoked following an injection as described above, among these, the first three stimuli gave fairly similar signals.

Experimental Setup for Flash Photolysis

For these experiments, cells should be cultured in polyornithine-coated Petriperm dishes to facilitate UV-light transmission. Photolysis of DM-nitrophen is accomplished using brief discharges from a high-pressure xenon flash lamp (35 S, Chadwick-Helmuth, El Monte, CA). The Chadwick-Helmuth lamp in combination with the power supply (Strobex model 238) allows one to elicit short and intensive light flashes which are triggered by capacitive discharge of between 50 and 230 W electrical energy. Eighty percent of the light is delivered within 200 μs (Fig. 4D). The flash lamp is placed in front of the epifluorescent port of an inverted microscope (IM, Zeiss). The light is directed via a dichroic mirror (FT 395, Zeiss) and a Zeiss 25× objective (Plan-Neofluar 0.8 Imm. Korr.) onto the cells (Fig. 4C). Furthermore, wavelengths most efficient for photolysis (300–400 nm) are selected from the spectrum of a UG11 filter (Zeiss,

[51] G. C. R. Ellis-Davies and J. H. Kaplan, *Proc. Natl. Acad. Sci. USA* **91,** 187 (1994).

3 mm). By using a first surface reflector instead of the dichroic mirror (FT 395), light intensities on the specimen can be increased. Our flash lamp housing and focusing optics were custom made, but similar equipment[52] can be obtained from Rapp Optoelectronic, Hamburg, Germany (Hi-Tech Scientific LTD, Salisbury, U.K. distributor).

Incorporation of a flash lamp into the electrophysiological setup requires electrostatic shielding in order to minimize artifacts arising from capacitor discharge. Furthermore, it is good practice to isolate the power supply of the lamps from the electrophysiological equipment and to use isolation circuits in its trigger pulse connection (TTL-compatible optoisolator). A photoelectric effect occurs when intensive UV light hits the Ag–AgCl wire. By covering the silver wires in both patch pipettes with a black polyethylene tubing, this artifact can be reduced.

Appendix

Animals

Experiments are performed on cells of adult leeches (*Hirudo medicinalis*) kept in laboratory aquaria. Leeches can be obtained from Ricarimpex (33980 Audenge, Gde, France) or from Leeches USA (Westbury, NY).

Materials and Solutions

Patch Pipettes and Microelectrodes. Patch pipettes for whole-cell recording are made from borosilicate glass capillaries, such as Kimax-51 (Witz Scientific, Maumee, OH). These thin-walled capillaries make it possible to pull sharply tapered pipettes, tapering being a function of the thickness of the capillary wall. We use patch pipettes heat-polished to a tip opening of 3 to 5 μm, having resistances of 0.4 to 1 MΩ, depending on the taper shape and the pipette solution. Microelectrodes are made from thick-walled borosilicate glass (GC150F-10, Clark Electromedical Instruments, Reading, U.K.), giving resistances of 30 to 50 MΩ.

Solutions

Recordings of 5-HT– and 5,7-DHT–induced uptake currents

External saline: 130 m*M* sodium gluconate, 6 m*M* potassium gluconate, 5 m*M* calcium acetate, 10 m*M* tetraethylammonium acetate, 10 m*M* HEPES (pH 7.3 with NMDG).

[52] G. Rapp and K. Güth, *Pfluegers Arch.* **411,** 200 (1988).

Patch pipette and microelectrode filling solution: 110 m*M* potassium gluconate, 10 m*M* sodium gluconate, 10 m*M* EGTA, 1 m*M* calcium acetate, 20 m*M* HEPES (pH 7.3 with NMDG). Microelectrodes used for single-electrode voltage clamp recordings are backfilled with 3 *M* cesium acetate.

Caged Ca^{2+} photolysis

External saline: 130 sodium chloride, 6 m*M* potassium chloride, 5 m*M* calcium chloride, 10 m*M* HEPES (pH 7.3 with NMDG), optional 10 m*M* tetraethylammonium chloride

Presynaptic patch pipette filling solution: 110 m*M* potassium gluconate, 10 m*M* potassium chloride, 10 m*M* sodium gluconate, 1 m*M* DM-nitrophen-Na_4 (Ca^{2+}-unloaded), 20 m*M* HEPES (pH 7.3 with NMDG)

Injection solution: 75 m*M* DM-nitrophen-Na_4, 25 m*M* calcium acetate, 75 m*M* HEPES (pH 7.3 with NaOH)

Postsynaptic patch pipette filling solution: 130 m*M* potassium chloride, 10 m*M* sodium chloride, 10 m*M* EGTA, 1 m*M* calcium chloride, 20 m*M* HEPES (pH 7.3 with KOH).

Osmolarity of all solutions was adjusted to 320 mosmol with glucose.

Chemicals

5-HT and 5,7-DHT (Sigma Chemical Corp., St. Louis, MO); reserpine (Research Biochemical, Natick, MA); DM-nitrophen (Calbiochem Corp., San Diego, CA).

Acknowledgments

I would like to thank C. Hahnel and G. Brändle for comments on the 5,7-DHT imaging section of this chapter. I am also grateful to Drs. R. Jahn and J. Avery for critically reading the manuscript.

[41] Glutamate Uptake in Purkinje Cells in Rat Cerebellar Slices

By MICHIKO TAKAHASHI, MONIQUE SARANTIS, and DAVID ATTWELL

Introduction

Although much information on the properties of plasma membrane neurotransmitter transporters can be gained from studying them expressed heterologously in oocytes or cell lines, or in cells isolated from the nervous system, it is important to have methods for studying the transporters *in situ*. Only in this way can the density of carrier expression and the relevance of the carriers to nervous system function be assessed. Furthermore, now that the localization of cloned transporters is increasingly being studied using antibodies, it is essential to have methods that assess whether the proteins detected are actually functional in the cell surface membrane. Finally, studying transporters *in situ* should allow a comparison of their properties in the cells that they are normally expressed in with their properties when expressed in oocytes or cell lines; this may be particularly important for assessing the importance of modulation by (for example) phosphorylation, which might differ in different cell types.

To this end we have whole-cell clamped cells in slices of rat central nervous system (CNS) to try to detect currents produced by glutamate transporters, choosing cells to study based on the localization of cloned transporters suggested by antibody labeling. The four cloned mammalian glutamate transporters reported to date are electrogenic,[1–4] and so their activity can be monitored from the membrane current they produce when active.[5,6]

[1] T. Storck, S. Schulte, K. Hofmann, and W. Stoffel, *Proc. Natl. Acad. Sci. USA* **89,** 10955 (1992).
[2] G. Pines, N. C. Danbolt, M. Bjoras, Y. Zhang, A. Bendahan, L. Eide, H. Koepsell, J. Storm-Mathisen, and B. I. Kanner, *Nature* **360,** 464 (1992).
[3] Y. Kanai and M. A. Hediger, *Nature* **360,** 467 (1992).
[4] W. A. Fairman, R. J. Vandenberg, J. L. Arriza, M. P. Kavanaugh, and S. G. Amara, *Nature* **375,** 599 (1995).
[5] H. Brew and D. Attwell, *Nature* **327,** 707 (1987).
[6] B. Billups, M. Szatkowski, D. Rossi, and D. Attwell, *Methods Enzymol.* **296,** [42], (1998) (this volume).

0076-6879/98 $25.00

Choice of Cell to Study

Glutamate uptake is usually thought of as being into presynaptic cells (to recycle glutamate to vesicles) or into glial cells (to keep the extracellular glutamate concentration below neurotoxic levels). However, immunohistochemical techniques have shown that two cloned glutamate transporters (EAAC1 and EAAT4) are expressed in the soma and dendritic tree of cerebellar Purkinje cells.[7,8] We thought it particularly important to assess whether these transporters were functional in the cell membrane, because they appear to be expressed in a location which is postsynaptic to the glutamatergic inputs to the Purkinje cell from the climbing fiber and parallel fibers, suggesting a possible role in terminating transmission at these synapses. Furthermore, such a location for transporters would challenge the generally held idea that neuronal glutamate transporters are in presynaptic terminals.

Preparation of Rat Cerebellar Slices

Cerebellar slices are prepared as described by Sakmann and Stuart.[9] In brief, 12-day-old Sprague-Dawley rats are killed by cervical dislocation followed by decapitation, and their whole brain is removed. After removal, the brain is immediately submerged in ice-cold oxygenated saline solution. Then the two side lobes of the cerebellum are cut off using a razor blade, leaving the central part of the cerebellum intact.

The slicing chamber is half-filled with the saline solution to a level below the stage of the chamber, and precooled in a freezer until the solution is frozen. A sagittal surface of the cerebellar block is then glued with a cyanoacrylate glue to the slicing chamber stage, and the chamber is immediately filled with ice-cold oxygenated saline solution. A vibrating tissue slicer is used to cut slices. Four to 10 200-μm thick slices are usually obtained from one cerebellum.

After slicing, each slice is immediately transferred to a holding chamber containing oxygenated saline solution, which is kept either in a water bath at a temperature of 37° or at room temperature for 1 hr to allow recovery of the cells from the slicing, then stored at room temperature until required. Sometimes 1 m*M* sodium kynurenate (which blocks the ionotropic receptors

[7] J. D. Rothstein, L. Martin, M. Dykes-Hoberg, L. Jin, D. Wu, N. Nash, and R. W. Kuncl, *Neuron* **13,** 713 (1994).

[8] K. Yamada, M. Watanabe, T. Shibata, K. Tanaka, K. Wada, and Y. Inoue, *Soc. Neurosci. Abstracts* **22,** 144.4 (1996).

[9] B. Sakmann and G. Stuart, Chapter 8 of "Single Channel Recording," 2nd ed. B. Sakmann and E. Neher, Eds. Plenum, New York, 1995.

of glutamate) is added to the holding solution to try to decrease cell death caused by glutamate released into the slice from cells damaged by the slicing procedure.

Choice of Agonist to Activate Uptake

To study glutamate uptake it is obviously desirable to use the naturally transported substrate, L-glutamate, as the activating substance. However, there are severe disadvantages to this approach when studying neurons in brain slices, because neurons express large numbers of glutamate-gated ionotropic and metabotropic receptors. Purkinje cells, for example, express a large number of AMPA (alpha-amino-3-hydroxy-5-isoxazole) and metabotropic receptors [but no NMDA (*N*-methyl-D-aspartate) receptors in rats of the age we use], which generate a current in response to glutamate application, making it difficult to isolate any current generated by transporters.

The following argument shows that it is not very feasible to isolate the transporter current by blocking glutamate's action on its receptors. Glutamate produces steady-state activation of AMPA receptors with[10] an EC_{50} of K_{Glu} of 19 μM and a Hill coefficient of 1.5, and CNQX (a commonly used competitive blocker of AMPA receptors) blocks these receptors with a K_I of 0.3 μM. If a glutamate concentration, [Glu], is applied in the presence of an inhibitory concentration, [I], of CNQX, therefore, it will activate an AMPA receptor current that is a fraction

$$[Glu]^{1.5}/\{[Glu]^{1.5} + K_{Glu}{}^{1.5}(1 + [I]/K_i)^{1.5}\}$$

of the maximum possible. For a [Glu] of 200 μM (say) this predicts a current in the absence of CNQX that is 97% of the maximum. Adding 10 μM CNQX (a dose often used to block synaptic currents) reduces the response by 85%, but because the AMPA receptor generated current can be several nanoamps in magnitude in the absence of blockers, this still leaves a large current generated by AMPA receptors. (In fact, 10 μM CNQX only blocks the synaptic current successfully because the affinity of AMPA receptors for glutamate is 10- to 30-fold lower when the glutamate is applied rapidly.[10]) Even increasing the CNQX level to 50 μM still leaves 1.5% of the AMPA–receptor current unblocked, which can be similar in size to the current generated in Purkinje cells by glutamate transporters. Higher levels of CNQX are undesirable both because of possible nonspecific effects and because of the cost. Thus, it is essential to activate uptake with a glutamate analog which has little effect on AMPA (and preferably also

[10] D. K. Patneau and M. L. Mayer, *J. Neurosci.* **10,** 2385 (1990).

on metabotropic) receptors. D-Aspartate, an analog of glutamate, fulfills these requirements because it is transported by glutamate uptake carriers,[11] but does not activate AMPA receptors.[12] In addition, it has not been reported to activate metabotropic receptors.

Advantages of Local Application of D-Aspartate

Because D-aspartate is transported by glutamate carriers,[11] it is a competitive blocker of glutamate uptake, and applying D-aspartate globally (for example, by superfusion) inhibits uptake into adjacent glial cells and presynaptic terminals and cause a rise in extracellular glutamate concentration (as has been seen previously[13–15] when applying the similarly acting transport blocker PDC). This rise of glutamate concentration could activate AMPA and metabotropic receptors in the Purkinje cell. We therefore have chosen to apply D-aspartate to a localized part of the Purkinje cell, and use iontophoresis to apply D-aspartate to the soma of the cell being whole-cell clamped. An iontophoresis electrode (resistance around 100 MΩ) filled with 100 m*M* sodium D-aspartate is placed 5–10 μm from the cell body in the flow of the bath solution. The experimental arrangement is shown in Fig. 1. D-Aspartate is ejected for around 4 sec, once every 30 sec, by switching from passing a holding current of +20 nA through the iontophoresis electrode, to passing an ejection current of −40 nA. The position of the iontophoresis electrode is monitored continuously using a TV camera on the microscope, and thus maintained constant relative to the cell.

Choice of Intracellular Solution to Fill Whole-Cell Pipette

The intracellular solution used contains cesium as a major cation instead of potassium to reduce the cell membrane's potassium currents and thus obtain better voltage-clamp uniformity. This approach of removing internal potassium is possible because, although glutamate transporters need to countertransport K^+ to cycle, cesium can substitute for potassium.[16] The internal solution used also contains calcium and pH buffers and comprises (in m*M*): CsF 110; CsCl 30; NaCl 4; $CaCl_2$ 0.5; *N*-methyl-D-glucamine$_2$-EGTA 5; HEPES 10, pH 7.3.

[11] M. Sarantis and D. Attwell, *Brain Res.* **516,** 322 (1990).
[12] D. T. Monaghan, D. Yao, and C. W. Cotman, *Brain Res.* **324,** 160 (1984).
[13] M. Sarantis, L. Ballerini, B. Miller, R. A. Silver, M. Edwards, and D. Attwell, *Neuron* **11,** 541 (1993).
[14] B. Barbour, B. U. Keller, I. Llano, and A. Marty, *Neuron* **12,** 1331 (1994).
[15] M. Takahashi, Y. Kovalchuk, and D. Attwell, *J. Neurosci.* **15,** 5693 (1995).
[16] B. Barbour, H. Brew, and D. Attwell, *J. Physiol.* **436,** 169 (1991).

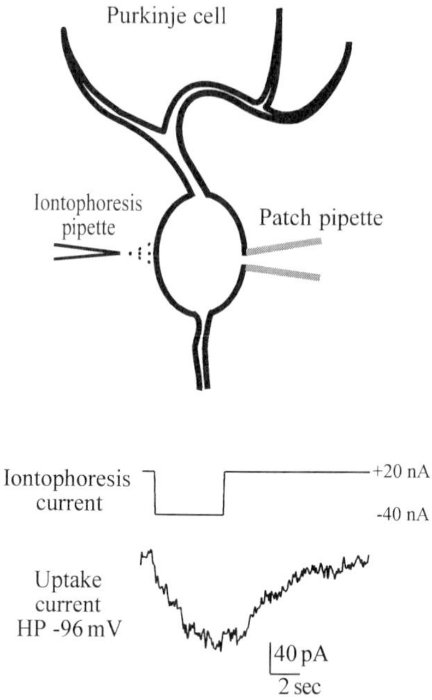

FIG. 1. *Top:* Schematic diagram showing how an iontophoresis electrode was placed near a whole-cell clamped Purkinje cell body in a cerebellar slice, and used to eject D-aspartate onto the soma to activate glutamate uptake carriers. *Bottom:* Specimen data showing the current change evoked in the Purkinje cell at −96 mV by the application of D-aspartate. [From the study of M. Takahashi, M. Sarantis, and D. Attwell, *J. Physiol.* **497,** 523 (1996).]

Response to D-Aspartate

As shown in Fig. 1, applying D-aspartate to Purkinje cells evoked an inward current at negative potentials. This current showed the pharmacology of glutamate uptake, being blocked by the transport blocker PDC but essentially unaffected by blockers of glutamate's ionotropic and metabotropic receptors.[17] In addition, the current showed a voltage- and ion-dependence consistent with it being generated by a sodium-dependent glutamate transporter with an associated anion channel.[17] Thus, it is possible to study glutamate transport *in situ* in neurons.

Interestingly, Renard and Crepel[18] found a different response to D-aspartate, dominated by an AMPA–receptor current, presumably activated by a rise of extracellular glutamate concentration following a block

[17] M. Takahashi, M. Sarantis, and D. Attwell, *J. Physiol.* **497,** 523 (1996).
[18] A. Renard and F. Crepel, *Eur. J. Neurosci.* **8,** 978 (1996).

of uptake by D-aspartate (as described above). This response may have dominated their recordings because they used superfusion of the D-aspartate–containing solution, which, although the superfusion was local rather than of the whole slice, may have resulted in more penetration of the D-aspartate into the slice where it could inhibit uptake into glial cells around the Purkinje cells.

Method to Block Uptake Selectively in Purkinje Cells

The EPSC produced in Purkinje cells by activating the climbing fiber input is prolonged when glutamate uptake is blocked by applying uptake blockers extracellularly[14,15] to slow the removal of glutamate from the synaptic cleft. However, extracellular uptake blockers block postsynaptic uptake as well as uptake into presynaptic terminals and glial cells. To determine whether postsynaptic uptake contributes significantly to removing glutamate from the synaptic cleft, it is desirable to be able to block glutamate uptake specifically in the Purkinje cells without changing glial or presynaptic uptake. One strategy to block Purkinje cell uptake alone is to include D-aspartate in the whole-cell pipette used to clamp the cell. It is known that internal glutamate slows the uptake of external glutamate,[16] presumably by binding to the glutamate-binding site of the carrier at the inner surface of the membrane, making it harder for the carrier to lose glutamate at that surface and reorient to the outer surface to pick up more glutamate. Since D-aspartate can also bind to the glutamate transport site, then internal D-aspartate also ought to slow glutamate uptake.

To test this approach, we compared the uptake currents produced by iontophoresing D-aspartate onto Purkinje cells clamped with pipettes either containing or lacking sodium D-aspartate (earlier work[16] showed that for internal glutamate to inhibit uptake it is essential to have sodium inside as well: a result that can be interpreted as being a consequence of transported glutamate dissociating from the carrier before the sodium). The intracellular solution used was (in mM): CsF 110; CsCl 14; NaCl 20 or 0; sodium-D-aspartate 0 or 20; $CaCl_2$ 0.5; $NMDG_2$-EGTA 5; HEPES 10, pH 7.3. Adding 20 mM sodium-D-aspartate to the pipette solution (replacing NaCl), to try to slow uptake, reduced the current produced by iontophoresed D-aspartate by 53% (Fig. 2), from 111 $\pm$ 24 pA in 14 cells studied with D-aspartate inside to 52 $\pm$ 14 pA in 11 cells without D-aspartate ($p < 0.06$, 2-tailed t-test).

Comparison of Climbing Fiber EPSC Decay with and without Uptake Inhibited in Different Cells

The fact that internal D-aspartate slows uptake provides a method for investigating the role of postsynaptic uptake in determining the duration

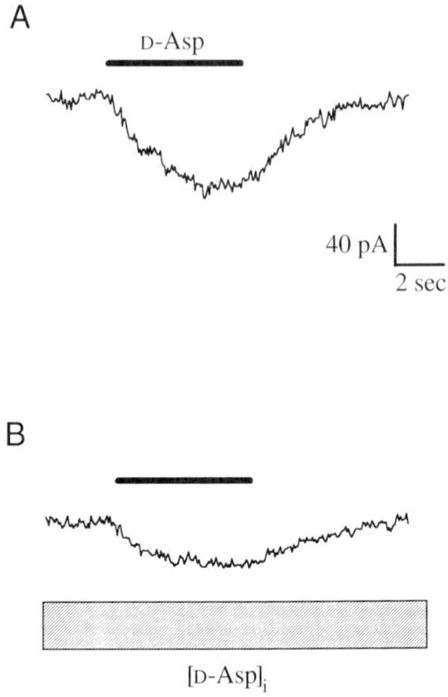

FIG. 2. Including 20 m*M* sodium D-aspartate in the whole-cell pipette (and hence inside the cell) reduces the uptake current evoked in Purkinje cell bodies by iontophoresed D-aspartate (black bar). *Top:* Specimen data from a cell clamped without D-aspartate inside. *Bottom:* Specimen data from a cell clamped with D-aspartate inside. [From the study of M. Takahashi, M. Sarantis, and D. Attwell, *J. Physiol.* **497,** 523 (1996).]

of synaptic currents in the Purkinje cell: the duration of the EPSC can be measured (as the time constant of the EPSC decay) in cells clamped with electrodes either containing or lacking D-aspartate.[17] The climbing fiber synapses were chosen for this analysis because, in the 12-day-old rats used, they impinge much closer to the soma where D-aspartate is introduced than do the parallel fiber synapses. With D-aspartate inside the EPSC decay time constant (8.89 ± 0.96 msec, measured from 90% to 10% of the current amplitude in 16 cells) was significantly longer ($p < 0.02$, 2-tailed *t*-test) than that with no D-aspartate inside (6.33 ± 0.58 msec, 24 cells: see Fig. 3A). Thus, it appears that postsynaptic uptake contributes to terminating the synaptic action of glutamate, and that altering the contents of the whole-cell pipette solution is a useful technique to block uptake selectively in one cell.

Double Patch-Clamping as Method of Comparing EPSC Decay with and without Uptake Inhibited in Same Cells

Although the experiments described above on cells with uptake functioning and cells with uptake blocked suggested a significant role for post-

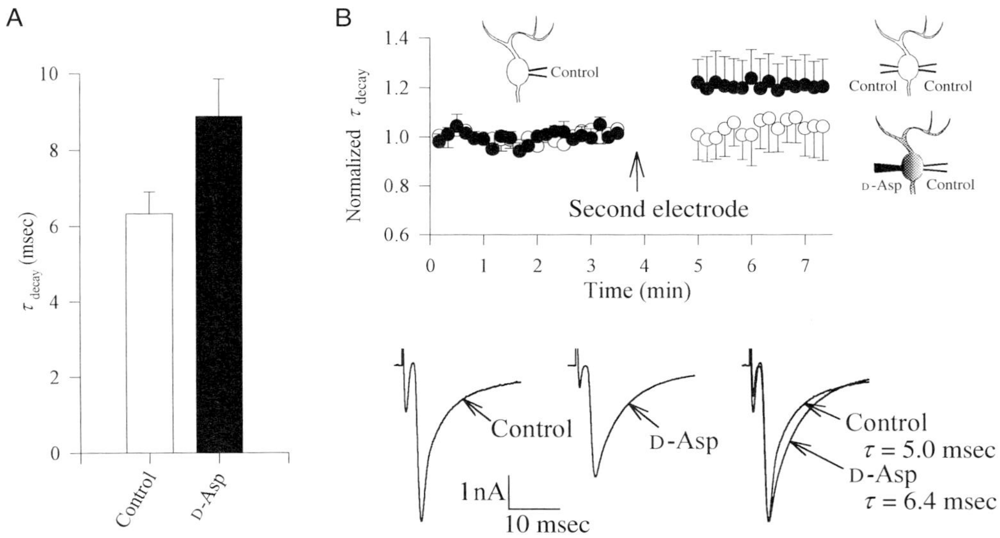

FIG. 3. Including D-aspartate inside the pipette used to whole-cell clamp Purkinje cells slows the decay of the EPSC evoked by climbing fiber stimulation. (A) EPSC decay time constant (± s.e.m.) in cells clamped without (control) or with (D-Asp) 20 m*M* sodium D-aspartate in the whole-cell pipette. (B) *Top:* Normalized EPSC decay time constant in cells clamped first with an electrode lacking sodium D-aspartate and then having a second electrode attached which either lacked (open symbols) or contained (closed symbols) 20 m*M* Na(D-Asp). *Bottom:* Specimen data for the closed symbols, showing the EPSC before (control) and after (D-Asp) attaching the second electrode, and these EPSCs normalized to the same peak. [From the study of M. Takahashi, M. Sarantis, and D. Attwell, *J. Physiol.* **497,** 523 (1996).]

synaptic uptake, ideally it would be better to measure the climbing fiber EPSC decay time constant without and with uptake blocked in the same cell. To do this,[17] a Purkinje cell was first whole-cell clamped with an electrode lacking D-aspartate (electrode 1) and the EPSC decay time constant was measured. Next, immediately prior to attaching a second patch electrode (electrode 2), which either contained or lacked D-aspartate, electrode 1 was switched from the voltage-clamp configuration to the current-clamp configuration, and a seal was obtained with electrode 2. Rupture of the patch of membrane under electrode 2 was then achieved to obtain the whole-cell configuration. Then electrode 2 was set to current-clamp mode while electrode 1 was switched back to voltage-clamp mode so that subsequent EPSC recordings were made with the same electrode as before electrode 2 was attached. To guard against an apparent change in EPSC waveform recorded with electrode 1, the series resistance of this electrode was continuously monitored by giving 2-mV steps prior to the stimuli used to evoke a synaptic current. The EPSC time constant just after the second electrode was attached was increased by about 20% ($p = 0.097$, 2-tailed paired t-test, $n = 4$) from its value before the electrode was attached (Fig. 3B) in cells to which a second electrode containing D-Asp was attached, while in cells to which a second electrode filled with control internal solution was attached there was no change in time constant ($p = 0.93$, $n = 3$; Fig. 3B). These results are similar to those in Fig. 3A obtained by comparing separate cells clamped with and without internal D-aspartate (although much less statistically significant because of the small number of cells we were able to do the experiment on). Thus, dialysis of the cell with an uptake inhibitor by attaching a second electrode is a promising method for testing the role of postsynaptic uptake in shaping EPSC kinetics.

Summary

We have described how whole-cell clamping of neurons in brain slices has allowed a characterization of postsynaptic transporters, probably a mixture of EAAC1 and EAAT4, in cerebellar Purkinje cells. Similar experiments have been carried out on transporters (mainly GLAST) in cerebellar Bergmann glia, and have revealed an uptake current occurring as these carriers remove glutamate released at the parallel fiber synapses.[19] As more transporters are cloned and their regulation is characterized in heterologous expression systems, it will be increasingly important to use methods similar

[19] B. Clark and B. Barbour, *J. Physiol.* **502,** 335 (1997).

to those outlined above to investigate to what extent the behavior of the carriers is similar *in situ* in the nervous system.

Acknowledgments

This work was supported by the Wellcome Trust, the M.R.C., INTAS, and the European Community (CT 95-571).

[42] Patch-Clamp, Ion-Sensing, and Glutamate-Sensing Techniques to Study Glutamate Transport in Isolated Retinal Glial Cells

By BRIAN BILLUPS, MAREK SZATKOWSKI, DAVID ROSSI, and DAVID ATTWELL

Introduction

It is becoming increasingly clear that no one technique can provide a complete understanding of the operation of neurotransmitter transporters. For years these transporters were studied with radiotracing techniques, which established that the main energy source for accumulation of neurotransmitter against an electrochemical gradient was the cotransport of Na^+ ions. These techniques also determined, through autoradiography, into which cells transmitters were transported. However, radiotracing experiments have several limitations. They do not provide good control of the membrane potential, which is a key determinant of the rate of transmitter uptake (complicating the interpretation of experiments using agents that may alter the membrane potential). Furthermore, they have poor temporal resolution, and they do not allow good control of the solution composition inside the cell.

To overcome some of these difficulties, more recent work has often monitored neurotransmitter transport electrically, taking advantage of the fact that a net charge movement accompanies the translocation of each transmitter molecule. For example, glutamate transporters cotransport an excess of Na^+ ions with each glutamate anion (as well as other ions moving, as described below), so an inward current is generated during glutamate uptake.[1] Using whole-cell clamping to monitor this current provides control of the membrane potential and intracellular solution (because the pipette

[1] H. Brew and D. Attwell, *Nature* **327,** 707 (1987).

0076-6879/98 $25.00

solution equilibrates with that in the cell[2]) as well as high temporal resolution of the transmitter-activated uptake current. Electrical recording is most useful if the current generated by a transporter is proportional to the rate at which transmitter is transported, as would be the case if the transporter stoichiometry were constant and the transporter could only generate current by transporting substrate. Recently, however, it has become clear that glutamate transporters can also generate current by activating an anion channel in their structure.[3,4] This, together with suggestions of an electroneutral mode of glutamate transport,[5] has spurred development of a more direct method to sense the amount of glutamate transported.[6,7]

Some agents, such as potassium,[8–10] alter the amount of glutamate transported and the current generated by the transporter even when the voltage is held constant, and it is useful to be able to determine whether these agents are transported substrates or just modulators of the rate of transport. Thus there has been increasing use of ion-sensitive dyes and electrodes to measure changes of the concentration of transported ions inside or outside the cells expressing the transporters.[11–13]

In this article we review the use of patch-clamping and ion- and glutamate-sensing methods to monitor glutamate transport by glial cells (Müller cells) of the salamander retina.

Studying Glutamate Transport by Patch Clamping

Choice of Cell to Study

We started to study glutamate transport in whole-cell clamped retinal glial cells as part of a program to look at transmitter-gated currents in isolated retinal cells. For several reasons this turns out to have been a highly fortuitous choice of transmitter and cell. First, radiotracing studies suggest that the density of glutamate transport in the nervous system is an

[2] O. P. Hamill, A. Marty, E. Neher, B. Sakmann, and F. Sigworth, *Pflügers Arch.* **391,** 85 (1981).
[3] J. I. Wadiche, S. G. Amara, and M. P. Kavanaugh, *Neuron* **15,** 721 (1995).
[4] W. A. Fairman, R. J. Vandenberg, J. L. Arriza, M. P. Kavanaugh, and S. G. Amara, *Nature* **375,** 599 (1995).
[5] E. A. Schwartz and M. Tachibana, *J. Physiol.* **426,** 43 (1990).
[6] B. Billups and D. Attwell, *Nature* **379,** 171 (1996).
[7] B. Billups, D. Rossi, and D. Attwell, *J. Neurosci.* **16,** 6722 (1996).
[8] B. I. Kanner and I. Sharon, *Biochemistry* **17,** 3949 (1978).
[9] B. Barbour, H. Brew, and D. Attwell, *Nature* **335,** 433 (1988).
[10] M. Sarantis and D. Attwell, *Brain Res.* **516,** 322 (1990).
[11] M. Bouvier, M. Szatkowski, A. Amato, and D. Attwell, *Nature* **360,** 471 (1992).
[12] A. Amato, L. Ballerini, and D. Attwell, *J. Neurophysiol.* **72,** 1686 (1994).
[13] A. Amato, B. Barbour, M. Szatkowski, and D. Attwell, *J. Physiol.* **479,** 371 (1994).

order of magnitude higher than that for other transmitters (possibly because of the need to prevent the extracellular glutamate concentration from rising to neurotoxic levels). This suggests that, for cells *in vivo* or freshly isolated cells, it might be easier to detect membrane currents associated with glutamate transport than it would be to detect γ-aminobutyric acid (GABA) or glycine transport currents (this consideration does not apply to transfected cells which may express different cloned carriers at similar densities). Second, because retinal neurons encode information as graded potentials rather than action potentials, the retinal photoreceptors and bipolar cells release glutamate continuously, necessitating a particularly powerful glutamate uptake system to control the extracellular glutamate concentration. Finally, salamander Müller cells express essentially no *N*-methyl D-aspartate (NMDA), (s)-α-amino-3-hydroxy-5-methyl-4-isoxazole propionic acid (AMPA), or kainate receptors,[1] and although they express a G-protein–linked receptor that gates potassium channels,[14] those potassium channels can be blocked with barium so that essentially all the current evoked by glutamate in these cells is generated by glutamate transporters.

Isolated Cells vs Cells in Slices

Studying cells isolated from the nervous system with enzymes offers several advantages. It is possible to control accurately the concentration of substances applied to the cells, it usually provides better voltage-clamp quality compared to the situation in slices where cells may retain long processes or gap junctions connecting them to other cells, and it is easier to attach the patch-clamp electrode to the cell in the first place. Nevertheless, it is possible to record glutamate uptake currents in cells in brain slices[15] as described in the chapter by Takahashi *et al.* in this volume,[16] and ion-sensing techniques can also be profitably applied in slices.[12]

Recording Forward Uptake Currents by Whole-Cell Clamping

A detailed description of the whole-cell patch-clamp technique is given by Hamill *et al.*[2] Applying glutamate to the outside of whole-cell clamped salamander Müller cells generates an inward current in the cell membrane (Fig. 1A: ref. 1). Applying glutamate analogs which are either substrates for transporters (D- and L-aspartate) or agonists at the ionotropic or metabotropic receptors (NMDA, AMPA, kainate, quisqualate, ACPD) of glutamate shows that the inward current has the pharmacology of glutamate

[14] E. A. Schwartz, *Neuron* **10,** 1141 (1993).
[15] M. Takahashi, M. Sarantis, and D. Attwell, *J. Physiol.* **497,** 523 (1996).
[16] M. Takahashi, M. Sarantis, and D. Attwell, *Methods Enzymol.* **296,** [41], (1998) (this volume).

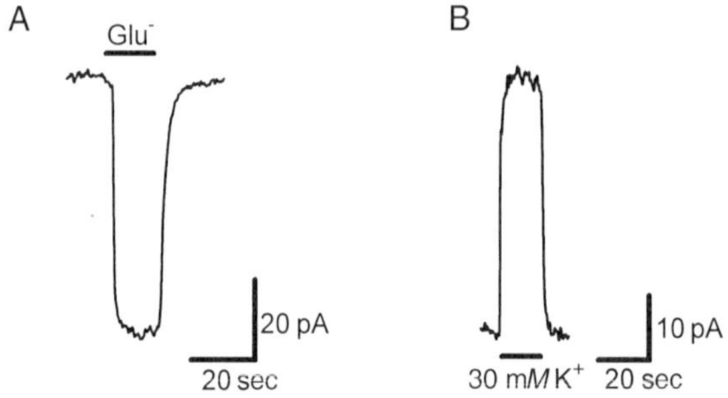

FIG. 1. Recording glutamate transport by whole-cell patch-clamping of isolated salamander retinal Müller cells. (A) Uptake of glutamate (200 μM, applied during bar) produces an inward membrane current in a cell voltage clamped to −40 mV. (B) Release of glutamate by reversed uptake produces an outward membrane current. An elevated potassium concentration (bar) was applied to a cell clamped to 0 mV (with a pipette containing 10 mM sodium glutamate) to stimulate reversed glutamate transport (with Na^+ and Glu^- leaving the cell, K^+ entering the cell, and either an OH^- entering the cell or an H^+ leaving the cell). [Data are from B. Billups and D. Attwell, *Nature* **379,** 171 (1996).]

uptake (only the transported substrates evoked a current). Removing external sodium ions abolishes the glutamate-evoked current, consistent with radiotracing work showing that uptake is driven partly by Na^+ cotransport.

Use of Intracellular Dialysis for Studying Countertransported Substrates

The control of the internal milieu provided by whole-cell clamping is extremely useful when studying the effect of intracellular ions on the rate of glutamate transport. For example, removing potassium from the pipette solution almost abolishes the glutamate-evoked current,[17] consistent with the idea that potassium binds to the glutamate transporter at the inner face of the membrane and is transported out of the cell.[8] Similarly, increasing the glutamate or sodium concentration in the whole-cell pipette solution, and hence in the cell, decreases the glutamate-evoked current,[17] as expected if it is more difficult for the transporter to lose glutamate and sodium at the inner membrane surface.

Importance of Electrode Series Resistance

When studying the effect of alterations in the pipette solution, it is important to use pipettes which are large enough to provide adequate exchange of the contents of the cell interior with those of the pipette. Because the geometrical factors determining diffusion into the cell are the

[17] B. Barbour, H. Brew, and D. Attwell, *J. Physiol.* **436,** 169 (1991).

same as those determining current flow, this requirement is essentially that of having a low series resistance (which is in any case desirable to provide good control of the cell voltage). An example of this is provided by experiments looking at the size of electrode needed to adequately control the potassium concentration within the cell.[18] With low series resistance electrodes (2 MΩ), removing potassium from the electrode greatly reduces the currents evoked by L-glutamate and D-aspartate. By contrast, when high series resistance electrodes are used (20 MΩ) the reduction of the currents is less clear, particularly for D-aspartate. Having a low series resistance is more important for the case of D-aspartate because the affinity of the transporters for K^+ binding to the inner face of the transporter is fivefold higher when the carrier is transporting D-aspartate than when it is transporting glutamate,[13] so that any residual K^+ not dialyzed out of the cell will still be able to activate a significant uptake current.

Recording Reversed Uptake Currents by Whole-Cell Clamping

Whole-cell clamping provides a relatively simple means of investigating reversed operation of glutamate transporters, based on the observation that, as described above, forward uptake requires glutamate and sodium outside the cell and potassium inside the cell. If this substrate distribution across the membrane is reversed, with glutamate and sodium being introduced into the cell via the whole-cell pipette, and with a high potassium concentration in the external solution, the transporters should be able to run backward. Consistent with this, raising $[K^+]_0$ outside cells clamped with glutamate- and Na^+-containing electrodes evokes an outward membrane current,[19] as expected if net positive charge is moved across the membrane with each glutamate anion (Fig. 1B).

Using Ion-Sensing Techniques to Identify Substrates Transported by Glutamate Transporters

To investigate which ions are directly involved in driving the transport of glutamate, we have found it useful to detect the movement of specific ions across the Müller cell membrane using two different approaches, either measuring intracellular ion concentration changes using fluorescent dyes, or measuring extracellular concentration changes at the outer surface of cells using microelectrodes.

[18] M. Szatkowski, B. Barbour, and D. Attwell, *Brain Res.* **555,** 343 (1991).
[19] M. Szatkowski, B. Barbour, and D. Attwell, *Nature* **348,** 443 (1990).

Measurement of Intracellular pH (pH) Changes using pH-Sensitive Fluorescent Dye BCECF

To investigate whether the transport of glutamate was associated with the transport of pH-changing ions,[20,21] we decided to see if we could measure any changes in pH_i associated with glutamate uptake. The small size of the Müller cells under study and their relative mechanical instability when isolated precludes the use of membrane-penetrating ion-sensitive microelectrodes to measure any such changes, and so a fluorometric method has been adopted using pH-sensitive dyes. Cells are loaded with the pH-sensitive indicator[22] BCECF (2′,7′-bis(carboxyethyl)carboxyfluoroescein) by dissolving the dye (96 μM) in the internal solution used to fill the patch pipettes (the concentration of pH buffer in the solution is reduced from our normal value of 5 mM to 0.5 mM to minimize the internal buffering capacity of the cells). In this way the indicator is allowed to equilibrate with the cell interior through the large tip of the pipette. Similar results are obtained[11] when the dye is loaded without the need to attach a pipette to the cell by bathing cells in the membrane-permeant acetoxymethyl ester of BCECF (the ester groups being cleaved from the dye inside the cell).

BCECF is a so-called dual-excitation wavelength, single-emission wavelength dye. This means that the ratio of fluorescence signal intensities evoked by different stimulation wavelengths can be used for the accurate measurement of concentration, because pH-independent factors that affect signal intensity (such as nonuniform intracellular dye concentrations, dye bleaching, or leakage out of the cell) cancel out in the ratio measurements. Whole-cell clamped Müller cells are illuminated with an epifluorescence attachment on the microscope which passes the excitation light onto the cells through the ×40 water immersion lens used to visualize the cells. The excitation beam passes through one of two manually switchable filters providing light of wavelength 490 nm (close to the most pH-sensitive part of the excitation spectrum) or 440 nm (close to the isosbestic or pH-independent point of the excitation spectrum). Light from the fluorescence illuminator is deflected toward the cells using a dichroic mirror splitting the light at 510 nm. The fluorescence signal emitted from the cells is collected by the ×40 lens and passes through the dichroic mirror and then through a 530 nm filter (the wavelength of maximum emission for BCECF) to be detected by a photomultiplier. Between the

[20] M. Erecinska, D. Wantorsky, and D. F. Wilson, *J. Biol. Chem.* **258,** 9069 (1983).
[21] P. J. Nelson, G. E. Dean, P. S. Aronson, and G. Rudnick, *Biochemistry* **22,** 5459 (1983).
[22] T. J. Rink, R. Y. Tsien, and T. Pozzan, *J. Cell Biol.* **95,** 189 (1982).

530 nm filter and the photomultiplier, an adjustable iris allows us to select the field from which the fluorescence signal is collected, so it covers only the cell and excludes the patch pipette (from which the BCECF would give too large a background signal). In practice, no changes in emitted fluorescence intensity are observed using 440 nm excitation of BCECF in our experiments.[11] To calibrate the changes in fluorescence intensity at 490 nm excitation with respect to pH, we use the method of Eisner *et al.*,[23] in which the pH changes produced by externally applied weak acids and bases are used to calculate the resting pH, the buffering power, and the fluorescence change per pH unit, or else use the H^+/K^+ exchanger nigericin to set the pH_i to defined values: when applied to the same cell, both methods give the same result.

Application of glutamate to whole-cell clamped Müller cells filled with BCECF results not only in an inward uptake current, but also in a decrease in fluorescence intensity excited by 490-nm light (Fig. 2A), implying an intracellular acidification.[11] Control experiments show that this pH change is not a secondary consequence of pH-regulating transporters (like Na^+/H^+ exchange) being activated by the influx of Na^+ on the uptake carriers. The size of the glutamate-induced intracellular acidification (around 0.2 units) and measurements of the buffering power, volume, and uptake current of the cells shows that the pH change is consistent with the movement of roughly one effective proton into the cell per unitary charge of uptake current. It can also be shown that the size of the pH_i change per uptake current does not change even if ligands that are transported slowly on the uptake carrier, such as D-aspartate, are applied to the cells. Use of the pH-sensitive fluorescent dye therefore leads to the tentative conclusion that either one H^+ is transported into the cell or one OH^- is transported out by the transporter for each net positive charge entering the cell, independent of which amino acid was transported. However, it could be argued that the intracellular acidification is a result of metabolism of glutamate inside the cell. Although this explanation would be hard to sustain for the pH change produced by the nonmetabolized D-aspartate, we have thought it worth testing whether glutamate uptake was associated with an extracellular alkalinization (consistent with transport of H^+ or OH^- on the carrier) or acidification (consistent with the intracellular pH change being produced by metabolism). To do this, and also to measure the movement of other ions on the uptake carrier, we have used ion-sensitive microelectrodes (ISMs) placed just outside the cell membrane.

[23] D. A. Eisner, N. A. Kenning, S. C. O'Neill, R. Pocock, C. D. Richards, and M. Valdeolmillos, *Pflügers Arch.* **413,** 553 (1989).

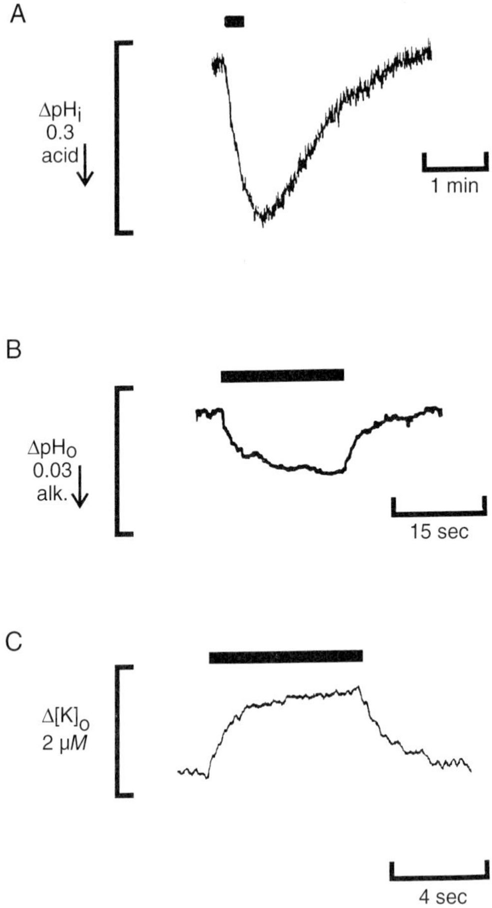

FIG. 2. Changes of ion concentration associated with glutamate uptake into salamander retinal Müller cells. (A) Change of intracellular pH (acid downwards), measured with the dye BCECF, associated with activation of glutamate uptake into the cell (black bar). (B) Change of extracellular pH (alkaline downwards), measured with a pH-sensitive microelectrode positioned just outside the cell, associated with the activation of glutamate uptake (bar). (C) Rise of extracellular potassium concentration, measured with a K^+-sensitive microelectrode positioned just outside the cell, associated with the activation of glutamate uptake (bar). [Data taken from M. Bouvier, M. Szatkowski, A. Amato, and D. Attwell, *Nature* **360,** 471 (1992), and A. Amato, B. Barbour, M. Szatkowski, and D. Attwell, *J. Physiol.* **479,** 371 (1994).]

Measuring Extracellular Ion Concentration Changes using Ion-Sensitive Microelectrodes

Fabrication of Ion-Sensitive Microelectrodes

We have used ISMs to detect changes of extracellular pH (pH_o), and also changes in the external concentrations of countertransported K^+ and of anions moving through what turns out to be an anion conductance present in the transporter structure.[11,13] For all of these measurements the microelectrodes are single barrel and pulled to a tip size of ~1 μm from 1.5 mm diameter, thin walled, filamented borosilicate glass tubing (Clark Electromedical, Reading, U.K., GC150TF10). They are silanized by dipping in a 10% solution of hexamethyldisilazane in 1-chloronaphthalene and baked at 40° for 5 hr. The tips are then filled with ion-sensitive resins bought from Fluka Chemicals (Ronkonkoma, NY) and backfilled appropriately (as described below). The ISMs are connected via chlorided silver wires to a high input-impedance electrometer (World Precision Instruments). The reference potential is measured with a chlorided silver wire positioned in the bulk solution at the edge of the experimental chamber. Since the voltage gradients next to the tips of the ISMs have been both calculated and measured[13] to be negligible, it is not necessary to use a reference voltage electrode positioned close to the ISM. To measure ion concentration changes just outside a Müller cell, an ISM is positioned within a few micrometers of the cell membrane (touching of the cell membrane often resulted in failure of the ISM, because the lipophilic resin inside the electrode seemed to be attracted out of the electrode and into contact with the cell membrane). Throughout the experiments the ISM and cell are observed on a TV monitor (linked to a TV camera on the microscope) to ensure that the distance between ISM and cell is constant.

Calculated predictions of the changes in the extracellular ion concentrations associated with uptake[13] have indicated them to be small. We therefore maximize the measured changes in two ways. First, the external solutions have been modified. For pH_o measurements the buffering capacity of the extracellular solution is reduced by lowering the HEPES concentration from its normal value of 5 m*M* to 0.05 m*M*, and for $[K]_o$ measurements potassium is omitted from the external solution since the micromolar predicted increases in $[K]_o$ generated by glutamate uptake[13] would be undetectable if superimposed on the normal millimolar external $[K]_o$. Second, instead of applying glutamate in the perfusing solution to activate uptake, which might dissipate local changes of pH_o or $[K]_o$, glutamate uptake is activated in a nonflowing solution by stepping the voltage of the cell from positive to negative values in the presence of glutamate, relying on the steep voltage depenence of the uptake[17] (the uptake current at +20 mV is only 15% of that at −80 mV).

pH-Sensitive Microelectrodes

For pH-sensitive microelectrodes, we fill the electrodes with Fluka ionophore 95291 and backfill them with extracellular solution. Activation of uptake leads to an extracellular alkalinization (of a few hundredths of a pH unit: Fig. 2B), proportional to the uptake current at different voltages, and (for a given uptake current) independent of whether L-glutamate or D-aspartate are transported. These data show conclusively that a pH-changing ion is transported by the glutamate transporter in these cells, and that the stoichiometry (ratio of movement of pH-changing ion to net charge movement) is voltage- and substrate-independent.

Potassium-Sensitive Microelectrodes

Removing potassium from inside cells or raising the external potassium concentration at constant voltage inhibits glutamate uptake.[8–10] We use potassium-sensitive electrodes to show definitively that these effects are a consequence of K^+ being transported out of the cell on the glutamate carrier.[13] The microelectrodes are filled with Fluka ionophore 60398, and backfilled with 1 m*M* KCl. Activating uptake leads to a small accumulation of potassium outside the cell (a few micromolar: Fig. 2C), which is proportional to the uptake current at different voltages, and (for a given uptake current) independent of whether L-glutamate or D-aspartate is the transported species. Thus, these data show that the glutamate transporter countertransports K^+ with a stoichiometry (K^+ movement per uptake current) that is voltage- and substrate-independent.

Anion-Sensitive Microelectrodes

The glutamate-evoked current in salamander Müller cells is increased by the presence of certain anions (SCN^-, ClO_4^-, NO_3^-) inside the cell,[11] and we wanted to know whether these anions actually moved out of the cell via the uptake carrier of if they just acted at an intracellular modulatory site. To study this, we have used microelectrodes filled with Fluka resin 24902 (and backfilled with 100 m*M* NaCl). This resin is nominally Cl^--sensitive, but in fact is many orders of magnitude more sensitive to SCN^- and ClO_4^-, allowing these ISMs to detect micromolar levels of SCN^- and ClO_4^- in the presence of a normal extracellular Cl^- concentration. Activating uptake is found to result in an efflux of ClO_4^- from cells clamped with a pipette solution containing this anion.[11] We originally attributed this efflux to ClO_4^- being transported out of the cell on a site that normally transports OH^-, but more recent work[7,24] has shown that the ClO_4^- actually leaves the cell through an anion channel in the carrier structure. Thus, it remains unclear whether the transporter carries H^+ into the cell or OH^- out of the cell.

Sensing Glutamate Released by Transporters using Glutamate-Gated Channels in Isolated Neurons

As described above, under certain conditions glutamate uptake carriers can be stimulated to run backwards, releasing glutamate from cells rather than taking it up. Release of glutamate by reversed glutamate uptake may have important implications during brain ischemia or anoxia, because it is a rise in extracellular glutamate concentration which triggers the death of neurons[25] and because the rise in extracellular potassium concentration and consequent membrane depolarization which occur during anoxia/ischemia will stimulate glutamate uptake carriers to run backward and release glutamate.[26]

Reversal of uptake carriers can be observed experimentally as an outward membrane current when potassium is applied to whole-cell voltage-clamped Müller cells, as long as sodium and glutamate are present inside the cell and the membrane is depolarized.[19] However, using the reversed glutamate uptake current as a way of accurately assessing the amount of glutamate released relies on the assumption that the carrier transports the same amount of net charge outward with each glutamate ion under all conditions. The validity of this assumption is brought into question by the recent discovery that glutamate transporters gate an anion conductance,[3,4,24] which is activated by both forward and reversed glutamate uptake.[7] Thus, contamination of the reversed uptake current by anion fluxes through this conductance may occur, and might be expected to represent a different proportion of the total measured current at different membrane voltages. There has also been a suggestion that glutamate release can be mediated by the glutamate transporter transporting only one sodium and one glutamate ion out of the cell.[5] This mode of glutamate release would be electroneutral and so would not be detectable as a membrane current. Thus, to accurately evaluate the glutamate release, a more direct method of detecting glutamate efflux is required: one that is independent of charge movements.

[24] S. Eliasof and C. Jahr, *Proc. Natl. Acad. Sci USA* **93,** 4153 (1996).
[25] D. W. Choi and S. M. Rothman, *Ann. Rev. Neurosci.* **13,** 171 (1990).
[26] M. Szatkowski and D. Attwell, *Trends Neurosci.* **17,** 359 (1994).

Sensing Neurotransmitter Release

Previous studies have demonstrated that release of neurotransmitters can be measured directly using cells or patches of cell membrane, expressing transmitter-gated ion channels, as sensors. This technique was first used to demonstrate that acetylcholine (ACh) is released from the growth cones of developing motor neurons of the chick[27] and the frog.[28] Outside-out membrane patches pulled from muscle cells contain ACh receptors which, when placed next to the growth cone of a motor neuron, are activated by the released ACh. Release of GABA by reversal of the GABA uptake carrier has similarly been demonstrated using whole-cell voltage-clamped goldfish bipolar cells to detect the release of GABA from catfish horizontal cells.[29] More recently, glutamate has been identified as the neurotransmitter released from turtle photoreceptors[30] and chick cochlea hair cells[31] by activation of NMDA channels in outside-out membrane patches pulled from rat hippocampal CA1 neurons and cerebellar granule cells, respectively. In our experiments, whole-cell voltage-clamped neurons isolated from the rat cerebellum have been used to sense the release of glutamate, mediated by reversed glutamate uptake, from salamander Müller cells.

Choice of Glutamate-Sensing Cells

Slices (200 μm thick) from the cerebellum of 12-day-old rats are dissociated by papain treatment followed by trituration with a fine glass pipette, in the same manner as was used for isolated Müller cells[1] but with the tonicity of the solution raised to 330 mOsm by the addition of NaCl. When plated onto coverslips, two cell types are easily identifiable. Granule cells can be identified by their small spherical appearance, and Purkinje cells by their larger size and remains of axons and dendritic trees. Both these cell types express glutamate gated ion channels. The granule cells express NMDA receptors, whereas the Purkinje cells do not express any functional NMDA receptors, but do express both the AMPA and kainate subtypes of non-NMDA receptors. Both cultured granule cells and freshly isolated Purkinje cells have been used successfully to detect glutamate release from Müller cells.[6] NMDA receptors have a higher affinity for glutamate, with an EC_{50} of 2.3 μM for the steady-state current seen during prolonged glutamate application, compared to 19 μM for AMPA receptors,[32] and also

[27] R. I. Hume, L. W. Role, and G. D. Fischbach, *Nature* **305,** 632 (1983).
[28] S. H. Young and M. M. Poo, *Nature* **305,** 634 (1983).
[29] E. A. Schwartz, *Science* **238,** 350 (1987).
[30] D. R. Copenhagen and C. E. Jahr, *Nature* **341,** 536 (1989).
[31] Y. Kataoka and H. Ohmori, *J. Physiol.* **477,** 403 (1994).
[32] D. K. Patneau and M. L. Mayer, *J. Neurosci.* **10,** 2385 (1990).

show less rapid desensitization in the maintained presence of glutamate than do AMPA receptors. For some purposes, then, granule cells will be the cell of choice as a glutamate sensor. However, the experiments we wished to perform[6] involved changing the pH to mimic the events of ischemia, so the pH sensitivity of the different receptor types was an important issue. The IC_{50} for protons blocking the NMDA receptor[33] is at pH 7.3, whereas for non-NMDA receptors the IC_{50} is at pH 6.3–5.7. For our experiments, therefore, it is more appropriate to use AMPA receptors in Purkinje cells as the glutamate sensors. To sense low levels of glutamate with these channels it is essential to block desensitization[34] of the receptors, using a class of drugs termed benzothiadiazides (e.g., cyclothiazide, diazoxide, or trichlormethiazide).

Calibration of Glutamate Sensing Cell

To quantify changes in glutamate release by transporters, calibration of the glutamate-sensing cell is essential. Isolated cerebellar Purkinje cells are whole-cell voltage clamped, and their response to external glutamate is assessed (with desensitization of the AMPA receptors expressed in the Purkinje cells blocked by adding 1 m*M* trichlormethiazide to the external medium). The dose–response curve for the glutamate response is shown in Fig. 3A. The data are fitted by a Hill equation with an EC_{50} of 23 μM and a Hill coefficient of 1.2.

Communication between Two Cells

With a Purkinje cell pushed up against a whole-cell voltage-clamped Müller cell, it is possible to detect glutamate released from the Müller cell (Fig. 3B). Reversed glutamate uptake is stimulated in the Müller cell by depolarizing the cell from a membrane potential of −60 mV to +20 mV, in the presence of 10 m*M* internal sodium and glutamate, and 30 m*M* external potassium. This method of stimulating reversed uptake is desirable because the voltage change can easily be applied to the Müller cell without affecting the glutamate-sensing cell. Stimulating reversed uptake by raising the external potassium concentration[19] would affect both the Müller cell and the membrane current through potassium channels in the membrane of the glutamate-sensing cell, making interpretation of the results more complicated. Electrical stimulation has the added bonus of accurate temporal regulation, enabling many responses to be easily averaged to give a better signal-to-noise ratio in the signal of the glutamate-sensing cell.

[33] S. F. Traynelis and S. G. Cull-Candy, *Nature* **345,** 347 (1990).
[34] K. A. Yamada and C.-M. Tang, *J. Neurosci.* **13,** 3904 (1993).

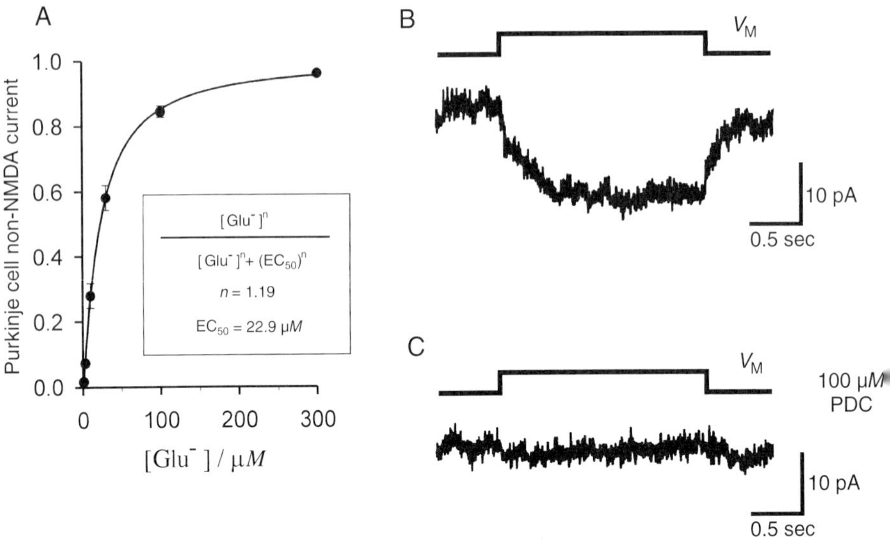

FIG. 3. Using non-NMDA receptor channels in isolated rat cerebellar Purkinje cells to detect glutamate released from isolated salamander retinal Müller cells. (A) Calibration of the Purkinje cell current response at −60 mV (ordinate, mean of data from four cells) to different glutamate concentrations (abscissa) in the presence of 1 m*M* trichlormethiazide to remove receptor desensitization. *Inset:* Hill equation fit to the data. (B) Current response in a Purkinje cell clamped to −60 mV, when the voltage V_M of an apposed Müller cell (clamped with an electrode containing 10 m*M* sodium glutamate) was depolarized from −60 to +20 mV (top trace): this voltage change activates glutamate release from the Müller cell by reversed uptake. (C) An experiment like that in (B) on the same cells, showing suppression of the current evoked in the Purkinje cell when glutamate transport in the Müller cell membrane was slowed with 100 μ*M* PDC. [Data are from B. Billups and D. Attwell, *Nature* **379,** 171 (1996).]

When the Müller cell is stimulated to release glutamate in this way, a membrane current is observed in the adjacent Purkinje cell (Fig. 3B). At a negative Purkinje cell membrane potential, an inward current is observed. The current becomes outward at a positive Purkinje cell membrane potential, reverses around 0 mV, and is blocked by the non-NMDA receptor blocker CNQX.[6] These results are consistent with the opening of non-NMDA receptors in the Purkinje cell. They also rule out the possibility that the response observed in the glutamate-sensing cell is a result of an artifactual electrical connection between the apposed membranes of the two cells. When granule cells are used to detect glutamate release, the response is similarly blocked by the specific NMDA receptor blocker D-APV.

Glutamate Release by Reversed Uptake

To demonstrate that the release of glutamate from the Müller cell is mediated by reversed glutamate uptake, the competitive glutamate uptake blocker PDC is used. With 100 μM PDC present in the external solution, stimulation of the Müller cell does not result in any glutamate release being detected by the adjacent Purkinje cell (Fig. 3C). Using a pharmacological agent in this way to determine the nature of the glutamate release relies on the assumption that the agent does not impair the Purkinje cell's ability to detect glutamate. This assumption has been tested by comparing the response of an isolated Purkinje cell to 3 μM external glutamate, in the presence and absence of PDC. No significant difference in the Purkinje cell glutamate-evoked current is produced by the addition of PDC. This indicates that PDC's inhibition of the Purkinje cell current induced by Müller cell stimulation must be as a result of PDC acting on the rate of glutamate release from the Müller cell.

Removal of the extracellular K^+ blocks glutamate release by reversed uptake, as expected from the stoichiometry of the transport process (a K^+ moves into the cell during reversed uptake, just as it moves out during forward uptake[6,19]).

Detecting Tonic Release of Glutamate

Stimulation of reversed uptake by depolarization as described above will only allow the Purkinje cell to detect a voltage-dependent change in glutamate transport. However, voltage-independent electroneutral glutamate release could also be occurring,[5] which will go undetected by the methods described above. To check for such tonic glutamate release, the Purkinje cell is moved from a position close to the Müller cell, where it would detect the released glutamate, to a distant position, out of range of the glutamate released. Any change in membrane current resulting from this withdrawal can be attributed to glutamate released from the Müller cell. Using this technique it has been possible to show that all the glutamate released by reversed operation of glutamate transporters is by simple reversal of the voltage-dependent electrogenic transport mode which normally transports glutamate into the cell.[6]

Summary

We have described how a combination of electrical, ion-sensing, and glutamate-sensing techniques has advanced our understanding of glutamate uptake into isolated salamander retinal glial cells. The next steps in understanding glutamate transport will inevitably depend strongly on molecular biological methods, as described elsewhere in this book, but will also require more detailed study of transporters in their normal environment, perhaps by using patch-clamping or imaging techniques to study cells *in situ*.

Acknowledgments

This work was supported by the Wellcome Trust, M.R.C., and the European Community (CT95-571).

[43] Measurement of Glial Transport Currents in Microcultures: Application to Excitatory Neurotransmission

By STEVEN MENNERICK and CHARLES F. ZORUMSKI

Introduction

Transporters are often ascribed the important role of clearing transmitter from the synaptic cleft following presynaptic release. This role has been compared to the role of the synaptic cleft enzyme acetylcholinesterase, which rapidly hydrolyzes acetylcholine following synaptic release at the vertebrate neuromuscular synapse. Nevertheless, clear evidence for a direct role of transporters in clearing the vertebrate central nervous system (CNS) transmitters glutamate and γ-aminobutyric acid (GABA) has been lacking until recently. In addition, although both neurons and glia possess transporters for fast neurotransmitters in the vertebrate CNS, the relative contributions of the two cell types to transporter function is typically difficult to discern. Pinpointing the role of transporters in vertebrate CNS synaptic physiology requires an experimental preparation that balances tight experimental control with the real-world validity of the intact nervous system.

Toward the goal of balancing experimental control with external validity, we have used patch-clamp technology and primary microcultures of postnatal rat hippocampal cells to help elucidate the role of glutamate transporters in clearing synaptically released glutamate. The first physiological studies using microcultures were performed on peripheral cells,[1,2] and

[1] E. J. Furshpan, P. R. MacLeish, P. H. O'Lague, and D. D. Potter, *Proc. Natl. Acad. Sci. USA* **73,** 4225 (1976).
[2] E. J. Furshpan, S. C. Landis, S. G. Matsumoto, and D. D. Potter, *J. Neurosci.* **6,** 1061 (1986).

0076-6879/98 $25.00

later studies extended the technique to the study of hippocampal neurons.[3,4] An individual hippocampal microculture consists of one or a few neurons and an associated complement of astrocytes. Each microculture is prevented from physical contact with other microcultures by growing the cells on small dots (20–1000 μm diameter) of adhesive substrate, such as collagen, in a surrounding sea of nonadhesive substrate, usually agarose. The physical boundaries of the microculture lend important advantages to the electrophysiological study of both neuron and glial cells and allow studies to be performed in this culture environment that would be difficult or impossible to achieve *in situ* or in conventional cultures. To eliminate potential polysynaptic contributions to the effects being studied, most of our experiments of neuron–glial interactions rely on microcultures of a single excitatory neuron and its surrounding astrocytes. In this chapter we focus on the advantages of these microcultures for the study of glial glutamate transporters in the context of excitatory synapses.

Advantages of Microcultures

Figure 1 shows examples of two microcultures prepared from postnatal day 1 rats. The microcultures were fixed following 12 days *in vitro* and were subsequently stained using antibodies directed against a peptide representing amino acid residues 510–524 of the neuronal EAAC1 transporter (Fig. 1A, B) or against a peptide representing amino acid residues 522–541 of the glial glutamate transporter GLAST (Fig. 1C, D).[5,6] The stains demonstrate microculture expression of both neuronal and glial transporters and suggest that microcultures may be a suitable environment for examining the role of transporters in synaptic transmission.

Microcultures provide a number of desirable features compared to other preparations (e.g., CNS slices and conventional primary cultures) that allow the controlled study of transporter operation during synaptic transmission. Neurons grown in synaptic isolation from neighboring neurons form numerous autaptic (self-synaptic) connections. Although autapses have been shown to exist *in vivo*,[7] the prevalence of autapses in microcultures appears greatly increased over that seen *in situ* or in conventional cultures prepared

[3] M. M. Segal and E. J. Furshpan, *J. Neurophysiol.* **64,** 1390 (1990).

[4] M. M. Segal, *J. Neurophysiol.* **65,** 761 (1991).

[5] K. P. Lehre, L. M. Levy, O. P. Ottersen, J. Storm-Mathisen, and N. C. Danbolt, *J. Neurosci.* **15,** 1835 (1995).

[6] Ø. Haugeto, K. Ullensvang, L. M. Levy, F. A. Chaudhry, T. Honoré, M. Nielsen, K. P. Lehre, and N. C. Danbolt, *J. Biol. Chem.* **271,** 27715 (1996).

[7] H. Van der Loos and E. M. Glaser, *Brain Res.* **48,** 355 (1972).

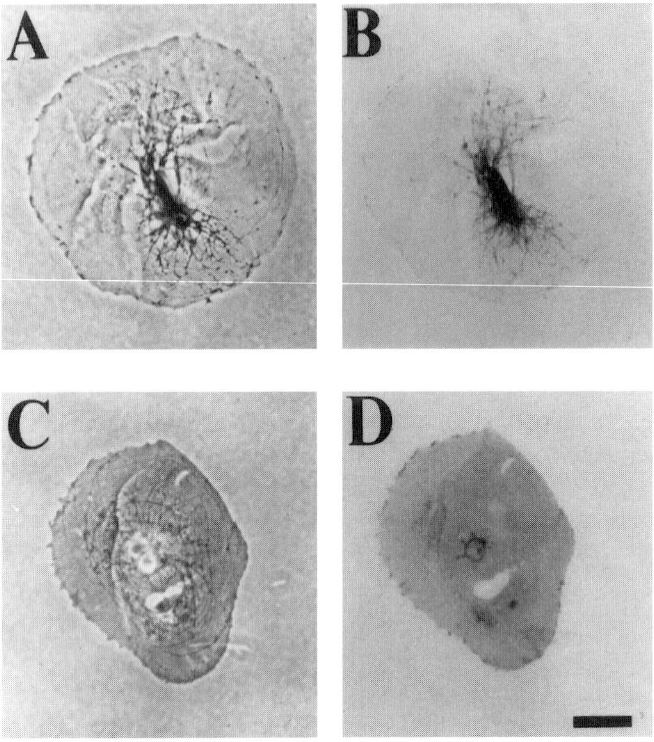

FIG. 1. Microcultures stained with antibodies against the neuronal glutamate transporter EAAC1 and the glial transporter GLAST. (A) Phase-contrast photomicrograph of a single-neuron microculture fixed 12 days following plating and stained with an antibody raised against the EAAC1 transporter. (B) Bright-field photomicrograph of the same microculture. (C) Phase-contrast photomicrograph of a two-neuron microculture from the same plating as the microculture in (A) and stained with an antibody raised against the glial transporter GLAST. Two phase-bright neuronal somata are visible near the center of the microculture, along with numerous thin neurites encircling the microculture. (D) Bright-field view of the same microculture. Note the complementary staining pattern compared with (B). The scale bar applies to A–D and represents 44 μm. Cultures were briefly fixed with 4% paraformaldehyde/0.2% glutaraldehyde in phosphate-buffered saline followed by permeabilization with 0.1% Triton X and block in 10% normal goat serum. Cultures were incubated in a 1:5000 dilution of antibody raised against EAAC1 and GLAST peptides, respectively (kind gift of Dr. N. C. Danbolt). Bound primary antibody was visualized using a commercially available nickel-intensified peroxidase reaction (Vectastain Elite ABC kit, Vector Laboratories, Burlingame, CA).

under the same conditions and from the same animals as microcultures.[8] For the study of the small electrical currents generated by glial transporters, the numerous (often several hundred to >1000) autaptic connections in a single microculture provide the requisite glutamate release to generate detectable transporter currents in underlying astrocytes. In addition, autaptic synapses allow both presynaptic and postsynaptic function of a neuron to be monitored with a single recording pipette, making it technically feasible to use another patch pipette to monitor glial membrane responses to neuronal stimulation.

The physical restrictions of microcultures also have advantages specific to the study of the Type I astrocytes found in these cultures. Type I astrocytes are known to form extensive gap junctions,[9] and individual cells can grow quite large if not physically constrained by the size of the microculture collagen dot. The large syncytium formed by astrocytes in conventional high-density cultures makes it difficult to voltage clamp the glia distal to the recording pipette. In contrast, if studies are constrained to microcultures of <300 μm in diameter, the entire glial bed of the microculture can be reasonably well voltage clamped. This allows experimental control over glial membrane potential, an important variable in the operation of transporters driven by electrochemical gradients.[10] In addition, the spatial restriction of microcultures allows confidence in complete local exchanges of the experimental bath solution surrounding the culture (see below). Therefore, ion and substrate concentrations can be precisely varied during an experiment.

Microcultures, like all primary culture systems, suffer from some limitations that are worth bearing in mind. Microcultures are prepared from young animals, and therefore may represent an immature physiological state. Although in many culture systems antigenic and physiological markers are preserved between the *in situ* and *in vitro* environments,[11] it also remains possible that culture environments alter cell phenotype in unexpected ways.

Microculture Preparation

Our general culture procedure for primary postnatal CNS cells is adapted from that described previously.[11] Our microculture procedure is adapted from that originally reported for use with peripheral neurons[1,2] and later extended to hippocampal cells.[3,4]

[8] S. Mennerick, J. Que, A. Benz, and C. F. Zorumski, *J. Neurophysiol.* **73,** 320 (1995).
[9] G. J. Stephens, M. B. A. Djamgoz, and G. P. Wilkin, *Receptors and Channels* **1,** 39 (1993).
[10] B. Barbour, H. Brew, and D. Attwell, *J. Physiol.* (*Lond.*) **436,** 169 (1991).
[11] J. E. Huettner and R. W. Baughman, *J. Neurosci.* **6,** 3044 (1986).

Chemicals and Ingredients

Supply	Supplier
Earle's minimal essential medium (MEM)	GIBCO (Grand Island, NY)
Leibovitz's L-15	GIBCO
Fetal bovine serum	GIBCO
Horse serum	GIBCO
L-Glutamine	GIBCO
Penicillin/Streptomycin	GIBCO
Agarose, Type IV-A	Sigma (St. Louis, MO)
Collagen, Type I, rat tail	Sigma
Papain	Sigma
Bovine serum albumin (BSA)	Sigma
Glucose	Sigma
Cytosine arabinoside (ARA-C)	Sigma

MEM, L-15, serum, glutamine, antibiotics, glucose (30% w/v stock), and ARA-C (1 m*M* stock) are aliquoted and stored at −80°. Collagen is prepared at 0.5 mg ml^{-1} in 0.1% acetic acid and stored at 4° for up to 3 weeks. Agarose, papain, and BSA solutions are prepared immediately before use (see below). All dish preparation and cell preparation procedures are performed in a tissue-culture hood.

Other Supplies

37° incubator with CO_2 supply [5% (v/v) CO_2]
Autoclaved 20 ml and 2 dram vials
35 mm plastic culture dishes (Falcon, Fisher Scientific, St. Louis, MO)
37° water bath
Microatomizer (Thomas Scientific, Swedesboro, NJ)
Miscellaneous dissection and surgical supplies

Dish Preparation

The day before plating cells, a thin layer of agarose is applied to the bottom of culture dishes. Although we typically apply the agarose substrate and perform platings directly on the bottom of 35 mm plastic culture dishes, others have successfully plated microcultures on glass coverslips,[3] which allows better visualization of cells and increased flexibility in performing optical recordings. A 0.15% (w/v) agarose solution is prepared in tissue-culture water and the solution is autoclaved for 20 min. While still warm, a small drop of agarose is spread uniformly across the bottom of a 35 mm plastic culture dish and allowed to dry overnight.

The day after agarose application, collagen solution is sprayed on the bottom of the dishes using a microatomizer. We hold dishes in the air at arm's length from the atomizer and spray each dish four times. This procedure usually results in small collagen dots (50–1000 μm diameter) of an appropriate density. To assess the size of collagen dots, one can visualize the dots on an inverted phase–contrast microscope immediately after spraying, before the droplets have dried. The dishes are sterilized for 30 min under ultraviolet irradiation. Plates are briefly rinsed with unsupplemented warm MEM before adding the cell suspension to remove residual acetic acid. Several adhesive substrates other than rat tail collagen have been used successfully for hippocampal microcultures. Others have used polylysine and palladium as adhesive substrates.[3] We have found that human placental collagen (Sigma) gives similar results to rat tail collagen. We have also had some limited success using Matrigel (Collaborative Research, Bedford, MA), a mixture of several adhesive substrates.

Cell Preparation

The growth medium for the cultures consists of supplemented MEM. The day of plating, supplements are added to warmed (37°) MEM at the following final concentrations: horse serum (5%), fetal calf serum (5%), D-glucose (17 mM), glutamine (400 μM), penicillin (50 U ml^{-1}), and streptomycin (50 μg ml^{-1}).

A solution of 0.02% BSA (w/v) in L-15 is prepared. Then, 9 ml of this solution is filter sterilized, warmed to 37°, and gently bubbled through a cotton-plugged, sterile pipette with 95% O_2/5% CO_2 (v/v) for later use as a dissection solution. Papain is dissolved (1 mg ml^{-1}) in a separate 3 ml aliquot of the BSA/L-15 solution. The papain solution is also filter sterilized and warmed to 37° with oxygenation.

Rat pups, 1–3 days postnatal, are anesthetized and decapitated. The skull is gently removed by cutting along the midline with a small surgical scissors and using forceps to pull the skull away laterally. The brain is gently removed with a spatula into a culture dish containing the warmed and oxygenated BSA/L-15 solution. Following hemisection of the brain along the midline, the cerebellum and midbrain are removed, and hippocampi are dissected away from surrounding cortex. Transverse hippocampal slices approximately 500-μm thick are prepared using a small scalpel, and the slices are transferred to the papain solution. The slices are occasionally gently agitated during a 25 min incubation in papain. Following enzymatic digestion, papain is removed and slices are placed in supplemented MEM. Pasteur pipettes of decreasing diameter are used to triturate the slices until a single-cell suspension is obtained. Six hippocampi (3 rat pups) yield approximately 5 ml of the cell suspension, which we subsequently dilute 15-fold in supplemented MEM. This yields an approximate cell density of 20,000 ml^{-1}, and 1.5 ml of this final suspension is added to each culture dish. We have found that with higher plating densities, microcultures form balls containing too many cells and are useless for physiology. Typically, we do not feed the cultures with fresh media following plating, but 72 hr after plating, cultures are treated with 10 μM ARA-C to inhibit glial overgrowth of the cultures. Also, although some authors have performed a second plating of fresh cells after allowing neurons in the original culture to die and glia to reach confluence,[3,4] we have performed most of our studies using a single plating of cells.

For synaptic physiology, we typically use cultures 8–15 days following plating, when functional synapses have formed. Up to 15 days *in vitro,* average neuronal synaptic responses increase in size, probably reflecting the continued addition of functional synapses. For unknown reasons, our microcultures do not typically survive as long as our conventional high-density mass cultures. Whereas conventional cultures can survive for several months, our microcultures typically survive 3–4 weeks. Others have reported longer survival of microcultures using a somewhat different culturing protocol.[3]

Following plating, glial cells, most of which exhibit properties of Type I astrocytes, adhere to the collagen droplet and expand over several days, often filling the entire collagen droplet. The glia exhibit a flattened morphology and large, prominent nuclei when viewed with phase-contrast optics. Both by electrophysiological and by dye-coupling criteria, we have observed extensive gap–junction coupling of microculture glia. In addition, the cells express glial fibrillary acidic protein immunoreactivity. These features are all typical of Type I astrocytes in other primary culture systems.[9]

Neuronal somata, in turn, adhere to the glial cells following plating. Having lost most of their processes during dissociation, neurons regrow processes throughout the culture period. Although some of the processes remain atop the glial bed, electron photomicrographs reveal many neurites between the layers of glial processes,[12] thus simulating the situation *in vivo,* where glial appendages surround neuronal processes and synapses.[13] By physiological, pharmacological, and immunohistochemical criteria, about half the neurons in microcultures appear to be GABAergic, and half appear to be glutamatergic. Cells with larger somata and thicker, more elaborate processes tend to be excitatory;[4,8] these visual characteristics allow us to

[12] S. Mennerick, A. Benz, and C. F. Zorumski, *J. Neurosci.* **16,** 55 (1996).

[13] A. Peters, S. L. Palay, and H. D. Webster, "The Fine Structure of the Nervous System." Oxford University Press, New York, 1991.

select excitatory cells for study *a priori* with a high degree of success. Typically, with systematic searching, we expect to find 5–10 single-neuron microcultures suitable for recording in a single culture plate. For dual recordings from glia and neurons, we exercise some bias toward choosing smaller microcultures with neurites that appear to be buried between glia.

Recording

Recording Apparatus

We use a typical patch-clamp electrophysiological setup.[14] An inverted microscope (Nikon, Boyce Scientific, St. Louis, MO) equipped with phase-contrast optics and a 40× phase-contrast objective is used to visualize cells. Two recording headstages are necessary for dual neuron/glia recordings. These are mounted onto remotely operated hydraulic micromanipulators (Narashige, East Meadow, NY) for positioning. For recordings, we use either an Axopatch 1D patch-clamp amplifier (Axon Instruments, Foster City, CA) or an Axoclamp 2A amplifier (Axon Instruments). Because of its low noise levels, we typically use the patch-clamp amplifier for glial recordings, where signals are small (2–200 pA), and we often use the Axoclamp 2A amplifier in the discontinuous single-electrode voltage-clamp mode for studying large (1–15 nA) autaptic currents generated by neurons. The outputs of the amplifiers are interfaced with a computer through an analog-to-digital converter. For dual neuron–glia recordings, the interface and acquisition software must be capable of multichannel data acquisition. Our data acquisition and analysis software is a combination of commercially available programs (Pclamp, Axon Instruments) and applications written in the Axobasic programming environment (Axon Instruments).

Pipettes and Pipette Solutions

Patch pipettes are pulled from thick-walled borosillicate glass (1.2 mm outer diameter, World Precision Instruments, Sarasota, FL) to a tip diameter of ~1.0 μm. Tips are lightly fire polished using a microfuge (Narishige), and the pipettes are filled with a solution consisting of (in m*M*) potassium gluconate (140), sodium chloride (4), EGTA (5), calcium chloride (0.5), HEPES (10), magnesium ATP (2), and GTP (0.5). The solution is adjusted to pH 7.25 using KOH. When filled with this pipette solution, pipettes typically have an open tip resistance of 3–6 MΩ, measured with small voltage-command pulses applied to the patch pipette with the pipette tip submerged in the bath solution.

[14] B. Rudy and L. E. Iverson, Eds. *Methods Enzymol.* **207** (1992).

When the whole-cell recording configuration is achieved, the contents of the patch pipette diffuse relatively quickly into the cell and replace the cytoplasmic contents.[15] Therefore, selection of the patch-pipette solution is important. For instance, potassium is used as the main monovalent cation in the neuronal pipette solution in order to preserve normal action potential propagation down the axon during neuronal stimulation (see below). Potassium is also used in the glial patch pipette because of the reported dependence of glutamate transport on intracellular potassium.[10]

Extracellular Solutions

Extracellular recording solution, which replaces culture medium at the time of recording, is composed of (in m*M*) sodium chloride (140), potassium chloride (4), calcium chloride (2.0), magnesium chloride (1.0), glucose (10), and HEPES (10). The solution is adjusted to pH 7.25 with NaOH. Experiments are performed at room temperature (22–24°).

Because characterization of glial and neuronal currents induced by neuronal stimulation relies heavily on pharmacological alteration of the extracellular bath solution, we sought a convenient means of quickly, completely, and gently changing the solution surrounding cells during an experiment. We achieve exchanges of experimental bath solution with a local microperfusion system. The perfusion system is a multibarrel pipette constructed of an SMI capillary (size E) pulled to a sharp tip, then cleanly broken and lightly fire polished in a flame to a diameter of 100 μm. Six or seven polyethylene tubes (PE 10) are collared by a larger polyethylene tube (PE 205, 1.5 cm long). Silicon sealant is allowed to run between the small tubes to seal gaps between tubes. The small tubes are cut flush with the end of the collar, and the collar is then inserted into the glass tube as far as possible toward the pipette tip, thereby minimizing the volume of the pipette tip that needs to be exchanged when switching solutions. Silicone sealant is used to make a water-tight seal between the collar and glass. The uncollared ends of the polyethylene tubes are attached to solution reservoirs (10 ml glass syringes), which are positioned 6–12 in above the microscope stage. Outflow from the reservoirs can be either manually controlled (XPERTEK miniature flow-through valve, P. J. Cobert Associates, St. Louis, MO), or electronically controlled by miniature solenoid flow-through valves (Lee Valve Co., Essex, CT). Solution flow rate can be adjusted by adjusting the height of the solution reservoirs or by altering flow resistance by changing the tubing diameter in a section of the flow lines.

[15] O. P. Hamill, A. Marty, E. Neher, B. Sakmann, and F. J. Sigworth, *Pflügers Arch.* **391,** 85 (1981).

We find that this multibarrel pipette is sufficiently gentle that it does not disrupt recordings, yet fast enough that multiple solutions can be tested on an individual microculture. The glass pipette is mounted on a micromanipulator and positioned ~500 μm from the microculture under study. During an experiment, outflow from one of the reservoirs is always on while outflow from the others is off; therefore, the microculture is perfused constantly (~100 μl min^{-1}) with either control bath solution or with one of up to six experimental solutions. Changes in junction currents at the tip of an open patch pipette suggest that local solution switches are achieved in 200–500 ms. Commercial versions of a multibarrel local perfusion system are also available (List-electronic, Darmstadt, Germany).

For study of responses of glial cells to exogenous glutamate applications, it is sometimes desirable to make faster solution exchanges than the above system allows. For faster drug applications, we use a linear array of 1.2 mm (outer diameter) glass tubes, which taper at their output to an outer diameter of ~400 μm. Solution flow is gravity driven, with outflow from a given tube controlled electronically with miniature solenoid flow-through valves (Lee Valve Co.). Solution exchanges on the order of 10 ms can be achieved on intact cells with this system, but the linear array of tubes requires repositioning to align the relevant tube with the recorded cell before each drug application. Also, the higher solution flow rate (up to 1 ml min^{-1}) with this system can disrupt recordings.

Recording Method

After the recording pipette is filled and mounted in the headstage pipette holder, slight positive pressure is applied through the side port of the pipette holder to prevent contamination of the pipette tip by bath debris. The pipette is then submerged in bath and lowered using the micromanipulator until the tip just touches the cell of interest. The positive pressure is then relieved, and subsequent application of slight negative pressure usually results in formation of a high-resistance (1–100 gigohms) seal between the pipette tip and cell membrane. For glial cells, there is a rather small margin for error in lowering and manipulating the patch pipette because of the very flat morphology of the cells. One can monitor membrane contact either visually, using a high-power microscope objective, or electrically, by looking for a slight increase in pipette resistance indicative of membrane contact. Following gigaseal formation, the whole-cell recording mode is achieved by rupturing the membrane beneath the patch pipette with slight suction.

For most experiments both cells are voltage clamped at −70 mV. Usually, this potential is slightly negative to the neuron's resting potential, but

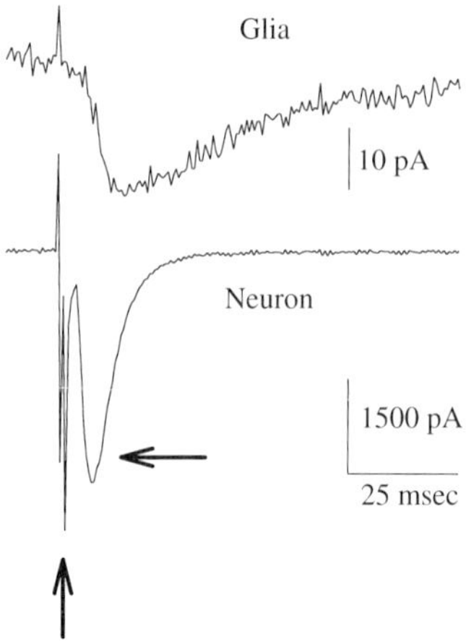

FIG. 2. Dual whole-cell voltage-clamp recording from an excitatory microculture neuron (lower trace) and surrounding glia (upper trace). The vertical arrow indicates the fast transient currents caused by briefly (1.5 msec) pulsing the neuronal command potential to +20 mV from the holding potential of −70 mV. This stimulus effectively triggers autaptic transmitter release, as indicated by the inward autaptic current in the neuron (horizontal arrow). Glia also respond to the stimulus with an inward current that we have found is largely due to glutamate transport [see S. Mennerick, A. Benz, and C. F. Zorumski, *J. Neurosci.* **16,** 55 (1996), and S. Mennerick and C. F. Zorumski, *Nature* **368,** 59 (1994)].

often slightly positive to the glial resting potential. Neurons are stimulated by a brief (1.5 ms) voltage pulse to +20 mV. The resulting neuronal currents include very fast transient currents, reflecting the capacitive and ionic currents that flow during a step in the command potential (Fig. 2). With single-neuron excitatory microcultures, these transient currents are followed by an inward excitatory autaptic current (Fig. 2). By a number of pharmacological and kinetic criteria, these currents can be shown to be mediated by both α-amino-3-hydroxy-5-methyl-4-isoxazolepropionic acid (AMPA) and *N*-methyl D-aspartate (NMDA) subclasses of glutamate receptors. Evoked autaptic transmitter release has been shown to be dependent on an action potential that escapes voltage clamp in the axon of the neuron, because stimulation no longer elicits an evoked autaptic response when the sodium-channel blocker tetrodotoxin is present in the extracellular bath.[16] We typically stimulate neurons at a frequency <0.1 Hz to avoid frequency-dependent facilitation and depression of synaptic responses.

[16] J. M. Bekkers and C. F. Stevens, *Proc. Natl. Acad. Sci. USA* **88,** 7834 (1991).

Results: Glutamate Transport during Synaptic Glutamate Release

To our initial surprise, we found that underlying glial cells also respond with inward currents to stimulation of the neuron (Fig. 2). Further characterization suggests that these currents are largely (~85% of the peak current)[12] due to the electrogenic transport of glutamate released synaptically from the neuron. The evidence that these glial currents are largely due to electrogenic glutamate uptake is as follows. First, the current–voltage relationship of the glial currents is similar to that reported with exogenous glutamate applications to cells expressing transporters.[10,17] Second, application of glutamate transporter substrates, such as *threo*-3-hydroxyaspartate (THA) or L-*trans*-pyrrolidine-2,4-dicarboxylic acid (PDC), in the extracellular bath induces inward currents in the glial cells and suppresses the glial responses to neuronal stimulation.[12,17] Third, the glial currents are sensitive to extracellular sodium removal, consistent with known sodium dependence of glutamate transport.[10,17] Fourth, when the glial patch pipette contains glutamate, depolarization of glia can induce slow, small glutamatergic responses in the overlying neuron, consistent with known reversal of transporter function at depolarized membrane potentials.[17,18] Finally, under certain experimental conditions, reduction of the glial currents by reversing the glial membrane potential can prolong neuronal autaptic currents, consistent with the idea that depressing the glial current prolongs the lifetime of synaptically released glutamate.[17,19] Pharmacological transport blockade or substitution of lithium for sodium in the extracellular bath can cause a similar prolongation of synaptic currents.[17]

To define the conditions under which glutamate transport is important in clearing synaptically released glutamate, we used changes in the decay of the AMPA–receptor-mediated autaptic current of the neuron to assay the time course of the glutamate transient following release, and we used pharmacological and nonpharmacological methods to inhibit glutamate transport. THA, PDC, and sodium removal were used as nonselective inhibitors of both neuronal and glial transporters.[20] In contrast, selective inhibition of glial glutamate transport was achieved by varying the glial cell membrane potential using the whole-cell recording method outlined above and by exploiting the known voltage dependence of glutamate transport.[10]

[17] S. Mennerick and C. F. Zorumski, *Nature* **368,** 59 (1994).

[18] M. Szatkowski, B. Barbour, and D. Attwell, *Nature* **348,** 443 (1990).

[19] S. Mennerick and C. F. Zorumski, *J. Neurosci.* **15,** 3178 (1995).

[20] J. L. Arriza, W. A. Fairman, J. I. Wadiche, G. H. Murdoch, M. P. Kavanaugh, and S. G. Amara, *J. Neurosci.* **14,** 5559 (1994).

The results of these studies suggest that glutamate transport plays a detectable role in clearing synaptic glutamate only when many transmitter quanta (transmitter vesicles) are released during a stimulus. Thus, glutamate transport inhibition has no detectable effect on the time course of miniature autaptic currents (believed to reflect the postsynaptic response to a single quantum of transmitter), or evoked autaptic currents made up of only a few quanta, even when postsynaptic receptors have been pharmacologically sensitized to residual glutamate.[19] These results contrast with those obtained at the neuromuscular junction, where cholinesterase inhibition prolongs endplate currents of small quantal content.[21] At microculture glutamate synapses following release of a single glutamate quantum, diffusion alone is presumably responsible for quickly clearing transmitter to levels below those detectable by postsynaptic receptors, although a very fast role for transporters in buffering transmitter has also been proposed.[22] In contrast to the lack of effect of transport inhibition on small autaptic currents, large evoked autaptic currents, made up of perhaps several thousand transmitter quanta, can be prolonged by nonselective (neuronal and glial) transport inhibition or by selective glial transport inhibition.[19] Furthermore, our data suggest that even in the absence of transport inhibition, the glutamate transient following the release of many quanta is significantly prolonged compared with the glutamate transient following a miniature autaptic current.[19] This suggests that glutamate transporters, again unlike endplate cholinesterase, are too slow to limit the glutamate transient of a large multiquantal autaptic current to that of a single glutamate quantum. This slowness may reflect slow transporter turnover kinetics and may also reflect the diffusion time required for synaptic glutamate to reach glial transporters. Interestingly, the similarity between the effects of broad-spectrum pharmacological transport inhibitors and selective glial transport inhibition suggests that glial cells in these cultures are the primary transporters involved in removing glutamate from the vicinity of postsynaptic receptors, despite the presumed presence of neuronal transporters (Fig. 1B).

We have also found evidence that microculture glia have functional AMPA-type glutamate receptors.[12] Although these receptors mediate only a small fraction of the response of glia to synaptic glutamate release, AMPA–receptor current dominates the response to exogenous applications of high glutamate concentrations.[12] The functional role of these receptors remains to be elucidated.

[21] H. C. Hartzell, S. W. Kuffler, and D. Yoshikami, *J. Physiol.* (*Lond.*) **251,** 427 (1975).
[22] G. Tong and C. E. Jahr, *Neuron* **13,** 1195 (1994).

Summary and Future Directions

We have found microcultures of neurons and astrocytes to provide a useful experimental preparation for studying the role of glutamate transporters in synaptic physiology. The preparation balances experimental control over a number of important variables with an environment that mimics many fundamental features of synapses *in situ*.

Several avenues exist for further use of microcultures in the study of transporters. Although our studies have focused primarily on the role of transporters in the few milliseconsds subsequent to individual synaptic events, it is likely that microcultures may prove useful for studying slower effects of transporters in maintaining or modulating tonic ambient glutamate levels. Another avenue for further research is to understand which of the several cloned glutamate transporters may be responsible for the observed physiological effects of transporters in microcultures. Finally, because a number of neuronal transmitter phenotypes have been shown to survive in the microculture environment, it may be possible to extend studies of transmitter transport to other transmitter phenotypes and to other nervous system regions.

Acknowledgments

The authors thank P. Dhond, N. C. Danbolt, and K. Isenberg for collaborating in the immunohistochemistry shown in Fig. 1, and thank A. Benz for help in development and maintenance of the microcultures. Work in the authors' laboratory is supported by a McDonnell fellowship (S.M.), by NIMH Research Scientist Development Award MH00964, grants MH45493 and AG11355, and a fellowship from the Bantly Foundation.

Section VII

Microdialysis and Electrochemical Measurements

Articles 44 through 49

[44] Voltammetric Studies on Kinetics of Uptake and Efflux at Catecholamine Transporters

By KRISSTINA DANEK and JAY B. JUSTICE, JR.

Introduction

The focus of this chapter is a description of an application of a well-established electrochemical method, rotating disk electrode voltammetry, to the measurement of transporter kinetics. The time course of uptake and efflux of catecholamines in stable cell lines expressing the human norepinephrine and dopamine transporters (LLC-NET and LLC-DAT cells) can be followed using this method. The induced efflux experiment, which allows for the kinetic analysis of substrates not oxidized at the applied potential and provides information on both inward and outward substrate transport in the same assay, will also be introduced.

The plasma membrane catecholamine transporters are integral membrane proteins whose function is to terminate synaptic transmission by removal of the respective neurotransmitter from the extracellular fluid. The transport of dopamine (DA) and norepinephrine (NE) are of particular interest because the dopamine transporter is the site of action of many drugs of abuse, including cocaine and amphetamine, and alterations in norepinephrine transmission have been associated with depressive disorders.[1,2] Typically, catecholamine transport studies have relied on radioactively labeled substrates. These experiments involve measuring the accumulation of the radiolabeled substrate in a cell or tissue preparation after a period of incubation. A single measurement is taken and is used to represent the average rate of transport over the incubation time period. This interval, however, may provide sufficient time for the substrate not only to collect intracellularly, but to undergo outward transport as well.

The rotating disk voltammetry (RDV) method, which has been developed by Schenk *et al.*[3–6] to study catecholamine kinetics, eliminates the use of radioactivity. This allows a greater range of substrates to be examined. All substrates with a catechol functionality are easily oxidized, and those

[1] W. E. Bunney, Jr., and J. M. Davis, *Arch. Gen. Psychiatry* **13,** 509 (1965).
[2] P. Willner, "Depression: a Psychobiological Synthesis." Wiley-Interscience, New York, 1985.
[3] J. S. McElvain and J. O. Schenk, *Biochem. Pharmacol.* **43,** 2189 (1992).
[4] S. M. Meiergerd, J. S. McElvain, and J. O. Schenk, *Biochem. Pharmacol.* **47,** 1627 (1994).
[5] S . M. Meiergerd and J. O. Schenk, *J. Neurochem.* **63,** 1683 (1994).
[6] S. M. Meiergerd and J. O. Schenk, *J. Neurochem.* **62,** 998 (1994).

0076-6879/98 $25.00

with a single ring hydroxyl group are also oxidizable, although at a somewhat higher potential. Rotating disk electrode voltammetry has been applied to study DA kinetics in tissue homogenates and synaptosomal preparations,[3–6] and to stable cell lines expressing hNET.[7] This method allows for initial rate measurements and also provides a time course of uptake. In these studies, in contrast to those which measure accumulated radioactivity, the change in DA concentration in the extracellular medium is measured. The rotating disk electrode continuously generates an oxidation current proportional to the concentration in the external medium as uptake or efflux proceeds. The rotating electrode also rapidly mixes the suspension, minimizing effects of diffusion on transport measurement.

Some complications are encountered when performing *in vitro* studies in tissue preparations. For example, more than one catecholamine transporter may be present, and cellular machinery for the accumulation, storage, and metabolism of neurotransmitters can alter results. These complications can be avoided by using a cell line transfected with the transporter DNA of interest. In the past few years, the hNET and hDAT have been cloned and their amino acid sequences determined.[8,9] They have also been transfected into LLC-PK_1 cells, a stable cell line derived from porcine kidney epithelium[10] (ATCC Rockville, MD, CL-101). In our lab, studies using the LLC-NET cells have been performed and the Michaelis–Menten parameters for zero-*trans* uptake of DA and NE obtained.[7] In order to be able to study the transport kinetics of a wider range of substrates, which are not electroactive, induced efflux measurements have been made. For example, *l*- and *d*-amphetamine are not electroactive compounds, but by studying their effect in inducing efflux of electroactive DA, information can be obtained about the transport kinetics of the effluxing agent. Comparisons of the efflux induced by a range of structural variations in these substrates at the NET and DAT can also provide information about the relative importance of different structural features in transport.

Methods

Apparatus

Rotating disk electrode voltammetric measurements are made using a 3 mm diameter glassy carbon working electrode driven by an AFMSRX

[7] W. B. Burnette, M. D. Bailey, S. Kukoyi, R. D. Blakely, C. G. Trowbridge, and J. B. Justice, Jr., *Anal. Chem.* **68,** 2932 (1996).

[8] T. Pacholczyk, R. D. Blakely, and S. G. Amara, *Nature* **350,** 350 (1991).

[9] B. Giros, S. El Mestikawy, N. Godinot, K. Zheng, H. Han, T. Yang-Feng, and M. G. Caron, *Mol. Pharmacol.* **42,** 383 (1992).

[10] R. N. Hull, W. R. Cherry, and G. W. Weaver, *In Vitro* **12,** 670 (1976).

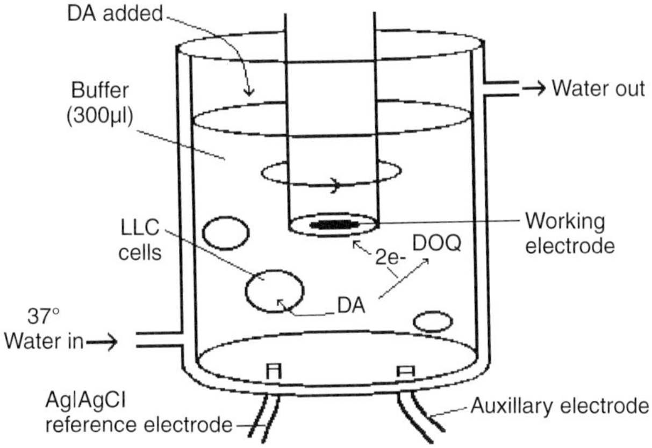

FIG. 1. The experimental apparatus. The working electrode is rotated at 4000 rpm and the potential is held at +450 mV vs Ag|AgCl. The volume is 300 μl.

Analytical Rotator System (Pine Instrument Company, Grove City, PA). The electrochemical cell is equipped with a Pt auxiliary electrode and an Ag|AgCl reference electrode (Fig. 1). An LC-4 amperometric detector (Bioanalytical Systems, Lafayette, IN) serves as a potentiostat and the output current is amplified by a Keithley Model 427 amplifier (Keithley Instruments, Cleveland, OH) using a 300 msec time constant. For all experiments, the electrode is rotated at 4000 rpm and the applied potential is +450 mV vs Ag|AgCl, allowing for the oxidation of DA or NE. The temperature is held at 37° and a 95% O_2/5% CO_2 mixture is directed over the cell suspension in the electrochemical cell. Data is collected at a frequency of 64 Hz, averaged, and recorded at a frequency of 4 Hz on a 486 PC by means of a DT 2801A interface board (Data Translation, Marlboro, MA) controlled by Origin data acquisition software (MicroCal Software, Northhampton, MA).

The working electrode is concentrically rotated in solution, creating convection paths flowing first perpendicular toward the electrode and then radially outward past its surface. This enables constant replenishment of the detected species to the electrode surface. The limiting current, I_L, produced by the oxidation or reduction of electroactive species at the electrode surface can be described by the Levich equation[11]:

$$I_L = 0.62nFACD^{2/3}\nu^{-1/6}\omega^{1/2} \tag{1}$$

[11] Y. V. Pleskov and V. Y. Fillinovski, "The Rotating Disk Electrode." Plenum, New York, 1976.

where I_L is in milliamperes, n is the number of electrons transferred per molecule, F is Faraday's constant (96,487 C/mol e^-), A is the electrode surface area (cm^2), C is the analyte concentration ($mmol/cm^3$), D is the analyte diffusion coefficient (cm^2/sec), ν is the kinematic viscosity of the solution (cm^2/sec), and ω is the angular velocity of the rotating disk (sec^{-1}). From Eq. (1) it is evident that RDEV can be used to measure concentration changes of the electroactive species at the electrode surface.

Cell Preparation

The LLC-NET and PK_1 cells, nontransfected parent cells, are propagated from cell lines obtained form Dr. Randy Blakely (Department of Pharmacology, Vanderbilt University, Nashville, TN) and the LLC-DAT cells were obtained from Dr. Michael Owens (Department of Psychiatry/Behavioral Sciences, Emory University, Atlanta, GA). The cells are maintained in Eagle's minimum essential medium (Fisher Scientific, Pittsburgh, PA) supplemented with 10% fetal bovine serum (Hyclone Laboratories, Logan, UT), 2 m*M* L-glutamine, and 10 ml 10,000 U/ml penicillin–streptomycin (Life Technologies, Grand Island, NY) to provide the essential amino acids, vitamins, and salt.[12] The cells are grown in an incubator which controls the 5% CO_2 (v/v) and 37° environment needed to maintain the pH for optimum cell growth, approximately 7.4.[12] An inverted microscope is used to monitor the daily cell growth and for detection of any morphological changes that could signal deterioration in the culture. All cell line maintenance and propagation occurs under a laminar flow hood. The cells grow exponentially so when their growth ceases or is reduced or when all the available growth space, the face of the flask, is occupied, they must be divided or subcultured. The common method involves removal of the old medium, dissociation of the cells with trypsin, and diluting as desired in new medium into Corning flasks, Falcon cell culture dishes, or Corning Cell Wells.

Procedure

The following typical procedure is performed when running an experiment. The temperature of the electrochemical cell is maintained at 37° by circulating water through silastic tubing surrounding the cell during all experiments. The working electrode is polished using alumina powder and the reference electrode is coated with silver chloride by connecting the Ag wire to the positive terminal of a 12-V lantern battery and the platinum wire auxiliary electrode to the negative terminal. Both wires are exposed

[12] R. I. Freshney, "Culture of Animal Cells." Wiley-Liss, New York, 1994.

to a 0.1*M* HCl solution and current starts to flow, with hydrogen evolving at the platinum wire or cathode and silver oxidizing to silver ion at the anode. The silver ions quickly deposit on the silver wire as silver chloride. Once the coating of the silver wire is sufficient, there is a visible slowing of the rate of evolution of the hydrogen gas at the cathode.

The old medium is aspirated from the dish of cells and they are washed with 20 ml of room temperature filtered Krebs–Ringer–HEPES (KRH) buffer.

Krebs–Ringer–HEPES Buffer

120 m*M* NaCl	7.01 g
4.7 m*M* KCl	0.35 g
2.2 m*M* $CaCl_2$	0.32 g
1.2 m*M* $MgSO_4$	0.14 g
1.2 m*M* KH_2PO_4	0.16 g
10 m*M* HEPES	2.4 g
Water to adjust to pH 7.4.	1 liter

The cells are scraped from the 150 mm dish with two 2.5 ml additions of KRH buffer into a 12 × 75 mm clear glass test tube. They are then centrifuged at 2500 rpm (1000*g*) for 2 min in a Dynac variable speed centrifuge (Becton, Dickinson and Co., Cockeysville, MD). Higher gravity centrifugation may cause cell damage.[12] The supernatant is aspirated from the resulting cell pellet, 1200 μl of new KRH buffer is added, and the suspension is transferred to a 1.5 ml centrifuge tube. It is then placed in a 37° water bath under O_2. A 300 μl aliquot of this suspension is placed into the electrochemical cell and the cell is raised so that the working electrode is positioned just below the surface of the solution. A constant stream of a 95% O_2/5% CO_2 (v/v) mixture is blown over the electrochemical cell. The working electrode rotation is started, the potential of 450 mV vs the Ag|AgCl reference electrode is applied, and the current is monitored on the computer. After waiting approximately 3–5 min for stable baseline acquisition, 6 μl of DA, NE, or any electroactive substrate dissolved in an unbuffered electrolyte solution, KR, is added via pipette to make a 1 μ*M* solution concentration in the electrochemical cell.

KR Solution

120 m*M* NaCl	7.01 g
4.7 m*M* KCl	0.35 g
2.2 m*M* $CaCl_2$	0.32 g
1.2 m*M* $MgSO_4$	0.14 g
Water to	1 liter

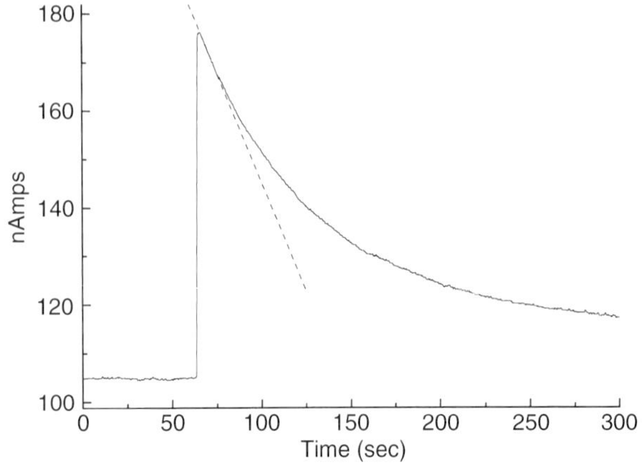

FIG. 2. Uptake of DA at the hNET. At 65 sec, 6 μl of a DA solution is added to 300 μl LLC-NET cells/KRH buffer to make a 1 μM solution. Oxidation of DA at the electrode surface is seen as the sharp rise in current. The subsequent decrease in current is due to uptake of DA at the hNET. The linear regression used to estimate initial rate of DA uptake is also shown (dotted line).

A sharp rise in the oxidiative current is observed (Fig. 2) as DA is oxidized to the *o*-quinone. The decreasing current is then recorded for an additional 7 min as uptake occurs until a new steady state is reached. For those experiments in which the initial rate of uptake is what is desired, the applied potential is turned off and the working electrode rotation stopped. For efflux experiments, 6 μl of a second, nonelectroactive substrate is added once steady state has been reached (a typical final concentration of 10 μM). The efflux of DA, observed as an increase in oxidative current, is monitored for an additional 5–7 min. The potential is then turned off, the rotation of the electrode is terminated, and the electrochemical cell is lowered so that the suspension can be transferred to a test tube. At this point, the cell suspension is diluted with KRH buffer and an aliquot is used to quantitate cell number with a hemocytometer under the inverted microscope.[12] Because the cell numbers can be inaccurate due to the formation of cell clumps, the data are expressed as the ratio of the initial rate of efflux to initial rate of uptake. This normalizes the data for the number of cells and level of transporter expression in each assay. Finally, the electrochemical cell is rinsed with KRH buffer and H_2O before the next 300 μl aliquot of cells is added. The working electrode is also polished, if needed, between each assay.

Analyzing Results

All data are analyzed using Origin software using a semiautomated graphical interface written in-house. The data is first converted from voltage vs time to current vs time. Baseline current from over the 10 sec before catecholamine addition is then subtracted from the raw current vs time record. A linear regression is next performed on the uptake profile using an interval of 10–20 sec beginning 1 sec after the point of maximum change in oxidative current (Fig. 2). Typically, an interval of 10 sec is used. However, there are particular assays where, in order to obtain an acceptable r^2 value, the regression is extended over a longer time. The data is converted from current vs time to concentration vs time using a calibration factor obtained from extrapolating the regression line back to the time of maximum rate of change in the oxidation current. The initial rate of substrate clearance, v_o, can then be calculated from the slope of the regression line and the calibration factor. These rates are normalized to the cell count and expressed in units of 10^{-18} mol/min/cell. Four experimental runs are averaged and expressed as mean ± SEM. To obtain Michaelis–Menten kinetic parameter estimates for the substrate of interest, initial rates are fit to the Eadie–Hofstee equation:

$$v_o = V_{max} - K_m(v_o/[N]) \tag{2}$$

where the y intercept of the line is V_{max}, the maximal rate of transport, the slope is K_m, the Michaelis–Menten constant, and N is the initial concentration of substrate. Then, with these estimated K_m and V_{max} values, a nonlinear least squares curve fitting algorithm in Origin is used to fit the experimental data to the Michaelis–Menten function:

$$v_o = (V_{max} \times [N])/(K_m + [N]) \tag{3}$$

Nontransfected parent LLC-PK_1 cells are used as controls. On addition of substrate, these cells typically show negligible decrease in the oxidative current. However, for some substrates, for example *m*-tyramine and *p*-tyramine, a measurable decrease in current is seen following substrate addition to an LLC-PK_1 cell suspension, and so the control rates are subtracted from the rates measured in the transfected cells. This decrease may be due to polymeric filming of the electrode surface following oxidation.[13] Figure 3 shows the initial rates of uptake for 1 μM solutions of various catecholamine analogs normalized to the rate of DA uptake.

For the efflux experiments, a baseline subtraction is first performed using 60 sec of raw voltage vs time data prior to the second substrate

[13] A. J. Fry, "Synthetic Organic Electrochemistry." John Wiley & Sons, New York, 1989.

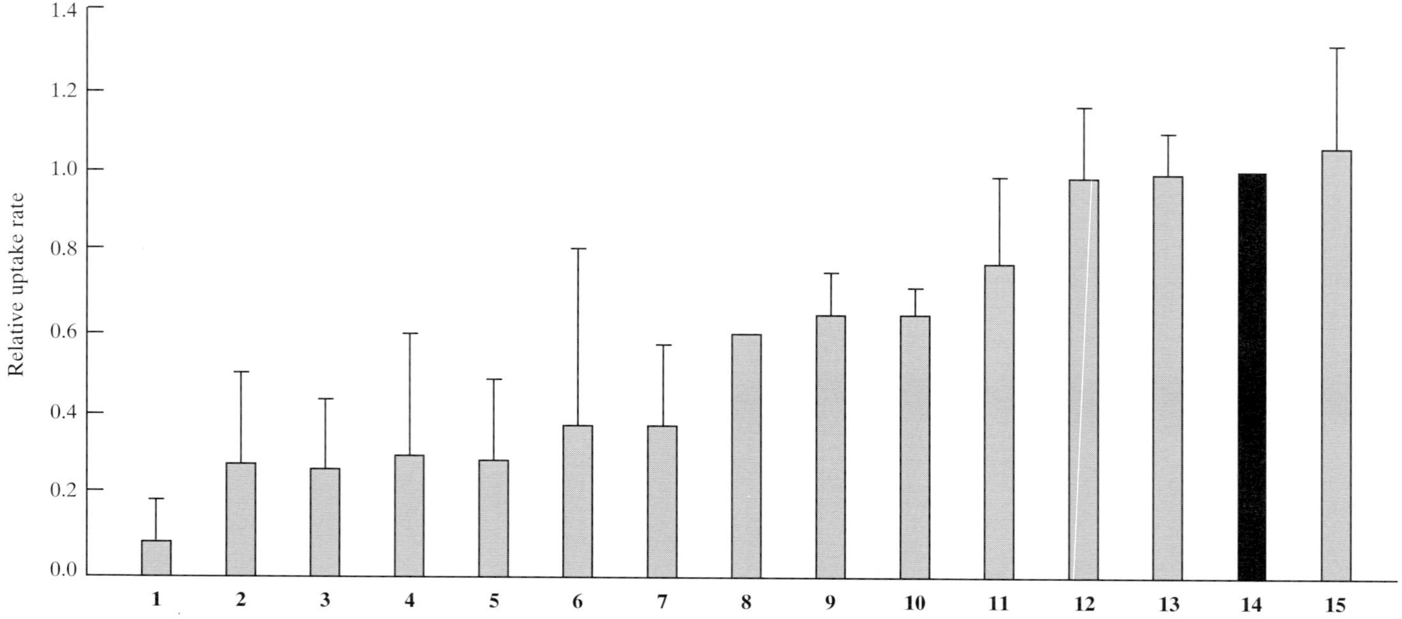

FIG. 3. Initial rates of transport for 1 μM solution of given structure (n = 2–5). Applied potentials varied from +.45 V to +.9 V vs Ag|AgCl. [Reprinted with permission from W. B. Burnette, M. D. Bailey, S. Kukoyi, R. D. Blakely, C. G. Trowbridge, and J. B. Justice, Jr., *Anal. Chem.* **68,** 2932 (1996). Copyright (1996) American Chemical Society.]

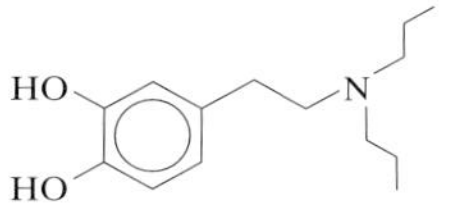

1

dipropyldopamine

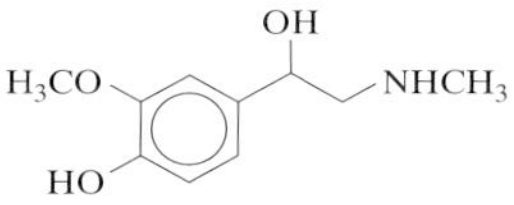

2

metanephrine

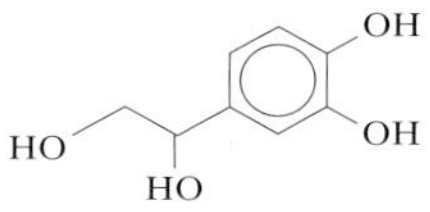

3

3,4-dihydroxyphenylglycol

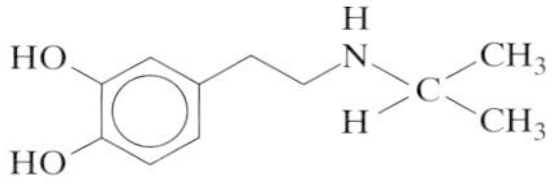

4

isoproterenol

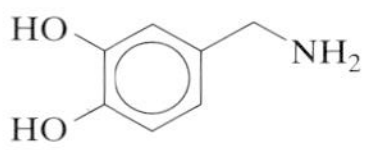

5

3,4-dihydroxybenzylamine

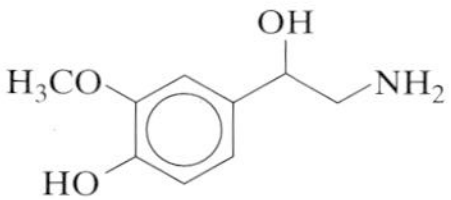

6

normetanephrine

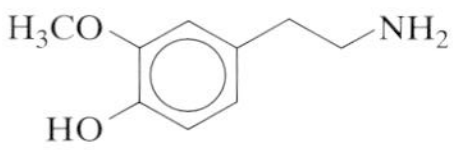

7

3-methoxy-4-hydroxyphenethylamine

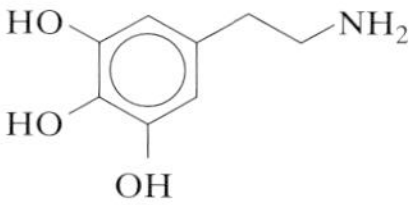

8

5-hydroxydopamine

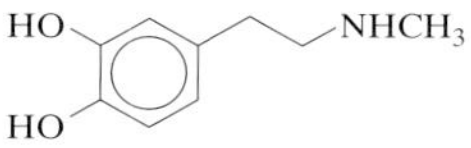

9

N-methyldopamine

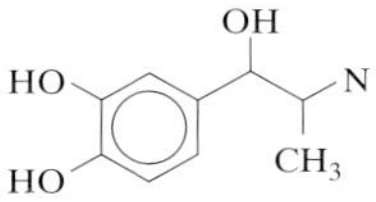

10

3,4-dihydroxynorephedrine

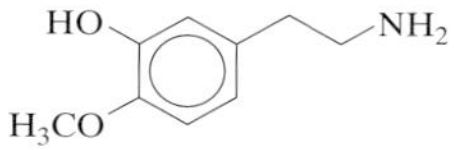

11

4-methoxy-3-hydroxyphenethylamine

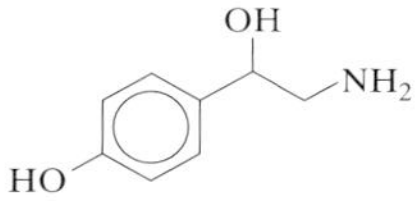

12

octopamine

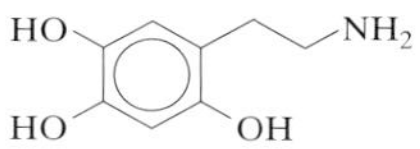

13

6-hydroxydopamine

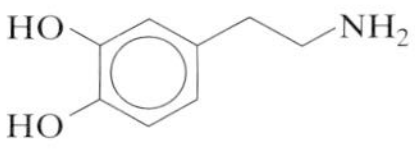

14

dopamine

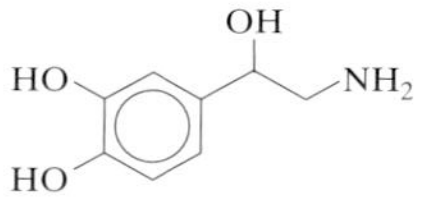

15

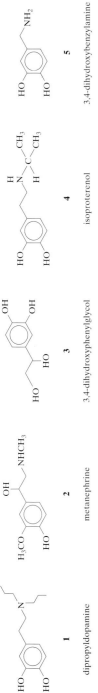

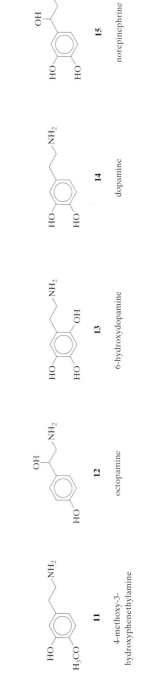

norepinephrine

addition. The calibration factor is used to convert voltage to concentration. An initial rate of efflux is then determined from a linear regression on the first seconds of data following the nonelectroactive substrate addition. As with the uptake data, the exact length of time included in the regression analysis depends on the value of r^2 but is typically 15 sec. The data are then expressed as the ratio of the rate of efflux to the rate of uptake for each individual run. Ratios from individual runs are averaged to produce a mean ± SEM.

Figure 4 shows the clearance of 1 μM DA followed by 10 μM *l*-amphetamine induced efflux of DA in hNET transfected cells. Although the transport of *l*-amphetamine or other effluxing agents is not directly observed in these experiments, their ability to stimulate the efflux of DA demonstrates that they are transported. The efflux curve can be best understood from a simple two-substrate model of a transporter (Fig. 5). Binding of cotransported sodium or chloride is not depicted. Transport is accomplished by a transporter, T, which can exist in empty or bound states facing inward or outward. Once transport of the preloaded substrate, DA, has reached a steady state, addition of a second substrate, S, on the external side of the membrane induces efflux by competing for the transporter, T_{out}, at the external face of the membrane. This competition reduces the rate of inward transport of DA while initially having no effect on outward DA transport,

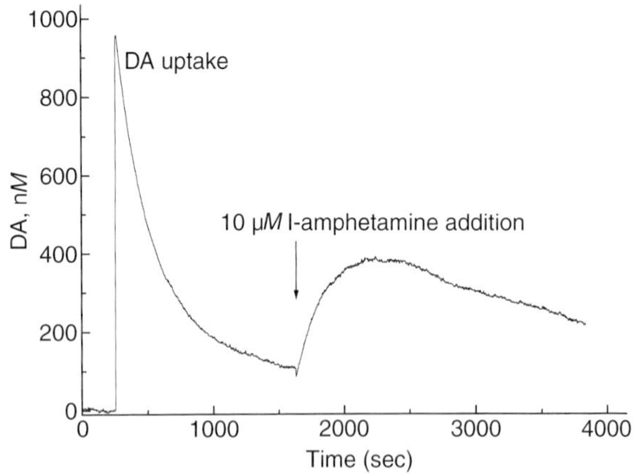

FIG. 4. Uptake of 1 μM DA at the NET followed by 10 μM *l*-amphetamine induced efflux. At 30 sec, 6 μl of DA solution is added to 300 μl LLC-NET cells/KRH buffer at 37° to make a 1 μM solution. Uptake occurs and a steady state DA level of approximately 120 nM is reached. Six microliters of an *l*-amphetamine solution is then added to make a solution concentration of 10 μM. The immediate efflux of DA is observed as an increase in oxidative current which continues through a maximum and then proceeds to a new steady state.

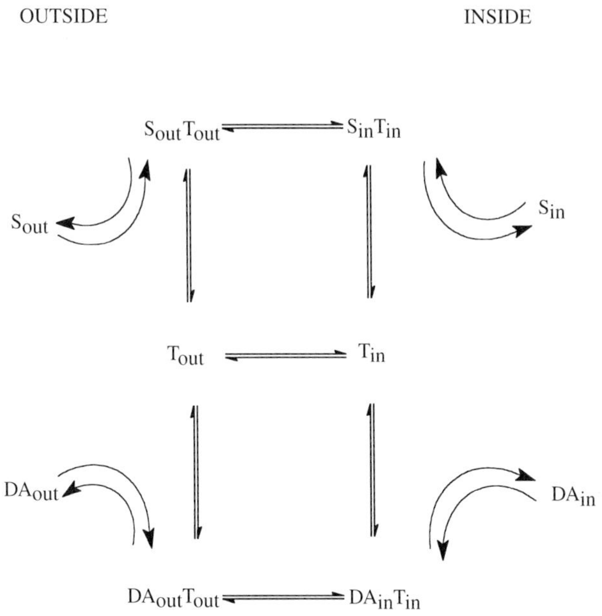

FIG. 5. The two-substrate model of the transporter.

since there is no S present at the inside face. A net outward flux of DA results. This is observed as an increase in external concentration in the early rising phase of the DA efflux curve induced by *l*-amphetamine (Fig. 4).

As *l*-amphetamine is transported inward it dissociates from the transporter, allowing the much higher concentration of DA to bind. At first, there is very little *l*-amphetamine on the inside of the cell to compete with DA for outward transport. However, as it accumulates, and the concentration of DA inside decreases, the *l*-amphetamine competes with DA for binding to the empty transporter, T_{in}, and DA efflux decreases. Simultaneously, DA concentration in the extracellular medium is increasing, shifting the extracellular competition for T_{out}, which increases the inward flux of DA. The result is a maximum in external DA concentration when the increasing inward and decreasing outward DA fluxes are equal. Beyond this maximum, net DA transport, driven by the sodium gradient, is inward as the system approaches a new steady state. In this particular example, the rate of uptake of DA is 2.17×10^8 amol/min and the rate of efflux of DA is 9.94×10^7 amol/min, resulting in an efflux/influx ratio of 45.8%.

In summary, the efflux experiment is the observation of the relaxation in substrate concentrations from an instantaneous perturbation away from their equilibria with respect to the steady state ion gradients. The ion gradients remain constant throughout and drive the direction of relaxation of substrate concentrations back toward equilibrium with respect to the ionic gradients. The mechanism of the transporter results in various substrate concentrations having maxima or minima over the time course of the relaxation. The RDE (rotating disk electrode) method allows the time course of these changes to be followed, which may eventually allow new insight into catecholamine transporter structure and function.

Acknowledgments

This research was supported by the National Science Foundation, Grant IBN 94-12703, and by RSDA K02-DA00179 from the National Institutes on Drug Abuse.

[45] Resolution of Biogenic Amine Transporter Kinetics by Rotating Disk Electrode Voltammetry: Methodology and Mechanistic Interpretations

By CYNTHIA EARLES, HOLLIE WAYMENT, MITCHELL GREEN, and JAMES O. SCHENK

Introduction

The movement of molecules and ions across biological membranes has been an object of study since at least the late 1930s, perhaps beginning with the studies reviewed and described in the monograph "The Permeability of Natural Membranes" by Davson and Danielli in 1943.[1] The main challenge of this area of research can be appreciated from the view that a biological membrane serves to isolate the interior of a cell from its external environment, but because the cell relies on external chemical constituents for life, it must be able to discriminate between what it needs from what it does not need and find a way to move the needed chemicals inside for use. At this writing there seem to be only a few processes that the cell uses to achieve the goal outlined above. That is, membranes evolved to contain within them two broad classes of "conduits" to allow inorganic ions and molecules to cross the membrane: pores (or channels) and transporters. A pore or channel can be viewed as a dynamic transmem-

[1] H. Davson and J. F. Danielli, "The Permeability of Natural Membranes." Cambridge University Press, Cambridge and New York, 1943.

0076-6879/98 $25.00

brane spanning hole which is sometimes open and sometimes closed, providing a path for ions to move across the membrane under diffusion control. The opening or closing of these pores is gated by complex biological and biophysical controls, but they are either in a state of conducting ions or in a state of not conducting ions. The energy force driving ion movements mediated by pores (or channels) is defined by the concentration gradient of that ion across the membrane, and the ion moves in the direction from high concentration to low concentration. Selectivity of the pore (or channel) for a given ion is defined in a complex way by the size of the solvated ion, the size of the naked ion, the energies required for and/or kinetics of removing waters of hydration from the solvated ion, and the energies required for and/or kinetics of replacing the solvating sites of the ion with ligands from the membrane constituents of the pore or channel.[2]

However, transporters are also membrane-spanning proteins, but appear to have evolved to exploit a more complex means of moving substrates across membranes. They serve to move larger chemical species, such as organic molecules (many with similar sizes and molecular weights such as glucose, the amino acids, and biogenic amine neurotransmitters), across the membrane. They cannot use molecular size, energy and geometry of hydration, or the kinetics of ligand exchange for selectivity. Instead they use molecular recognition, similar to enzymes and antibodies, to obtain chemical selectivity and employ more complex energy sources such as coupling the flow of one substrate to that of another and/or to the hydrolysis of ATP. The energy of activation of transporter activity is greater than that for diffusion. It has been observed also that transport of the natural substrate is inhibited competitively by structural analogs of the substrate, whereas protein-modifying agents noncompetitively inhibit transporter activity. In addition, transporters have properties quite distinct from those of pores or channels in that many of them concentrate the substrate inside the cell by transporting it against a concentration gradient. That is, the movement of substrate at a given concentration on one side of the membrane is not influenced by the concentration of substrate on the other side of the membrane. In addition, they undergo conformational changes during the course of the translocation event. Because of the observation of the induction of counterflow by substrate, they are thought to exist in at least two states of activity, an inwardly biased action and an outwardly biased action. This implies that the recognition site of a transporter has associations with the inside as well as the outside of the membrane. The image of a

[2] G. Eisenman, *J. Membrane Biol.* **76,** 197, 1983.

"gated pore" is used for the transporter[3-5] where the transport of substrate associated with a recognition site on one side of the membrane is not influenced by the concentration gradient of substrate across the membrane. Once on the other side of the membrane, the substrate is again shielded from the transmembrane substrate concentration gradient. Thermodynamic and kinetic analyses preclude a ferry-boat mechanism, whereby a transporter complexes with the substrate on side of the membrane and shuttles it to the other side of the membrane by diffusion.[5] Thus, instead of the model of the transmembrane-spanning hole as has been used for the pore or channel, the image of the "boa constrictor's lunch" applies here, where the substrate appears as a bolus in a tube constricted at both ends. However, the movement of the bolus is directional.

The differences between the mass transport functioning of pores (or channels) and transporters and their kinetic mechanistic characteristics have been under study at least since the 1950s to 1960s, and three monographs on the subject have appeared since 1967.[3-5] It has generally been found that pores or channels have two kinetic conditions, open or closed, and that the velocity of movement of the ions as a function of ion concentration ([ion]) follows a hyperbolic kinetic description of the same form as Michaelis–Menten kinetics, but including terms for [ion] which result in a diminution of velocity toward the membrane side of increasing ion concentration. In contrast, transporters can be described by three or four state models in which the transporter itself is included as a "reactant" in the transport process (see Stein[4]), and transmembrane concentrations of substrate influence transport kinetics, not as a function of the transmembrane concentration gradient, but by direct binding reactions with the transporter. Although the study of the function and mechanism of transporters is still a challenging area of research, much progress has been made in the understanding of the mechanistic differences between pores or channels and transporters and more than 150 different transporter systems had been characterized by 1986. Kinetic mechanistic studies of the transporters have also been the subject of a previous volume in this series in 1989.[6]

In the neurosciences transporters are viewed as important entities in neuronal communication because they frequently operate to clear neurotransmitters from the synaptic cleft and other extracellular regions following

[3] W. D. Stein, "The Movement of Molecules across Cell Membrane." Academic Press, New York, 1967.

[4] W. D. Stein, "Transport and Diffusion across Cell Membranes." Academic Press, San Diego, 1986.

[5] W. D. Stein, "Channels, Carriers, and Pumps: An Introduction to Membrane Transport," Academic Press, San Diego, 1990.

[6] S. Fleischer and B. Fleischer, Eds. *Methods Enzymol.* **171** (1989).

neurotransmitter release during an action potential. Thus, an understanding of how transporters for neurotransmitters operate is important in understanding the timing and intensity of neuronal chemical signaling. In addition, these transporters have been shown to be important as targets for drugs treating depression and for some drugs of abuse, most notably the amphetamines and cocaine.

Transporter Kinetics and Their Unique Challenges

Transporter activity can be viewed as "catalyzing" the movement of substrates across membranes. The cycle of events includes the binding of the substrate to the transporter protein, followed by the substrate–transporter complex undergoing a change in conformation analogous to a transition state in enzymes, except that covalent bonds are probably not made or broken. As a result of these interactions the substrate is exposed to the opposite side of the membrane. The transporter protein then relaxes to its former state, ready to undergo the cycle again with another substrate molecule. The kinetic behavior of this activity can be described by various forms of the Michaelis–Menten expression but must include all of the events described. This condition is more complicated than the cases of pores (or channels) for ions or enzymes because the theory has to account for transport activity in two directions (for example, counterflow) and include terms for the reorientation (relaxation) of the transporter. As a result, the description of the kinetic activity of the transporter must account for real binding and transport events in two directions. In enzymology, the action of enzymes in the reverse direction is generally ignored because enzymes frequently catalyze reactions in only one direction or are experimentally set up to ensure this condition. Thus, the simple kinetic description of the action of a simple transporter is given by[4,5]

$$\nu_{o\to i} = \frac{K[\mathrm{S}]_o + [\mathrm{S}]_o[\mathrm{S}]_i}{K^2R_{oo} + KR_{o\to i}[\mathrm{S}]_o + KR_{i\to o}[\mathrm{S}]_i + R_{ee}[\mathrm{S}]_o[\mathrm{S}]_i} \tag{1}$$

where [S] is the substrate concentration, R values are resistances defined as the reciprocal of experimentally observed maximal velocity (ν) (the $V_{\max}$), i and o indicate inward and outward, and the arrows indicate direction of transport. The R values describe the kinetics of the roundtrip of the transporter under four conditions: loaded inwardly—unloaded outwardly ($R_{i\to o}$), unloaded inwardly–loaded outwardly ($R_{o\to i}$), loaded in both directions (R_{ee}), and unloaded in both directions (R_{oo}). The affinity constant (K) is a term defined by a collection of the rate constants of all of the transporter-related movements.

To obtain constants for Eq. (1) or propose alternative models so that predictive models of transporter activity can be developed, an investigator has to make measurements of the transporter activity in the kinetic domain. That is, the observed changes in [S] must be due to transporter activity and not diffusion or mixing within the experimental medium. The kinetic activity of the biogenic amine transporters has never been fully characterized in the context of Eq. (1) (as have the glucose and some amino acid transporters), and to our knowledge no alternative transporter kinetic description has been proposed. In order to fully characterize the biogenic amine transporters, rapid measurements of kinetic events should be employed. The rotating disk electrode (RDE) approach described here has been found to be a simple and useful approach to measuring the kinetics of biogenic amine transport because these neurotransmitters are easy to oxidize. The following sections provide a brief overview of rapid chemical measurements and how they may be applied to studies of biogenic amine transporter systems, background on the theory of the RDE, practical descriptions of the RDE experimental setup, and a brief summary of how the RDE technique has been used to study biogenic amine transporters.

Analytical Challenge

In order to kinetically resolve transporter activity, it is necessary to make measurements under experimental conditions in which the observed changes in concentration of substrate over time represent transporter activity and not other factors, such as mixing or diffusion times within the experimental preparation. Some rapid measurement methods that are commonly used to make rapid chemical measurements include flow injection analysis (FIA), temperature-jump, rapid filtration, and electroanalytical techniques such as RDE. There are limitations in the practical value of some of these techniques due to the nature of the transporter-containing sample (tissue slices, homogenates, synaptosomes, or free cells).

FIA involves the combination of two separate analytes, such as enzyme and substrate in a flowing stream. The mixing times define the limits on the determination of rate values; however, rapid mixing times of fractions of a second can readily be obtained, providing kinetic information on substrate turnover numbers at $\leq 1000 \text{ sec}^{-1}$. Two common methods of FIA are the continuous flow and the stop flow methods. The continuous flow method involves rapid mixing of enzyme (or transporter-containing preparation) and substrate via separate syringes compressed at a constant rate or a single pump and a single flow stream equipped with injection ports along the flow path. The solutions are rapidly combined in a mixing chamber or even more rapidly by utilizing a mixing jet that employs turbulent flow to mix the two solutions. The combined solutions then pass through a flow tube where detectors measure the progress of the reaction. Although many detection schemes may be employed, the most common include spectrophotometric detection. In continuous flow methods, the detectors are at fixed positions along the flow tube and the flow rate is fixed, generating a series of measurements of which the postmixing time at a given detector is constant and is simply a function of the flow rate divided by the distance of the detector from the mixing apparatus. Continuous flow analysis can generate a set of values for product formation or changes in [substrate] at different times, allowing for the calculation of an initial rate. The R. M. Wightman group has used a variation of this approach to study release and reuptake of biogenic amines in tissue slices, homogenates, and synaptosomes.[7] In stopped flow analysis the two samples are mechanically injected into a common chamber where they are allowed to mix. The detector is placed in the small mixing chamber, and the solution in the chamber is continuously monitored as the reaction proceeds.

Although the time resolution of flow injection experiments would allow for the measurements of transport rates in biogenic amine systems, there are problems that limit FIA as a tool for transporter kinetic analyses: Tissue preparations are suspensions of cells (and cell parts such as synaptosomes), and physiological buffer measurements in these samples are difficult with the conventional spectroscopic methods employed by FIA because of the opaque and light-scattering properties of the preparation. The Wightman group used a thin layer electrochemical flowthrough detector for their work.[7] It may be difficult to perform successive analyses of these preparations by stop flow spectrometry because the previous solution is merely replaced by the next injection by displacing it into the stop syringe. With tissue preparations a rinse would be necessary to remove any tissue remaining in the observation chamber that might interfere with subsequent runs, making stop flow spectrometry labor intensive in comparison to the RDE voltammetry experiments conducted in this lab.

Rate data for a transport system may be obtained by establishing an equilibrium, perturbing the equilibrium, and measuring the time it takes to establish a new equilibrium. A common example of this experimental method is the temperature jump method. In temperature jump experiments, an equilibrium is established at a given temperature and concentration is measured, typically by spectroscopic means. The temperature is then increased by 5 to 10° very rapidly (over a 0.01–1 μsec interval) by means of a capacitor discharge or laser pulse. If the equilibrium involves an enthalpic change, a new equilibrium will be established. A series of relaxation times,

[7] R. M. Wightman, C. E. Bright, and J. N. Caviness, *Life Sci.* **28,** 1279 (1981).

or τ, defined as the inverse of the rate constnat, will precede the establishment of a new equilibrium.[8] Although temperature jump experiments can give rapid measures of rates, they have not been employed in the studies of transporters because of the analytical difficulties in detection as described for the other methods above. To our knowledge, electroanalytical techniques have not been used in combination with temperature jump methodology.

Finally, kinetic rate data for biogenic amine transport can be obtained using rapid filtration techniques. In this case, the tissue is homogenized and suspended in physiological buffer. Substrate labeled with ^3H or ^{14}C is then added to the preparation and transport is measured by rapidly filtering the preparation at various time intervals followed by rinsing the filters, dissolving the filters, and determining the amount of substrate taken up or released by using liquid scintillation counting. Concentration vs time profiles are obtained from the radiolabel counts from experiments at different incubation times. These experiments are repeated at different values of initial [substrate].

RDE voltammetric experimental methods are more facile at obtaining the same data as the methods outlined above. In addition, it is unnecessary to work with radioactivity to obtain transport profiles, and the RDE approach can provide a continuous record of the transport profile in a single experimental preparation. Using the RDE, initial rate information is easily obtained directly from the continuous record of transport profiles measured in "real time."

Measurements using Rotating Disk Electrode Voltammetry

Brief History and Theoretical Background

Little work was done using RDE until the 1950s,[9,10] although the first mass transport equation describing this technique,

$$I_L = 0.62nFACD^{2/3}\nu^{-1/6}\omega^{1/2} \tag{2}$$

was given by Levich in 1942, where I_L is the limiting current, n is the number of electrons transferred, F is Faraday's constant (96,485 coulombs/equivalent of electrons), A is the area of the electrode in cm^2, D is the

[8] K. Hiromi, "Kinetics of Fast Enzyme Reactions: Theory and Practice," John Wiley & Sons, New York, 1979.

[9] R. N. Adams, "Electrochemistry at Solid Electrodes," p. 67. Marcel Dekker, 1969.

[10] A. J. Bard and L. R. Faulkner, "Electrochemical Methods: Fundamentals and Applications," p. 280. John Wiley & Sons, New York, 1980.

diffusion coefficient of the electroactive species in cm^2/sec, C is the concentration of the electroactive species in the bulk solution in mol/liter, ν is the kinematic viscosity of the solution in cm^2/sec, ω is the angular velocity of rotation in rad/sec, where $\omega = 2\pi N$, and N is the number of rotations per second.

The RDE theory is based on considering an infinitely thin plane, the electrode, rotating in solution at a constant rate about its axis. The motion of the liquid, caused by the drag (stirring) of the electrode, pulls the solution to the electrode surface from a field perpendicular to the electrode. The analyte, dissolved in solution, is brought to the electrode and then spun radially away horizontally by centrifugal force (Fig. 1). If a potential sufficient to oxidize (or reduce) the dissolved analyte is applied to the electrode, then electron flow to (or from) the electrode is detected as current. This electrolysis at the electrode surface drives the concentration of the analyte to zero, producing a thin diffusion layer. Diffusion across this layer is driven by the concentration gradient between the solution at the electrode surface and the bulk solution, and therefore must remain constant since the concentration gradient across the diffusion layer is constant. Thus, it has been shown experimentally that the observed current is dependent on both the diffusion coefficient and the motion of convection currents caused by the

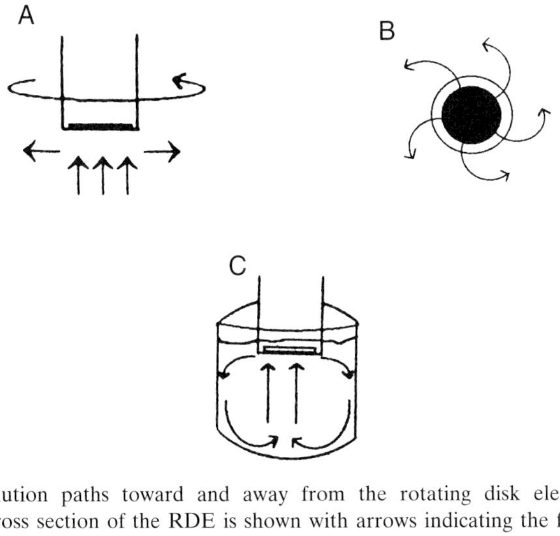

FIG. 1. Solution paths toward and away from the rotating disk electrode (RDE). (A) Vertical cross section of the RDE is shown with arrows indicating the flow of solution from a field perpendicular to the electrode surface and exit of solution from the electrode in a field parallel to the plane of the electrode. (B) Radial diffusion paths (due to centrifugal force) across the electrode surface, shown in an end-on view of the electrode. (C) Diffusion path occurring within the RDE/incubation chamber used in the studies of transporter functioning.

stirring of the solution. As the RDE became a more commonly used technique, experiments were conducted to better define the parameters for the disk size as well as the proportion of volume, or bulk solution, relative to the size of the electrode. These experiments showed that these parameters in fact did not limit the practical use of the RDE.[9]

Historically, RDE has been used to investigate rates and mechanisms of electrode processes and has been shown to be useful in the neurosciences to study fast kinetics of dopamine depolarization-stimulated release with coupled uptake as well as uptake into resting preparations.[11] The only requirement for such measurements is that the neurotransmitter be electroactive. Currently, the RDE can be used to measure release and/or transport rates of the catecholamine family in which the oxidation reaction yields two electrons and two protons,[11,12] and serotonin where the overall oxidation reaction yields two electrons and one proton.[13] Uric acid (a possible mirror of adenosine biochemistry) and ascorbic acid may also be studied using the RDE.

Experimental Setup

The experimental setup, as illustrated in Fig. 2, consists of a thermostatted, glass incubation chamber; a water circulator; an RDE; a precision rotator; a potentiostat; and a signal output device. Each component is described in turn.

The Pyrex glass incubation chamber is custom-made by a glass blower. The approximate dimensions of the inner chamber are 1 cm in diameter by 1.5 cm deep. The compartment is water-jacketed so that the temperature of the inner chamber can be held constant. Reference and auxiliary electrodes are heat annealed into the glass walls of the chamber. Both of these electrodes are composed of a platinum (Pt) wire. In our experience, thicker Pt wire (up to 1 mm diameter) annealed better than thin 100 μm diameter wire. The portion of the wires that span the water-jacketed region must be encased in glass to prevent electrical contact with the water circulating heater. The Ag|AgCl reference electrode is completed by placing a drop of molten Ag on the inner-chamber end of one of the two Pt wires. The silver ball is then anodized in 0.1 *M* HCl to produce a coating of AgCl on its surface. The remaining bare Pt wire is used as the auxiliary electrode. The ends of the wires that extend outside are soldered to copper (Cu)

[11] S. M. Meiergerd and J. O. Schenk, in "Voltammetric Methods in Brain Systems" (A. A. Boulton, G. B. Baker, and R. N. Adams, Eds.), p. 305. Humana Press, Totowa, NJ, 1995.

[12] G. Dryhurst, K. M. Kadish, F. Scheller, and R. Renneberg, "Biological Electrochemistry," Vol. 1, p. 116. Academic Press, New York, 1982.

[13] Z. Wrona and G. Dryhurst, *Bioorganic Chem.* **18,** 291 (1990).

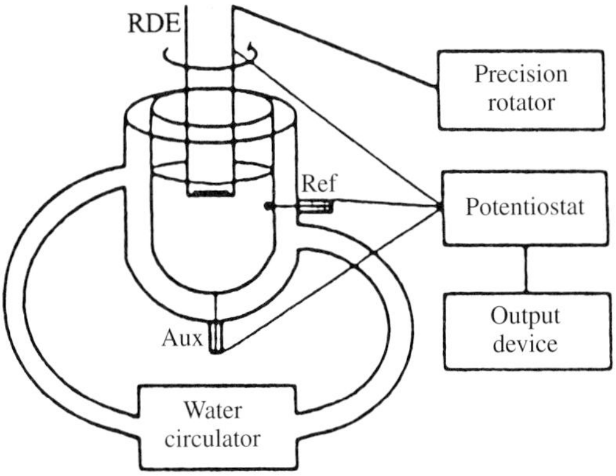

FIG. 2. Glass RDE/incubation chamber and peripheral connections to the devices required for thermostatting its contents, applying a potential to the electrode and measuring the current for detection, and recording the output signal. Details of construction of the chamber and each attached component are described in the text.

tubing which is glued with Torr Seal (Varian Associates, Lexington, MA) to the outside of the chamber. The Cu tubing provides a location structurally strong enough to make electrical contact between the potentiostat and the reference and auxiliary electrodes.

The commonly available water circulating heater/cooler devices by Lauda or others work quite well, holding temperatures within the incubation chamber to within $\pm 0.1°$. An inexpensive alternative for a fixed temperature water recirculating device is the Aqamatic K-Module sold by American Hospital Supply (Cincinnati, OH) for use in physical therapy.

The RDE consists of a 3.0 mm glassy carbon (GC) disk surrounded by Teflon cladding (overall diameter, GC disk + cladding = 5 mm) and is commercially available from Pine Instruments, Inc. (Grove City, PA). The electrode is attached to a 6 in stainless steel shaft which is designed to fit into a Pine Instrument Model MSRX high-precision rotator. Pine Instruments silver–carbon brushes make contact between the electrode and the potentiostat. The Pine Instruments high-precision rotator can rotate the electrode at a constant rotation rate between 1 and 10,000 rpm, or any wave form can be applied by a wave generator to alter rotation speed.

A potentiostat is used to apply a constant potential to the electrode and monitor the resulting current. The most important requirements for the potentiostat are that it have a current range of 0.1–100 nA/full scale detection, an adjustable full-scale offset, and a $\leq$100 millisecond response time. Our laboratory modifies the response time of commercial potentiostats by changing the capacitors in the feedback loop of the current follower. The commercially available potentiostats which we have found to be useful are the Bioanalytical Systems Inc. (West Lafayette, IN) line, including the old LC4A and the current LC4B and the Petite Ampere LC3D models. We have also built potentiostats which have the characteristics above plus the capability of a continuously variable time constant. The details of these circuits are available upon request.

A strip chart recorder or digital oscilloscope are both suitable output devices. However, a digital oscilloscope capable of storing data on computer disks is more convenient for subsequent data analyses because the digitally stored data can be analyzed using a variety of commercially available software such as GraphPad PRISM (San Diego, CA), Jandel Sigma Plot (San Rafael, CA), or Microsoft Excel (see Microsoft Press, Redmond, WA). The commercially available oscilloscopes that we have found to work well include the Nicolet and Tektronix lines.

Striatal or Other Biogenic Amine Transporter-Containing Suspensions and Synaptosomes

Our laboratory prepares striatal suspensions as described previously by Meiergerd and Schenk.[11] A freshly dissected striatum (or other tissue) is hand chopped on an ice-cold watch glass using a sharp razor blade. Immediately after chopping, the tissue is transferred to the incubation chamber containing 500 μl physiological buffer [composition: 124 mM NaCl, 1.80 mM KCl, 1.24 KH_2PO_4, 1.30 mM $MgSO_4$, 2.50 mM $CaCl_2$, 26 mM $NaHCO_3$, and 10.0 mM glucose, saturated with a gas mixture of 95% O_2 and 5% CO_2 (v/v)]. Saturation of the buffer throughout the experiment is maintained by gently directing a stream of the gas mixture with a Pasteur pipette across the surface of the contents of the incubation chamber. The tissue is suspended by drawing 250 μl of the tissue–buffer solution into and expelling 250 μl of the tissue–buffer solution out of an Eppendorf pipette tip 10 times. We use a 200–1000 μl size Universal Fit Pipette tip (VWR) with an orifice diameter of ca. 1.0 mm. The suspended tissue is then washed 6 to 10 times by adding 250 μl of fresh buffer to the suspension and, after the tissue has settled, removing this volume. Once the tissue is washed, the working electrode (RDE) is lowered into the tissue–buffer suspension to a level approximately 3 mm below the liquid surface. A potential of +450 mV (or other value sufficient to detect the compound under study) with respect to the Ag|AgCl reference electrode is applied and the rotation of the electrode is set at a constant 2000 rpm (mixing time of *ca.* 25 msec). After a 15-min incubation period, a stable baseline is obtained.

Synaptosomes to be studied by the RDE method are prepared by a slight modification of the procedure described by Booth and Clark,[14] and a detailed protocol can be found in McElvain and Schenk.[15]

HEK hDAT Cells

The cloned human transporter hDAT has been expressed in a variety of different cell lines. Our laboratory, in collaboration with Drs. Eshleman, Janoswsky, and Neve (at the VA Medical Center, Portland, OR) have used RDE voltammetry to study hDAT transfected into human embryonic kidney (HEK) cells.[16] These cells are easy to use and require much less preparation time than rat striatal tissue suspensions. The simple protocol is as follows. A volume of physiological buffer is added to a rinsed (with physiological buffer) confluent cell culture plate. We have found that 1.2 ml buffer/100 × 20 cm plate works well. The cells can be removed from the bottom of the culture plate by gently scraping with a cell scraper. A suspension is made with a Pasteur pipette, and now 300 to 500 μl (constant and defined for a given set of experiments) of cell suspension is transferred directly to the glass incubation chamber. The electrode is lowered to a depth of approximately 3 mm below the liquid surface, the suitable potential is applied, and the rotation rate is set at 2000 rpm. Unlike the tissue preparations described above, a stable baseline is achieved almost immediately. Transport velocities are normalized to protein concentration or cell number. Transporter activity in these preparations is sharply temperature dependent with apparent optimal performance at 37°.

^{3}[H]Dopamine release and uptake experiments are often performed in well plates where the transfected cells are growing on the bottom. Transfected cells growing in these wells may be suitable for study using RDE voltammetry. A AgCl-coated Ag wire and a Pt wire, acting as the reference and auxiliary electrodes, respectively, could be draped down the side of the well. A glassy carbon electrode such as that described above could be lowered directly into the well as long as the dimensions of the diameter of the well are not less than about 12 mm.

Data Acquisition and Analysis of Data

Transport velocities are measured in the following manner. After a baseline is obtained at a given RDE rate of rotation (optimized for a given experiment) in a transporter-containing preparation, a known concentra-

[14] R. F. Booth and J. B. Clark, *Biochem. J.* **176,** 365 (1978).
[15] J. S. McElvain and J. O. Schenk, *Biochem. Pharmacol.* **43,** 2189 (1992).
[16] A. J. Eshleman, C. Earles, K. Neve, A. Janowsky, and J. O. Schenk, *Neurosci. Abstr.* **26,** 702.4.

tion of S (for example, dopamine), other biogenic amine, or structural analog is added. A continuous record of the inwardly directed transport is then monitored over ~80 sec (with 50–100 msec resolution) and stored on a floppy disk. Transport velocities are determined from linear regression analyses of the $[S]_o$ vs time data over the apparent linear region of the transport kinetics. This is the linear region of the data where the value of the *y* intercept is not significantly different from the initial [S] added. Velocities are expressed conveniently in picomoles of substrate/sec/gm wet weight of tissue and in our hands have values of precision within ±5% relative standard errors. Values for V_{max} and K_m are then obtained by fitting the observed velocities of transport at various $[substrate]_o$ values to the Michaelis–Menten expression. This is generally accomplished by nonlinear curve fitting using estimates of V_{max} and K_m from Eadie–Hofstee analyses as the initial guesses. A simple "analysis of residuals" is performed by using the kinetic parameters from the curve fit to calculate a hypothetical velocity–[substrate] curve and comparing it point by point with the experimental curve. In our hands the fits evaluated in this way using a minimum of five pairs of data are within 5 to 10% with no systematic errors observable at the two ends of the hyperbolic curves. Because of propagation of errors, values at this level of precision are needed in order to perform detailed kinetic analyses.

To eliminate intracellular biochemical processes of metabolism and sequestration from influences on kinetic results in rat tissues, reserpine is injected intraperitoneally (5 to 12 mg/kg) 8 hr prior to experimentation to destroy vesicles[17] and 100 μM pargyline is added to the tissue suspension after tissue washing, but before the 15 min incubation, to inhibit monoamine oxidase.[18] In HEK cells tropolone is used to inhibit catechol *O*-methyltransferase.

Overview of Applications

Multisubstrate Analyses

The RDE technique has been used mostly to investigate the mechanism(s) of the Na^+ and Cl^- dependent dopaminergic transporter. Reaction orders of Na^+ and Cl^- involvement in transport were determined by substituting NaCl in the physiological buffer with choline chloride and sodium isethionate, respectively, and monitoring the result on the transport of

[17] A. J. Bean, T. E. Adrian, I. M. Modlin, and R. H. Roth, *J. Pharmacol. Exp. Thera.* **249,** 681 (1989).
[18] N. Y. Liang and C. O. Rutledge, *Eur. J. Pharmacol.* **89,** 153 (1982).

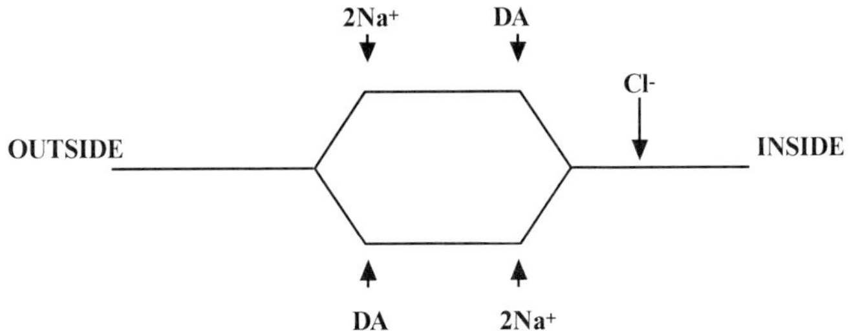

FIG. 3. Reaction pathway for the multisubstrate mechanism of transport of dopamine and Na^+ and Cl^- cosubstrates by the dopaminergic transporter (modified from ref. 15). The line from left to right is the transport reaction pathway as depicted for enzyme reactions by H. J. Fromm, "Initial Rate Enzyme Kinetics." Springer-Verlag, New York, 1975. It illustrates a random AB terreactant model in which dopamine and 2 Na^+ bind to the transporter randomly prior to the binding of Cl^- before substrates are transported inwardly into or across the membrane. This mechanism has been observed in tissues from both rat striatum and nucleus accumbens. The constants for the involvement of the substrates have been shown to distinguish transporter functioning in these two anatomic areas.

dopamine. Any given ion substitute may have some effects of its own,[19] and the choice of substitute should be considered in this light. However, exact integer reaction orders were obtained using the substitutes above with kinetically resolved kinetics of transport by the RDE. It should be noted that *N*-methyl-D-glucamine is also sometimes used[20] as a substitute for Na^+. It is commercially available (Sigma Chemical Co., St. Louis, MO) as the free base, and thus an investigator should take care in the neutralization of the base (usually with HCl) in the physiological buffer so as not to affect the final concentration of Cl. It is suggested that the neutralization procedure utilize a standard solution of HCl in order to provide accurate knowledge of the amount of Cl^- added.

A multisubstrate mechanism has been determined for the transport of dopamine by the dopamine transporter (see Fig. 3).[15] Our laboratory has used methods described by Fromm[21] and/or Segel[22] to study the involvement of Na^+ and Cl^- in dopamine uptake. In each experimental trial, initial dopamine uptake velocities are measured in a buffer whose Na^+ and Cl^- conentrations are held at a constant ratio of 1.15 to 1.00, to mimic the ratio in physiological buffer, whereas the initial initial dopamine concentration,

[19] R. P. Shank, C. R. Schneider, and J. J. Tighe, *J. Neurochem.* **49,** 381 (1987).
[20] D. D. Wheeler, A. M. Edward, B. M. Chapman, and J. G. Ondo, *Neurochem. Res.* **18,** 927 (1993).
[21] H. J. Fromm, "Initial Rate Enzyme Kinetics." Springer-Verlag, New York, 1975.
[22] I. H. Segel, "Enzyme Kinetics Behavior and Analysis of Rapid Equilibrium and Steady State Enzyme Systems." John Wiley & Sons, Wiley Classics Library Edition, New York, 1993.

$[DA]_o$, is varied. The $[Na^+]_o$ and $[Cl^-]_o$ are decreased in each trial, whereas their ratio is held constant. The data is plotted as 1/velocity vs $1/[DA]_o$. The slopes of these plots are replotted with respect to $1/[Cl^-]_o$ and/or $1/[Na^+]_o$. The graphic shape of this replot is compared to the models of Fromm[21] to determine multisubstrate reaction mechanism with respect to Na^+, Cl^-, and dopamine. Some 10 different multisubstrate mechanisms can be identified by these procedures (see Fromm[21]).

Other Studies

In addition to the multisubstrate mechanisms described above, the RDE method has been used to conduct struture activity studies for the transport of dopamine to determine what structural features of dopamine are important in the recognition and translocation of substrate.[23] Analyses of the results of these studies suggest that the catechol moiety of dopamine mediates transporter recognition of the molecule, whereas catechol and the ethylamine side chain mediate translocation. In addition, apparent transacceleration, a result of the general simple transporter theory of Eq. (1),[4,5] has been demonstrated for the striatal transporter,[24] and it is worth noting that this phenomenon has also been observed in an expression system of cloned norepinephrine transporter.[25] Inhibition of the transport of dopamine has also been investigated using Michaelis–Menten analyses[15,26] and theory developed by Devés and Krupka[27–29] for investigating the effects of inhibitions by pairs of inhibitors. These latter studies[30] were used to compare the binding domains of different inhibitors of dopamine transport relative to the binding of cocaine. It should be noted here that the models used in the cited work were only those applicable to comparing external inhibitor binding sites, and the original literature cited provides solutions for many other combinations of binding site relationships. The RDE method has been used to investigate the changes (kinetic up-regulation) in dopaminergic transporter function following withdrawal from repeated systemic treatments with cocaine in investigator-administered treatment paradigms[26] and in self-administration paradigms.[24] In addition, the RDE technique has been used to identify a feedback mechanism in which extra-

[23] S. M. Meiergerd and J. O. Schenk, *J. Neurochem.* **62,** 998 (1994).

[24] S. M. Meiergerd, S. M. Hooks, and J. O. Schenk, *J. Neurochem.* **63,** 1277 (1994).

[25] W. B. Burnette, M. D. Bailey, S. Kukoyi, R. D. Blakely, C. G. Trowbridge, and J. B. Justice, Jr. *Analyt. Chem.* **68,** 2932 (1996).

[26] S. M. Meiergerd, J. S. McElvain, and J. O. Schenk, *Biochem. Pharmacol.* **47,** 1627 (1994).

[27] R. Devés and R. M. Krupka, *J. Biol. Chem.* **255,** 11870 (1980).

[28] M. Krupka and R. Devés, *Int. Rev. Cytol.* **84,** 303 (1983).

[29] R. Devés and R. M. Krupka, *Methods Enzymol.* **171** (1989).

[30] S. M. Meiergerd and J. O. Schenk, *J. Neurochem.* **63,** 1683 (1994).

cellular dopamine feeds back on the function of its transporter via a D2-like receptor to upregulate transporter activity.[31] Finally, an overview of other investigations of the mechanism of dopamine transport and how the RDE results fit into the body of work appears in Povlock *et al.*[32]

In summary, the RDE method appears to provide chemical information about transport in the kinetic domain. It is expected that similar studies can be conducted in other biogenic amine systems.

Acknowledgments

This work was supported by a grant to J.O.S. from the National Institute on Drug Abuse (NIDA), DA 07384, and the State of Washington. J.O.S. is also the recipient of a NIDA Research Scientist Development Award (KO 2 DA 00184) and C.E. is a recipient of a NIDA predoctoral fellowship, DA 05762.

[31] S. M. Meiergerd, T. A. Patterson, and J. O. Schenk, *J. Neurochem.* **61,** 764 (1993).
[32] S. L. Povlock, S. M. Meiergerd, and J. O. Schenk, *in* "CNS Neurotransmitters and Neuromodulators Dopamine" (T. W. Stone, Ed.), p. 21. CRC Press, Boca Raton, FL, 1996.

[46] Electrochemical Detection of Reverse Transport from *Planorbis* Giant Dopamine Neuron

By BRIAN B. ANDERSON, ANDREW G. EWING, and DAVID SULZER

Introduction

Large identified neurons from invertebrates have proved valuable for the elucidation of basic mechanisms of neurotransmission. Early insights into exocytic transmission resulted from studies in preparations including the squid giant neuron and various identified neurons in the leech and *Aplysia californica*. In addition, alternate forms of neurotransmission have been clearly demonstrated first from invertebrate preparations; the initial demonstration of synaptic electrical transmission by gap junctions was in the giant axons of the abdominal nerve cord of the crayfish *Astacus fluviatilis*.[1,2]

Another form of nonexocytic neurotransmission may be due to reverse transport across uptake transporters. This action was initially suggested by studies in which uptake blockers inhibited increased extracellular transmit-

[1] E. J. Furshpan and D. D. Potter, *J. Physiol.* **145,** 289 (1959).
[2] E. J. Furshpan and D. D. Potter, *Nature* **180,** 342 (1957).

0076-6879/98 $25.00

ter levels.[3-5] Electrophysiological recordings from postsynaptic cells also indicated the existence of reverse transport,[6] but it was difficult to rule out possible alternate forms of neurotransmission, and observation of reverse transport directly from the secretory cell was possible only in glial cells, in which potassium currents mediated by flux across the transporter were measured.[7] Reverse transport was suggested to underlie the mechanism of monoamine transmitter release by amphetamines (ref. 8 and references therein), but the slow time course of such studies, which typically take tens of minutes, have made it difficult to isolate effects such as reuptake blockade from reverse transport.

Here we show how studies of the giant dopamine neuron (GDN) located in the left pedal ganglion of the pond snail *Planorbis corneus* have provided direct demonstration of reverse transport from a neuron in real time. The GDN is advantageous for these studies since it has nomifensine-sensitive plasma membrane dopamine (DA) uptake, reserpine-sensitive vesicular DA uptake, metabolizes DA to dihydroxyphenylacetic acid,[9] and lacks ascorbic acid, facilitating the unambiguous amperometric detection of DA.[10] Exocytic quantal release events can easily be distinguished from release by reverse transport using carbon fiber electrodes. Moreover, the large size of the neuron (approximately 150 μm in diameter) has allowed estimation of cytosolic levels of DA with intracellular carbon electrodes.[11] The large size of the cell, the presence of the DA uptake system, and presence of many synaptic vesicles provides an enormous transmitter pool, proving the means to detect reverse transport from a single cell.

Dissection Procedure

Planorbic corneus obtained from NASCO (Fort Atkinson, WI) are maintained in aquaria at room temperature until used. The snails are dissected under a snail Ringer's solution (39.5 m*M* NaCl, 1.3 m*M* KCl, 4.5 m*M* $CaCl_2$, and 6.9 m*M* $NaHCO_3$ adjusted to pH 7.4). The dissection procedure and the identification of the DA containing neuron have been described

[3] V. Adam-Vizi, *J. Neurochem.* **58,** 395 (1992).
[4] G. Levi and M. Raiteri, *Trends Neurosci.* **16,** 415 (1993).
[5] D, Attwell, B. Barbour, and M. Szatkowski, *Neuron* **11,** 401 (1993).
[6] E. A. Schwartz, *Science* **238,** 350 (1987).
[7] M. Szatkowski, B. Barbour, and D. Attwell, *Nature* **348,** 443 (1990).
[8] D. Sulzer, N. T. Maidment, and S. Rayport, *J. Neurochem.* **60,** 527 (1993).
[9] N. N. Osborne, E. Priggemeier, and V. Neuhoff, *Brain Res.* **90,** 261 (1975).
[10] A. Barber and B. Kempter, *Comp. Biochem. Physiol.* **84C,** 171 (1986).
[11] J. B. Chien, R. A. Wallingford, and A. G. Ewing, *J. Neurochem.* **54,** 633 (1990).

previously.[12,13] The snail shell is broken and removed from the body before the snail is pinned with the dorsal side uppermost in a wax-filled petri dish. The whole snail is initially held in place by three pins. Two pins hold the front of the head of the snail, one through each side of the head, near the antennae. The third pin is used to expose the skin that covers the brain by pulling back the body of the snail and pinning through the body. The skin of the head part is then cut open by making a medial incision from the caudal base of the head in the rostral direction along the axis of the body, to reveal the brain and esophagus. The esophagus is cut and pulled through the ganglion ring and pinned rostrally to keep it out of the way. Next, the tissue connecting the left and right cerebral ganglia (for diagram, see ref. 14) is cut exposing the pedal ganglia. The connectives attached to the visceral and left parietal ganglia are then cut as long as possible and pinned rostrally through the animal, exposing the ventral side of the ganglia ring. The dissection process leaves the intact ganglia ring pinned to the wax by the connective tissue with 7–10 pins in a position so that the left pedal ganglion is posteriormost and to the left of midline. Individual cells, including the GDN (Fig. 1, rightmost arrow) can be seen beneath the outer connective tissue surrounding the brain using a binocular microscope (70× magnification). However, the outer connective tissue is removed from the surface of the pedal ganglion with a home-made fine pin with a 90° hook, thus allowing free diffusion of released substances. The DA neuron is usually the largest cell (approximately 150 μm in diameter) in the left pedal ganglion and is easily identified[15] by its specific location close to the statocyst (Fig. 1, leftmost arrow).

Electrode Preparation

Ring Electrodes

Carbon ring electrodes are fabricated in pulled quartz capillary tubes by pyrolysis of methane on the interior surface.[16] Quartz capillary tubes (0.06 inch × 6 in, Pyromatics, Willoughby, OH) are pulled manually in a methane/oxygen flame to tip diameters of 1–4 μm. Carbon is deposited inside the pulled quartz tip by passing methane through the whole capillary while holding it over a Bunsen burner flame. The Bunsen burner flame

[12] M. S. Berry, *J. Exp. Biol.* **56,** 621 (1972).
[13] G. Chen, P. F. Gavin, G. Luo, and A. G. Ewing, *J. Neurosci.* **15,** 7747 (1995).
[14] B. A. Ger and E. V. Zeimal, *Nature* **259,** 681 (1976).
[15] C. Marsden and G. A. Kerkut, *Comp. Gen. Pharmacol.* **1,** 101 (1970).
[16] Y. Kim, T., D. M. Scarnulis, and A. G. Ewing, *Anal. Chem.* **58,** 1782 (1986).

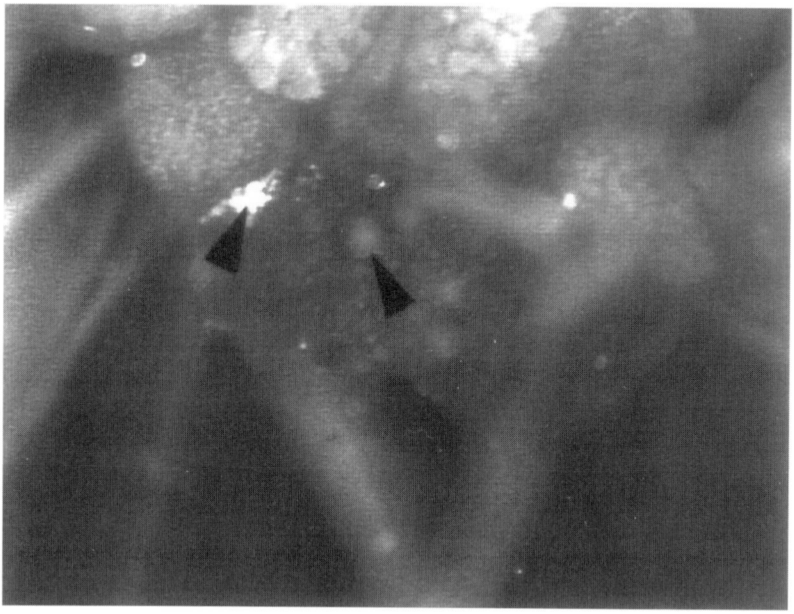

FIG. 1. Photomicrograph of the ventral side of the pedal ganglion of *P. corneus*. Typically 7 to 10 pins are used to hold the ganglia ring against the dissection wax (not visible in the photomicrograph). The rightmost black arrow is pointing at the GDN, which can be seen beneath the transparent tissue surrounding the ganglion. The GDN is easily identifiable based on its large size and its location near the statocyst (leftmost arrow). The horizontal field of view of the photomicrograph is about 1.6 mm.

often softens the capillary tip, resulting in a bent electrode tip, but heating the pulled capillary inside another quartz tube alleviates this problem. The pyrolysis results in a shiny black carbon coating inside the pulled quartz capillary. The tips are then filled by dipping them in freshly prepared epoxy (Epon 828 with 14% by weight *m*-phenylenediamine chloride as hardener, Miller-Stephenson Chemical Co., Danbury, CT) for 3 min and cured in a 100° oven for 1 hr. The ring-shaped conductive carbon surface is then exposed by cleaving the tip to the desired diameter with a scalpel blade. Electrical contact with the pyrolyzed carbon coating is made by filling the capillary tube with mercury [colloidal graphite (Energy Beam Sciences, Agawam, MA) can alternatively be used] and securing a nickel–chromium wire inside.

Disk Electrodes

Carbon disk electrodes are fabricated by securing a carbon fiber inside a pulled gas capillary.[17] Ten micrometer diameter carbon fibers (Amoco, Greenville, SC) are aspirated into glass capillaries (1.2 mm o.d. × 0.68 mm i.d. × 4 in, A-M Systems, Inc., Everett, WA). The capillaries are pulled with a commercial microelectrode puller (Ealing, Harvard Apparatus, Edenbridge, KY) and dipped in fresh Epon 828 epoxy for 30 sec. The epoxy is then cured in a 100° oven for 1 hr. Prior to use, the carbon fiber is cut at the fiber–glass interface with a scalpel blade to expose a fresh conductive carbon surface. The electrode surface can also be replenished by polishing. Polishing, or beveling, is performed by lowering the electrode at an angle of approximately 45° onto a rotating surface that contains a thin layer of diamond polishing compound. After the electrode has been in contact with the diamond polishing surface for more than 30 sec, a smooth carbon surface is obtained.

An alternative preparation of disk electrodes is used for Axopatch 200 series amplifier headstages (Axon Instruments, Foster City, CA). Carbon fibers (5 μm, Amoco) are aspirated into glass capillaries. The glass capillaries are pulled in a micropipette puller (Sutter Instruments, Novato, CA) so that the tip fits tightly around approximately 1 mm of the fiber and the tip is cut with a scalpel under a dissecting microscope so that only the cut face of the fiber is exposed. The tips are then sealed by dipping into epoxy (Epo-Tek 301, Epoxy Technology, Billerica, MA) for 5 min and then cured in a 100° oven for 24 hr. The tips are polished at a 45° angle for 5–15 min using a beveler (World Precision Instruments, New Haven, CT) fitted with a diamond abrasive surface (Sutter Intruments) and backfilled with 3 *M* KCl.

Electrode Testing

The electrochemical response of both the ring- and disk-shaped electrodes is tested prior to use in an experiment. Electrode testing is typically performed in freshly prepared 100 μM DA in Ringer's solution. A saturated sodium calomel electrode (SSCE) or silver | silver chloride (Ag | AgCl) reference electrode and a platinum wire auxiliary electrode are used to complete the three-electrode electrochemical cell for testing. Typically, a slow scan (100 mV/sec) triangle potential waveform ranging from −0.2 to 1.0 V is applied to the reference electrode using an analog triangle wave generator (Ensman EI-400, Bloomington, IN). The resulting redox current is amplified by the EI-400 potentiostat and monitored on an *X–Y* recorder or on a

[17] R. S. Kelly and R. W. Wightman, *Anal. Chim. Acta.* **187,** 79 (1986).

computer. This process allows the experimenter to calibrate the electrode for quantitation and check the electrode response for correct sigmoidal wave shape and half wave potential. The half wave potential for DA is typically in the range of 0.15 to 0.25 V vs SSCE for carbon microelectrodes. In addition, the stability of the electrode is observed by the shape of the *X–Y* chart between the rising and falling voltage slopes, so that electrodes with wide or irregular responses are rejected. The working surface of the electrode can be estimated by the relationship

$$r = \frac{i_{\text{lim}}}{2\pi nFDC} \tag{1}$$

where r is the electrode radius, i_{lim} is the limiting current, n is the number of electrons transferred in the electrochemical reaction (2 for catecholamines), C is the DA concentration in mol/cm^3, F is the Faraday constant (96,485 coulomb/equivalent), and D is the diffusion constant ($\sim 6 \times 10^{-6}$ cm^2/sec for DA in aqueous solution). Finally, electrode performance can be tested with the Axopatch 200B by observing the rms (root mean square) noise. With the electrodes in buffered saline, the rms can be measured on the Axopatch 200B front panel. If the rms noise output of the Axopatch 200B is greater than 0.9 pA after filtering with a 5 kHz 4-pole lowpass Butterworth filter, the electrode is discarded.

Measuring Reverse Transport

Immediately after dissection, the petri dish containing the dissected snail is placed on the stage of a stereomicroscope (Nikon Corporation, Tokyo Japan). Figure 2 is a schematic of the region of interest as viewed through the microscope. The working electrode is positioned over the top of the cell body of the GDN with a three-dimensional micromanipulator (Mertzhauser, Zeiss, Jena, Germany). Then, the electrode is manipulated so that it lightly touches the center of the cell body. The electrode potential is typically held constant at 0.8 V, or scanned from −0.2 to 0.8 V for voltammetry. With the electrode positioned over the cell, it is in position to measure reverse transport of DA to the extracellular medium.

Release of DA from the cell body of the DA neuron by reverse transport is induced by intracellular injection of either DA or amphetamine. Intracellular injections are performed with glass capillaries pulled to small tips (0.5 μm tip diameter) attached to a two-channel pressure application device (Picospritzer, General Valve Corporation, Fairfield, NJ). An injector is positioned with a micropositioner (Medical Systems Corporation, Greenvale, NY) until it just touches the outer membrane of the cell body and the injector is then forced through the membrane. The injection appara-

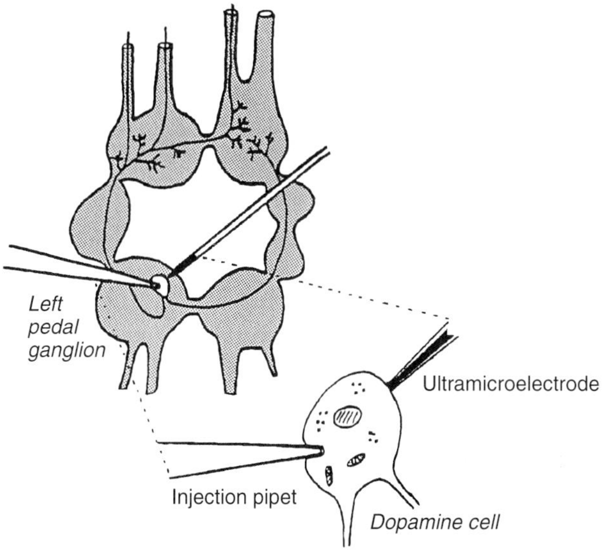

FIG. 2. Experimental setup for measuring reverse transport at the identified GDN of *P. corneus*. Single GDNs in the left pedal ganglion of adult *P. corneus* were impaled with an injection pipette. A 2 μm carbon ring ultamicroelectrode was positioned on the surface of the cell so that oxidizable species released from the cell could be detected. Detection was performed either by constant potential electrochemistry (amperometry) at 0.8 V or by scanning electrochemistry (cyclic voltammetry) by ramping the potential from −0.2 to 0.8 V. [Reproduced with permission from D. Sulzer, T. K. Chen, Y. Y. Lau, H. Kristensen, S. Rayport, and A. Ewing, *J. Neurosci.* **15,** 4102 (1995).]

tus is calibrated so that a timed injection delivers a known volume of solution. Calibration is performed by measuring the radius of a drop of aqueous solution created by a series of timed injections into mineral oil. The drop volume is then calculated from the drop radius for the different injection times and a calibration curve is created. This technique allows for the routine injection of picoliter volumes of solution into the cell body without harming the cell. For the GDN (approximately 150 μm in diameter), 4 to 8 picoliters are typically injected.

On injection of DA or amphetamine into the cell, released DA is measured as an oxidation current using a homemade low-current three-electrode potentiostat. The output of the potentiostat is digitized at 820 Hz (Labmaster AD, Scientific Solutions, Solon, OH) and then saved to a computer and analyzed by software written in-house.[18] For constant potential (amperometry) experiments, the data are saved and displayed (in real

[18] T. K. Chen, G. Luo, and A. G. Ewing, *Anal. Chem.* **66,** 3031 (1994).

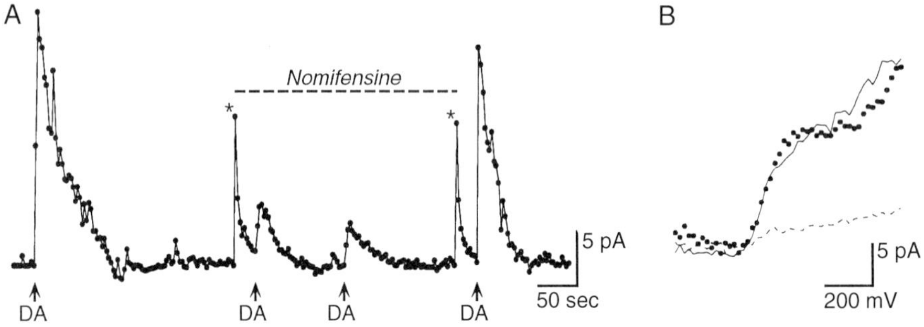

FIG. 3. Intracellular injection of DA induces reverse transport. (A) Intracellular injection of 4 pl of 0.5 m*M* DA (*arrows*) reliably increases extracellular DA. During extracellular perfusion with 10 μM nomifensine (dashed line), the DA release due to the same DA injections was markedly attenuated. Perfusion with nomifensine-containing medium and its replacement by control medium produce current spikes (asterisks). These data have been baseline corrected by subtracting a curve fitted to the background signal. (B) Background subtracted voltammograms (from −0.2 to 0.8 V) for the first (filled circles), second (dashed line), and fourth (solid line) intracellular DA injections shown in (A). The voltammograms monitored at the face of the neuron are similar to DA. Although the shapes of the voltammograms without nomifensine differ slightly at potentials above 0.5 V, all three voltammograms are sigmoidal with a half-wave potential of approximately 0.2 V. [Reproduced with permission from D. Sulzer, T. K. Chen, Y. Y. Lau, H. Kristensen, S. Rayport, and A. Ewing, *J. Neurosci.* **15,** 4102 (1995).]

time) as a current vs time plot. For voltammetry experiments, the data are saved as current vs potential but displayed as current vs time by plotting the current value at a potential in the cyclic waveform capable of oxidizing DA (typically 0.75 V). Voltammetry is useful for measuring reverse transport because it provides qualitative structural as well as kinetic information. Although it is difficult to distinguish between catecholamines by voltammetry, the method can discriminate catecholamines from their metabolites or metabolic cofactors such as ascorbate. Amperometry at a constant voltage is advantageous because it provides a better signal-to-noise (*S*/N) ratio and faster time resolution, a useful characteristic for measuring rapid events such as vesicle exocytosis.

Reverse transport of DA from the cell body of the GDN can be measured by cyclic voltammetry. In the resting state, cytosolic DA is about 2.2 μM and the extracellular level is below the detection limit of the microelectrodes used.[19] Injection of DA into the cell body results in the release of DA[20]; the data are shown in Fig. 3A. The observation that the flux of

[19] A. G. Ewing, T. S. Stein, and Y. Y. Lau, *Accts, Chem. Res.* **25,** 440 (1992).

[20] D. Sulzer, T. K. Chen, Y. Y. Lau, H. Kristensen, S. Rayport, and A. Ewing, *J. Neurosci.* **15,** 4102 (1995).

DA is attenuated (by 87 ± 4%) by nomifensine, a DA transporter antagonist, suggests that most of the DA is released by reverse transport. Removal of nomifensine from the environment then leads to an increase in extracellular DA due to intracellular injections (20.4 ± 1.5 μM peak concentration), and, in time, the signal returns to near control levels. Intracellular injection of control solutions lacking DA induces no DA release. Difference voltammograms obtained after the first, second, and fourth DA injections shown in Fig. 3A are plotted in Fig. 3B. These voltammograms have shapes and half-wave potentials that closely resemble voltammetry runs obtained by *in vitro* calibrations. Therefore, the evidence points to the released substance as being DA. The current traces observed for reverse transport of DA by this procedure provide a means to obtain both kinetic and quantitative information about this process in a simple model system.

Measurement of DA Released by Exocytosis

The intact GDN releases DA by exocytosis at the cell body.[13] The experimental setup is different from that used to measure reverse transport in that a pipette is used to apply a secretagogue close to the cell and a carbon disk electrode is used instead of the carbon ring electrode. Depolarization-induced release following application of 1 M KCl can be observed as current transients by amperometry. Figure 4A is an example of the data obtained from such an exocytosis experiment. Figure 4B shows an expansion of the time axis so that the shape of individual transients can be seen. The current transients are consistent with fusion of individual vesicles to the cell membrane and release of their contents into the space between the cell surface and the electrode.

The detection of individual release events allows the amount of neurotransmitter to be calculated on a per vesicle basis, assuming that DA is quantitatively converted to a quinone, with the transfer of two electrons. The area under each current transient is equivalent to the charge passed at the electrode surface and is related to the number of moles of DA detected by Faraday's law,

$$Q = nNF \tag{2}$$

where Q is the integrated charge for a single current transient and N is the number of moles of DA detected. The average number of molecules measured per current transient (quantal size) is 1.3 amol (1 amol ≈ 6 × 10^5 molecules); kinetic models of such events suggest that essentially all DA is oxidized from release that occurs directly under the electrode,

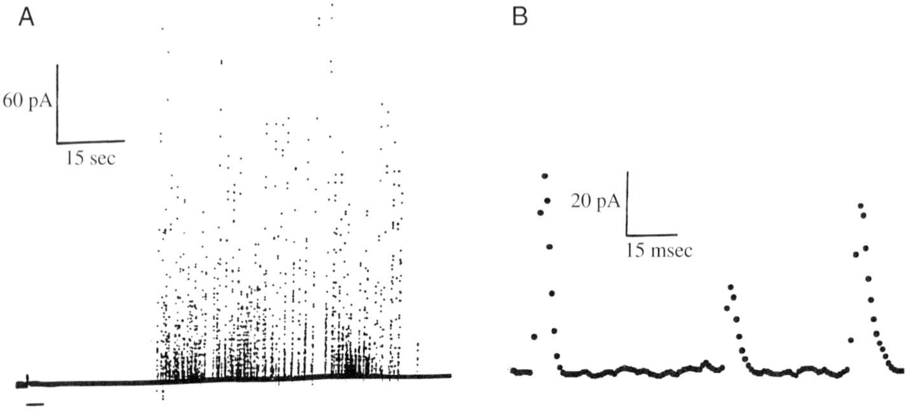

FIG. 4. Dopamine current transients detected by carbon fiber disk electrodes following cellular stimulation. (A) An example of current transients recorded with the amperometric constant-voltage method. A GDN (diameter about 100 μm) was stimulated with a 4 sec potassium chloride (1 *M*) pulse (87 nl) delivered from a glass pipette that was placed about 15 μm from the cell body. The stimulation is shown by the horizontal bar below the trace. (B) Examples of the expanded secretory events with base widths ranging from 4 to 40 msec and an average width of 14 msec (13 cells and 12,324 transients). Data points were acquired every 1.0 msec and typical rise times were 2 to 4 msec. (C) Frequency vs attomoles of DA released (16 cells, 18,456 release events). Only release events with base widths less than 40 msec were considered. (D) Frequency vs the cube root of attomoles of DA released (16 cells and 18,456 events). Based on the data from all 16 cells, the average vesicle content was 1.36 $\pm$ 0.53 amol (mean $\pm$ SD). [Reproduced with permission from G. Chen, P. F. Gavin, G. Luo, and A. G. Ewing, *J. Neurosci.* **15,** 7747 (1995).]

whereas more distal events contribute only to the overall background current, if at all.[21]

A histogram of quanta constructed by binning *N* values is shown in Fig. 4C. The histogram reveals more than one distribution of quanta. The bimodal distribution could result from a bimodal distribution of vesicle radii with a constant DA concentration. Evidence supporting this hypothesis is shown in Fig. 4D, in which the cube root of the number of molecules released per transient is used as the ordinate of the histogram. If the vesicles in the GDN are spherical with a near Gaussian distribution of radii, and the vesicular DA concentration is either constant or in a uniform distribution, the cube root of the amount of DA histogram should also be a near Gaussian distribution. In fact, two apparent Gaussian distributions are

[21] T. J. Schroeder, J. A. Jankowski, K. T. Kawagoe, R. M. Wightman, C. Lefrou, and C. Amatore, *Anal. Chem.* **64,** 3077 (1992).

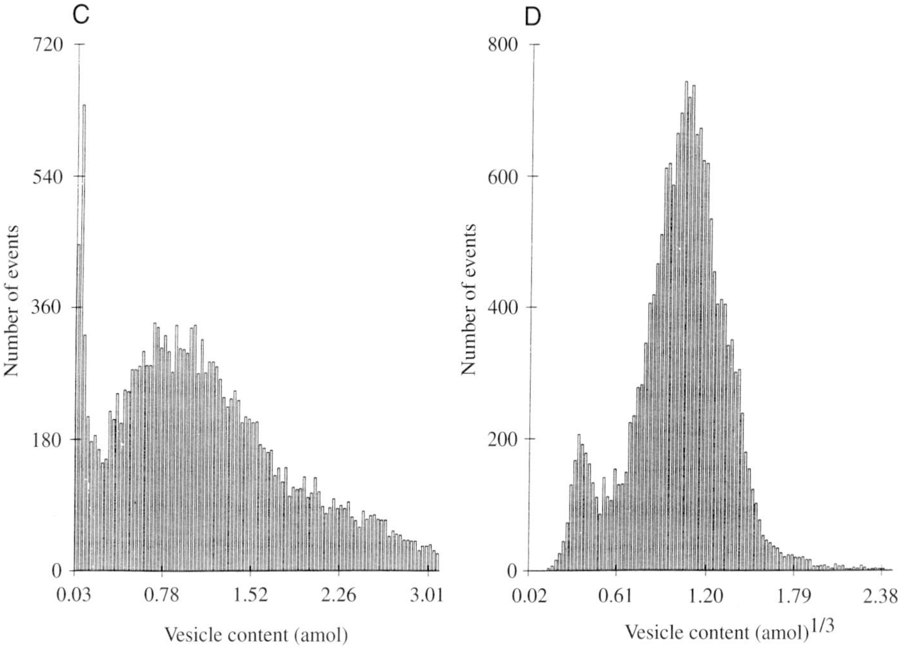

FIG. 4. (*continued*)

present, suggesting that there are at least two distinct classes of vesicles in this cell and that the variability is due to vesicle radius.

Effect of Amphetamine on Quantal Size

It has been suggested that amphetamine acts on vesicles to displace DA by competing for intravesicular free protons and/or by an action at the vesicular monoamine transporter. Measurements of exocytosis have provided evidence for this hypothesis. Incubation of the GDN with 10 μM *d*-amphetamine shifts the quantal distribution to the left.[20] Figure 5A shows a cube root of DA content histogram for control data with the bimodal distribution described for the GDN. After incubation with amphetamine, a third peak becomes apparent (Fig. 5B). An overlay of the control and amphetamine treated data is shown in Fig. 5C. These data suggest that DA is displaced from some of the vesicles to yield a third vesicle class after treatment with amphetamine.

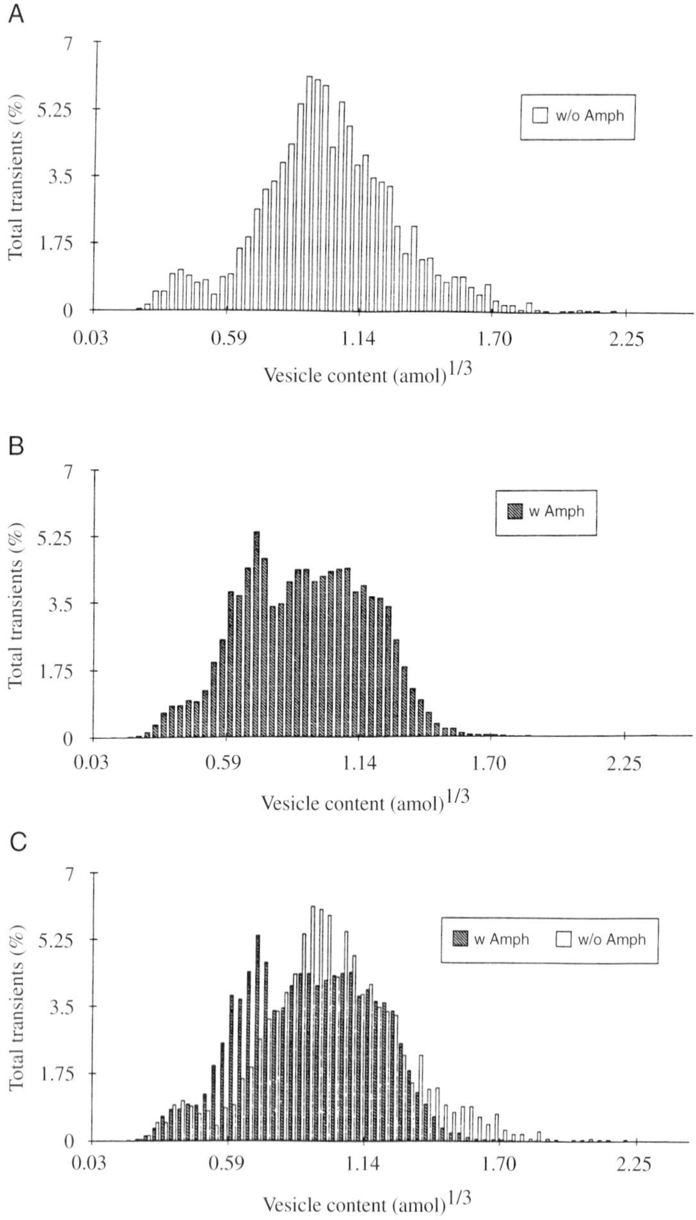

FIG. 5. Histograms of percentage of release transients vs the cube root of attomoles of DA released per vesicle. These data were obtained by amperometric measurements of KCl (1 *M*) stimulated release of vesicular DA from the cell body of the GDN before (5 cells,

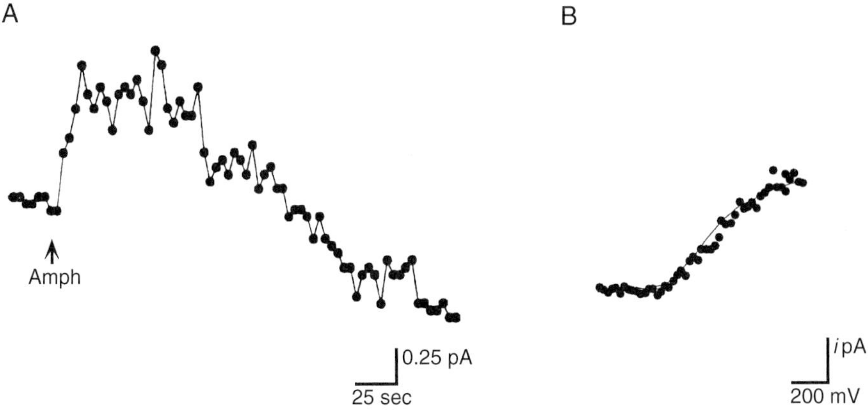

FIG. 6. Intracellular injection of amphetamine promotes DA release. (A) Amphetamine (8 pl of 100 mM) was injected intracellularly (arrow), thereby bypassing the uptake transporter, and the increase in extracellular DA measured. Similar levels of DA release were found in three replications of the experiment using DC amperometry. The small, slow baseline decrease during these experiments results from adsorption of electrode oxidation products to the electrode surface [Y. Y. Lau, J. B. Chien, D. K. Y. Wong, and A. G. Ewing, *Electroanalysis* **3,** 87 (1991)]. (B) The shape of the voltammogram (0.2 initial potential to 0.8 V final potential) taken at the peak current value 14 sec postinjection (filled circles) closely matched the voltammogram for a DA standard (solid line). Calibration: *i* equals 0.5 for the experiment and 31.5 for the standard. [Reproduced with permission from D. Sulzer, T. K. Chen, Y. Y. Lau, H. Kristensen, S. Rayport, and A. Ewing, *J. Neurosci.* **15,** 4102 (1995).]

Amphetamine-Induced Reverse Transport

In addition to changing quantal size as observed by exocytosis experiments, these techniques indicate that amphetamine induces reverse transport of DA from the cell body via the DA transporter.[20] Figure 6 shows the data obtained from measuring reverse transport by voltammetry at a carbon ring electrode. There is a rapid increase in DA concentration detected outside the cell to a maximum concentration of 1.43 ± 0.58 μM (mean $\pm$ SEM, $n = 3$) following intracellular injection of 8 picoliters of 100 mM amphetamine (Fig. 6A). The injection results in an intracellular amphetamine concentration of 100 μM based on a cytosolic volume estimation of 8 nl. Voltammetry of the released substance, shown in Fig. 6B,

2,669 release events) and after (5 cells, 8,987 release events) amphetamine incubation. (A) Response obtained before amphetamine treatment. (B) Response obtained from a 20 min incubation with 10 μM amphetamine. (C) Histograms from (A) and (B) superimposed on the same x axis. [Reproduced from G. Chen and A. G. Ewing, *Brain Res.* **701,** 167 (1996) with kind permission from Elsevier Science—NL, Sara Burgerhartstraat 25, 1055 KV Amsterdam, The Netherlands.]

reveals a wave shape similar to that of an *in vitro* DA voltammogram, indicating that DA is the detected substance. Amphetamine, which lacks hydroxyl groups, is not electrochemically reactive at the relatively low voltage used and does not interfere with these measurements.

Resolving DA Release by Reverse Transport and Exocytosis

These studies directly demonstrate reverse transport from a neuron in real time and that the DA uptake blocker nomifensine blocks reverse transport. The observations are consistent with studies (e.g., ref. 22) in which cloned neurotransmitter transporters are expressed in nonneuronal cells and actions can be examined without an intact neurosecretory system. However, neurons such as the GDN are well suited for analysis of mechanisms that involve both cytosolic and vesicle DA sequestration, such as studies of amphetamine action.

Elevation of extracellular DA due to uptake blockade, vesicle exocytosis, and reverse transport can clearly be resolved in the GDN. Because the cell is the only DA-secreting cell in the ganglion and extracellular DA is undetectable if the cell is not stimulated, uptake blockade does not underlie the present results. Following application of a secretagogue, tens of thousands of quantal events can be recorded amperometrically.[23] Dopamine (DA) release stimulated by intracellular injection of DA or amphetamine occurs on a much slower time scale than exocytosis, so current transients are not observed. The lack of apparent exocytic release following injection of DA along with inhibition of release by nomifensine indicates that reverse transport is mostly through the transporter and that altering the equilibrium across the transporter can induce reverse transport. The ability of intracellularly injected amphetamine to induce reverse transport suggests that redistribution of transmitter from the vesicles to the cytosol can promote reverse transport even in the absence of amphetamine uptake by the plasma membrane transporter, although additional actions of amphetamine as a plasma membrane transporter substrate appear to contribute to the action of this drug. The ability of electrochemical techniques to resolve different pools of DA in the GDN, including cytosolic levels, vesicular levels, and transmitter released by reverse transport, appears to make it an excellent preparation for determining the relationship between transmembrane ionic, voltage, and chemical gradients, as well as vesicle/cytosolic gradients, and the kinetics of flux through the transporter.

[22] C. Pifl, H. Drobny, H. Reither, O. Hornykiewicz, and E. A. Singer, *Molec. Pharmacol.* **47,** 368 (1995).

[23] G. Chen and A. G. Ewing, *Brain Res.* **701,** 167 (1996).

Acknowledgments

The work reviewed was supported, in part, by grants from the National Science Foundation (AGE) and the National Institute of Drug Abuse and the Parkinson's Disease Foundation (DS). We gratefully acknowledge the efforts of co-workers who have contributed to the work referenced in this chapter.

[47] Measuring Uptake Rates in Intact Tissue

By MELISSA A. BUNIN and R. MARK WIGHTMAN

Introduction

Measurements of the uptake of neurotransmitters have been performed in a multitude of experimental preparations, from synaptosomes to brain slices to the living brain. Comparison of the uptake parameters determined from these different experiments is not always straightforward, as the results depend on the tissue preparation and the methods of quantification, as well as the experimental conditions. Characterizing uptake from experiments in intact tissue can be complicated due to the convolution of diffusion and uptake. This laboratory uses electrical stimulation to evoke the release of endogenous dopamine and utilizes fast scan cyclic voltammetry (FSCV) at carbon fiber microelectrodes to detect the resulting concentration in the extracellular fluid and the changes in concentration due to uptake. This analytical technique provides a method for the determination of uptake kinetics in intact brain tissue that avoids the problems typically encountered in intact tissue. Simulation of the experimental response reveals values for V_{max} and K_m that closely resemble those previously obtained in synaptosomes. Through the use of this technique, the effects of various pharmacological agents on the rates of dopamine (DA) uptake have been determined and compared in various brain regions. Furthermore, the kinetic parameters for uptake have been determined and compared in many brain regions, demonstrating the heterogeneous nature of the uptake process and the complexity of dopaminergic neurotransmission.

Traditional Methods for Measuring Neurotransmitter Uptake

Studies in Synaptosomes and Homogenates

The classic method for measuring neurotransmitter uptake kinetics is to incubate membrane fractions (synaptosomes) in varying concentrations

0076-6879/98 $25.00

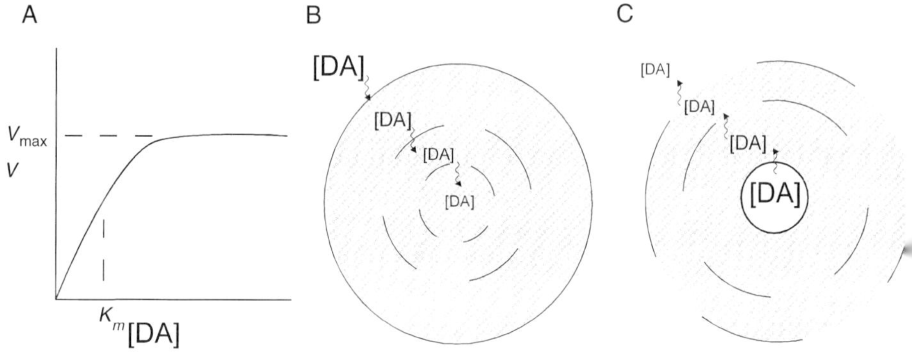

FIG. 1. (A) The classic-Michaelis–Menten plot used to determine uptake kinetics. (B) Schematic diagram of the problems inherent in determining uptake kinetics in intact tissue preparations when the substrate is applied to the solution surrounding the tissue. Diffusion and uptake couple, causing a concentration gradient within the tissue, such that the DA concentration at the center of the tissue is much smaller than that applied to the solution. The concentration is thus overestimated. (C) Schematic diagram of the problems inherent in determining uptake kinetics in intact tissue preparations when the substrate is applied as a bolus within the tissue. As the substrate moves outward, it decreases in concentration due to the processes of diffusion and uptake. The concentration is thus underestimated.

of radioactive substrate for various periods of time and measure the accumulation of the substrate by the homogenized tissue. Michaelis–Menten plots such as that shown in Fig. 1A can be constructed to determine the apparent values of the kinetic parameters, K_m and V_{max}, for various experimental conditions.[1] Then presumed uptake inhibitors or other pharmacological agents can be applied to examine their effects on the rate of substrate accumulation.[2]

Although well established, the use of homogenized tissue has its drawbacks. First, the uptake processes occurring *in vivo* depend on the local concentration of the neurotransmitter; the concentrations employed in experiments using synaptosomes may differ from the physiological concentrations occurring in the intact brain. Second, homogenization destroys the compartmentalization and organization that occurs *in vivo*.[3] The use of homogenized tissue results in a microscopic view of the uptake process which may significantly differ from the macroscopic picture seen *in vivo*. The environment of the uptake transporter in intact tissue is much more

[1] L. Annunziato, P. Leblanc, C. Kordon, and R. I. Weiner, *Neuroendocrinology* **31,** 316 (1980).

[2] A. S. Horn, *Prog. Neurobiol.* **34,** 387 (1990).

[3] A. L. Green, *J. Pharm. Pharmac.* **28,** 265 (1976).

complex than the *in vitro* environment, and the effects of the complete system on the transport process (i.e., regulatory mechanisms, ionic fluctuations, etc.) may be overlooked or oversimplified in such *in vitro* experiments.

Studies in Intact Brain Tissue

For the reasons listed above, investigation of the uptake process has moved toward the study of intact brain tissue. Brain slices have been used in much of the same manner as synaptosomes have been used for the classic experiments described above.[4] Alternatively, the substrate has been applied through the perfusion buffer and its permeation through the slice has been measured by a probe located within the slice tissue.[5] In addition, chopped rather than homogenized tissue has been used to obtain kinetic information in relatively intact tissue in order to compare these results with those obtained from synaptosomes.[6]

Although studies in intact brain tissue appear to remove many of the problems associated with homogenized tissue, they, too, have their disadvantages. Most important of these is that, with regard to the uptake site as well as the actual uptake process, diffusion affects the data obtained in these experiments. If this is not recognized, it can lead to a misinterpretation of the results. For example, the uptake kinetics obtained in chopped tissue were profoundly different from those obtained with synaptosomes, implying the existence of both high affinity and low affinity uptake.[6] In intact tissue, the effects of diffusion and uptake or metabolism couple and lead to a concentration gradient for the substrate inside the tissue. This is shown schematically in Fig. 1B: the concentration of substrate applied to the solution surrounding the tissue (in this case, [DA]) gets smaller and smaller in size as it diffuses through the tissue, having the lowest size (concentration) at the very center of the tissue. This is because the substrate continually encounters uptake sites as it diffuses through tissue containing the same transporter that removes it from solution. Therefore, the concentration within the tissue is nonuniform and significantly lower than that in the surrounding solution. Because kinetic analysis requires that the concentration at the transporter be known (i.e., for the ordinate as in Fig. 1A), this causes a distortion of the kinetic values that are determined.[3] If corrections were made to this data to account for diffusion, then accurate measurements of uptake could presumably be obtained. However, such corrections are complex[3] and are usually not attempted. For this reason, it was concluded

[4] C. Missale, L. Castelletti, S. Govoni, P. F. Spano, M. Trabucchi, and I. Hanbauer, *J. Neurochem.* **45,** 51 (1985).
[5] R. S. Kelly and R. M. Wightman, *Brain Res.* **423,** 79 (1987).
[6] J. A. Near, J. C. Bigelow, and R. M. Wightman, *J. Pharm. Exp. Ther.* **245,** 921 (1988).

that the "low affinity component" observed in chopped tissue was an artifact due to hindered diffusion in such preparations.

An alternate route for measuring uptake rates in intact tissue involves applying the substrate to tissue preparations in a small bolus that then diffuses through tissue to a concentration sensor.[5] Such experiments have been performed in brain slices and in the intact brain of anesthetized animals through the use of pressure ejection or microiontophoresis from micropipettes containing the substrate of interest.[7] In this type of experiment, the substrate is applied at a fixed distance (usually ~100 μm) from a probe detecting the substrate concentration within the tissue. The maximum amplitude measured (the substrate concentration at the probe) and the rate of disappearance of the substrate after its application can then be examined. Drugs which may affect uptake can be applied (with a fixed concentration of substrate) to examine their effects on the amplitude and time course of the signal obtained by the probe.[8]

This type of method also leads to complex results because of the coupling of diffusion with uptake. In other words, with techniques such as pressure ejection and microiontophoresis, the substrate must diffuse outward for detection, being taken up by transporters in the process (Fig. 1C). It has been shown that the declining phase of the curves obtained with pressure ejection *in vivo* contains elements of diffusion of the substrate as well as uptake processes.[7] Thus, the determination of uptake parameters from these curves is quite complex.[9]

Idealized Sensor for Detection of Neurotransmitter Dynamics in Intact Tissue

In order to characterize uptake kinetics in intact brain tissue while overcoming the problems discussed above, a method which utilizes endogenous changes in neurotransmitter concentration is desirable. Then, differences between the exogenous concentrations applied experimentally and the typical endogenous concentrations occurring in reality would be irrelevant. In addition, in regions of high neurotransmitter density, where endogenous changes in neurotransmitter concentration are large, the disappearance of the endogenous substance would correlate directly to its uptake, without any complication by the effects of diffusion, if (1) the probe used to detect the substance is small enough, (2) detection of the substance is fast enough, and (3) uptake rates are sufficiently fast. For instance, in the

[7] G. A. Gerhardt and M. R. Palmer, *J. Neurosci. Meth.* **22,** 147 (1987).

[8] W. A. Cass, N. R. Zahniser, K. A. Flach, and G. A. Gerhardt, *J. Neurochem.* **61,** 2269 (1993).

[9] C. Nicholson, *Biophys. J.* **68,** 1699 (1995).

caudate nucleus, dopaminergic nerve terminals are known to be an average of 4 μm apart.[10] If these terminals were to release their neurotransmitter simultaneously, the concentration between each neuron would become homogeneous in about 3 msec.[11] Thus, a concentration gradient does not exist in such a local environment and diffusion is not a factor. If the probe used to measure this change in neurotransmitter concentration were small enough so that it only sampled this active and homogeneous region of the brain, and if the probe sampled the extracellular fluid composition with a sampling time longer than 3 msec but not so long that the rapid, local process of uptake removed all of the neurotransmitter, then the probe would measure a disappearance rate that is similar to that at the neuronal surface.

Characterizing Uptake Kinetics with Fast Scan Cyclic Voltammetry

Advantages of Measuring Uptake with FSCV

We have used FSCV to detect dopamine release evoked by electrical stimulation in the intact brain of the anesthetized rat and in rat brain slices.[12,13] Because it elicits the release of endogenous dopamine, electrical stimulation reduces concern over the differences between the local substrate concentrations occurring naturally in the intact brain and those applied exogenously in experiments employing autoradiography, perfusion, microiontophoretic, and pressure injection techniques. Furthermore, the maximal stimulus intensities used in our experiments activate the neurons (within the range of the stimulating electrode) uniformly, such that a maximal number of terminals are forced to release simultaneously.

FSCV with carbon-fiber microelectrodes (CFE) is a well-established technique for the measurement of DA in intact brain tissue.[14,15] The superior spatial (micron) and temporal (millisecond) resolution of FSCV enable site-specific, real-time detection of electrically stimulated neurotransmitter release and subsequent disappearance.[12] As discussed earlier, the capability to make spatially and temporally resolved measurements is a necessity for the meaningful evaluation of extracellular neurotransmitter dynamics; thus, FSCV is well suited for this purpose. This technique causes minimal damage

[10] V. M. Pickel, S. C. Beckley, T. K. Joh, and B. J. Reis, *Brain Res.* **225,** 373 (1981).

[11] P. A. Garris, E. L. Ciolkowski, P. Pastore, and R. M. Wightman, *J. Neurosci.* **14,** 6084 (1994).

[12] W. G. Kuhr and R. M. Wightman, *Brain Res.* **381,** 168 (1986).

[13] R. T. Kennedy, S. R. Jones, and R. M. Wightman, *Neuroscience* **47,** 603 (1992).

[14] K. T. Kawagoe, J. B. Zimmerman, and R. M. Wightman, *J. Neurosci. Methods* **48,** 225 (1993).

[15] J. A. Stamford, P. Palij, C. Davidson, C. M. Jorm, and P. E. M. Phillips, *in* "Voltammetric Methods in Brain Systems" (A. A. Boulton, G. B. Baker, and R. N. Adams, Eds.), p. 81. Humana Press, Clifton, NJ, 1995.

to the brain tissue due to the small tip size (~15 μm) of the probe employed for measurement. In addition, there is minimal perturbation of the concentration of DA being measured because of the short time scale of the measurements and the return of the oxidation product generated during the detection of DA to its original form.[16,17] Finally, the small size of the electrode means that it can be placed well within the boundaries of brain regions activated by electrical stimulation, such that one of the requirements for an ideal sensor (listed above) is fulfilled.

Because the time resolution of FSCV is 100 msec or less, when we position our electrode in a region where there is dense dopaminergic innervation, the use of electrical stimulation and FSCV fulfills the second of the requirements for an ideal sensor, thereby eliminating concern over the effects of diffusion on our measurements. Thus, the use of electrical stimulation and FSCV generates data displaying real-time temporal changes in DA concentrations within intact brain tissue. Because a concentration gradient does not exist, the measured concentration is the same as that at the release site.

Simulating Experimental Data with Michaelis–Menten Model

Dopamine uptake is a saturable process that is described by Michaelis–Menten kinetics:

$$V = -\{d[\mathrm{DA}]/dt\}_{\mathrm{uptake}} = V_{\max}/(\{K_m/[\mathrm{DA}] + 1\}) \quad (1)$$

The classic Michaelis–Menten plot used to characterize enzyme kinetics, such as uptake by the dopamine transporter, is shown in Fig. 1. Uptake of radiolabeled DA into synaptosomes and brain homogenate has revealed regionally distinct values of $V_{\max}$ for the uptake of DA in various brain regions.[2] In most regions, the K_m value is similar.

We have shown that the temporal response curves obtained by FSCV for the electrically stimulated release of DA contain information on both uptake and release. Figure 2 displays a simulated curve for the temporal response obtained by FSCV after a one-pulse stimulation. During the stimulation, marked by the open squares, the concentration of DA in the extracellular fluid rises. The release process has been characterized elsewhere and is not discussed in this chapter.[18,19] After the stimulation pulse has ceased,

[16] A. G. Ewing, M. A. Dayton, and R. M. Wightman, *Anal. Chem.* **53,** 1842 (1981).

[17] R. M. Wightman, L. J. May and A. C. Michael, *Anal. Chem.* **60,** 769A (1988).

[18] R. M. Wightman, C. Amatore, R. C. Engstrom, P. D. Hale, E. W. Kristensen, W. G. Kuhr, and L. J. May, *Neuroscience* **25,** 513 (1988).

[19] R. M. Wightman and J. B. Zimmerman, *Brain Res. Rev.* **15,** 135 (1990).

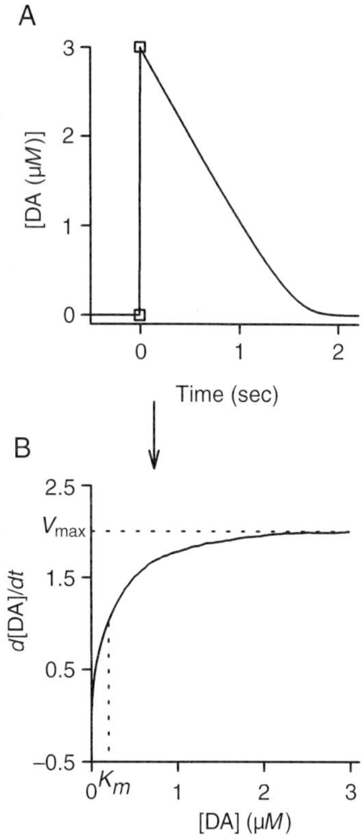

FIG. 2. (A) Simulated curve for release evoked by a single stimulation pulse followed by uptake with $V_{max} = 2\ \mu M$/sec and $K_m = 0.16\ \mu M$. (B) Michaelis–Menten plot obtained by taking the derivative of the DA concentration in (A) with respect to time and plotting it vs the measured concentration. The values for V_{max} and K_m are exactly those used for the simulation.

the DA concentration rapidly returns to baseline. This descending half of the curve relies solely on cellular uptake, because diffusion is far too slow to account for the observed rate of clearance of evoked DA from the extracellular fluid,[20] and because no significant release is occurring during this time because the electrical stimulation and associated neural activity have ceased.[21] Therefore, the descending portion of the curve can be modeled by the Michaelis–Menten equation described above.

Investigation of the Michaelis–Menten equation for DA uptake reveals a very useful situation when the concentration of DA is much greater than

[20] A. G. Ewing and R. M. Wightman, *J. Neurochem.* **43,** 570 (1984).
[21] W. G. Kuhr, R. M. Wightman, and G. V. Rebec, *Brain Res.* **418,** 122 (1987).

K_m. In this case, the right side of the equation simplifies to merely V_{max}, such that the rate of DA disappearance is equal to the maximal rate of uptake. Thus, after stimulations which result in the release of large amounts of DA, the descending linear slope of the response curve yields V_{max} as long as the concentration of DA is greater than or equal to 10 times the value of K_m. Once V_{max} has been calculated in this manner, it can be substituted into the Michaelis–Menten expression above, along with the concentration of DA and its rate of disappearance (obtained from the temporal response curve) to solve for K_m.

As shown in Fig. 2, this approach is not far from the classic use of a Michaelis–Menten plot to estimate the values of V_{max} and K_m from autoradiography experiments employed to characterize enzyme rates. If the response curve obtained by electrical stimulation and FSCV is differentiated and the clearance of [DA] with respect to time is plotted vs DA concentration, a Michaelis–Menten plot exactly like that shown in Fig. 1 is generated. From this plot, V_{max} and K_m can be obtained in the same manner as they are obtained in classical Michaelis–Menten experiments. In Fig. 1, the Michaelis–Menten plot generated gives the same values for V_{max} and K_m as those that were used to simulate the response curve. Similarly, Lineweaver–Burk, Scatchard, Hofstee, and any number of plots based on rearrangement of the Michaelis–Menten equation can be generated from our FSCV data.

Rather than performing the tedious calculations necessary to convert FSCV data to Michaelis–Menten-type plots, we have chosen an alternate way to use this model. Based on the estimates of V_{max} and K_m, Eq. (1) is numerically integrated with locally written software, and the simulated curve is then compared to the data. Typically, the value for K_m is taken as the literature value from synaptosome studies for the brain region of interest, V_{max} is determined from the maximum slope of the descending side of the curve, and the release term is optimized based on the maximum correlation coefficient (or best fit) obtained when compared to the actual data. More recently, kinetic parameters for DA release and uptake have been calculated from efflux curves utilizing nonlinear regression by simplex optimization. In this way, the parameters are obtained without experimenter bias and K_m can be determined directly from the experiment. Such values obtained for K_m are in good agreement with the values obtained in synaptosomes for various brain regions.[22] Likewise, the values obtained for V_{max} agree reasonably well with those previously obtained in synaptosomes or homogenates.[9]

[22] S. R. Jones, P. A. Garris, C. D. Kilts, and R. M. Wightman, *J. Neurochem.* **64**, 2581 (1995).

It is important to note that the units for V_{max} obtained by FSCV differ from those obtained in radiography studies. Measurements of V_{max} in synaptosome, tissue, or membrane preparations in which DA is applied exogenously are reported in pmol/mg tissue/min (or similar units). This is because such experimental procedures vary considerably. In particular, the amount of tissue used in the assay depends on the given experiment, and there is the need to account for dilution. However, the value for V_{max} obtained by FSCV with electrical stimulation has units of μM/sec. This is because FSCV detects the real concentration in the extracellular fluid at any given moment. This concentration depends neither on the amount of tissue present nor on the effects of dilution.

Deconvoluting Effects of Nafion

A schematic view of the events to be considered during detection of electrically evoked release and uptake is shown in Fig. 3. On stimulation,

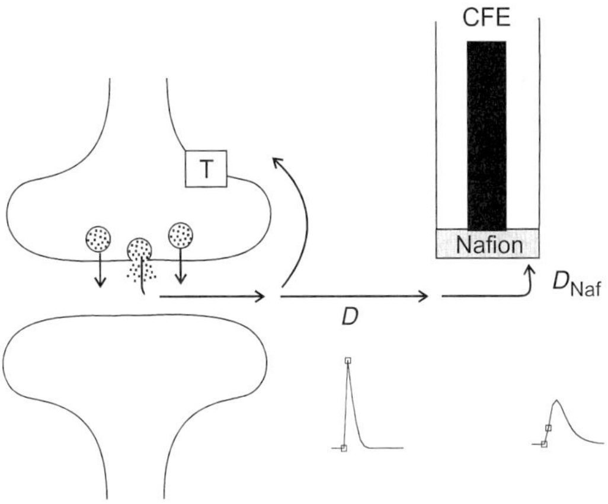

FIG. 3. Schematic diagram portraying the events to be considered during detection of electrically evoked release and uptake. During the stimulation, DA is released from presynaptic vesicles into the synapse, from which it overflows into the extracellular space. Once at the synaptic–extracellular interface, DA diffuses through the extracellular space with a speed determined by *D*, the diffusion coefficient in the tissue. Termination of DA neurotransmission is accomplished by DA transporters. The carbon fiber electrode (CFE) used to detect the changes in DA concentration is larger than the neuronal components and samples the synaptic overflow from many neurons. The electrode is coated with Nafion, which has a diffusion coefficient (D_{Naf}) that is much smaller than *D*. The ideal response for a given change in DA concentration after a one-pulse stimulation is shown beneath "*D*"; this is the response that would be obtained if there were no temporal delay inherent in the detection system. The response obtained with a Nafion-coated CFE is shown beneath "D_{Naf}"; the temporal delay caused by the Nafion film is readily apparent.

DA is released from vesicles in the presynaptic terminal into the synapse, from which it can overflow into the extracellular fluid (ECF). Either within the synapse or at extrasynaptic uptake sites,[11,23] termination of dopaminergic transmission is primarily accomplished by DA transporters, which reaccumulate DA into the terminal region where it can be subsequently repackaged into vesicles or eliminated by metabolism. The electrode used to measure the changes in DA concentration is typically about 15 μm in total diameter. Thus, the electrode tip is larger than the majority of cellular components and is unable to sample concentrations within an individual synapse. Instead, it samples the synaptic overflow from many different structures. Nonetheless, the electrode can provide information about the concentration of released substances at the synaptic–ECF interface, as described earlier in this chapter. It should be noted that with disk electrodes, the kinetic values obtained apply only to the brain area within a few micrometers of the electrode; however, when cylinder electrodes are employed, the signal is an average over the length of the cylinder, such that a much larger brain region is sampled with each measurement and heterogeneity is not so easily observed. Because the sites of greatest dopamine release ("hotspots") are very localized,[24] the averaged response obtained by cylinders is often lower than that of disk electrodes. Thus, different results are obtained with different sizes of electrodes, a point that should be considered when comparing results.

Our electrodes are coated with Nafion (Aldrich, Milwaukee, WI), a perfluorinated ion-exchange membrane, in order to increase sensitivity and selectivity for catecholamines.[25] Many easily oxidized substances exist throughout the brain and can be detected with FSCV. Ascorbate has the highest concentration of all of these substances throughout the brain. In addition, DOPAC, the major metabolite of DA, is present at relatively high concentrations where DA neurotransmission occurs. Detection of these substances can complicate the accurate detection of changes in DA concentration. The use of Nafion prevents anions from reaching the electrode surface, virtually eliminating concern over interference from anions such as ascorbate and DOPAC. In addition, the Nafion film accumulates cations such as DA in the time between scans, increasing the electrode sensitivity for DA to 16 times the theoretically calculated value.[26] Another advantage

[23] M. J. Nirenberg, R. A. Vaughan, G. R. Uhl, M. J. Kuhar, and V. M. Pickel, *J. Neurosci.* **16,** 436 (1996).

[24] S. D. Young and A. C. Michael, *Brain Res.* **600,** 305 (1993).

[25] G. A. Gerhardt, A. F. Oke, G. Nagy, B. Mohaddam, and R. N. Adams, *Brain Res.* **290,** 390 (1984).

[26] J. E. Baur, E. W. Kristensen, L. J. May, D. J. Wiedemann, and R. M. Wightman, *Anal. Chem.* **60,** 1268 (1988).

is that the use of Nafion protects the electrode surface from protein adsorption, which can decrease sensitivity, resulting from the implantation process. Furthermore, the restricted diffusion within the Nafion film ensures that, for the time scale of our measurements, we are sampling only the composition within the Nafion; therefore, comparison of the amplitudes of efflux curves in brain tissue and postcalibrations need not consider the differences between the complex milieu that exists in brain tissue and the free solution used for postcalibrations.[27]

In spite of these advantages, Nafion slows the temporal response of the electrode, such that the DA efflux curves obtained with Nafion-coated electrodes are temporally distorted. Figure 3 shows actual signals obtained for release of DA in the caudate nucleus after a one-pulse electrical stimulation. The effects of the Nafion film can be readily seen by comparing the two overflow curves, where the overflow curve on the left is that which would be observed if no Nafion were present, and the curve on the right is that which is obtained with a Nafion-coated electrode. At the synaptic–extracellular interface the change in DA concentration appears as a curve similar to that shown in Fig. 2, where the effects of Nafion are not taken into account. The DA then diffuses through the brain tissue to the electrode, where it is accumulated by the Nafion film. Because the diffusion coefficient in Nafion (D_{Naf}) is three orders of magnitude less than in free solution,[28] it takes much longer for DA to reach the electrode through the Nafion film than through the same volume of free solution. This greatly slows the response of the electrode to a concentration change. In addition, the restricted diffusion in the Nafion film results in an apparent rise in DA concentration even after the stimulation has ceased (indicated by the second open square).

To minimize the temporal delay caused by Nafion, the films applied to our electrodes are kept at a minimum thickness (200–500 μm).[27] However, slowed temporal responses are still apparent, and it is necessary to account for the presence of the Nafion film when analyzing the measured responses in brain tissue. Postcalibrations can be used to do just this. In a flow injection analysis system (FIA), a concentration step of DA is applied to the electrode *in vitro*. The observed response to the concentration step reveals the temporal delay caused by the Nafion film. This delay is described by $t_{1/2}$, the time required for the electrode response to reach half-maximal amplitude. Typically, it takes <500 msec for the maximum amplitude of the concentra-

[27] E. W. Kristensen, W. G. Kuhr, and R. M. Wightman, *Anal. Chem.* **59,** 1752 (1987).

[28] G. Nagy, G. A. Gerhardt, A. F. Oke, M. E. Rice, R. N. Adams, R. B. Moore, M. N. Szentimay, and C. R. J. Martin, *J. Electroanal. Chem. Interfacial Electrochem.* **188,** 85 (1985).

tion pulse to be detected at a Nafion-coated electrode that has previously been used for measurements in brain tissue.

The effect of diffusion through a thin layer, such as Nafion, when exposed to a concentration step is exactly known.[29] From this equation, the established diffusion coefficient for DA in Nafion ($D_{Naf} = 1 \times 10^{-9}$ cm^2/sec),[27] and the measured response time of the electrode, the apparent thickness of the Nafion film can be calculated.[29,30] The following empirical expression, obtained by evaluation of the diffusion coefficient equation, provides an adequate estimate of the Nafion film thickness:

$$l[\text{thickness (nm)}] = 521.9[t_{1/2}(s)]^{1/2} - 3.8 \tag{2}$$

Once the apparent film thickness is known, the temporal effects of Nafion on the *in vivo* response can be removed by the mathematical process of deconvolution.[30] In the regions of dense dopaminergic innervation studied in this laboratory, deconvoluting the effects of the Nafion film results in a DA efflux curve that begins to return to baseline immediately after the stimulation has been terminated. Because after deconvolution there is no continued rise in DA concentration after the stimulation has ceased, it is clear that there is no measurable effect of diffusion through brain tissue. This result shows that the concentration of DA measured at the electrode tip is, in fact, equal to that at the synaptic–extracellular interface, as claimed earlier.

In actual practice, the simulated signal obtained using Michaelis–Menten kinetics is convoluted with the effects of the Nafion film. This is to avoid noise artifacts that occur during the deconvolution of real data. In addition, it allows the primary experimental data to be compared directly to the simulation.

It is imperative to account for the effects of the Nafion film (or for the temporal response inherent in any biosensor) when simulating curves to determine the kinetic parameters for uptake, or else the values determined will be erroneous. In fact, some of the first experiments performed in this laboratory made this error. In initial experiments measuring electrically evoked DA efflux in the caudate nucleus of anesthetized rats, a high capacity, low affinity uptake system seemed apparent, while there was failure to observe any evidence of a high affinity system.[31,32] An example is shown

[29] R. C. Engstrom, R. M. Wightman, and E. W. Kristensen, *Anal. Chem.* **60,** 652 (1988).
[30] K. T. Kawagoe, P. A. Garris, D. J. Wiedeman, and R. M. Wightman, *Neuroscience* **51,** 55 (1992).
[31] J. A. Stamford, Z. L. Kruk, J. Millar, and R. M. Wightman, *Neurosci. Letts.* **51,** 133 (1984).
[32] R. M. Wightman, W. G. Kuhr, and A. G. Ewing, *Ann. NY Acad. Sci.* **473,** 92 (1987).

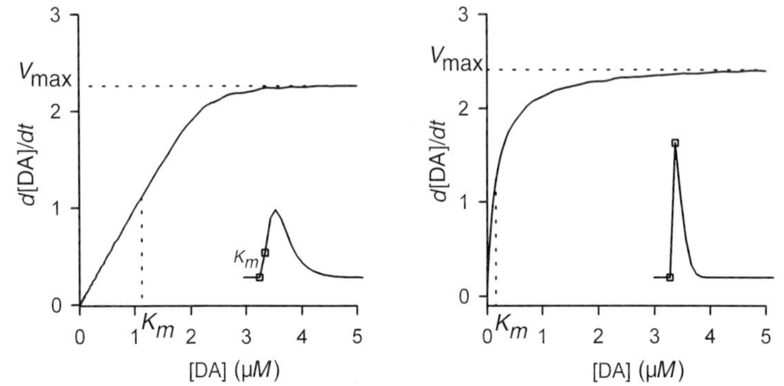

FIG. 4. Michaelis–Menten plots obtained in the manner of Fig. 2. (A) Data obtained by a Nafion-coated CFE (after a 60 Hz, 10 sec stimulation) without taking the temporal effects of the Nafion film into account. (B) Data in (A) after deconvolution of the effects of the Nafion film. The value obtained for K_m when the effects of Nafion are not deconvoluted is severely overestimated.

in Fig. 4, in which the differential (d[DA]/dt) of the DA overflow curve obtained in the presence of Nafion is plotted as a Michaelis–Menten plot. The values for V_{max} and K_m differ immensely from the actual values. After the response time of the electrode was accounted for, however, values for V_{max} and K_m which were more consistent with a high affinity system were obtained.[18,30] Thus, once again the effects of diffusion emerged in data from this laboratory, resulting in the initial false conclusion that a low affinity system exists in this dopaminergic region. To further emphasize this point, the DA efflux curve obtained after deconvolution of diffusion through the Nafion was differentiated and displayed as a Michaelis–Menten plot in Fig. 4. When the curve is extrapolated to DA concentrations that are greater than 10 times K_m, appropriate values for V_{max} and K_m are obtained. These are the same values obtained by the simulation procedure outlined above when the simulation based on Michaelis–Menten kinetics is convoluted with the effects of the Nafion film.

Determining Uptake

To summarize, there are many lines of evidence indicating that what we are measuring is indeed uptake. First of all, the rate of disappearance of DA from the ECF after cessation of the stimulation is far too rapid to be due to diffusion through brain tissue alone. Furthermore, no delay is observed in the disappearance after deconvolution of the effects of the Nafion film, implying that diffusion through brain tissue plays a negligible role in the measured DA disappearance. Another piece of evidence comes from Marc Caron's laboratory at Duke University. FSCV detection of electrically stimulated DA efflux in brain slices of DA transporter knockout mice revealed that DA persists in the extracellular space for a prolonged period of time, the time scale of DA clearance being consistent with that predicted for diffusion in the slice.[33] Comparison of the time for DA clearance in the wild-type (~1 sec) and knockout mice (100 sec) unequivocally demonstrates that the disappearance of DA in wild-type animals is due to a mechanism that is much more efficient than diffusion. Finally, the rate of disappearance is sensitive to well-characterized inhibitors of dopamine uptake. Nomifensine, bupropion, and cocaine significantly slow the rate of DA disappearance in the caudate putamen of anesthetized rats and in rat brain slices.[34,35] Application of fluoxetine and desipramine, potent and selective uptake inhibitors for 5-hydroxytryptamine and norepinephrine, respectively, do not significantly affect the uptake of DA in the caudate nucleus,[22] a result that is not surprising given the sparce serotonergic and noradrenergic innervation of this brain region. As stated above, the values for K_m and V_{max} obtained by simulation of uptake by Michaelis–Menten kinetics and deconvolution of the effects of the Nafion film, are in agreement with those obtained in synaptosomes. These results indicate that the predominant cause of DA disappearance after stimulation cessation is, in fact, uptake by DA transporters.

An alternate route for the clearance of DA from the extracellular fluid is metabolism by catechol *O*-methyltransferase (COMT) or monoamine oxidase (MAO). If metabolism is a source of DA clearance on the time scale of our measurements, the use of metabolic inhibitors should increase the residence time of DA in the extracellular space following electrical stimulation. Administration of pargyline, an MAO inhibitor, does not affect the time course of the DA efflux curves obtained by our methods. In addition, the COMT inhibitors tropolone and Ro 40-7592 have no significant effects on the rate of DA disappearance.[36] These results confirm that the extracellular clearance of evoked DA is not the result of metabolic degradation. Thus, FSCV measurements of electrically evoked DA efflux and disappearance reflect clearance for which the DA transporter is solely responsible.

[33] B. Giros, M. Jaber, S. R. Jones, R. M. Wightman, and M. Caron, *Nature* **379,** 606 (1996).
[34] L. J. May, W. G. Kuhr, and R. M. Wightman, *J. Neurochem.* **51,** 1060 (1988).
[35] S. R. Jones, P. A. Garris, and R. M. Wightman, *J. Pharm. Exp. Ther.* **274,** 396 (1995).
[36] P. A. Garris and R. M. Wightman, *Synapse* **20,** 269 (1995).

Elucidating Brain Function through Characterization of Uptake Kinetics

Assessment of Action of Pharmacological Agents

The kinetic results obtained in experiments with rat brain slices are consistent with those obtained in anesthetized rats.[22] In light of this fact, many of the experiments performed in this laboratory utilize the rat brain slice because the slice preparation is a more convenient way to examine the effects of pharmacological agents. Agents can be applied to the slice through the perfusion buffer, their effects can be rapidly evaluated and compared, and their reversibility can be established without concern over the different routes or rates of transport of the drugs to their sites of action. An added benefit is the ability to obtain many slices containing the same region (i.e., the caudate) or slices from different regions, such that numerous experiments can be performed on the same animal and on the same day; the *in vivo* experiment requires considerable time for setup, and typically only a single drug application is possible with a single anesthetized rat.

Numerous pharmacological agents have been applied to rat brain slices to assess their effects on the uptake of DA. As described above, the inhibition of DA uptake by agents such as nomifensine and bupropion has been affirmed in our experiments. In addition to affirming the action of these agents as DA uptake inhibitors, assessment of their actions as *competitive* uptake inhibitors and determination of apparent K_m values in their presence are possible. Figure 5 shows the efflux curves obtained by electrical stimulation (1 sec, 10 Hz) and FSCV detection in the caudate nucleus of a rat brain slice in the absence and presence of 2 μM nomifensine. Values for the release term, $V_{\max}$, and K_m were determined from the control experiment as described above. No significant difference was observed between the value of $V_{\max}$ measured prior to and after the administration of nomifensine, nor was the amount of DA release altered. Only a change in the apparent K_m resulted in a reasonable fit of the simulation to the experimental efflux curve. This amounted to an increase of K_m from 0.2 μM (control) to 4 μM. (Thus, the change in amplitude of the nomifensine signal is a result of decreased uptake during the time between stimulation pulses, and not the result of an increase in the amount of DA released.) The value of K_m in the presence of an uptake inhibitor, $(K_m)_{\text{app}}$, is related to the value for the dissociation constant for the reaction, K_i, by the following equation[35]:

$$(K_m)_{\text{app}} = K_m(1 + [\text{I}]/K_i) \tag{3}$$

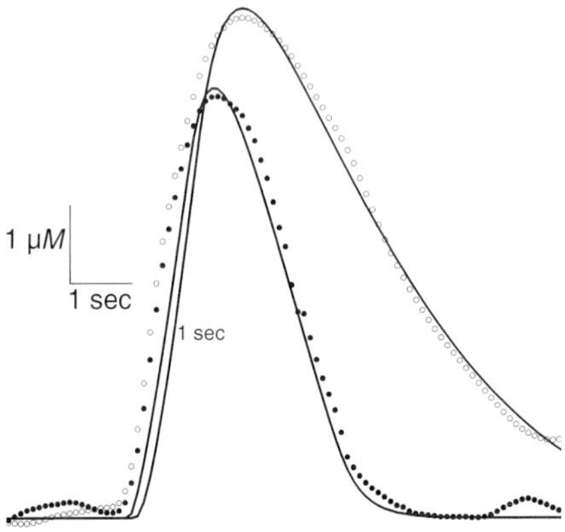

FIG. 5. DA response curves obtained in the caudate putamen after a 10 Hz, 1 sc (350 μA, 2 msec/phase) electrical stimulation. Solid circles: control; solid line is the simulation with $DA_p = 0.9\ \mu M$, $V_{max} = 3.0\ \mu M$/sec, $K_m = 0.16\ \mu M$. Open circles: after the application of 2 μM Nomifensine; solid line is the simulation with $DA_p = 0.9\ \mu M$, $V_{max} = 3.0\ \mu M$/sec, $K_m = 4.0\ \mu M$.

Thus, a plot of $(K_m)_{app}$ vs inhibitor concentration, [I], should be linear with a slope of K_m/K_i. From such a plot, the value of K_i for a given inhibitor can, therefore, be obtained. In this manner, K_i values for cocaine and nomifensine have been determined in the caudate, and these values are similar to those obtained in synaptosomes prepared from the caudate.[35]

Similar experiments have been performed in other brain regions and with other pharmacological agents. From such studies it was found that cocaine inhibited DA uptake with similar kinetics in the caudate and the nucleus accumbens, whereas nomifensine exhibited twice the potency as a DA uptake inhibitor in the caudate when compared to the nucleus accumbens.[35] These differences provide evidence that the transporter, itself, differs in the caudate and nucleus accumbens. In particular, the results support the existence of separate binding sites on the DA transporter for cocaine and nomifensine and suggest that, whereas the binding site for cocaine may be identical in the caudate and nucleus accumbens, that for nomifensine is different in the two regions. In another study, it was found that nomifensine had no observable effect and cocaine had only a minimal effect on DA uptake in the baseolateral amygdaloid nucleus.[22] This provides further evidence that DA systems do not exhibit a generalized pharmacology and suggests that different anatomic areas may implement DA neurotransmission in different ways.

Meaning of Uptake Rate Constants

This laboratory has investigated the rate of DA uptake in a variety of brain regions, including the caudate putamen (CP) and the nucleus accumbens (NAc).[22] It has been found that there is a large degree of variability in the rate of DA clearance between regions. For instance, the value for V_{max} obtained in the CPu is about twice that obtained in the core of the NAc. In addition, heterogeneity of DA uptake has been demonstrated within brain regions. Uptake in the NAc shell is threefold less than in the NAc core, a result that is opposite of that expected, given that the shell NAc contains a significantly greater tissue concentration of DA than the core region.[37] Furthermore, a significant, negative relationship was found between the rate of DA disappearance and depth within the striatum, showing that there are not just two distinct rates for DA uptake in the CP and NAc, but a gradual decrease in V_{max} on moving from the CP to the dorsal striatum, tail, and NAc.[38]

In an attempt to determine the basis of the heterogeneity observed between brain regions, we have compared the rates of uptake to the number of binding sites for the DA transporter in the CP, NAc core, and NAc shell. The Michaelis–Menten kinetics to describe uptake implies the following reaction scheme:

$$\mathrm{DA_o} + \mathrm{U} \underset{k_{-1}}{\overset{k_1}{\rightleftharpoons}} [\mathrm{DA} - \mathrm{U}] \overset{k_2}{\rightarrow} \mathrm{DA_i} + \mathrm{U}, \tag{4}$$

where DA_o and DA_i are extracellular and intracellular dopamine, respectively, and U is the dopamine transporter. The experimentally determined V_{max} is equal to $k_2[\mathrm{U}]_{tot}$, where $[\mathrm{U}]_{tot}$ is the total concentration of uptake sites (often referred to as B_{max} in binding studies).[11] For a V_{max} of 3.8 μM/sec[22] and a $[\mathrm{U}]_{tot}$ of 0.6 μM,[39] k_2 is calculated to be 6 $\sec^{-1}$ in the striatum. The turnover rate, k_2, is the mean number of DA molecules transported per second. The affinity constant, K_m, is given by:

$$K_m = (k_2 + k_{-1})/k_1 \tag{5}$$

Thus, setting $k_{-1} = 0$, the lower limit for k_1 can be estimated. With $K_m = 0.16$ μM for the striatum,[6] the lower limit for $k_1 = 4 \times 10^7 M^{-1} \sec^{-1}$. This value is comparable to that reported for the initial binding of acetylcholine to acetylcholinesterase,[40] one of the fastest biochemical reactions known.

[37] S. R. Jones, S. J. O'Dell, J. F. Marshall, and R. M. Wightman, *Synapse* **23,** 224 (1996).
[38] P. A. Garris, E. L. Ciolkowski, and R. M. Wightman, *Neuroscience* **59,** 417 (1994).
[39] E. K. Richfield, *Brain Res.* **540,** 1 (1991).
[40] T. Rosenberry, *Adv. Enzymol.* **43,** 103 (1975).

TABLE I
VALUES OF V_{max}, B_{max}, k_2, AND k_1 IN DIFFERENT REGIONS

Region[a]	V_{max} (μM sec^{-1})	B_{max} (μM)[b,c]	k_2 (sec^{-1})	k_1 (M^{-1} sec^{-1})
CP	3.8[d]	0.6	6	4×10^7
NAc core	2.4[d]	0.4	6	4×10^7
NAc shell	0.63[e]	0.2	3	2×10^7

[a] CP, caudate putamen; NAc, nucleus accumbens.

[b] For CPu: average of values from P. H. Anderson, *Eur. J. Pharmacol.* **166,** 493 (1989); J. W. Boja, W. M. Mitchell, A. Patel, T. A. Kopajtic, F. I. Carroll, A. H. Lewin, P. Abraham, and M. J. Kuhar, *Synapse* **12,** 27 (1992); B. Scatton, A. Dubois, M. L. Dubocovich, N. R. Zahniser, and D. Fage, *Life Sci.* **36,** 815 (1985); A. S. Horn, *Prog. Neurobiol.* **34,** 387 (1990); M. L. Dubocovich and N. R. Zahniser, *Biochem. Pharmacol.* **34,** 1137 (1985). For CPu and NAc core: J. F. Marshall, S. J. O'Dell, R. Navarrete, and A. J. Rosenstein, *Neuroscience* **37,** 11 (1990); J. A. Javitch, S. M. Strittmatter, and S. H. Snyder, *J. Neurosci.* **5,** 1513 (1985); F. Mennicken, *J. Neural Transm.* **87,** 1 (1992); P. Berger, J. D. Elsworth, J. Arroyo, and R. H. Roth, *Eur. J. Pharmacol.* **177,** 91 (1990); S. Izenwasser, L. L. Werling, J. G. Rosenberger, and B. M. Cox, *Neuropharmacology* **29,** 1017 (1990). For NAc shell: taken as 0.5× that for NAc core based on results from L. G. Sharpe, N. S. Pilotte, W. M. Mitchell, and E. B. De Souza, *Eur. J. Pharmacol.* **203,** 141 (1991); S. R. Boulay, I. Leroux-Nicollet, D. Duterte-Boucher, L. Naudon, and J. Costentin, *J. Neural. Transm.* **98,** 209 (1994); S. R. Jones, S. J. O'Dell, J. F. Marshall, and R. M. Wightman, *Synapse* **23,** 224 (1996).

[c] Original units are pmol/mg protein. Converted by dividing by 10.

[d] Values from S. R. Jones, P. A. Garris, C. D. Kilts, and R. M. Wightman, *J. Neurochem.* **64,** 2581 (1995).

[e] Values from P. A. Garris, E. L. Ciolkowski, and R. M. Wightman, *Neuroscience* **59,** 417 (1994).

The large value obtained for k_1 indicates that nearly every collision of DA with its transporter results in binding.

The results of these same calculations for the NAc core and NAc shell are shown in Table I. All three regions had a calculated turnover number of about 3 to 6 sec^{-1}. Because the value of K_m appears to be the same for these regions,[37] the value for k_1 is around 3×10^7 for all three regions as well. These similarities indicate that the DA carrier is the same between regions. Furthermore, the DA innervation of the dorsal striatum has associated with it a higher density of uptake sites than the ventral striatum, and the gradient uptake sites extends throughout the dorsoventral axis of the striatum.[41,42] This should account for the negative relationship between the rate of DA disappearance and depth within the striatum. Thus, the turnover rate for the DA carrier is most likely conserved throughout the striatum.

[41] J. F. Marshall, S. J. O'Dell, R. Navarrete, and A. J. Rosenstein, *Neuroscience* **37,** 11 (1990).

[42] L. G. Sharpe, N. S. Pilotte, W. M. Mitchell, and E. B. De Souza, *Eur. J. Pharmacol.* **203,** 141 (1991).

These similarities are not surprising, given that only one gene for the DA transporter has been identified[43,44] and that the mRNA for the DA transporter has been found in every midbrain neuron.[45] Indeed, the similarities in structure and amino acid sequence for the dopamine, norepinephrine, and serotonin transporters suggest that the uptake mechanism (and, thus, the turnover number) may be conserved between all three neurotransmitter carriers.[46,47]

Conclusions

The detection of electrically evoked DA overflow by FSCV avoids many of the problems encountered with experiments involving measurements in intact tissue. Because there is no concentration gradient resulting from the coupling of diffusion and the active uptake process, measurements of electrically evoked changes in DA concentration by FSCV directly reflect the actual concentration of DA at the release sites. Thus, uptake can be characterized directly from the data obtained. When diffusion through the Nafion film is accounted for, comparison of the data obtained by FSCV with simulations based on the Michaelis–Menten model for enzyme kinetics have revealed $V_{\max}$ and K_m values for the transport process that are comparable to those obtained with the classical method of characterizing uptake kinetics in synaptosomes. Using this technique, the effects of pharmacological agents on the uptake process have been determined. These experiments revealed surprising differences between brain regions and implied the existence of separate binding sites for nomifensine and cocaine. Furthermore, K_i values for nomifensine and cocaine have been assessed directly from the FSCV data, and these values agree well with previous literature values. Investigation of the heterogeneous values for $V_{\max}$ obtained in various dopaminergic brain regions reveals the differences in uptake to be due to differences in transporter density. A turnover rate of 6 $\sec^{-1}$ was determined for the dopamine transporter in the CP and NAc; a similar turnover rate is anticipated for all dopaminergic regions.

Acknowledgments

Research in this area was supported by the National Institutes of Health.

[43] J. E. Kilty, D. Lorang, and S. G. Amara, *Science* **254,** 578 (1991).

[44] S. Shimada, S. Kitayama, C. L. Lin, A. Patel, E. Nanthakumar, P. Gregor, M. Kuhar, and G. Uhl, *Science* **254,** 576 (1991).

[45] D. Lorang, S. G. Amara, and R. B. Simerly, *Neuroscience* **14,** 4903 (1994).

[46] S. G. Amara and M. J. Kuhar, *Annu. Rev. Neurosci.* **16,** 73 (1993).

[47] B. Giros, Y. Wang, S. Suter, S. B. McLeskey, C. Pifl, and M. G. Caron, *J. Biol. Chem.* **269,** 15985 (1994).

[48] High-Speed Chronoamperometric Electrochemical Measurements of Dopamine Clearance

By NANCY R. ZAHNISER, SHELLY D. DICKINSON, and GREG A. GERHARDT

Introduction

The dopamine transporter (DAT), a member of the superfamily of Na^+- and Cl^--dependent neurotransmitter transporters, is localized on the plasma membrane of dopaminergic neurons where it removes dopamine (DA) from the extracellular space. DAT is the primary mechanism by which the action of extracellular DA is terminated. Thus, DAT activity is critical in determining the level of dopaminergic neurotransmission in the brain and is an important parameter to measure. Several methods are available for measuring the activity of DAT. In this chapter we discuss the use of high-speed chronoamperometric electrochemistry to measure DA clearance. This has been shown to be a reliable index of DAT activity, including being sensitive to drugs and other perturbations that alter DAT function.[1–3] This technique provides for a high degree of temporal and spatial resolution and is useful with both *in vitro* and *in vivo* preparations. However, in this chapter we focus primarily on its use to measure *in vivo* DAT activity in circumscribed regions of rodent and nonhuman primate brains.

Overview of Method

An electrochemical electrode is placed directly into the tissue containing DATs and used to measure the disappearance, or "clearance," of extracellular DA signals. The response of the Nafion-coated, carbon fiber electrode is first calibrated to DA in solution, and then the electrochemical electrode/micropipette assembly is constructed by attaching a glass micropipette a fixed distance from the electrode. The assembly is lowered under stereotaxic control into the brain region of interest. Using high-speed chronoamperometry, oxidation and reduction signals are sequentially recorded at a speed of 5 Hz. To measure exogenous DA clearance, the baseline signal is zeroed

[1] M. N. Friedemann and G. A. Gerhardt, *Neurobiol. Aging* **13,** 325 (1992).
[2] C. van Horne, B. J. Hoffer, I. Stromberg, and G. A. Gerhardt, *J. Pharmacol. Exp. Ther.* **263,** 1285 (1992).
[3] W. A. Cass, N. R. Zahniser, K. A. Flach, and G. A. Gerhardt, *J. Neurochem.* **61,** 2269 (1993).

0076-6879/98 $25.00

and then a finite, nanoliter volume of DA is pressure-ejected from the micropipette. DA ejection is repeated at a fixed time interval (usually 5 min) to ensure reproducible responses. Once a reproducible signal is obtained, the kinetics of DAT can be determined by generating a DA concentration–response curve. Changes in DA clearance in response to either systemic or local injection of drugs can also be investigated. It is important to determine the potential contribution of other monoamine transporters—particularly the norepinephrine transporter (NET) and possibly the serotonin transporter (SERT)—to DA clearance by using inhibitors selective for DAT, NET, and SERT.[4] Endogenous DA clearance can be determined by ejecting KCl, rather than DA, from the micropipette to stimulate release of endogenous DA.[1] Parameters obtained from the DA oxidation signal includes signal rise time, maximal signal amplitude, signal decay time, and clearance rate. The ratio of the amplitudes of the reduction signal : oxidation signal is characteristic of each monoamine and, thus, provides a means of identifying the species being measured.

Electrode/Micropipette Assemblies

Carbon fiber electrochemical electrodes can either be purchased (Quanteon, Denver, CO) or made.[1,5] To make them, a vertical electrode puller is used to pull 4 mm o.d. glass (Schott Scientific Glass, Parkersburg, WV) to obtain a 3–4 cm long capillary with a uniformly tapered end. The capillary tip is cut back to an opening of 35–40 μm (i.e., slightly larger than the carbon fiber to be used) and excess glass is cut off the back to facilitate the packing of graphite paste in a later step. The carbon fibers (30 μm diameter; AVCO Specialty Materials, Textron, Lowell, MA) are cut to 5–6 cm lengths and are threaded through the glass capillaries. The tip of the capillary is then filled with epoxy (Epoxylite Corp., Westerville, OH) using a syringe and 30-gauge needle. The epoxy is cured for 12–16 hr at 120°. After the curing step, graphite epoxy paste (Dylon Industries, Inc., Berea, OH) is packed into the back of the capillary shaft with a sharpened wooden stick. The graphite paste should be packed far enough down the shaft to make contact with the carbon fiber. The glass is then cut so that only the narrow shaft of the capillary remains. Lacquer-coated copper wire (28-gauge) is cut to lengths of 10 cm and stripped (~3 mm) at both ends. One end is inserted into the paste. At this point, the electrodes are baked in a 125° oven for 12–16 hr. Finally, the carbon fibers are cut back to yield an exposed length of ~100 μm, and amphenols are soldered to the end of the copper wire.

[4] W. A. Cass and G. A. Gerhardt, *J. Neurochem.* **65,** 201 (1995).
[5] G. A. Gerhardt, K. Pang, and G. M. Rose, *J. Pharmacol. Exp. Ther.* **241,** 714 (1987).

The electrode is coated with Nafion [Aldrich, Milwaukee, WI; 5% solution (v/v)] to give a selectivity for DA over ascorbic acid (AA) of ≥1000 : 1.[3,6] First, the electrode is rinsed with distilled water and dried in a 200° oven for 5 min. The tip of each electrode is then immersed in Nafion, the tip is swirled for 2–3 sec, and the electrode is returned to the 200° oven for an additional 5–10 min. The coating and drying process is repeated 4–6 times. Immediately prior to use, each electrode is calibrated in solution containing DA over the range of interest (1–40 μM) and should exhibit linearity (correlation coefficients of 0.998 to 1.000) over the desired range.[4,7] Electrodes are also tested with solutions of AA or dihydroxyphenylacetic acid (DOPAC) to determine the selectivity of the probes for DA vs the potential anionic inferences, AA and DOPAC, in tissues.

Glass micropipettes (1 mm o.d.; 0.58 mm i.d.: World Precision Instruments, Inc., Sarasota, FL) are pulled and bumped to a tip diameter of 10–15 μm. For multibarrel pipettes, two or more glass capillaries are attached to each other with epoxy and/or brass collars. Alternatively, multibarrel fused glass micropipettes can be purchased (World Precision Instruments, Inc.). The micropipettes are rotated 360° while they are being pulled to twist the tips together and are then bumped to a diameter of 10–15 μm for each tip. Micropipettes are filled using syringes fitted with 30-gauge needles with the desired solution, e.g., 200 μM DA and 100 μM AA in 154 mM NaCl (pH 7.4, adjusted with NaOH). For K^+ stimulation, the solution contains 70 mM KCl, 79 mM NaCl, 2.5 mM $CaCl_2$ (pH 7.4).

Electrode/micropipette assemblies are constructed by attaching (with wax) a single- or multiple-barrel glass micropipette to each Nafion-coated carbon fiber electrode. The tips of the micropipette(s) and electrode should be even and positioned 270–330 μm apart (Fig. 1). The distance should be verified with a microscope. It is also critical to position the micropipette(s) and electrode parallel to one another and in the same plane, i.e., all should be simultaneously in focus under the microscope. If they are not parallel, the distance between them may change as the assembly is lowered through the tissue and aberrant signal shapes or no signals will result. In addition, the electrode/micropipette assembly should be positioned in a plane parallel to the midline suture of the skull before being lowered into the brain.

Animal Surgery

Rats have been used most often for clearance measurements. For nonsurvival surgery, rats are anesthetized with an intraperitoneal (i.p.) injection

[6] W. A. Cass, G. A. Gerhardt, R. D. Mayfield, P. Curella, and N. R. Zahniser, *J. Neurochem.* **59,** 259 (1992).

[7] A. Gratton, B. J. Hoffer, and G. A. Gerhardt, *Neuroscience* **29,** 57 (1989).

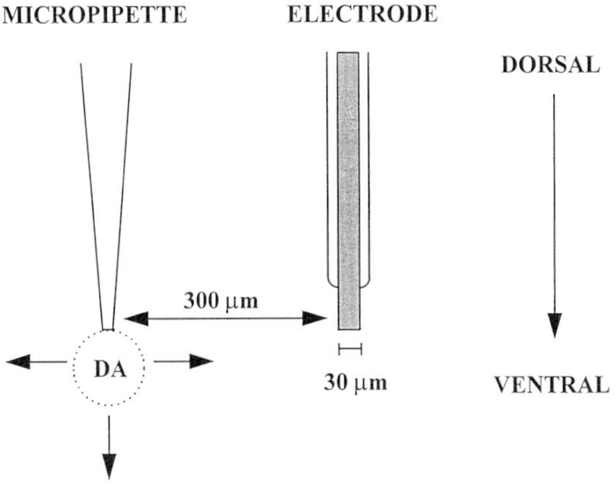

FIG. 1. Diagram of a single carbon fiber electrochemical electrode/micropipette assembly. Note that the electrode and micropipette are glued together so that they are parallel, their tips are even, and they are approximately 300 μm apart.

of urethane (1.25–1.5 g/kg). After the animal has reached a deep surgical level of anesthesia, it is placed in a stereotaxic frame and prepared for electrochemical recording as previously described.[1,7] The rat rests on a heating pad. Body temperature, as monitored by a rectal thermometer, is maintained at 37°. Using aseptic technique, a 2 cm incision is made in the scalp, and the scalp is retracted. Using a variable speed drill (Dremel, Racine, WI), the skull overlying the anterior cortex is removed bilaterally for recordings in the dorsal striatum, nucleus accumbens, and medial prefrontal cortex. After removal of the dura, the electrode/micropipette assembly is lowered slowly under stereotaxic control into the area of interest. The general coordinates in 250–400 g Sprague-Dawley rats are as follows: dorsal striatum (1–1.5 mm anterior to bregma, 2.2 mm lateral from the midline, and 4–5.0 mm below the surface of the brain), nucleus accumbens (1.5 mm anterior, 2.2 mm lateral, and 6.5–7.5 mm below the surface of the brain), or medial prefrontal cortex (2.8–3.0 mm anterior or bregma, 0.8–1.2 mm lateral from the midline, and 0.5–5.0 mm below the surface of the brain).[8] A silver|silver chloride (Ag|AgCl) reference electrode, prepared fresh daily, is positioned via a small hole drilled in the skull at a distance from the working electrode, usually in the posterior cerebral cortex.

DA clearance can also be measured in mice, using procedures very similar to those used in rats, with some exceptions. As with rats, mice are anesthesized with urethane (1.25 to 1.5 g/kg, i.p.). However, an injection

[8] G. Paxinos and C. Watson, "The Rat Brain in Stereotaxic Coordinates," 2nd ed. Academic Press, San Diego, 1986.

volume of 10 ml/kg results in a very narrow volume range, and it is fairly easy to kill the mice by overdose. To overcome this problem, we dilute the concentration of urethane by half and then inject twice the original volume. After reaching deep surgical anesthesia, the mouse is placed in a stereotaxic frame fitted with a mouse adapter. Because ear bars are not generally used with mice, a molded foam bed is often used to further stabilize the mouse. An isothermal heating pad under the stereotaxic adapter platform and a low wattage lamp are used to maintain the body temperature of the mouse at 36–37°. Using surgical scissors, a 2 cm incision is made in the scalp, and the skull overlying anterior cortex is removed using a drill (Dremel) fitted with a small engraver's bit. The dura is then removed, and the electrode/micropipette assembly is lowered slowly under stereotaxic control into the area of interest. General coordinates in 25–50 g C57BL/6 mice are as follows: dorsal striatum (1–1.4 mm anterior to bregma, 1.2–1.5 mm lateral from the midline, and 2.2–2.8 mm below dura) and nucleus accumbens (1.2–1.5 mm anterior, 1.4 mm lateral, and 3.5–4.5 mm below dura).[9] A Ag|AgCl reference electrode is placed in posterior cortex, as in a rat, with a small amount of petroleum jelly used to cover the hole. When measuring DA clearance in mice, the barrel concentration of DA is routinely doubled (400 μM). In addition, the electrode and micropipette tips are positioned slightly closer together than in rats (approximately 280–290 μm).

Rhesus monkeys have also been used for measurements of DA clearance.[10,11] The monkeys are treated preoperatively with ketamine HCl [10–30 mg/kg, intramuscular (i.m.)] and atropine (0.04 mg/kg, i.m.) and then are anesthetized with isoflurane gas (1–3%). This anesthetic is used to maintain the animals throughout the experiment. Surgery is performed under sterile field conditions. Animals are intubated and placed in a special stereotaxic apparatus designed to be compatible with magnetic resonance imaging (MRI). Areas of the skull overlying the brain areas of interest are removed (approximately 1.5 × 2 cm), and the overlying meninges are removed for penetration of the electrode/micropipette assembly. The placements of the assemblies are performed using MRI-guided stereotaxic coordinates. The general coordinates for the head of the caudate nucleus are as follows: 17.7–24.5 mm anterior to the ear bars, 4–6 mm lateral from the midline, and 13–22 mm below the surface of the cortex. A standard Ag|AgCl reference electrode (RE-5, Bioanalytical Systems, Lafeyette, IN) is placed

[9] K. B. J. Franklin and G. Paxinos, "The Mouse Brain in Stereotaxic Coordinates." Academic Press, San Diego, 1997.

[10] W. A. Cass, G. A. Gerhardt, Z. Zhang, A. Ovadia, and D. M. Gash, *Neurosci. Lett.* **185,** 52 (1995).

[11] G. A. Gerhardt, W. A. Cass, J. Hudson, M. Henson, Z. Zhang, A. Ovadia, B. J. Hoffer, and D. M. Gash, *J. Neurochem.* **66,** 579 (1996).

under the skin flap adjacent to the skull, sutured in place, and irrigated with saline. At the end of the experiment, gel foam is placed in the skull cavity and the remaining skin is sutured over the exposed areas to seal the wound. An antibiotic (Ambipen; 40,000 units/kg) is administered prophylactically every other day for 10 days.

In Vivo Electrochemical Measurements

After implantation of the electrode/micropipette assembly, the electrode is left undisturbed for ~10 min. High-speed chronoamperometric measurements are continuously made at 5 Hz and averaged to 1 Hz using either an IVEC-5 or IVEC-10 recording system (Medical Systems Corp., Greenvale, NY), and the recordings are allowed to come to a stable baseline before local application of DA or other drugs. For DA measures, an oxidation potential of +0.55 V (versus the reference electrode) is applied for 100 msec and then a resting potential of 0.0 V is applied for 100 msec. Oxidation and reduction currents are digitally integrated during the last 70–90 msec (steady-state) of each 100 msec pulse. To measure exogenous DA clearance, DA is pressure-ejected (6–125 nl, 5–30 pounds per square inch (psi) for 2–9 sec) at 5-min intervals (Fig. 2). The amount ejected is monitored by using a dissection microscope to measure the volume of fluid displaced from the micropipette. [We have calculated that a 1-mm segment of our single- or double-barrel micropipettes (0.58 mm i.d. glass) contains

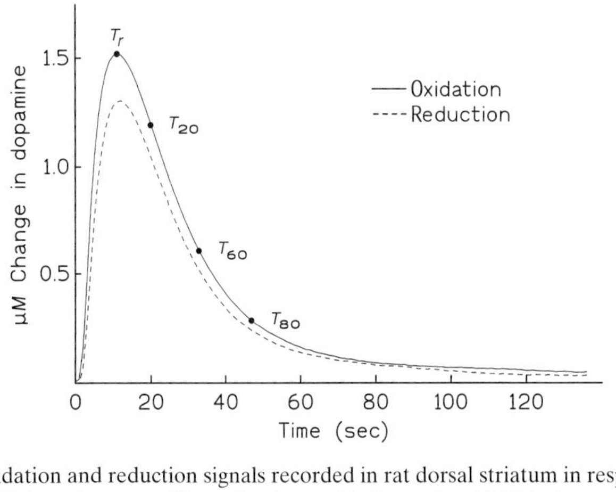

FIG. 2. Oxidation and reduction signals recorded in rat dorsal striatum in response to local application of DA at time zero. The reduction : oxidation ratio of 0.86 is characteristic of DA. The signal rise time (T_r) and the time for the signal to decay from its maximum by 20% (T_{20}), 60% (T_{60}), and 80% (T_{80}) are indicated.

246 nl.] Generally, an amplitude in the range of 1–2 μM DA is desirable. Once a steady baseline is achieved and the DA signals are reproducible at the location of interest, experimental manipulations such as drug application or DA concentration–response studies are carried out. A new baseline must be established at each location in the brain.

Drugs may be injected systemically or applied locally. For systemic injection, after two or three reproducible baseline DA signals ($\leq$10% change in amplitude) have been obtained, the drug or vehicle (control solution) is injected into the animal.[6] The experimenter continues to pressure-eject DA at 5-min intervals for an appropriate time period in order to obtain the maximal drug effect and return to baseline values (if appropriate). For local application, after two or three reproducible baseline DA signals are obtained, the drug or vehicle solution is pressure ejected 30–90 sec before the next ejection of DA.[3] The drug or vehicle solution is applied slowly over 20–40 sec to minimize disturbances to the baseline signal. The volume of drug applied is routinely twice that of DA. Again, the experimenter continues to pressure-eject DA at 5-min intervals for an appropriate time period in order to obtain the maximal drug effect and drug washout (if appropriate, Fig. 3).

DA clearance can also be studied by using KCl stimulation to release DA into the extracellular space. For these studies, the micropipette is filled with KCl solution (see above), and the electrochemical recordings are allowed to stabilize in a selected brain region for approximately 5–10 min to obtain a stable baseline. Once a stable baseline is achieved, local applica-

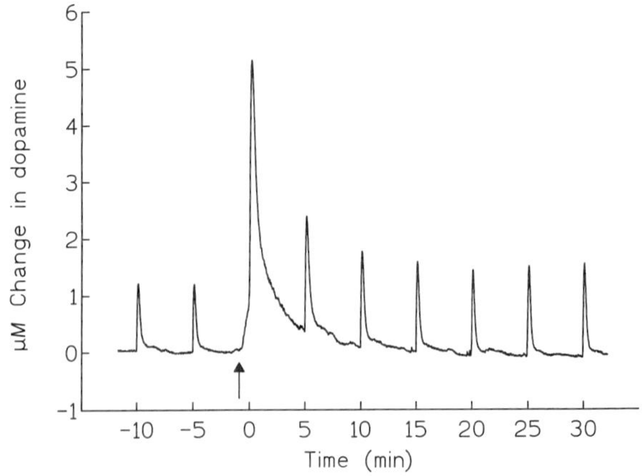

FIG. 3. Effect of local application of the DAT inhibitor cocaine on electrochemical signals recorded in the rat dorsal striatum. DA (10 pmol, 50 nl of 200 μM solution) was pressure-ejected at 5-min intervals and resulted in reproducible signals. Cocaine (80 pmol; 100 nl of 800 μM solution) was applied (arrow) 45 sec before DA.

tions of K^+ (100–500 nl, 6–25 psi for 2–9 sec) are used to release DA into the extracellular space. The recordings are continued at 15 min intervals until the signals return to the pre-KCl application baseline values. The procedure can cause the release of NE and serotonin in certain brain regions. Signals that are "DA-like" are selected based on the reduction : oxidation signal ratios to ensure the calculation of DA clearance values.[11] To verify electrode placement, brains are fixed, sectioned, stained, and examined with a microscope (Fig. 4). Because the electrode track produces relatively little tissue damage, it can sometimes be difficult to visualize at the light microscopic level. In order to facilitate determination of the position of the electrode tip more precisely, at the end of the recording session a small amount of current (≤10 μA/unit brain area) can be passed through the electrode tip. This produces a small electrolytic lesion. After dissection from the skull, the brain is fixed by immersion in 10% neutral buffered formalin. The fixed brains are frozen, sectioned with a cryostat at a thickness of 10–40 μm, stained with cresyl violet acetate, and examined by light (4×) microscopy.

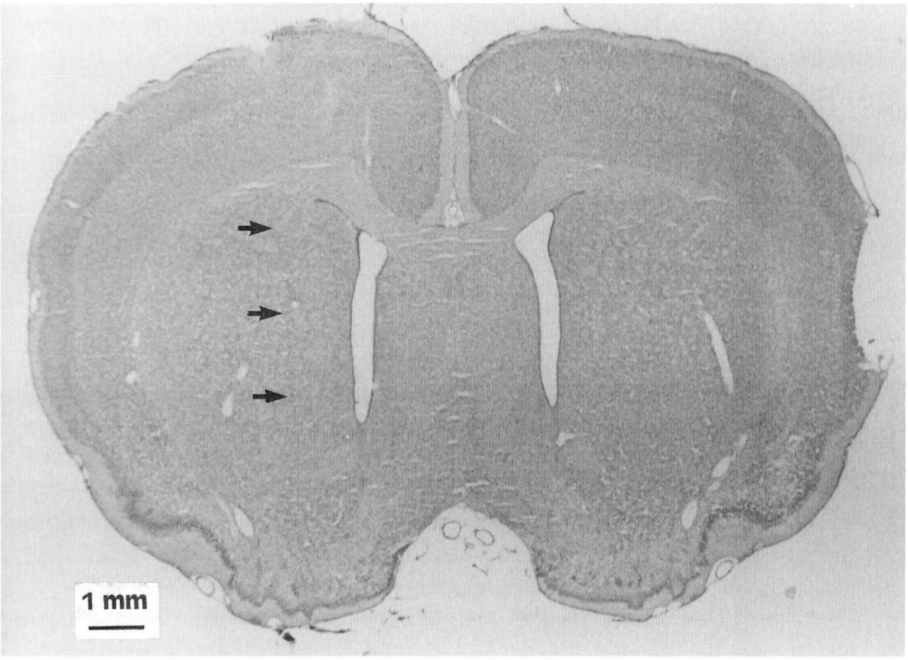

FIG. 4. Localization of a single fiber electrochemical electrode tract (arrows) terminating in nucleus accumbens. Shown is a cresyl violet–stained coronal section (40 μm) of rat brain. Note the minimal tissue damage produced by the electrode/micropipette assembly.

Data Analysis

Several parameters are routinely examined from these types of experiments (Fig. 2). The rise time of the signal is the time that it takes the signal to reach maximal amplitude. This reflects the diffusion of the pressure-ejected DA from the micropipette to the electrode or diffusion of released DA to the probe. The DA signal amplitude and time course of the signal reflect transporter activity. The maximal amplitude is measured from the maximal height of the oxidation signal. The time course is measured as the oxidation signal rise time plus the time to decay by a certain percentage, e.g., the T_{80} is the time for the signal to rise and then decay by 80% from the maximal amplitude. The early pseudolinear portion of the decay curve (T_{20-60}) is used in the clearance rate measurement, whereas the later curvilinear portion of the curve (T_{40-80} or T_{80}) is more sensitive to changes by uptake blockers. Clearance rate considers changes in both the amplitude and time course of the signal.[12] It is determined from the slope of the initial linear portion of the decay curve between the T_{20} and T_{60} points. Drug effects are often measured as a percent change from baseline value, with baseline being the mean of the two values just prior to local drug ejection or systemic drug injection.

Experimental Findings

High-speed *in vivo* chronoamperometric recordings of DA clearance with a particular electrode placement are quite stable. Thus, ejection of a finite amount of DA at 5-min intervals into rodent or primate brain produces transient and reproducible signals (Fig. 3). For example, in rat striatum or nucleus accumbens, the amplitudes of the signals decrease by approximately 30% during a 90-min recording session, whereas the signal rise times and T_{40-80} intervals are constant.[6]

This technique has been used to study the mechanisms contributing to the clearance of locally applied DA and have confirmed that DAT is the major determinant of DA clearance in dorsal striatum and nucleus accumbens of the anesthetized rat.[2,3] Thus, local application of DAT inhibitors (DA uptake blockers; Fig. 3) or 6-hydroxydopamine–induced loss of DA terminals significantly alters the DA signal. In contrast, neither metabolism by monoamine oxidase nor metabolism by catechol-*O*-methyltransferase appears to contribute significantly to the decline in the DA signal. In addition, local anesthetics do not appear to alter DAT function. Moreover, diffusion of DA appears to play a minimal role in the striatum and nucleus accumbens. However, it may play a greater role in brain areas, such as the

[12] R. M. Wightman and J. B. Zimmerman, *Brain Res. Rev.* **15,** 135 (1990).

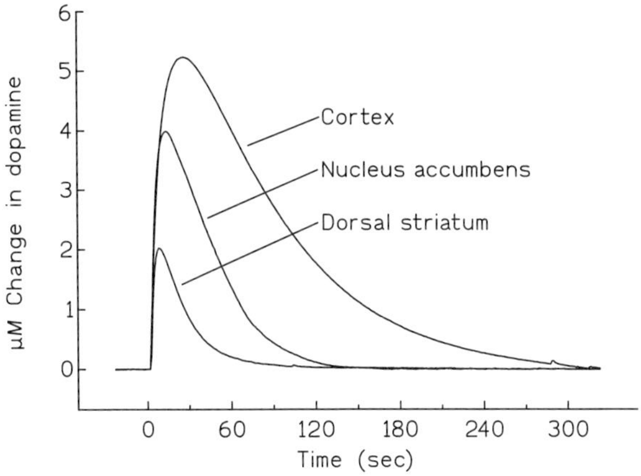

FIG. 5. Comparison of the signals recorded using the same electrochemical electrode/micropipette assembly in the cerebral cortex, nucleus accumbens, and dorsal striatum of a single rat. In each region 20 pmol (100 nl of 200 μM solution) of DA was pressure-ejected. The assembly was positioned anterior to bregma: 1.5 mm; lateral to the midline: 2.2 mm; and below the surface of the brain: 1.5 mm (cortex), 4.9 mm (striatum) and 6.7 mm (nucleus accumbens).

frontal cortex, which contain much lower levels of DAT. These results suggest that DA clearance, measured as outlined in this chapter, primarily reflects the activity of DAT.

The idea that DA clearance reflects DAT activity is further strengthened by the finding that the magnitude of the DA signal depends on the number of DATs expressed in the brain region (Fig. 5). In general, to obtain signals of equivalent amplitude in two brain regions, less DA needs to be applied in the brain region with fewer DAT levels (e.g., nucleus accumbens) than in the brain region with higher DAT levels (e.g., dorsal striatum).[4,5] The clearance time is longer and the clearance rate is slower in the brain region with fewer DATs. This is exemplified by the 6-hydroxydopamine–lesioned striatum (>95% depletion of DA), in which the clearance rate appears to be dominated by diffusion.[2,3] We have also observed that, compared with dorsal striatum, DA clearance in the nucleus accumbens is more sensitive to inhibition by low doses of DAT antagonists such as cocaine.[3,6] This is true whether cocaine is injected systemically or applied locally. These data, taken together with other observations, suggest that the striatum may possess "spare" DATs, whereas the nucleus accumbens does not. However, it should be noted that even within a region such as the striatum, DAT activity is heterogeneous.[13]

[13] E. J. Cline, C. E. Adams, G. A. Larson, G. A. Gerhardt, and N. R. Zahniser, *Exp. Neurol.* **134,** 135 (1995).

Results of experiments with acute administration of DAT inhibitors suggest that as DAT function is antagonized, the clearance rate of the remaining active transporters increases until it reaches a maximal rate (V_{max}). Thus, depending on the concentration of DA applied, DAT inhibitors may alter the amplitude and/or the time course of the DA signal.[3,13,14] An increased DA clearance rate is reflected as an increase in the amplitude of the signal without prolongation of the time course. Measuring changes in DA clearance has been useful to study the regulation of DAT activity in response to aging, neuronal lesioning, or chronic cocaine administration.[1,2,11,14]

In addition to studying the effects of DAT inhibitors on DA clearance, this approach has been used to investigate the modulation of DAT activity by receptors.[15] Activation of presynaptic D_2 DA autoreceptors stimulates DAT activity, and blockade of these receptors diminishes DAT activity. Consistent with this notion and with inhibition of DAT activity, local application of a D_2 DA receptor antagonist increases the DA signal amplitude and time course in rat striatum, nucleus accumbens, and medial prefrontal cortex.[15] In contrast, local application of a D_1 DA receptor antagonist or saline does not alter DA clearance. Although D_2 DA receptor antagonists are effective in this paradigm, D_2 DA receptor agonists are not. This is presumably because the receptor is already occupied by DA, particularly the DA locally applied to measure clearance.

Additional Applications/Future Directions

To address some questions, it is advantageous to measure DA clearance in *in vitro* brain slices. Advantages include the fact that drug concentrations and electrode placement can be more precisely controlled. For example, placement of the electrode/micropipette assembly in the substantia nigra pars compacta vs the substantia nigra pars reticulata can be easily accomplished. Electrochemical and electrophysiological measurements can also be made concurrently in the slice. The electrode/micropipette assembly inserted approximately 80 μm into a 400 μm brain slice yields reproducible results.

Another avenue of interest involves defolliculated stage V or VI *Xenopus lavis* oocytes, which have proved to be a useful system for expressing the cloned DAT. When a DAT-expressing oocyte is bathed in a DA solu-

[14] W. A. Cass, G. A. Gerhardt, K. Gillespie, P. Curella, R. D. Mayfield, and N. R. Zahniser, *J. Neurochem.* **61,** 273 (1993).

[15] W. A. Cass and G. A. Gerhardt, *Neurosci. Lett.* **176,** 259 (1994).

tion, DA clearance can theoretically be measured by an electrode inserted into the oocyte and the initial rate of the appearance of the DA signal. However, this approach has proved to be challenging. Preliminary studies have shown that the oocytes contain an oxidizable substance (not DA) that produces a background signal and the electrodes often lose sensitivity when inserted into oocytes. Using electrochemical recording in this system remains one of the technical challenges for the future.

Finally, most measures of DA clearance have been made in anesthetized animals. However, DA clearance can also be measured in awake, behaving rats. This offers the obvious advantage of being able to measure DAT activity and behavior concurrently. It also eliminates the potential confound of anesthesia on DAT activity. The only change is that the electrodes and micropipettes are assembled using fused silica tubing rather than glass micropipettes. Under surgical anesthesia, 1 mm holes are drilled in the rat's skull for electrode injector assembly implantation and for screws to fasten a miniature connector to the skull. The animals are allowed to recover from surgery for 2 days. On the day of recording, the assembly is connected to a miniature amplifier/commutator system. The animal is unrestrained and free to move in the recording chamber. We have found that the small amounts of locally applied DA do not produce behavioral effects. Furthermore, DA clearance is not affected by doses of urethane or choral hydrate that produce complete anesthesia. The long-term stability of DA clearance measurements in chronically implanted rats has been determined to be at least 2 weeks. Future studies will investigate the functional properties of DAT in freely moving animals.

[49] *In Vivo* Microdialysis for Measurement of Extracellular Monoamine Levels Following Inhibition of Monoamine Transporters

By NIAN-HANG CHEN and MAARTEN E. A. REITH

Introduction

Intracerebral microdialysis, developed by Ungerstedt *et al.*,[1] is a technique to sample endogenous substances from and to introduce exogenous

[1] U. Ungerstedt, *in* "Measurement of Neurotransmitter Release *in Vivo* (C. A. Marsden, Ed.), p. 81. John Wiley and Sons, New York, 1984.

METHODS IN ENZYMOLOGY, VOL. 296
0076-6879/98 $25.00

substances into the brain extracellular fluid by implanting a dialysis fiber in discrete brain areas. Once removed by dialysis from the brain, the endogenous substances may be analyzed with various analytic techniques. Many reviews on the theory and applications of microdialysis are available[2,3] containing pertinent, basic information. This technique is now providing insights correlating the activity of monoamine transporters with monoamine neurotransmission in live, conscious, and freely moving animals.[4] Various monoamine uptake blockers, which are either selective (interacting with only one type of monoamine transporter) or nonselective (interacting with more than one type of monoamine transporter) according to the *in vitro* studies, have been used as tools to explore the *in vivo* relationship between monoamine transporter function and monoamine neurotransmission.[4] We describe here the use of microdialysis to identify the monoamine profile arising from inhibition of monoamine transporters in the ventral tegmental area (VTA)–nucleus accumbens (NAC) DA pathway. The utility of this method is documented by the results obtained in our laboratory.[4–9]

Principle of Method

Monoamine transporters utilizing norepinephrine (NE), dopamine (DA), and serotonin (5-HT) as substrates are referred to as NE, DA, and 5-HT transporters, respectively. Neuronal monoamine transporters function to regulate monoamine neurotransmission in the central nervous system (CNS) by transporting monoamine neurotransmitters, released by nerve impulses, from inside and outside the synaptic clefts into monoamine neurons.[10,11] Uptake blockers directly interact with monoamine transporters and prevent the transporting process of monoamine neurotransmitters, resulting in increased monoamine levels in the synaptic clefts.[3,12] In the presence of uptake inhibition, this increased monoamine level in the synaptic cleft is presumed to be in equilibrium with the monoamine levels in the

[2] H. Benveniste and P. C. Hüttemeier, *Prog. Neurobiol.* **35,** 195 (1990).

[3] T. E. Robinson and J. B. Justice, Jr., Eds., "Microdialysis in the Neurosciences." Elsevier Science Publishers BV, Netherlands, 1991.

[4] N. H. Chen and M. E. A. Reith, *in* "Neurotransmitter Transporters: Structure, Function and Regulation" (M. E. A. Reith, Ed.), p. 345. Humana Press, Clifton, NJ, 1997.

[5] N.-H. Chen and M. E. A. Reith, *J. Neurochem.* **63,** 1701 (1994).

[6] N.-H. Chen and M. E. A. Reith, *J. Pharmacol. Exp. Ther.* **271,** 1597 (1994).

[7] N.-H. Chen and M. E. A. Reith, *J. Neurochem.* **64,** 1585 (1995).

[8] M.-Y. Li, Q.-S. Yan, L. L. Coffey, and M. E. A. Reith, *J. Neurochem.* **66,** 559 (1996).

[9] S.-Y. Chen, R. L. Burger, and M. E. A. Reith, *Brain Res.* **729,** 294 (1996).

[10] S. G. Amara and M. J. Kuhar, *Annu. Rev. Neurosci.* **16,** 73 (1993).

[11] G. R. Uhl and P. S. Johnson, *J. Exp. Biol.* **196,** 229 (1994).

[12] G. D. Chiara, *Trends Pharmacol. Sci.* **11,** 116 (1990).

brain extracellular fluid surrounding the dialysis probe.[3] The monoamines diffuse down their concentration gradients from the brain extracellular fluid into the perfusion solution flowing through the microdialysis probe. Subsequently, microdialysis carries monoamines out of the brain (without removing extracellular fluid) and makes them accessible to analysis by high performance liquid chromatography with electrochemical detection (HPLC–EC). It should always be kept in mind that there are substantial differences between *in vivo* microdialysis and the *in vitro* procedures used to determine transporter activity. In addition to the autoregulation of monoamine release and the activation of other adaptive cellular responses, there are many other possible ways in which NE, DA, and 5-HT could interact to change each other's release. Therefore, the dialyzate monoamine levels measured after administration of uptake blockers reflect a kinetic balance among many processes resulting from inhibition of transporters that occur on a short time scale that can be measured with the current microdialysis technique. The term *efflux* is used to describe this effect of uptake blockers on dialyzate monoamines. Dialysis can also be used to introduce a compound into the brain: When present in the perfusate, a compound will diffuse down its concentration gradient into the brain extracellular fluid surrounding the probe. Simultaneous measurement of NE, DA, and 5-HT levels in the dialyzate from a brain region during focal infusion of an uptake blocker into the same brain region allows the evaluation of the *in vivo* apparent selectivity of the uptake blocker. With multiple routes for administration of uptake blockers and/or sampling of DA, the different changes induced by an uptake blocker in the extracellular DA level between DA cell body and terminal fields can be understood by considering the impacts of DA transporter density, impulse-dependent release, and interconnecting neuronal circuits impinging on the VTA–NAC DA pathway.

Experimental Variables

Anesthesia

Because monoamine transporters remove monoamine transmitters from the extracellular space after their release by nerve impulses, the monoamine effect of uptake blockers is contingent on the activity of monoamine neurons. Inherent in dialysis studies using anesthetized animals is the concern that the anesthetic itself may compromise the activity of monoamine neurons[13–16] and/or interact with uptake blockers,[17] resulting in misleading

[13] M. D. Kelland, A. S. Freeman, and L. A. Chiodo, *Synapse* **3,** 30 (1989).

[14] M. D. Kelland, L. A. Chiodo, and A. S. Freeman, *Synapse* **6,** 207 (1990).

[15] M. E. Hamilton, A. Mele, and A. Pert, *Brain Res.* **597,** 1 (1992).

interpretation of findings. Therefore, it is preferable to conduct microdialysis in conscious animals.

Perfusion Solution

In principle, the composition of a perfusion solution should be as close as possible to that of the brain extracellular fluid.[2] However, for the simultaneous determination of NE, DA, and 5-HT in dialyzates, we also use a Ringer's solution. The Ca^{2+} and K^{+} concentrations in Ringer's solution are close to those in plasma but higher than those in brain extracellular fluid. Choice of such a perfusion solution is guided by the specific characteristics of the experimental conditions to be studied. First, in the case of the mesolimbic dopamine pathways in the brain, the levels of extracellular NE and 5-HT in both the VTA and NAC are very low and barely determined with a perfusion solution containing concentrations of Ca^{2+} and K^{+} mimicking those in the brain extracellular fluid. The higher concentrations of Ca^{2+} and K^{+} stimulate release and allow analysis of the basal levels of monoamines with a greater precision. This problem might be circumvented by adding uptake blockers to the perfusion solution, but such a manipulation is certainly not suitable for a study to investigate the monoamine effect of uptake blockers. Second, the HPLC–EC system used is highly sensitive to impurities in perfusion solution. Thus, use of Ringer's solution devoid of Mg^{2+} and phosphate ions reduces the presence of electroactive contaminants and causes less interference with the chromatographic analysis. Although it has been reported that a higher concentration (3.4 m*M*) of calcium can affect the changes in DA induced by certain pharmacological stimuli in striatal dialyzates,[18,19] in our experience the use of Ringer's solution containing 2.3 m*M* Ca^{2+} does not affect the efflux of monoamines in dialyzates in response to uptake blockers. The observed effects of uptake blockers fulfill the classic criteria (see below).

Microdialysis Probes

Three types of probes have been used for the dialysis of monoamines: concentric, U-shaped, and transcerebral. The smaller size of concentric probes permits them to be applied to deep and small brain structures such

[16] A. Fink-Jensen, S. H. Ingwersen, P. G. Nielsen, L. Hansen, E. B. Nielsen, and A. J. Hansen, *Naunyn-Schmiedeberg's Arch. Pharmacol.* **350,** 239 (1994).

[17] W. H. T. Pan, Y.-J. Lai, and N.-H. Chen, *J. Neurochem.* **64,** 2653 (1995).

[18] B. H. C. Westerink, H. M. Hofsteede, G. Damsma, J. B. de Vries, *Naunyn-Schmiedeberg's Arch. Pharmacol.* **337,** 373 (1988).

[19] B. Moghaddam and B. S. Bunney, *J. Neurochem.* **53,** 652 (1989).

as the VTA and NAC, and also permits less damage to the brain tissue.[20] We usually use the commercially available concentric probes equipped with Cuprophan membranes. Generally, more than 70% of monoamine efflux is impulse-dependent 3–4 hr after implantation of the probe. In addition, a rapid and unobstructed exchange of the three monoamines between the dialysis perfusion fluid and the outside medium is achieved with the Cuprophan membrane, which may not be met by some other membranes.[21] It is important to calibrate the *in vitro* recovery of probes before and after an experiment. Although the relative recovery of a probe *in vitro* does not directly predict their diffusion efficiency *in vivo*,[22] it helps to identify and thereby avoid use of "bad" probes *in vivo*. Moreover, use of probes with appreciable variations in their *in vitro* performance will add noise to the results. For example, the reduction in magnitude of the probe recovery may vary with the monoamines, being usually most severe for NE and least serious for 5-HT. A probe with unusually low recovery (<5%) for catecholamines often shows a greater relative change, compared with one with normal recovery (on the average 12%), in dialyzate NE and DA in response to an uptake blocker. In addition, perfusion of an uptake blocker into the extracellular fluid (dialysis into the brain) may be affected by a probe with lower *in vitro* diffusion efficiency. Therefore, keeping the *in vitro* recovery of probes relatively constant in each experiment is necessary for the correct understanding of the *in vivo* monoamine profile of uptake blockers. If maintained properly, the probe can be reused more than 10 times with little change in its *in vitro* recovery.

Contaminations

The monoamines are very easily destroyed by bacteria and metal ions present in the perfusion system. This is a particularly serious problem in our HPLC–EC system which handles minute amounts of dialyzate monoamines. When the HPLC signals for tested monoamines are abnormally low even when a new probe is used, one should consider these possibilities. An approach to keep bacteria away from the dialysis system is to flush the system with 1:200 Kathon (Bioanalytical Systems) solution at 1 μl/min overnight after each experiment. In addition, metal-made components should be accommodated into the microdialysis system as rarely as possible. Never assemble any aluminum-coated/made parts into this dialysis system. Another type of contamination occurs when low amounts of neuronal

[20] M. Santiago and B. H. C. Westerink, *Naunyn-Schmiedeberg's Arch. Pharmacol.* **342,** 407 (1990).
[21] R. Tao and S. Hjorth, *J. Neurochem.* **59,** 1778 (1992).
[22] S. D. Glick, N. Dong, R. W. Keller, Jr., and J. N. Carlson, *J. Neurochem.* **62,** 2017 (1994).

5-hydroxytryptamine (5-HT) are contaminated with 5-HT of blood origin. Both the VTA and NAC are deeply located, with the VTA close to the dura. Penetration of the dura by the probe tip can cause bleeding. In this context it should be recalled that 5-HT transporters in platelets can also be a target for uptake blockers. If the 5-HT originating from blood accounts for a high proportion of dialyzate 5-HT, it can outweigh the change in brain extracellular 5-HT resulting from the inhibition of neuronal 5-HT transporters. In this context, the Ringer's solution with a higher concentration of Ca^{2+} may attenuate this interference by promoting an impulse-dependent 5-HT efflux. Clearly, optimal placement of the dialysis probe is critical in minimizing the contamination of 5-HT of blood origin. Other potential contaminations are often caused by drugs sticking to the probe and the tubing. This can be a substantial problem, especially when perfusing uptake blockers such as GBR analogs, desipramine, and fluoxetine. The adsorption can be so strong that repeated washing may not provide complete remedy. A contamination of this type can shift the concentration–response curve for the same compound to the right in the next experiment. The only way to solve this problem is to change the probe and tubing before a new experiment.

Criteria for Evaluating Monoamine Effect of Uptake Blockers

Because uptake blockers simply prevent removal from the extracellular space of the transmitter released by nerve impulses, Ca^{2+} dependence and tetrodotoxin sensitivity are regarded as fundamental criteria for an uptake blocker to increase monoamine efflux in dialyzates.[5,12] For the same reason, an increase in dialyzate monoamine efflux by inhibition of transporters can be modified with agents acting at monoamine autoreceptors, which modify impulse-dependent monoamine release. Thus, an autoreceptor agonist reduces the uptake blocker-induced monoamine efflux,[23] whereas an autoreceptor antagonist enhances it.[6] These criteria are currently used to determine whether the observed effect of an uptake blocker is due to inhibition of monoamine transporters. If the observed effect of a putative uptake blocker stands in contrast to these criteria, it indicates contamination with blood, a nonspecific effect of the compound, or methodological problems. These criteria also differentiate inhibition of monoamine transporters from reversed transport of monoamines. The latter is Ca^{2+} independent and tetrodotoxin insensitive because a releaser displaces monoamines from vesicular pools into the cytoplasm, making them available for active transport out of the terminal by the transporter.

[23] S. M. Florin, R. Kuczenski, and D. S. Segal, *Brain Res.* **654,** 53 (1994).

Routes for Administration of Uptake Blockers and Sampling of Monoamines

Focal application of an uptake blocker into a brain region via dialysis probes can circumvent pharmacokinetic factors and minimize the effect of the compound on other brain structures which could indirectly influence monoamine transmission in the target region. It allows the evaluation of the local effect of transporters on neurotransmission. By using different loci of focal application, we can target either the DA transporters localized at DA somatodendrites or the monoamine transporters localized at monoamine terminals. Unlike focal infusion, systemic administration of an uptake blocker takes into account the impact of brain circuitries on the overall effect of both somatodendritic and axonal transporter inhibition. Thus, inhibition of somatodendritic transporters can produce an inhibitory effect on the monoamine efflux induced by inhibition of terminal transporters.[4] With dual probes, one in the VTA and the other ipsilateral in the NAC, focal infusion of a DA uptake blocker into the VTA allows one to examine the effect of inhibiting somatodendritic DA transporters on the extracellular DA level in the nucleus accumbens,[9] whereas systemic application of uptake blockers can be used to investigate regional differences in monoamine efflux following inhibition of monoamine transporters. Multiple combinations of these approaches provides an enormous amount of information about the role of monoamine transporters in the modulation of monoamine neurotransmission in the brain.

Data Presentation

If an external calibration indicates that a probe is functioning normally, no attempt should be made to correct dialyzate concentration for *in vitro* probe recovery because of the reported lack of relationship between *in vitro* and *in vivo* recoveries.[22] It is common practice to express the microdialysis data as either absolute values or relative values (percent of baseline). The former has the appeal of being more quantitative, but the latter can compensate for variations in probe recovery and animals. Two main factors should be considered when people evaluate experimental data on the effects of uptake blockers. One factor is the rate of ongoing monoamine release. Because uptake blocker–induced increase in extracellular monoamines is dependent on exocytotic release, the region with a higher basal release would be expected to show the larger absolute increase in dialyzate monoamine efflux after application of uptake blockers. The other factor is the density of monoamine transporters. Apart from considering the interrelationships between brain regions, at higher transporter density, more uptake sites are occupied by a given synaptic concentration of uptake blocker, causing more monoamine efflux in dialyzates. When expressed as absolute values, microdialysis data include the impact of basal release and correlate more with actual monoamine levels in brain extracellular fluid after administration of uptake blockers. When expressed as relative values, the data exclude the impact of basal release and hence correlate better with the combined effect of transporter inhibition on uptake, release, and other processes. Obviously, the combination of both data presentations is important in addressing the functional significance of a certain modification of monoamine transporter activity. However, the relevance of both data presentations regarding transporter activity should also be considered in the light of the potential effect of application of an uptake blocker on the *in vivo* probe recovery for DA.[24,25]

Methods

Implantation of Guide Cannula

Anesthetize the rat (we use male Sprague-Dawley rats, 250–300 g) with a combination of sodium pentobarbital [30 mg/kg, intraperitoneal (i.p.)] and halothane (5% in oxygen). Place the rat in a stereotaxic apparatus and adjust the incisor bar so that the skull is flat. Shave the head of animal and make a saggittal midline incision through the skin to expose the skull. Drill a hole where the guide cannula is to be implanted. Drill holes for the anchor screws in different bones of the skull. Open the dura with the tip of a syringe needle. Lower the guide cannula (CMA/11, 0.38 mm o.d, Bioanalytical Systems, W. Lafayette, IN) slowly to 2 mm above the ventral surface of the target brain regions. Fix the guide cannula to the skull of the rat with dental acrylic and anchor screws, and keep it patent during a 7-day recovery period with a stainless steel stylet. The coordinates for the ventral tegmental area are 5.2 mm posterior to bregma, 2 mm right from the midline directed towards the ventral tegmental area at an angle of 10° to avoid damaging the sagittal sinus, and 6.8 mm below the surface of the skull. The coordinates for the nucleus accumbens are 1.8 mm anterior to bregma, 1.5 mm right from the midline, and 5.8 mm below the surface of the skull. All the surgery procedures should be carried out under sterile conditions.

Microdialysis

Apparatus. Concentric microdialysis probes (CMA/11, Cuprophan, 2.0 mm length, 0.21 mm i.d., 0.24 mm o.d.); inlet and outlet tubing (FEP

[24] R. J. Olson and J. B. Justice, Jr., *Anal. Chem.* **65,** 1017 (1993).
[25] A. D. Smith and J. B. Justice, Jr., *J. Neurosci. Meth.* **54,** 75 (1994).

polymer, 0.12 mm i.d., 0.68 mm o.d., Bioanalytical Systems); a two-channel liquid swivel (Harvard Apparatus, Inc., South Natick, MA) located on a balance beam to allow free movement of the rat within the experimental chamber; a spring tether (ESA Inc., Bedford, MA) with one end attached to the liquid swivel and the other end to be attached to a jacket on animals to protect the tubing from the damage of animals; a syringe pump (model 22, Harvard); gas-tight syringes equipped with 0.2 μm filters; a liquid switch (CMA/110, Bioanalytical Systems); Plexiglas animal chamber.

Preparation of Perfusion Solution. Ringer's solution (in m*M*, NaCl 147, KCl 4, $CaCl_2$ 2.3, pH 6–6.3): dissolve 4.3 g of NaCl, 0.15 g of KCl, and 0.169 g of $CaCl_2 \cdot 2H_2O$ in 500 ml fresh purified water; it can be stored at 4° for 1 week. Artificial cerebrospinal fluid (in m*M*, NaCl 145, KCl 3, $CaCl_2$ 1.4, $MgCl_2$ 1, Na_2HPO_4 2, pH 7.4): prepare 10× stock solutions of NaCl (1450 m*M*), KCl (30 m*M*), $MgCl_2$ (10 m*M*), $CaCl_2$ (14 m*M*), and Na_2HPO_4 (20 m*M*, pH 7.4 with 85% H_3PO_4) with fresh purified water and store them at 4°; on the day of use, 1 ml of each stock is combined in a volumetric flask and diluted to 10 ml. Both types of perfusion solution should be filtered through 0.2 μm filters before using.

Calibration of Probes. In vitro recovery of a probe is determined daily prior to implantation, to verify probe consistency and system stability. Flush the probe with the perfusion solution at 20–30 μl/min for 5 min to remove air bubbles and glycerol. Immerse the probe in a beaker containing 10 ml of perfusion solution to which NE, DA, and 5-HT are added at a concentration of 10 n*M*. Perfuse the probe with the perfusion solution without monoamines at a flow rate of 2 μl/min. Collect at least two 20-min perfusate samples. Determine the monoamine amount in the perfusate obtained from the probe and that in an equal volume obtained from the beaker. The relative recovery is expressed as a percentage of a monoamine concentration in perfusate to that in the beaker.

Microdialysis Procedure. On the day before the test day, have the rat wear a jacket. On the test day, place the rat into the Plexiglas chamber. Fill a gas-tight syringe with filtered perfusion solution. Make sure the fluid is at ambient temperature and there is no air bubble in the syringe. Mount the syringe in the syringe pump. Connect the syringe in the syringe pump to the liquid switch and the switch to the liquid swivel with FEB tubing (Teflon, 0.65 mm OD, 0.12 mm ID; Bioanalytical Systems). Connect the inlet tubing to the outlet of the liquid swivel. Start the pump at 2 μl/min to perfuse the calibrated probe (see above) at a flow rate of 2 μl/min. Slowly insert the probe through the guide cannula by gently restraining the awake rat with a wrap technique. The tip of the probe extends 2 mm below the end of the guide cannula to reach the ventral surface of the target brain region. Fix both inlet and outlet tubing of the probe to the spring tether with taps. Attach the other end of the spring tether to the jacket on the rat. Approximately 4 hr after probe insertion, collect dialyzate samples from the outlet tubing of the probe into polyethylene microvials every 20 min. After obtaining three to four consecutive stable baseline dialyzate samples, administer uptake blockers focally or systemically (i.p.) For focal application of uptake blockers, dissolve the compounds in water or perfusion solution at 1 m*M*; sonicate the solutions for 10 min; further dilute them with the perfusion solution to the desired concentrations; and perfuse the drug solutions via the probe. For the determination of the concentration–response relationships, perfuse progressively higher concentrations of an uptake blocker through the probe, with each concentration lasting 80 min measured over four collected samples of 20 min. Use the liquid switch for changing the perfusion solution. For the systemic administration of uptake blockers, set up a control group in which animals receive injections of saline (the vehicle used for uptake blockers). Drug delivery and sample collection time are corrected for the lag time resulting from the dead volume of the inlet and outlet tubes.

HPLC–EC Analysis

Apparatus. An ESA solvent dietary system (model 580, ESA) including a low stroke volume (10 μl) dual piston HPLC pump equipped with polyether ether ketone (PEEK) tubing and two in-line serial PEEK pulse dampeners; an ESA HR-80 column (3 μm, ODS, 80 $\times$ 4.6 mm); an inert Rheodyne 9125 injector with a 100-μl PEEK injector loop; an ESA Coulochem II electrochemical detector equipped with a dual electrode analysis cell (M5014 Microdialysis) and a guard cell (M5020).

Preparation of Mobile Phase. Prepare a phosphate buffer by mixing 14.01 g of Na_2HPO_4 and 0.57 g of sodium dodecyl sulfate with approximately 960 ml fresh purified water, adding 132 μl of triethylamine and 1.32 ml of 0.02 *M* EDTA-2Na, adjusting the pH to 5.6 with 85% H_3PO_4, and adjusting the volume to 1000 ml. Filter the phosphate buffer through 0.2-μm filter paper. Add 120 ml methanol and 120 ml acetonitrile to 760 ml phosphate buffer and mix together. The mobile phase used is designed to selectively retain the monoamines while eluting the acid metabolites in the solvent front.[26] The final concentrations are 75 m*M* NaH_2PO_4, 1.5 m*M* sodium dodecyl sulfate, 20 μ*M* EDTA-2Na, 100 μl/l triethylamine, 12% methanol, and 12% acetonitrile (pH 5.6). The concentration of organic may be changed based on a good separation of peaks and the desired retention time. After being degassed, the mobile phase can be introduced to the HPLC system.

[26] K. C. Gariepy, B. Bailey, J. Yu, T. Maher, and I. N. Ackworth, *J. Liquid Chromat.* **17,** 1541 (1994).

The mobile phase is always prepared freshly and circulated through the HPLC system no longer than 5–6 days.

Preparation of Monoamine Standard Solutions. Separate NE, DA, or 5-HT stock standard solutions (1 m*M*): dissolve 31.93 mg of NE bitartrate, 18.96 mg of DA HCl, or 38.74 mg of 5-HT in 0.1 m*M* HCl and adjust the volume to 100 ml with 0.1 m*M* HCl. Mixed monoamine stock standard solution (1 μM): mix 0.1 ml of each of the above concentrated stock standard solution with 0.01 m*M* HCl to 100 ml. Working monoamine standard solutions: dilute the mixed monoamine stock standard solution with the perfusion solution to desired concentrations. The stock standard solution can be stored at 4° for up to 4–6 months. The working standard solutions should

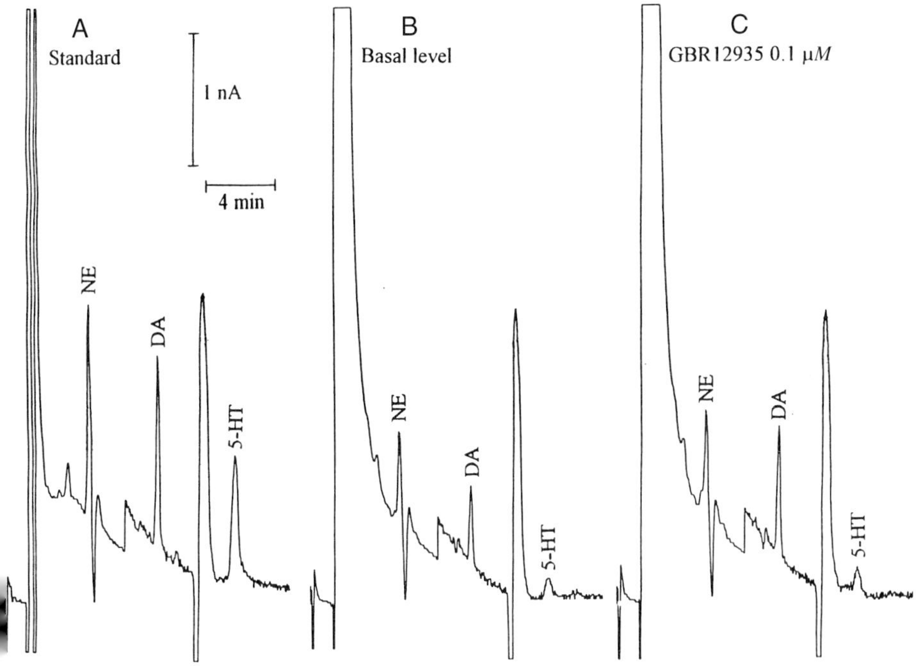

FIG. 1. Chromatograms achieved following injections at 40 μl standard solution or 40 μl microdialyzate from the ventral tegmental area of the freely moving rat. (A): Standard; the on-column amount of NE, DA, and 5-HT was 40 fmol each. (B): Basal microdialyzate; the computed on-column amount of NE, DA, and 5-HT was 14.9, 15.3, and 5.3 fmol, respectively. (C): Microdialyzate following focal infusion of the selective DA uptake blocker GBR 12935 (0.1 μM) through the probe; the computed on-column amount of NE, DA, and 5-HT was 16.7, 32.2, and 5.7 fmol, respectively. The microdialysis probe (CMA/11) was perfused with Ringer's solution at a flow rate of 2 μl/min. Chromatographic conditions are described in the text.

be freshly prepared just before each injection for peak identification. The injection volume of the working standard solution is 40 μl.

Working Conditions. This system is dedicated solely to the simultaneous analysis of NE, DA, and 5-HT in one run to allow evaluation of the *in vivo* monoamine profile of uptake blockers. The guard cell is placed immediately before the injector and is set at 400 mV. The dual electrode analysis cell is set with electrode 1 at -100 mV and electrode 2 at 225 mV with respect to palladium reference electrodes. The mobile phase is pumped through the HPLC system at 1 ml/in. Chromatograms are integrated, compared with standards run separately before, during, and after each experiment, and analyzed using a computer-based data acquisition system (we have used, among others, Maxima 820, Waters, Milford, MA). The retention time for elution of all three monoamines is less than 15 min. The detection limit for NE, DA, and 5-HT is 1 fmol at a 2 : 1 signal-to-noise ratio. Standard curves ranging from 2 to 60 fmol on column give a linear increase in peak area with an R^2 of >0.99. Representive chromatograms are shown in Fig. 1.

Assay of Dialyzate Samples. Typically, the dialyzate samples are collected at 20-min intervals and injected directly into the HPLC–EC system. For experiments with dual probes, samples from the two regions are analyzed in two ways that are alternated between regions: dialyzates from one region are analyzed immediately, whereas dialyzates from the other region are frozen at $-20°$ and analyzed within 24 hr. Control experiments do not reveal a change in monoamine content over this period in the frozen aliquots. There is no difference between samples collected in acid (final concentration of 0.05 *M* $HClO_4$) and nonacidified samples.

Histology

At the end of each experiment, anesthetize the rat with an overdose of sodium pentobarbital. Perfuse the rat transcardially with saline followed by 10% formalin. Decapitate the rat, remove the brain, and fix it in 10% formalin for at least 3 days. Cut coronal sections (40 μm) with a cryostat microtome and stain them with cresyl violet. Verify the placement of the probe under a microscope. Only use the data obtained from an experiment with correct placement of the probe.

Author Index

Numbers in parentheses are footnote reference numbers and indicate that an author's work is referred to although the name is not cited in the text.

A

C

D

E

F

H

K

M

N

O

P

Q

R

S

T

U

V

W

X

Y

Z

Subject Index

A

B

C

D

F

G

K

L

M

T

U

V

Y

ISBN 0-12-182197-8